The Finite Element Method in Engineering

The Finite Element Method in Engineering

FOURTH EDITION

Singiresu S. Rao
Professor and Chairman
Department of Mechanical and Aerospace Engineering
University of Miami, Coral Gables, Florida, USA

ELSEVIER
BUTTERWORTH HEINEMANN

Amsterdam • Boston • Heidelberg • London • New York • Oxford
Paris • San Diego • San Francisco • Singapore • Sydney • Tokyo

Elsevier Butterworth–Heinemann
30 Corporate Drive, Suite 400, Burlington, MA 01803, USA
Linacre House, Jordan Hill, Oxford OX2 8DP, UK

∞ Recognizing the importance of preserving what has been written, Elsevier prints its books on acid-free paper whenever possible.

Library of Congress Cataloging-in-Publication Data
APPLICATION SUBMITTED

British Library Cataloguing-in-Publication Data
A catalogue record for this book is available from the British Library.

ISBN: 0-7506-7828-3

For information on all Butterworth-Heinemann publications
visit our Web site at www.books.elsevier.com.

04 05 06 07 08 09 10 10 9 8 7 6 5 4 3 2 1

Printed in the United States of America

CONTENTS

PREFACE

The finite element method is a numerical method that can be used for the accurate solution of complex engineering problems. The method was first developed in 1956 for the analysis of aircraft structural problems. Thereafter, within a decade, the potentialities of the method for the solution of different types of applied science and engineering problems were recognized. Over the years, the finite element technique has been so well established that today it is considered to be one of the best methods for solving a wide variety of practical problems efficiently. In fact, the method has become one of the active research areas for applied mathematicians. One of the main reasons for the popularity of the method in different fields of engineering is that once a general computer program is written, it can be used for the solution of any problem simply by changing the input data.

The objective of this book is to introduce the various aspects of finite element method as applied to engineering problems in a systematic manner. It is attempted to give details of development of each of the techniques and ideas from basic principles. New concepts are illustrated with simple examples wherever possible. Several Fortran computer programs are given with example applications to serve the following purposes:

- to enable the student to understand the computer implementation of the theory developed;
- to solve specific problems;
- to indicate procedure for the development of computer programs for solving any other problem in the same area.

The source codes of all the Fortran computer programs can be found at the Web site for the book, www.books.elsevier.com. Note that the computer programs are intended for use by students in solving simple problems. Although the programs have been tested, no warranty of any kind is implied as to their accuracy.

After studying the material presented in the book, a reader will not only be able to understand the current literature of the finite element method but also be in a position to develop short computer programs for the solution of engineering problems. In addition, the reader will be in a position to use the commercial software, such as ABAQUS, NASTRAN, and ANSYS, more intelligently.

The book is divided into 22 chapters and an appendix. Chapter 1 gives an introduction and overview of the finite element method. The basic approach and the generality of the method are illustrated through simple examples. Chapters 2 through 7 describe the basic finite element procedure and the solution of the resulting equations. The finite element discretization and modeling, including considerations in selecting the number and types of elements, is discussed in Chapter 2. The interpolation models in terms of Cartesian and natural coordinate systems are given in Chapter 3. Chapter 4 describes the higher order and isoparametric elements. The use of Lagrange and Hermite polynomials is also discussed in this chapter. The derivation of element characteristic matrices and vectors using direct, variational, and weighted residual approaches is given in Chapter 5.

The assembly of element characteristic matrices and vectors and the derivation of system equations, including the various methods of incorporating the boundary conditions, are indicated in Chapter 6. The solutions of finite element equations arising in equilibrium, eigenvalue, and propagation (transient or unsteady) problems, along with their computer implementation, are briefly outlined in Chapter 7.

The application of the finite element method to solid and structural mechanics problems is considered in Chapters 8 through 12. The basic equations of solid mechanics — namely, the internal and external equilibrium equations, stress–strain relations, strain–displacement relations and compatibility conditions — are summarized in Chapter 8. The analysis of trusses, beams, and frames is the topic of Chapter 9. The development of inplane and bending plate elements is discussed in Chapter 10. The analysis of axisymmetric and three-dimensional solid bodies is considered in Chapter 11. The dynamic analysis, including the free and forced vibration, of solid and structural mechanics problems is outlined in Chapter 12.

Chapters 13 through 16 are devoted to heat transfer applications. The basic equations of conduction, convection, and radiation heat transfer are summarized and the finite element equations are formulated in Chapter 13. The solutions of one-, two-, and three-dimensional heat transfer problems are discussed in Chapters 14–16, respectively. Both the steady state and transient problems are considered. The application of the finite element method to fluid mechanics problems is discussed in Chapters 17–19. Chapter 17 gives a brief outline of the basic equations of fluid mechanics. The analysis of inviscid incompressible flows is considered in Chapter 18. The solution of incompressible viscous flows as well as non-Newtonian fluid flows is considered in Chapter 19. Chapters 20–22 present additional applications of the finite element method. In particular, Chapters 20–22 discuss the solution of quasi-harmonic (Poisson), Helmholtz, and Reynolds equations, respectively. Finally, Green–Gauss theorem, which deals with integration by parts in two and three dimensions, is given in Appendix A.

This book is based on the author's experience in teaching the course to engineering students during the past several years. A basic knowledge of matrix theory is required in understanding the various topics presented in the book. More than enough material is included for a first course at the senior or graduate level. Different parts of the book can be covered depending on the background of students and also on the emphasis to be given on specific areas, such as solid mechanics, heat transfer, and fluid mechanics. The student can be assigned a term project in which he/she is required to either modify some of the established elements or develop new finite elements, and use them for the solution of a problem of his/her choice. The material of the book is also useful for self study by practicing engineers who would like to learn the method and/or use the computer programs given for solving practical problems.

I express my appreciation to the students who took my courses on the finite element method and helped me improve the presentation of the material. Finally, I thank my wife Kamala for her tolerance and understanding while preparing the manuscript.

Miami
May 2004

S. S. Rao
srao@miami.edu

PRINCIPAL NOTATION

a	length of a rectangular element
a_x, a_y, a_z	components of acceleration along x, y, z directions of a fluid
A	area of cross section of a one-dimensional element; area of a triangular (plate) element
$A^{(e)}$	cross-sectional area of one-dimensional element e
$A_i(A_j)$	cross-sectional area of a tapered one-dimensional element at node $i(j)$
b	width of a rectangular element
$\vec{B}$	body force vector in a fluid $= \{B_x, B_y, B_z\}^T$
c	specific heat
c_v	specific heat at constant volume
$C_1, C_2, \ldots$	constants
$[C]$	compliance matrix; damping matrix
D	flexural rigidity of a plate
$[D]$	elasticity matrix (matrix relating stresses and strains)
E	Young's modulus; total number of elements
$E^{(e)}$	Young's modulus of element e
E_{ii}	Young's modulus in a plane defined by axis i
$f_1(x), f_2(x), \ldots$	functions of x
F	shear force in a beam
g	acceleration due to gravity
G	shear modulus
G_{ij}	shear modulus in plane ij
h	convection heat transfer coefficient
$H_{oi}^{(0)}(x)$	Lagrange polynomial associated with node i
$H_{ki}^{(j)}$	jth order Hermite polynomial
i	$(-1)^{1/2}$
I	functional to be extremized; potential energy; area moment of inertia of a beam
$\vec{i}$, $(\vec{I})$	unit vector parallel to $x(X)$ axis
$I^{(e)}$	contribution of element e to the functional I
I_{zz}	area moment of inertia of a cross section about z axis
J	polar moment of inertia of a cross section
$\vec{j}$ $(\vec{J})$	unit vector parallel to $y(Y)$ axis
$[J]$	Jacobian matrix

k	thermal conductivity
k_x, k_y, k_z	thermal conductivities along x, y, z axes
k_r, k_θ, k_z	thermal conductivities along r, θ, z axes
$\vec{k}$ $(\vec{K})$	unit vector parallel to $z(Z)$ axis
$[k^{(e)}]$	stiffness matrix of element e in local coordinate system
$[K^{(e)}] = [K_{ij}^{(e)}]$	stiffness matrix of element e in global coordinate system
$[K] = [K_{ij}]$	stiffness (characteristic) matrix of complete body after incorporation of boundary conditions
$[\underset{\sim}{K}] = [\underset{\sim}{K}_{ij}]$	stiffness (characteristic) matrix of complete body before incorporation of boundary conditions
l	length of one-dimensional element
$l^{(e)}$	length of the one-dimensional element e
l_x, l_y, l_z	direction cosines of a line
l_{ox}, m_{ox}, n_{ox}	direction cosines of x axis
l_{ij}, m_{ij}, n_{ij}	direction cosines of a bar element with nodes i and j
L	total length of a bar or fin; Lagrangian
L_1, L_2	natural coordinates of a line element
L_1, L_2, L_3	natural coordinates of a triangular element
L_1, L_2, L_3, L_4	natural coordinates of a tetrahedron element
$\pounds$	distance between two nodes
m	mass of beam per unit length
M	bending moment in a beam; total number of degrees of freedom in a body
M_x, M_y, M_{xy}	bending moments in a plate
M_z	torque acting about z axis on a prismatic shaft
$[m^{(e)}]$	mass matrix of element e in local coordinate system
$[M^{(e)}]$	mass matrix of element e in global coordinate system
$[M]$	mass matrix of complete body after incorporation of boundary conditions
$[\underset{\sim}{M}]$	mass matrix of complete body before incorporation of boundary conditions
n	normal direction
N_i	interpolation function associated with the ith nodal degree of freedom
$[N]$	matrix of shape (nodal interpolation) functions
p	distributed load on a beam or plate; fluid pressure
P	perimeter of a fin
$\vec{P}_c$	vector of concentrated nodal forces
$P_i(P_j)$	perimeter of a tapered fin at node $i(j)$
P_x, P_y, P_z	external concentrated loads parallel to x, y, z axes
$\vec{p}^{(e)}$	load vector of element e in local coordinate system
$\vec{p}_b^{(e)}(\vec{P}_b^{(e)})$	load vector due to body forces of element e in local (global) coordinate system

$\vec{p}_i^{(e)}(\vec{P}_i^{(e)})$	load vector due to initial strains of element e in local (global) coordinate system
$\vec{p}_s^{(e)}(\vec{P}_s^{(e)})$	load vector due to surface forces of element e in local (global) coordinate system
$\vec{P}^{(e)} = \{P_i^{(e)}\}$	vector of nodal forces (characteristic vector) of element e in global coordinate system
$\vec{P} = \{P_i\}$	vector of nodal forces of body after incorporation of boundary conditions
$\underset{\sim}{\vec{P}} = \{\underset{\sim}{P}_i\}$	vector of nodal forces of body before incorporation of boundary conditions
q	rate of heat flow
$\dot{q}$	rate of heat generation per unit volume
q_x	rate of heat flow in x direction
Q_i	mass flow rate of fluid across section i
Q_x, Q_y	vertical shear forces in a plate
Q_x, Q_y, Q_z	external concentrated moments parallel to x, y, z axes
$\vec{q}^{(e)}(\vec{Q}^{(e)})$	vector of nodal displacements (field variables) of element e in local (global) coordinate system
$\underset{\sim}{\vec{Q}}$	vector of nodal displacements of body before incorporation of boundary conditions
$\underline{\vec{Q}}_j$	mode shape corresponding to the frequency ω_j
r, s	natural coordinates of a quadrilateral element
r, s, t	natural coordinates of a hexahedron element
r, θ, z	radial, tangential, and axial directions
(r_i, s_i, t_i)	values of (r, s, t) at node i
R	radius of curvature of a deflected beam; residual; region of integration; dissipation function
S	surface of a body
S_1, S_2	part of surface of a body
$S^{(e)}$	surface of element e
$S_1^{(e)}$, $S_2^{(e)}$	part of surface of element e
t	time; thickness of a plate element
T	temperature; temperature change; kinetic energy of an elastic body
T_i	temperature at node i
T_0	temperature at the root of fin
T_∞	surrounding temperature
$T_i^{(e)}$	temperature at node i of element e
$\vec{T}^{(e)}$	vector of nodal temperatures of element e

$\vec{\underset{\sim}{T}}$	vector of nodal temperatures of the body before incorporation of boundary conditions
u	flow velocity along x direction; axial displacement
u, v, w	components of displacement parallel to x, y, z axes; components of velocity along x, y, z directions in a fluid (Chapter 17)
$\vec{U}$	vector of displacements $= \{u, v, w\}^T$
V	volume of a body
$\vec{V}$	velocity vector $= \{u, v, w\}^T$ (Chapter 17)
w	transverse deflection of a beam
W	amplitude of vibration of a beam
W_i	value of W at node i
W_p	work done by external forces
$\vec{W}^{(e)}$	vector of nodal displacements of element e
x	x coordinate; axial direction
(x_c, y_c)	coordinates of the centroid of a triangular element
(x_i, y_i, z_i)	(x, y, z) coordinates of node i
(X_i, Y_i, Z_i)	global coordinates (X, Y, Z) of node i
α	coefficient of thermal expansion
α_i	ith generalized coordinate
δ	variation operator
ε_{ii}	normal strain parallel to ith axis
ε_{ij}	shear strain in ij plane
$\varepsilon^{(e)}$	strain in element e
$\vec{\varepsilon}$	strain vector $= \{\varepsilon_{xx}, \varepsilon_{yy}, \varepsilon_{zz}, \varepsilon_{xy}, \varepsilon_{yz}, \varepsilon_{zx}\}^T$ for a three-dimensional body; $= \{\varepsilon_{rr}, \varepsilon_{\theta\theta}, \varepsilon_{zz}, \varepsilon_{rz}\}^T$ for an axisymmetric body
$\vec{\varepsilon_0}$	initial strain vector
θ	torsional displacement or twist
$[\lambda^{(e)}]$	coordinate transformation matrix of element e
$\eta_j^{(t)}$	jth generalized coordinate
μ	dynamic viscosity
ν	Poisson's ratio
ν_{ij}	Poisson's ratio in plane ij
π	potential energy of a beam; strain energy of a solid body
π_c	complementary energy of an elastic body
π_p	potential energy of an elastic body
π_R	Reissner energy of an elastic body
$\pi^{(e)}$	strain energy of element e
ρ	density of a solid or fluid
σ_{ii}	normal stress parallel to ith axis

σ_{ij}	shear stress in ij plane
$\sigma^{(e)}$	stress in element e
$\vec{\sigma}$	stress vector $= \{\sigma_{xx}, \sigma_{yy}, \sigma_{zz}, \sigma_{xy}, \sigma_{yz}, \sigma_{zx}\}^T$ for a three-dimensional body; $= \{\sigma_{rr}, \sigma_{\theta\theta}, \sigma_{zz}, \sigma_{rz}\}^T$ for an axisymmetric body
τ	shear stress in a fluid
ϕ	field variable; axial displacement; potential function in fluid flow
ϕ_x, ϕ_y, ϕ_z	body force per unit volume parallel to x, y, z axes
$\vec{\phi}$	vector valued field variable with components u, v, and w
$\vec{\bar{\phi}}$	vector of prescribed body forces
Φ	dissipation function for a fluid
Φ_x, Φ_y, Φ_z	surface (distributed) forces parallel to x, y, z axes
Φ_i	ith field variable
Φ_i^*	prescribed value of ϕ_i
$\Phi_i^{(e)}$	value of the field variable ϕ at node i of element e
$\vec{\Phi}^{(e)}$	vector of nodal values of the field variable of element e
$\vec{\Phi}$	vector of nodal values of the field variables of complete body after incorporation of boundary conditions
$\underset{\sim}{\vec{\Phi}}$	vector of nodal values of the field variables of complete body before incorporation of boundary conditions
ψ	stream function in fluid flow
ω	frequency of vibration
ω_j	jth natural frequency of a body
ω_x	rate of rotation of fluid about x axis
$\tilde{\omega}_i$	approximate value of ith natural frequency
Ω	body force potential in fluid flow
superscript e	element e
arrow over a symbol ($\vec{X}$)	column vector $\vec{X} = \begin{Bmatrix} X_1 \\ X_2 \\ \vdots \end{Bmatrix}$
$\vec{X}^T$([ ]^T)	transpose of $\vec{X}$([ ])
dot over a symbol ($\dot{x}$)	derivative with respect to time $\left(\dot{x} = \dfrac{\mathrm{d}x}{\mathrm{d}t}\right)$

INTRODUCTION

1

OVERVIEW OF FINITE ELEMENT METHOD

1.1 BASIC CONCEPT

The basic idea in the finite element method is to find the solution of a complicated problem by replacing it by a simpler one. Since the actual problem is replaced by a simpler one in finding the solution, we will be able to find only an approximate solution rather than the exact solution. The existing mathematical tools will not be sufficient to find the exact solution (and sometimes, even an approximate solution) of most of the practical problems. Thus, in the absence of any other convenient method to find even the approximate solution of a given problem, we have to prefer the finite element method. Moreover, in the finite element method, it will often be possible to improve or refine the approximate solution by spending more computational effort.

In the finite element method, the solution region is considered as built up of many small, interconnected subregions called finite elements. As an example of how a finite element model might be used to represent a complex geometrical shape, consider the milling machine structure shown in Figure 1.1(a). Since it is very difficult to find the exact response (like stresses and displacements) of the machine under any specified cutting (loading) condition, this structure is approximated as composed of several pieces as shown in Figure 1.1(b) in the finite element method. In each piece or element, a convenient approximate solution is assumed and the conditions of overall equilibrium of the structure are derived. The satisfaction of these conditions will yield an approximate solution for the displacements and stresses. Figure 1.2 shows the finite element idealization of a fighter aircraft.

1.2 HISTORICAL BACKGROUND

Although the name of the finite element method was given recently, the concept dates back for several centuries. For example, ancient mathematicians found the circumference of a circle by approximating it by the perimeter of a polygon as shown in Figure 1.3.

In terms of the present-day notation, each side of the polygon can be called a "finite element." By considering the approximating polygon inscribed or circumscribed, one can obtain a lower bound $S^{(l)}$ or an upper bound $S^{(u)}$ for the true circumference S. Furthermore, as the number of sides of the polygon is increased, the approximate values

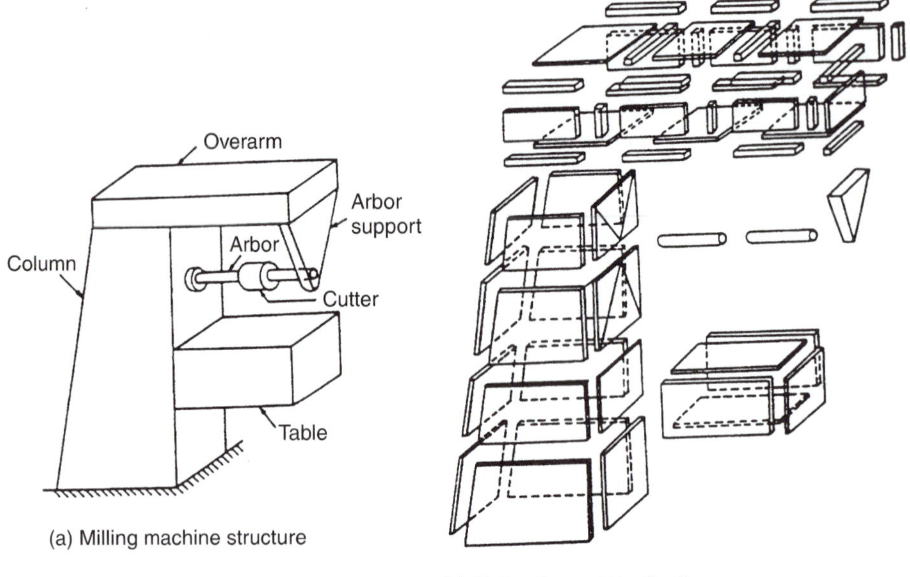

Figure 1.1. Representation of a Milling Machine Structure by Finite Elements.

Figure 1.2. Finite Element Mesh of a Fighter Aircraft (Reprinted with Permission from Anamet Laboratories, Inc.).

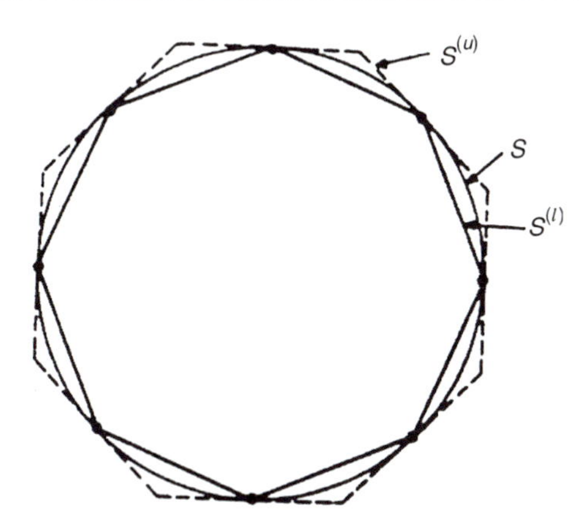

Figure 1.3. Lower and Upper Bounds to the Circumference of a Circle.

converge to the true value. These characteristics, as will be seen later, will hold true in any general finite element application. In recent times, an approach similar to the finite element method, involving the use of piecewise continuous functions defined over triangular regions, was first suggested by Courant [1.1] in 1943 in the literature of applied mathematics.

The basic ideas of the finite element method as known today were presented in the papers of Turner, Clough, Martin, and Topp [1.2] and Argyris and Kelsey [1.3]. The name *finite element* was coined by Clough [1.4]. Reference [1.2] presents the application of simple finite elements (pin-jointed bar and triangular plate with inplane loads) for the analysis of aircraft structure and is considered as one of the key contributions in the development of the finite element method. The digital computer provided a rapid means of performing the many calculations involved in the finite element analysis and made the method practically viable. Along with the development of high-speed digital computers, the application of the finite element method also progressed at a very impressive rate. The book by Przemieniecki [1.33] presents the finite element method as applied to the solution of stress analysis problems. Zienkiewicz and Cheung [1.5] presented the broad interpretation of the method and its applicability to any general field problem. With this broad interpretation of the finite element method, it has been found that the finite element equations can also be derived by using a weighted residual method such as Galerkin method or the least squares approach. This led to widespread interest among applied mathematicians in applying the finite element method for the solution of linear and nonlinear differential equations. Over the years, several papers, conference proceedings, and books have been published on this method.

A brief history of the beginning of the finite element method was presented by Gupta and Meek [1.6]. Books that deal with the basic theory, mathematical foundations, mechanical design, structural, fluid flow, heat transfer, electromagnetics and manufacturing applications, and computer programming aspects are given at the end of the chapter [1.10–1.32]. With all the progress, today the finite element method is considered one of the well-established and convenient analysis tools by engineers and applied scientists.

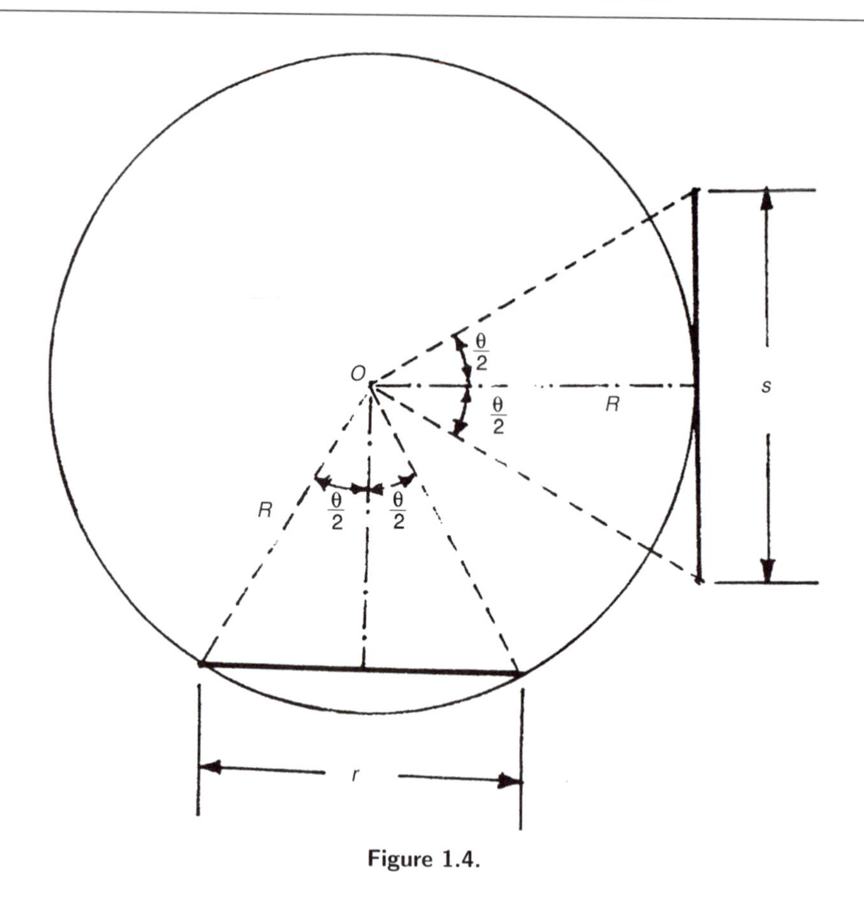

Figure 1.4.

Example 1.1 The circumference of a circle (S) is approximated by the perimeters of inscribed and circumscribed n-sided polygons as shown in Figure 1.3. Prove the following:

$$\lim_{n\to\infty} S^{(l)} = S \quad \text{and} \quad \lim_{n\to\infty} S^{(u)} = S$$

where $S^{(l)}$ and $S^{(u)}$ denote the perimeters of the inscribed and circumscribed polygons, respectively.

Solution If the radius of the circle is R, each side of the inscribed and the circumscribed polygon can be expressed as (Figure 1.4)

$$r = 2R\sin\frac{\pi}{n}, \qquad s = 2R\tan\frac{\pi}{n} \tag{E$_1$}$$

Thus, the perimeters of the inscribed and circumscribed polygons are given by

$$S^{(l)} = nr = 2nR\sin\frac{\pi}{n}, \qquad S^{(u)} = ns = 2nR\tan\frac{\pi}{n} \tag{E$_2$}$$

which can be rewritten as

$$S^{(l)} = 2\pi R \left[\frac{\sin\dfrac{\pi}{n}}{\dfrac{\pi}{n}}\right], \qquad S^{(u)} = 2\pi R \left[\frac{\tan\dfrac{\pi}{n}}{\dfrac{\pi}{n}}\right] \tag{E$_3$}$$

As $n \to \infty$, $\dfrac{\pi}{n} \to 0$, and hence

$$S^{(l)} \to 2\pi R = S, \qquad S^{(u)} \to 2\pi R = S \tag{E$_4$}$$

1.3 GENERAL APPLICABILITY OF THE METHOD

Although the method has been extensively used in the field of structural mechanics, it has been successfully applied to solve several other types of engineering problems, such as heat conduction, fluid dynamics, seepage flow, and electric and magnetic fields. These applications prompted mathematicians to use this technique for the solution of complicated boundary value and other problems. In fact, it has been established that the method can be used for the numerical solution of ordinary and partial differential equations. The general applicability of the finite element method can be seen by observing the strong similarities that exist between various types of engineering problems. For illustration, let us consider the following phenomena.

1.3.1 One-Dimensional Heat Transfer

Consider the thermal equilibrium of an element of a heated one-dimensional body as shown in Figure 1.5(a). The rate at which heat enters the left face can be written as [1.7]

$$q_x = -kA\frac{\partial T}{\partial x} \tag{1.1}$$

where k is the thermal conductivity of the material, A is the area of cross section through which heat flows (measured perpendicular to the direction of heat flow), and $\partial T/\partial x$ is the rate of change of temperature T with respect to the axial direction.

The rate at which heat leaves the right face can be expressed as (by retaining only two terms in the Taylor's series expansion)

$$q_{x+\mathrm{d}x} = q_x + \frac{\partial q_x}{\partial x}\,\mathrm{d}x = -kA\frac{\partial T}{\partial x} + \frac{\partial}{\partial x}\left(-kA\frac{\partial T}{\partial x}\right)\mathrm{d}x \tag{1.2}$$

The energy balance for the element for a small time dt is given by

Heat inflow in time dt	+	Heat generated by internal sources in time dt	=	Heat outflow in time dt	+	Change in internal energy during time dt

That is,

$$q_x\,\mathrm{d}t + \dot{q}A\,\mathrm{d}x\,\mathrm{d}t = q_{x+\mathrm{d}x}\,\mathrm{d}t + c\rho\frac{\partial T}{\partial t}\,\mathrm{d}x\,\mathrm{d}t \tag{1.3}$$

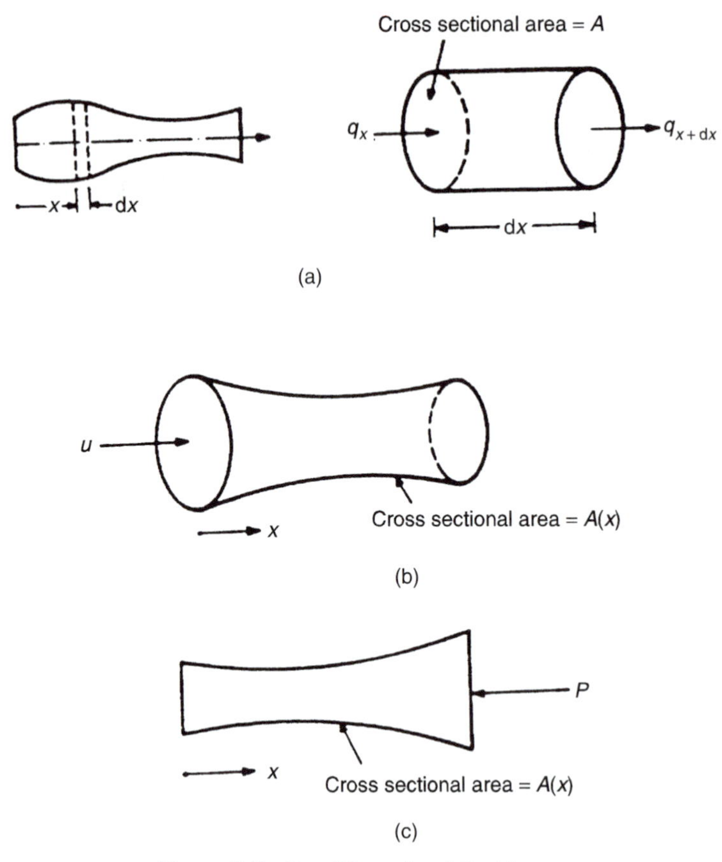

Figure 1.5. One-Dimensional Problems.

where $\dot{q}$ is the rate of heat generation per unit volume (by the heat source), c is the specific heat, ρ is the density, and $\partial T/\partial t\,\mathrm{d}t = \mathrm{d}T$ is the temperature change of the element in time $\mathrm{d}t$. Equation (1.3) can be simplified to obtain

$$\frac{\partial}{\partial x}\left(kA\frac{\partial T}{\partial x}\right) + \dot{q}A = c\rho\frac{\partial T}{\partial t} \tag{1.4}$$

Special cases

If the heat source $\dot{q} = 0$, we get the Fourier equation

$$\frac{\partial}{\partial x}\left(kA\frac{\partial T}{\partial x}\right) = c\rho\frac{\partial T}{\partial t} \tag{1.5}$$

If the system is in a steady state, we obtain the Poisson equation

$$\frac{\partial}{\partial x}\left(kA\frac{\partial T}{\partial x}\right) + \dot{q}A = 0 \tag{1.6}$$

If the heat source is zero and the system is in steady state, we get the Laplace equation

$$\frac{\partial}{\partial x}\left(kA\frac{\partial T}{\partial x}\right) = 0 \tag{1.7}$$

If the thermal conductivity and area of cross section are constant, Eq. (1.7) reduces to

$$\frac{\partial^2 T}{\partial x^2} = 0 \tag{1.8}$$

1.3.2 One-Dimensional Fluid Flow

In the case of one-dimensional fluid flow (Figure 1.5(b)), we have the net mass flow the same at every cross section; that is,

$$\rho A u = \text{constant} \tag{1.9}$$

where ρ is the density, A is the cross-sectional area, and u is the flow velocity. Equation (1.9) can also be written as

$$\frac{\mathrm{d}}{\mathrm{d}x}(\rho A u) = 0 \tag{1.10}$$

If the fluid is inviscid, there exists a potential function $\phi(x)$ such that [1.8]

$$u = \frac{\mathrm{d}\phi}{\mathrm{d}x} \tag{1.11}$$

and hence Eq. (1.10) becomes

$$\frac{\mathrm{d}}{\mathrm{d}x}\left(\rho A\frac{\mathrm{d}\phi}{\mathrm{d}x}\right) = 0 \tag{1.12}$$

1.3.3 Solid Bar under Axial Load

For the solid rod shown in Figure 1.5(c), we have at any section x,

$$\begin{aligned}\text{Reaction force} &= (\text{area})(\text{stress}) = (\text{area})(E)(\text{strain})\\ &= AE\frac{\partial u}{\partial x} = \text{applied force}\end{aligned} \tag{1.13}$$

where E is the Young's modulus, u is the axial displacement, and A is the cross-sectional area. If the applied load is constant, we can write Eq. (1.13) as

$$\frac{\partial}{\partial x}\left(AE\frac{\partial u}{\partial x}\right) = 0 \tag{1.14}$$

A comparison of Eqs. (1.7), (1.12), and (1.14) indicates that a solution procedure applicable to any one of the problems can be used to solve the others also. We shall see how the finite element method can be used to solve Eqs. (1.7), (1.12), and (1.14) with appropriate boundary conditions in Section 1.5 and also in subsequent chapters.

1.4 ENGINEERING APPLICATIONS OF THE FINITE ELEMENT METHOD

As stated earlier, the finite element method was developed originally for the analysis of aircraft structures. However, the general nature of its theory makes it applicable to a wide variety of boundary value problems in engineering. A boundary value problem is one in which a solution is sought in the domain (or region) of a body subject to the satisfaction of prescribed boundary (edge) conditions on the dependent variables or their derivatives. Table 1.1 gives specific applications of the finite element method in the three major categories of boundary value problems, namely, (i) equilibrium or steady-state or time-independent problems, (ii) eigenvalue problems, and (iii) propagation or transient problems.

In an equilibrium problem, we need to find the steady-state displacement or stress distribution if it is a solid mechanics problem, temperature or heat flux distribution if it is a heat transfer problem, and pressure or velocity distribution if it is a fluid mechanics problem.

In eigenvalue problems also, time will not appear explicitly. They may be considered as extensions of equilibrium problems in which critical values of certain parameters are to be determined in addition to the corresponding steady-state configurations. In these problems, we need to find the natural frequencies or buckling loads and mode shapes if it is a solid mechanics or structures problem, stability of laminar flows if it is a fluid mechanics problem, and resonance characteristics if it is an electrical circuit problem.

The propagation or transient problems are time-dependent problems. This type of problem arises, for example, whenever we are interested in finding the response of a body under time-varying force in the area of solid mechanics and under sudden heating or cooling in the field of heat transfer.

1.5 GENERAL DESCRIPTION OF THE FINITE ELEMENT METHOD

In the finite element method, the actual continuum or body of matter, such as a solid, liquid, or gas, is represented as an assemblage of subdivisions called finite elements. These elements are considered to be interconnected at specified joints called nodes or nodal points. The nodes usually lie on the element boundaries where adjacent elements are considered to be connected. Since the actual variation of the field variable (e.g., displacement, stress, temperature, pressure, or velocity) inside the continuum is not known, we assume that the variation of the field variable inside a finite element can be approximated by a simple function. These approximating functions (also called interpolation models) are defined in terms of the values of the field variables at the nodes. When field equations (like equilibrium equations) for the whole continuum are written, the new unknowns will be the nodal values of the field variable. By solving the field equations, which are generally in the form of matrix equations, the nodal values of the field variable will be known. Once these are known, the approximating functions define the field variable throughout the assemblage of elements.

The solution of a general continuum problem by the finite element method always follows an orderly step-by-step process. With reference to static structural problems, the step-by-step procedure can be stated as follows:

Step (i): Discretization of the structure

The first step in the finite element method is to divide the structure or solution region into subdivisions or elements. Hence, the structure is to be modeled with suitable finite elements. The number, type, size, and arrangement of the elements are to be decided.

Table 1.1. Engineering Applications of the Finite Element Method

Area of study	Equilibrium problems	Eigenvalue problems	Propagation problems
1. Civil engineering structures	Static analysis of trusses, frames, folded plates, shell roofs, shear walls, bridges, and prestressed concrete structures	Natural frequencies and modes of structures; stability of structures	Propagation of stress waves; response of structures to aperiodic loads
2. Aircraft structures	Static analysis of aircraft wings, fuselages, fins, rockets, spacecraft, and missile structures	Natural frequencies, flutter, and stability of aircraft, rocket, spacecraft, and missile structures	Response of aircraft structures to random loads; dynamic response of aircraft and spacecraft to aperiodic loads
3. Heat conduction	Steady-state temperature distribution in solids and fluids	—	Transient heat flow in rocket nozzles, internal combustion engines, turbine blades, fins, and building structures
4. Geomechanics	Analysis of excavations, retaining walls, underground openings, rock joints and soil–structure interaction problems; stress analysis in soils, dams, layered piles, and machine foundations	Natural frequencies and modes of dam–reservoir systems and soil–structure interaction problems	Time-dependent soil–structure interaction problems; transient seepage in soils and rocks; stress wave propagation in soils and rocks
5. Hydraulic and water resources engineering; hydrodynamics	Analysis of potential flows, free surface flows, boundary layer flows, viscous flows, transonic aerodynamic problems; analysis of hydraulic structures and dams	Natural periods and modes of shallow basins, lakes, and harbors; sloshing of liquids in rigid and flexible containers	Analysis of unsteady fluid flow and wave propagation problems; transient seepage in aquifers and porous media; rarefied gas dynamics; magneto-hydrodynamic flows

(continued)

Table 1.1 (continued). Engineering Applications of the Finite Element Method

Area of study	Equilibrium problems	Eigenvalue problems	Propagation problems
6. Nuclear engineering	Analysis of nuclear pressure vessels and containment structures; steady-state temperature distribution in reactor components	Natural frequencies and stability of containment structures; neutron flux distribution	Response of reactor containment structures to dynamic loads; unsteady temperature distribution in reactor components; thermal and viscoelastic analysis of reactor structures
7. Biomedical engineering	Stress analysis of eyeballs, bones, and teeth; load-bearing capacity of implant and prosthetic systems; mechanics of heart valves	—	Impact analysis of skull; dynamics of anatomical structures
8. Mechanical design	Stress concentration problems; stress analysis of pressure vessels, pistons, composite materials, linkages, and gears	Natural frequencies and stability of linkages, gears, and machine tools	Crack and fracture problems under dynamic loads
9. Electrical machines and electromagnetics	Steady-state analysis of synchronous and induction machines, eddy current, and core losses in electric machines, magnetostatics	—	Transient behavior of electromechanical devices such as motors and actuators, magnetodynamics

Step (ii): Selection of a proper interpolation or displacement model

Since the displacement solution of a complex structure under any specified load conditions cannot be predicted exactly, we assume some suitable solution within an element to approximate the unknown solution. The assumed solution must be simple from a computational standpoint, but it should satisfy certain convergence requirements. In general, the solution or the interpolation model is taken in the form of a polynomial.

Step (iii): Derivation of element stiffness matrices and load vectors

From the assumed displacement model, the stiffness matrix $[K^{(e)}]$ and the load vector $\vec{P}^{(e)}$ of element e are to be derived by using either equilibrium conditions or a suitable variational principle.

Step (iv): Assemblage of element equations to obtain the overall equilibrium equations

Since the structure is composed of several finite elements, the individual element stiffness matrices and load vectors are to be assembled in a suitable manner and the overall equilibrium equations have to be formulated as

$$[\underset{\sim}{K}]\vec{\underset{\sim}{\Phi}} = \vec{\underset{\sim}{P}} \tag{1.15}$$

where $[\underset{\sim}{K}]$ is the assembled stiffness matrix, $\vec{\underset{\sim}{\Phi}}$ is the vector of nodal displacements, and $\vec{\underset{\sim}{P}}$ is the vector of nodal forces for the complete structure.

Step (v): Solution for the unknown nodal displacements

The overall equilibrium equations have to be modified to account for the boundary conditions of the problem. After the incorporation of the boundary conditions, the equilibrium equations can be expressed as

$$[K]\vec{\Phi} = \vec{P} \tag{1.16}$$

For linear problems, the vector $\vec{\phi}$ can be solved very easily. However, for nonlinear problems, the solution has to be obtained in a sequence of steps, with each step involving the modification of the stiffness matrix $[K]$ and/or the load vector $\vec{P}$.

Step (vi): Computation of element strains and stresses

From the known nodal displacements $\vec{\Phi}$, if required, the element strains and stresses can be computed by using the necessary equations of solid or structural mechanics.

The terminology used in the previous six steps has to be modified if we want to extend the concept to other fields. For example, we have to use the term continuum or domain in place of structure, field variable in place of displacement, characteristic matrix in place of stiffness matrix, and element resultants in place of element strains. The application of the six steps of the finite element analysis is illustrated with the help of the following examples.

Example 1.2 (*Stress analysis of a stepped bar*) Find the stresses induced in the axially loaded stepped bar shown in Figure 1.6(a). The bar has cross-sectional areas of $A^{(1)}$ and $A^{(2)}$ over the lengths $l^{(1)}$ and $l^{(2)}$, respectively. Assume the following data: $A^{(1)} = 2\ \text{cm}^2$, $A^{(2)} = 1\ \text{cm}^2$; $l^{(1)} = l^{(2)} = 10$ cm; $E^{(1)} = E^{(2)} = E = 2 \times 10^7\ \text{N/cm}^2$; $P_3 = 1$ N.

Solution

(i) Idealization

Let the bar be considered as an assemblage of two elements as shown in Figure 1.6(b). By assuming the bar to be a one-dimensional structure, we have only axial displacement at any point in the element. As there are three nodes, the axial displacements of the nodes, namely, Φ_1, Φ_2, and Φ_3, will be taken as unknowns.

(ii) Displacement model

In each of the elements, we assume a linear variation of axial displacement ϕ so that (Figure 1.6(c))

$$\phi(x) = a + bx \tag{E$_1$}$$

where a and b are constants. If we consider the end displacements $\Phi_1^{(e)}$ (ϕ at $x = 0$) and $\Phi_2^{(e)}$ (ϕ at $x = l^{(e)}$) as unknowns, we obtain

$$a = \Phi_1^{(e)} \quad \text{and} \quad b = (\Phi_2^{(e)} - \Phi_1^{(e)})/l^{(e)}$$

where the superscript e denotes the element number. Thus,

$$\phi(x) = \Phi_1^{(e)} + (\Phi_2^{(e)} - \Phi_1^{(e)})\frac{x}{l^{(e)}} \tag{E$_2$}$$

(iii) Element stiffness matrix

The element stiffness matrices can be derived from the principle of minimum potential energy. For this, we write the potential energy of the bar (I) under axial deformation as

$$\begin{aligned} I &= \text{strain energy} - \text{work done by external forces} \\ &= \pi^{(1)} + \pi^{(2)} - W_p \end{aligned} \tag{E$_3$}$$

where $\pi^{(e)}$ represents the strain energy of element e, and W_p denotes the work done by external forces. For the element shown in Figure 1.6(c),

$$\pi^{(e)} = A^{(e)} \int_0^{l^{(e)}} \frac{1}{2}\sigma^{(e)} \cdot \varepsilon^{(e)} \cdot \mathrm{d}x = \frac{A^{(e)}E^{(e)}}{2} \int_0^{l^{(e)}} \varepsilon^{(e)^2}\, \mathrm{d}x \tag{E$_4$}$$

where $A^{(e)}$ is the cross-sectional area of element e, $l^{(e)}$ is the length of element $e = L/2$, $\sigma^{(e)}$ is the stress in element e, $\varepsilon^{(e)}$ is the strain in element e, and $E^{(e)}$ is the Young's modulus of element $e = E$. From the expression of $\phi(x)$, we can write

$$\varepsilon^{(e)} = \frac{\partial \phi}{\partial x} = \frac{\Phi_2^{(e)} - \Phi_1^{(e)}}{l^{(e)}} \tag{E$_5$}$$

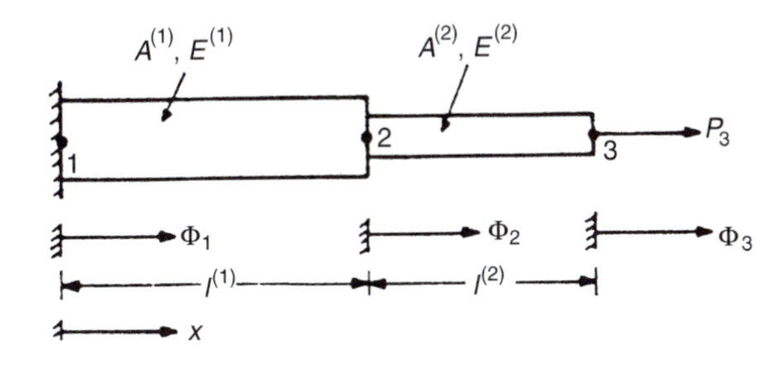

(a) Element characteristics

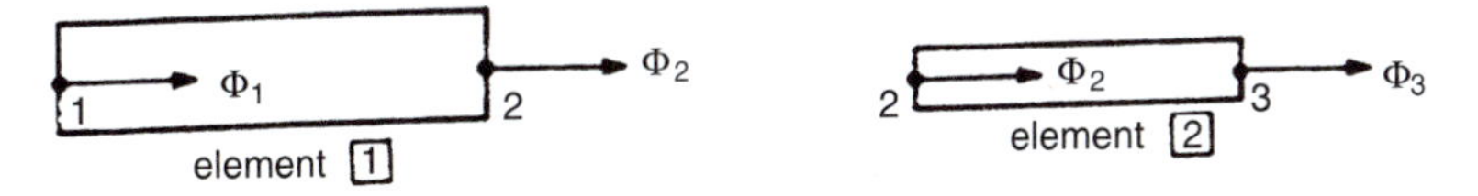

(b) Element degrees of freedom

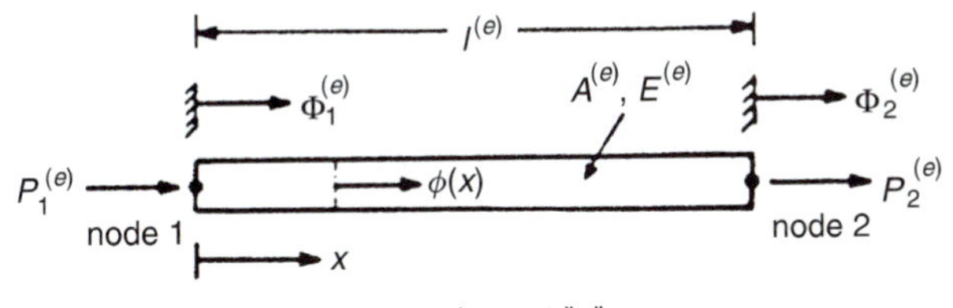

$$\vec{\Phi}^{(e)} = \begin{Bmatrix} \Phi_1 \\ \Phi_2 \end{Bmatrix}^{(e)}, \quad \vec{P}^{(e)} = \begin{Bmatrix} P_1 \\ P_2 \end{Bmatrix}^{(e)}$$

(c) Displacements and loads for element e

Figure 1.6. A Stepped Bar under Axial Load.

and hence

$$\pi^{(e)} = \frac{A^{(e)}E^{(e)}}{2} \int_0^{l^{(e)}} \left\{ \frac{\Phi_2^{(e)^2} + \Phi_1^{(e)^2} - 2\Phi_1^{(e)}\Phi_2^{(e)}}{l^{(e)2}} \right\} \mathrm{d}x$$

$$= \frac{A^{(e)}E^{(e)}}{2l^{(e)}} \left(\Phi_1^{(e)^2} + \Phi_2^{(e)^2} - 2\Phi_1^{(e)}\Phi_2^{(e)} \right) \tag{E$_6$}$$

This expression for $\pi^{(e)}$ can be written in matrix form as

$$\pi^{(e)} = \frac{1}{2}\vec{\Phi}^{(e)^T}[K^{(e)}]\vec{\Phi}^{(e)} \tag{E$_7$}$$

where $\vec{\Phi}^{(e)} = \begin{Bmatrix} \Phi_1^{(e)} \\ \Phi_2^{(e)} \end{Bmatrix}$ is the vector of nodal displacements of element e

$\equiv \begin{Bmatrix} \Phi_1 \\ \Phi_2 \end{Bmatrix}$ for $e = 1$ and $\begin{Bmatrix} \Phi_2 \\ \Phi_3 \end{Bmatrix}$ for $e = 2$, and

$[K^{(e)}] = \dfrac{A^{(e)}E^{(e)}}{l^{(e)}}\begin{bmatrix} 1 & -1 \\ -1 & 1 \end{bmatrix}$ is called the stiffness matrix of element e.

Since there are only concentrated loads acting at the nodes of the bar (and no distributed load acts on the bar), the work done by external forces can be expressed as

$$W_p = \Phi_1 P_1 + \Phi_2 P_2 + \Phi_3 P_3 \equiv \underset{\sim}{\vec{\Phi}}^T \underset{\sim}{\vec{P}_c} \tag{E$_8$}$$

where P_i denotes the force applied in the direction of the displacement Φ_i $(i = 1, 2, 3)$. In this example, P_1 = reaction at fixed node, $P_2 = 0$, and $P_3 = 1.0$.

If external distributed loads act on the elements, the corresponding element load vectors, $\vec{P}_d^{(e)}$, will be generated for each element and the individual load vectors will be assembled to generate the global load vector of the system due to distributed load, $\underset{\sim}{\vec{P}_d}$. This load vector is to be added to the global load vector due to concentrated loads, $\underset{\sim}{\vec{P}_c}$, to generate the total global nodal load vector of the system, $\underset{\sim}{\vec{P}} = \underset{\sim}{\vec{P}_d} + \underset{\sim}{\vec{P}_c}$. In the present example, there are no distributed loads on the element; external load acts only at one node and hence the global load vector $\underset{\sim}{\vec{P}}$ is taken to be same as the vector of concentrated loads acting at the nodes of the system.

If the bar as a whole is in equilibrium under the loads $\underset{\sim}{\vec{P}} = \begin{Bmatrix} P_1 \\ P_2 \\ P_3 \end{Bmatrix}$, the principle of minimum potential energy gives

$$\frac{\partial I}{\partial \Phi_i} = 0, \qquad i = 1, 2, 3 \tag{E$_9$}$$

This equation can be rewritten as

$$\frac{\partial I}{\partial \Phi_i} = \frac{\partial}{\partial \Phi_i}\left(\sum_{e=1}^{2}\pi^{(e)} - W_p\right) = 0, \qquad i = 1, 2, 3 \tag{E$_{10}$}$$

where the summation sign indicates the addition of the strain energies (scalars) of the elements. Equation (E$_{10}$) can be written as

$$\sum_{e=1}^{2}\left([K^{(e)}]\vec{\Phi}^{(e)} - \vec{P}^{(e)}\right) = \vec{0} \tag{E$_{11}$}$$

where the summation sign indicates the assembly of vectors (not the addition of vectors) in which only the elements corresponding to a particular degree of freedom in different vectors are added.

(iv) Assembly of element stiffness matrices and element load vectors

This step includes the assembly of element stiffness matrices $[K^{(e)}]$ and element load vectors $\vec{P}^{(e)}$ to obtain the overall or global equilibrium equations. Equation (E_{11}) can be rewritten as

$$[\underset{\sim}{K}]\underset{\sim}{\vec{\Phi}} - \underset{\sim}{\vec{P}} = \vec{0} \tag{E$_{12}$}$$

where $[\underset{\sim}{K}]$ is the assembled or global stiffness matrix $= \Sigma_{e=1}^{2}[K^{(e)}]$, and $\underset{\sim}{\vec{\Phi}} = \begin{Bmatrix} \Phi_1 \\ \Phi_2 \\ \Phi_3 \end{Bmatrix}$ is the vector of global displacements. For the data given, the element matrices would be

$$[K^{(1)}] = \frac{A^{(1)}E^{(1)}}{l^{(1)}}\begin{bmatrix} 1 & -1 \\ -1 & 1 \end{bmatrix} = 10^6 \begin{array}{c} \begin{array}{cc} \Phi_1 & \Phi_2 \end{array} \\ \begin{bmatrix} 4 & -4 \\ -4 & 4 \end{bmatrix} \end{array} \begin{array}{c} \Phi_1 \\ \Phi_2 \end{array} \tag{E$_{13}$}$$

$$[K^{(2)}] = \frac{A^{(2)}E^{(2)}}{l^{(2)}}\begin{bmatrix} 1 & -1 \\ -1 & 1 \end{bmatrix} = 10^6 \begin{array}{c} \begin{array}{cc} \Phi_2 & \Phi_3 \end{array} \\ \begin{bmatrix} 2 & -2 \\ -2 & 2 \end{bmatrix} \end{array} \begin{array}{c} \Phi_2 \\ \Phi_3 \end{array} \tag{E$_{14}$}$$

Since the displacements of the left and right nodes of the first element are Φ_1 and Φ_2, the rows and columns of the stiffness matrix corresponding to these unknowns are identified as indicated in Eq. (E_{13}). Similarly, the rows and columns of the stiffness matrix of the second element corresponding to its nodal unknowns Φ_2 and Φ_3 are also identified as indicated in Eq. (E_{14}).

The overall stiffness matrix of the bar can be obtained by assembling the two element stiffness matrices. Since there are three nodal displacement unknowns (Φ_1, Φ_2, and Φ_3), the global stiffness matrix, $[\underset{\sim}{K}]$, will be of order three. To obtain $[\underset{\sim}{K}]$, the elements of $[K^{(1)}]$ and $[K^{(2)}]$ corresponding to the unknowns Φ_1, Φ_2, and Φ_3 are added as shown below:

$$[\underset{\sim}{K}] = 10^6 \begin{array}{c} \begin{array}{ccc} \Phi_1 & \Phi_2 & \Phi_3 \end{array} \\ \begin{bmatrix} 4 & -4 & 0 \\ -4 & 4+2 & -2 \\ 0 & -2 & 2 \end{bmatrix} \end{array} \begin{array}{c} \Phi_1 \\ \Phi_2 \\ \Phi_3 \end{array}$$

$$= 2 \times 10^6 \begin{bmatrix} 2 & -2 & 0 \\ -2 & 3 & -1 \\ 0 & -1 & 1 \end{bmatrix} \tag{E$_{15}$}$$

The overall or global load vector can be written as

$$\underset{\sim}{\vec{P}} = \begin{Bmatrix} P_1 \\ P_2 \\ P_3 \end{Bmatrix} = \begin{Bmatrix} P_1 \\ 0 \\ 1 \end{Bmatrix}$$

where P_1 denotes the reaction at node 1. Thus, the overall equilibrium equations, Eqs. (E_{12}), become

$$2 \times 10^6 \begin{bmatrix} 2 & -2 & 0 \\ -2 & 3 & -1 \\ 0 & -1 & 1 \end{bmatrix} \begin{Bmatrix} \Phi_1 \\ \Phi_2 \\ \Phi_3 \end{Bmatrix} = \begin{Bmatrix} P_1 \\ 0 \\ 1 \end{Bmatrix} \tag{E$_{16}$}$$

Note that a systematic step-by-step finite element procedure has been used to derive Eq. (E_{16}). If a step-by-step procedure is not followed, Eq. (E_{16}) can be derived in a much simpler way, in this example, as follows:

The potential energy of the stepped bar, Eq. (E_3), can be expressed using Eqs. (E_6) and (E_8) as

$$\begin{aligned} I &= \pi^{(1)} + \pi^{(2)} - W_p \\ &= \frac{1}{2}\frac{A^{(1)}E^{(1)}}{l^{(1)}}\left(\Phi_1^2 + \Phi_2^2 - 2\Phi_1\Phi_2\right) + \frac{1}{2}\frac{A^{(2)}E^{(2)}}{l^{(2)}}\left(\Phi_2^2 + \Phi_3^2 - 2\Phi_2\Phi_3\right) \\ &\quad - P_1\Phi_1 - P_2\Phi_2 - P_3\Phi_3 \end{aligned} \tag{E$_{17}$}$$

Equations (E_9) and (E_{17}) yield

$$\frac{\partial I}{\partial \Phi_1} = \frac{A^{(1)}E^{(1)}}{l^{(1)}}(\Phi_1 - \Phi_2) - P_1 = 0 \tag{E$_{18}$}$$

$$\frac{\partial I}{\partial \Phi_2} = \frac{A^{(1)}E^{(1)}}{l^{(1)}}(\Phi_2 - \Phi_1) + \frac{A^{(2)}E^{(2)}}{l^{(2)}}(\Phi_2 - \Phi_3) - P_2 = 0 \tag{E$_{19}$}$$

$$\frac{\partial I}{\partial \Phi_3} = \frac{A^{(2)}E^{(2)}}{l^{(2)}}(\Phi_3 - \Phi_2) - P_3 = 0 \tag{E$_{20}$}$$

For the given data, Eqs. (E_{18})–(E_{20}) can be seen to reduce to Eq. (E_{16}).

(v) Solution for displacements

If we try to solve Eq. (E_{16}) for the unknowns Φ_1, Φ_2, and Φ_3, we will not be able to do it since the matrix

$$[\underset{\sim}{K}] = 2 \times 10^6 \begin{bmatrix} 2 & -2 & 0 \\ -2 & 3 & -1 \\ 0 & -1 & 1 \end{bmatrix}$$

is singular. This is because we have not incorporated the known geometric boundary condition, namely $\Phi_1 = 0$. We can incorporate this by setting $\Phi_1 = 0$ or by deleting the row and column corresponding to Φ_1 in Eq. (E_{16}). The final equilibrium equations can be written as

$$[K]\vec{\Phi} = \vec{P}$$

or

$$2 \times 10^6 \begin{bmatrix} 3 & -1 \\ -1 & 1 \end{bmatrix} \begin{Bmatrix} \Phi_2 \\ \Phi_3 \end{Bmatrix} = \begin{Bmatrix} 0 \\ 1 \end{Bmatrix} \tag{E$_{21}$}$$

The solution of Eq. (E_{21}) gives

$$\Phi_2 = 0.25 \times 10^{-6} \text{ cm} \quad \text{and} \quad \Phi_3 = 0.75 \times 10^{-6} \text{ cm}$$

(vi) Element strains and stresses

Once the displacements are computed, the strains in the elements can be found as

$$\varepsilon^{(1)} = \frac{\partial \phi}{\partial x} \text{ for element } 1 = \frac{\Phi_2^{(1)} - \Phi_1^{(1)}}{l^{(1)}} \equiv \frac{\Phi_2 - \Phi_1}{l^{(1)}} = 0.25 \times 10^{-7}$$

and $$\varepsilon^{(2)} = \frac{\partial \phi}{\partial x} \text{ for element } 2 = \frac{\Phi_2^{(2)} - \Phi_1^{(2)}}{l^{(2)}} \equiv \frac{\Phi_3 - \Phi_2}{l^{(2)}} = 0.50 \times 10^{-7}$$

The stresses in the elements are given by

$$\sigma^{(1)} = E^{(1)}\varepsilon^{(1)} = (2 \times 10^7)(0.25 \times 10^{-7}) = 0.5 \text{ N/cm}^2$$

and $$\sigma^{(2)} = E^{(2)}\varepsilon^{(2)} = (2 \times 10^7)(0.50 \times 10^{-7}) = 1.0 \text{ N/cm}^2$$

Example 1.3 (*Temperature distribution in a fin*) Find the distribution of temperature in the one-dimensional fin shown in Figure 1.7(a).

The differential equation governing the steady-state temperature distribution $T(x)$ along a uniform fin is given by [1.7]

$$kA\frac{d^2T}{dx^2} - hp(T - T_\infty) = 0$$

or $$\frac{d^2T}{dx^2} - \frac{hp}{kA}(T - T_\infty) = 0 \qquad (E_1)$$

with the boundary condition $T(x = 0) = T_0$

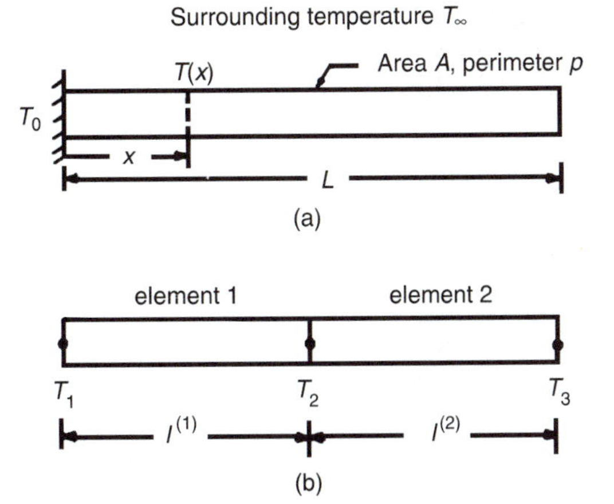

Figure 1.7. A One-Dimensional Fin.

where h is the convection heat transfer coefficient, p is the perimeter, k is the thermal conductivity, A is the cross-sectional area, T_∞ is the surrounding temperature, and T_o is the temperature at the root of the fin. The derivation of Eq. (E_1) is similar to that of Eq. (1.4) except that convection term is also included in the derivation of Eq. (E_1) along with the assumption of $\dot{q} = \partial T/\partial t = 0$. The problem stated in Eq. (E_1) is equivalent to [1.11]

$$\left.\begin{aligned} &\text{Minimize} \quad I = \frac{1}{2}\int_{x=0}^{L}\left[\left(\frac{dT}{dx}\right)^2 + \frac{hp}{kA}(T^2 - 2T\,T_\infty)\right]dx \\ &\text{with the boundary condition } T(x=0) = T_0. \end{aligned}\right\} \quad (E_2)$$

Assume the following data: $h = 10$ W/cm^2–°C, $k = 70$ W/cm–°C, $T_\infty = 40$°C, $T_0 =$ 140°C, and $L = 5$ cm, and the cross section of fin is circular with a radius of 1 cm.

Solution

Note: Since the present problem is a heat transfer problem, the terms used in the case of solid mechanics problems, such as solid body, displacement, strain, stiffness matrix, load vector, and equilibrium equations, have to be replaced by terms such as body, temperature, gradient of temperature, characteristic matrix, characteristic vector, and governing equations, respectively.

(i) Idealization

Let the fin be idealized into two finite elements as shown in Figure 1.7(b). If the temperatures of the nodes are taken as the unknowns, there will be three nodal temperature unknowns, namely T_1, T_2, and T_3, in the problem.

(ii) Interpolation (temperature distribution) model

In each element e ($e = 1, 2$), the temperature (T) is assumed to vary linearly as

$$T(x) = a + bx \quad (E_3)$$

where a and b are constants. If the nodal temperatures $T_1^{(e)}$ (T at $x = 0$) and $T_2^{(e)}$ (T at $x = l^{(e)}$) of element e are taken as unknowns, the constants a and b can be expressed as $a = T_1^{(e)}$ and $b = \left(T_2^{(e)} - T_1^{(e)}\right)/l^{(e)}$, where $l^{(e)}$ is the length of element e. Thus,

$$T(x) = T_1^{(e)} + \left(T_2^{(e)} - T_1^{(e)}\right)\frac{x}{l^{(e)}} \quad (E_4)$$

(iii) Element characteristic matrices and vectors

The element characteristic matrices and vectors can be identified by expressing the functional I in matrix form. When the integral in I is evaluated over the length of element e, we obtain

$$I^{(e)} = \frac{1}{2}\int_{x=0}^{l^{(e)}}\left[\left(\frac{dT}{dx}\right)^2 + \frac{hp}{kA}(T^2 - 2T_\infty T)\right]dx \quad (E_5)$$

Substitution of Eq. (E_4) into (E_5) leads to

$$I^{(e)} = \frac{1}{2} \int_{x=0}^{l^{(e)}} \left[\left(\frac{T_2^{(e)} - T_1^{(e)}}{l^{(e)}} \right)^2 + \frac{hp}{kA} \left\{ T_1^{(e)} + \left(T_2^{(e)} - T_1^{(e)} \right) \frac{x}{l^{(e)}} \right\}^2 - \frac{2hpT_\infty}{kA} \left\{ T_1^{(e)} + \left(T_2^{(e)} - T_1^{(e)} \right) \frac{x}{l^{(e)}} \right\} \right] dx \tag{E$_6$}$$

Equation (E_6) can be expressed, after evaluating the integral, in matrix notation as

$$I^{(e)} = \frac{1}{2} \vec{T}^{(e)^T} [K^{(e)}] \vec{T}^{(e)} - \vec{T}^{(e)^T} \vec{P}^{(e)} \tag{E$_7$}$$

where $\vec{T}^{(e)} = \begin{Bmatrix} T_1^{(e)} \\ T_2^{(e)} \end{Bmatrix}$ is the vector of nodal temperatures of

element $e = \begin{Bmatrix} T_1 \\ T_2 \end{Bmatrix}$ for $e = 1$ and $\begin{Bmatrix} T_2 \\ T_3 \end{Bmatrix}$ for $e = 2$,

$[K^{(e)}]$ is the characteristic matrix of element e

$$= \frac{1}{l^{(e)}} \begin{bmatrix} 1 & -1 \\ -1 & 1 \end{bmatrix} + \frac{hpl^{(e)}}{6kA} \begin{bmatrix} 2 & 1 \\ 1 & 2 \end{bmatrix} \tag{E$_8$}$$

and $\vec{P}^{(e)} = \begin{Bmatrix} P_1^{(e)} \\ P_2^{(e)} \end{Bmatrix}$ is the characteristic vector of element e

$= \begin{Bmatrix} P_1 \\ P_2 \end{Bmatrix}$ for $e = 1$ and $\begin{Bmatrix} P_2 \\ P_3 \end{Bmatrix}$ for $e = 2$

$$= \frac{hpT_\infty l^{(e)}}{2kA} \begin{Bmatrix} 1 \\ 1 \end{Bmatrix} \tag{E$_9$}$$

(iv) Assembly of element matrices and vectors and derivation of governing equations

As stated in Eq. (E_2), the nodal temperatures can be determined by minimizing the functional I. The conditions for the minimum of I are given by

$$\frac{\partial I}{\partial T_i} = \frac{\partial \left(\sum_{e=1}^{2} I^{(e)} \right)}{\partial T_i} = \sum_{e=1}^{2} \frac{\partial I^{(e)}}{\partial T_i} = 0, \qquad i = 1, 2, 3 \tag{E$_{10}$}$$

where I has been replaced by the sum of elemental contributions, $I^{(e)}$. Equation (E_{10}) can also be stated as

$$\sum_{e=1}^{2} \frac{\partial I^{(e)}}{\partial \vec{T}^{(e)}} = \sum_{e=1}^{2} \left(\left[K^{(e)} \right] \vec{T}^{(e)} - \vec{P}^{(e)} \right) = [\underset{\sim}{K}] \underset{\sim}{\vec{T}} - \underset{\sim}{\vec{P}} = \vec{0} \tag{E$_{11}$}$$

where $[\underset{\sim}{K}] = \Sigma_{e=1}^{2}[K^{(e)}]$ is the assembled characteristic matrix, $\underset{\sim}{\vec{P}} = \Sigma_{e=1}^{2}\vec{P}^{(e)}$ is the assembled characteristic vector, and $\underset{\sim}{\vec{T}}$ is the assembled or overall nodal temperature vector $= \begin{Bmatrix} T_1 \\ T_2 \\ T_3 \end{Bmatrix}$. Equation ($E_{11}$) gives the governing matrix equations as

$$[\underset{\sim}{K}]\underset{\sim}{\vec{T}} = \underset{\sim}{\vec{P}} \tag{E_{12}}$$

From the given data we can obtain

$$[K^{(1)}] = \frac{1}{2.5}\begin{bmatrix} 1 & -1 \\ -1 & 1 \end{bmatrix} + \frac{10 \times 2\pi \times 2.5}{6 \times 70 \times \pi}\begin{bmatrix} 2 & 1 \\ 1 & 2 \end{bmatrix}$$

$$= \begin{array}{c} \begin{array}{cc} T_1 & \quad T_2 \end{array} \\ \begin{bmatrix} 0.6382 & -0.2809 \\ -0.2809 & 0.6382 \end{bmatrix} \end{array} \begin{array}{c} \\ T_1 \\ T_2 \end{array} \tag{E_{13}}$$

$$[K^{(2)}] = \begin{array}{c} \begin{array}{cc} T_2 & \quad T_3 \end{array} \\ \begin{bmatrix} 0.6382 & -0.2809 \\ -0.2809 & 0.6382 \end{bmatrix} \end{array} \begin{array}{c} \\ T_2 \\ T_3 \end{array} \tag{E_{14}}$$

$$\vec{P}^{(1)} = \frac{10 \times 2\pi \times 40 \times 2.5}{2 \times 70 \times \pi}\begin{Bmatrix} 1 \\ 1 \end{Bmatrix} = 14.29\begin{Bmatrix} 1 \\ 1 \end{Bmatrix}\begin{matrix} T_1 \\ T_2 \end{matrix} \tag{E_{15}}$$

$$\vec{P}^{(2)} = 14.29\begin{Bmatrix} 1 \\ 1 \end{Bmatrix}\begin{matrix} T_2 \\ T_3 \end{matrix} \tag{E_{16}}$$

where the nodal unknowns associated with each row and column of the element matrices and vectors were also indicated in Eqs. (E_{13})–(E_{16}). The overall characteristic matrix of the fin can be obtained by adding the elements of $[K^{(1)}]$ and $[K^{(2)}]$ corresponding to the unknowns T_1, T_2, and T_3:

$$[\underset{\sim}{K}] = \begin{array}{c} \begin{array}{ccc} T_1 & \qquad T_2 & \qquad T_3 \end{array} \\ \begin{bmatrix} 0.6382 & -0.2809 & 0 \\ -0.2809 & (0.6382 + 0.6382) & -0.2809 \\ 0 & -0.2809 & 0.6382 \end{bmatrix} \end{array} \begin{array}{c} \\ T_1 \\ T_2 \\ T_3 \end{array} \tag{E_{17}}$$

Similarly, the overall characteristic vector of the fin can be obtained as

$$\underset{\sim}{\vec{P}} = \begin{Bmatrix} 14.29 \\ (14.29 + 14.29) \\ 14.29 \end{Bmatrix}\begin{matrix} T_1 \\ T_2 \\ T_3 \end{matrix} \tag{E_{18}}$$

Thus, the governing finite element equation of the fin, Eq. (E_{12}), becomes

$$\begin{bmatrix} 0.6382 & -0.2809 & 0 \\ -0.2809 & 1.2764 & -0.2809 \\ 0 & -0.2809 & 0.6382 \end{bmatrix}\begin{Bmatrix} T_1 \\ T_2 \\ T_3 \end{Bmatrix} = \begin{Bmatrix} 14.29 \\ 28.58 \\ 14.29 \end{Bmatrix} \tag{E_{19}}$$

(v) Solution for nodal temperatures

Equation (E_{19}) has to be solved after applying the boundary condition, namely, T (at node 1) $= T_1 = T_0 = 140\,°C$. For this, the first equation of (E_{19}) is replaced by $T_1 = T_0 = 140$ and the remaining two equations are written in scalar form as

$$-0.2809T_1 + 1.2764T_2 - 0.2809T_3 = 28.58$$
$$-0.2809T_2 + 0.6382T_3 = 14.29$$

or

$$\left.\begin{aligned} 1.2764T_2 - 0.2809T_3 &= 28.58 + 0.2809 \times 140 = 67.906 \\ -0.2809T_2 + 0.6382T_3 &= 14.29 \end{aligned}\right\} \qquad (E_{20})$$

The solution of Eq. (E_{20}) gives the nodal temperatures as

$$T_2 = 64.39\,°C \quad \text{and} \quad T_3 = 50.76\,°C.$$

Example 1.4 (*Inviscid fluid flow in a tube*) Find the velocity distribution of an inviscid fluid flowing through the tube shown in Figure 1.8(a). The differential equation governing the velocity distribution $u(x)$ is given by Eq. (1. 12) with the boundary condition $u(x = 0) = u_0$. This problem is equivalent to

$$\left.\begin{aligned} &\text{Minimize} \quad I = \frac{1}{2}\int_{x=0}^{L} \rho A \left(\frac{d\phi}{dx}\right)^2 \cdot dx \\ &\text{with the boundary condition } u(x = 0) = u_0 \end{aligned}\right\} \qquad (E_1)$$

Assume the area of cross section of the tube as $A(x) = A_0 \cdot e^{-(x/L)}$.

Solution

Note: In this case the terminology of solid mechanics, such as solid body, displacement, stiffness matrix, load vector, and equilibrium equations, has to be replaced by the terms continuum, potential function, characteristic matrix, characteristic vector, and governing equations.

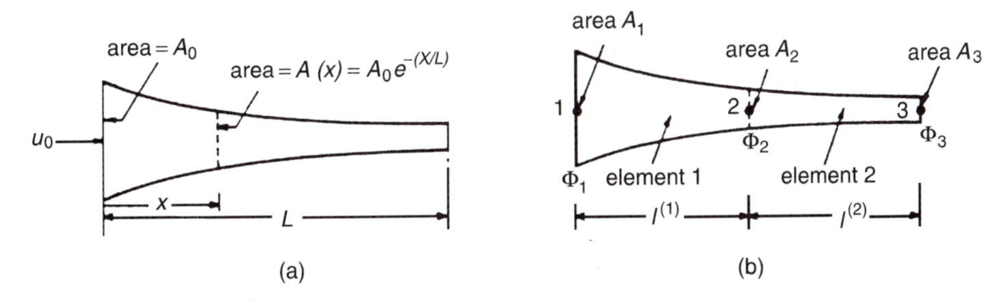

Figure 1.8. A One-Dimensional Tube of Varying Cross Section.

(i) Idealization

Divide the continuum into two finite elements as shown in Figure 1.8(b). If the values of the potential function at the various nodes are taken as the unknowns, there will be three quantities, namely Φ_1, Φ_2, and Φ_3, to be determined in the problem.

(ii) Interpolation (potential function) model

The potential function, $\phi(x)$, is assumed to vary linearly within an element e $(e = 1, 2)$ as

$$\phi(x) = a + bx \tag{E$_2$}$$

where the constants a and b can be evaluated using the nodal conditions $\phi(x = 0) = \Phi_1^{(e)}$ and $\phi(x = l^{(e)}) = \Phi_2^{(e)}$ to obtain

$$\phi(x) = \Phi_1^{(e)} + \left(\Phi_2^{(e)} - \Phi_1^{(e)}\right) \frac{x}{l^{(e)}} \tag{E$_3$}$$

where $l^{(e)}$ is the length of element e.

(iii) Element characteristic matrices

The functional I corresponding to element e can be expressed as

$$\begin{aligned} I^{(e)} &= \frac{1}{2} \int\limits_{x=0}^{l^{(e)}} \rho A \left(\frac{\mathrm{d}\phi}{\mathrm{d}x}\right)^2 \mathrm{d}x = \frac{1}{2} \int\limits_{x=0}^{l^{(e)}} \rho A \left(\frac{\Phi_2^{(e)} - \Phi_1^{(e)}}{l^{(e)}}\right)^2 \mathrm{d}x \\ &= \frac{1}{2} \vec{\Phi}^{(e)^T} [K^{(e)}] \vec{\Phi}^{(e)} \end{aligned} \tag{E$_4$}$$

where

$[K^{(e)}]$ is the characteristic matrix of element e

$$= \frac{\rho A^{(e)}}{l^{(e)}} \begin{bmatrix} 1 & -1 \\ -1 & 1 \end{bmatrix}, \tag{E$_5$}$$

$A^{(e)}$ is the cross-sectional area of element e (which can be taken as $(A_1 + A_2)/2$ for $e = 1$ and $(A_2 + A_3)/2$ for $e = 2$ for simplicity),

and $\vec{\Phi}^{(e)}$ is the vector of nodal unknowns of element e

$$= \begin{Bmatrix} \vec{\Phi}_1^{(e)} \\ \vec{\Phi}_2^{(e)} \end{Bmatrix} = \begin{Bmatrix} \Phi_1 \\ \Phi_2 \end{Bmatrix} \text{ for } e = 1 \quad \text{and} \quad \begin{Bmatrix} \Phi_2 \\ \Phi_3 \end{Bmatrix} \text{ for } e = 2.$$

(iv) Governing equations

The overall equations can be written as

$$\begin{bmatrix} \frac{\rho A^{(1)}}{l^{(1)}} & -\frac{\rho A^{(1)}}{l^{(1)}} & 0 \\ -\frac{\rho A^{(1)}}{l^{(1)}} & \left(\frac{\rho A^{(1)}}{l^{(1)}} + \frac{\rho A^{(2)}}{l^{(2)}}\right) & -\frac{\rho A^{(2)}}{l^{(2)}} \\ 0 & -\frac{\rho A^{(2)}}{l^{(2)}} & \frac{\rho A^{(2)}}{l^{(2)}} \end{bmatrix} \begin{Bmatrix} \Phi_1 \\ \Phi_2 \\ \Phi_3 \end{Bmatrix} = \begin{Bmatrix} Q_1 = -\rho A_1 u_0 \\ Q_2 = 0 \\ Q_3 = \rho A_3 u_3 \end{Bmatrix} \tag{E$_6$}$$

where Q_i is the mass flow rate across section i ($i = 1, 2, 3$) and is nonzero when fluid is either added to or subtracted from the tube with $Q_1 = -\rho A_1 u_1$ (negative since u_1 is opposite to the outward normal to section 1), $Q_2 = 0$, and $Q_3 = \rho A_3 u_3$. Since $u_1 = u_0$ is given, Q_1 is known, whereas Q_3 is unknown.

(v) Solution of governing equations

In the third equation of (E_6), both Φ_3 and Q_3 are unknowns and thus the given system of equations cannot be solved. Hence, we set $\Phi_3 = 0$ as a reference value and try to find the values of Φ_1 and Φ_2 with respect to this value. The first two equations of (E_6) can be expressed in scalar form as

$$\frac{\rho A^{(1)}}{l^{(1)}}\Phi_1 - \frac{\rho A^{(1)}}{l^{(1)}}\Phi_2 = Q_1 = -\rho A_1 u_0 \tag{E$_7$}$$

and

$$-\frac{\rho A^{(1)}}{l^{(1)}}\Phi_1 + \left(\frac{\rho A^{(1)}}{l^{(1)}} + \frac{\rho A^{(2)}}{l^{(2)}}\right)\Phi_2 = \frac{\rho A^{(2)}}{l^{(2)}}\Phi_3 = 0 \tag{E$_8$}$$

By substituting $A^{(1)} \simeq (A_1 + A_2)/2 = 0.8032A_0$, $A^{(2)} \simeq (A_2 + A_3)/2 = 0.4872A_0$, and $l^{(1)} = l^{(2)} = L/2$, Eqs. (E_7) and (E_8) can be written as

$$0.8032\Phi_1 - 0.8032\Phi_2 = -u_0 L/2 \tag{E$_9$}$$

and

$$-0.8032\Phi_1 + 1.2904\Phi_2 = 0 \tag{E$_{10}$}$$

The solution of Eqs. (E_9) and (E_{10}) is given by

$$\Phi_1 = -1.650\ u_0 L \quad \text{and} \quad \Phi_2 = -1.027\ u_0 L$$

(vi) Computation of velocities of the fluid

The velocities of the fluid in elements 1 and 2 can be found as

$$\begin{aligned} u \text{ in element } 1 = u^{(1)} &= \frac{\mathrm{d}\phi}{\mathrm{d}x} \text{ (element 1)} \\ &= \frac{\Phi_2 - \Phi_1}{l^{(1)}} = 1.246u_0 \end{aligned}$$

and
$$u \text{ in element } 2 = u^{(2)} = \frac{\mathrm{d}\phi}{\mathrm{d}x} \text{ (element 2)}$$
$$= \frac{\Phi_3 - \Phi_2}{l^{(2)}} = 2.054u_0$$

These velocities will be constant along the elements in view of the linear relationship assumed for $\phi(x)$ within each element. The velocity of the fluid at node 2 can be approximated as

$$u_2 = (u^{(1)} + u^{(2)})/2 = 1.660u_0.$$

The third equation of (E_6) can be written as

$$-\frac{\rho A^{(2)}}{l^{(2)}}\Phi_2 + \frac{\rho A^{(2)}}{l^{(2)}}\Phi_3 = Q_3$$

or
$$\frac{\rho(0.4872A_0)}{(L/2)}(-\Phi_2 + \Phi_3) = Q_3$$

or
$$Q_3 = \rho A_0 u_0.$$

This shows that the mass flow rate is the same at nodes 1 and 3, which proves the principle of conservation of mass.

1.6 COMPARISON OF FINITE ELEMENT METHOD WITH OTHER METHODS OF ANALYSIS[†]

The common analysis methods available for the solution of a general field problem (e.g., elasticity, fluid flow, and heat transfer problems) can be classified as follows:

Methods of analysis (solution of differential equations)

- Analytical methods
  - Exact methods (e.g., separation of variables and Laplace transformation methods)
  - Approximate methods (e.g., Rayleigh–Ritz and Galerkin methods)
- Numerical methods
  - Numerical solution of differential equations
    - Numerical integration
    - Finite differences
  - Finite or discrete element method

The finite element method will be compared with some of the other analysis methods in this section by considering the beam vibration problem as an example.

[†]This section may be omitted without loss of continuity in the text material.

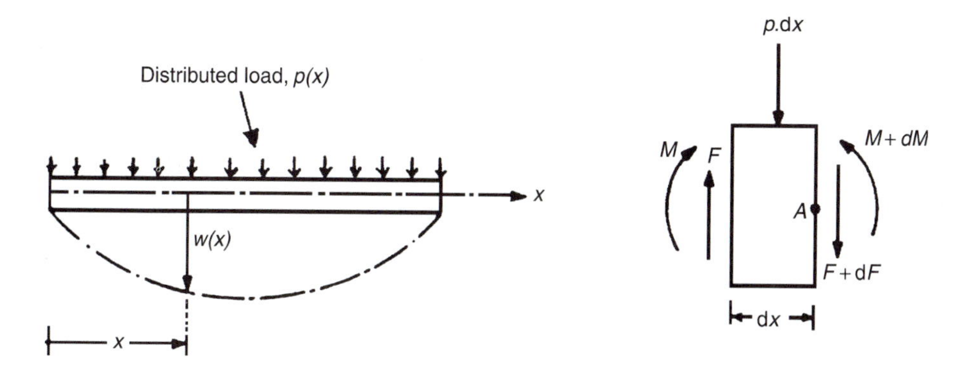

Figure 1.9. Free Body Diagram of an Element of Beam.

1.6.1 Derivation of the Equation of Motion for the Vibration of a Beam [1.9]

By considering the dynamic equilibrium of an element of the beam shown in Figure 1.9, we have

$$F - (F + \mathrm{d}F) - p\,\mathrm{d}x = 0 \text{ (vertical force equilibrium)}$$

or

$$-\frac{\mathrm{d}F}{\mathrm{d}x} = p \tag{1.17}$$

where F and M denote the shear force and bending moment at a distance x, $\mathrm{d}F$ and $\mathrm{d}M$ indicate their increments over an elemental distance $\mathrm{d}x$, p represents the external force acting on the beam per unit length, and

$$M - (M + \mathrm{d}M) + F\,\mathrm{d}x - p\,\mathrm{d}x\frac{\mathrm{d}x}{2} = 0 \text{ (moment equilibrium about point } A)$$

which can be written, after neglecting the term involving $(\mathrm{d}x)^2$, as

$$\frac{\mathrm{d}M}{\mathrm{d}x} = F \tag{1.18}$$

Combining Eqs. (1.17) and (1.18) we obtain

$$-\frac{\mathrm{d}^2 M}{\mathrm{d}x^2} = p(x) \tag{1.19}$$

The curvature of the deflected center line of the beam, $w(x)$, is given by

$$\frac{1}{R} = -\frac{(\mathrm{d}^2 w/\mathrm{d}x^2)}{[1 + (\mathrm{d}w/\mathrm{d}x)^2]^{3/2}} \tag{1.20}$$

For small deflections, Eq. (1.20) can be approximated as

$$\frac{1}{R} \simeq -\frac{\mathrm{d}^2 w}{\mathrm{d}x^2} \tag{1.21}$$

From strength of materials, we know the relation

$$\frac{1}{R} = \frac{M}{E \cdot I(x)} \tag{1.22}$$

where R is the radius of curvature, I is the moment of inertia of the cross section of the beam, and E is the Young's modulus of material. By combining Eqs. (1.19), (1.21), and (1.22), we have

$$M(x) = -\frac{\mathrm{d}^2 w}{\mathrm{d}x^2} EI(x) \tag{1.23}$$

and

$$p(x) = \frac{\mathrm{d}^2}{\mathrm{d}x^2}\left[EI(x)\frac{\mathrm{d}^2 w}{\mathrm{d}x^2}\right] \tag{1.24}$$

According to D'Alembert's rule, for free vibrations, we have

$$p(x) = \text{inertia force} = -m\frac{\mathrm{d}^2 w}{\mathrm{d}t^2} \tag{1.25}$$

where m is the mass of beam per unit length. If the cross section of the beam is constant throughout its length, we have the final beam vibration equation

$$EI\frac{\partial^4 w}{\partial x^4} + m\frac{\partial^2 w}{\partial t^2} = 0 \tag{1.26}$$

Equation (1.26) has to be solved for any given beam by satisfying the associated boundary conditions. For example, if we consider a fixed–fixed beam, the boundary conditions to be satisfied are

$$\left.\begin{aligned} w &= 0 \\ \frac{\partial w}{\partial x} &= 0 \end{aligned}\right\} \text{ at } x = 0 \quad \text{and} \quad x = L \tag{1.27}$$

where L is the length of the beam.

1.6.2 Exact Analytical Solution (Separation of Variables Technique)

For free vibrations, we assume harmonic motion and hence

$$w(x,t) = W(x) \cdot e^{i\omega t} \tag{1.28}$$

where $W(x)$ is purely a function of x, and ω is the circular natural frequency of vibration. Substituting Eq. (1.28) into Eq. (1.26) we get

$$\frac{d^4W}{dx^4} - \lambda W = 0 \tag{1.29}$$

where

$$\lambda = \frac{m\omega^2}{EI} \equiv \beta^4 \tag{1.30}$$

The general solution of Eq. (1.29) can be written as

$$W(x) = C_1 \sin\beta x + C_2 \cos\beta x + C_3 \sinh\beta x + C_4 \cosh\beta x \tag{1.31}$$

where C_1–C_4 are constants to be determined from the boundary conditions. In view of Eq. (1.28), the boundary conditions for a fixed–fixed beam can be written as

$$\left.\begin{aligned} W(x=0) = W(x=L) = 0 \\ \frac{dW}{dx}(x=0) = \frac{dW}{dx}(x=L) = 0 \end{aligned}\right\} \tag{1.32}$$

If we substitute Eq. (1.31) into Eqs. (1.32), we get four linear homogeneous equations in the unknowns C_1–C_4. For a nontrivial solution, the determinant of the coefficient matrix must be zero. This gives the condition [1.9]

$$\cos\beta L \cosh\beta L = 1 \tag{1.33}$$

Equation (1.33) is called the frequency equation and there will be infinite number of solutions to it. Let us call the nth solution of Eq. (1.33) as $\beta_n L$. If we use this solution in Eqs. (1.30) and (1.31) we obtain the natural frequencies as

$$\omega_n^2 = \frac{\beta_n^4 EI}{m} \tag{1.34}$$

where $\beta_1 L = 4.73$, $\beta_2 L = 7.85$, $\beta_3 L = 11.00$, and so on, and mode shapes as

$$W_n(x) = B_n \left\{ \cosh\beta_n x - \cos\beta_n x - \left(\frac{\cos\beta_n L - \cosh\beta_n L}{\sin\beta_n L - \sinh\beta_n L} \right) (\sinh\beta_n x - \sin\beta_n x) \right\} \tag{1.35}$$

where B_n is a constant. Finally, the solution of w is given by

$$w_n(x,t) = W_n(x) \cdot e^{i\omega_n t}, \qquad n = 1, 2, \ldots \tag{1.36}$$

1.6.3 Approximate Analytical Solution (Rayleigh's Method)

To obtain an approximate solution to the first natural frequency and the mode shape, the Rayleigh's method can be used. In this method we equate the maximum kinetic energy during motion to the maximum potential energy. For a beam, the potential energy π is given by [1.9]

$$\pi = \frac{1}{2}\int_{o}^{L} EI\left(\frac{\partial^2 w}{\partial x^2}\right)^2 \mathrm{d}x \tag{1.37}$$

and the kinetic energy T by

$$T = \frac{1}{2}\int_{o}^{L} m(x)\left(\frac{\partial w}{\partial t}\right)^2 \mathrm{d}x \tag{1.38}$$

By assuming harmonic variation of $w(x,t)$ as

$$w(x,t) = W(x)\cos\omega t, \tag{1.39}$$

the maximum values of π and T can be found as

$$(\pi)_{\max} = \frac{1}{2}\int_{o}^{L} EI\left[\frac{\mathrm{d}^2 W}{\mathrm{d}x^2}\right]^2 \mathrm{d}x \tag{1.40}$$

and

$$(T)_{\max} = \frac{\omega^2}{2}\int_{o}^{L} mW^2\,\mathrm{d}x \tag{1.41}$$

By equation $(\pi)_{\max}$ to $(T)_{\max}$, we obtain

$$\omega^2 = \frac{\int_{o}^{L} EI(\mathrm{d}^2W/\mathrm{d}x^2)^2\,\mathrm{d}x}{\int_{o}^{L} m(x)W^2\,\mathrm{d}x} \tag{1.42}$$

To find the value of ω^2 from Eq. (1.42), we assume a certain deflection or mode shape $W(x)$ known as admissible function that satisfies the geometric boundary conditions but not necessarily the governing equilibrium equation, Eq. (1.29), and substitute it in Eq. (1.42). Let us take

$$W(x) = \left(1 - \cos\frac{2\pi x}{L}\right) \tag{1.43}$$

This satisfies the boundary conditions stated in Eq. (1.32) and not the equation of motion, Eq. (1.29). By substituting Eq. (1.43) into Eq. (1.42), we obtain the approximate value of the first natural frequency ($\tilde{\omega}_1$) as

$$\tilde{\omega}_1 = \frac{22.792}{L^2}\sqrt{\frac{EI}{m}} \tag{1.44}$$

which can be seen to be 1.87% greater than the exact value of

$$\omega_1 = \frac{22.3729}{L^2}\sqrt{\frac{EI}{m}}$$

1.6.4 Approximate Analytical Solution (Rayleigh–Ritz Method)

If we want many natural frequencies, we have to substitute a solution, made up of a series of admissible functions that satisfy the forced boundary conditions, in Eq. (1.42). For example, if we want n frequencies, we take

$$W(x) = C_1 f_1(x) + C_2 f_2(x) + \cdots + C_n f_n(x) \tag{1.45}$$

where $C_1, C_2, \ldots, C_n$ are constants and $f_1, f_2, \ldots, f_n$ are admissible functions.

If we substitute Eq. (1.45) into Eq. (1.42), we obtain $\tilde{\omega}^2$ as a function of $C_1, C_2, \ldots, C_n$. Since the actual frequency ω will be smaller than $\tilde{\omega}$ [1.9], we want to choose $C_1, C_2, \ldots, C_n$ such that they make $\tilde{\omega}$ a minimum. For this, we set

$$\frac{\partial(\tilde{\omega}^2)}{\partial C_1} = \frac{\partial(\tilde{\omega}^2)}{\partial C_2} = \cdots = \frac{\partial(\tilde{\omega}^2)}{\partial C_n} = 0 \tag{1.46}$$

This gives a set of n equations in the n unknowns $C_1, C_2, \ldots, C_n$, and hence we will be able to solve them. A typical equation in Eq. (1.46) is given by

$$\frac{\partial}{\partial C_j}\int_0^L EI\left(\frac{\mathrm{d}^2W}{\mathrm{d}x^2}\right)^2 \mathrm{d}x - \omega_j^2 \frac{\partial}{\partial C_j}\int_0^L m(x)W^2\,\mathrm{d}x = 0 \tag{1.47}$$

As an example, consider the following two-term solution for a fixed–fixed beam:

$$W(x) = C_1\left(1 - \cos\frac{2\pi x}{L}\right) + C_2\left(1 - \cos\frac{4\pi x}{L}\right) \tag{1.48}$$

Substitution of Eq. (1.48) into Eq. (1.42) gives

$$\tilde{\omega}^2 = \frac{\dfrac{8\pi^4 EI}{L^3}(C_1^2 + 16C_2^2)}{\dfrac{mL}{2}(3C_1^2 + 3C_2^2 + 4C_1C_2)} \tag{1.49}$$

The conditions for the minimum of $\tilde{\omega}^2$, namely,

$$\frac{\partial(\tilde{\omega}^2)}{\partial C_1} = \frac{\partial(\tilde{\omega}^2)}{\partial C_2} = 0$$

lead to the following algebraic eigenvalue problem:

$$\frac{16\pi^4 EI}{L^3}\begin{bmatrix}1 & 0\\0 & 16\end{bmatrix}\begin{Bmatrix}C_1\\C_2\end{Bmatrix} = \omega^2 mL\begin{bmatrix}3 & 2\\2 & 3\end{bmatrix}\begin{Bmatrix}C_1\\C_2\end{Bmatrix} \tag{1.50}$$

The solution of Eq. (1.50) is given by

$$\omega_1 = \frac{22.35}{L^2}\sqrt{\frac{EI}{m}} \quad \text{with} \quad \begin{Bmatrix}C_1\\C_2\end{Bmatrix} = \begin{Bmatrix}1.0\\0.5750\end{Bmatrix}$$

and

$$\omega_2 = \frac{124.0}{L^2}\sqrt{\frac{EI}{m}} \quad \text{with} \quad \begin{Bmatrix}C_1\\C_2\end{Bmatrix} = \begin{Bmatrix}1.0\\-1.4488\end{Bmatrix} \tag{1.51}$$

1.6.5 Approximate Analytical Solution (Galerkin Method)

To find the approximate solution of Eq. (1.29) with the boundary conditions stated in Eq. (1.32), using the Galerkin method, we assume the solution to be of the form

$$W(x) = C_1 f_1(x) + C_2 f_2(x) + \cdots + C_n f_n(x) \tag{1.52}$$

where $C_1,\ C_2,\ldots,C_n$ are constants and $f_1,\ f_2,\ldots,f_n$ are functions that satisfy all the specified boundary conditions. Since the solution assumed, Eq. (1.52), is not an exact one, it will not satisfy Eq. (1.29), and we will obtain, upon substitution into the left-hand side of the equation, a quantity different from zero (known as the residual, R). The values of the constants $C_1, C_2, \ldots, C_n$ are obtained by setting the integral of the residual multiplied by each of the functions f_i over the length of the beam equal to zero; that is,

$$\int_{x=0}^{L} f_i R\,\mathrm{d}x = 0, \qquad i = 1, 2, \ldots, n \tag{1.53}$$

Equation (1.53) represents a system of linear homogeneous equations (since the problem is an eigenvalue problem) in the unknowns $C_1,\ C_2,\ldots,C_n$. These equations can be solved to find the natural frequencies and mode shapes of the problem. For illustration, let us consider a two-term solution as

$$W(x) = C_1 f_1(x) + C_2 f_2(x) \tag{1.54}$$

where

$$f_1(x) = \cos\frac{2\pi x}{L} - 1 \tag{1.55}$$

and

$$f_2(x) = \cos\frac{4\pi x}{L} - 1 \tag{1.56}$$

Substitution of Eq. (1.54) into Eq. (1.29) gives the residue R as

$$R = C_1\left\{\left(\frac{2\pi}{L}\right)^4 - \beta^4\right\}\cos\frac{2\pi x}{L} + C_1\beta^4 + C_2\left\{\left(\frac{4\pi}{L}\right)^4 - \beta^4\right\}\cos\frac{4\pi x}{L} + C_2\beta^4 \tag{1.57}$$

Thus, the application of Eq. (1.53) leads to

$$\int_{x=0}^{L}\left(\cos\frac{2\pi x}{L}-1\right)\left[C_1\left\{\left(\frac{2\pi}{L}\right)^4-\beta^4\right\}\cos\frac{2\pi x}{L}+C_1\beta^4\right.$$
$$\left.+C_2\left\{\left(\frac{4\pi}{L}\right)^4-\beta^4\right\}\cos\frac{4\pi x}{L}+C_2\beta^4\right]dx=0 \tag{1.58}$$

and

$$\int_{x=0}^{L}\left(\cos\frac{4\pi x}{L}-1\right)\left[C_1\left\{\left(\frac{2\pi}{L}\right)^4-\beta^4\right\}\cos\frac{2\pi x}{L}+C_1\beta^4\right.$$
$$\left.+C_2\left\{\left(\frac{4\pi}{L}\right)^4-\beta^4\right\}\cos\frac{4\pi x}{L}+C_2\beta^4\right]dx=0 \tag{1.59}$$

or

$$C_1\left[\frac{1}{2}\left\{\left(\frac{2\pi}{L}\right)^4-\beta^4\right\}-\beta^4\right]-C_2\beta^4=0 \tag{1.60}$$

and

$$-C_1\beta^4+C_2\left[\frac{1}{2}\left\{\left(\frac{4\pi}{L}\right)^4-\beta^4\right\}-\beta^4\right]=0 \tag{1.61}$$

For a nontrivial solution of Eqs. (1.60) and (1.61), the determinant of the coefficient matrix of C_1 and C_2 must be zero. This gives

$$\begin{vmatrix}\frac{1}{2}\left\{\left(\frac{2\pi}{L}\right)^4-\beta^4\right\}-\beta^4 & -\beta^4\\ -\beta^4 & \frac{1}{2}\left\{\left(\frac{4\pi}{L}\right)^4-\beta^4\right\}-\beta^4\end{vmatrix}=0$$

or

$$(\beta L)^8-15900(\beta L)^4+7771000=0 \tag{1.62}$$

The solution of Eq. (1.62) is

$$\beta L=4.741 \text{ or } 11.140.$$

Thus, the first two natural frequencies of the beam are given by

$$\omega_1=\frac{22.48}{L^2}\sqrt{\frac{EI}{m}},\qquad \omega_2=\frac{124.1}{L^2}\sqrt{\frac{EI}{m}} \tag{1.63}$$

The eigenvectors corresponding to ω_1 and ω_2 can be obtained by solving Eq. (1.60) or (1.61) with the appropriate value of β. The results are

$$\text{For } \omega_1:\ \begin{Bmatrix}C_1\\C_2\end{Bmatrix}=\begin{Bmatrix}23.0\\1.0\end{Bmatrix},\qquad \text{For } \omega_2:\ \begin{Bmatrix}C_1\\C_2\end{Bmatrix}=\begin{Bmatrix}-0.69\\1.00\end{Bmatrix} \tag{1.64}$$

1.6.6 Finite Difference Method of Numerical Solution

The main idea in the finite difference method is to use approximations to derivatives. We will derive finite difference approximations (similar to central difference method) to various order derivatives below. Let $f = f(x)$ be any given function. The Taylor's series expansion of f around any point x gives us

$$f(x+\Delta x) \simeq f(x) + \left.\frac{\mathrm{d}f}{\mathrm{d}x}\right|_x \Delta x + \left.\frac{\mathrm{d}^2 f}{\mathrm{d}x^2}\right|_x \frac{(\Delta x)^2}{2!} + \left.\frac{\mathrm{d}^3 f}{\mathrm{d}x^3}\right|_x \frac{(\Delta x)^3}{3!} + \left.\frac{\mathrm{d}^4 f}{\mathrm{d}x^4}\right|_x \frac{(\Delta x)^4}{4!} \tag{1.65}$$

$$f(x-\Delta x) \simeq f(x) - \left.\frac{\mathrm{d}f}{\mathrm{d}x}\right|_x \Delta x + \left.\frac{\mathrm{d}^2 f}{\mathrm{d}x^2}\right|_x \frac{(\Delta x)^2}{2!} - \left.\frac{\mathrm{d}^3 f}{\mathrm{d}x^3}\right|_x \frac{(\Delta x)^3}{3!} + \left.\frac{\mathrm{d}^4 f}{\mathrm{d}x^4}\right|_x \frac{(\Delta x)^4}{4!} \tag{1.66}$$

By taking two terms only and subtracting Eq. (1.66) from (1.65), we get

$$f(x+\Delta x) - f(x-\Delta x) = \left(\frac{\mathrm{d}f}{\mathrm{d}x}\Delta x\right) - \left(-\frac{\mathrm{d}f}{\mathrm{d}x}\Delta x\right)$$

or

$$\left.\frac{\mathrm{d}f}{\mathrm{d}x}\right|_x = \frac{f(x+\Delta x) - f(x-\Delta x)}{2\Delta x} \tag{1.67}$$

Take terms up to second derivative and add Eqs. (1.65) and (1.66) to obtain

$$f(x+\Delta x) + f(x-\Delta x) = 2f(x) + \left.\frac{\mathrm{d}^2 f}{\mathrm{d}x^2}\right|_x (\Delta x)^2$$

or

$$\left.\frac{\mathrm{d}^2 f}{\mathrm{d}x^2}\right|_x = \frac{f(x+\Delta x) - 2f(x) + f(x-\Delta x)}{(\Delta x)^2} \tag{1.68}$$

Using $\left.(\mathrm{d}^2 f/\mathrm{d}x^2)\right|_x$ in place of $f(x)$, Eq. (1.68) can be expressed as

$$\left.\frac{\mathrm{d}^4 f}{\mathrm{d}x^4}\right|_x = \frac{\left.\frac{\mathrm{d}^2 f}{\mathrm{d}x^2}\right|_{x+\Delta x} - 2\left.\frac{\mathrm{d}^2 f}{\mathrm{d}x^2}\right|_x + \left.\frac{\mathrm{d}^2 f}{\mathrm{d}x^2}\right|_{x-\Delta x}}{(\Delta x)^2} \tag{1.69}$$

By substituting Eq. (1.68) on the right-hand side of Eq. (1.69), we obtain

$$\begin{aligned}\left.\frac{\mathrm{d}^4 f}{\mathrm{d}x^4}\right|_x = \Bigg[&\left\{\frac{f(x+2\Delta x) - 2f(x+\Delta x) + f(x)}{(\Delta x)^2}\right\} \\ &- 2\left\{\frac{f(x+\Delta x) - 2f(x) + f(x-\Delta x)}{(\Delta x)^2}\right\} \\ &+ \left\{\frac{f(x) - 2f(x-\Delta x) + f(x-2\Delta x)}{(\Delta x)^2}\right\}\Bigg] \Big/ (\Delta x)^2\end{aligned} \tag{1.70}$$

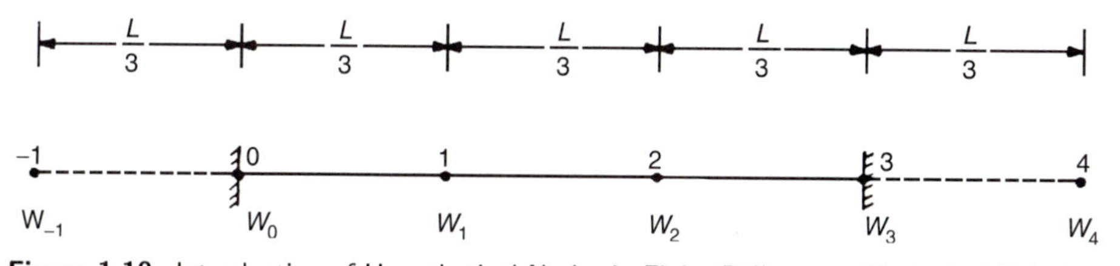

Figure 1.10. Introduction of Hypothetical Nodes in Finite Differences Method of Solution.

Equation (1.70) can be simplified to obtain the formula for the fourth derivative as

$$\left.\frac{\mathrm{d}^4 f}{\mathrm{d}x^4}\right|_x = \frac{1}{(\Delta x)^4}[f(x+2\Delta x) + f(x-2\Delta x) + 6f(x) - 4f(x+\Delta x) - 4f(x-\Delta x)] \tag{1.71}$$

To find the approximate solution of

$$\frac{\mathrm{d}^4 W}{\mathrm{d}x^4} - \beta^4 W = 0 \tag{1.72}$$

we approximate this equation around points 1 and 2 in Figure 1.10 by finite differences. For this, we need to imagine two hypothetical node points -1 and 4 as shown in Figure 1.10. We obtain, by approximating Eq. (1.72) at points 1 and 2,

$$\left.\begin{aligned}(W_{-1} - 4W_0 + 6W_1 - 4W_2 + W_3) - \beta_1^4 W_1 = 0\\ (W_0 - 4W_1 + 6W_2 - 4W_3 + W_4) - \beta_1^4 W_2 = 0\end{aligned}\right\} \tag{1.73}$$

where

$$\beta_1^4 = \left(\frac{L}{3}\right)^4 \beta^4 \tag{1.74}$$

The boundary conditions are

$$\left.\begin{aligned}&W_0 = W_3 = 0\\ &\frac{\mathrm{d}W}{\mathrm{d}x} = 0 \text{ at nodes 0 and 3, or } W_{-1} = W_1 \text{ and } W_2 = W_4\end{aligned}\right\} \tag{1.75}$$

By substituting Eqs. (1.75), Eqs. (1.73) reduce to

$$7W_1 - 4W_2 = \beta_1^4 \cdot W_1$$

$$-4W_1 + 7W_2 = \beta_1^4 \cdot W_2$$

or

$$\begin{bmatrix} 7 & -4 \\ -4 & 7 \end{bmatrix} \begin{Bmatrix} W_1 \\ W_2 \end{Bmatrix} = \beta_1^4 \begin{bmatrix} 1 & 0 \\ 0 & 1 \end{bmatrix} \begin{Bmatrix} W_1 \\ W_2 \end{Bmatrix} \tag{1.76}$$

By solving this standard eigenvalue problem, Eq. (1.76), we can obtain the approximate first two natural frequencies and mode shapes of the given beam as

$$\omega_1 = \frac{15.59}{L^2}\sqrt{\frac{EI}{m}} \quad \text{with} \quad \begin{Bmatrix} W_1 \\ W_2 \end{Bmatrix} = \begin{Bmatrix} 1 \\ 1 \end{Bmatrix}$$

and

$$\omega_2 = \frac{29.85}{L^2}\sqrt{\frac{EI}{m}} \quad \text{with} \quad \begin{Bmatrix} W_1 \\ W_2 \end{Bmatrix} = \begin{Bmatrix} 1 \\ -1 \end{Bmatrix} \tag{1.77}$$

The accuracy of the solution can be improved by taking more node points in the beam.

1.6.7 Finite Element Method of Numerical Solution (Displacement Method)

In the finite element method, we divide the given structure (beam) into several elements and assume a suitable solution within each of the elements. From this we formulate the necessary equations from which the approximate solution can be obtained easily. Figure 1.11 shows the beam divided into two elements of equal length. Within each element, let us assume a solution of the type*

$$\begin{aligned} w(x) = {} & W_1^{(e)} \cdot \frac{1}{l^3}(2x^3 - 3lx^2 + l^3) + W_3^{(e)} \cdot \frac{1}{l^3}(3lx^2 - 2x^3) \\ & + W_2^{(e)} \cdot \frac{1}{l^2}(x^3 - 2lx^2 + l^2 x) + W_4^{(e)} \cdot \frac{1}{l^2}(x^3 - lx^2) \end{aligned} \tag{1.78}$$

where $W_1^{(e)}$ to $W_4^{(e)}$ denote the displacements at the ends of the element e (to be found), and l indicates the length of the element. By writing the expressions for strain and kinetic

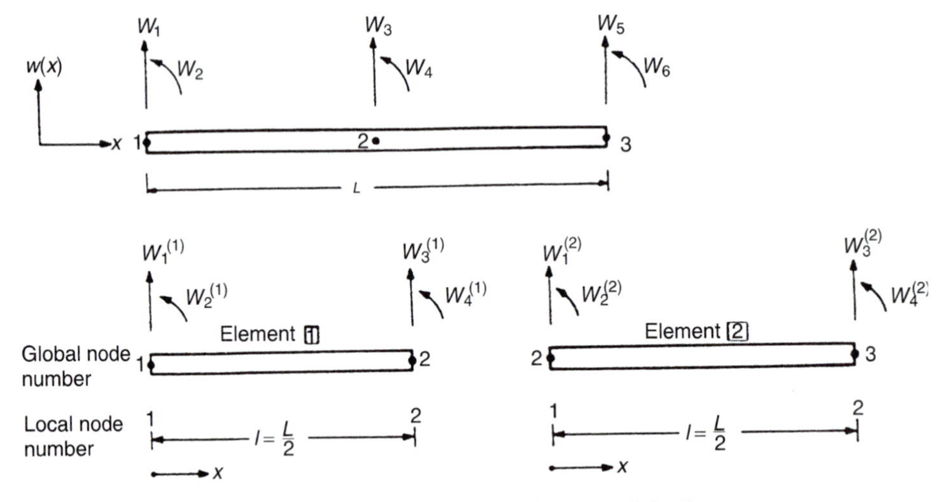

Figure 1.11. Finite Element Idealization of the Beam.

* Equation (1.78) implies that the interpolation model is a cubic equation.

energies of the elements as

$$\pi^{(e)} = \frac{1}{2} \int_0^l EI \left(\frac{\partial^2 w}{\partial x^2} \right)^2 dx = \frac{1}{2} \vec{W}^{(e)^T} [K^{(e)}] \vec{W}^{(e)} \tag{1.79}$$

and

$$T^{(e)} = \frac{1}{2} \int_0^l \rho A \left(\frac{\partial w}{\partial t} \right)^2 dx = \frac{1}{2} \dot{\vec{W}}^{(e)^T} [M^{(e)}] \dot{\vec{W}}^{(e)} \tag{1.80}$$

where ρ is the mass density, A is the cross-sectional area of the element, and a dot over $\vec{W}^{(e)}$ represents time derivative of $\vec{W}^{(e)}$. One can obtain, after substituting Eq. (1.78) into Eqs. (1.79) and (1.80), the stiffness matrix $[K^{(e)}]$ and mass matrix $[M^{(e)}]$ as

$$[K^{(e)}] = \frac{2EI}{l^3} \begin{bmatrix} 6 & 3l & -6 & 3l \\ 3l & 2l^2 & -3l & l^2 \\ -6 & -3l & 6 & -3l \\ 3l & l^2 & -3l & 2l^2 \end{bmatrix} \tag{1.81}$$

$$[M^{(e)}] = \frac{\rho A l}{420} \begin{bmatrix} 156 & 22l & 54 & -13l \\ 22l & 4l^2 & 13l & -3l^2 \\ 54 & 13l & 156 & -22l \\ -13l & -3l^2 & -22l & 4l^2 \end{bmatrix} \tag{1.82}$$

Figure 1.11 shows that

$$\vec{W}^{(1)} = \text{vector of unknown displacements for element 1}$$

$$\begin{Bmatrix} W_1^{(1)} \\ W_2^{(1)} \\ W_3^{(1)} \\ W_4^{(1)} \end{Bmatrix} = \begin{Bmatrix} W_1 \\ W_2 \\ W_3 \\ W_4 \end{Bmatrix}$$

and

$$\vec{W}^{(2)} = \text{vector of unknown displacements for element 2}$$

$$\begin{Bmatrix} W_1^{(2)} \\ W_2^{(2)} \\ W_3^{(2)} \\ W_4^{(2)} \end{Bmatrix} = \begin{Bmatrix} W_3 \\ W_4 \\ W_5 \\ W_6 \end{Bmatrix}$$

By assembling the stiffness matrices of the two elements (details are given in Chapter 6), one obtains the assembled stiffness matrix as

$$\begin{array}{c} \\ [\underset{\sim}{K}] = \dfrac{2EI}{l^3} \end{array} \begin{array}{c} \begin{array}{cccccc} W_1 = W_1^{(1)} & W_2 = W_2^{(1)} & W_3 = W_3^{(1)} = W_1^{(2)} & W_4 = W_4^{(1)} = W_2^{(2)} & W_5 = W_3^{(2)} & W_6 = W_4^{(2)} \end{array} \\ \begin{bmatrix} 6 & 3l & -6 & 3l & 0 & 0 \\ 3l & 2l^2 & -3l & l^2 & 0 & 0 \\ -6 & -3l & 6+6 & -3l+3l & -6 & 3l \\ 3l & l^2 & -3l+3l & 2l^2+2l^2 & -3l & l^2 \\ 0 & 0 & -6 & -3l & 6 & -3l \\ 0 & 0 & 3l & l^2 & -3l & 2l^2 \end{bmatrix} \end{array} \begin{array}{l} W_1 = W_1^{(1)} \\ W_2 = W_2^{(1)} \\ W_3 = W_3^{(1)} = W_1^{(2)} \\ W_4 = W_4^{(1)} = W_2^{(2)} \\ W_5 = W_3^{(2)} \\ W_6 = W_4^{(2)} \end{array}$$

After deleting the rows and columns corresponding to the degrees of freedom W_1, W_2, W_5, and W_6 (since $W_1 = W_2 = W_5 = W_6 = 0$ are the boundary conditions), we obtain the stiffness matrix $[K]$ of the beam as

$$[K] = \frac{2EI}{l^3}\begin{bmatrix} 12 & 0 \\ 0 & 4l^2 \end{bmatrix} = \frac{16EI}{L^3}\begin{bmatrix} 12 & 0 \\ 0 & L^2 \end{bmatrix} \tag{1.83}$$

Similarly, the assembled mass matrix is given by

$$\begin{array}{c} \\ [\underset{\sim}{M}] = \dfrac{\rho A l}{420} \end{array}
\begin{array}{c}
\begin{array}{cccccc} W_1 = W_1^{(1)} & W_2 = W_2^{(1)} & W_3 = W_3^{(1)} = W_1^{(2)} & W_4 = W_4^{(1)} = W_2^{(2)} & W_5 = W_3^{(2)} & W_6 = W_4^{(2)} \end{array} \\
\begin{bmatrix} 156 & 22l & 54 & -13l & 0 & 0 \\ 22l & 4l^2 & 13l & -3l^2 & 0 & 0 \\ 54 & 13l & 156+156 & -22l+22l & 54 & -13l \\ -13l & -3l^2 & -22l+22l & 4l^2+4l^2 & 13l & -3l^2 \\ 0 & 0 & 54 & 13l & 156 & -22l \\ 0 & 0 & -13l & -3l^2 & -22l & 4l^2 \end{bmatrix}
\end{array}
\begin{array}{l} W_1 = W_1^{(1)} \\ W_2 = W_2^{(1)} \\ W_3 = W_3^{(1)} = W_1^{(2)} \\ W_4 = W_4^{(1)} = W_2^{(2)} \\ W_5 = W_3^{(2)} \\ W_6 = W_4^{(2)} \end{array}$$

By reducing this matrix (by deleting the rows and columns corresponding to fixed degrees of freedom), we obtain the matrix $[M]$ of the beam as

$$[M] = \frac{\rho A l}{420}\begin{bmatrix} 312 & 0 \\ 0 & 8l^2 \end{bmatrix} = \frac{\rho A L}{840}\begin{bmatrix} 312 & 0 \\ 0 & 2L^2 \end{bmatrix} \tag{1.84}$$

Once the stiffness and mass matrices of the complete beam are available, we formulate the eigenvalue problem as

$$[K]\vec{W} = \lambda[M]\vec{W} \tag{1.85}$$

where $\vec{W} = \begin{Bmatrix} W_3 \\ W_4 \end{Bmatrix}$ is the eigenvector, and λ is the eigenvalue. The solution of Eq. (1.85) gives us two natural frequencies and the corresponding mode shapes of the beam as

$$\omega_1 = \frac{22.7}{L^2}\sqrt{\frac{EI}{m}} \quad \text{with} \quad \begin{Bmatrix} W_3 \\ W_4 \end{Bmatrix} = \begin{Bmatrix} 1 \\ 0 \end{Bmatrix}$$

and

$$\omega_2 = \frac{82.0}{L^2}\sqrt{\frac{EI}{m}} \quad \text{with} \quad \begin{Bmatrix} W_3 \\ W_4 \end{Bmatrix} = \begin{Bmatrix} 0 \\ 1 \end{Bmatrix} \tag{1.86}$$

1.6.8 Stress Analysis of Beams Using Finite Element Method

The displacement model of Eq. (1.78) and the resulting element stiffness matrix of the beam, given by Eq. (1.81), can be used for the stress analysis of beams subjected to loads. The procedure is illustrated with the following example.

Example 1.5

A beam of uniform rectangular cross section with width 1 cm and depth 2 cm and length 60 cm is subjected to a vertical concentrated load of 1000 N as shown in Figure 1.12(a). If the beam is fixed at both the ends, find the stresses in the beam using a two-element idealization. Assume the Young's modulus of the beam as $E = 10^7$ N/cm^2.

Solution By assuming the two fixed ends of the beam as the end nodes and introducing an additional node at the point of application of the load, the beam can be replaced by a two-element idealization as shown in Figures 1.12(a) and 1.12(b). The global degrees of freedom of the beam are indicated in Figure 1.12(a) so that the vector of displacement degrees of freedom of the system (beam) is given by

$$\vec{W} = \begin{Bmatrix} W_1 \\ W_2 \\ W_3 \\ W_4 \\ W_5 \\ W_6 \end{Bmatrix} \tag{E$_1$}$$

The element nodal degrees of freedom can be identified from Fig. 1.12(b) as

$$\vec{W}^{(1)} = \begin{Bmatrix} W_1^{(1)} \\ W_2^{(1)} \\ W_3^{(1)} \\ W_4^{(1)} \end{Bmatrix} = \begin{Bmatrix} W_1 \\ W_2 \\ W_3 \\ W_4 \end{Bmatrix} \tag{E$_2$}$$

$$\vec{W}^{(2)} = \begin{Bmatrix} W_1^{(2)} \\ W_2^{(2)} \\ W_3^{(2)} \\ W_4^{(2)} \end{Bmatrix} = \begin{Bmatrix} W_3 \\ W_4 \\ W_5 \\ W_6 \end{Bmatrix} \tag{E$_3$}$$

The element stiffness matrices are given by Eq. (1.81):

$$\left[K^{(e)}\right] = \frac{2E^{(e)}I^{(e)}}{l^{(e)3}} \begin{array}{c} \begin{matrix} W_1^{(e)} & W_2^{(e)} & W_3^{(e)} & W_4^{(e)} \end{matrix} \\ \begin{bmatrix} 6 & 3l^{(e)} & -6 & 3l^{(e)} \\ 3l^{(e)} & 2l^{(e)^2} & -3l^{(e)} & l^{(e)^2} \\ -6 & -3l^{(e)} & 6 & -3l^{(e)} \\ 3l^{(e)} & l^{(e)^2} & -3l^{(e)} & 2l^{(e)^2} \end{bmatrix} \end{array} \begin{matrix} W_1^{(e)} \\ W_2^{(e)} \\ W_3^{(e)} \\ W_4^{(e)} \end{matrix} \tag{E$_4$}$$

where $E^{(e)}$ is the Young's modulus, $I^{(e)}$ is the area moment of inertia of the cross section, and $l^{(e)}$ is the length of element e. Using $E^{(1)} = E^{(2)} = 10^7$ N/cm^2, $I^{(1)} = I^{(2)} = (1)(2^3)/12 = 2/3$ cm^4, $l^{(1)} = 20$ cm, and $l^{(2)} = 40$ cm in Eq. (E$_4$), we obtain

$$\left[K^{(1)}\right] = 10^4 \begin{array}{c} \begin{matrix} W_1 & W_2 & W_3 & W_4 \end{matrix} \\ \begin{bmatrix} 1 & 10 & -1 & 10 \\ 10 & \frac{400}{3} & -10 & \frac{200}{3} \\ -1 & 10 & 1 & -10 \\ 10 & \frac{200}{3} & -10 & \frac{300}{3} \end{bmatrix} \end{array} \begin{matrix} W_1 \\ W_2 \\ W_3 \\ W_4 \end{matrix} \tag{E$_5$}$$

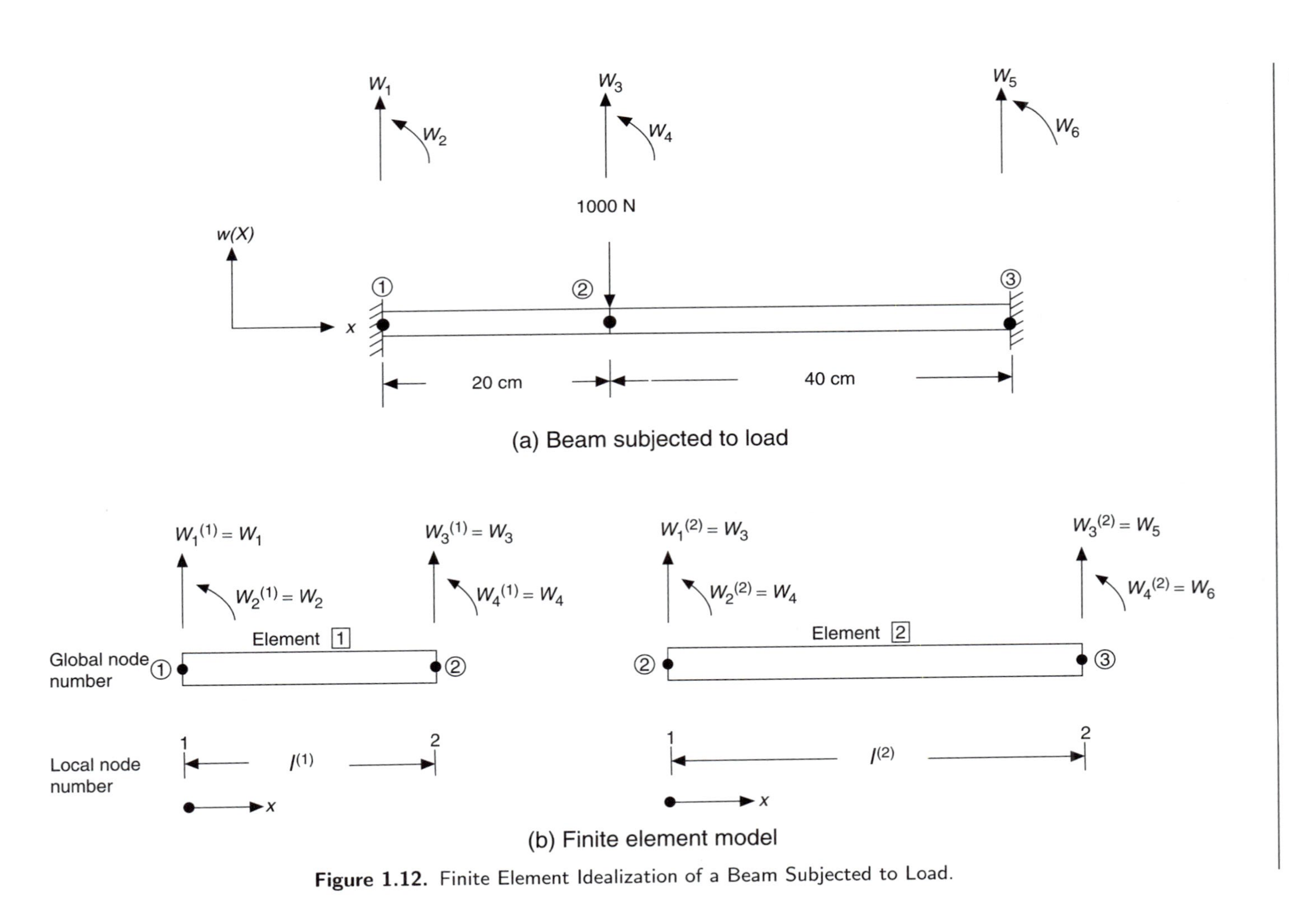

Figure 1.12. Finite Element Idealization of a Beam Subjected to Load.

$$\left[K^{(2)}\right] = 10^4 \begin{bmatrix} \frac{1}{8} & \frac{5}{2} & \frac{-1}{8} & \frac{5}{2} \\ \frac{5}{2} & \frac{200}{3} & \frac{-5}{2} & \frac{100}{3} \\ \frac{-1}{8} & \frac{-5}{2} & \frac{1}{8} & \frac{-5}{2} \\ \frac{5}{2} & \frac{100}{3} & \frac{-5}{2} & \frac{200}{3} \end{bmatrix} \begin{matrix} W_3 \\ W_4 \\ W_5 \\ W_6 \end{matrix} \quad \text{(columns } W_3, W_4, W_5, W_6\text{)} \tag{E$_6$}$$

Note that the nodal degrees of freedom associated with the various rows and columns of the matrices $[K^{(e)}]$ are also indicated in Eqs. (E_4)–(E_6). The assembled stiffness matrix of the beam can be obtained as

$$[\underset{\sim}{K}] = 10^4 \begin{bmatrix} 1 & 10 & -1 & 10 & & \\ 10 & \frac{400}{3} & -10 & \frac{200}{3} & & \\ -1 & 10 & \left(1+\frac{1}{8}\right) & \left(-10+\frac{5}{2}\right) & -\frac{1}{8} & \frac{5}{2} \\ 10 & \frac{200}{3} & \left(-10+\frac{5}{2}\right) & \left(\frac{400}{3}+\frac{200}{3}\right) & -\frac{5}{2} & \frac{100}{3} \\ & & -\frac{1}{8} & -\frac{5}{2} & \frac{1}{8} & -\frac{5}{2} \\ & & \frac{5}{2} & \frac{100}{3} & -\frac{5}{2} & \frac{200}{3} \end{bmatrix} \begin{matrix} W_1 \\ W_2 \\ W_3 \\ W_4 \\ W_5 \\ W_6 \end{matrix} \quad \text{(columns } W_1, \ldots, W_6\text{)} \tag{E$_7$}$$

Noting that nodes 1 and 3 are fixed, we have $W_1 = W_2 = W_5 = W_6 = 0$ and hence by deleting the rows and columns corresponding to these degrees of freedom in Eq. (E_7), we obtain the final stiffness matrix of the beam as

$$[K] = 10^4 \begin{bmatrix} \frac{9}{8} & -\frac{15}{2} \\ -\frac{15}{2} & 200 \end{bmatrix} \begin{matrix} W_3 \\ W_4 \end{matrix} \quad \text{(columns } W_3, W_4\text{)} \tag{E$_8$}$$

Since the externally applied vertical load at node 2 (in the direction of W_3) is -1000 N and the rotational load (bending moment) at node 2 (in the direction of W_4) is 0, the load vector of the beam corresponding to the degrees of freedom W_3 and W_4 can be expressed as

$$\vec{P} = \begin{Bmatrix} P_3 \\ P_4 \end{Bmatrix} = \begin{Bmatrix} -1000 \\ 0 \end{Bmatrix} \tag{E$_9$}$$

Thus the equilibrium equations of the beam are given by

$$[K]\vec{W} = \vec{P}$$

or

$$10^4 \begin{bmatrix} \frac{9}{8} & -\frac{15}{2} \\ -\frac{15}{2} & 200 \end{bmatrix} \begin{Bmatrix} W_3 \\ W_4 \end{Bmatrix} = \begin{Bmatrix} -1000 \\ 0 \end{Bmatrix} \tag{E$_{10}$}$$

The solution of Eq. (E_{10}) gives

$$W_3 = -0.11851852 \text{ cm}, \qquad W_4 = -0.00444444 \text{ radian} \tag{E$_{11}$}$$

The bending stress, $\sigma_{xx}^{(e)}(x, y)$, induced at a point (fiber) located at a distance x from node 1 (left side node) in horizontal direction and y from the neutral axis in vertical direction in element e is given by (see Section 9.3):

$$\sigma_{xx}^{(e)} = \sigma_{xx}^{(e)}(x, y) = [E][B]\vec{W}^{(e)} \equiv -yE^{(e)} \frac{d^2 w^{(e)}(x)}{dx^2} \tag{E$_{12}$}$$

where $w^{(e)}(x)$ is given by Eq. (1.78) so that

$$\begin{aligned} \frac{d^2 w^{(e)}(x)}{dx^2} &= \frac{1}{l^{(e)3}}(12x - 6l^{(e)})W_1^{(e)} + \frac{1}{l^{(e)2}}(6x - 4l^{(e)})W_2^{(e)} \\ &+ \frac{1}{l^{(e)3}}(6l^{(e)} - 12x)W_3^{(e)} + \frac{1}{l^{(e)2}}(6x - 2l^{(e)})W_4^{(e)} \end{aligned} \tag{E$_{13}$}$$

Thus the stresses in the elements can be determined as follows:

Element 1: Using $E^{(1)} = 10^7$, $l^{(1)} = 20$, $W_1^{(1)} = W_1 = 0$, $W_2^{(1)} = W_2 = 0$, $W_3^{(1)} = W_3 = -0.11851852$, and $W_4^{(1)} = W_4 = -0.00444444$, Eq. ($E_{12}$) gives

$$\sigma_{xx}^{(1)}(x, y) = 1777.7778(10 - x)y + 222.2222(3x - 20)y \tag{E$_{14}$}$$

The stress induced in the top fibers of element 1 is given by (with $y = +1$ cm)

$$\sigma_{xx}^{(1)}(x) = 1777.7778(10 - x) - 222.2222(20 - 2x) \tag{E$_{15}$}$$

For instance, the maximum stresses induced at $x = 0$ (fixed end) and $x = 20$ cm (point of load application) will be

$$\sigma_{xx}^{(1)}(0) = 13{,}333.334 \text{ N/cm}^2 \quad \text{and} \quad \sigma_{xx}^{(1)}(20) = -8{,}888.890 \text{ N/cm}^2$$

Element 2: Using $E^{(2)} = 10^7$, $l^{(2)} = 40$, $W_1^{(2)} = W_3 = -0.11851852$, $W_2^{(2)} = W_4 = -0.00444444$, $W_3^{(2)} = W_5 = 0$, and $W_4^{(2)} = W_6 = 0$, Eq. (E_{12}) gives

$$\sigma_{xx}^{(2)}(x, y) = 222.2222(x - 20)y + 55.5550(3x - 80)y \tag{E$_{16}$}$$

The stress induced in the top fibers of element 2 is given by (with $y = +1$ cm)

$$\sigma_{xx}^{(2)}(x) = 222.2222(x - 20) + 55.5550(3x - 80) \qquad (E_{17})$$

For instance, the maximum stresses induced at $x = 0$ (point of load application) and $x = 40$ cm (fixed end) will be

$$\sigma_{xx}^{(2)}(0) = -8{,}888.844 \text{ N/cm}^2 \quad \text{and} \quad \sigma_{xx}^{(2)}(40) = 6{,}666.644 \text{ N/cm}^2$$

1.7 FINITE ELEMENT PROGRAM PACKAGES

The general applicability of the finite element method makes it a powerful and versatile tool for a wide range of problems. Hence, a number of computer program packages have been developed for the solution of a variety of structural and solid mechanics problems. Some of the programs have been developed in such a general manner that the same program can be used for the solution of problems belonging to different branches of engineering with little or no modification.

Many of these packages represent large programs that can be used for solving real complex problems. For example, the NASTRAN (National Aeronautics and Space Administration Structural Analysis) program package contains approximately 150,000 Fortran statements and can be used to analyze physical problems of practically any size, such as a complete aircraft or an automobile structure.

The availability of supercomputers (e.g., the Cray-1 and the Cyber 205) has made a strong impact on the finite element technology. In order to realize the full potential of these supercomputers in finite element computation, special parallel numerical algorithms, programming strategies, and programming languages are being developed. The use of personal computers and workstations in engineering analysis and design is becoming increasingly popular as the price of hardware is decreasing dramatically. Many finite element programs, specially suitable for the personal computer and workstation environment, have been developed. Among the main advantages are a user-friendly environment and inexpensive graphics.

REFERENCES

1.1 R. Courant: Variational methods for the solution of problems of equilibrium and vibrations, *Bulletin of American Mathematical Society, 49*, 1–23, 1943.

1.2 M.J. Turner, R.W. Clough, H.C. Martin, and L.J. Topp: Stiffness and deflection analysis of complex structures, *Journal of Aeronautical Sciences, 23*, 805–824, 1956.

1.3 J.H. Argyris and S. Kelsey: "Energy theorems and structural analysis," *Aircraft Engineering*, Vols. 26 and 27, October 1954 to May 1955. Part I by J.H. Argyris and Part II by J.H. Argyris and S. Kelsey.

1.4 R.W. Clough: "The finite element method in plane stress analysis," *Proceedings, Second ASCE Conference on Electronic Computation*, Pittsburgh, PA, pp. 345–378, September 1960.

1.5 O.C. Zienkiewicz and Y.K. Cheung: *The Finite Element Method in Structural and Continuum Mechanics*, McGraw-Hill, London, 1967.

1.6 K.K. Gupta and J.L. Meek: A brief history of the beginning of the finite element method, *International Journal for Numerical Methods in Engineering, 39*, 3761–3774, 1996.

1.7 F.P. Incropera and D.P. DeWitt: *Introduction to Heat Transfer*, 3rd Ed., Wiley, New York, 1996.
1.8 R.W. Fox and A.T. McDonald: *Introduction to Fluid Mechanics*, 4th Ed., Wiley, New York, 1992.
1.9 S.S. Rao: *Mechanical Vibrations*, 3rd Ed., Addison-Wesley, Reading, MA, 1995.
1.10 K.J. Bathe: *Finite Element Procedures*, Prentice Hall, Englewood Cliffs, NJ, 1996.
1.11 O.C. Zienkiewicz: *The Finite Element Method*, McGraw-Hill, London, 1989.
1.12 Z.H. Zhong: *Finite Element Procedures for Contact-Impact Problems*, Oxford University Press, Oxford, 1993.
1.13 W.B. Bickford: *A First Course in the Finite Element Method*, Irwin, Burr Ridge, IL, 1994.
1.14 S. Kobayashi, S.I. Oh, and T. Altan: *Metal Forming and the Finite Element Method*, Oxford University Press, New York, 1989.
1.15 M. Kleiber and T.D. Hien: *The Stochastic Finite Element Method: Basic Perturbation Technique and Computer Implementation*, Wiley, Chichester, UK, 1992.
1.16 R.J. Melosh: *Structural Engineering Analysis by Finite Elements*, Prentice Hall, Englewood Cliffs, NJ, 1990.
1.17 O. Pironneau: *Finite Element Method for Fluids*, Wiley, Chichester, UK, 1989.
1.18 J.M. Jin: *The Finite Element Method in Electromagnetics*, Wiley, New York, 1993.
1.19 E. Zahavi: *The Finite Element Method in Machine Design*, Prentice Hall, Englewood Cliffs, NJ, 1992.
1.20 C.E. Knight: *The Finite Element Method in Mechanical Design*, PWS-Kent, Boston, 1993.
1.21 T.J.R. Hughs: *The Finite Element Method: Linear Static and Dynamic Finite Element Analysis*, Prentice Hall, Englewood Cliffs, NJ, 1987.
1.22 H.C. Huang: *Finite Element Analysis for Heat Transfer: Theory and Software*, Springer-Verlag, London, 1994.
1.23 E.R. Champion, Jr.: *Finite Element Analysis in Manufacturing Engineering*, McGraw-Hill, New York, 1992.
1.24 O.O. Ochoa: *Finite Element Analysis of Composite Laminates*, Kluwer, Dordrecht, The Netherlands, 1992.
1.25 S.J. Salon: *Finite Element Analysis of Electrical Machines*, Kluwer, Boston, 1995.
1.26 Y.K. Cheung: *Finite Element Implementation*, Blackwell Science, Oxford, 1996.
1.27 P.P. Silvester and R.L. Ferrari: *Finite Elements for Electrical Engineers*, 2nd Edition, Cambridge University Press, Cambridge, UK, 1990.
1.28 R.D. Cook, D.S. Malkus, and M.E. Plesha: *Concepts and Applications of Finite Element Analysis*, 3rd Ed., Wiley, New York, 1989.
1.29 J.N. Reddy: *An Introduction to the Finite Element Method*, 2nd Ed., McGraw-Hill, New York, 1993.
1.30 I.M. Smith and D.V. Griffiths: *Programming the Finite Element Method*, 2nd Ed., Wiley, Chichester, UK, 1988.
1.31 W. Weaver, Jr., and P.R. Johnston: *Structural Dynamics by Finite Elements*, Prentice Hall, Englewood Cliffs, NJ, 1987.
1.32 G.R. Buchanan: *Schaum's Outline of Theory and Problems of Finite Element Analysis*, McGraw-Hill, New York, 1995.
1.33 J.S. Przemieniecki: *Theory of Matrix Structural Analysis*, McGraw-Hill, New York, 1968.

PROBLEMS

1.1 If $S^{(l)}$ and $S^{(u)}$ denote the perimeters of the inscribed and circumscribed polygons, respectively, as shown in Figure 1.3, prove that

$$S^{(l)} \leq S \leq S^{(u)}$$

where S is the circumference of the circle.

1.2 Find the stress distribution in the tapered bar shown in Figure 1.13 using two finite elements under an axial load of $P = 1$ N.

1.3 Find the temperature distribution in the stepped fin shown in Figure 1.14 using two finite elements.

1.4 Using a one-beam element idealization, find the stress distribution under a load of P for the uniform cantilever beam shown in Figure 1.15. (Hint: Use the displacement model of Eq. (1.78), the strain–displacement relation given in Eq. (9.25), and the stress–strain relation $\sigma_{xx} = E\varepsilon_{xx}$, where E is the Young's modulus).

1.5 Find the stress distribution in the cantilever beam shown in Figure 1.16 using one beam element. (Hint: Use the displacement model of Eq. 1.78).

1.6 Find the stress distribution in the beam shown in Figure 1.17 using two beam elements.

1.7 Find the stress distribution in the beam shown in Figure 1.18 using two beam elements.

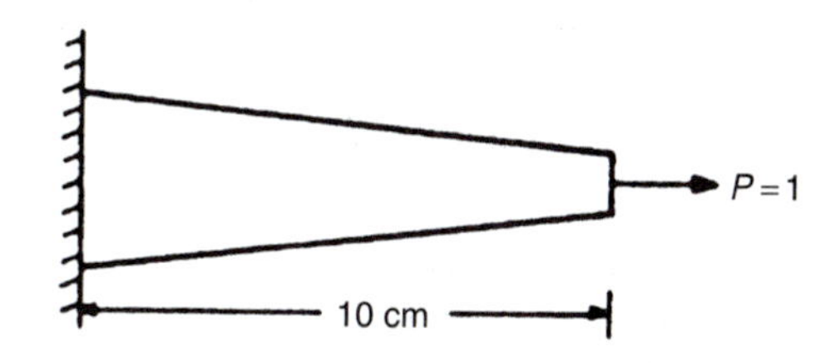

Cross sectional area at root = 2 cm^2
Cross sectional area at end = 1 cm^2
Young's modulus = 2 x 10^7 N/cm^2

Figure 1.13.

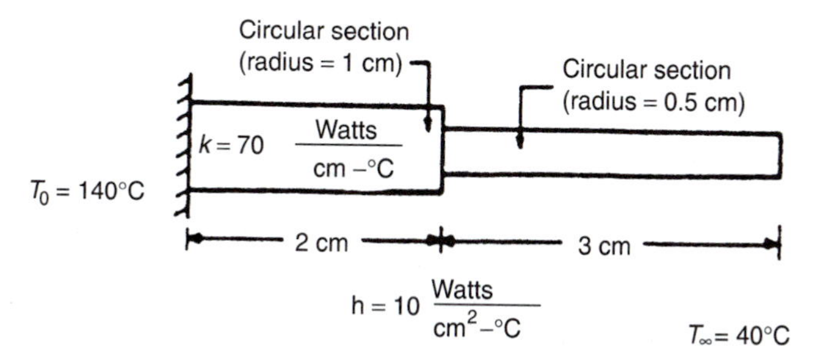

Figure 1.14.

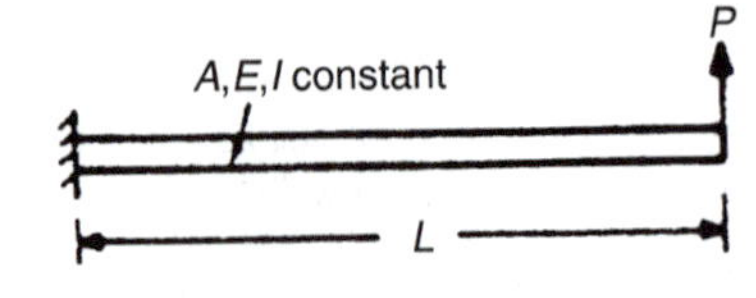

Figure 1.15.

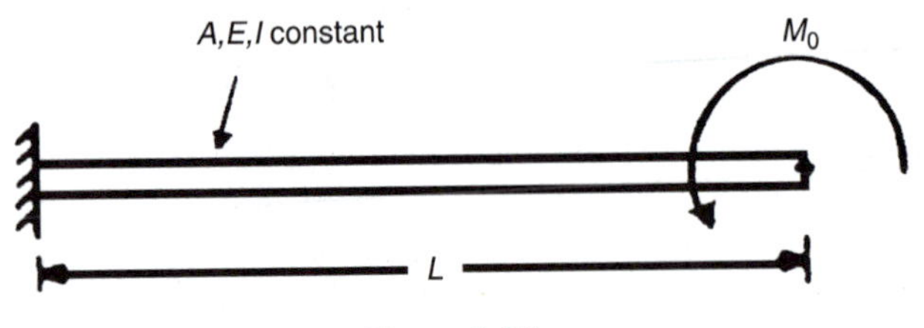

Figure 1.16.

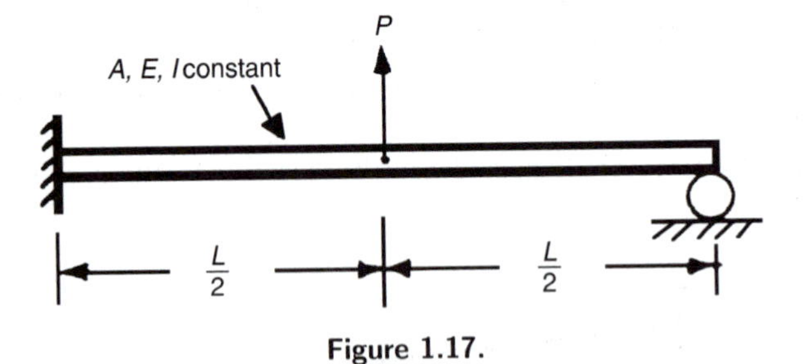

Figure 1.17.

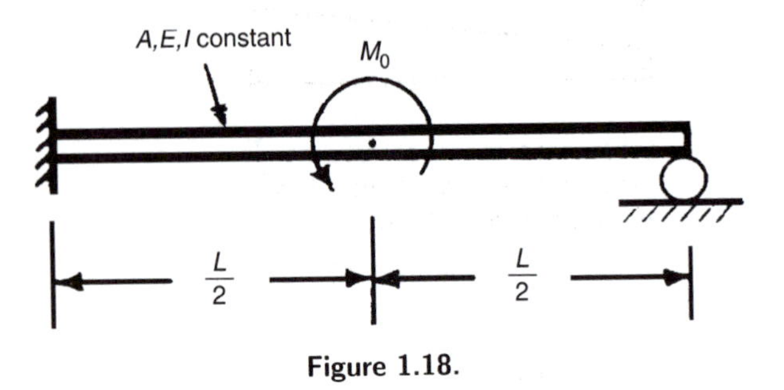

Figure 1.18.

1.8 Find the stress distribution in the beam shown in Figure 1.19 using two beam elements.

1.9 For the tapered bar shown in Figure 1.20, the area of cross section changes along the length as $A(x) = A_0 e^{-(x/l)}$, where A_0 is the cross-sectional area at $x = 0$, and l is the length of the bar. By expressing the strain and kinetic energies of the bar in matrix forms, identify the stiffness and mass matrices of a typical element. Assume a linear model for the axial displacement of the bar element.

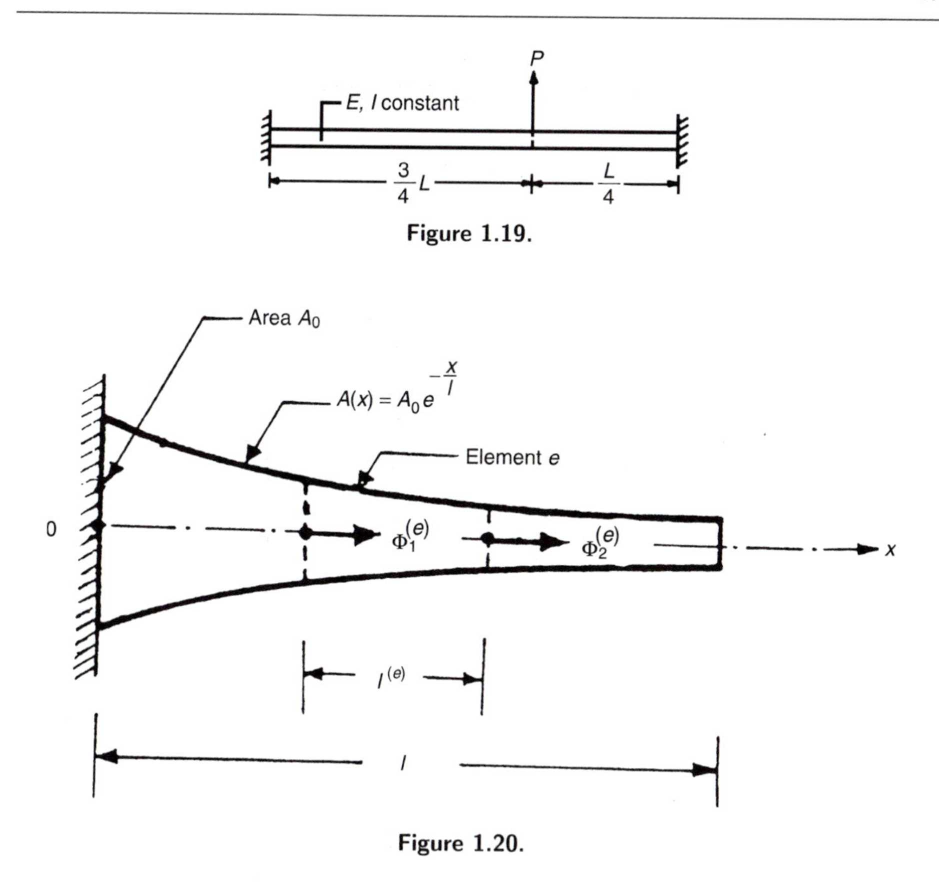

Figure 1.19.

Figure 1.20.

1.10 Find the fundamental natural frequency of axial vibration of the bar shown in Figure 1.20 using Rayleigh's method.

1.11 Find two natural frequencies of axial vibration of the bar shown in Figure 1.20 using the Rayleigh–Ritz method.

1.12 Find two natural frequencies of axial vibration of the bar shown in Figure 1.20 using the Galerkin method.

1.13 Find two natural frequencies of axial vibration of the bar shown in Figure 1.20 using the finite difference method.

1.14 Find two natural frequencies of axial vibration of the bar shown in Figure 1.20 using the finite element method. (Use a two-element idealization.)

1.15 For the cantilever beam shown in Figure 1.21, find the fundamental natural frequency using Rayleigh's method.

1.16 For the cantilever beam shown in Figure 1.21, find two natural frequencies using the Rayleigh–Ritz method.

1.17 For the cantilever beam shown in Figure 1.21, find two natural frequencies using the Galerkin method.

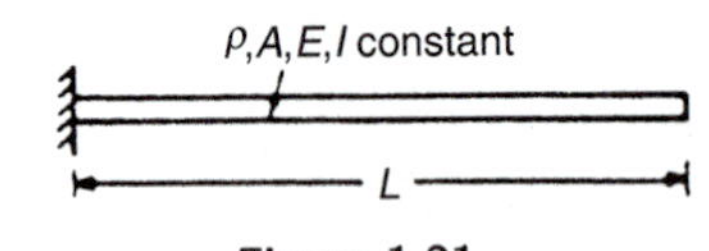

Figure 1.21.

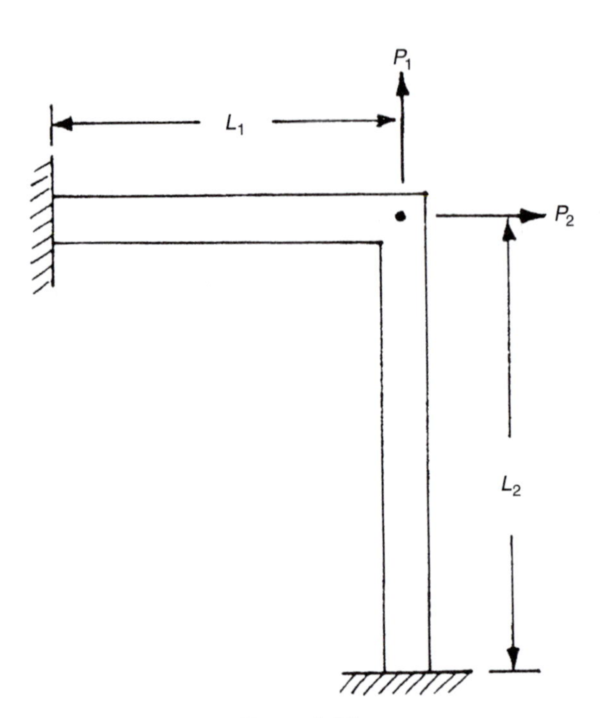

Figure 1.22.

1.18 Determine two natural frequencies of the cantilever beam shown in Figure 1.21 using the finite difference method.

1.19 Determine two natural frequencies of the cantilever beam shown in Figure 1.21 using the finite element method using a one-element idealization.

1.20 The differential equation governing the free longitudinal vibrations of a uniform bar is given by

$$EA\frac{\partial^2 u}{\partial x^2} - m\frac{\partial^2 u}{\partial t^2} = 0$$

where E is Young's modulus, A is the area of cross section, m is the mass per unit length, u is the axial displacement (in the direction of x), and t is time. If the bar is fixed at $x = 0$, find the first two natural frequencies of the bar using the finite difference method.

1.21 Find the first two natural frequencies of the bar described in Problem 1.20 using the finite element method with a linear displacement model.

1.22 Find the fundamental natural frequency of longitudinal vibration of the bar described in Problem 1.20 using Rayleigh's method.

1.23 Find two natural frequencies of longitudinal vibration of the bar described in Problem 1.20 using the Rayleigh–Ritz method.

1.24 Find two natural frequencies of longitudinal vibration of the bar described in Problem 1.20 using the Galerkin method.

1.25 Suggest a method of finding the stresses in the frame shown in Figure 1.22 using the finite element method.

1.26 The stiffness matrix of a spring (Figure 1.23(a)) is given by

$$[K^{(e)}] = k \begin{array}{c} \begin{array}{cc} \Phi_1^{(e)} & \Phi_2^{(e)} \end{array} \\ \begin{bmatrix} 1 & -1 \\ -1 & 1 \end{bmatrix} \end{array} \begin{array}{c} \\ \Phi_1^{(e)} \\ \Phi_2^{(e)} \end{array} \tag{1.87}$$

where k denotes the stiffness of the spring. Using this, determine the displacements of nodes 1 and 2 of the system shown in Figure 1.23(b).

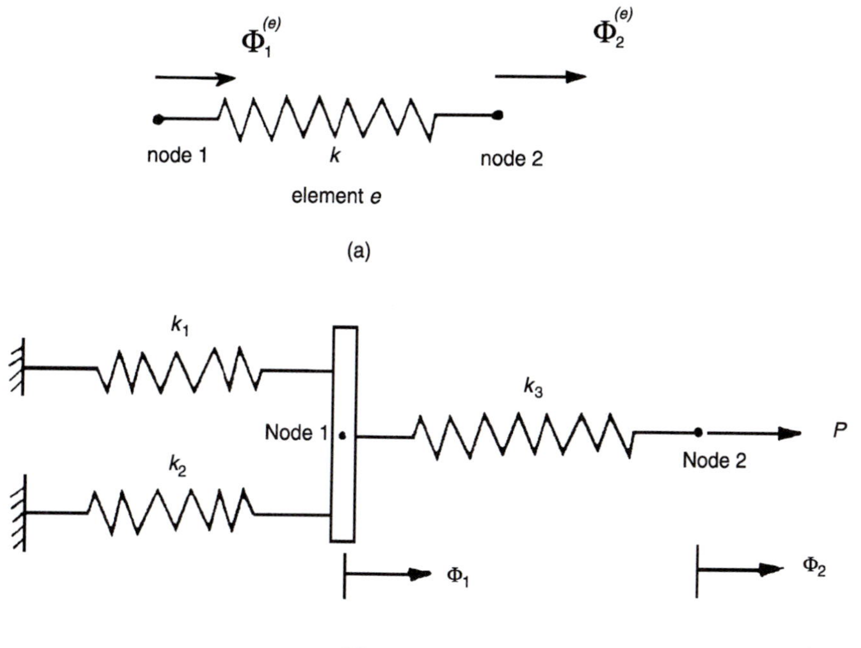

(b)

Figure 1.23.

BASIC PROCEDURE

2

DISCRETIZATION OF THE DOMAIN

2.1 INTRODUCTION

In most engineering problems, we need to find the values of a field variable such as displacement, stress, temperature, pressure, and velocity as a function of spatial coordinates (x, y, z). In the case of transient or unsteady state problems, the field variable has to be found as a function of not only the spatial coordinates (x, y, z) but also time (t). The geometry (domain or solution region) of the problem is often irregular. The first step of the finite element analysis involves the discretization of the irregular domain into smaller and regular subdomains, known as finite elements. This is equivalent to replacing the domain having an infinite number of degrees of freedom by a system having finite number of degrees of freedom.

A variety of methods can be used to model a domain with finite elements. Different methods of dividing the domain into finite elements involve different amounts of computational time and often lead to different approximations to the solution of the physical problem. The process of discretization is essentially an exercise of engineering judgment. Efficient methods of finite element idealization require some experience and a knowledge of simple guidelines. For large problems involving complex geometries, finite element idealization based on manual procedures requires considerable effort and time on the part of the analyst. Some programs have been developed for the *automatic mesh generation* for the efficient idealization of complex domains with minimal interface with the analyst.

2.2 BASIC ELEMENT SHAPES

The shapes, sizes, number, and configurations of the elements have to be chosen carefully such that the original body or domain is simulated as closely as possible without increasing the computational effort needed for the solution. Mostly the choice of the type of element is dictated by the geometry of the body and the number of independent coordinates necessary to describe the system. If the geometry, material properties, and the field variable of the problem can be described in terms of only one spatial coordinate, we can use the one-dimensional or line elements shown in Figure 2.1(a). The temperature distribution in a rod (or fin), the pressure distribution in a pipe flow, and the deformation of a bar under axial load, for example, can be determined using these elements. Although these elements have cross-sectional area, they are generally shown schematically as a line element (Figure 2.1(b)). In some cases, the cross-sectional area of the element may be nonuniform.

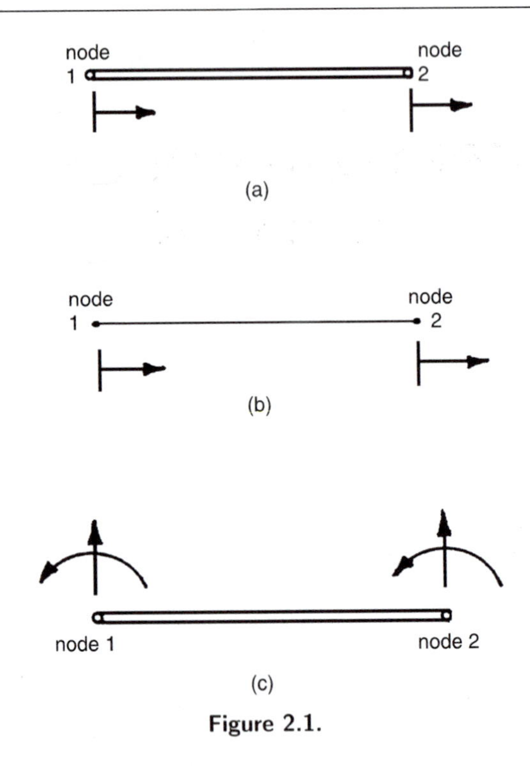

Figure 2.1.

For a simple analysis, one-dimensional elements are assumed to have two nodes, one at each end, with the corresponding value of the field variable chosen as the unknown (degree of freedom). However, for the analysis of beams, the values of the field variable (transverse displacement) and its derivative (slope) are chosen as the unknowns (degrees of freedom) at each node as shown in Figure 2.1(c).

When the configuration and other details of the problem can be described in terms of two independent spatial coordinates, we can use the two-dimensional elements shown in Figure 2.2. The basic element useful for two-dimensional analysis is the triangular element. Although a quadrilateral (or its special forms, rectangle and parallelogram) element can be obtained by assembling two or four triangular elements, as shown in Figure 2.3, in some cases the use of quadrilateral (or rectangle or parallelogram) elements proves to be advantageous. For the bending analysis of plates, multiple degrees of freedom (transverse displacement and its derivatives) are used at each node.

If the geometry, material properties, and other parameters of the body can be described by three independent spatial coordinates, we can idealize the body by using the three-dimensional elements shown in Figure 2.4. The basic three-dimensional element, analogous to the triangular element in the case of two-dimensional problems, is the tetrahedron element. In some cases the hexahedron element, which can be obtained by assembling five tetrahedrons as indicated in Figure 2.5, can be used advantageously. Some problems, which are actually three-dimensional, can be described by only one or two independent coordinates. Such problems can be idealized by using an axisymmetric or ring type of elements shown in Figure 2.6. The problems that possess axial symmetry, such as pistons, storage tanks, valves, rocket nozzles, and reentry vehicle heat shields, fall into this category.

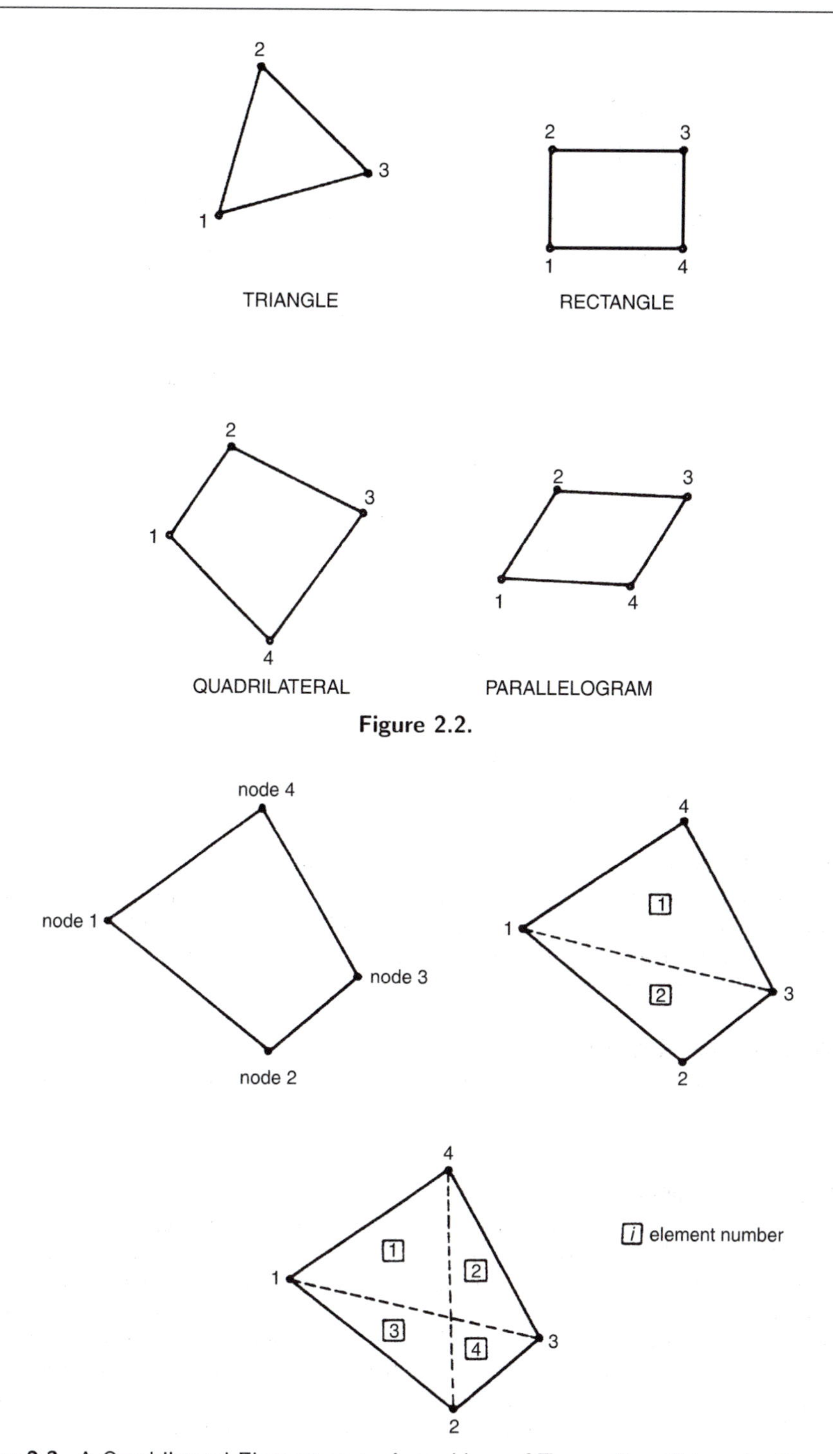

Figure 2.2.

Figure 2.3. A Quadrilateral Element as an Assemblage of Two or Four Triangular Elements.

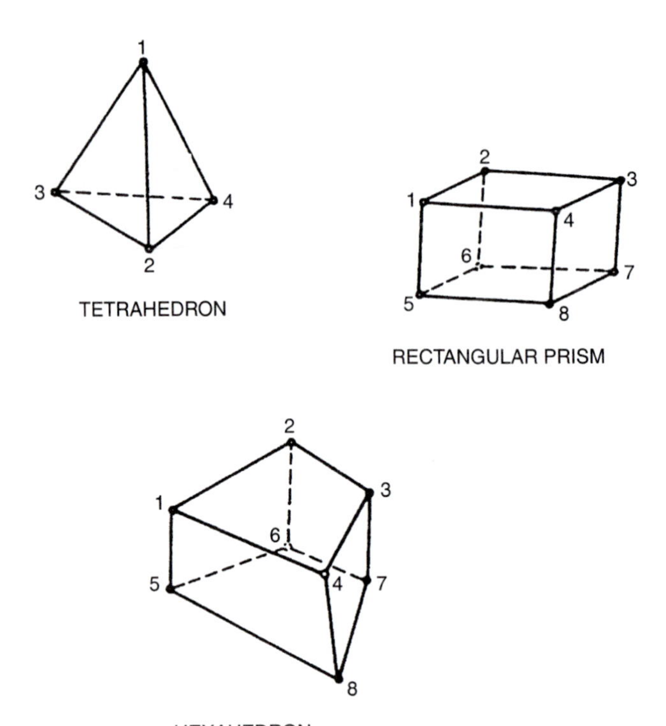

Figure 2.4. Three-Dimensional Finite Elements.

For the discretization of problems involving curved geometries, finite elements with curved sides are useful. Typical elements having curved boundaries are shown in Figure 2.7. The ability to model curved boundaries has been made possible by the addition of midside nodes. Finite elements with straight sides are known as linear elements, whereas those with curved sides are called higher order elements.

2.3 DISCRETIZATION PROCESS

The various considerations to be taken in the discretization process are given in the following sections [2.1].

2.3.1 Type of Elements

Often, the type of elements to be used will be evident from the physical problem. For example, if the problem involves the analysis of a truss structure under a given set of load conditions (Figure 2.8(a)), the type of elements to be used for idealization is obviously the "bar or line elements" as shown in Figure 2.8(b). Similarly, in the case of stress analysis of the short beam shown in Figure 2.9(a), the finite element idealization can be done using three-dimensional solid elements as shown in Figure 2.9(b). However, the type of elements to be used for idealization may not be apparent, and in such cases one has to choose the type of elements judicially. As an example, consider the problem of analysis of the thin-walled shell shown in Figure 2.10(a). In this case, the shell can be idealized

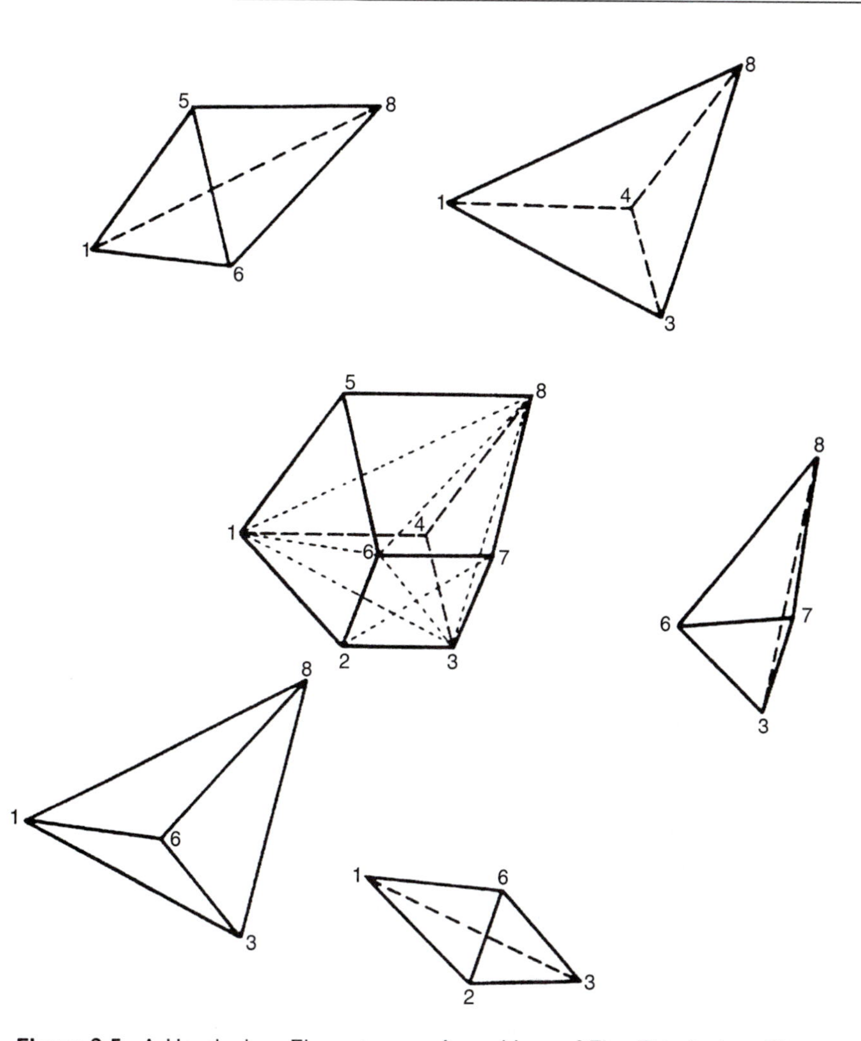

Figure 2.5. A Hexahedron Element as an Assemblage of Five Tetrahedron Elements.

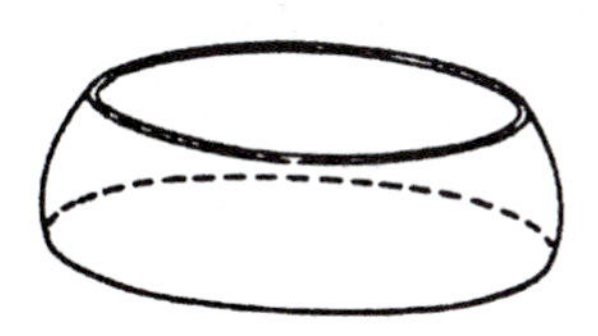

(a) An one dimensional axisymmetric (shell) element

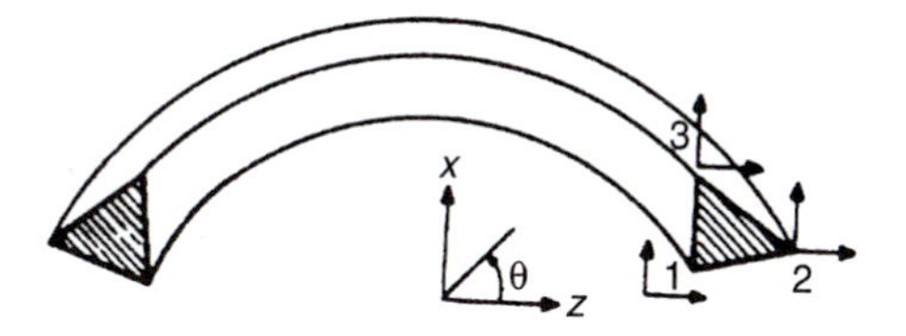

(b) A two dimensional axisymmetric (toroidal) element

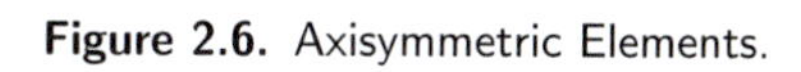

Figure 2.6. Axisymmetric Elements.

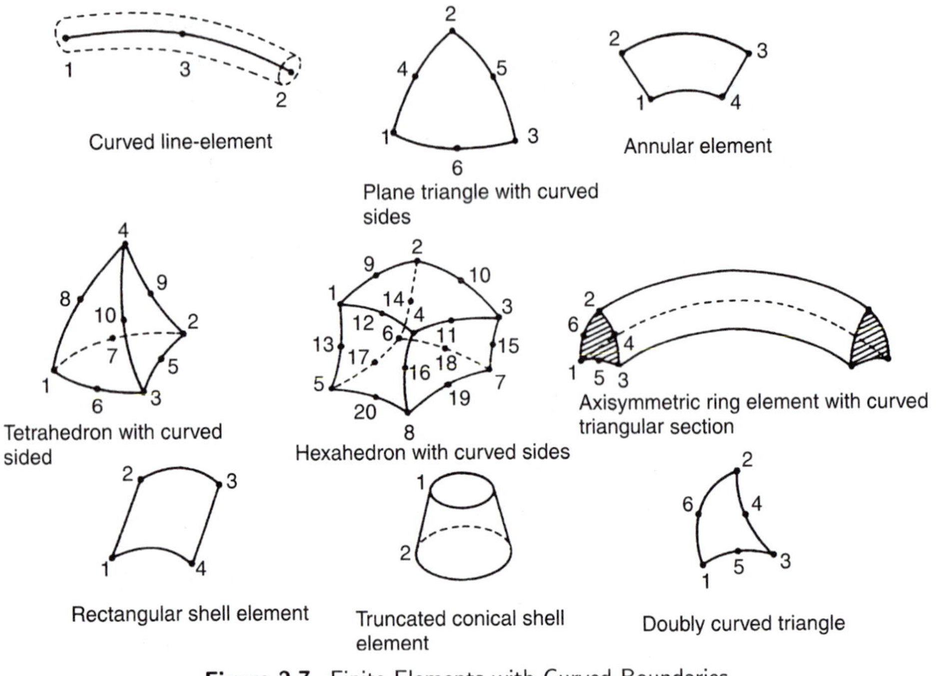

Figure 2.7. Finite Elements with Curved Boundaries.

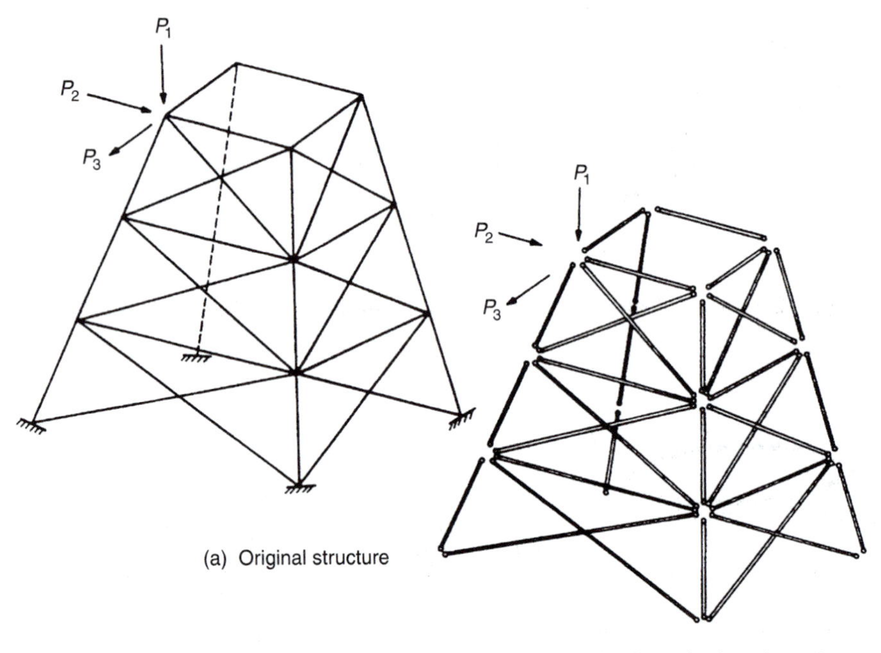

Figure 2.8.

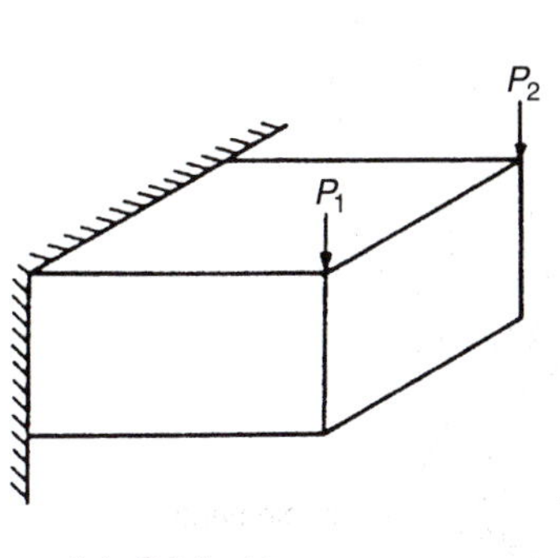

(a) Original beam

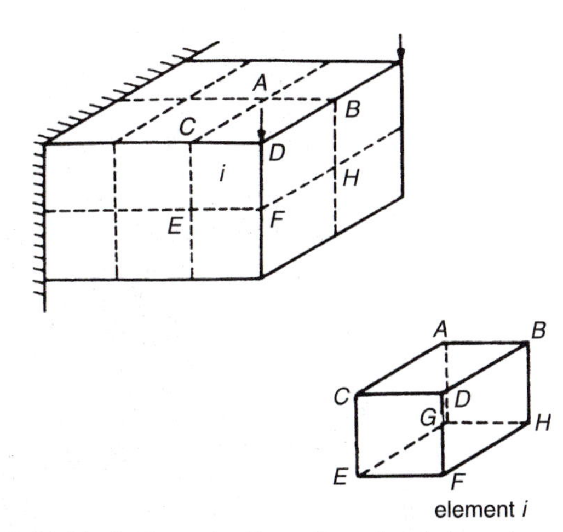

element i

(b) Idealization using three-dimensional elements

Figure 2.9.

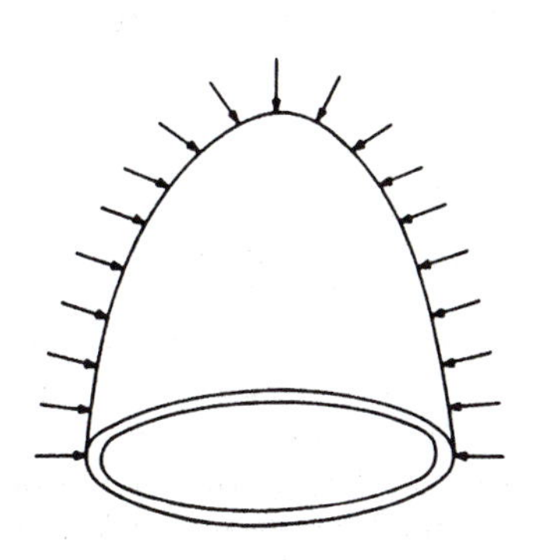

(a) Original shell

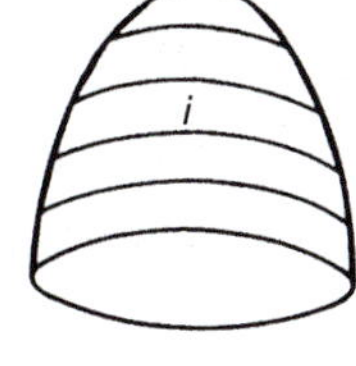

Using conical ring elements

Using axisymmetric ring elements

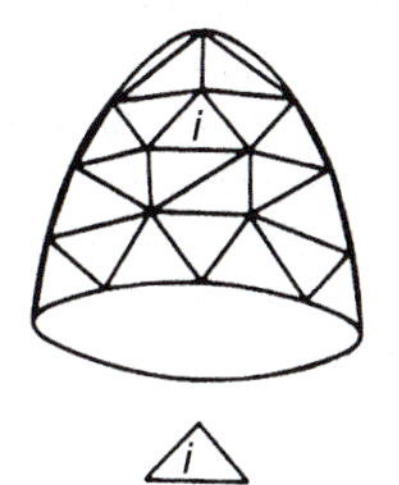

Using flat triangular plate elements

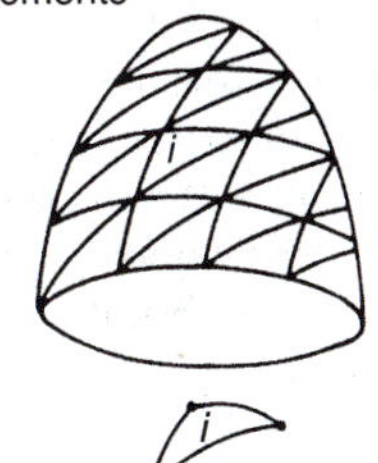

Using curved triangular plate elements

(b) Idealization using different types of elements

Figure 2.10. A Thin-Walled Shell under Pressure.

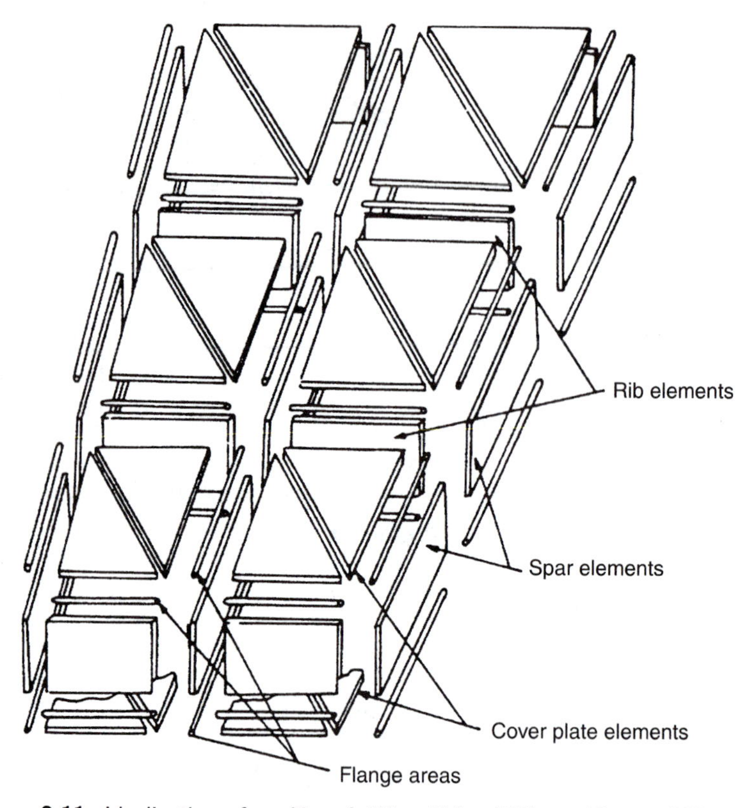

Figure 2.11. Idealization of an Aircraft Wing Using Different Types of Elements.

by several types of elements as shown in Figure 2.10(b). Here, the number of degrees of freedom needed, the expected accuracy, the ease with which the necessary equations can be derived, and the degree to which the physical structure can be modeled without approximation will dictate the choice of the element type to be used for idealization. In certain problems, the given body cannot be represented as an assemblage of only one type of elements. In such cases, we may have to use two or more types of elements for idealization. An example of this would be the analysis of an aircraft wing. Since the wing consists of top and bottom covers, stiffening webs, and flanges, three types of elements, namely, triangular plate elements (for covers), rectangular shear panels (for webs), and frame elements (for flanges), have been used in the idealization shown in Figure 2.11.

2.3.2 Size of Elements

The size of elements influences the convergence of the solution directly and hence it has to be chosen with care. If the size of the elements is small, the final solution is expected to be more accurate. However, we have to remember that the use of elements of smaller size will also mean more computational time. Sometimes, we may have to use elements of different sizes in the same body. For example, in the case of stress analysis of the box beam shown

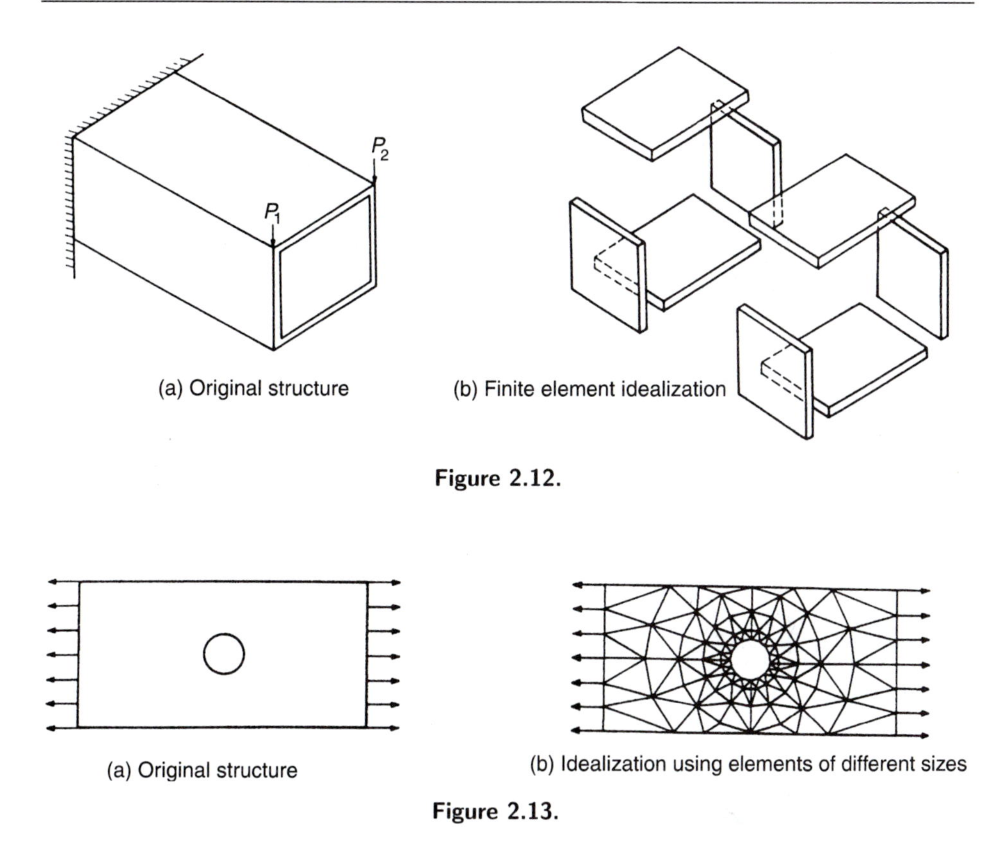

(a) Original structure (b) Finite element idealization

Figure 2.12.

(a) Original structure (b) Idealization using elements of different sizes

Figure 2.13.

in Figure 2.12(a), the size of all the elements can be approximately the same, as shown in Figure 2.12(b). However, in the case of stress analysis of a plate with a hole shown in Figure 2.13(a), elements of different sizes have to be used, as shown in Figure 2.13(b). The size of elements has to be very small near the hole (where stress concentration is expected) compared to far away places. In general, whenever steep gradients of the field variable are expected, we have to use a finer mesh in those regions. Another characteristic related to the size of elements that affects the finite element solution is the aspect ratio of the elements. The aspect ratio describes the shape of the element in the assemblage of elements. For two-dimensional elements, the aspect ratio is taken as the ratio of the largest dimension of the element to the smallest dimension. Elements with an aspect ratio of nearly unity generally yield best results [2.2].

2.3.3 Location of Nodes

If the body has no abrupt changes in geometry, material properties, and external conditions (e.g., load and temperature), the body can be divided into equal subdivisions and hence the spacing of the nodes can be uniform. On the other hand, if there are any discontinuities in the problem, nodes have to be introduced at these discontinuities, as shown in Figure 2.14.

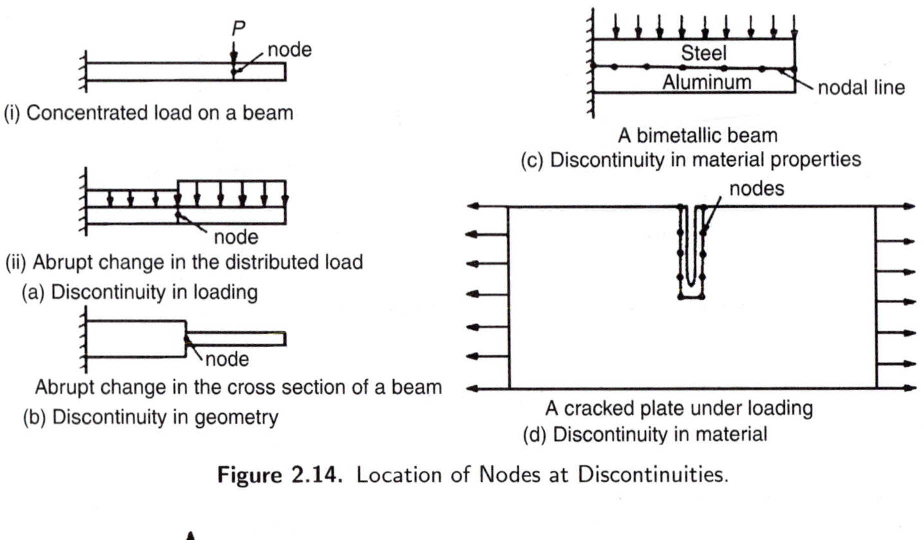

Figure 2.14. Location of Nodes at Discontinuities.

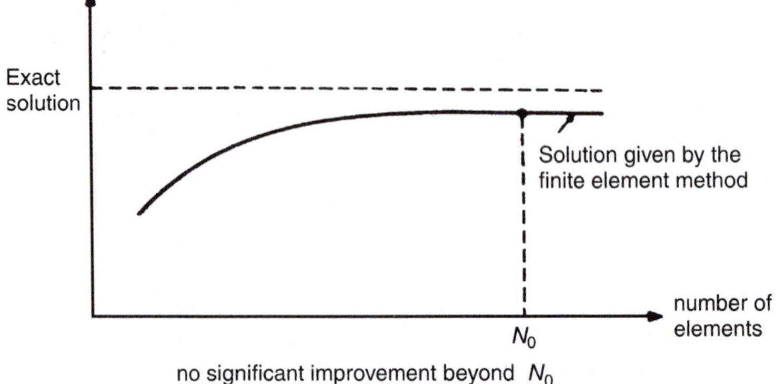

Figure 2.15. Effect of Varying the Number of Elements.

2.3.4 Number of Elements

The number of elements to be chosen for idealization is related to the accuracy desired, size of elements, and the number of degrees of freedom involved. Although an increase in the number of elements generally means more accurate results, for any given problem, there will be a certain number of elements beyond which the accuracy cannot be improved by any significant amount. This behavior is shown graphically in Figure 2.15. Moreover, since the use of large number of elements involves a large number of degrees of freedom, we may not be able to store the resulting matrices in the available computer memory.

2.3.5 Simplifications Afforded by the Physical Configuration of the Body

If the configuration of the body as well as the external conditions are symmetric, we may consider only half of the body for finite element idealization. The symmetry conditions,

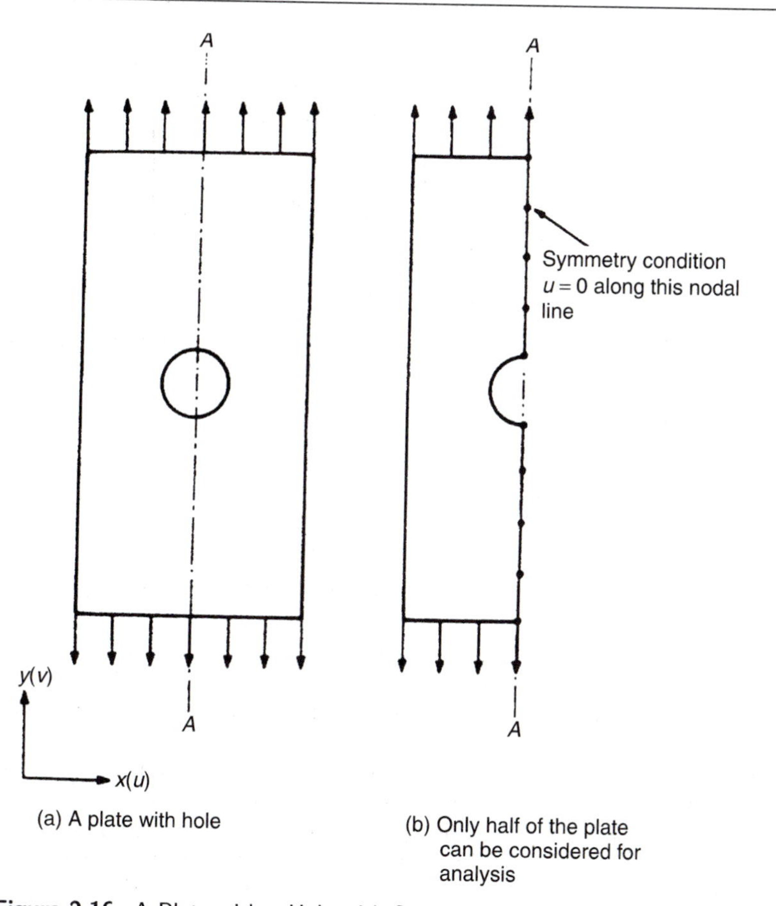

Figure 2.16. A Plate with a Hole with Symmetric Geometry and Loading.

however, have to be incorporated in the solution procedure. This is illustrated in Figure 2.16, where only half of the plate with hole, having symmetry in both geometry and loading, is considered for analysis.* Since there cannot be a horizontal displacement along the line of symmetry AA, the condition that $u = 0$ has to be incorporated while finding the solution.

2.3.6 Finite Representation of Infinite Bodies

In most of the problems, like in the case of analysis of beams, plates, and shells, the boundaries of the body or continuum are clearly defined. Hence, the entire body can be considered for the element idealization. However, in some cases, as in the case of analysis of dams, foundations, and semiinfinite bodies, the boundaries are not clearly defined. In the case of dams (Figure 2.17), since the geometry is uniform and the

---

* In this example, even one-fourth of the plate can be considered for analysis due to symmetry about both horizontal and vertical center lines.

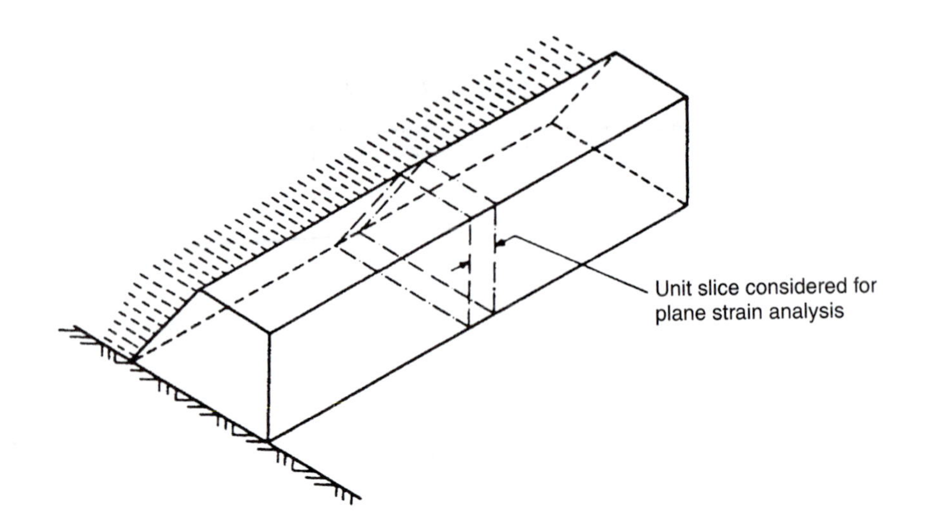

Figure 2.17. A Dam with Uniform Geometry and Loading.

loading does not change in the length direction, a unit slice of the dam can be considered for idealization and analyzed as a plane strain problem. However, in the case of the foundation problem shown in Figure 2.18(a), we cannot idealize the complete semiinfinite soil by finite elements. Fortunately, it is not really necessary to idealize the infinite body. Since the effect of loading decreases gradually with increasing distance from the point of loading, we can consider only that much of the continuum in which the loading is expected to have significant effect as shown in Figure 2.18(b). Once the significant extent of the infinite body is identified as shown in Figure 2.18(b), the boundary conditions for this finite body have to be incorporated in the solution. For example, if the horizontal movement only has to be restrained for sides AB and CD (i.e., $u = 0$), these sides are supposed to be on rollers as shown in Figure 2.18(b). In this case, the bottom boundary can be either completely fixed ($u = v = 0$) or constrained only against vertical movement ($v = 0$). The fixed conditions ($u = v = 0$ along BC) are often used if the lower boundary is taken at the known location of a bedrock surface.

In Figure 2.18 the semiinfinite soil has been simulated by considering only a finite portion of the soil. In some applications, the determination of the size of the finite domain may pose a problem. In such cases, one can use infinite elements for modeling [2.3–2.5]. As an example, Figure 2.19 shows a four-node element that is infinitely long in the x direction. The coordinates of the nodes of this infinite element can be transformed to the natural coordinate system (s, t) as [see Section 4.3.3 for the definition of natural coordinate system]

$$s = 1 - 2\left\{\frac{1}{x} \cdot \frac{(y_3 - y)x_1 + (y - y_1)x_4}{(y_3 - y_1)}\right\}^m; \quad m \geq 1$$

$$t = 1 - 2\left\{\frac{y_3 - y}{y_3 - y_1}\right\}$$

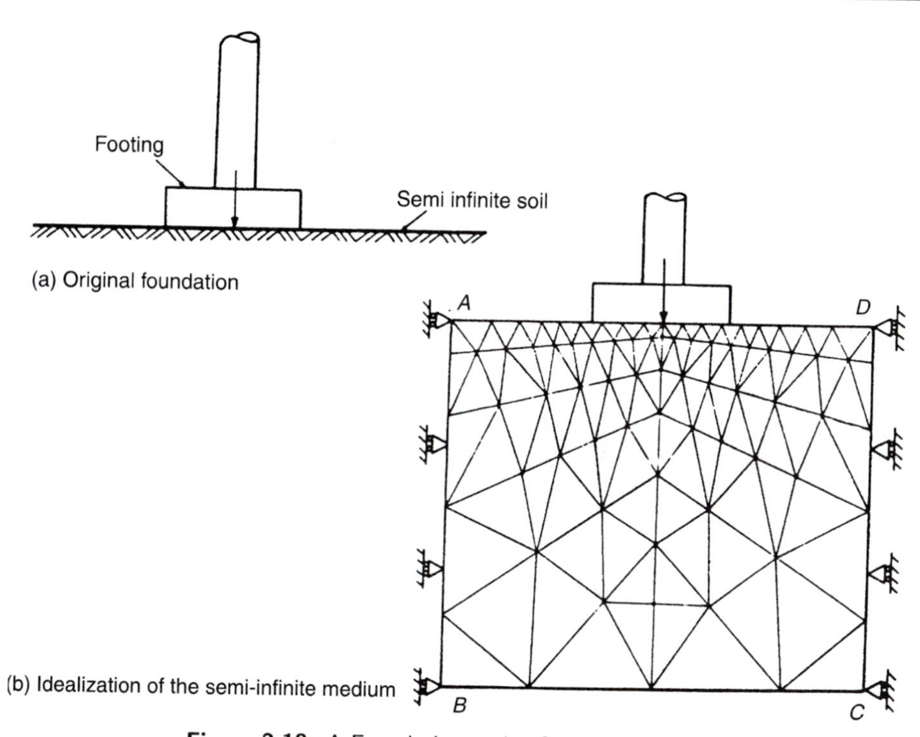

Figure 2.18. A Foundation under Concentrated Load.

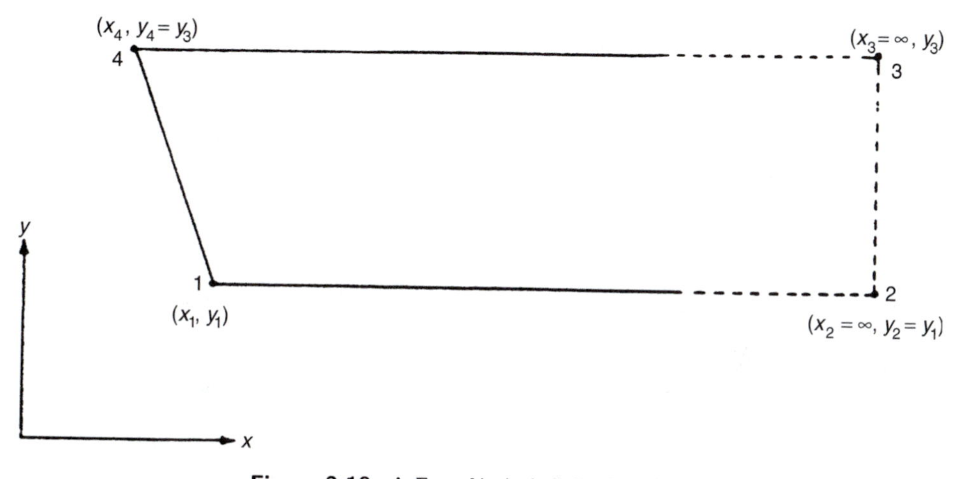

Figure 2.19. A Four-Node Infinite Element.

2.4 NODE NUMBERING SCHEME

As seen in Chapter 1, the finite element analysis of practical problems often leads to matrix equations in which the matrices involved will be banded. The advances in the

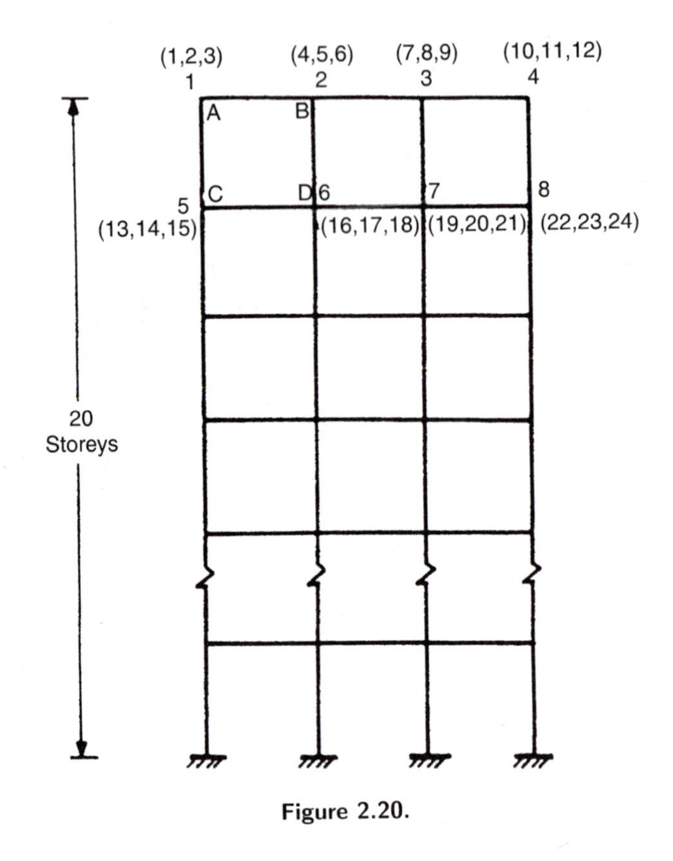

Figure 2.20.

finite element analysis of large practical systems have been made possible largely due to the banded nature of the matrices. Furthermore, since most of the matrices involved (e.g., stiffness matrices) are symmetric, the demands on the computer storage can be substantially reduced by storing only the elements involved in half bandwidth instead of storing the whole matrix.

The bandwidth of the overall or global characteristic matrix depends on the node numbering scheme and the number of degrees of freedom considered per node [2.6]. If we can minimize the bandwidth, the storage requirements as well as solution time can also be minimized. Since the number of degrees of freedom per node is generally fixed for any given type of problem, the bandwidth can be minimized by using a proper node numbering scheme. As an example, consider a three-bay frame with rigid joints, 20 storeys high, shown in Figure 2.20. Assuming that there are three degrees of freedom per node, there are 240 unknowns in the final equations (excluding the degrees of freedom corresponding to the fixed nodes) and if the entire stiffness matrix is stored in the computer it will require $240^2 = 57{,}600$ locations. The bandwidth (strictly speaking, half bandwidth) of the overall stiffness matrix is 15 and thus the storage required for the upper half band is only $15 \times 240 = 3600$ locations.

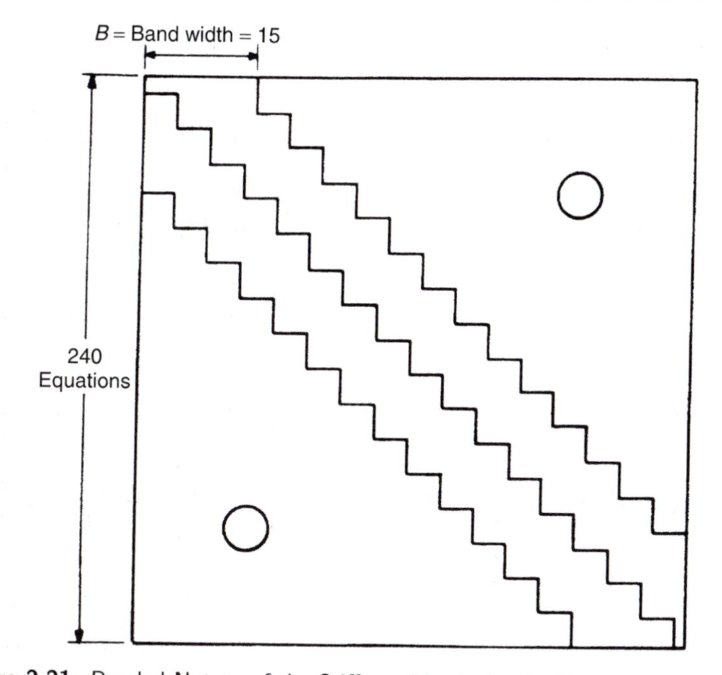

Figure 2.21. Banded Nature of the Stiffness Matrix for the Frame of Figure 2.20.

Before we attempt to minimize the bandwidth, we discuss the method of calculating the bandwidth. For this, we consider again the rigid jointed frame shown in Figure 2.20. By applying constraints to all the nodal degrees of freedom except number 1 at node 1 (joint A), it is clear that an imposed unit displacement in the direction of 1 will require constraining forces at the nodes directly connected to node A—that is, B and C. These constraining forces are nothing but the cross-stiffnesses appearing in the stiffness matrix and these forces are confined to the nodes B and C. Thus, the nonzero terms in the first row of the global stiffness matrix (Figure 2.21) will be confined to the first 15 positions. This defines the bandwidth (B) as

$$\text{Bandwidth } (B) = (\text{maximum difference between the numbered degrees of freedom at the ends of any member} + 1)$$

This definition can be generalized so as to be applicable for any type of finite element as

$$\text{Bandwidth } (B) = (D + 1) \cdot f \tag{2.1}$$

where D is the maximum largest difference in the node numbers occurring for all elements of the assemblage, and f is the number of degrees of freedom at each node.

The previous equation indicates that D has to be minimized in order to minimize the bandwidth. Thus, a shorter bandwidth can be obtained simply by numbering the nodes across the shortest dimension of the body. This is clear from Figure 2.22 also,

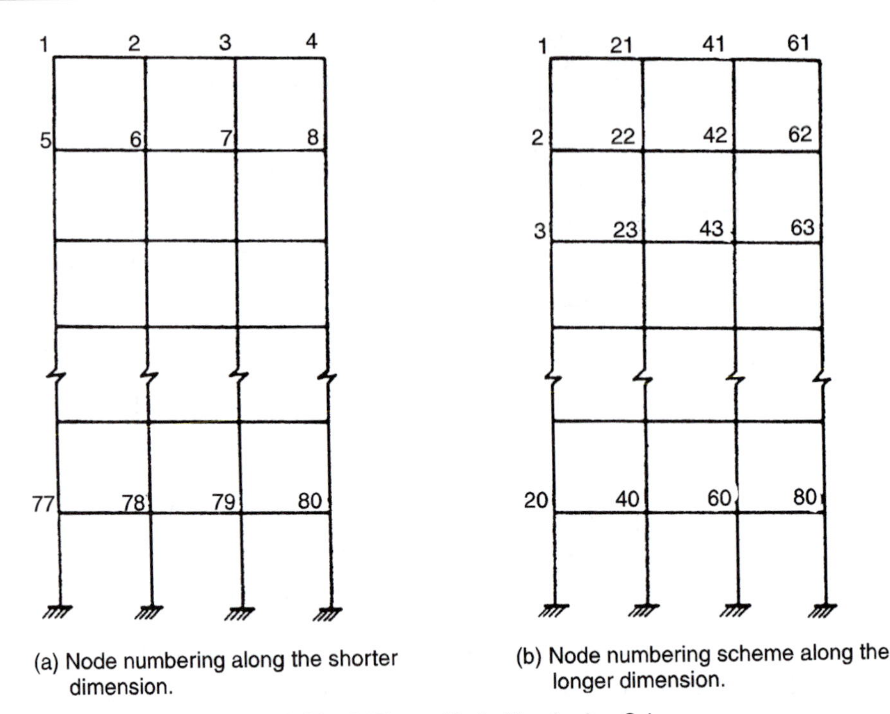

(a) Node numbering along the shorter dimension.

(b) Node numbering scheme along the longer dimension.

Figure 2.22. Different Node Numbering Schemes.

where the numbering of nodes along the shorter dimension produces a bandwidth of $B = 15$ $(D = 4)$, whereas the numbering along the longer dimension produces a bandwidth of $B = 63$ $(D = 20)$.

2.5 AUTOMATIC MESH GENERATION*

As indicated in the previous section, the bandwidth of the overall system matrix depends on the manner in which the nodes are numbered. For simple systems or regions, it is easy to label the nodes so as to minimize the bandwidth. But for large systems, the procedure becomes nearly impossible. Hence, automatic mesh generation algorithms, capable of discretizing any geometry into an efficient finite element mesh without user intervention, have been developed [2.7, 2.8]. Most commercial finite element software has built-in automatic mesh generation codes. An automatic mesh generation program generates the locations of the node points and elements, labels the nodes and elements, and provides the element–node connectivity relationships. The automatic mesh generation schemes are usually tied to solid modeling and computer-aided design schemes. When the user supplies information on the surfaces and volumes of the material domains that make up the object or system, an automatic mesh generator generates the nodes and elements in the object. The user can also specify the minimum permissible element sizes for different regions of the object.

*This section may be omitted without loss of continuity in the text material.

The most common methods used in the development of automatic mesh generators are the tesselation and octree methods [2.9, 2.10]. In the tesselation method, the user gives a collection of node points and also an arbitrary starting node. The method then creates the first simplex element using the neighboring nodes. Then a subsequent or neighboring element is generated by selecting the node point that gives the least distorted element shape. The procedure is continued until all the elements are generated. The step-by-step procedure involved in this method is illustrated in Figure 2.23 for a two-dimensional example. Alternately, the user can define the boundary of the object by a series of nodes. Then the tesselation method connects selected boundary nodes to generate simplex elements. The stepwise procedure used[†] in this approach is shown in Figure 2.24.

The octree methods belong to a class of mesh generation schemes known as tree structure methods, which are extensively used in solid modeling and computer graphics display methods. In the octree method, the object is first considered enclosed in a three-dimensional cube. If the object does not completely (uniformly) cover the cube, the cube is subdivided into eight equal parts. In the two-dimensional analog of the octree method, known as the quadtree method, the object is first considered enclosed in a square region. If the object does not completely cover the square, the square is subdivided into four equal quadrants. If any one of the resulting quadrants is full (completely occupied by the object) or empty (not occupied by the object), then it is not subdivided further. On the other hand, if any one of the resulting quadrants is partially full (partially occupied by the object), it is subdivided into four quadrants. This procedure of subdividing partially full quadrants is continued until all the resulting regions are either full or empty or until some predetermined level of resolution is achieved. At the final stage, the partially full quadrants are assumed to be either full or empty arbitrarily based on a prespecified criterion.

Example 2.1 Generate the finite element mesh for the two-dimensional object (region) shown by the crossed lines in Figure 2.25(a) using the quadtree method.

Solution First, the object is enclosed in a square region as shown by the dotted lines in Figure 2.25(a). Since the object does not occupy the complete square, the square is divided into four parts as shown in Figure 2.25(b). Since none of these parts are fully occupied by the object, each part is subdivided into four parts as shown in Figure 2.25(c). It can be seen that parts 1, 3, and 4 of A, part 3 of B, parts 2–4 of C, and parts 1–3 of D are completely occupied by the object, whereas parts 1, 2, and 4 of B and part 1 of C are empty (not occupied by the object). In addition, part 2 of A and part 4 of D are partially occupied by the object; hence, they are further subdivided into four parts each as shown in Figure 2.25(d). It can be noted that parts α and γ of part 2 (of A) and parts α and β of part 4 (of D) are completely occupied while the remaining parts, namely β and δ of part 2 (of A) and γ and δ of part 4 (of D), are empty. Since all the parts at this stage are either completely occupied or completely empty, no further subdivision is necessary. The corresponding quadtree representation is shown in Figure 2.25(e). Note that the shape of the finite elements is assumed to be square in this example.

[†] A simplex in an n-dimensional space is defined as a geometric figure having $n + 1$ nodes or corners. Thus, the simplex will be a triangle in a two-dimensional space and a tetrahedron in three-dimensional space.

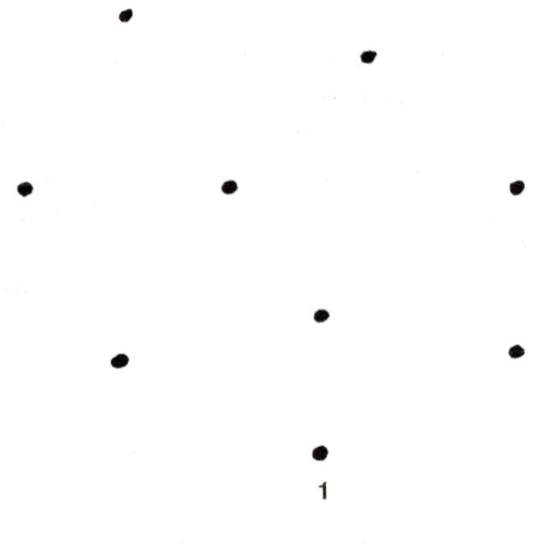

(a) Nodes in the object or region

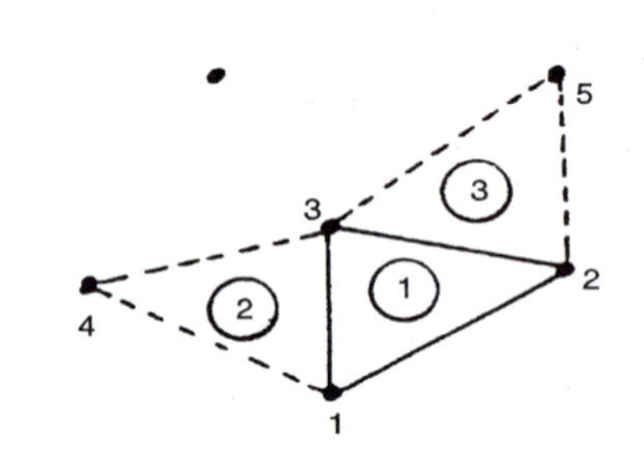

(b) Generation of simplex elements

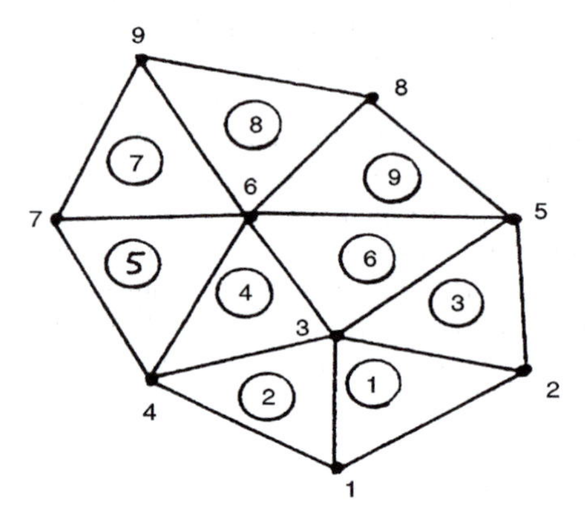

(c) Complete set of nodes and elements

Figure 2.23. Mesh Generation Using Tesselation Method.

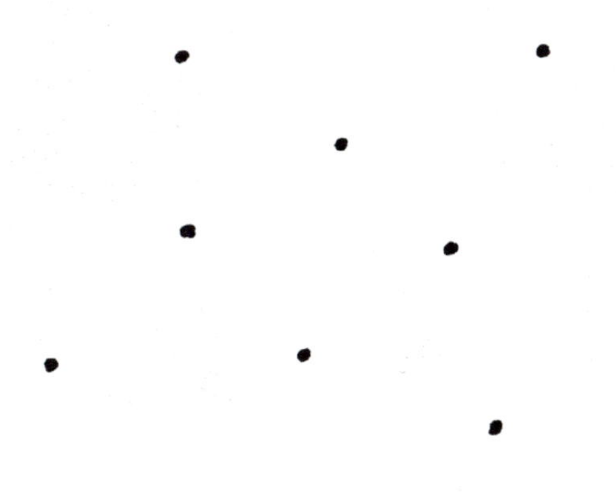

(a) Nodes on the boundary of the object or region

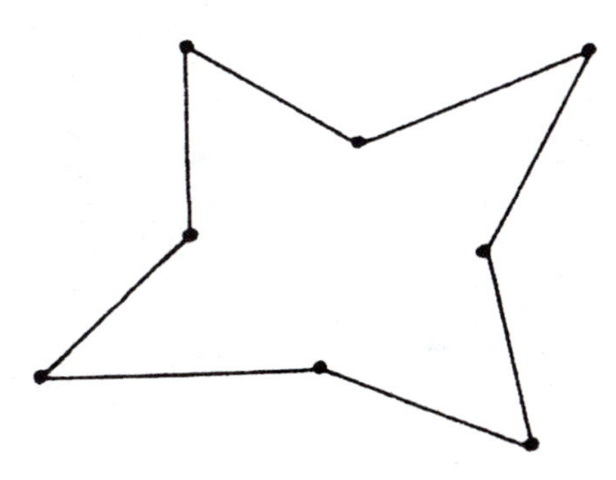

(b) Geometry of the object or region

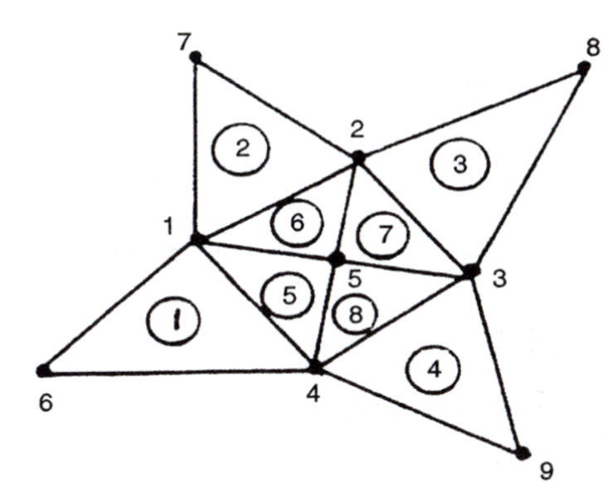

(c) Complete set of nodes and elements

Figure 2.24. Tesselation Method with Nodes Defined on the Boundary.

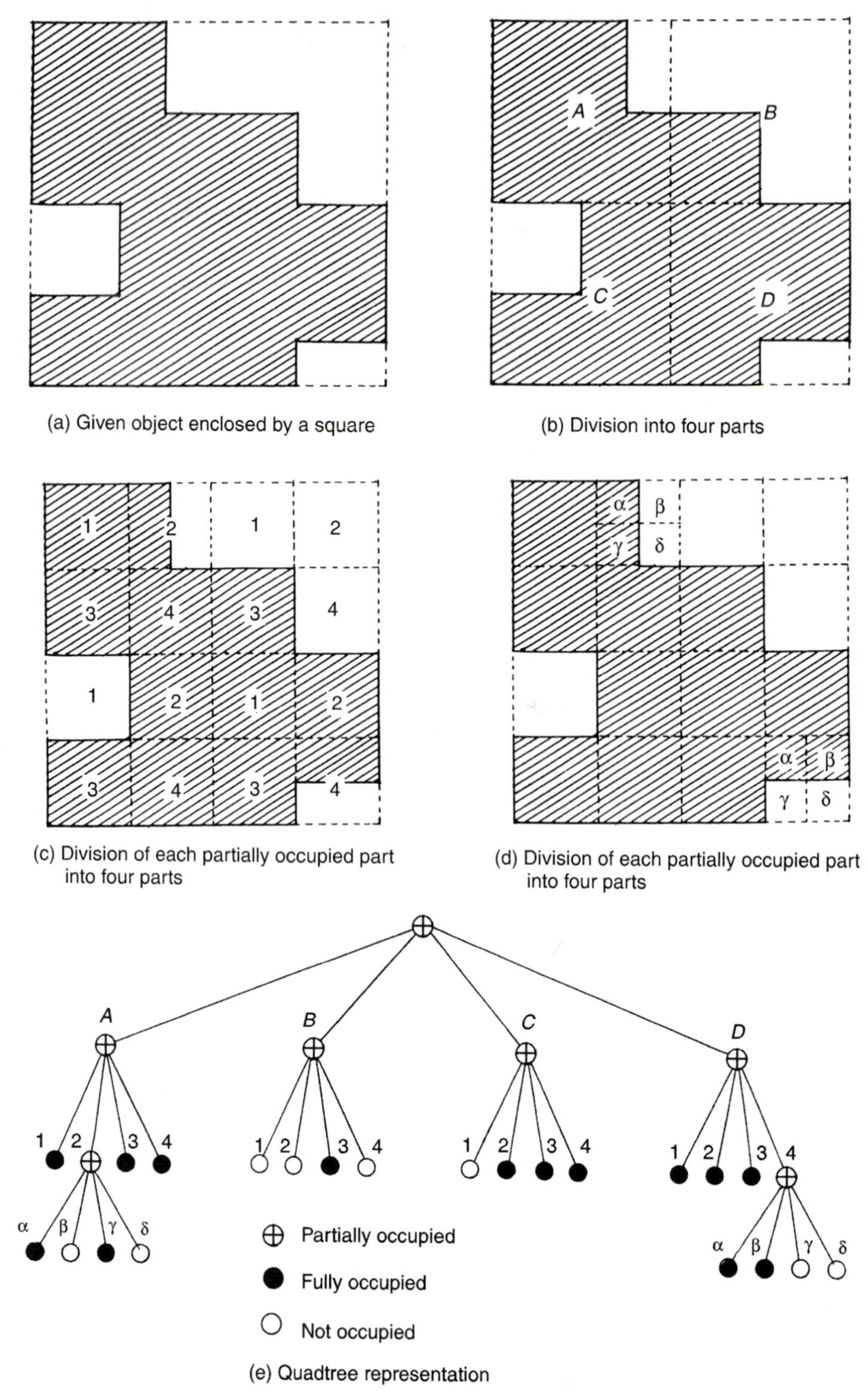

Figure 2.25. Mesh Generation Using Quadtree Method.

REFERENCES

2.1 O.C. Zienkiewicz: The finite element method: From intuition to generality, *Applied Mechanics Reviews*, *23*, 249–256, 1970.

2.2 R.W. Clough: Comparison of three dimensional finite elements, *Proceedings of the Symposium on Application of Finite Element Methods in Civil Engineering*. Vanderbilt University, Nashville, pp. 1–26, November 1969.

2.3 P. Bettess: Infinite elements, *International Journal for Numerical Methods in Engineering*, *11*, 53–64, 1977.

2.4 F. Medina and R.L. Taylor: Finite element techniques for problems of unbounded domains, *International Journal for Numerical Methods in Engineering*, *19*, 1209–1226, 1983.

2.5 S. Pissanetzky: An infinite element and a formula for numerical quadrature over an infinite interval, *International Journal for Numerical Methods in Engineering*, *19*, 913–927, 1983.

2.6 R.J. Collins: Bandwidth reduction by automatic renumbering, *International Journal for Numerical Methods in Engineering*, *6*, 345–356, 1973.

2.7 J.E. Akin: *Finite Elements for Analysis and Design*, Academic Press, London, 1994.

2.8 K. Baldwin (ed.): *Modern Methods for Automatic Finite Element Mesh Generation*, American Society of Civil Engineers, New York, 1986.

2.9 P.L. George: *Automatic Generation of Meshes*, Wiley, New York, 1991.

2.10 C.G. Armstrong: Special issue: Automatic mesh generation, *Advances in Engineering Software*, *13*, 217–337, 1991.

PROBLEMS

2.1 A thick-walled pressure vessel is subjected to an internal pressure as shown in Figure 2.26. Model the cross section of the pressure vessel by taking advantage of the symmetry of the geometry and load condition.

2.2 A helical spring is subjected to a compressive load as shown in Figure 2.27. Suggest different methods of modeling the spring using one-dimensional elements.

2.3 A rectangular plate with a v-notch is shown in Figure 2.28. Model the plate using triangular elements by taking advantage of the symmetry of the system.

2.4 A drilling machine is modeled using one-dimensional beam elements as shown in Figure 2.29. If two degrees of freedom are associated with each node, label the node numbers for minimizing the bandwidth of the stiffness matrix of the system.

2.5 The plate shown in Figure 2.30 is modeled using 13 triangular and 2 quadrilateral elements. Label the nodes such that the bandwidth of the system

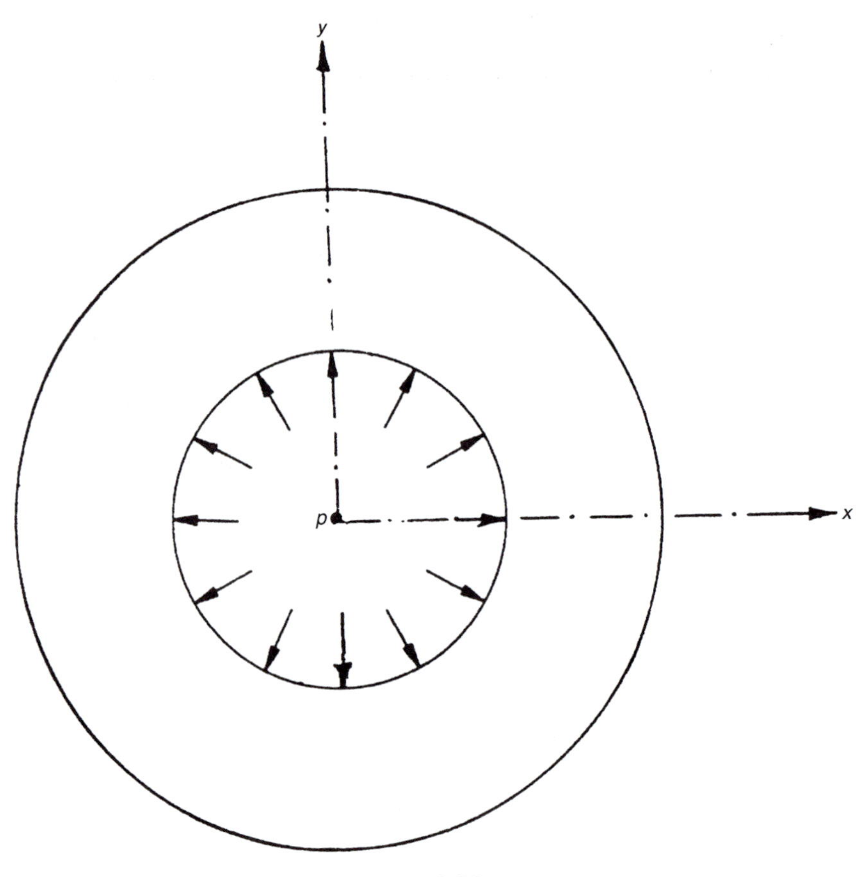

Figure 2.26.

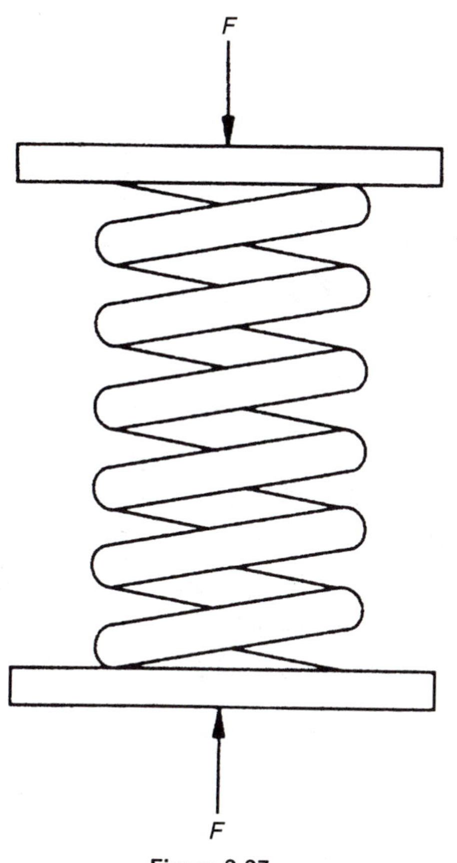

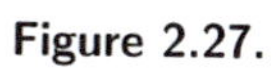
Figure 2.27.

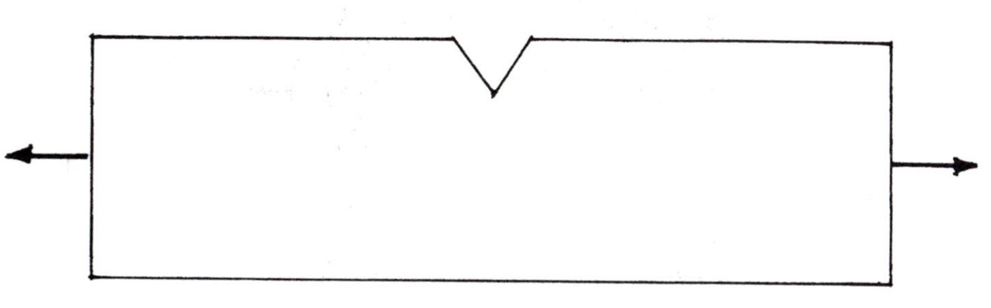

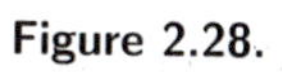
Figure 2.28.

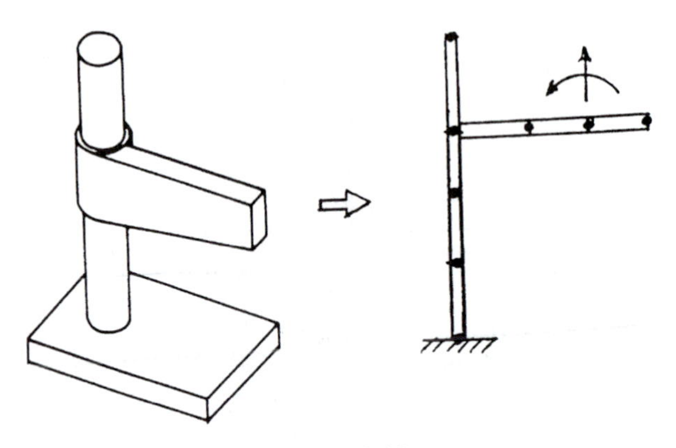

Figure 2.29.

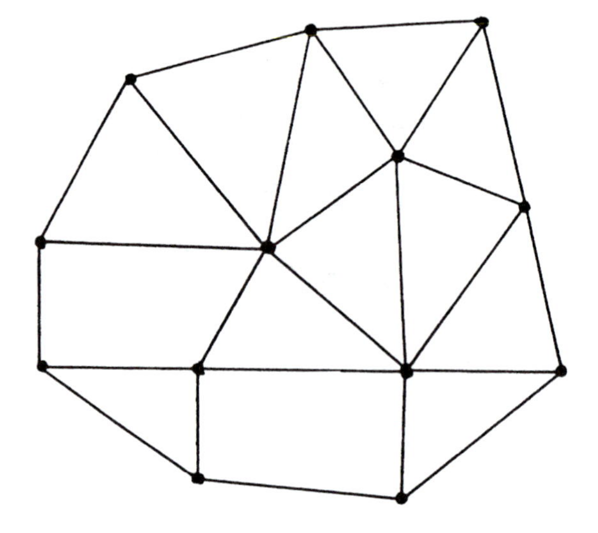

Figure 2.30.

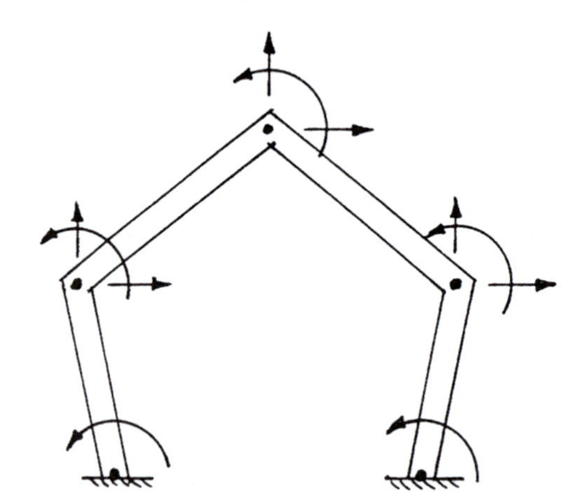

Figure 2.31.

matrix is minimum. Compute the resulting bandwidth assuming one degree of freedom at each node.

2.6–2.10 Label the elements and nodes for each of the systems shown in Figures 2.31–2.35 to produce a minimum bandwidth.

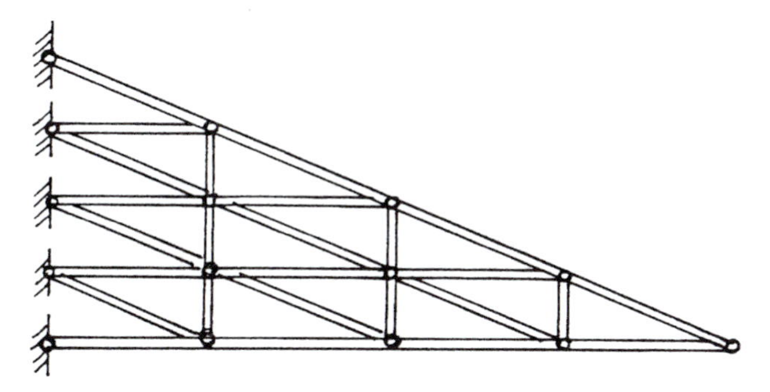

Figure 2.32.

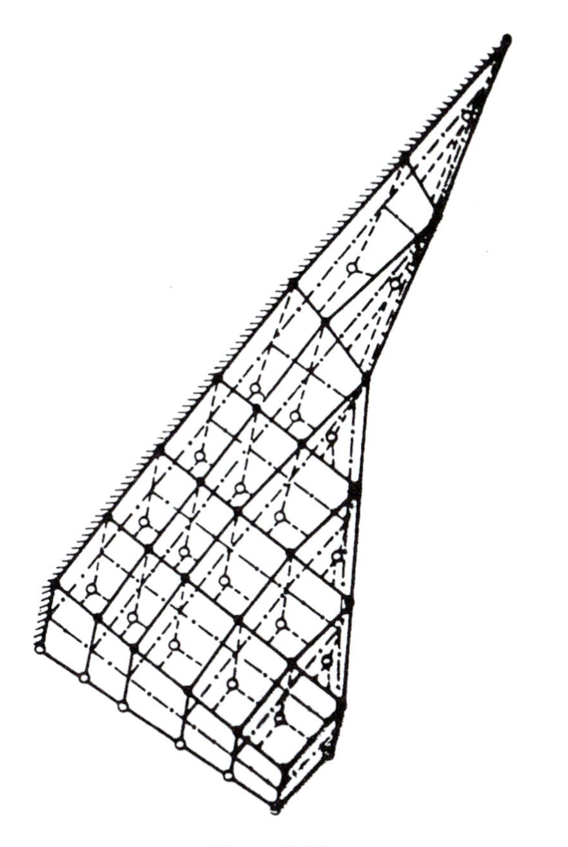

Figure 2.33.

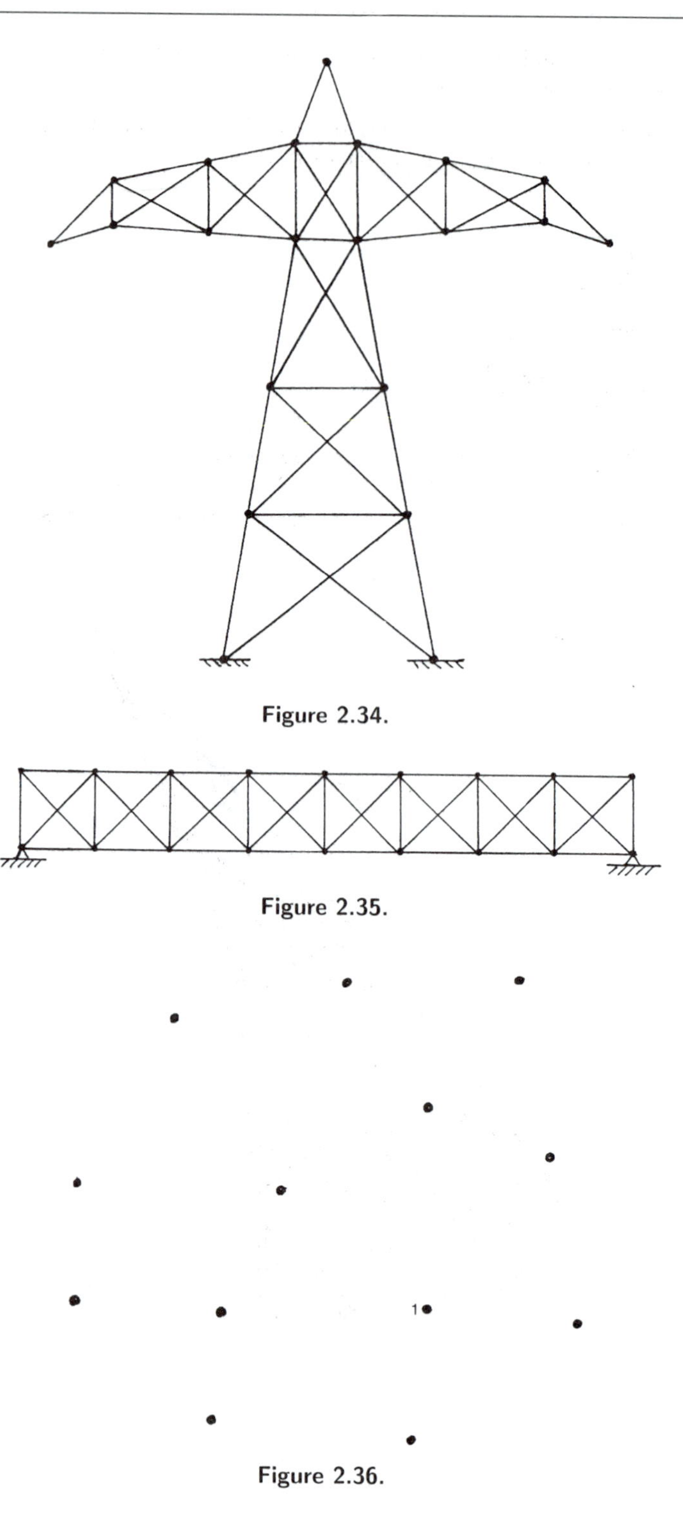

Figure 2.34.

Figure 2.35.

Figure 2.36.

Figure 2.37.

2.11 Consider the collection of node points shown in Figure 2.36 for a two-dimensional object. Generate the finite element mesh using the tesselation method.

2.12 Generate the finite element mesh for the two-dimensional object shown in Figure 2.37 using the quadtree method.

3

INTERPOLATION MODELS

3.1 INTRODUCTION

As stated earlier, the basic idea of the finite element method is piecewise approximation—that is, the solution of a complicated problem is obtained by dividing the region of interest into small regions (finite elements) and approximating the solution over each subregion by a simple function. Thus, a necessary and important step is that of choosing a simple function for the solution in each element. The functions used to represent the behavior of the solution within an element are called interpolation functions or approximating functions or interpolation models. Polynomial-type interpolation functions have been most widely used in the literature due to the following reasons:

(i) It is easier to formulate and computerize the finite element equations with polynomial-type interpolation functions. Specifically, it is easier to perform differentiation or integration with polynomials.
(ii) It is possible to improve the accuracy of the results by increasing the order of the polynomial, as shown in Figure 3.1. Theoretically, a polynomial of infinite order corresponds to the exact solution. But in practice we use polynomials of finite order only as an approximation.

Although trigonometric functions also possess some of these properties, they are seldom used in the finite element analysis [3.1]. We shall consider only polynomial-type interpolation functions in this book.

When the interpolation polynomial is of order one, the element is termed a linear element. A linear element is called a simplex element if the number of nodes in the element is 2, 3, and 4 in 1, 2, and 3 dimensions, respectively. If the interpolation polynomial is of order two or more, the element is known as a higher order element. In higher order elements, some secondary (midside and/or interior) nodes are introduced in addition to the primary (corner) nodes in order to match the number of nodal degrees of freedom with the number of constants (generalized coordinates) in the interpolation polynomial.

In general, fewer higher order elements are needed to achieve the same degree of accuracy in the final results. Although it does not reduce the computational time, the reduction in the number of elements generally reduces the effort needed in the preparation of data and hence the chances of errors in the input data. The higher order elements are especially useful in cases in which the gradient of the field variable is expected to

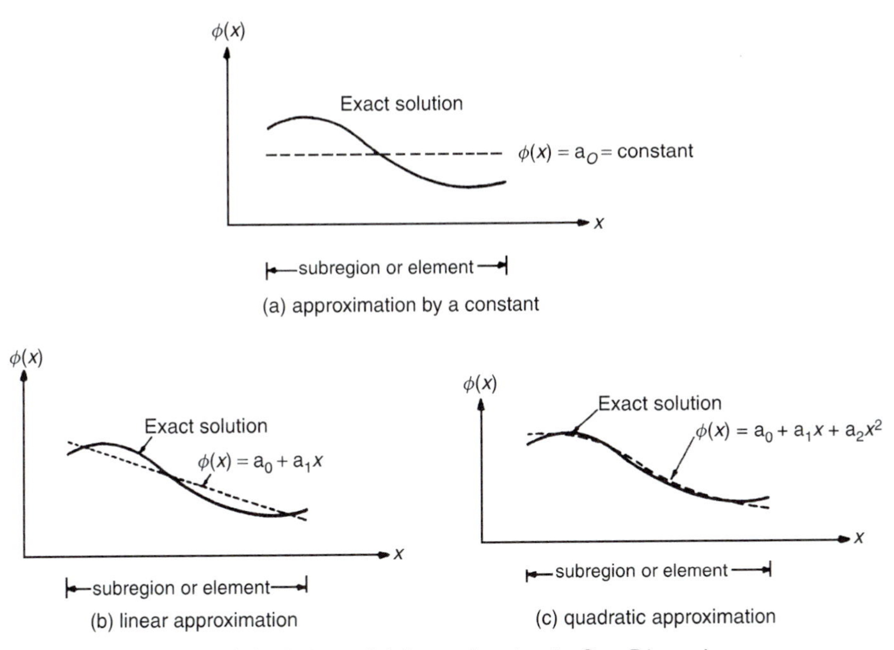

Figure 3.1. Polynomial Approximation in One Dimension.

vary rapidly. In these cases the simplex elements, which approximate the gradient by a set of constant values, do not yield good results. The combination of greater accuracy and a reduction in the data preparation effort has resulted in the widespread use of higher order elements in several practical applications. We shall consider mostly linear elements in this chapter.

If the order of the interpolation polynomial is fixed, the discretization of the region (or domain) can be improved by two methods. In the first method, known as the r-method, the locations of the nodes are altered without changing the total number of elements. In the second method, known as the h-method, the number of elements is increased. On the other hand, if improvement in accuracy is sought by increasing the order of the interpolation of polynomial, the method is known as the p-method.

Problems involving curved boundaries cannot be modeled satisfactorily by using straight-sided elements. The family of elements known as "isoparametric" elements has been developed for this purpose. The basic idea underlying the isoparametric elements is to use the same interpolation functions to define the element shape or geometry as well as the variation of the field variable within the element. To derive the isoparametric element equations, we first introduce a local or natural coordinate system for each element shape. Then the interpolation or shape functions are expressed in terms of the natural coordinates. The representation of geometry in terms of (nonlinear) shape functions can be considered as a mapping procedure that transforms a regular shape, such as a straight-sided triangle or rectangle in the local coordinate system, into a distorted shape, such as a curved-sided triangle or rectangle in the global Cartesian coordinate system. This concept can be used in representing problems with curved boundaries with the help of curved-sided isoparametric elements. Today, isoparametric elements are extensively used in three-dimensional and shell analysis problems. The formulation of isoparametric elements, along with the aspect of numerical integration that is essential for compuatations with isoparametric elements, is considered in the next chapter.

3.2 POLYNOMIAL FORM OF INTERPOLATION FUNCTIONS

If a polynomial type of variation is assumed for the field variable $\phi(x)$ in a one-dimensional element, $\phi(x)$ can be expressed as

$$\phi(x) = \alpha_1 + \alpha_2 x + \alpha_3 x^2 + \cdots + \alpha_m x^n \tag{3.1}$$

Similarly, in two- and three-dimensional finite elements the polynomial form of interpolation functions can be expressed as

$$\phi(x, y) = \alpha_1 + \alpha_2 x + \alpha_3 y + \alpha_4 x^2 + \alpha_5 y^2 + \alpha_6 xy + \cdots + \alpha_m y^n \tag{3.2}$$

$$\begin{aligned}\phi(x, y, z) = \alpha_1 + \alpha_2 x + \alpha_3 y + \alpha_4 z + \alpha_5 x^2 + \alpha_6 y^2 + \alpha_7 z^2 \\ + \alpha_8 xy + \alpha_9 yz + \alpha_{10} zx + \cdots + \alpha_m z^n\end{aligned} \tag{3.3}$$

where $\alpha_1, \alpha_2, \ldots, \alpha_m$ are the coefficients of the polynomial, also known as generalized coordinates; n is the degree of the polynomial; and the number of polynomial coefficients m is given by

$$m = n + 1 \text{ for one-dimensional elements (Eq. 3.1)} \tag{3.4}$$

$$m = \sum_{j=1}^{n+1} j \text{ for two-dimensional elements (Eq. 3.2)} \tag{3.5}$$

$$m = \sum_{j=1}^{n+1} j(n + 2 - j) \text{ for three-dimensional elements (Eq. 3.3)} \tag{3.6}$$

In most of the practical applications the order of the polynomial in the interpolation functions is taken as one, two, or three. Thus, Eqs. (3.1)–(3.3) reduce to the following equations for various cases of practical interest.

For $n = 1$ (linear model)

One-dimensional case:

$$\phi(x) = \alpha_1 + \alpha_2 x \tag{3.7}$$

Two-dimensional case:

$$\phi(x, y) = \alpha_1 + \alpha_2 x + \alpha_3 y \tag{3.8}$$

Three-dimensional case:

$$\phi(x, y, z) = \alpha_1 + \alpha_2 x + \alpha_3 y + \alpha_4 z \tag{3.9}$$

For $n = 2$ (quadratic model)
One-dimensional case:

$$\phi(x) = \alpha_1 + \alpha_2 x + \alpha_3 x^2 \tag{3.10}$$

Two-dimensional case:

$$\phi(x, y) = \alpha_1 + \alpha_2 x + \alpha_3 y + \alpha_4 x^2 + \alpha_5 y^2 + \alpha_6 xy \tag{3.11}$$

Three-dimensional case:

$$\begin{aligned}\phi(x, y, z) = {} & \alpha_1 + \alpha_2 x + \alpha_3 y + \alpha_4 z + \alpha_5 x^2 + \alpha_6 y^2 + \alpha_7 z^2 \\ & + \alpha_8 xy + \alpha_9 yz + \alpha_{10} xz\end{aligned} \tag{3.12}$$

For $n = 3$ (cubic model)
One-dimensional case:

$$\phi(x) = \alpha_1 + \alpha_2 x + \alpha_3 x^2 + \alpha_4 x^3 \tag{3.13}$$

Two-dimensional case:

$$\begin{aligned}\phi(x, y) = {} & \alpha_1 + \alpha_2 x + \alpha_3 y + \alpha_4 x^2 + \alpha_5 y^2 + \alpha_6 xy \\ & + \alpha_7 x^3 + \alpha_8 y^3 + \alpha_9 x^2 y + \alpha_{10} xy^2\end{aligned} \tag{3.14}$$

Three-dimensional case:

$$\begin{aligned}\phi(x, y, z) = {} & \alpha_1 + \alpha_2 x + \alpha_3 y + \alpha_4 z + \alpha_5 x^2 + \alpha_6 y^2 + \alpha_7 z^2 + \alpha_8 xy + \alpha_9 yz \\ & + \alpha_{10} xz + \alpha_{11} x^3 + \alpha_{12} y^3 + \alpha_{13} z^3 + \alpha_{14} x^2 y + \alpha_{15} x^2 z \\ & + \alpha_{16} y^2 z + \alpha_{17} xy^2 + \alpha_{18} xz^2 + \alpha_{19} yz^2 + \alpha_{20} xyz\end{aligned} \tag{3.15}$$

3.3 SIMPLEX, COMPLEX, AND MULTIPLEX ELEMENTS

Finite elements can be classified into three categories as simplex, complex, and multiplex elements depending on the geometry of the element and the order of the polynomial used in the interpolation function [3.2]. The simplex elements are those for which the approximating polynomial consists of constant and linear terms. Thus, the polynomials given by Eqs. (3.7)–(3.9) represent the simplex functions for one-, two-, and three-dimensional elements. Noting that a simplex is defined as a geometric figure obtained by joining $n+1$ joints (nodes) in an n-dimensional space, we can consider the corners of the elements as nodes in simplex elements. For example, the simplex element in two dimensions is a triangle with three nodes (corners). The three polynomial coefficients α_1, α_2, and α_3 of Eq. (3.8) can thus be expressed in terms of the nodal values of the field variable ϕ. The complex elements are those for which the approximating polynomial consists of quadratic, cubic, and higher order terms, according to the need, in addition to the constant and linear terms. Thus, the polynomials given by Eqs. (3.10)–(3.15) denote complex functions.

Figure 3.2. Example of a Multiplex Element.

The complex elements may have the same shapes as the simplex elements but will have additional boundary nodes and, sometimes, internal nodes. For example, the interpolating polynomial for a two-dimensional complex element (including terms up to quadratic terms) is given by Eq. (3.11). Since this equation has six unknown coefficients α_i, the corresponding complex element must have six nodes. Thus, a triangular element with three corner nodes and three midside nodes satisfies this requirement. The multiplex elements are those whose boundaries are parallel to the coordinate axes to achieve interelement continuity, and whose approximating polynomials contain higher order terms. The rectangular element shown in Figure 3.2 is an example of a multiplex element in two dimensions. Note that the boundaries of the simplex and complex elements need not be parallel to the coordinate axes.

3.4 INTERPOLATION POLYNOMIAL IN TERMS OF NODAL DEGREES OF FREEDOM

The basic idea of the finite element method is to consider a body as composed of several elements (or subdivisions) that are connected at specified node points. The unknown solution or the field variable (e.g., displacement, pressure, or temperature) inside any finite element is assumed to be given by a simple function in terms of the nodal values of that element. The nodal values of the solution, also known as nodal degrees of freedom, are treated as unknowns in formulating the system or overall equations. The solution of the system equations (e.g., force equilibrium equations or thermal equilibrium equations or continuity equations) gives the values of the unknown nodal degrees of freedom. Once the nodal degrees of freedom are known, the solution within any finite element (and hence within the complete body) will also be known to us.

Thus, we need to express the approximating polynomial in terms of the nodal degrees of freedom of a typical finite element e. For this, let the finite element have M nodes. We can evaluate the values of the field variable at the nodes by substituting the nodal coordinates into the polynomial equation given by Eqs. (3.1)–(3.3). For example, Eq. (3.1) can be expressed as

$$\vec{\phi}(x) = \vec{\eta}^T \vec{\alpha} \tag{3.16}$$

where $\vec{\phi}(x) = \phi(x)$,

$$\vec{\eta}^T = \{1 \quad x \quad x^2 \quad \dots \quad x^n\},$$

and

$$\vec{\alpha} = \begin{Bmatrix} \alpha_1 \\ \alpha_2 \\ \vdots \\ \alpha_{n+1} \end{Bmatrix}$$

The evaluation of Eq. (3.16) at the various nodes of element e gives

$$\begin{Bmatrix} \vec{\phi} \text{ (at node 1)} \\ \vec{\phi} \text{ (at node 2)} \\ \vdots \\ \vec{\phi} \text{ (at node } M) \end{Bmatrix}^{(e)} = \vec{\Phi}^{(e)} = \begin{bmatrix} \vec{\eta}^T \text{ (at node 1)} \\ \vec{\eta}^T \text{ (at node 2)} \\ \vdots \\ \vec{\eta}^T \text{ (at node } M) \end{bmatrix} \vec{\alpha} \equiv [\underset{\sim}{\eta}]\vec{\alpha} \tag{3.17}$$

where $\vec{\Phi}^{(e)}$ is the vector of nodal values of the field variable corresponding to element e, and the square matrix $[\underset{\sim}{\eta}]$ can be identified from Eq. (3.17). By inverting Eq. (3.17), we obtain

$$\vec{\alpha} = [\underset{\sim}{\eta}]^{-1}\vec{\Phi}^{(e)} \tag{3.18}$$

Substitution of Eq. (3.18) into Eqs. (3.1)–(3.3) gives

$$\vec{\phi} = \vec{\eta}^T\vec{\alpha} = \vec{\eta}^T[\underset{\sim}{\eta}]^{-1}\vec{\Phi}^{(e)} = [N]\vec{\Phi}^{(e)} \tag{3.19}$$

where

$$[N] = \vec{\eta}^T[\underset{\sim}{\eta}]^{-1} \tag{3.20}$$

Equation (3.19) now expresses the interpolating polynomial inside any finite element in terms of the nodal unknowns of that element, $\vec{\Phi}^{(e)}$. A major limitation of polynomial-type interpolation functions is that one has to invert the matrix $[\underset{\sim}{\eta}]$ to find $\vec{\phi}$, and $[\underset{\sim}{\eta}]^{-1}$ may become singular in some cases [3.3]. The latter difficulty can be avoided by using other types of interpolation functions discussed in Chapter 4.

3.5 SELECTION OF THE ORDER OF THE INTERPOLATION POLYNOMIAL

While choosing the order of the polynomial in a polynomial-type interpolation function, the following considerations have to be taken into account:

(i) The interpolation polynomial should satisfy, as far as possible, the convergence requirements stated in Section 3.6.
(ii) The pattern of variation of the field variable resulting from the polynomial model should be independent of the local coordinate system.
(iii) The number of generalized coordinates (α_i) should be equal to the number of nodal degrees of freedom of the element (Φ_i).

A discussion on the first consideration, namely, the convergence requirements to be satisfied by the interpolation polynomial, is given in the next section. According to the second consideration, as can be felt intuitively also, it is undesirable to have a preferential coordinate direction. That is, the field variable representation within an element, and hence the polynomial, should not change with a change in the local coordinate system (when a linear transformation is made from one Cartesian coordinate system to another). This property is called geometric isotropy or geometric invariance or spatial isotropy [3.4]. In order to achieve geometric isotropy, the polynomial should contain terms that do not violate symmetry in Figure 3.3, which is known as Pascal triangle in the case of two dimensions and Pascal tetrahedron in the case of three dimensions.

Thus, in the case of a two-dimensional simplex element (triangle), the interpolation polynomial should include terms containing both x and y, but not only one of them, in addition to the constant term. In the case of a two-dimensional complex element (triangle), if we neglect the term x^3 (or x^2y) for any reason, we should not include y^3 (or xy^2) also in order to maintain goemetric isotropy of the model. Similarly, in the case of a three-dimensional simplex element (tetrahedron), the approximating polynomial should contain terms involving x, y, and z in addition to the constant term.

The final consideration in selecting the order of the interpolation polynomial is to make the total number of terms involved in the polynomial equal to the number of nodal degrees of freedom of the element. The satisfaction of this requirement enables us to express the polynomial coefficients in terms of the nodal unknowns of the element as indicated in Section 3.4.

3.6 CONVERGENCE REQUIREMENTS

Since the finite element method is a numerical technique, we obtain a sequence of approximate solutions as the element size is reduced successively. This sequence will converge to the exact solution if the interpolation polynomial satisfies the following convergence requirements [3.5–3.8]:

(i) The field variable must be continuous within the elements. This requirement is easily satisfied by choosing continuous functions as interpolation models. Since polynomials are inherently continuous, the polynomial type of interpolation models discussed in Section 3.2 satisfy this requirement.

(ii) All uniform states of the field variable ϕ and its partial derivatives up to the highest order appearing in the functional $I(\phi)$ must have representation in the interpolation polynomial when, in the limit, the element size reduces to zero.

The necessity of this requirement can be explained physically. The uniform or constant value of the field variable is the most elementary type of variation. Thus, the interpolation polynomial must be able to give a constant value of the field variable within the element when the nodal values are numerically identical. Similarly, when the body is subdivided into smaller and smaller elements, the partial derivatives of the field variable up to the highest order appearing in the functional* $I(\phi)$ approach a constant value within each element. Thus, we

* The finite element method can be considered as an approximate method of minimizing a functional $I(\phi)$ in the form of an integral of the type

$$I(\phi) = I\left(\phi, \frac{\mathrm{d}\phi}{\mathrm{d}x}, \frac{\mathrm{d}^2\phi}{\mathrm{d}x^2}, \ldots, \frac{\mathrm{d}^r\phi}{\mathrm{d}x^r}\right)$$

The functionals for simple one-dimensional problems were given in Examples 1.2–1.4.

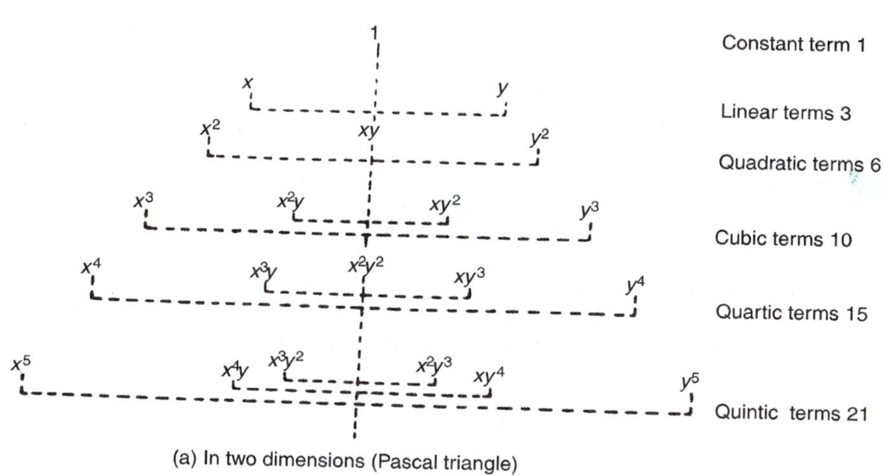

(a) In two dimensions (Pascal triangle)

(b) In three dimensions (Pascal tetrahedron)

Figure 3.3. Array of Terms in Complete Polynomials of Various Orders.

cannot hope to obtain convergence to the extact solution unless the interpolation polynomial permits this constant derivative state.

In the case of solid mechanics and structural problems, this requirement states that the assumed displacement model must permit the rigid body (zero strain) and the constant strain states of the element.

(iii) The field variable ϕ and its partial derivatives up to one order less than the highest order derivative appearing in the functional $I(\phi)$ must be continuous at element boundaries or interfaces.

We know that in the finite element method the discrete model for the continuous function ϕ is taken as a set of piecewise continuous functions, each defined over a single element. As seen in Examples 1.2–1.4, we need to evaluate integrals of the form

$$\int \frac{\mathrm{d}^r \phi}{\mathrm{d}x^r}\, \mathrm{d}x$$

to derive the element characteristic matrices and vectors. We know that the integral of a stepwise continuous function, say $f(x)$, is defined if $f(x)$ remains bounded in the interval of integration. Thus, for the integral

$$\int \frac{\mathrm{d}^r \phi}{\mathrm{d}x^r}\, \mathrm{d}x$$

to be defined, ϕ must be continuous to the order $(r-1)$ to ensure that only finite jump discontinuities occur in the rth derivative of ϕ. This is precisely the requirement stated previously.

The elements whose interpolation polynomials satisfy the requirements (i) and (iii) are called "compatible" or "conforming" elements and those satisfying condition (ii) are called "complete" elements. If rth derivative of the field variable ϕ is continuous, then ϕ is said to have C^r continuity. In terms of this notation, the completeness requirement implies that ϕ must have C^r continuity within an element, whereas the compatibility requirement implies that ϕ must have C^{r-1} continuity at element interfaces.†

In the case of general solid and structural mechanics problems, this requirement implies that the element must deform without causing openings, overlaps, or discontinuities between adjacent elements. In the case of beam, plate, and shell elements, the first derivative of the displacement (slope) across interelement boundaries also must be continuous.

Although it is desirable to satisfy all the convergence requirements, several interpolation polynomials that do not meet all the requirements have been used in the finite element literature. In some cases, acceptable convergence or convergence to an incorrect solution has been obtained. In particular, the interpolation polynomials that are complete but not conforming have been found to give satisfactory results.

If the interpolation polynomial satisfies all three requirements, the approximate solution converges to the correct solution when we refine the mesh and use an increasing number of smaller elements. In order to prove the convergence mathematically, the

† This statement assumes that the functional (I) corresponding to the problem contains derivatives of ϕ up to the rth order.

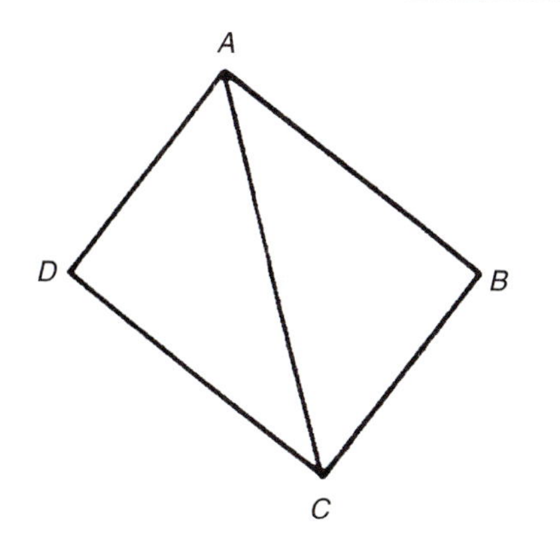

(a) Idealization with 2 elements

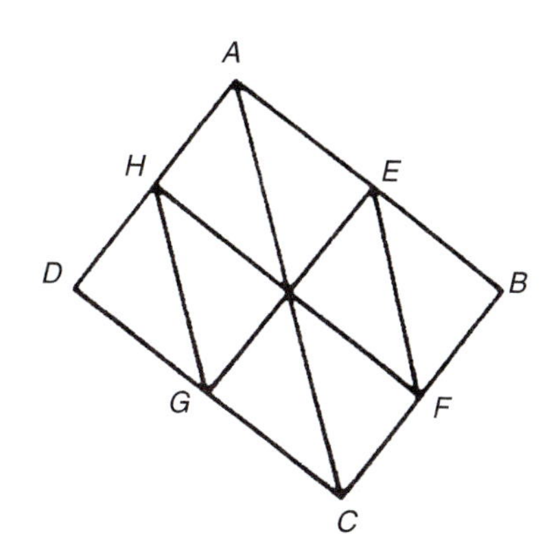

(b) Idealization with 8 elements

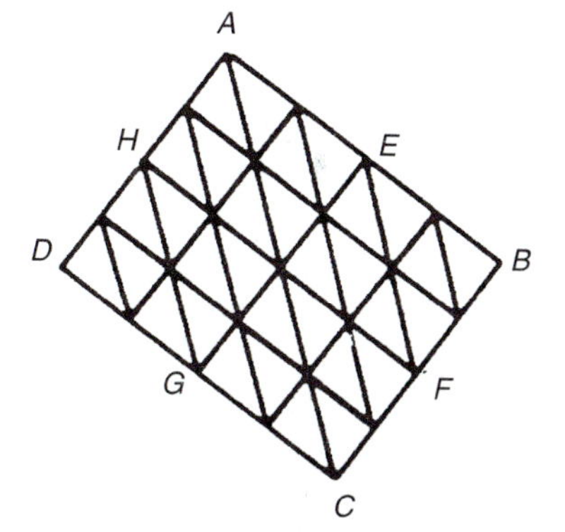

(c) Idealization with 32 elements

Figure 3.4. All Previous Meshes Contained in Refined Meshes.

mesh refinement has to be made in a regular fashion so as to satisfy the following conditions:

(i) All previous (coarse) meshes must be contained in the refined meshes.
(ii) The elements must be made smaller in such a way that every point of the solution region can always be within an element.
(iii) The form of the interpolation polynomial must remain unchanged during the process of mesh refinement.

Conditions (i) and (ii) are illustrated in Figure 3.4, in which a two-dimensional region (in the form of a parallelogram) is discretized with an increasing number of triangular elements. From Figure 3.5, in which the solution region is assumed to have a curved boundary, it can be seen that conditions (i) and (ii) are not satisfied if we use elements with straight boundaries. In structural problems, interpolation polynomials satisfying all the convergence requirements always lead to the convergence of the displacement solution from below while nonconforming elements may converge either from below or from above.

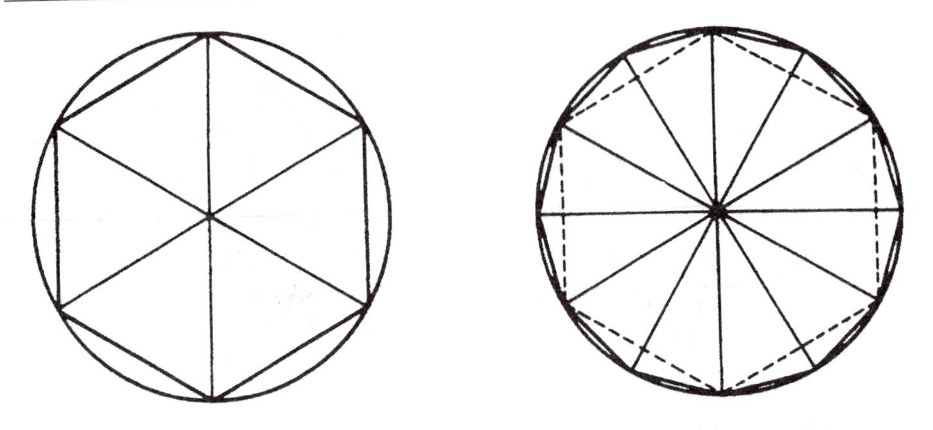

(a) Idealization with 6 elements

(b) Idealization with 12 elements

Figure 3.5. Previous Mesh Is Not Contained in the Refined Mesh.

Notes:

1. For any physical problem, the selection of finite elements and interpolation polynomials to achieve C^0 continuity is not very difficult. However, the difficulty increases rapidly when higher order continuity is required. In general, the construction of finite elements to achieve specified continuity of order $C^0, C^1, C^2, \ldots,$ requires skill, ingenuity, and experience. Fortunately, most of the time, we would be able to use the elements already developed in an established area such as stress analysis for solving new problems.
2. The construction of an efficient finite element model involves (a) representing the geometry of the problem accurately, (b) developing a finite element mesh to reduce the bandwidth, and (c) choosing a proper interpolation model to obtain the desired accuracy in the solution. Unfortunately, there is no a priori method of creating a reasonably efficient finite element model that can ensure a specified degree of accuracy. Several numerical tests are available for assessing the convergence of a finite element model [3.9, 3.10].

   Some adaptive finite element methods have been developed to employ the results from previous meshes to estimate the magnitude and distribution of solution errors and to adaptively improve the finite element model [3.11–3.15]. There are four basic approaches to adaptively improve a finite element model:

   (a) Subdivide selected elements (called h-method)
   (b) Increase the order of the polynomial of selected elements (called p-refinement)
   (c) Move node points in fixed element topology (called r-refinement)
   (d) Define a new mesh having a better distribution of elements

   Various combinations of these approaches are also possible. Determining which of these approaches is the best for a particular class of problems is a complex problem that must consider the cost of the entire solution process.

3.7 LINEAR INTERPOLATION POLYNOMIALS IN TERMS OF GLOBAL COORDINATES

The linear interpolation polynomials correspond to simplex elements. In this section, we derive the linear interpolation polynomials for the basic one-, two-, and three-dimensional elements in terms of the global coordinates that are defined for the entire domain or body.

3.7.1 One-Dimensional Simplex Element

Consider a one-dimensional element (line segment) of length l with two nodes, one at each end, as shown in Figure 3.6. Let the nodes be denoted as i and j and the nodal values of the field variable ϕ as Φ_i and Φ_j. The variation of ϕ inside the element is assumed to be linear as

$$\phi(x) = \alpha_1 + \alpha_2 x \tag{3.21}$$

where α_1 and α_2 are the unknown coefficients. By using the nodal conditions

$$\phi(x) = \Phi_i \quad \text{at} \quad x = x_i$$
$$\phi(x) = \Phi_j \quad \text{at} \quad x = x_j$$

and Eq. (3.21), we obtain

$$\Phi_i = \alpha_1 + \alpha_2 x_i$$
$$\Phi_j = \alpha_1 + \alpha_2 x_j$$

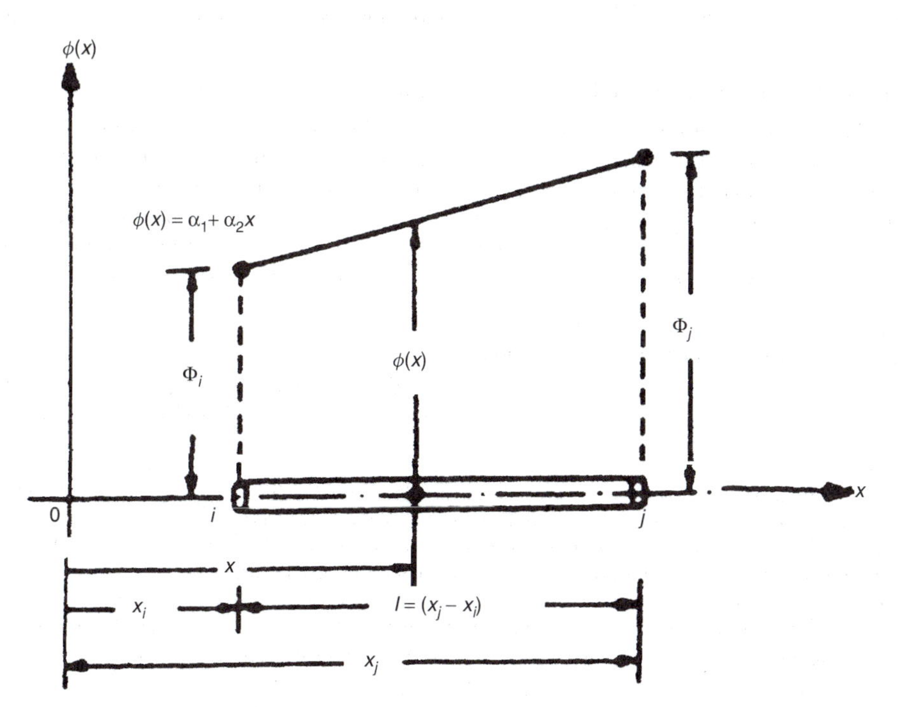

Figure 3.6.

The solution of these equations gives

$$\left.\begin{aligned} \alpha_1 &= \frac{\Phi_i x_j - \Phi_j x_i}{l} \\ \text{and} \qquad \alpha_2 &= \frac{\Phi_j - \Phi_i}{l} \end{aligned}\right\} \tag{3.22}$$

where x_i and x_j denote the global coordinates of nodes i and j, respectively. By substituting Eq. (3.22) into Eq. (3.21), we obtain

$$\phi(x) = \left(\frac{\Phi_i x_j - \Phi_j x_i}{l}\right) + \left(\frac{\Phi_j - \Phi_i}{l}\right) x \tag{3.23}$$

This equation can be written, after rearrangement of terms, as

$$\phi(x) = N_i(x)\Phi_i + N_j(x)\Phi_j = [N(x)]\vec{\Phi}^{(e)} \tag{3.24}$$

where

$$[N(x)] = [N_i(x)\ N_j(x)], \tag{3.25}$$

$$\left.\begin{aligned} N_i(x) &= \frac{x_j - x}{l} \\ N_j(x) &= \frac{x - x_i}{l} \end{aligned}\right\} \tag{3.26}$$

and

$$\vec{\Phi}^{(e)} = \begin{Bmatrix} \Phi_i \\ \Phi_j \end{Bmatrix} = \text{vector of nodal unknowns of element } e \tag{3.27}$$

Notice that the superscript e is not used for Φ_i and Φ_j for simplicity.

The linear functions of x defined in Eq. (3.26) are called interpolation or shape functions.* Notice that each interpolation function has a subscript to denote the node to which it is associated. Furthermore, the value of $N_i(x)$ can be seen to be 1 at node i ($x = x_i$) and 0 at node j ($x = x_j$). Likewise, the value of $N_j(x)$ will be 0 at node i and 1 at node j. These represent the common characteristics of interpolation functions. They will be equal to 1 at one node and 0 at each of the other nodes of the element.

3.7.2 Two-Dimensional Simplex Element

The two-dimensional simplex element is a straight-sided triangle with three nodes, one at each corner, as indicated in Figure 3.7. Let the nodes be labeled as i, j, and k by

---

* The original polynomial type of interpolation model $\phi = \vec{\eta}^T \vec{\alpha}$ (which is often called the *interpolation polynomial* or interpolation model of the element) should not be confused with the *interpolation functions* N_i associated with the nodal degrees of freedom. There is a clear difference between the two. The expression $\vec{\eta}^T \vec{\alpha}$ denotes an interpolation polynomial that applies to the entire element and expresses the variation of the field variable inside the element in terms of the generalized coordinates α_i. The interpolation function N_i corresponds to the ith nodal degree of freedom $\Phi_i^{(e)}$ and only the sum $\Sigma_i N_i \Phi i^{(e)}$ represents the variation of the field variable inside the element in terms of the nodal degrees of freedom $\Phi_i^{(e)}$. In fact, the interpolation function corresponding to the ith nodal degree of freedom (N_i) assumes a value of 1 at node i and 0 at all the other nodes of the element.

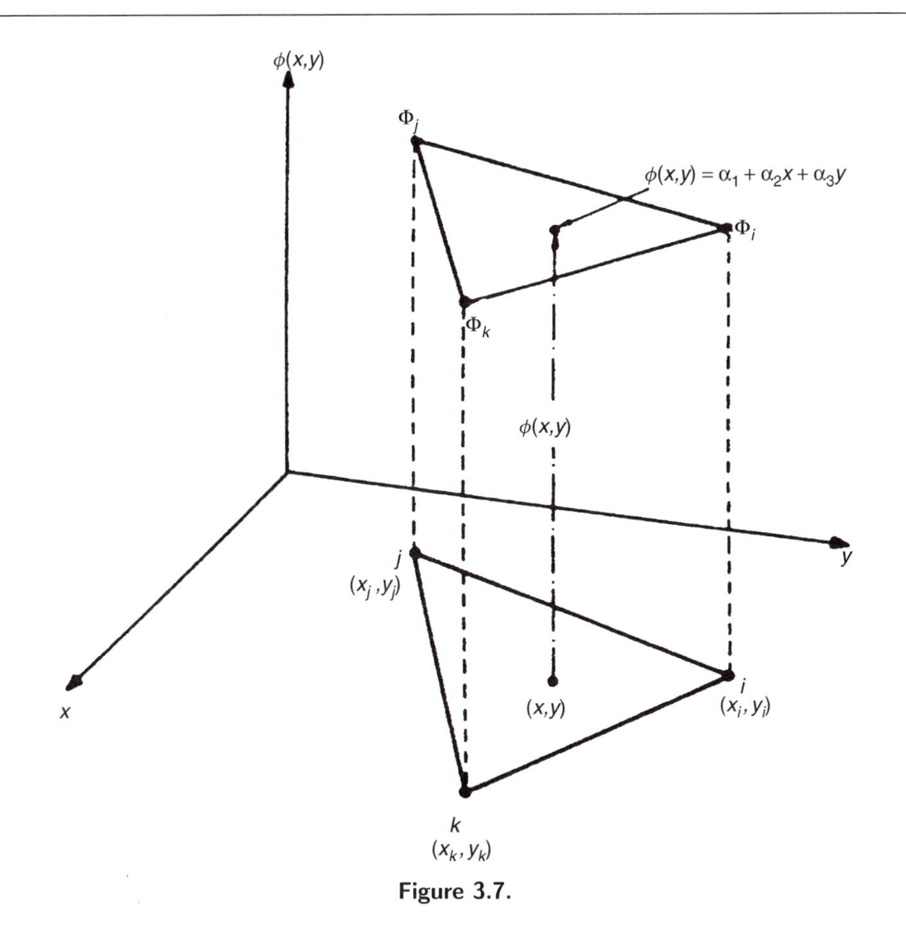

Figure 3.7.

proceeding counterclockwise from node i, which is arbitrarily specified. Let the global coordinates of the nodes i, j, and k be given by (x_i, y_i), (x_j, y_j), and (x_k, y_k) and the nodal values of the field variable $\phi(x, y)$ by Φ_i, Φ_j, and Φ_k, respectively. The variation of ϕ inside the element is assumed to be linear as

$$\phi(x, y) = \alpha_1 + \alpha_2 x + \alpha_3 y \tag{3.28}$$

The nodal conditions

$$\begin{aligned}
\phi(x, y) &= \Phi_i \quad \text{at} \quad (x = x_i, y = y_i) \\
\phi(x, y) &= \Phi_j \quad \text{at} \quad (x = x_j, y = y_j) \\
\phi(x, y) &= \Phi_k \quad \text{at} \quad (x = x_k, y = y_k)
\end{aligned}$$

lead to the system of equations

$$\begin{aligned}
\Phi_i &= \alpha_1 + \alpha_2 x_i + \alpha_3 y_i \\
\Phi_j &= \alpha_1 + \alpha_2 x_j + \alpha_3 y_j \\
\Phi_k &= \alpha_1 + \alpha_2 x_k + \alpha_3 y_k
\end{aligned} \tag{3.29}$$

The solution of Eqs. (3.29) yields

$$\begin{aligned}\alpha_1 &= \frac{1}{2A}(a_i\Phi_i + a_j\Phi_j + a_k\Phi_k)\\ \alpha_2 &= \frac{1}{2A}(b_i\Phi_i + b_j\Phi_j + b_k\Phi_k)\\ \alpha_3 &= \frac{1}{2A}(c_i\Phi_i + c_j\Phi_j + c_k\Phi_k)\end{aligned} \tag{3.30}$$

where A is the area of the triangle ijk given by

$$A = \frac{1}{2}\begin{bmatrix}1 & x_i & y_i\\ 1 & x_j & y_j\\ 1 & x_k & y_k\end{bmatrix} = \frac{1}{2}(x_iy_j + x_jy_k + x_ky_i - x_iy_k - x_jy_i - x_ky_j) \tag{3.31}$$

$$\begin{aligned}a_i &= x_jy_k - x_ky_j\\ a_j &= x_ky_i - x_iy_k\\ a_k &= x_iy_j - x_jy_i\\ b_i &= y_j - y_k\\ b_j &= y_k - y_i\\ b_k &= y_i - y_j\\ c_i &= x_k - x_j\\ c_j &= x_i - x_k\\ c_k &= x_j - x_i\end{aligned} \tag{3.32}$$

Substitution of Eqs. (3.30) into Eq. (3.28) and rearrangement yields the equation

$$\phi(x,y) = N_i(x,y)\Phi_i + N_j(x,y)\Phi_j + N_k(x,y)\Phi_k = [N(x,y)]\vec{\Phi}^{(e)} \tag{3.33}$$

where

$$[N(x,y)] = [N_i(x,y)\ N_j(x,y)\ N_k(x,y)], \tag{3.34}$$

$$\begin{aligned}N_i(x,y) &= \frac{1}{2A}(a_i + b_ix + c_iy)\\ N_j(x,y) &= \frac{1}{2A}(a_j + b_jx + c_jy)\\ N_k(x,y) &= \frac{1}{2A}(a_k + b_kx + c_ky)\end{aligned} \tag{3.35}$$

and

$$\vec{\Phi}^{(e)} = \begin{Bmatrix}\Phi_i\\ \Phi_j\\ \Phi_k\end{Bmatrix} = \text{vector of nodal unknowns of element } e. \tag{3.36}$$

Notes:

1. The shape function $N_i(x, y)$ when evaluated at node i (x_i, y_i) gives

$$\begin{aligned} N_i(x_i, y_i) &= \frac{1}{2A}(a_i + b_i x_i + c_i y_i) \\ &= \frac{1}{2A}(x_j y_k - x_k y_j + x_i y_j - x_i y_k + x_k y_i - x_j y_i) = 1 \end{aligned} \tag{3.37}$$

It can be shown that $N_i(x, y) = 0$ at nodes j and k, and at all points on the line passing through these nodes. Similarly, the shape functions N_j and N_k have a value of 1 at nodes j and k, respectively, and 0 at other nodes.

2. Since the interpolation functions are linear in x and y, the gradient of the field variable in x or y direction will be a constant. For example,

$$\frac{\partial \phi(x, y)}{\partial x} = \frac{\partial}{\partial x}[N(x, y)]\vec{\Phi}^{(e)} = (b_i \Phi_i + b_j \Phi_j + b_k \Phi_k)/2A \tag{3.38}$$

Since Φ_i, Φ_j, and Φ_k are the nodal values of ϕ (independent of x and y) and b_i, b_j, and b_k are constants whose values are fixed once the nodal coordinates are specified, $(\partial \phi / \partial x)$ will be a constant. A constant value of the gradient of ϕ within an element means that many small elements have to be used in locations where rapid changes are expected in the value of ϕ.

3.7.3 Three-Dimensional Simplex Element

The three-dimensional simplex element is a flat-faced tetrahedron with four nodes, one at each corner, as shown in Figure 3.8. Let the nodes be labeled as i, j, k, and l, where i, j, and k are labeled in a counterclockwise sequence on any face as viewed from the vertex opposite this face, which is labeled as l. Let the values of the field variable be Φ_i, Φ_j, Φ_k, and Φ_l and the global coordinates be (x_i, y_i, z_i), (x_j, y_j, z_j), (x_k, y_k, z_k), and (x_l, y_l, z_l) at nodes i, j, k, and l, respectively. If the variation of $\phi(x, y, z)$ is assumed to be linear,

$$\phi(x, y, z) = \alpha_1 + \alpha_2 x + \alpha_3 y + \alpha_4 z \tag{3.39}$$

the nodal conditions $\phi = \Phi_i$ at (x_i, y_i, z_i), $\phi = \Phi_j$ at (x_j, y_j, z_j), $\phi = \Phi_k$ at (x_k, y_k, z_k), and $\phi = \Phi_l$ at (x_l, y_l, z_l) produce the system of equations

$$\begin{aligned} \Phi_i &= \alpha_1 + \alpha_2 x_i + \alpha_3 y_i + \alpha_4 z_i \\ \Phi_j &= \alpha_1 + \alpha_2 x_j + \alpha_3 y_j + \alpha_4 z_j \\ \Phi_k &= \alpha_1 + \alpha_2 x_k + \alpha_3 y_k + \alpha_4 z_k \\ \Phi_l &= \alpha_1 + \alpha_2 x_l + \alpha_3 y_l + \alpha_4 z_l \end{aligned} \tag{3.40}$$

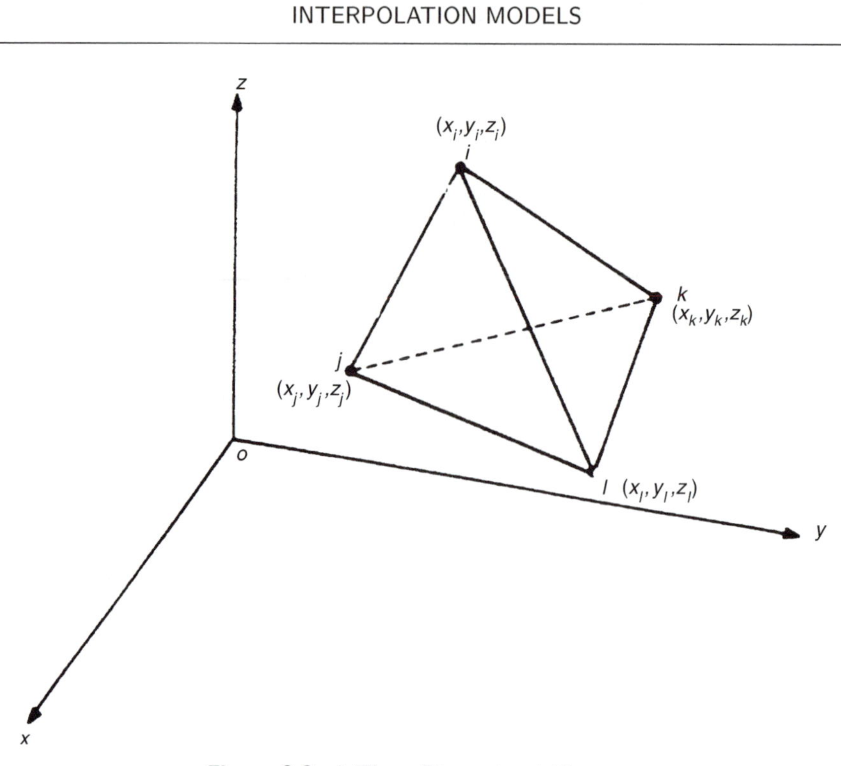

Figure 3.8. A Three-Dimensional Element.

Equations (3.40) can be solved and the coefficients α_1, α_2, α_3, and α_4 can be expressed as

$$\begin{aligned} \alpha_1 &= \frac{1}{6V}(a_i\Phi_i + a_j\Phi_j + a_k\Phi_k + a_l\Phi_l) \\ \alpha_2 &= \frac{1}{6V}(b_i\Phi_i + b_j\Phi_j + b_k\Phi_k + b_l\Phi_l) \\ \alpha_3 &= \frac{1}{6V}(c_i\Phi_i + c_j\Phi_j + c_k\Phi_k + c_l\Phi_l) \\ \alpha_4 &= \frac{1}{6V}(d_i\Phi_i + d_j\Phi_j + d_k\Phi_k + d_l\Phi_l) \end{aligned} \tag{3.41}$$

where V is the volume of the tetrahedron $i\ j\ k\ l$ given by

$$V = \frac{1}{6}\begin{vmatrix} 1 & x_i & y_i & z_i \\ 1 & x_j & y_j & z_j \\ 1 & x_k & y_k & z_k \\ 1 & x_l & y_l & z_l \end{vmatrix}, \tag{3.42}$$

$$a_i = \begin{vmatrix} x_j & y_j & z_j \\ x_k & y_k & z_k \\ x_l & y_l & z_l \end{vmatrix}, \tag{3.43}$$

$$b_i = -\begin{vmatrix} 1 & y_j & z_j \\ 1 & y_k & z_k \\ 1 & y_l & z_l \end{vmatrix}, \tag{3.44}$$

$$c_i = -\begin{vmatrix} x_j & 1 & z_j \\ x_k & 1 & z_k \\ x_l & 1 & z_l \end{vmatrix}, \tag{3.45}$$

and

$$d_i = -\begin{vmatrix} x_j & y_j & 1 \\ x_k & y_k & 1 \\ x_l & y_l & 1 \end{vmatrix}, \tag{3.46}$$

with the other constants defined by cyclic interchange of the subscripts in the order l, i, j, k. The signs in front of determinants in Eqs. (3.43)–(3.46) are to be reversed when generating a_j, b_j, c_j, d_j and a_l, b_l, c_l, d_l. By substituting Eqs. (3.41) into Eq. (3.39), we obtain

$$\begin{aligned}\phi(x,y,z) &= N_i(x,y,z)\Phi_i + N_j(x,y,z)\Phi_j + N_k(x,y,z)\Phi_k + N_l(x,y,z)\Phi_l \\ &= [N(x,y,z)]\vec{\Phi}^{(e)}\end{aligned} \tag{3.47}$$

where

$$\begin{aligned}
[N(x,y,z)] &= [N_i(x,y,z) \quad N_j(x,y,z) \quad N_k(x,y,z) \quad N_l(x,y,z)] \\
N_i(x,y,z) &= \frac{1}{6V}(a_i + b_i x + c_i y + d_i z) \\
N_j(x,y,z) &= \frac{1}{6V}(a_j + b_j x + c_j y + d_j z) \\
N_k(x,y,z) &= \frac{1}{6V}(a_k + b_k x + c_k y + d_k z) \\
N_l(x,y,z) &= \frac{1}{6V}(a_l + b_l x + c_l y + d_l z)
\end{aligned} \tag{3.48}$$

and

$$\vec{\Phi}^{(e)} = \begin{Bmatrix} \Phi_i \\ \Phi_j \\ \Phi_k \\ \Phi_l \end{Bmatrix} \tag{3.49}$$

3.8 INTERPOLATION POLYNOMIALS FOR VECTOR QUANTITIES

In Eqs. (3.21), (3.28), and (3.39), the field variable ϕ has been assumed to be a scalar quantity. In some problems the field variable may be a vector quantity having both magnitude and direction (e.g., displacement in solid mechanics problems). In such cases, the usual procedure is to resolve the vector into components parallel to the coordinate axes and treat these components as the unknown quantities. Thus, there will be more than one unknown (degree of freedom) at a node in such problems. The number of degrees of freedom at a node will be one, two, or three depending on whether the problem is one-, two-, or three-dimensional. The notation used in this book for the vector components is shown in Figure 3.9. All the components are designated by the same symbol, Φ, with a subscript denoting the individual components. The subscripts, at any node, are ordered in the sequence x, y, z starting with the x component. The x, y, and z components of the vector quantity (field variable) ϕ are denoted by u, v, and w, respectively.

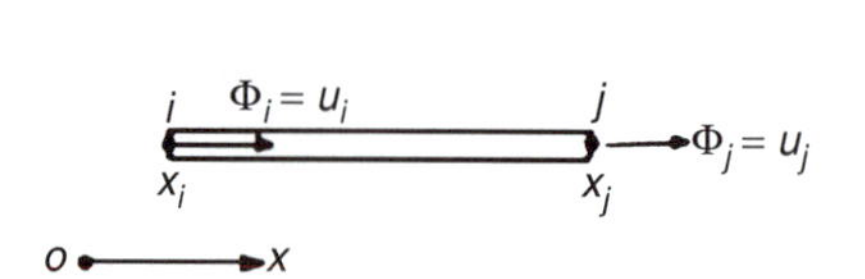

(a) One-dimensional problem

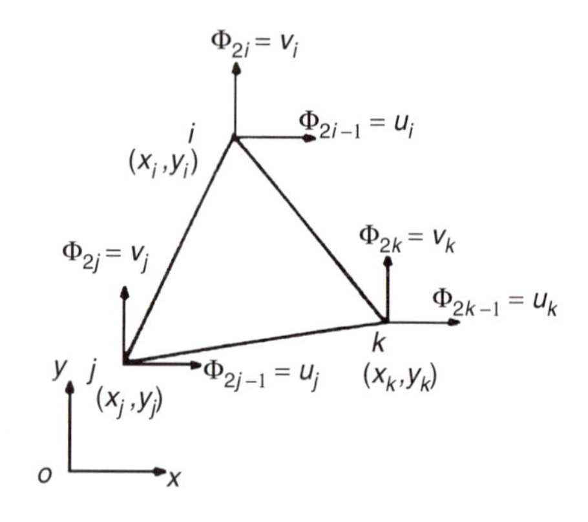

(b) Two-dimensional problem

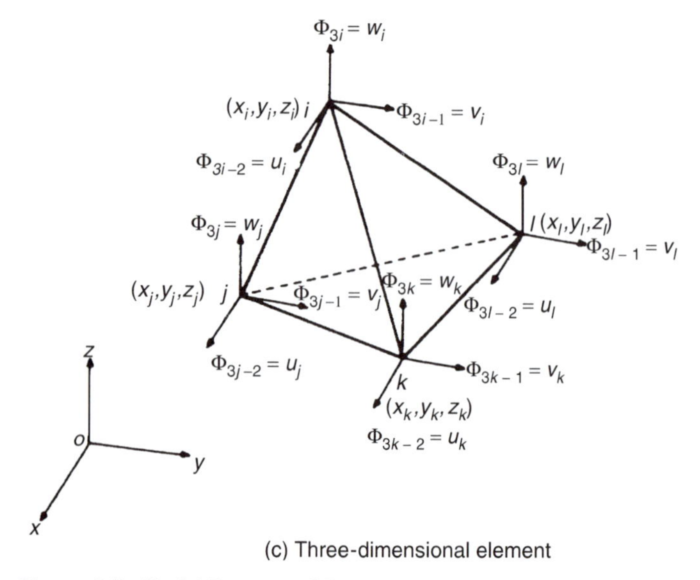

(c) Three-dimensional element

Figure 3.9. Nodal Degrees of Freedom When the Field Variable Is a Vector.

The interpolation function for a vector quantity in a one-dimensional element will be same as that of a scalar quantity since there is only one unknown at each node. Thus,

$$u(x) = N_i(x)\Phi_i + N_j(x)\Phi_j = [N(x)]\vec{\Phi}^{(e)} \tag{3.50}$$

where

$$[N(x)] = [N_i(x) \quad N_j(x)],$$

$$\vec{\Phi}^{(e)} = \begin{Bmatrix} \Phi_i \\ \Phi_j \end{Bmatrix},$$

and u is the component of ϕ (e.g., displacement) parallel to the axis of the element that is assumed to coincide with the x axis. The shape functions $N_i(x)$ and $N_j(x)$ are the same as those given in Eq. (3.26).

For a two-dimensional triangular (simplex) element, the linear interpolation model of Eq. (3.33) will be valid for each of the components of ϕ, namely, u and v. Thus,

$$u(x, y) = N_i(x, y)\Phi_{2i-1} + N_j(x, y)\Phi_{2j-1} + N_k(x, y)\Phi_{2k-1} \tag{3.51}$$

and

$$v(x, y) = N_i(x, y)\Phi_{2i} + N_j(x, y)\Phi_{2j} + N_k(x, y)\Phi_{2k} \tag{3.52}$$

where N_i, N_j, and N_k are the same as those defined in Eq. (3.35); Φ_{2i-1}, Φ_{2j-1}, and Φ_{2k-1} are the nodal values of u (component of ϕ parallel to the x axis); and Φ_{2i}, Φ_{2j}, and Φ_{2k} are the nodal values of v (component of ϕ parallel to the y axis). Equations (3.51) and (3.52) can be written in matrix form as

$$\vec{\phi}(x, y) = \begin{Bmatrix} u(x, y) \\ v(x, y) \end{Bmatrix} = [N(x, y)]\vec{\Phi}^{(e)} \tag{3.53}$$

where

$$[N(x, y)] = \begin{bmatrix} N_i(x, y) & 0 & N_j(x, y) & 0 & N_k(x, y) & 0 \\ 0 & N_i(x, y) & 0 & N_j(x, y) & 0 & N_k(x, y) \end{bmatrix} \tag{3.54}$$

and

$$\vec{\Phi}^{(e)} = \begin{Bmatrix} \Phi_{2i-1} \\ \Phi_{2i} \\ \Phi_{2j-1} \\ \Phi_{2j} \\ \Phi_{2k-1} \\ \Phi_{2k} \end{Bmatrix} = \text{vector of nodal degrees of freedom} \tag{3.55}$$

Extending this procedure to three dimensions, we obtain for a tetrahedron (simplex) element,

$$\vec{\phi}(x, y, z) = \begin{Bmatrix} u(x, y, z) \\ v(x, y, z) \\ w(x, y, z) \end{Bmatrix} = [N(x, y, z)]\vec{\Phi}^{(e)} \tag{3.56}$$

where

$$[N(x,y,z)] = \begin{bmatrix} N_i(x,y,z) & 0 & 0 & N_j(x,y,z) & 0 & 0 & N_k(x,y,z) & 0 & 0 & N_l(x,y,z) & 0 & 0 \\ 0 & N_i(x,y,z) & 0 & 0 & N_j(x,y,z) & 0 & 0 & N_k(x,y,z) & 0 & 0 & N_l(x,y,z) & 0 \\ 0 & 0 & N_i(x,y,z) & 0 & 0 & N_j(x,y,z) & 0 & 0 & N_k(x,y,z) & 0 & 0 & N_l(x,y,z) \end{bmatrix} \tag{3.57}$$

$$\vec{\Phi}^{(e)} = \begin{Bmatrix} \Phi_{3i-2} \\ \Phi_{3i-1} \\ \Phi_{3i} \\ \Phi_{3j-2} \\ \Phi_{3j-1} \\ \Phi_{3j} \\ \Phi_{3k-2} \\ \Phi_{3k-1} \\ \Phi_{3k} \\ \Phi_{3l-2} \\ \Phi_{3l-1} \\ \Phi_{3l} \end{Bmatrix} \tag{3.58}$$

and the shape functions N_i, N_j, N_k, and N_l are the same as those defined in Eq. (3.48).

3.9 LINEAR INTERPOLATION POLYNOMIALS IN TERMS OF LOCAL COORDINATES

The derivation of element characteristic matrices and vectors involves the integration of the shape functions or their derivatives or both over the element. These integrals can be evaluated easily if the interpolation functions are written in terms of a local coordinate system that is defined separately for each element.

In this section, we derive the interpolation functions of simplex elements in terms of a particular type of local coordinate systems, known as natural coordinate systems. A natural coordinate system is a local coordinate system that permits the specification of any point inside the element by a set of nondimensional numbers whose magnitude lies between 0 and 1. Usually, natural coordinate systems are chosen such that some of the natural coordinates will have unit magnitude at primary* or corner nodes of the element.

3.9.1 One-Dimensional Element

The natural coordinates for a one-dimensional (line) element are shown in Figure 3.10. Any point P inside the element is identified by two natural coordinates L_1 and L_2, which

* The nodes located at places other than at corners (e.g., midside nodes and interior nodes) are called secondary nodes.

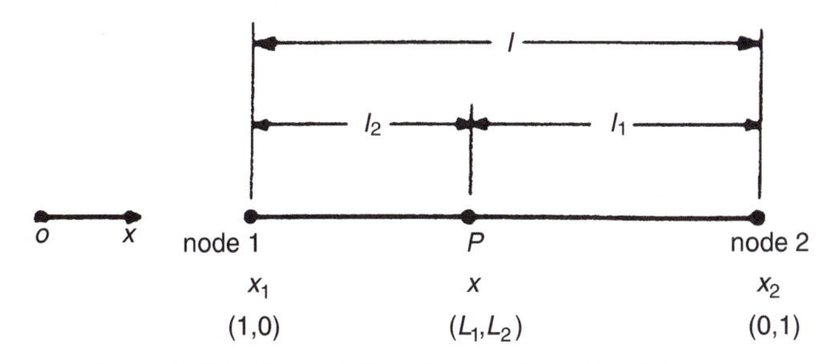

Figure 3.10. Natural Coordinates for a Line Element.

are defined as

$$L_1 = \frac{l_1}{l} = \frac{x_2 - x}{x_2 - x_1}$$

$$L_2 = \frac{l_2}{l} = \frac{x - x_1}{x_2 - x_1} \tag{3.59}$$

where l_1 and l_2 are the distances shown in Figure 3.10, and l is the length of the element. Since it is a one-dimensional element, there should be only one independent coordinate to define any point P. This is true even with natural coordinates because the two natural coordinates L_1 and L_2 are not independent but are related as

$$L_1 + L_2 = \frac{l_1}{l} + \frac{l_2}{l} = 1 \tag{3.60}$$

A study of the properties of L_1 and L_2 reveals something quite interesting. The natural coordinates L_1 and L_2 are also the shape functions for the line element [compare Eqs. (3.59) with Eqs. (3.26)]. Thus,

$$N_i = L_1, \qquad N_j = L_2 \tag{3.61}$$

Any point x within the element can be expressed as a linear combination of the nodal coordinates of nodes 1 and 2 as

$$x = x_1 L_1 + x_2 L_2 \tag{3.62}$$

where L_1 and L_2 may be interpreted as weighting functions. Thus, the relationship between the natural and the Cartesian coordinates of any point P can be written in matrix form as

$$\begin{Bmatrix} 1 \\ x \end{Bmatrix} = \begin{bmatrix} 1 & 1 \\ x_1 & x_2 \end{bmatrix} \begin{Bmatrix} L_1 \\ L_2 \end{Bmatrix} \tag{3.63}$$

or

$$\begin{Bmatrix} L_1 \\ L_2 \end{Bmatrix} = \frac{1}{(x_2 - x_1)} \begin{bmatrix} x_2 & -1 \\ -x_1 & 1 \end{bmatrix} \begin{Bmatrix} 1 \\ x \end{Bmatrix} = \frac{1}{l} \begin{Bmatrix} x_2 & -1 \\ -x_1 & 1 \end{Bmatrix} \begin{Bmatrix} 1 \\ x \end{Bmatrix} \tag{3.64}$$

If f is a function of L_1 and L_2, differentiation of f with respect to x can be performed, using the chain rule, as

$$\frac{\mathrm{d}f}{\mathrm{d}x} = \frac{\partial f}{\partial L_1}\frac{\partial L_1}{\partial x} + \frac{\partial f}{\partial L_2}\frac{\partial L_2}{\partial x} \tag{3.65}$$

where, from Eq. (3.59),

$$\frac{\partial L_1}{\partial x} = -\frac{1}{x_2 - x_1} \quad \text{and} \quad \frac{\partial L_2}{\partial x} = \frac{1}{x_2 - x_1} \tag{3.66}$$

Integration of polynomial terms in natural coordinates can be performed by using the simple formula

$$\int_{x_1}^{x_2} L_1^{\alpha} L_2^{\beta}\,\mathrm{d}x = \frac{\alpha!\beta!}{(\alpha+\beta+1)!}l \tag{3.67}$$

where $\alpha!$ is the factorial of α given by $\alpha! = \alpha(\alpha-1)(\alpha-2)\ldots(1)$. The value of the integral in Eq. (3.67) is given for certain combinations of α and β in Table 3.1.

3.9.2 Two-Dimensional (Triangular) Element

A natural coordinate system for a triangular element (also known as the triangular coordinate system) is shown in Figure 3.11(a). Although three coordinates L_1, L_2, and L_3 are used to define a point P, only two of them are independent. The natural coordinates are defined as

$$L_1 = \frac{A_1}{A}, \qquad L_2 = \frac{A_2}{A}, \qquad L_3 = \frac{A_3}{A} \tag{3.68}$$

Table 3.1.

Value of α	β	Value of the integral in Eq. (3.67)/l
0	0	1
1	0	1/2
1	1	1/6
2	0	1/3
1	2	1/12
3	0	1/4
4	0	1/5
2	2	1/30
3	1	1/20
1	4	1/30
3	2	1/60
5	0	1/6

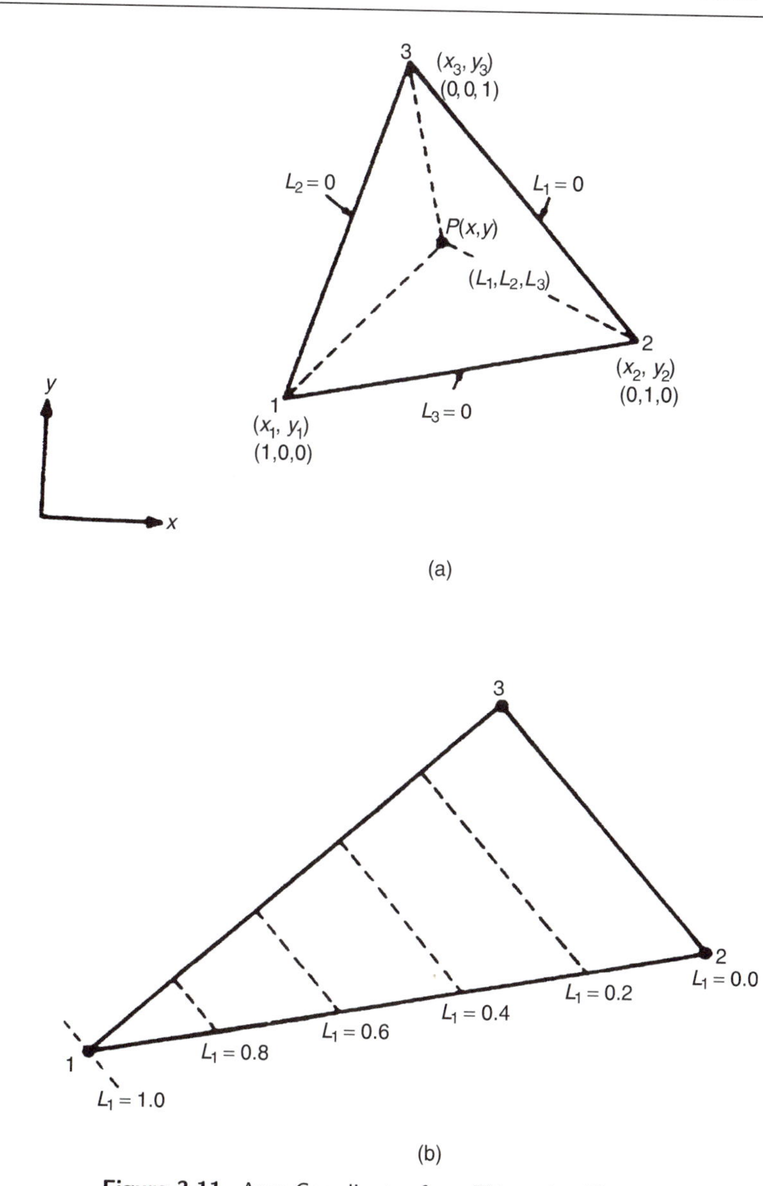

Figure 3.11. Area Coordinates for a Triangular Element.

where A_1 is the area of the triangle formed by the points P, 2 and 3; A_2 is the area of the triangle formed by the points P, 1 and 3; A_3 is the area of the triangle formed by the points P, 1 and 2; and A is the area of the triangle 123 in Figure 3.11. Because L_i are defined in terms of areas, they are also known as area coordinates. Since

$$A_1 + A_2 + A_3 = A,$$

we have

$$\frac{A_1}{A} + \frac{A_2}{A} + \frac{A_3}{A} = L_1 + L_2 + L_3 = 1 \tag{3.69}$$

A study of the properties of L_1, L_2, and L_3 shows that they are also the shape functions for the two-dimensional simplex (triangular) element:

$$N_i = L_1, \qquad N_j = L_2, \qquad N_K = L_3 \tag{3.70}$$

The relation between the natural and Cartesian coordinates is given by (see problem 3.8)

$$\left.\begin{aligned} x &= x_1 L_1 + x_2 L_2 + x_3 L_3 \\ y &= y_1 L_1 + y_2 L_2 + y_3 L_3 \end{aligned}\right\} \tag{3.71}$$

To every set of natural coordinates (L_1, L_2, L_3) [which are not independent but are related by Eq. (3.69)], there corresponds a unique set of Cartesian coordinates (x, y). At node 1, $L_1 = 1$ and $L_2 = L_3 = 0$, etc. The linear relationship between L_i $(i = 1, 2, 3)$ and (x, y) implies that the contours of L_1 are equally placed straight lines parallel to the side 2, 3 of the triangle (on which $L_1 = 0$), etc. as shown in Figure 3.11(b).

Equations (3.69) and (3.71) can be expressed in matrix form as

$$\begin{Bmatrix} 1 \\ x \\ y \end{Bmatrix} = \begin{bmatrix} 1 & 1 & 1 \\ x_1 & x_2 & x_3 \\ y_1 & y_2 & y_3 \end{bmatrix} \begin{Bmatrix} L_1 \\ L_2 \\ L_3 \end{Bmatrix} \tag{3.72}$$

Equation (3.72) can be inverted to obtain

$$\begin{Bmatrix} L_1 \\ L_2 \\ L_3 \end{Bmatrix} = \frac{1}{2A} \begin{bmatrix} (x_2 y_3 - x_3 y_2) & (y_2 - y_3) & (x_3 - x_2) \\ (x_3 y_1 - x_1 y_3) & (y_3 - y_1) & (x_1 - x_3) \\ (x_1 y_2 - x_2 y_1) & (y_1 - y_2) & (x_2 - x_1) \end{bmatrix} \begin{Bmatrix} 1 \\ x \\ y \end{Bmatrix} \tag{3.73}$$

where A is the area of the triangle 1, 2, 3 given by

$$A = \frac{1}{2} \begin{vmatrix} 1 & x_1 & y_1 \\ 1 & x_2 & y_2 \\ 1 & x_3 & y_3 \end{vmatrix} \tag{3.74}$$

Notice that Eq. (3.73) is identical to Eq. (3.35).

If f is a function of L_1, L_2, and L_3, the differentiation with respect to x and y can be performed as

$$\left.\begin{aligned} \frac{\partial f}{\partial x} &= \sum_{i=1}^{3} \frac{\partial f}{\partial L_i} \frac{\partial L_i}{\partial x} \\ \frac{\partial f}{\partial y} &= \sum_{i=1}^{3} \frac{\partial f}{\partial L_i} \frac{\partial L_i}{\partial y} \end{aligned}\right\} \tag{3.75}$$

where

$$\left.\begin{aligned}\frac{\partial L_1}{\partial x} &= \frac{y_2 - y_3}{2A}, & \frac{\partial L_1}{\partial y} &= \frac{x_3 - x_2}{2A}\\ \frac{\partial L_2}{\partial x} &= \frac{y_3 - y_1}{2A}, & \frac{\partial L_2}{\partial y} &= \frac{x_1 - x_3}{2A}\\ \frac{\partial L_3}{\partial x} &= \frac{y_1 - y_2}{2A}, & \frac{\partial L_3}{\partial y} &= \frac{x_2 - x_1}{2A}\end{aligned}\right\} \tag{3.76}$$

For integrating polynomial terms in natural coordinates, we can use the relations

$$\int_L L_1^\alpha L_2^\beta \cdot \mathrm{d}\mathcal{L} = \frac{\alpha!\beta!}{(\alpha+\beta+1)!}\mathcal{L} \tag{3.77}$$

and

$$\iint_A L_1^\alpha L_2^\beta L_3^\gamma \cdot \mathrm{d}A = \frac{\alpha!\beta!\gamma!}{(\alpha+\beta+\gamma+2)!}2A \tag{3.78}$$

Equation (3.77) is used to evaluate an integral that is a function of the length along an edge of the element. Thus, the quantity $\mathcal{L}$ denotes the distance between the two nodes that define the edge under consideration. Equation (3.78) is used to evaluate area integrals. Table 3.2 gives the values of the integral for various combinations of α, β, and γ.

Table 3.2.

Value of α	β	γ	Value of the integral in Eq. (3.77)/$\mathcal{L}$	Value of the integral in Eq. (3.78)/A
0	0	0	1	1
1	0	0	1/2	1/3
2	0	0	1/3	1/6
1	1	0	1/6	1/12
3	0	0	1/4	1/10
2	1	0	1/12	1/30
1	1	1	—	1/60
4	0	0	1/5	1/15
3	1	0	1/20	1/60
2	2	0	1/30	1/90
2	1	1	—	1/180
5	0	0	1/6	1/21
4	1	0	1/30	1/105
3	2	0	1/60	1/210
3	1	1	—	1/420
2	2	1	—	1/630

3.9.3 Three-Dimensional (Tetrahedron) Element

The natural coordinates for a tetrahedron element can be defined analogous to those of a triangular element. Thus, four coordinates L_1, L_2, L_3, and L_4 will be used to define a point P, although only three of them are independent. These natural coordinates are defined as

$$L_1 = \frac{V_1}{V}, \qquad L_2 = \frac{V_2}{V}, \qquad L_3 = \frac{V_3}{V}, \qquad L_4 = \frac{V_4}{V} \tag{3.79}$$

where V_i is the volume of the tetrahedron formed by the points P and the vertices other than the vertex i ($i = 1, 2, 3, 4$), and V is the volume of the tetrahedron element defined by the vertices $1, 2, 3$, and 4 (Figure 3.12). Because the natural coordinates are defined in terms of volumes, they are also known as volume or tetrahedral coordinates. Since

$$V_1 + V_2 + V_3 + V_4 = V$$

we obtain

$$\frac{V_1}{V} + \frac{V_2}{V} + \frac{V_3}{V} + \frac{V_4}{V} = L_1 + L_2 + L_3 + L_4 = 1 \tag{3.80}$$

The volume coordinates L_1, L_2, L_3, and L_4 are also the shape functions for a three-dimensional simplex element:

$$N_i = L_1, \qquad N_j = L_2, \qquad N_k = L_3, \qquad N_l = L_4 \tag{3.81}$$

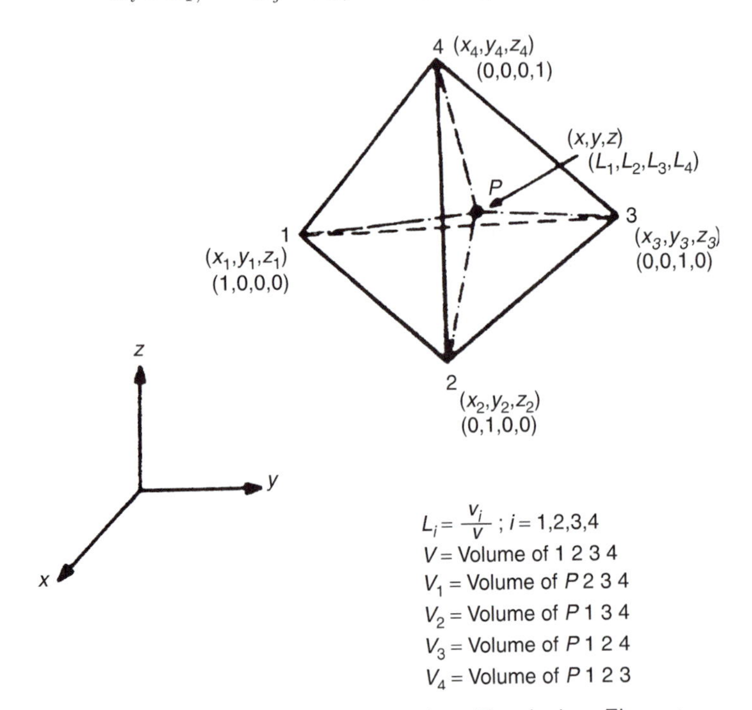

Figure 3.12. Volume Coordinates for a Tetrahedron Element.

The Cartesian and natural coordinates are related as

$$\left.\begin{aligned} x &= L_1x_1 + L_2x_2 + L_3x_3 + L_4x_4 \\ y &= L_1y_1 + L_2y_2 + L_3y_3 + L_4y_4 \\ z &= L_1z_1 + L_2z_2 + L_3z_3 + L_4z_4 \end{aligned}\right\} \tag{3.82}$$

Equations (3.80) and (3.82) can be expressed in matrix form as

$$\begin{Bmatrix} 1 \\ x \\ y \\ z \end{Bmatrix} = \begin{bmatrix} 1 & 1 & 1 & 1 \\ x_1 & x_2 & x_3 & x_4 \\ y_1 & y_2 & y_3 & y_4 \\ z_1 & z_2 & z_3 & z_4 \end{bmatrix} \begin{Bmatrix} L_1 \\ L_2 \\ L_3 \\ L_4 \end{Bmatrix} \tag{3.83}$$

The inverse relations can be expressed as

$$\begin{Bmatrix} L_1 \\ L_2 \\ L_3 \\ L_4 \end{Bmatrix} = \frac{1}{6V} \begin{bmatrix} a_1 & b_1 & c_1 & d_1 \\ a_2 & b_2 & c_2 & d_2 \\ a_3 & b_3 & c_3 & d_3 \\ a_4 & b_4 & c_4 & d_4 \end{bmatrix} \begin{Bmatrix} 1 \\ x \\ y \\ z \end{Bmatrix} \tag{3.84}$$

where

$$V = \frac{1}{6} \begin{vmatrix} 1 & x_1 & y_1 & z_1 \\ 1 & x_2 & y_2 & z_2 \\ 1 & x_3 & y_3 & z_3 \\ 1 & x_4 & y_4 & z_4 \end{vmatrix} = \text{volume of the tetrahedron 1, 2, 3, 4} \tag{3.85}$$

$$a_1 = \begin{vmatrix} x_2 & y_2 & z_2 \\ x_3 & y_3 & z_3 \\ x_4 & y_4 & z_4 \end{vmatrix} \tag{3.86}$$

$$b_1 = - \begin{vmatrix} 1 & y_2 & z_2 \\ 1 & y_3 & z_3 \\ 1 & y_4 & z_4 \end{vmatrix} \tag{3.87}$$

$$c_1 = - \begin{vmatrix} x_2 & 1 & z_2 \\ x_3 & 1 & z_3 \\ x_4 & 1 & z_4 \end{vmatrix} \tag{3.88}$$

$$d_1 = - \begin{vmatrix} x_2 & y_2 & 1 \\ x_3 & y_3 & 1 \\ x_4 & y_4 & 1 \end{vmatrix} \tag{3.89}$$

and the other constants are obtained through a cyclic permutation of subscripts 1, 2, 3, and 4. These constants are the cofactors of the terms in the determinant of Eq. (3.85) and hence it is necessary to give proper signs to them. If the tetrahedron element is defined in a right-handed Cartesian coordinate system as shown in Figure 3.12, Eqs. (3.86)–(3.89) are valid only when the nodes 1, 2, and 3 are numbered in a counterclockwise manner when viewed from node 4.

If f is a function of the natural coordinates, it can be differentiated with respect to cartesian coordinates as

$$\left.\begin{aligned} \frac{\partial f}{\partial x} &= \sum_{i=1}^{4} \frac{\partial f}{\partial L_i} \frac{\partial L_i}{\partial x} \\ \frac{\partial f}{\partial y} &= \sum_{i=1}^{4} \frac{\partial f}{\partial L_i} \frac{\partial L_i}{\partial y} \\ \frac{\partial f}{\partial z} &= \sum_{i=1}^{4} \frac{\partial f}{\partial L_i} \frac{\partial L_i}{\partial z} \end{aligned}\right\} \tag{3.90}$$

where

$$\frac{\partial L_i}{\partial x} = \frac{b_i}{6V}, \qquad \frac{\partial L_i}{\partial y} = \frac{c_i}{6V}, \qquad \frac{\partial L_i}{\partial z} = \frac{d_i}{6V} \tag{3.91}$$

The integration of polynomial terms in natural coordinates can be performed using the relation

$$\iiint_V L_1^\alpha L_2^\beta L_3^\gamma L_4^\delta \, \mathrm{d}V = \frac{\alpha!\beta!\gamma!\delta!}{(\alpha+\beta+\gamma+\delta+3)!} 6V \tag{3.92}$$

The values of this integral for different values of α, β, γ, and δ are given in Table 3.3.

Table 3.3.

Value of				Value of the integral
α	β	γ	δ	in Eq. (3.92)/V
0	0	0	0	1
1	0	0	0	1/4
2	0	0	0	1/10
1	1	0	0	1/20
3	0	0	0	1/20
2	1	0	0	1/60
1	1	1	0	1/120
4	0	0	0	1/35
3	1	0	0	1/140
2	2	0	0	1/210
2	1	1	0	1/420
1	1	1	1	1/840
5	0	0	0	1/56
4	1	0	0	1/280
3	2	0	0	1/560
3	1	1	0	1/1120
2	2	1	0	1/1680
2	1	1	1	1/3360

REFERENCES

3.1 J. Krahula and J. Polhemus: Use of Fourier series in the finite element method, *AIAA Journal, 6*, 726–728, 1968.

3.2 J.T. Oden: *Finite Elements of Nonlinear Continua*, McGraw-Hill, New York, 1972.

3.3 P. Dunne: Complete polynomial displacement fields for the finite element method, *Aeronautical Journal, 72*, 245–246, 1968 (Discussion: *72*, 709–711, 1968).

3.4 R.H. Gallagher: Analysis of plate and shell structures, *Proceedings of the Symposium on Application of Finite Element Methods in Civil Engineering*, Vanderbilt University, Nashville, pp. 155–206, November 1969.

3.5 R.J. Melosh: Basis for derivation of matrices for the direct stiffness method, *AIAA Journal, 1*, 1631–1637, 1963.

3.6 E.R. Arantes e Oliveira: Theoretical foundations of the finite element method, *International Journal of Solids and Structures, 4*, 929–952, 1968.

3.7 P.M. Mebane and J.A. Stricklin: Implicit rigid body motion in curved finite elements, *AIAA Journal, 9*, 344–345, 1971.

3.8 R.L. Taylor: On completeness of shape functions for finite element analysis, *International Journal for Numerical Methods in Engineering, 4*, 17–22, 1972.

3.9 I. Babuska and B. Szabo: On the rates of convergence of the finite element method, *International Journal for Numerical Methods in Engineering, 18*, 323–341, 1982.

3.10 A. Verma and R.J. Melosh: Numerical tests for assessing finite element modal convergence, *International Journal for Numerical Methods in Engineering, 24*, 843–857, 1987.

3.11 D.W. Kelly, J.P. Gago, O.C. Zienkiewicz, and I. Babuska: A posteriori error analysis and adaptive processes in the finite element method: Part I—Error anlaysis, *International Journal for Numerical Methods in Engineering, 19*, 1593–1619, 1983.

3.12 J.P. Gago, D.W. Kelly, O.C. Zienkiewicz, and I. Babuska: A posteriori error anlaysis and adaptive processes in the finite element method: Part II—Adaptive mesh refinement, *International Journal for Numerical Methods in Engineering, 19*, 1621–1656, 1983.

3.13 G.F. Carey and M. Seager: Projection and iteration in adaptive finite element refinement, *International Journal for Numerical Methods in Engineering, 21*, 1681–1695, 1985.

3.14 R.E. Ewing: Adaptive grid refinement methods for time-dependent flow problems, *Communications in Applied Numerical Methods, 3*, 351, 1987.

3.15 P. Roberti and M.A. Melkanoff: Self-adaptive stress analysis based on stress convergence, *International Journal for Numerical Methods in Engineering, 24*, 1973–1992, 1987.

PROBLEMS

3.1 What kind of interpolation model would you propose for the field variable ϕ for the six-node rectangular element shown in Figure 3.13. Discuss how the various considerations given in Section 3.5 are satisfied.

3.2 A one-dimensional simplex element has been used to find the temperature distribution in a straight fin. It is found that the nodal temperatures of the element are 140 and 100 °C at nodes i and j, respectively. If the nodes i and j are located 2 and 8 cm from the origin, find the temperature at a point 5 cm from the origin. Also find the temperature gradient inside the element.

3.3 Two-dimensional simplex elements have been used for modeling a heated flat plate. The (x, y) coordinates of nodes i, j, and k of an interior element are given by (5,4), (8,6) and (4,8) cm, respectively. If the nodal temperatures are found to be $T_i = 100\,°\text{C}$, $T_j = 80\,°\text{C}$, and $T_k = 110\,°\text{C}$, find (i) the temperature gradients inside the element and (ii) the temperature at point P located at $(x_p, y_p) = (6, 5)$ cm.

3.4 Three-dimensional simplex elements are used to find the pressure distribution in a fluid medium. The (x, y, z) coordinates of nodes i, j, k, and l of an element are given by (2,4,2), (0,0,0), (4,0,0), and (2,0,6) in. Find the shape functions N_i, N_j, N_k, and N_l of the element.

3.5 Show that the condition to be satisfied for constant value of the field variable is $\Sigma_{i=1}^{r} N_i = 1$, where N_i denotes the shape function corresponding to node i and r represents the number of nodes in the element.

3.6 Triangular elements are used for the stress analysis of a plate subjected to inplane loads. The components of displacement parallel to (x, y) axes at the nodes i, j, and k of an element are found to be $(-0.001, 0.01)$, $(-0.002, 0.01)$, and $(-0.002, 0.02)$ cm, respectively. If the (x, y) coordinates of the nodes shown in Figure 3.14 are in centimeters, find (i) the distribution of the (x, y) displacement components inside the element and (ii) the components of displacement of the point $(x_p, y_p) = (30, 25)$ cm.

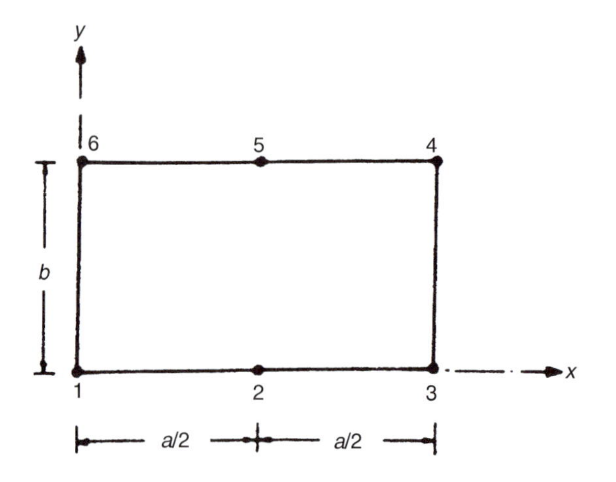

Figure 3.13. Six-Node Rectangular Element.

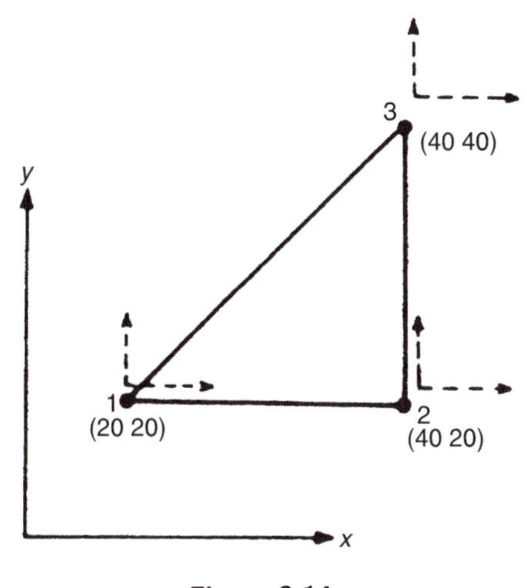

Figure 3.14.

3.7 The temperatures at the corner nodes of a rectangular element, in °C, are given by $T_i = 90$, $T_j = 84$, $T_k = 75$, and $T_l = 85$. If the length and width of the element are $x_{ij} = 15$ mm and $y_{il} = 10$ mm and the conduction coefficient of the material is $k = 42.5$ W/m-C, determine the following:

(a) Temperature distribution in the element

(b) Heat flow rates in x and y directions (q_x and q_y) using the relation

$$\begin{Bmatrix} q_x \\ q_y \end{Bmatrix} = -k \begin{Bmatrix} \dfrac{\partial T}{\partial x} \\ \dfrac{\partial T}{\partial y} \end{Bmatrix}$$

3.8 Derive the relationship between the natural (area) and Cartesian coordinates of a triangular element (Eq. 3.71).

3.9 The quadratic interpolation function of a one-dimensional element with three nodes is given by

$$\phi(x) = \alpha_1 + \alpha_2 x + \alpha_3 x^2$$

If the x coordinates of nodes 1, 2, and 3 are given by 1, 3, and 5 in., respectively, determine the matrices $[\underset{\sim}{\eta}]$, $[\underset{\sim}{\eta}]^{-1}$, and $[N]$ of Eqs. (3.17), (3.18), and (3.20).

3.10 The cubic interpolation function for the displacement of a beam element is expressed as

$$\phi(x) = \alpha_1 + \alpha_2 x + \alpha_3 x^2 + \alpha_4 x^3$$

with the nodal degrees of freedom defined as $\phi_1 = \phi(x = x_1)$, $\phi_2 = (\mathrm{d}\phi/\mathrm{d}x)(x = x_1)$, $\phi_3 = \phi(x = x_2)$, and $\phi_4 = (\mathrm{d}\phi/\mathrm{d}x)(x = x_2)$, where x_1 and x_2 denote the x coordinates of nodes 1 and 2 of the element. If $x_1 = 1.0$ in. and $x_2 = 6$ in., determine the matrices $[\underset{\sim}{\eta}]$, $[\underset{\sim}{\eta}]^{-1}$, and $[N]$ of Eqs. (3.17), (3.18), and (3.20).

3.11 The transverse displacement of a triangular bending element (w) is expressed as

$$w(x) = \alpha_1 + \alpha_2 x + \alpha_3 y + \alpha_4 x^2 + \alpha_5 xy + \alpha_6 y^2$$
$$+ \alpha_7 x^3 + \alpha_8 (x^2 y + xy^2) + \alpha_9 y^3$$

The nodal degrees of freedom are defined as $\phi_i = w(x_i, y_i)$, $\phi_{i+3} = (\partial w/\partial y)(x_i, y_i)$, $\phi_{i+6} = (\partial w/\partial x)(x_i, y_i)$; $i = 1, 2, 3$, where (x_i, y_i) denote the coordinates of node i. If the (x, y) coordinates of nodes 1, 2, and 3 are given by $(0, 0)$, $(0, 5)$, and $(10, 0)$, respectively, determine the matrices $[\underset{\sim}{\eta}]$, $[\underset{\sim}{\eta}]^{-1}$, and $[N]$ of Eqs. (3.17), (3.18), and (3.20).

3.12 Consider the displacement model of a triangular bending element given in Problem 3.11. Determine whether the convergence requirements of Section 3.6 are satisfied by this model.

Note: The expression for the functional I (potential energy) of a plate in bending is given by

$$I = \frac{1}{2} \iint\limits_A D \left\{ \left[\frac{\partial^2 w}{\partial x^2} + \frac{\partial^2 w}{\partial y^2} \right]^2 - 2(1 - v) \left[\frac{\partial^2 w}{\partial x^2} \frac{\partial^2 w}{\partial y^2} - \left[\frac{\partial^2 w}{\partial x \partial y} \right]^2 \right] \right\} \mathrm{d}x \, \mathrm{d}y$$
$$- \iint\limits_A (pw) \, \mathrm{d}x \, \mathrm{d}y$$

where p is the distributed transverse load per unit area, D is the flexural rigidity, v is the Poisson's ratio, and A is the surface area of the plate.

3.13 The coordinates of the nodes of a three-dimensional simplex element are given below:

Node number	Coordinates of the node		
	x	y	z
i	0	0	0
j	10	0	0
k	0	15	0
l	0	0	20

Determine the shape functions of the element.

3.14 The shape function matrix of a uniform one-dimensional simplex element is given by $[N] = [N_i \, N_j]$, with $N_i = 1 - (x/l)$ and $N_j = (x/l)$. Evaluate the integral: $\iiint_V [N]^T [N] \mathrm{d}V$, where $V = A \, \mathrm{d}x$, A is the cross-sectional area, and l is the length of the element.

3.15 Evaluate the integral $\int_{\mathcal{L}_{ij}} \vec{N} \vec{N}^T \mathrm{d}\mathcal{L}$ along the edge ij of a simplex triangle, where $\mathcal{L}_{ij}$ denotes the distance between the nodes i and j, and the vector of shape functions $\vec{N}$ is given by $\vec{N}^T = (N_i \, N_j \, N_k)$.

3.16 Evaluate the integral $\int_{S_{ijk}} \vec{N} \vec{N}^T \mathrm{d}S$ on the face ijk of a simplex tetrahedron, where S_{ijk} denotes the surface area bounded by the nodes i, j, and k, and the vector of shape functions is given by $\vec{N}^T = (N_i \, N_j \, N_k \, N_l)$.

4

HIGHER ORDER AND ISOPARAMETRIC ELEMENTS

4.1 INTRODUCTION

As stated earlier, if the interpolation polynomial is of order two or more, the element is known as a higher order element. A higher order element can be either a complex or a multiplex element. In higher order elements, some secondary (midside and/or interior) nodes are introduced in addition to the primary (corner) nodes in order to match the number of nodal degrees of freedom with the number of constants (also known as generalized coordinates) in the interpolation polynomial.

For problems involving curved boundaries, a family of elements known as "isoparametric" elements can be used. In isoparametric elements, the same interpolation functions used to define the element geometry are also used to describe the variation of the field variable within the element. Both higher order and isoparametric elements are considered in this chapter.

4.2 HIGHER ORDER ONE-DIMENSIONAL ELEMENTS

4.2.1 Quadratic Element

The quadratic interpolation model for a one-dimensional element can be expressed as

$$\phi(x) = \alpha_1 + \alpha_2 x + \alpha_3 x^2 \tag{4.1}$$

Since there are three constants α_1, α_2, and α_3 in Eq. (4.1), the element is assumed to have three degrees of freedom, one at each of the ends and one at the middle point as shown in Figure 4.1(b). By requiring that

$$\begin{aligned} \phi(x) &= \Phi_i \quad \text{at} \quad x = 0 \\ \phi(x) &= \Phi_j \quad \text{at} \quad x = l/2 \\ \phi(x) &= \Phi_k \quad \text{at} \quad x = l \end{aligned} \tag{4.2}$$

we can evaluate the constants α_1, α_2, and α_3 as

$$\begin{aligned} &\alpha_1 = \Phi_i, \qquad \alpha_2 = (4\Phi_j - 3\Phi_i - \Phi_k)/l, \\ &\alpha_3 = 2(\Phi_i - 2\Phi_j + \Phi_k)/l^2 \end{aligned} \tag{4.3}$$

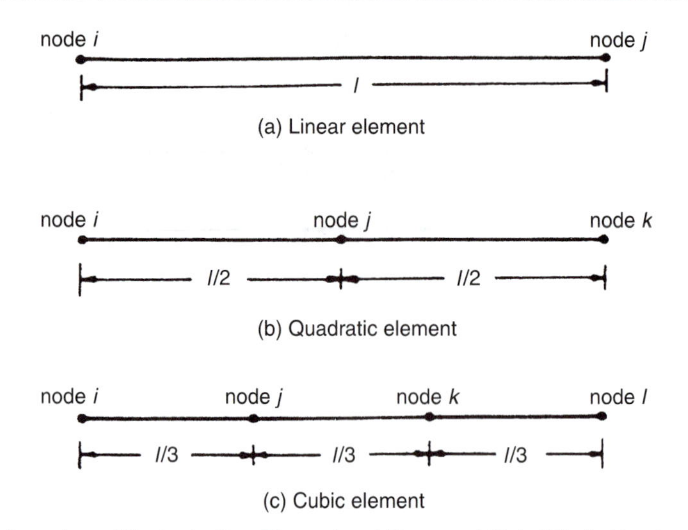

Figure 4.1. Location of Nodes in One-Dimensional Element (Global Node Numbers Indicated).

With the help of Eq. (4.2), Eq. (4.1) can be expressed after rearrangement as

$$\phi(x) = [N(x)]\vec{\Phi}^{(e)} \tag{4.4}$$

where

$$\begin{aligned}
[N(x)] &= \begin{bmatrix} N_i(x) & N_j(x) & N_k(x) \end{bmatrix}, \\
N_i(x) &= \left(1 - 2\frac{x}{l}\right)\left(1 - \frac{x}{l}\right), \\
N_j(x) &= 4\frac{x}{l}\left(1 - \frac{x}{l}\right), \\
N_k(x) &= -\frac{x}{l}\left(1 - 2\frac{x}{l}\right),
\end{aligned} \tag{4.5}$$

and

$$\vec{\Phi}^{(e)} = \begin{Bmatrix} \Phi_i \\ \Phi_j \\ \Phi_k \end{Bmatrix} \tag{4.6}$$

4.2.2 Cubic Element

The cubic interpolation model can be expressed as

$$\phi(x) = \alpha_1 + \alpha_2 x + \alpha_3 x^2 + \alpha_4 x^3 \tag{4.7}$$

Because there are four unknown coefficients α_1, α_2, α_3, and α_4, the element is assumed to have four degrees of freedom, one at each of the four nodes shown in Figure 4.1(c).

By requiring that

$$\begin{aligned} \phi(x) &= \Phi_i \quad \text{at} \quad x = 0 \\ \phi(x) &= \Phi_j \quad \text{at} \quad x = l/3 \\ \phi(x) &= \Phi_k \quad \text{at} \quad x = 2l/3 \\ \phi(x) &= \Phi_l \quad \text{at} \quad x = l \end{aligned} \tag{4.8}$$

the constants α_1, α_2, α_3, and α_4 can be evaluated. The substitution of values of these constants into Eq. (4.7) leads to

$$\phi(x) = [N(x)]\vec{\Phi}^{(e)} \tag{4.9}$$

where

$$\begin{aligned} [N(x)] &= [N_i(x) \quad N_j(x) \quad N_k(x) \quad N_l(x)], \\ N_i(x) &= \left(1 - \frac{3x}{l}\right)\left(1 - \frac{3x}{2l}\right)\left(1 - \frac{x}{l}\right), \\ N_j(x) &= 9\frac{x}{l}\left(1 - \frac{3x}{2l}\right)\left(1 - \frac{x}{l}\right), \\ N_k(x) &= -\frac{9}{2}\frac{x}{l}\left(1 - \frac{3x}{l}\right)\left(1 - \frac{x}{l}\right), \\ N_l(x) &= \frac{x}{l}\left(1 - \frac{3x}{l}\right)\left(1 - \frac{3x}{2l}\right), \end{aligned} \tag{4.10}$$

and

$$\vec{\Phi}^{(e)} = \begin{Bmatrix} \Phi_i \\ \Phi_j \\ \Phi_k \\ \Phi_l \end{Bmatrix} \tag{4.11}$$

It can be observed that the application of the previous procedure for determining the coefficients α_i and the nodal interpolation functions $N_i(x)$ becomes more tedius as the order of the interpolation polynomial increases. The nodal interpolation functions $N_i(x)$ can be constructed in a simpler manner by employing either natural coordinates or classical interpolation polynomials.

4.3 HIGHER ORDER ELEMENTS IN TERMS OF NATURAL COORDINATES

4.3.1 One-Dimensional Element

(i) Quadratic element

The normalized or natural coordinates L_1 and L_2 for a one-dimensional element were shown in Figure 3.10. If the values of ϕ at three stations x_1, $(x_1 + x_2)/2$, and x_2 are taken as nodal unknowns, the quadratic model for $\phi(x)$ can be expressed as

$$\phi(x) = [N]\vec{\Phi}^{(e)} = [N_1 \quad N_2 \quad N_3]\vec{\Phi}^{(e)} \tag{4.12}$$

where

$$\vec{\Phi}^{(e)} = \begin{Bmatrix} \Phi_1 \\ \Phi_2 \\ \Phi_3 \end{Bmatrix}^{(e)} = \begin{Bmatrix} \phi(x_1) \\ \phi(x_2) \\ \phi(x_3) \end{Bmatrix}^{(e)} = \begin{Bmatrix} \phi \text{ (at } L_1 = 1, L_2 = 0) \\ \phi \text{ (at } L_1 = \frac{1}{2}, L_2 = \frac{1}{2}) \\ \phi \text{ (at } L_1 = 0, L_2 = 1) \end{Bmatrix}^{(e)} \tag{4.13}$$

and the quadratic nodal interpolation functions N_i can be expressed in general form as

$$N_i = a_1^{(i)} L_1 + a_2^{(i)} L_2 + a_3^{(i)} L_1 L_2; \qquad i = 1, 2, 3 \tag{4.14}$$

For N_1 we impose the requirements

$$N_1 = \begin{cases} 1 \text{ at node } 1 & (L_1 = 1, L_2 = 0) \\ 0 \text{ at node } 2 & (L_1 = L_2 = \frac{1}{2}) \\ 0 \text{ at node } 3 & (L_1 = 0, L_2 = 1) \end{cases}$$

and find the values of the constants $a_1^{(1)}$, $a_2^{(1)}$, and $a_3^{(1)}$ as

$$a_1^{(1)} = 1, \qquad a_2^{(1)} = 0, \qquad a_3^{(1)} = -2$$

so that Eq. (4.14) becomes

$$N_1 = L_1 - 2L_1 L_2$$

By using the condition $L_1 + L_2 = 1$, we obtain

$$N_1 = L_1(2L_1 - 1) \tag{4.15}$$

Similarly, the other two nodal interpolation functions can be derived as

$$N_2 = 4L_1 L_2 \tag{4.16}$$

and

$$N_3 = L_2(2L_2 - 1) \tag{4.17}$$

The nodal interpolation functions N_i appearing in Eqs. (4.15)–(4.17) are shown in Figure 4.2.

(ii) Cubic element

For a cubic element, we consider four nodal degrees of freedom, one at each of the nodes shown in Figure 4.1(c). The cubic interpolation model can be written as

$$\phi(x) = [N]\vec{\Phi}^{(e)} = [N_1 \quad N_2 \quad N_3 \quad N_4]\vec{\Phi}^{(e)} \tag{4.18}$$

where

$$\vec{\Phi}^{(e)} = \begin{Bmatrix} \Phi_1 \\ \Phi_2 \\ \Phi_3 \\ \Phi_4 \end{Bmatrix}^{(e)} = \begin{Bmatrix} \phi(x_1) \\ \phi(x_2) \\ \phi(x_3) \\ \phi(x_4) \end{Bmatrix}^{(e)} = \begin{Bmatrix} \phi \text{ (at } L_1 = 1, L_2 = 0) \\ \phi \text{ (at } L_1 = 2/3, L_2 = 1/3) \\ \phi \text{ (at } L_1 = 1/3, L_2 = 2/3) \\ \phi \text{ (at } L_1 = 0, L_2 = 1) \end{Bmatrix}^{(e)}$$

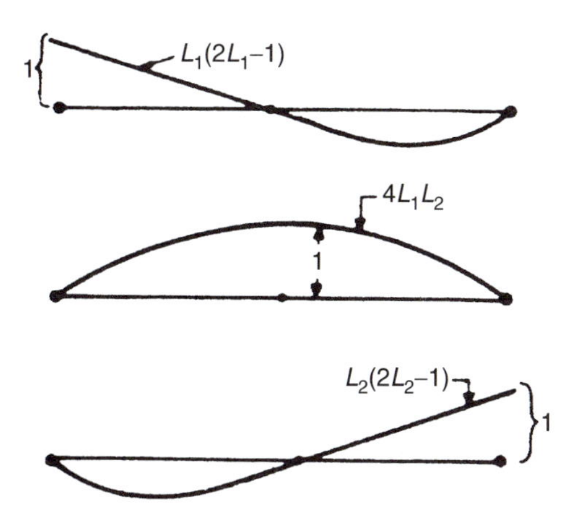

Figure 4.2. Nodal Interpolation (or Shape) Functions for a Line Element.

and the nodal interpolation functions N_i appearing in Eq. (4.18) can be expressed in terms of the natural coordinates as

$$N_i = a_1^{(i)} L_1 + a_2^{(i)} L_2 + a_3^{(i)} L_1 L_2 + a_4^{(i)} L_1^2 L_2 \tag{4.19}$$

By requiring that N_i be equal to one at node i and zero at each of the other nodes, we find that

$$N_1 = L_1 \left(1 - \tfrac{9}{2} L_1 L_2\right) \tag{4.20}$$

$$N_2 = -\tfrac{9}{2} L_1 L_2 (1 - 3L_1) \tag{4.21}$$

$$N_3 = 9 L_1 L_2 \left(1 - \tfrac{3}{2} L_1\right) \tag{4.22}$$

$$N_4 = L_2 - \tfrac{9}{2} L_1 L_2 (1 - L_1) \tag{4.23}$$

4.3.2 Two-Dimensional (Triangular) Element

(i) Quadratic element

The natural or triangular coordinates L_1, L_2, and L_3 of a triangular element were shown in Figure 3.11(a). For a quadratic interpolation model, the values of the field variable at three corner nodes and three midside nodes (Figure 4.3(a)) are taken as the nodal unknowns and $\phi(x, y)$ is expressed as

$$\phi(x, y) = [N]\vec{\Phi}^{(e)} = [N_1 \quad N_2 \quad \cdots \quad N_6]\vec{\Phi}^{(e)} \tag{4.24}$$

where N_i can be derived from the general quadratic relationship

$$N_i = a_1^{(i)} L_1 + a_2^{(i)} L_2 + a_3^{(i)} L_3 + a_4^{(i)} L_1 L_2 + a_5^{(i)} L_2 L_3 + a_6^{(i)} L_1 L_3 \tag{4.25}$$

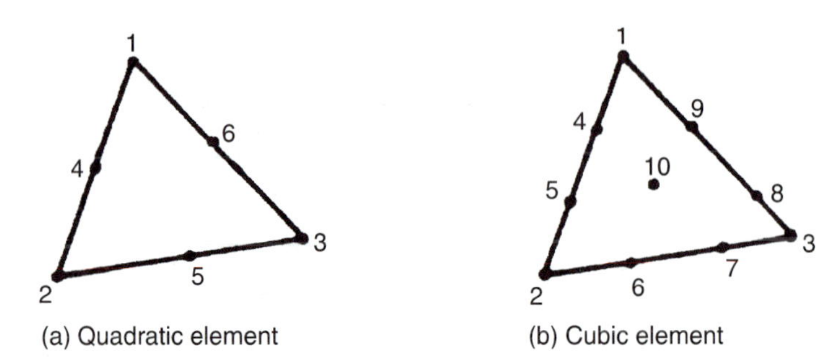

Figure 4.3. Location of Nodes in a Triangular Element.

as

$$\begin{aligned} N_i &= L_i(2L_i - 1), \qquad i = 1, 2, 3 \\ N_4 &= 4L_1 L_2 \\ N_5 &= 4L_2 L_3 \\ N_6 &= 4L_1 L_3 \end{aligned} \tag{4.26}$$

and

$$\vec{\Phi}^{(e)} = \begin{Bmatrix} \Phi_1 \\ \Phi_2 \\ \vdots \\ \Phi_6 \end{Bmatrix}^{(e)} = \begin{Bmatrix} \phi(x_1, y_1) \\ \phi(x_2, y_2) \\ \vdots \\ \phi(x_6, y_6) \end{Bmatrix}^{(e)} = \begin{Bmatrix} \phi \text{ (at } L_1 = 1, L_2 = L_3 = 0) \\ \phi \text{ (at } L_2 = 1, L_1 = L_3 = 0) \\ \vdots \\ \phi \text{ (at } L_1 = L_3 = \frac{1}{2}, L_2 = 0) \end{Bmatrix}^{(e)} \tag{4.27}$$

The nodal interpolation or shape functions of Eq. (4.26) are shown in Figure 4.4.

(ii) Cubic element

If a cubic interpolation model is used, 10 nodal unknowns are required. The location of the nodes is shown in Figure 4.3(b), in which the nodes 4 and 5 are located at one-third points along the edge 12 with similar locations for the nodes 6 and 7, and 8 and 9 along the edges 23 and 31, respectively. The node 10 is located at the centroid of the triangle 123. In this case, the interpolation model is given by

$$\phi(x, y) = [N]\vec{\Phi}^{(e)} = [N_1 \quad N_2 \quad \ldots \quad N_{10}]\vec{\Phi}^{(e)} \tag{4.28}$$

where the general form of the nodal interpolation function can be assumed as

$$\begin{aligned} N_i = {} & a_1^{(i)} L_1 + a_2^{(i)} L_2 + a_3^{(i)} L_3 + a_4^{(i)} L_1 L_2 + a_5^{(i)} L_2 L_3 + a_6^{(i)} L_1 L_3 \\ & + a_7^{(i)} L_1^2 L_2 + a_8^{(i)} L_2^2 L_3 + a_9^{(i)} L_3^2 L_1 + a_{10}^{(i)} L_1 L_2 L_3 \end{aligned} \tag{4.29}$$

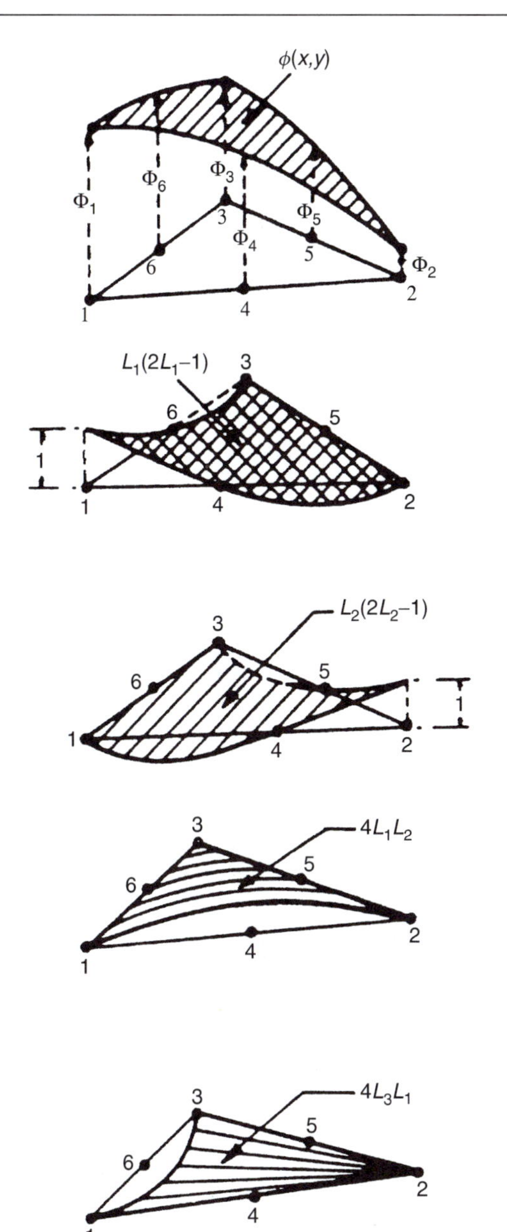

Figure 4.4. Quadratic Interpolation or Shape Functions for a Triangular Element.

By imposing the conditions that N_i be equal to one at node i and zero at each of the remaining nine nodes, we can obtain

$$\begin{aligned}
N_i &= \tfrac{1}{2}L_i(3L_i-1)(3L_i-2), \qquad i=1,2,3\\
N_4 &= \tfrac{9}{2}L_1L_2(3L_1-1)\\
N_5 &= \tfrac{9}{2}L_1L_2(3L_2-1)\\
N_6 &= \tfrac{9}{2}L_2L_3(3L_2-1)\\
N_7 &= \tfrac{9}{2}L_2L_3(3L_3-1)\\
N_8 &= \tfrac{9}{2}L_1L_3(3L_3-1)\\
N_9 &= \tfrac{9}{2}L_1L_3(3L_1-1)\\
N_{10} &= 27L_1L_2L_3
\end{aligned} \tag{4.30}$$

and

$$\vec{\Phi}^{(e)} = \begin{Bmatrix} \Phi_1 \\ \Phi_2 \\ \vdots \\ \Phi_{10} \end{Bmatrix}^{(e)} = \begin{Bmatrix} \phi(x_1,y_1) \\ \phi(x_2,y_2) \\ \vdots \\ \phi(x_{10},y_{10}) \end{Bmatrix}^{(e)} = \begin{Bmatrix} \phi\ (\text{at } L_1=1, L_2=L_3=0) \\ \phi\ (\text{at } L_2=1, L_1=L_3=0) \\ \vdots \\ \phi\ (\text{at } L_1=L_2=L_3=\tfrac{1}{3}) \end{Bmatrix}^{(e)} \tag{4.31}$$

4.3.3 Two-Dimensional (Quadrilateral) Element

Natural Coordinates

A different type of natural coordinate system can be established for a quadrilateral element in two dimensions as shown in Figure 4.5. For the local r, s (natural) coordinate system,

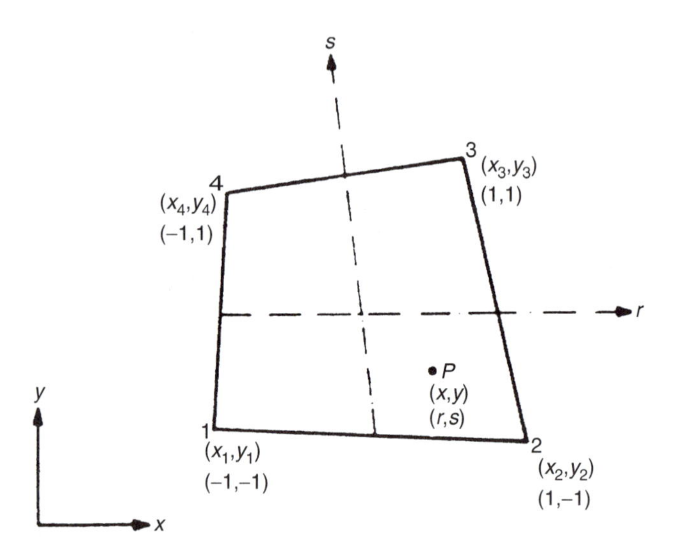

Figure 4.5. Natural Coordinates for a Quadrilateral Element.

the origin is taken as the intersection of lines joining the midpoints of opposite sides and the sides are defined by $r = \pm 1$ and $s = \pm 1$. The natural and Cartesian coordinates are related by the following equation:

$$\begin{Bmatrix} x \\ y \end{Bmatrix} = \begin{bmatrix} N_1 & N_2 & N_3 & N_4 & 0 & 0 & 0 & 0 \\ 0 & 0 & 0 & 0 & N_1 & N_2 & N_3 & N_4 \end{bmatrix} \begin{Bmatrix} x_1 \\ x_2 \\ x_3 \\ x_4 \\ y_1 \\ y_2 \\ y_3 \\ y_4 \end{Bmatrix} \tag{4.32}$$

where (x_i, y_i) are the (x, y) coordinates of node i $(i = 1, 2, 3, 4)$,

$$N_i = \tfrac{1}{4}(1 + rr_i)(1 + ss_i), \qquad i = 1, 2, 3, 4 \tag{4.33}$$

and the natural coordinates of the four nodes of the quadrilateral are given by

$$\begin{aligned} &(r_1, s_1) = (-1, -1), \qquad (r_2, s_2) = (1, -1), \\ &(r_3, s_3) = (1, 1), \quad \text{and} \quad (r_4, s_4) = (-1, 1). \end{aligned} \tag{4.34}$$

If ϕ is a function of the natural coordinates r and s, its derivatives with respect to x and y can be obtained as

$$\begin{Bmatrix} \dfrac{\partial \phi}{\partial x} \\ \dfrac{\partial \phi}{\partial y} \end{Bmatrix} = [J]^{-1} \begin{Bmatrix} \dfrac{\partial \phi}{\partial r} \\ \dfrac{\partial \phi}{\partial s} \end{Bmatrix} \tag{4.35}$$

where $[J]$ is a 2×2 matrix, called the Jacobian matrix, given by

$$\begin{aligned} [J] &= \begin{bmatrix} \partial x/\partial r & \partial y/\partial r \\ \partial x/\partial s & \partial y/\partial s \end{bmatrix} \\ &= \frac{1}{4} \begin{bmatrix} -(1-s) & (1-s) & (1+s) & -(1+s) \\ -(1-r) & -(1+r) & (1+r) & (1-r) \end{bmatrix} \begin{bmatrix} x_1 & y_1 \\ x_2 & y_2 \\ x_3 & y_3 \\ x_4 & y_4 \end{bmatrix} \end{aligned} \tag{4.36}$$

The integration of functions of r and s has to be performed numerically with

$$\mathrm{d}A = \mathrm{d}x\,\mathrm{d}y = \det\,[J] \cdot \mathrm{d}r\,\mathrm{d}s \tag{4.37}$$

and the limits of both r and s will be -1 and 1.

(i) Linear element

For a quadrilateral element, it is not possible to have linear variation of the field variable (in terms of two independent coordinates) if one degree of freedom is chosen at each of the four corner nodes. Hence, we take the interpolation model as

$$\phi(x,y) = [N]\vec{\Phi}^{(e)} = [N_1 \quad N_2 \quad N_3 \quad N_4]\vec{\Phi}^{(e)} \tag{4.38}$$

where

$$N_i = (1 + rr_i)(1 + ss_i)/4, \qquad i = 1,2,3,4 \tag{4.39}$$

and

$$\vec{\Phi}^{(e)} = \begin{Bmatrix} \Phi_1 \\ \Phi_2 \\ \Phi_3 \\ \Phi_4 \end{Bmatrix}^{(e)} = \begin{Bmatrix} \phi(x_1,y_1) \\ \phi(x_2,y_2) \\ \phi(x_3,y_3) \\ \phi(x_4,y_4) \end{Bmatrix}^{(e)} \equiv \begin{Bmatrix} \phi \text{ (at } r=-1, s=-1) \\ \phi \text{ (at } r=1, s=-1) \\ \phi \text{ (at } r=1, s=1) \\ \phi \text{ (at } r=-1, s=1) \end{Bmatrix}^{(e)} \tag{4.40}$$

The nodal shape functions represented by Eq. (4.39) are shown in Figure 4.6(a). It can be seen that the variation of the field variable along the edges of the quadrilateral is linear. Hence, this element is often called a linear element.

(ii) Quadratic element

If the values of $\phi(x,y)$ at the four corner nodes and four midside nodes are taken as the nodal unknowns, we get a "quadratic" element for which the variation of the field variable along any edge is given by a quadratic equation. In this case, the interpolation model can be expressed as

$$\phi(x,y) = [N]\vec{\Phi}^{(e)} = [N_1 \quad N_2 \quad \cdots \quad N_8]\vec{\Phi}^{(e)} \tag{4.41}$$

where

$$\begin{aligned} N_i &= \tfrac{1}{4}(1+rr_i)(1+ss_i)(rr_i+ss_i-1), \qquad i=1,2,3,4 \\ N_5 &= \tfrac{1}{2}(1-r^2)(1+ss_5) \\ N_6 &= \tfrac{1}{2}(1+rr_6)(1-s^2) \\ N_7 &= \tfrac{1}{2}(1-r^2)(1+ss_7) \\ N_8 &= \tfrac{1}{2}(1+rr_8)(1-s^2) \end{aligned} \tag{4.42}$$

(r_i, s_i) are the natural coordinates of node i $(i = 1, 2, \ldots, 8)$, and

$$\vec{\Phi}^{(e)} = \begin{Bmatrix} \Phi_1 \\ \Phi_2 \\ \vdots \\ \Phi_8 \end{Bmatrix}^{(e)} = \begin{Bmatrix} \phi(x_1,y_1) \\ \phi(x_2,y_2) \\ \vdots \\ \phi(x_8,y_8) \end{Bmatrix}^{(e)} \equiv \begin{Bmatrix} \phi \text{ (at } r=-1, s=-1) \\ \phi \text{ (at } r=1, s=-1) \\ \vdots \\ \phi \text{ (at } r=-1, s=0) \end{Bmatrix}^{(e)} \tag{4.43}$$

Typical quadratic interpolation or shape functions used in Eq. (4.41) are shown in Figure 4.6(b).

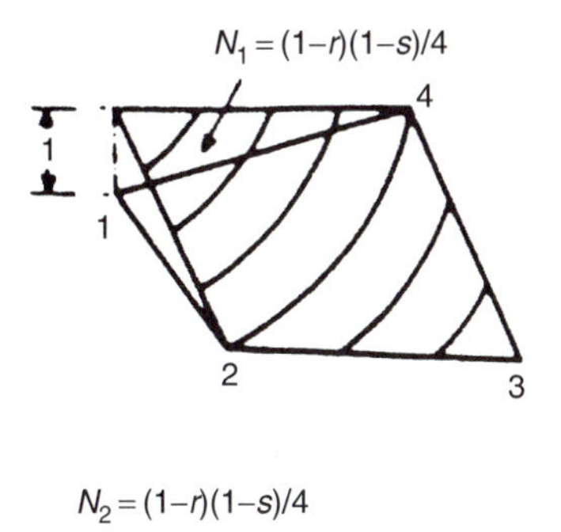

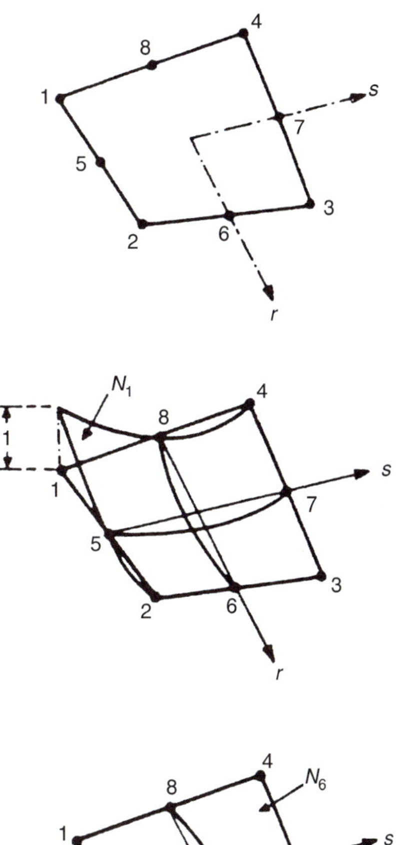

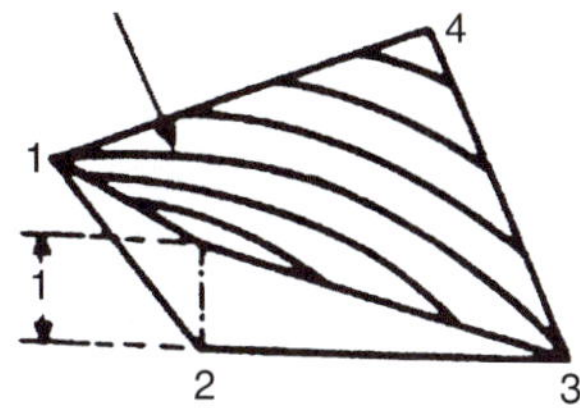

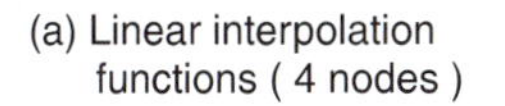

(a) Linear interpolation functions (4 nodes)

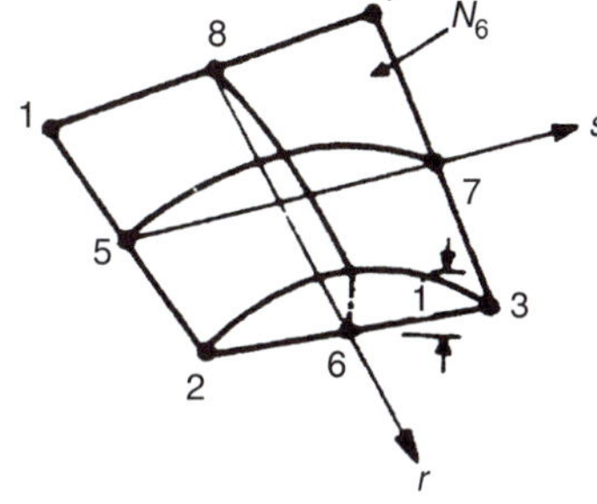

(b) Quadratic interpolation functions (8 nodes)

Figure 4.6. Interpolation Functions for a Quadrilateral Element.

4.3.4 Three-Dimensional (Tetrahedron) Element

(i) Quadratic element

The natural or tetrahedral coordinates L_1, L_2, L_3, and L_4 of a tetrahedron element were shown in Figure 3.12.

For a quadratic interpolation model, there will be 10 nodal unknowns, 1 at each of the nodes indicated in Figure 4.7(a). Here, the nodes 1, 2, 3, and 4 correspond to the corners,

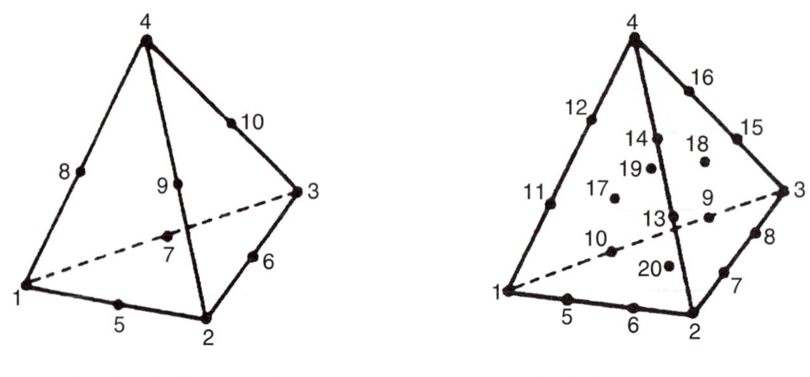

(a) Quadratic element (b) Cubic element

Figure 4.7. Location of Nodes in Tetrahedron Element.

whereas the nodes 5–10 are located at the midpoints of the edges of the tetrahedron. The variation of the field variable is given by

$$\phi(x,y,z) = [N]\vec{\Phi}^{(e)} = [N_1 \quad N_2 \quad \cdots \quad N_{10}]\vec{\Phi}^{(e)} \tag{4.44}$$

where N_i can be found as

$$\begin{aligned} N_i &= L_i(2L_i - 1), \qquad i = 1,2,3,4 \\ N_5 &= 4L_1L_2, \\ N_6 &= 4L_2L_3, \\ N_7 &= 4L_1L_3, \\ N_8 &= 4L_1L_4, \\ N_9 &= 4L_2L_4, \\ N_{10} &= 4L_3L_4, \end{aligned} \tag{4.45}$$

and

$$\vec{\Phi}^{(e)} = \begin{Bmatrix} \Phi_1 \\ \Phi_2 \\ \vdots \\ \Phi_{10} \end{Bmatrix}^{(e)} = \begin{Bmatrix} \phi(x_1,y_1,z_1) \\ \phi(x_2,y_2,z_2) \\ \vdots \\ \phi(x_{10},y_{10},z_{10}) \end{Bmatrix}^{(e)}$$

$$= \begin{Bmatrix} \phi \text{ (at } L_1 = 1, L_2 = L_3 = L_4 = 0) \\ \phi \text{ (at } L_2 = 1, L_1 = L_3 = L_4 = 0) \\ \vdots \\ \phi \text{ (at } L_3 = L_4 = \frac{1}{2}, L_1 = L_2 = 0) \end{Bmatrix}^{(e)} \tag{4.46}$$

(ii) Cubic element

The cubic interpolation model involves 20 nodal unknowns (nodes are shown in Figure 4.7(b)) and can be expressed as

$$\phi(x, y, z) = [N]\vec{\Phi}^{(e)} = [N_1 \quad N_2 \quad \cdots \quad N_{20}]\vec{\Phi}^{(e)} \tag{4.47}$$

where the nodal shape functions can be determined as follows:

For corner nodes: $N_i = \frac{1}{2}L_i(3L_i - 1)(3L_i - 2), \qquad i = 1, 2, 3, 4$ (4.48)

For one-third points of edges:
$$\begin{aligned} N_5 &= \tfrac{9}{2}L_1L_2(3L_1 - 1) \\ N_6 &= \tfrac{9}{2}L_1L_2(3L_2 - 1) \\ N_7 &= \tfrac{9}{2}L_2L_3(3L_2 - 1) \\ N_8 &= \tfrac{9}{2}L_2L_3(3L_3 - 1), \text{ etc.} \end{aligned} \tag{4.49}$$

For midface nodes:
$$\begin{aligned} N_{17} &= 27L_1L_2L_4 \\ N_{18} &= 27L_2L_3L_4 \\ N_{19} &= 27L_1L_3L_4 \\ N_{20} &= 27L_1L_2L_3 \end{aligned} \tag{4.50}$$

and

$$\vec{\Phi}^{(e)} = \begin{Bmatrix} \Phi_1 \\ \Phi_2 \\ \vdots \\ \Phi_{20} \end{Bmatrix}^{(e)} = \begin{Bmatrix} \phi(x_1, y_1, z_1) \\ \phi(x_2, y_2, z_2) \\ \vdots \\ \phi(x_{20}, y_{20}, z_{20}) \end{Bmatrix}^{(e)}$$
$$\equiv \begin{Bmatrix} \phi \text{ (at } L_1 = 1, L_2 = 0, L_3 = 0, L_4 = 0) \\ \phi \text{ (at } L_1 = 0, L_2 = 1, L_3 = 0, L_4 = 0) \\ \vdots \\ \phi\left(\text{at } L_1 = L_2 = L_3 = L_4 = \tfrac{1}{3}\right) \end{Bmatrix}^{(e)} \tag{4.51}$$

4.4 HIGHER ORDER ELEMENTS IN TERMS OF CLASSICAL INTERPOLATION POLYNOMIALS

It is possible to construct the nodal interpolation functions N_i by employing classical interpolation polynomials (instead of natural coordinates). We consider the use of Lagrange and Hermite interpolation polynomials in this section.

4.4.1 Classical Interpolation Functions

In numerical mathematics, an approximation polynomial that is equal to the function it approximates at a number of specified stations or points is called an interpolation function. A generalization of the interpolation function is obtained by requiring agreement with not only the function value $\phi(x)$ but also the first N derivatives of $\phi(x)$ at any number of distinct points $x_i, i = 1, 2, \ldots, n+1$. When $N = 0$—that is, when only the function values are required to match (agree) at each point of interpolation—the (classical) interpolation polynomial is called Lagrange interpolation formula. For the case of $N = 1$—that is, when the function and its first derivative are to be assigned at each point of interpolation—the (classical) interpolation polynomial is called the Hermite or osculatory interpolation formula. If higher derivatives of $\phi(x)$ are assigned (i.e., when $N > 1$), we obtain the hyperosculatory interpolation formula.

4.4.2 Lagrange Interpolation Functions for n Stations

The Lagrange interpolation polynomials are defined as [4.1]

$$L_k(x) = \prod_{\substack{i=0,\\ i\neq k}}^{n} \frac{(x - x_i)}{(x_k - x_i)} = \frac{(x - x_0)(x - x_1)\cdots(x - x_{k-1})(x - x_{k+1})\cdots(x - x_n)}{(x_k - x_0)(x_k - x_1)\cdots(x_k - x_{k-1})(x_k - x_{k+1})\cdots(x_k - x_n)} \tag{4.52}$$

It can be seen that $L_k(x)$ is an nth degree polynomial because it is given by the product of n linear factors. It can be seen that if $x = x_k$, the numerator would be equal to the denominator in Eq. (4.52) and hence $L_k(x)$ will have a value of unity. On the other hand, if $x = x_i$ and $i \neq k$, the numerator and hence $L_k(x)$ will be zero. This property of $L_k(x)$ can be used to represent any arbitrary function $\phi(x)$ over an interval on the x axis approximately.

For example, if the values of $\phi(x)$ are known only at the discrete points x_0, x_1, x_2, and x_3, the approximating polynomial $\underset{\sim}{\phi}(x)$ can be written as

$$\phi(x) = \underset{\sim}{\phi}(x) = \sum_{i=0}^{3} \Phi_i L_i(x) \tag{4.53}$$

where Φ_i is the value of ϕ at $x = x_i$, $i = 0, 1, 2, 3$. Figure 4.8 shows the typical shape of $L_i(x)$. Here, the function $\underset{\sim}{\phi}(x)$ is called the Lagrange interpolation formula. Thus, Lagrange interpolation functions can be used if the matching of only the function values (not derivatives) is involved for a line element.

4.4.3 General Two-Station Interpolation Functions

We denote a general one-dimensional interpolation polynomial as $H_{ki}^{(N)}(x)$, where N is the number of derivatives to be interpolated, k is an index varying from 0 to N, and i corresponds to the station index (i.e., the ith point of the discrete set of points of interpolation). For simplicity we consider the case in which there are only two points of interpolation (as in the case of one-dimensional elements). We denote the first point of interpolation as $i = 1(x_1 = 0)$ and the second point as $i = 2(x_2 = l)$, where l is the distance between the two points.

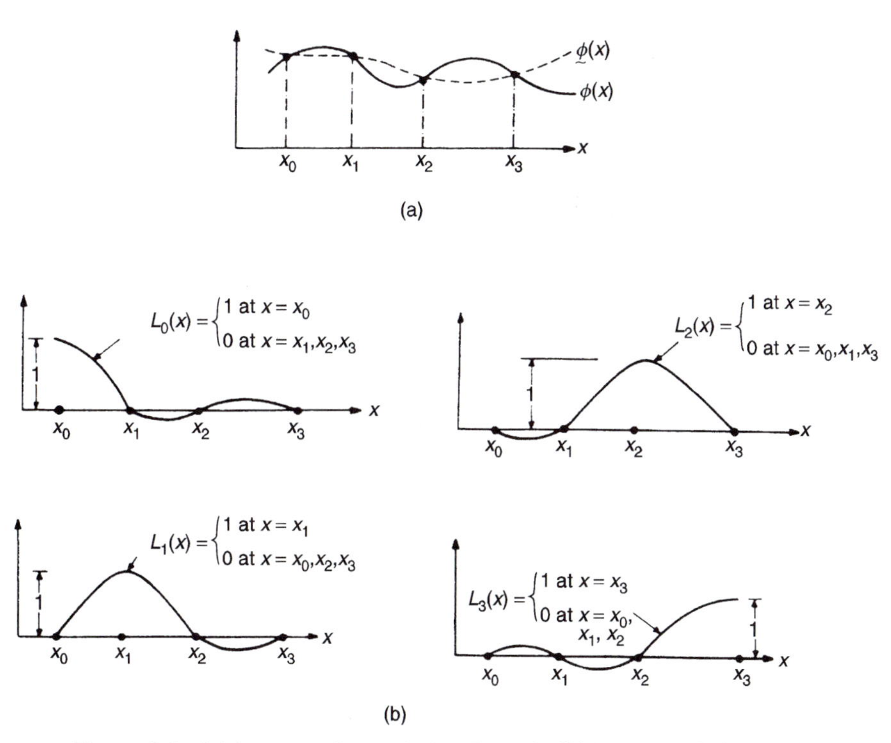

Figure 4.8. (a) Lagrange Interpolation Formula (b) Lagrange Polynomials.

Any function $\phi(x)$ shown in Figure 4.9 can be approximated by using Hermite functions as

$$
\begin{aligned}
\phi(x) &= \sum_{i=1}^{2} \sum_{k=0}^{N} H_{ki}^{(N)}(x)\Phi_i^k \\
&= \sum_{i=1}^{2} \left[H_{0i}^{(N)}(x)\Phi_i^{(0)} + H_{1i}^{(N)}(x)\Phi_i^{(1)} + H_{2i}^{(N)}(x)\Phi_i^{(2)} + \cdots + H_{Ni}^{(N)}(x)\Phi_i^{(N)} \right]
\end{aligned} \tag{4.54}
$$

where $\Phi_i^{(k)}$ are undetermined parameters. The Hermite polynomials have the following property:

$$
\left. \frac{\mathrm{d}^r H_{ki}^{(N)}}{\mathrm{d}x^r}(x_p) = \delta_{ip}\delta_{kr} \quad \text{for} \quad \begin{aligned} i, p &= 1, 2, \text{ and} \\ k, r &= 0, 1, 2, \ldots, N \end{aligned} \right\} \tag{4.55}
$$

where x_p is the value of x at pth station, and δ_{mn} is the Kronecker delta having the property

$$
\delta_{mn} = \begin{cases} 0 & \text{if } m \neq n \\ 1 & \text{if } m = n \end{cases} \tag{4.56}
$$

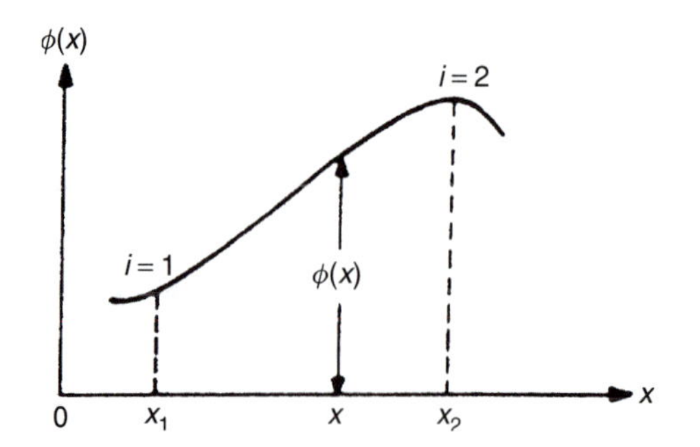

Figure 4.9. A One-Dimensional Function to Be Interpolated between Stations x_1 and x_2.

By using the property of Eq. (4.55) the undetermined parameters $\Phi_i^{(k)}$ appearing in Eq. (4.54) can be shown to have certain physical meaning. The rth derivative of $\phi(x)$ at $x = x_p$ can be written as, from Eq. (4.54),

$$\frac{\mathrm{d}^r \phi}{\mathrm{d}x^r}(x_p) = \sum_{i=1}^{2} \sum_{k=0}^{N} \frac{\mathrm{d}^r H_{ki}^{(N)}}{\mathrm{d}x^r}(x_p)\Phi_i^{(k)} \tag{4.57}$$

Using Eq. (4.55), Eq. (4.57) can be reduced to

$$\frac{\mathrm{d}^r \phi}{\mathrm{d}x^r}(x_p) = \sum_{i=1}^{2} \sum_{k=0}^{N} \delta_{ip}\delta_{kr}\Phi_i^{(k)} = \Phi_p^{(r)} \tag{4.58}$$

Thus, $\Phi_p^{(r)}$ indicates the value of rth derivative of $\phi(x)$ station p. For $r = 0$ and 1, the parameters $\Phi_p^{(r)}$ are shown in Figure 4.10. From Eqs. (4.58) and (4.54) the function $\phi(x)$ can be expressed as

$$\phi(x) = \sum_{i=1}^{2} \sum_{k=0}^{N} H_{ki}^{(N)}(x)\frac{\mathrm{d}^k \phi}{\mathrm{d}x^k}(x_i) \tag{4.59}$$

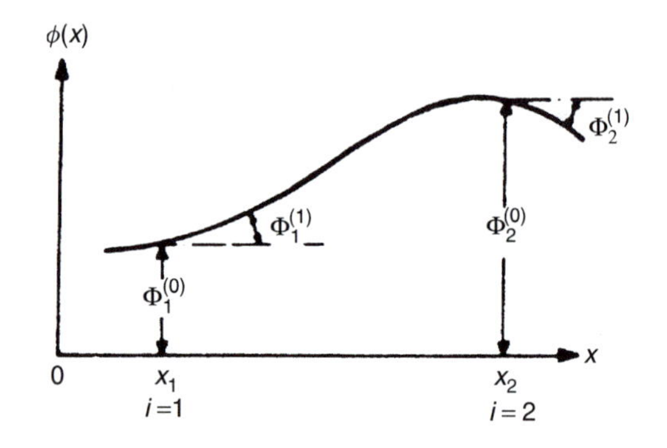

Figure 4.10. Physical Meaning of the Parameter $\Phi_p^{(r)}$.

Hermite interpolation functions find application in certain one- and two-dimensional (structural beam- and plate-bending) problems in which continuity of derivatives across element interfaces is important.

4.4.4 Zeroth-Order Hermite Interpolation Function

The general expression given in Eq. (4.54) can be specialized to the case of two-station zeroth-order Hermite (Lagrange) interpolation formula as

$$\phi(x) = \sum_{i=1}^{2} H_{0i}^{(0)} \Phi_i^{(0)} = \sum_{i=1}^{2} H_{0i}^{(0)}(x)\phi(x_i) \tag{4.60}$$

To find the polynomials $H_{01}^{(0)}(x)$ and $H_{02}^{(0)}(x)$, we use the property given by Eq. (4.55). For the polynomial $H_{01}^{(0)}(x)$, we have

$$\frac{\mathrm{d}^{(0)} H_{01}^{(0)}(x_p)}{\mathrm{d}x^{(0)}} = \delta_{1p}\delta_{00} = \delta_{1p} \equiv H_{01}^{(0)}(x_p)$$

or

$$H_{01}^{(0)}(x_1) = 1 \quad \text{and} \quad H_{01}^{(0)}(x_2) = 0 \tag{4.61}$$

Since two conditions are known (Eq. 4.61), we assume $H_{01}^{(0)}(x)$ as a polynomial involving two unknown coefficients as

$$H_{01}^{(0)}(x) = a_1 + a_2 x \tag{4.62}$$

By using Eq. (4.61), we find that

$$a_1 = 1 \quad \text{and} \quad a_2 = -1/l$$

by assuming that $x_1 = 0$ and $x_2 = l$. Thus, we have

$$H_{01}^{(0)}(x) = 1 - \frac{x}{l} \tag{4.63}$$

Similarly, the polynomial $H_{02}^{(0)}(x)$ can be found by using the conditions

$$H_{02}^{(0)}(x_1) = 0 \quad \text{and} \quad H_{02}^{(0)}(x_2) = 1 \tag{4.64}$$

as

$$H_{02}^{(0)}(x) = \frac{x}{l} \tag{4.65}$$

The shape of the Lagrange polynomials $H_{01}^{(0)}(x)$ and $H_{02}^{(0)}(x)$ and the variation of the function $\phi(x)$ approximated by Eq. (4.60) between the two stations are shown in Figures 4.11 and 4.12, respectively. Note that the Lagrange polynomials given by Eqs. (4.63) and (4.65) are special cases (two-station formulas) of the more general (n-station) polynomial given by Eq. (4.52).

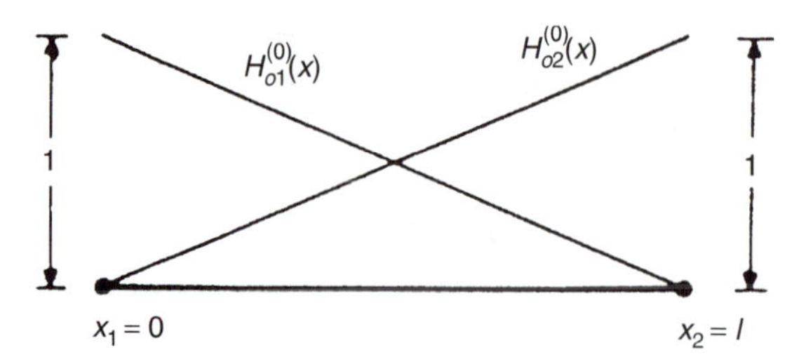

Figure 4.11. Variation of Lagrange Polynomials between the Two Stations.

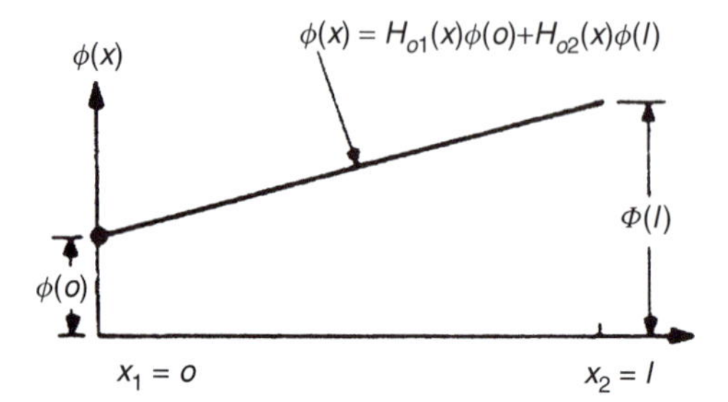

Figure 4.12. Variation of $\phi(x)$ Approximated by Lagrange Polynomials between the Two Stations.

4.4.5 First-Order Hermite Interpolation Function

If the function values as well as the first derivatives of the function are required to match with their true values, the two-station interpolation function is known as first-order Hermite (or osculatory) interpolation and is given by

$$\phi(x) = \sum_{i=1}^{2} \sum_{k=0}^{1} H_{ki}^{(1)}(x)\Phi_i^{(k)} = \sum_{i=1}^{2} \sum_{k=0}^{1} H_{ki}^{(1)}(x) \frac{\mathrm{d}^k \phi}{\mathrm{d}x^k}(x_i) \tag{4.66}$$

To determine the polynomials $H_{01}^{(1)}(x)$, $H_{02}^{(1)}(x)$, $H_{11}^{(1)}(x)$, and $H_{12}^{(1)}(x)$, four conditions are known from Eq. (4.55) for each of the polynomials. Thus, to find $H_{01}^{(1)}(x)$, we have

$$H_{01}^{(1)}(x_1) = 1, \qquad H_{01}^{(1)}(x_2) = 0, \qquad \frac{\mathrm{d}H_{01}^{(1)}}{\mathrm{d}x}(x_1) = 0, \quad \text{and} \quad \frac{\mathrm{d}H_{01}^{(1)}}{\mathrm{d}x}(x_2) = 0 \tag{4.67}$$

Since four conditions are known, we assume a cubic equation, which involves four unknown coefficients, for $H_{01}^{(1)}(x)$ as

$$H_{01}^{(1)}(x) = a_1 + a_2 x + a_3 x^2 + a_4 x^3 \tag{4.68}$$

By using Eqs. (4.67), the constants can be found as

$$a_1 = 1, \qquad a_2 = 0, \qquad a_3 = -\frac{3}{l^2}, \quad \text{and} \quad a_4 = \frac{2}{l^3}$$

Thus, the Hermite polynomial $H_{01}^{(1)}(x)$ becomes

$$H_{01}^{(1)}(x) = \frac{1}{l^3}(2x^3 - 3lx^2 + l^3) \tag{4.69}$$

Similarly, the other first-order Hermite polynomials can be obtained as

$$H_{02}^{(1)}(x) = -\frac{1}{l^3}(2x^3 - 3lx^2) \tag{4.70}$$

$$H_{11}^{(1)}(x) = \frac{1}{l^2}(x^3 - 2lx^2 + l^2x) \tag{4.71}$$

$$H_{12}^{(1)}(x) = \frac{1}{l^2}(x^3 - lx^2) \tag{4.72}$$

The variations of the first-order Hermite polynomials between the two stations are shown in Figure 4.13. The variation of the function approximated by these polynomials, namely,

$$\phi(x) = H_{01}^{(1)}(x)\phi(0) + H_{02}^{(1)}(x)\phi(l) + H_{11}^{(1)}(x)\frac{\mathrm{d}\phi}{\mathrm{d}x}(0) + H_{12}^{(1)}(x)\frac{\mathrm{d}\phi}{\mathrm{d}x}(l) \tag{4.73}$$

is shown in Figure 4.14.

4.5 ONE-DIMENSIONAL ELEMENTS USING CLASSICAL INTERPOLATION POLYNOMIALS

4.5.1 Linear Element

If the field variable $\phi(x)$ varies linearly along the length of a one-dimensional element and if the nodal values of the field variable, namely $\Phi_1 = \phi(x = x_1 = 0)$ and $\Phi_2 = \phi(x = x_2 = l)$, are taken as the nodal unknowns, we can use zeroth-order Hermite polynomials to express $\phi(x)$ as

$$\phi(x) = [N]\vec{\Phi}^{(e)} = [N_1 \quad N_2]\vec{\Phi}^{(e)} \tag{4.74}$$

where

$$N_1 = H_{01}^{(0)}(x) = 1 - \frac{x}{l},$$

$$N_2 = H_{02}^{(0)}(x) = \frac{x}{l},$$

$$\vec{\Phi}^{(e)} = \begin{Bmatrix} \Phi_1 \\ \Phi_2 \end{Bmatrix}^{(e)},$$

and l is the length of the element e.

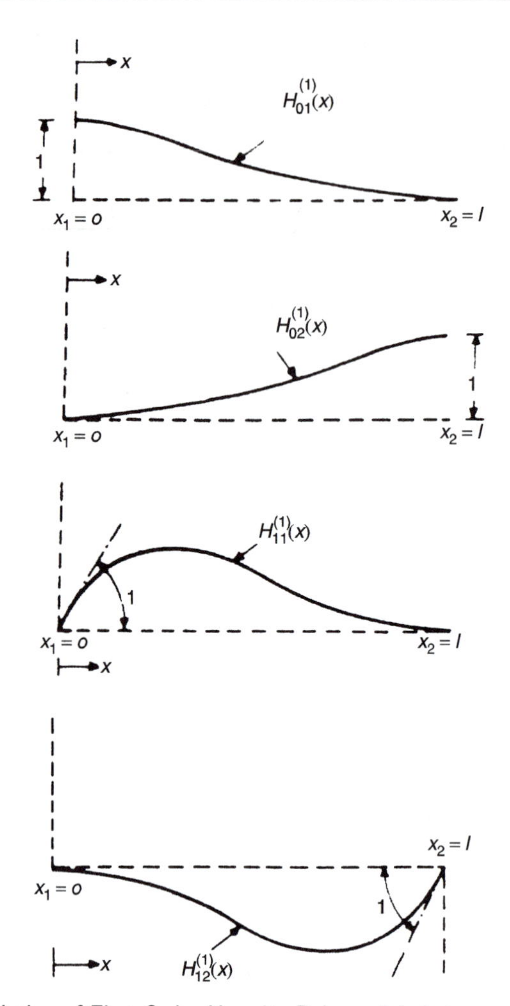

Figure 4.13. Variation of First-Order Hermite Polynomials between the Two Stations.

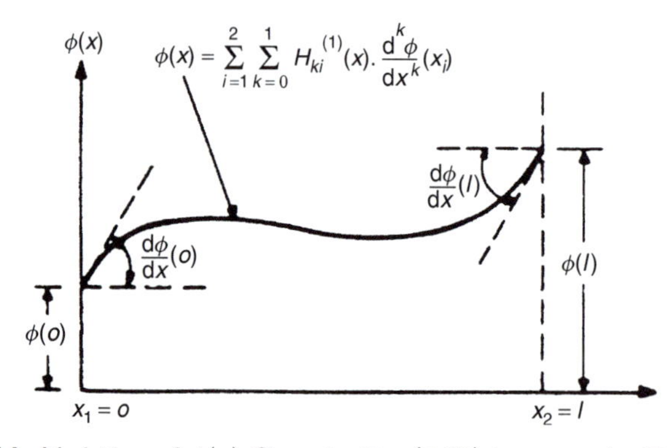

Figure 4.14. Variation of $\phi(x)$ Given by Eq. (4.73) between the Two Stations.

4.5.2 Quadratic Element

If $\phi(x)$ is assumed to vary quadratically along x and the values of $\phi(x)$ at three points x_1, x_2, and x_3 are taken as nodal unknowns, $\phi(x)$ can be expressed in terms of three-station Lagrange interpolation polynomials as

$$\phi(x) = [N]\vec{\Phi}^{(e)} = [N_1 \quad N_2 \quad N_3]\vec{\Phi}^{(e)} \tag{4.75}$$

where

$$N_1 = L_1(x) = \frac{(x - x_2)(x - x_3)}{(x_1 - x_2)(x_1 - x_3)},$$
$$N_2 = L_2(x) = \frac{(x - x_1)(x - x_3)}{(x_2 - x_1)(x_2 - x_3)},$$
$$N_3 = L_3(x) = \frac{(x - x_1)(x - x_2)}{(x_3 - x_1)(x_3 - x_2)},$$

and

$$\vec{\Phi}^{(e)} = \begin{Bmatrix} \Phi_1 \\ \Phi_2 \\ \Phi_3 \end{Bmatrix}^{(e)} = \begin{Bmatrix} \phi(x = x_1) \\ \phi(x = x_2) \\ \phi(x = x_3) \end{Bmatrix}^{(e)}$$

4.5.3 Cubic Element

If $\phi(x)$ is to be taken as a cubic polynomial and if the values of $\phi(x)$ and $(\mathrm{d}\phi/\mathrm{d}x)(x)$ at two nodes are taken as nodal unknowns, the first-order Hermite polynomials can be used to express $\phi(x)$ as

$$\phi(x) = [N]\vec{\Phi}^{(e)} = [N_1 \quad N_2 \quad N_3 \quad N_4]\vec{\Phi}^{(e)} \tag{4.76}$$

where

$$N_1(x) = H_{01}^{(1)}(x), \quad N_2(x) = H_{11}^{(1)}(x), \quad N_3(x) = H_{02}^{(1)}(x), \quad N_4(x) = H_{12}^{(1)}(x)$$

and

$$\vec{\Phi}^{(e)} = \begin{Bmatrix} \Phi_1 \\ \Phi_2 \\ \Phi_3 \\ \Phi_4 \end{Bmatrix}^{(e)} \equiv \begin{Bmatrix} \phi(x = x_1) \\ \dfrac{\mathrm{d}\phi}{\mathrm{d}x}(x = x_1) \\ \phi(x = x_2) \\ \dfrac{\mathrm{d}\phi}{\mathrm{d}x}(x = x_2) \end{Bmatrix}^{(e)}$$

4.6 TWO-DIMENSIONAL (RECTANGULAR) ELEMENTS USING CLASSICAL INTERPOLATION POLYNOMIALS

4.6.1 Using Lagrange Interpolation Polynomials

The Lagrange interpolation polynomials defined in Eq. (4.52) for one-dimensional problems can be used to construct interpolation functions for two- or higher

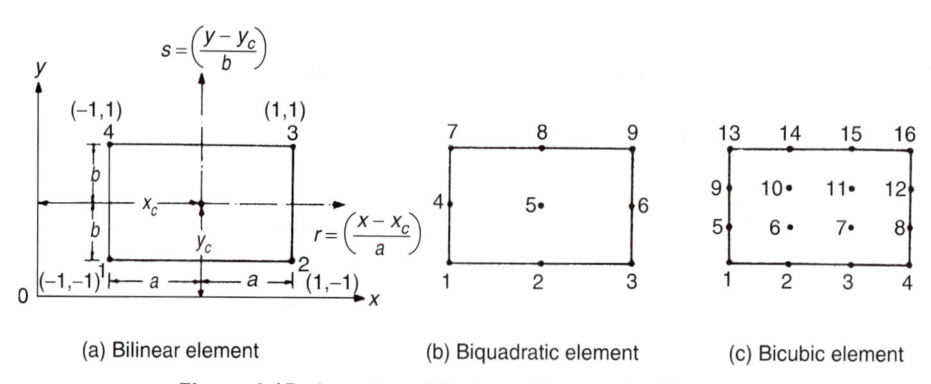

(a) Bilinear element (b) Biquadratic element (c) Bicubic element

Figure 4.15. Location of Nodes in Rectangular Elements.

dimensional problems. For example, in two dimensions, the product of Lagrange interpolation polynomials in x and y directions can be used to represent the interpolation functions of a rectangular element as [see Figure 4.15(a)]

$$\phi(r,s) = [N]\vec{\Phi}^{(e)} = [N_1 \quad N_2 \quad N_3 \quad N_4]\vec{\Phi}^{(e)} \tag{4.77}$$

where

$$N_i(r,s) = L_i(r) \cdot L_i(s), \qquad i = 1,2,3,4 \tag{4.78}$$

and

$$\vec{\Phi}^{(e)} = \begin{Bmatrix} \Phi_1 \\ \Phi_2 \\ \Phi_3 \\ \Phi_4 \end{Bmatrix}^{(e)} = \begin{Bmatrix} \phi(r=-1, s=-1) \\ \phi(r=1, s=-1) \\ \phi(r=1, s=1) \\ \phi(r=-1, s=1) \end{Bmatrix}^{(e)} \tag{4.79}$$

$L_i(r)$ and $L_i(s)$ denote Lagrange interpolation polynomials in r and s directions corresponding to node i and are defined, with reference to Figure 4.15(a), as

$$L_1(r) = \frac{r-r_2}{r_1-r_2}, \quad L_2(r) = \frac{r-r_1}{r_2-r_1}, \quad L_3(r) = \frac{r-r_4}{r_3-r_4}, \quad L_4(r) = \frac{r-r_3}{r_4-r_3}$$

$$L_1(s) = \frac{s-s_4}{s_1-s_4}, \quad L_2(s) = \frac{s-s_3}{s_2-s_3}, \quad L_3(s) = \frac{s-s_2}{s_3-s_2}, \quad L_4(s) = \frac{s-s_1}{s_4-s_1} \tag{4.80}$$

The nodal interpolation functions N_i given by Eq. (4.78) are called "bilinear" since they are defined as products of two linear functions.

The higher order elements, such as biquadratic and bicubic elements, can be formulated precisely the same way by taking products of Lagrange interpolation polynomials of degree two and three, respectively, as

$$N_i(r,s) = L_i(r) \cdot L_i(s) \tag{4.81}$$

where $L_i(r)$ and $L_i(s)$ can be obtained with the help of Eq. (4.52) and Figures 4.15(b) and 4.15(c). For example, in the case of the biquadratic element shown in Figure 4.15(b), the Lagrange interpolation polynomials are defined as follows:

$$L_1(r) = \frac{(r-r_2)(r-r_3)}{(r_1-r_2)(r_1-r_3)}, \qquad L_1(s) = \frac{(s-s_4)(s-s_7)}{(s_1-s_4)(s_1-s_7)} \tag{4.82}$$

$$L_2(r) = \frac{(r-r_1)(r-r_3)}{(r_2-r_1)(r_2-r_3)}, \qquad L_2(s) = \frac{(s-s_5)(s-s_8)}{(s_2-s_5)(s_2-s_8)} \tag{4.83}$$

etc. In this case, node 5 represents an interior node. It can be observed that the higher order Lagrangian elements contain a large number of interior nodes and this limits the usefulness of these elements. Of course, a technique known as "static condensation" can be used to suppress the degrees of freedom associated with the internal nodes in the final computation (see problem 12.7).

4.6.2 Using Hermite Interpolation Polynomials

Just as we have done with Lagrange interpolation polynomials, we can form products of one-dimensional Hermite polynomials and derive the nodal interpolation functions N_i for rectangular elements. If we use first-order Hermite polynomials for this purpose, we have to take the values of ϕ, $(\partial\phi/\partial x)$, $(\partial\phi/\partial y)$, and $(\partial^2\phi/\partial x\partial y)$ as nodal degrees of freedom at each of the four corner nodes. Thus, by using a two-number scheme for identifying the nodes of the rectangle as shown in Figure 4.16, the interpolation model for $\phi(x, y)$ can be expressed as

$$\begin{aligned}\phi(x,y) = \sum_{i=1}^{2}\sum_{j=1}^{2}\Bigg[& H_{0i}^{(1)}(x)\cdot H_{0j}^{(1)}(y)\cdot \phi_{ij} + H_{1i}^{(1)}(x)\cdot H_{0j}^{(1)}(y)\cdot\left(\frac{\partial\phi}{\partial x}\right)_{ij} \\ & + H_{0i}^{(1)}(x)\cdot H_{1j}^{(1)}(y)\cdot\left(\frac{\partial\phi}{\partial y}\right)_{ij} + H_{1i}^{(1)}(x)\cdot H_{1j}^{(1)}(y)\cdot\left(\frac{\partial^2\phi}{\partial x\partial y}\right)_{ij}\Bigg]\end{aligned} \tag{4.84}$$

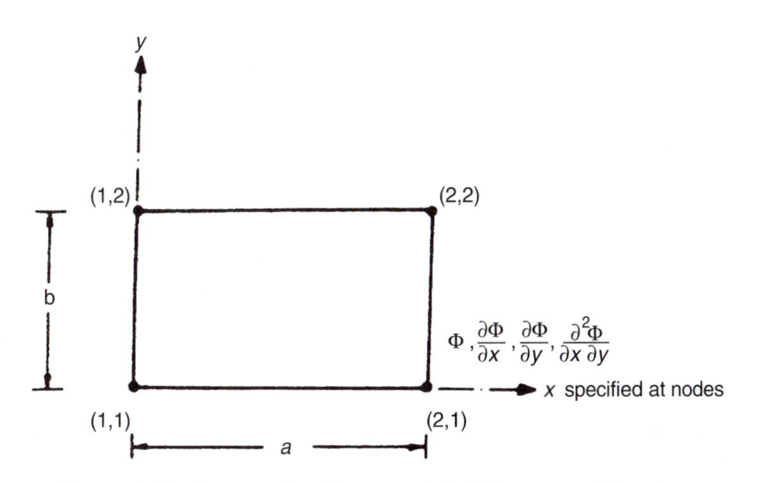

Figure 4.16. Rectangular Element with 16 Degrees of Freedom.

where ϕ_{ij}, $(\partial\phi/\partial x)_{ij}$, $(\partial\phi/\partial y)_{ij}$, and $(\partial^2\phi/\partial x\partial y)_{ij}$ denote the values of ϕ, $(\partial\phi/\partial x)$, $(\partial\phi/\partial y)$, and $(\partial^2\phi/\partial x\partial y)$, respectively, at node (i, j). Equation (4.84) can be rewritten in the familiar form as

$$\phi(x, y) = [N(x, y)]\vec{\Phi}^{(e)} = [N_1(x, y) \quad N_2(x, y) \quad \dots \quad N_{16}(x, y)]\vec{\Phi}^{(e)} \tag{4.85}$$

where

$$\begin{aligned}
N_1(x, y) &= H_{01}^{(1)}(x)H_{01}^{(1)}(y),\\
N_2(x, y) &= H_{11}^{(1)}(x)H_{01}^{(1)}(y),\\
N_3(x, y) &= H_{01}^{(1)}(x)H_{11}^{(1)}(y),\\
N_4(x, y) &= H_{11}^{(1)}(x)H_{11}^{(1)}(y),\\
N_5(x, y) &= H_{02}^{(1)}(x)H_{01}^{(1)}(y),\\
&\ \ \vdots\\
N_{16}(x, y) &= H_{11}^{(1)}(x)H_{12}^{(1)}(y),
\end{aligned} \tag{4.86}$$

and

$$\vec{\Phi}^{(e)} = \begin{Bmatrix} \Phi_1 \\ \Phi_2 \\ \Phi_3 \\ \Phi_4 \\ \Phi_5 \\ \vdots \\ \Phi_{16} \end{Bmatrix}^{(e)} = \begin{Bmatrix} \phi_{11} \\ \left(\dfrac{\partial\phi}{\partial x}\right)_{11} \\ \left(\dfrac{\partial\phi}{\partial y}\right)_{11} \\ \left(\dfrac{\partial^2\phi}{\partial x\partial y}\right)_{11} \\ \phi_{21} \\ \vdots \\ \left(\dfrac{\partial^2\phi}{\partial x\partial y}\right)_{12} \end{Bmatrix}^{(e)} \tag{4.87}$$

4.7 CONTINUITY CONDITIONS

We saw in Section 3.6 that the interpolation model assumed for the field variable ϕ has to satisfy the following conditions:

1. It has to be continuous inside and between the elements up to order $r-1$, where r is the order of the highest derivative in the functional I. For example, if the governing differential equation is quasi-harmonic as in the case of Example 1.3, ϕ have to be continuous (i.e., C^0 continuity is required). On the other hand, if the governing differential equation is biharmonic ($\nabla^4\phi = 0$), ϕ as well as its derivative $(\partial\phi/\partial n)$ have to be continuous inside and between elements (i.e., C^1 continuity is required). The continuity of the higher order derivatives associated with the free or natural boundary conditions need not be imposed because their eventual satisfaction is implied in the variational statement of the problem.

2. As the size of the elements decreases, the derivatives appearing in the functional of the variational statement will tend to have constant values. Thus, it is necessary to include terms that represent these conditions in the interpolation model of ϕ.

For elements requiring C^0 continuity (i.e., continuity of only the field variable ϕ at element interfaces), we usually take the nodal values of ϕ only as the degrees of freedom. To satisfy the interelement continuity condition, we have to take the number of nodes along a side of the element (and hence the number of nodal values of ϕ) to be sufficient to determine the variation of ϕ along that side uniquely. Thus, if a cubic interpolation model is assumed within the element and retains its cubic behavior along the element sides, then we have to take four nodes (and hence four nodal values of ϕ) along each side.

It can be observed that the number of elements (of a given shape) capable of satisfying C^0 continuity is infinite. This is because we can continue to add nodes and degrees of freedom to the elements to form ever increasing higher order elements. All such elements will satisfy the C^0 continuity. In general, higher order elements can be derived by increasing the number of nodes and hence the nodal degrees of freedom and assuming a higher order interpolation model for the field variable ϕ. As stated earlier, in general, smaller numbers of higher order elements yield more accurate results compared to larger numbers of simpler elements for the same overall effort. But this does not mean that we should always favor elements of very high order. Although there are no general guidelines available for choosing the order of the element for a given problem, elements that require polynomials of order greater than three have seldom been used for problems requiring C^0 continuity. The main reason for this is that the computational effort saved with fewer numbers of higher order elements will become overshadowed by the increased effort required in formulating and evaluating the element characteristic matrices and vectors.

4.7.1 Elements with C^0 Continuity

All simplex elements considered in Section 3.7 satisfy C^0 continuity because their interpolation models are linear. Furthermore, all higher order one-, two-, and three-dimensional elements considered in this chapter also satisfy the C^0 continuity. For example, each of the triangular elements shown in Figure 4.3 has a sufficient number of nodes (and hence the nodal degrees of freedom) to uniquely specify a complete polynomial of the order necessary to give C^0 continuity. Thus, the corresponding interpolation models satisfy the requirements of compatibility, completeness, and geometric isotropy. In general, for a triangular element, a complete polynomial of order n requires $(1/2)(n+1)(n+2)$ nodes for its specification. Similarly, a tetrahedron element requires $(1/6)(n+1)(n+2)(n+3)$ nodes in order to have the interpolation model in the form of a complete polynomial of order n. For such elements, if the nodal values of ϕ only are taken as degrees of freedom, the conditions of compatibility, completeness, and geometric isotropy will be satisfied.

The quadrilateral element discussed in Section 4.3.3 considers only the nodal values of ϕ as the degrees of freedom and satisfies C^0 continuity. For rectangular elements, if the nodal interpolation functions are defined by products of Lagrange interpolation polynomials (Figure 4.15), then the C^0 continuity is satisfied.

4.7.2 Elements with C^1 Continuity

The construction of elements that satisfy C^1 continuity of the field variable ϕ is much more difficult than constructing elements for C^0 continuity. To satisfy the C^1 continuity, we have to ensure continuity of ϕ as well as its normal derivative $\partial\phi/\partial n$ along the element boundaries. The one-dimensional cubic element considered in Section 4.5.3 guarantees the continuity of both ϕ and $d\phi/dx$ at the nodes and hence it satisfies the C^1 continuity.

For two-dimensional elements, we have to ensure that ϕ and $\partial\phi/\partial n$ are specified uniquely along an element boundary by the nodal degrees of freedom associated with the nodes of that particular boundary. The rectangular element considered in Figure 4.16 (Eq. 4.84) considers ϕ, $\partial\phi/\partial x$, $\partial\phi/\partial y$, and $\partial^2\phi/\partial x\partial y$ as nodal degrees of freedom and satisfies the C^1 continuity. In the case of a triangular element, some authors have treated the values of ϕ, $(\partial\phi/\partial x)$, $(\partial\phi/\partial y)$, $(\partial^2\phi/\partial x\partial y)$, $(\partial^2\phi/\partial x^2)$, and $(\partial^2\phi/\partial y^2)$ at the three corner nodes and the values of $(\partial\phi/\partial n)$ at the three midside nodes (Figure 4.17) as degrees of freedom and represented the interpolation model of ϕ by a complete quintic polynomial. If s denotes the linear coordinate along any boundary of the element, then ϕ varies along s as a fifth-degree polynomial. This fifth-degree polynomial is uniquely determined by the six nodal degrees of freedom, namely ϕ, $(\partial\phi/\partial s)$, and $(\partial^2\phi/\partial s^2)$ at each of the two end nodes. Hence, ϕ will be continuous along the element boundaries. Similarly, the normal slope $(\partial\phi/\partial n)$ can be seen to vary as a fourth-degree polynomial in s along the element boundary. There are five nodal degrees of freedom to determine this quartic polynomial uniquely. These are the values of $(\partial\phi/\partial n)$ and $(\partial^2\phi/\partial n^2)$ at each of the end nodes and $(\partial\phi/\partial n)$ at the midside node. Hence, the normal slope $(\partial\phi/\partial n)$ will also be continuous along the element boundaries. In the case of three-dimensional elements, the satisfaction of C^1 continuity is quite difficult and practically no such element has been used in the literature.

Note:
Since the satisfaction of C^1 continuity is difficult to achieve, many investigators have used finite elements that satisfy slope continuity at the nodes and other requirements but violate slope continuity along the element boundaries. Such elements are known as "incompatible" or "nonconforming" elements and have been used with surprising success in plate-bending (two-dimensional structural) problems.

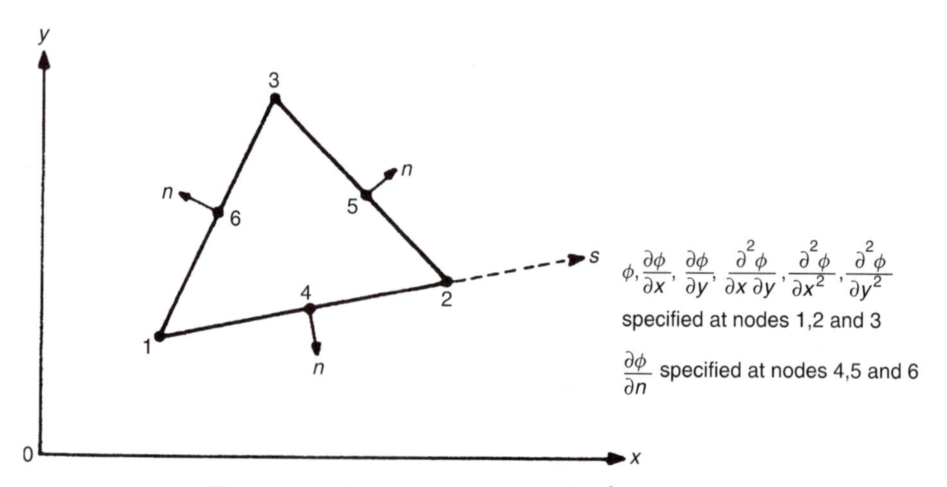

Figure 4.17. Triangular Element with C^1 Continuity.

4.8 COMPARATIVE STUDY OF ELEMENTS

The relative accuracy of the results obtained by using interpolation polynomials of different orders was studied by Emery and Carson [4.2]. They considered the solution of a one-dimensional steady-state diffusion equation as a test case. The governing equation is

$$\frac{d^2\phi}{dx^2} = \psi(x), \qquad 0 \le x \le 1 \tag{4.88}$$

with $\psi(x) = x^5$ and the boundary conditions are

$$\phi(x=0) = 0 \quad \text{and} \quad \frac{d\phi}{dx}(x=1) = 0 \tag{4.89}$$

By dividing the region ($x = 0$ to 1) into different numbers of finite elements, they obtained the results using linear, quadratic, and cubic interpolation models. The results are shown in Figure 4.18 along with those given by the finite difference method. The ordinate in Figure 4.18 denotes the error in the temperature (ϕ) at the point $x = 1$. The exact solution obtained by integrating Eq. (4.88) with the boundary conditions, Eq. (4.89), gives the value of ϕ at $x = 1$ as 0.1429.

The results indicate that the higher order models yield better results in this case. This characteristic has been found to be true even for higher dimensional problems. If the overall computational effort involved and the accuracy achieved are compared,

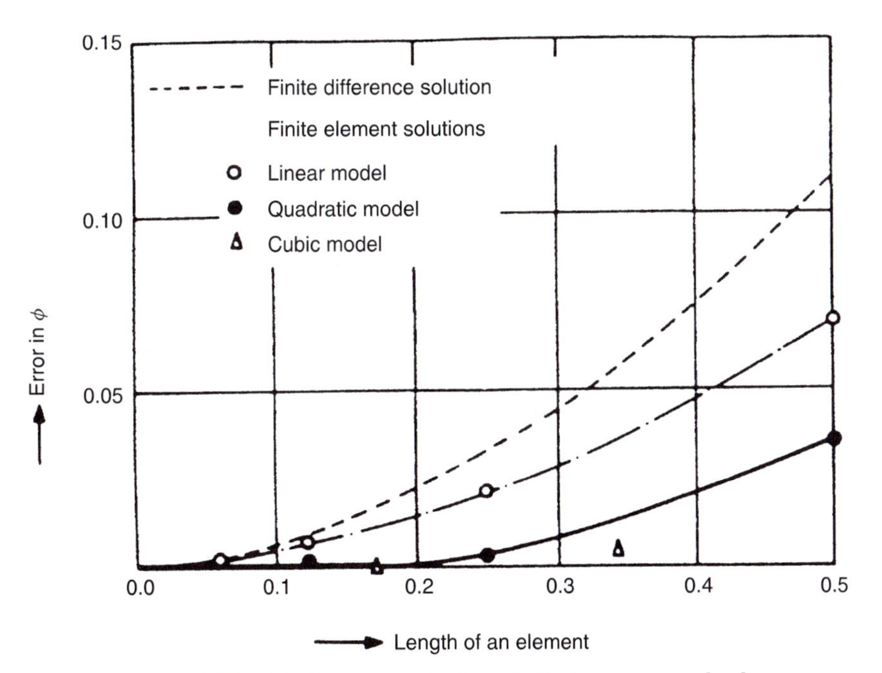

Figure 4.18. Solution of Steady-State Diffusion Equation [4.2].

we might find the quadratic model to be the most efficient one for use in complex practical problems.

4.9 ISOPARAMETRIC ELEMENTS

4.9.1 Definitions

In the case of one-dimensional elements, Eqs. (3.62) and (3.24) give

$$x = [N_1 \quad N_2] \begin{Bmatrix} x_1 \\ x_2 \end{Bmatrix} \tag{4.90}$$

and

$$\phi = [N_1 \quad N_2] \begin{Bmatrix} \Phi_1 \\ \Phi_2 \end{Bmatrix}^{(e)} \tag{4.91}$$

where $N_1 = L_1$ and $N_2 = L_2$. In the case of a triangular element, if we consider ϕ as a vector quantity with components $u(x, y)$ and $v(x, y)$, Eqs. (3.71) and (3.33) give

$$\begin{Bmatrix} x \\ y \end{Bmatrix} = \begin{bmatrix} N_1 & N_2 & N_3 & 0 & 0 & 0 \\ 0 & 0 & 0 & N_1 & N_2 & N_3 \end{bmatrix} \begin{Bmatrix} x_1 \\ x_2 \\ x_3 \\ y_1 \\ y_2 \\ y_3 \end{Bmatrix} \tag{4.92}$$

and

$$\begin{Bmatrix} u \\ v \end{Bmatrix} = \begin{bmatrix} N_1 & N_2 & N_3 & 0 & 0 & 0 \\ 0 & 0 & 0 & N_1 & N_2 & N_3 \end{bmatrix} \begin{Bmatrix} u_1 \\ u_2 \\ u_3 \\ v_1 \\ v_2 \\ v_3 \end{Bmatrix} \tag{4.93}$$

where $N_1 = L_1$, $N_2 = L_2$, $N_3 = L_3$, and $(x_i,\ y_i)$ are the Cartesian coordinates of node i, and u_i and v_i are the values of u and v, respectively, at node i $(i = 1, 2, 3)$. Similarly, for a quadrilateral element, the geometry and field variable are given by Eqs. (4.32) and (4.38) as (assuming ϕ to be a vector with components u and v)

$$\begin{Bmatrix} x \\ y \end{Bmatrix} = \begin{bmatrix} N_1 & N_2 & N_3 & N_4 & 0 & 0 & 0 & 0 \\ 0 & 0 & 0 & 0 & N_1 & N_2 & N_3 & N_4 \end{bmatrix} \begin{Bmatrix} x_1 \\ x_2 \\ x_3 \\ x_4 \\ y_1 \\ y_2 \\ y_3 \\ y_4 \end{Bmatrix} \tag{4.94}$$

and

$$\begin{Bmatrix} u \\ v \end{Bmatrix} = \begin{bmatrix} N_1 & N_2 & N_3 & N_4 & 0 & 0 & 0 & 0 \\ 0 & 0 & 0 & 0 & N_1 & N_2 & N_3 & N_4 \end{bmatrix} \begin{Bmatrix} u_1 \\ u_2 \\ u_3 \\ u_4 \\ v_1 \\ v_2 \\ v_3 \\ v_4 \end{Bmatrix} \tag{4.95}$$

where N_i $(i = 1, 2, 3, 4)$ are given by Eq. (4.33), $(x_i,\ y_i)$ are the Cartesian coordinates of node i, and $(u_i,\ v_i)$ are the components of $\phi(u, v)$ at node i. A comparison of Eqs. (4.90) and (4.91) or (4.92) and (4.93) or (4.94) and (4.95) shows that the geometry and field variables of these elements are described in terms of the same parameters and of the same order.

Such elements whose shape (or geometry) and field variables are described by the same interpolation functions of the same order are known as "isoparametric" elements. These elements have been used with great success in solving two- and three-dimensional elasticity problems, including those involving plates and shells [4.3]. These elements have become popular for the following reasons:

(i) If one element is understood, the same concepts can be extended for understanding all isoparametric elements.
(ii) Although linear elements have straight sides, quadratic and higher order isoparametric elements may have either straight or curved sides. Hence, these elements can be used for idealizing regions having curved boundaries.

It is not necessary to use interpolation functions of the same order for describing both geometry and the field variable of an element. If geometry is described by a lower order model compared to the field variable, the element is called a "subparametric" element. On the other hand, if the geometry is described by a higher order interpolation model than the field variable, the element is termed a "superparametric" element.

4.9.2 Shape Functions in Coordinate Transformation

The equations that describe the geometry of the element, namely,

$$\begin{Bmatrix} x \\ y \\ z \end{Bmatrix} = \begin{bmatrix} N_1 & N_2 \ldots N_p & 0 & 0 \ldots 0 & 0 & 0 \ldots 0 \\ 0 & 0 \ldots 0 & N_1 & N_2 \ldots N_P & 0 & 0 \ldots 0 \\ 0 & 0 \ldots 0 & 0 & 0 \ldots 0 & N_1 & N_2 \ldots N_p \end{bmatrix} \begin{Bmatrix} x_1 \\ x_2 \\ \vdots \\ x_p \\ y_1 \\ y_2 \\ \vdots \\ y_p \\ z_1 \\ z_2 \\ \vdots \\ z_p \end{Bmatrix} \tag{4.96}$$

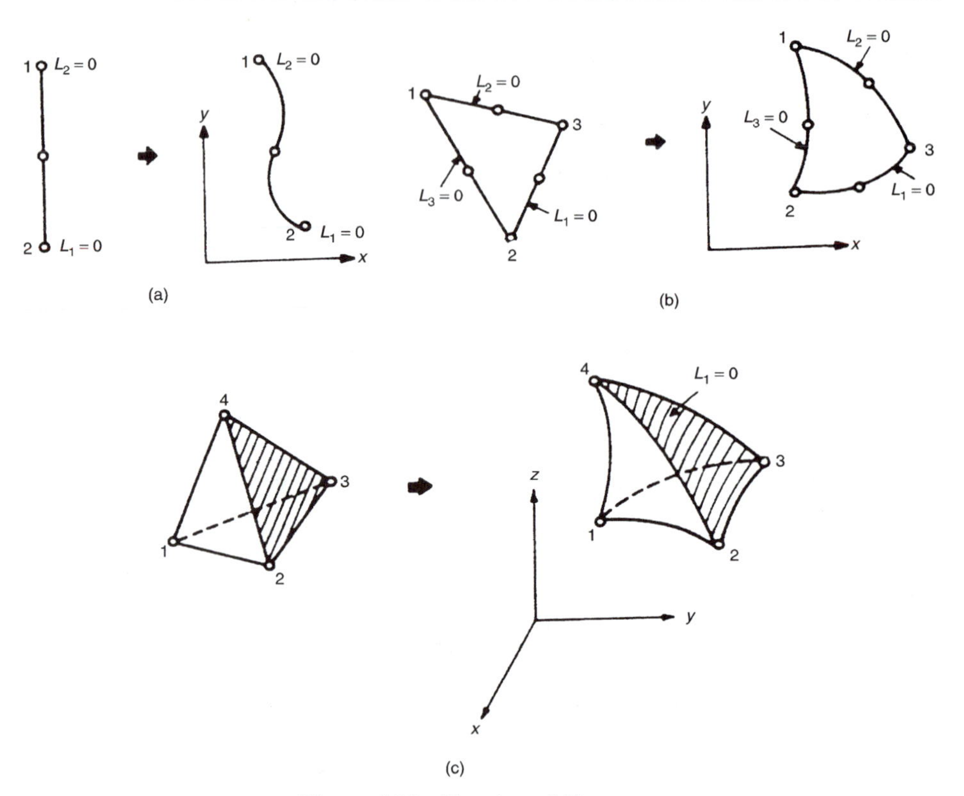

Figure 4.19. Mapping of Elements.

(p = number of nodes of the element) can be considered as a transformation relation between the Cartesian (x, y, z) coordinates and curvilinear (r, s, t or L_1, L_2, L_3, L_4) coordinates if the shape functions N_i are nonlinear in terms of the natural coordinates of the element. Equation (4.96) can also be considered as the mapping of a straight-sided element in local coordinates into a curved-sided element in the global Cartesian coordinate system. Thus, for any set of coordinates L_1, L_2, L_3, and L_4, or r, s, and t, there corresponds a set of x, y, and z. Such mapping permits elements of one-, two-, and three-dimensional types to be "mapped" into distorted forms in the manner illustrated in Figure 4.19.

To each set of local coordinates, there will be, in general, only one set of Cartesian coordinates. However, in some cases, a nonuniqueness may arise with violent distortion. In order to have unique mapping of elements, the number of coordinates (L_1, L_2, L_3, L_4) or (r, s, t) and (x, y, z) must be identical and the Jacobian, defined by,

$$|[J]| = \frac{\partial(x, y, \ldots)}{\partial(L_1, L_2, \ldots)} = \left| \begin{bmatrix} \dfrac{\partial x}{\partial L_1} & \dfrac{\partial x}{\partial L_2} & \cdots \\ \dfrac{\partial y}{\partial L_1} & \dfrac{\partial y}{\partial L_2} & \cdots \\ \vdots & \vdots & \vdots & \vdots \end{bmatrix} \right| \tag{4.97}$$

must not change sign in the domain.

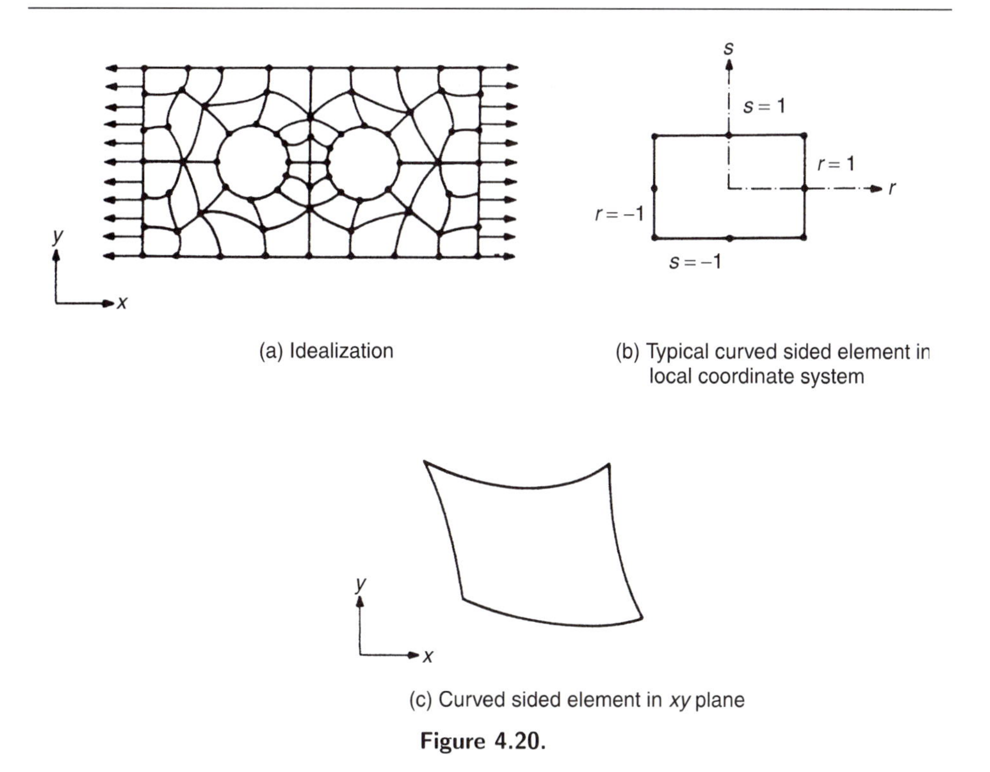

(a) Idealization

(b) Typical curved sided element in local coordinate system

(c) Curved sided element in *xy* plane

Figure 4.20.

4.9.3 Curved-Sided Elements

The main idea underlying the development of curved-sided elements centers on mapping or transforming simple geometric shapes (with straight edges or flat surfaces) in some local coordinate system into distorted shapes (with curved edges or surfaces) in the global Cartesian coordinate system and then evaluating the element equations for the resulting curved-sided elements.

To clarify the idea, we shall consider a two-dimensional example. The extension of the idea to one- and three-dimensional problems will be straightforward. Let the problem to be analyzed in a two-dimensional (x, y) space be as shown in Figure 4.20(a) and let the finite element mesh consist of curved-sided quadrilateral elements as indicated in Figure 4.20(a). Let the field variable ϕ (e.g., displacement) be taken to vary quadratically within each element. According to the discussion of Section 4.3.3, if we want to take only the nodal values of ϕ (but not the derivatives of ϕ) as degrees of freedom of the element, we need to take three nodes on each side of the quadrilateral element. In order to derive the finite element equations, we consider one typical element in the assemblage of Figure 4.20(a) and focus our attention on the simpler "parent" element in the local (r, s) coordinate system as shown in Figure 4.20(b).

We find from Section 4.3.3 that the quadratic variation of ϕ within this parent element can be expressed as

$$\phi(r, s) = \sum_{i=1}^{8} N_i(r, s)\Phi_i \tag{4.98}$$

where N_i are the quadratic shape or interpolation functions used in Eq. (4.41). The eight nodes in the (r, s) plane may be mapped into corresponding nodes in the (x, y) plane by defining the relations

$$\left.\begin{aligned} x &= \sum_{i=1}^{8} f_i(r, s) x_i \\ y &= \sum_{i=1}^{8} f_i(r, s)\, y_i \end{aligned}\right\} \tag{4.99}$$

where $f_i(r, s)$ are the mapping functions. These functions, in this case, must be at least quadratic since the curved boundaries of the element in the (x, y) plane need at least three points for their specification and the f_i should take the proper values of 0 and 1 when evaluated at the corner nodes in the (r, s) plane.

If we take the quadratic shape functions N_i given in Eq. (4.41) for this purpose, we can write

$$\left.\begin{aligned} x &= \sum_{i=1}^{8} N_i(r, s) x_i \\ y &= \sum_{i=1}^{8} N_i(r, s) y_i \end{aligned}\right\} \tag{4.100}$$

The mapping defined by Eq. (4.100) results in a curved-sided quadrilateral element as shown in Figure 4.20(c). Thus, for this element, the functional description of the field variable ϕ as well as its curved boundaries are expressed by interpolation functions of the same order. According to the definition, this element is an isoparametric element. Similarly, the element is called subparametric or superparametric if the functional representation of geometry of the element $[f_i(r, s)]$ is expressed in terms of a lower or a higher order polynomial than the one used for representing the field variable ϕ.

4.9.4 Continuity and Compatibility

We need to preserve the continuity and compatibility conditions in the global (x, y) coordinate system while constructing isoparametric elements using the following observations [4.4]:

(i) If the interpolation functions in natural (local) coordinates satisfy continuity of geometry and field variable both within the element and between adjacent elements, the compatibility requirement will be satisfied in the global coordinates.

The polynomial interpolation models discussed in Section 4.3 are inherently continuous within the element. Furthermore, we can notice that the field variable along any edge of the element depends only on the nodal degrees of freedom occurring on that edge when interpolation functions in natural coordinates are used. This can also be seen from Figures 4.4 and 4.6, where, for example, the field variable along the edge 2-6-3 of Figure 4.6(b) depends only on the values of the field variable at nodes 2, 6, and 3.

(ii) If the interpolation model provides constant values of ϕ in the local coordinate system, the conditions of both constant values of ϕ and its derivatives will be satisfied in the global coordinates.

Let the functional relation for the components of the vector-valued field variable in an isoparametric element be given by

$$\begin{Bmatrix} u \\ v \\ w \end{Bmatrix} = [N] \begin{Bmatrix} u_1 \\ \vdots \\ u_p \\ v_1 \\ \vdots \\ v_p \\ w_1 \\ \vdots \\ w_p \end{Bmatrix} \tag{4.101}$$

where

$$[N] = \begin{bmatrix} N_1 \ldots N_p & 0 \ldots 0 & 0 \ldots 0 \\ 0 \ldots 0 & N_1 \ldots N_p & 0 \ldots 0 \\ 0 \ldots 0 & 0 \ldots 0 & N_1 \ldots N_p \end{bmatrix} \tag{4.102}$$

and p is the number of nodes in the element. Thus, the u component of ϕ is given by

$$u = \sum_{i=1}^{p} N_i u_i \tag{4.103}$$

Let the geometry be given by

$$\begin{Bmatrix} x \\ y \\ z \end{Bmatrix} = [N] \begin{Bmatrix} x_1 \\ \vdots \\ x_p \\ y_1 \\ \vdots \\ y_p \\ z_1 \\ \vdots \\ z_p \end{Bmatrix} \tag{4.104}$$

where (x_i, y_i, z_i) are the coordinates of node i ($i = 1, 2, \ldots$). For constant u, all points on the element must have the same value of u, for example, u_0; hence, Eq. (4.103) becomes

$$u_0 = \left(\sum_{i=1}^{p} N_i \right) u_0 \tag{4.105}$$

Thus, we obtain the following necessary condition to be satisfied for constant values of ϕ in local coordinates:

$$\sum_{i=1}^{p} N_i = 1 \tag{4.106}$$

4.9.5 Derivation of Element Equations

In the case of structural and solid mechanics problems, the element characteristic (stiffness) matrix is given by*

$$[K^{(e)}] = \iiint_{V^{(e)}} [B]^T [D][B] \cdot \mathrm{d}V \tag{4.107}$$

and the element characteristic (load) vector by

$$\vec{p}^{(e)} = \iiint_{V^{(e)}} [B]^T [D]\vec{\varepsilon}_0 \cdot \mathrm{d}V + \iiint_{V^{(e)}} [N]^T \vec{\phi} \cdot \mathrm{d}V + \iint_{S_1^{(e)}} [N]^T \vec{\Phi} \cdot \mathrm{d}S_1 \tag{4.108}$$

where [B] is the matrix relating strains and nodal displacements, [D] is the elasticity matrix, $\vec{\Phi}$ is the vector of distributed surface forces, $\vec{\phi}$ is the vector of body forces, and $\vec{\varepsilon}_0$ is the initial strain vector.

For a plane stress or plane strain problem, we have

$$\begin{Bmatrix} u(x,y) \\ v(x,y) \end{Bmatrix} = [N]\vec{\Phi}^{(e)} \equiv \begin{bmatrix} N_1 & N_2 \dots N_p & 0 & 0 \dots 0 \\ 0 & 0 \dots 0 & N_1 & N_2 \dots N_p \end{bmatrix} \begin{Bmatrix} u_1 \\ u_2 \\ \vdots \\ u_p \\ v_1 \\ v_2 \\ \vdots \\ v_p \end{Bmatrix}, \tag{4.109}$$

$$\vec{\varepsilon} = \begin{Bmatrix} \varepsilon_{xx} \\ \varepsilon_{yy} \\ \varepsilon_{xy} \end{Bmatrix} = [B]\vec{Q}^{(e)} \tag{4.110}$$

and

$$[B] = \begin{bmatrix} \dfrac{\partial N_1}{\partial x} & \dfrac{\partial N_2}{\partial x} \dots \dfrac{\partial N_p}{\partial x} & 0 & 0 & \dots & 0 \\ 0 & 0 \quad \dots \quad 0 & \dfrac{\partial N_1}{\partial y} & \dfrac{\partial N_2}{\partial y} & \dots & \dfrac{\partial N_p}{\partial y} \\ \dfrac{\partial N_1}{\partial y} & \dfrac{\partial N_2}{\partial y} \dots \dfrac{\partial N_p}{\partial y} & \dfrac{\partial N_1}{\partial x} & \dfrac{\partial N_2}{\partial x} & \dots & \dfrac{\partial N_p}{\partial x} \end{bmatrix} \tag{4.111}$$

* Equations (4.107) and (4.108) are derived in Chapter 8.

where p denotes the number of nodes of the element, $(u_i,\ v_i)$ denote the values of (u, v) at node i, and N_i is the shape function associated with node i expressed in terms of natural coordinates (r, s) or $(L_1,\ L_2,\ L_3)$. Thus, in order to evaluate $[K^{(e)}]$ and $\vec{p}^{(e)}$, two transformations are necessary. First, the shape functions N_i are defined in terms of local curvilinear coordinates (e.g., r and s) and hence the derivatives of N_i with respect to the global coordinates x and y have to be expressed in terms of derivatives of N_i with respect to the local coordinates. Second, the volume and surface integrals needed in Eqs. (4.107) and (4.108) have to be expressed in terms of local coordinates with appropriate change of limits of integration.

For the first transformation, let us consider the differential of N_i with respect to the local coordinate r. Then, by the chain rule of differentiation, we have

$$\frac{\partial N_i}{\partial r} = \frac{\partial N_i}{\partial x} \cdot \frac{\partial x}{\partial r} + \frac{\partial N_i}{\partial y} \cdot \frac{\partial y}{\partial r} \tag{4.112}$$

Similarly,

$$\frac{\partial N_i}{\partial s} = \frac{\partial N_i}{\partial x} \cdot \frac{\partial x}{\partial s} + \frac{\partial N_i}{\partial y} \cdot \frac{\partial y}{\partial s}$$

Thus, we can express

$$\begin{Bmatrix} \partial N_i/\partial r \\ \partial N_i/\partial s \end{Bmatrix} = \begin{bmatrix} \partial x/\partial r & \partial y/\partial r \\ \partial x/\partial s & \partial y/\partial s \end{bmatrix} \begin{Bmatrix} \partial N_i/\partial x \\ \partial N_i/\partial y \end{Bmatrix} \equiv [J] \begin{Bmatrix} \partial N_i/\partial x \\ \partial N_i/\partial y \end{Bmatrix} \tag{4.113}$$

where the matrix $[J]$, called the Jacobian matrix, is given by

$$[J] = \begin{bmatrix} \partial x/\partial r & \partial y/\partial r \\ \partial x/\partial s & \partial y/\partial s \end{bmatrix} \tag{4.114}$$

Since x and y (geometry) are expressed as

$$\begin{Bmatrix} x \\ y \end{Bmatrix} = [N] \begin{Bmatrix} x_1 \\ x_2 \\ \vdots \\ x_p \\ y_1 \\ y_2 \\ \vdots \\ y_p \end{Bmatrix} \tag{4.115}$$

we can obtain the derivatives of x and y with respect to the local coordinates directly and hence the Jacobian matrix can be expressed as

$$[J] = \begin{bmatrix} \sum_{i=1}^{p} \left(\frac{\partial N_i}{\partial r} \cdot x_i \right) & \sum_{i=1}^{p} \left(\frac{\partial N_i}{\partial r} \cdot y_i \right) \\ \sum_{i=1}^{p} \left(\frac{\partial N_i}{\partial s} \cdot x_i \right) & \sum_{i=1}^{p} \left(\frac{\partial N_i}{\partial s} \cdot y_i \right) \end{bmatrix} \tag{4.116}$$

Thus, we can find the global derivatives needed in Eq. (4.111) as

$$\begin{Bmatrix} \partial N_i/\partial x \\ \partial N_i/\partial y \end{Bmatrix} = [J]^{-1} \begin{Bmatrix} \partial N_i/\partial r \\ \partial N_i/\partial s \end{Bmatrix} \tag{4.117}$$

For the second transformation, we use the relation $\mathrm{d}V = t\,\mathrm{d}x\,\mathrm{d}y = t \det [J]\,\mathrm{d}r\,\mathrm{d}s$ (for plane problems), where t is the thickness of the plate element, and $\mathrm{d}V = \mathrm{d}x\,\mathrm{d}y\,\mathrm{d}z = \det [J]\,\mathrm{d}r\,\mathrm{d}s\,dt^*$ (for three-dimensional problems). Assuming that the inverse of $[J]$ can be found, the volume integration implied in Eq. (4.107) can be performed as

$$[K^{(e)}] = t \int_{-1}^{1} \int_{-1}^{1} [B]^T [D][B] \cdot \det [J]\,\mathrm{d}r\,\mathrm{d}s \tag{4.118}$$

and a similar expression can be written for Eq. (4.108).

Notes:

1. Although a two-dimensional (plane) problem is considered for explanation, a similar procedure can be adopted in the case of one- and three-dimensional isoparametric elements.
2. If the order of the shape functions used is different in describing the geometry of the element compared to the displacements (i.e., for subparametric or superparametric elements), the shape functions used for describing the geometry would be used in Eqs. (4.115) and (4.116), whereas the shape functions used for describing the displacements would be used in Eqs. (4.112).
3. Although the limits of integration in Eq. (4.118) appear to be very simple, unfortunately, the explicit form of the matrix product $[B]^T[D][B]$ is not very easy to express in closed form. Hence, it is necessary to resort to numerical integration. However, this is not a severe restriction because general computer programs, not tied to a particular element, can be written for carrying out the numerical integration.

---

* For carrying out the volume integration, we assume that r, s, and t are the local coordinates and x, y, and z are the global coordinates so that the Jacobian matrix is given by

$$[J] = \begin{bmatrix} \partial x/\partial r & \partial y/\partial r & \partial z/\partial r \\ \partial x/\partial s & \partial y/\partial s & \partial z/\partial s \\ \partial x/\partial t & \partial y/\partial t & \partial z/\partial t \end{bmatrix}$$

4.10 NUMERICAL INTEGRATION

4.10.1 In One Dimension

There are several schemes available for the numerical evaluation of definite integrals. Because Gauss quadrature method has been proved to be most useful in finite element applications, we shall consider only this method in this section.

Let the one-dimensional integral to be evaluated be

$$I = \int_{-1}^{1} f(r)\,\mathrm{d}r \tag{4.119}$$

The simplest and crudest way of evaluating I is to sample (evaluate) f at the middle point and multiply by the length of the interval as shown in Figure 4.21(a) to obtain

$$I = 2f_1 \tag{4.120}$$

This result would be exact only if the curve happens to be a straight line. Generalization of this relation gives

$$I = \int_{-1}^{1} f(r)\,\mathrm{d}r \simeq \sum_{i=1}^{n} w_i f_i = \sum_{i=1}^{n} w_i f(r_i) \tag{4.121}$$

where w_i is called the "weight" associated with the ith point, and n is the number of sampling points. This means that in order to evaluate the integral, we evaluate the function at several sampling points, multiply each value f_i by an appropriate weight w_i, and add. Figure 4.21 illustrates sampling at one, two, and three points.

In the Gauss method, the location of sampling points is such that for a given number of points, greatest accuracy is obtained. The sampling points are located symmetrically about the center of the interval. The weight would be the same for symmetrically located points. Table 4.1 shows the locations and weights for Gaussian integration up to six points.

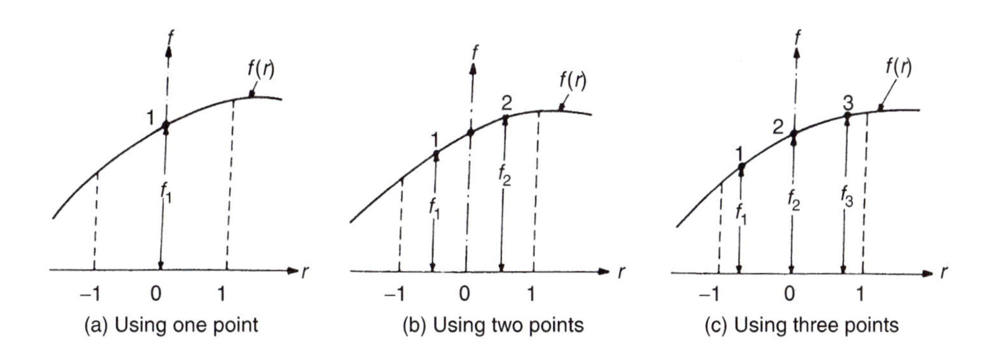

Figure 4.21.

Thus, for example, if we use the three-point Gaussian formula, we get

$$I \simeq 0.555556 f_1 + 0.888889 f_2 + 0.555556 f_3$$

which is the exact result if $f(r)$ is a polynomial of order less than or equal to 5. In general, Gaussian quadrature using n points is exact if the integrand is a polynomial of degree $2n-1$ or less.

The principle involved in deriving the Gauss quadrature formula can be illustrated by considering a simple function,

$$f(r) = a_1 + a_2 r + a_3 r^2 + a_4 r^3$$

If $f(r)$ is integrated between -1 and 1, the area under the curve $f(r)$ is

$$I = 2a_1 + \tfrac{2}{3} a_3$$

By using two symmetrically located points $r = \pm r_i$, we propose to calculate the area as

$$\underset{\sim}{I} = w \cdot f(-r_i) + w \cdot f(r_i) = 2w(a_1 + a_3 r_i^2)$$

If we want to minimize the error $e = I - \underset{\sim}{I}$ for any values of a_1 and a_3, we must have

$$\frac{\partial e}{\partial a_1} = \frac{\partial e}{\partial a_3} = 0$$

These equations give

$$w = 1$$

Table 4.1. Locations (r_i) and Weights (w_i) in Gaussian Integration (Eq. 4.121)

Number of points (n)	Location (r_i)	Weight (w_i)
1	$r_1 = 0.00000\ 00000\ 00000$	2.00000 00000 00000
2	$r_1, r_2 = \pm 0.57735\ 02691\ 89626$	1.00000 00000 00000
3	$r_1, r_3 = \pm 0.77459\ 66692\ 41483$	0.55555 55555 55555
	$r_2 = 0.00000\ 00000\ 00000$	0.88888 88888 88889
4	$r_1, r_4 = \pm 0.86113\ 63115\ 94053$	0.34785 48451 47454
	$r_2, r_3 = \pm 0.33998\ 10435\ 84856$	0.65214 51548 62546
5	$r_1, r_5 = \pm 0.90617\ 98459\ 38664$	0.23692 68850 56189
	$r_2, r_4 = \pm 0.53846\ 93101\ 05683$	0.47862 86704 99366
	$r_3 = 0.00000\ 00000\ 00000$	0.56888 88888 88889
6	$r_1, r_6 = \pm 0.93246\ 95142\ 03152$	0.17132 44923 79170
	$r_2, r_5 = \pm 0.66120\ 93864\ 66265$	0.36076 15730 48139
	$r_3, r_4 = \pm 0.23861\ 91860\ 83197$	0.46791 39345 72691

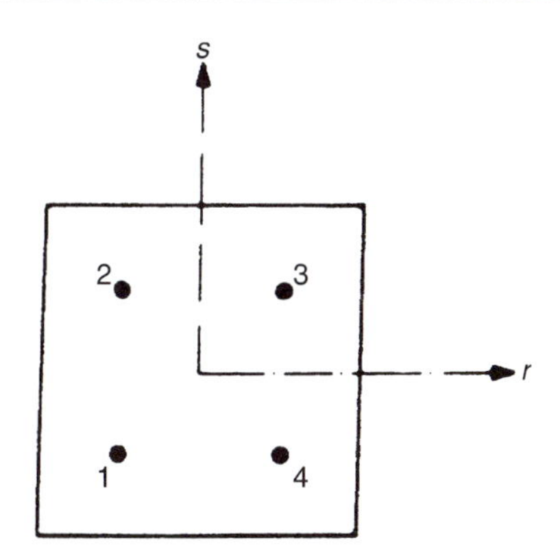

Figure 4.22. Four-Point Gaussian Quadrature Rule.

and

$$r_i = \frac{1}{\sqrt{3}} = 0.577350\ldots$$

4.10.2 In Two Dimensions

(i) In rectangular regions

In two-dimensional (rectangular) regions, we obtain the Gauss quadrature formula by integrating first with respect to one coordinate and then with respect to the second as

$$I = \int_{-1}^{1}\int_{-1}^{1} f(r,s)\,\mathrm{d}r\,\mathrm{d}s = \int_{-1}^{1}\left[\sum_{i=1}^{n} w_i f(r_i,s)\right]\mathrm{d}s$$
$$= \sum_{j=1}^{n} w_j\left[\sum_{i=1}^{n} w_i f(r_i,s_j)\right] = \sum_{i=1}^{n}\sum_{j=1}^{n} w_i w_j f(r_i,s_j) \tag{4.122}$$

Thus, for example, a four-point Gaussian rule (Figure 4.22) gives

$$I \simeq (1.000000)(1.000000)[f(r_1,s_1) + f(r_2,s_2) + f(r_3,s_3) + f(r_4,s_4)] \tag{4.123}$$

where the four sampling points are located at r_i, s_i $= \pm 0.577350$. In Eq. (4.122), the number of integration points in each direction was assumed to be the same. Clearly, it is not necessary and sometimes it may be advantageous to use different numbers in each direction.

(ii) In triangular regions

The integrals involved for triangular elements would be in terms of triangular or area coordinates and the following Gauss-type formula has been developed by Hammer and

Stroud [4.5]:

$$I = \iint_A f(L_1, L_2, L_3)\,\mathrm{d}A \simeq \sum_{i=1}^{n} w_i f(L_1^{(i)}, L_2^{(i)}, L_3^{(i)}) \tag{4.124}$$

where

for $n = 1$ (linear triangle):

$$w_1 = 1; \qquad L_1^{(1)} = L_2^{(1)} = L_3^{(1)} = \tfrac{1}{3}$$

for $n = 3$ (quadratic triangle):

$$\begin{aligned}
w_1 &= \tfrac{1}{3}; & L_1^{(1)} &= L_2^{(1)} = \tfrac{1}{2}, L_3^{(1)} = 0 \\
w_2 &= \tfrac{1}{3}; & L_1^{(2)} &= 0, L_2^{(2)} = L_3^{(2)} = \tfrac{1}{2} \\
w_3 &= \tfrac{1}{3}; & L_1^{(3)} &= L_3^{(3)} = \tfrac{1}{2}, L_2^{(3)} = 0
\end{aligned}$$

for $n = 7$ (cubic triangle):

$$\begin{aligned}
w_1 &= \tfrac{27}{60}; & L_1^{(1)} &= L_2^{(1)} = L_3^{(1)} = \tfrac{1}{3} \\
w_2 &= \tfrac{8}{60}; & L_1^{(2)} &= L_2^{(2)} = \tfrac{1}{2}, L_3^{(2)} = 0 \\
w_3 &= \tfrac{8}{60}; & L_1^{(3)} &= 0, L_2^{(3)} = L_3^{(3)} = \tfrac{1}{2} \\
w_4 &= \tfrac{8}{60}; & L_1^{(4)} &= L_3^{(4)} = \tfrac{1}{2}, L_2^{(4)} = 0 \\
w_5 &= \tfrac{3}{60}; & L_1^{(5)} &= 1, L_2^{(5)} = L_3^{(5)} = 0 \\
w_6 &= \tfrac{3}{60}; & L_1^{(6)} &= L_3^{(6)} = 0, L_2^{(6)} = 1 \\
w_7 &= \tfrac{3}{60}; & L_1^{(7)} &= L_2^{(7)} = 0, L_3^{(7)} = 1
\end{aligned}$$

The locations of the integration points are shown in Figure 4.23.

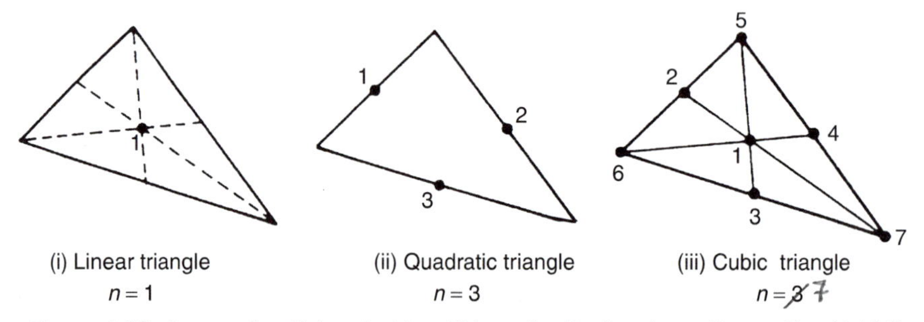

Figure 4.23. Integration Points inside a Triangular Region According to Eq. (4.124).

4.10.3 In Three Dimensions

(i) In rectangular prism-type regions

For a right prism, we can obtain an integration formula similar to Eq. (4.122) as

$$I = \int_{-1}^{1}\int_{-1}^{1}\int_{-1}^{1} f(r,s,t)\,\mathrm{d}r\,\mathrm{d}s\,\mathrm{d}t = \sum_{i=1}^{n}\sum_{j=1}^{n}\sum_{k=1}^{n} w_i w_j w_k f(r_i, s_j, t_k) \tag{4.125}$$

where an equal number of integration points (n) in each direction is taken only for convenience.

(ii) In tetrahedral regions

For tetrahedral regions, four volume coordinates are involved and the integral can be evaluated as in Eq. (4.125):

$$I \simeq \sum_{i=1}^{n} w_i f(L_1^{(i)}, L_2^{(i)}, L_3^{(i)}, L_4^{(i)}) \tag{4.126}$$

where

for $n = 1$ (linear tetrahedron):

$$w_1 = 1; \qquad L_1^{(1)} = L_2^{(1)} = L_3^{(1)} = L_4^{(1)} = \tfrac{1}{4}$$

for $n = 4$ (quadratic tetrahedron):

$$\begin{aligned}
w_1 &= \tfrac{1}{4}; & L_1^{(1)} &= a, L_2^{(1)} = L_3^{(1)} = L_4^{(1)} = b \\
w_2 &= \tfrac{1}{4}; & L_2^{(2)} &= a, L_1^{(2)} = L_3^{(2)} = L_4^{(2)} = b \\
w_3 &= \tfrac{1}{4}; & L_3^{(3)} &= a, L_1^{(3)} = L_2^{(3)} = L_4^{(3)} = b \\
w_4 &= \tfrac{1}{4}; & L_4^{(4)} &= a, L_1^{(4)} = L_2^{(4)} = L_3^{(4)} = b
\end{aligned}$$

with

$$a = 0.58541020,$$

and

$$b = 0.13819660,$$

for $n = 4$ (cubic tetrahedron):

$$\begin{aligned}
w_1 &= -\tfrac{4}{5}; & L_1^{(1)} &= L_2^{(1)} = L_3^{(1)} = L_4^{(1)} = \tfrac{1}{4} \\
w_2 &= \tfrac{9}{20}; & L_1^{(2)} &= \tfrac{1}{3}, L_2^{(2)} = L_3^{(2)} = L_4^{(2)} = \tfrac{1}{6}
\end{aligned}$$

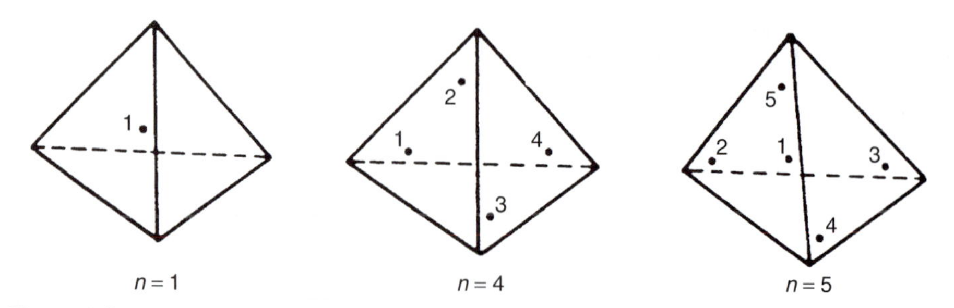

Figure 4.24. Location of Integration Points inside a Tetrahedron According to Eq. (4.126).

$$
\begin{aligned}
w_3 &= \tfrac{9}{20}; \qquad & L_2^{(3)} &= \tfrac{1}{3}, L_1^{(3)} = L_3^{(3)} = L_4^{(3)} = \tfrac{1}{6} \\
w_4 &= \tfrac{9}{20}; \qquad & L_3^{(4)} &= \tfrac{1}{3}, L_1^{(4)} = L_2^{(4)} = L_4^{(4)} = \tfrac{1}{6} \\
w_5 &= \tfrac{9}{20}; \qquad & L_4^{(5)} &= \tfrac{1}{3}, L_1^{(5)} = L_2^{(5)} = L_3^{(5)} = \tfrac{1}{6}
\end{aligned}
$$

The locations of the integration points used in Eq. (4.126) are shown in Figure 4.24.

REFERENCES

4.1 K.E. Atkinson: *An Introduction to Numerical Analysis*, 2nd Ed., Wiley, New York, 1989.

4.2 A.F. Emery and W.W. Carson: An evaluation of the use of the finite element method in the computation of temperature, *Journal of Heat Transfer, Transactions of ASME, 93,* 136–145, 1971.

4.3 V. Hoppe: Higher order polynomial elements with isoparametric mapping, *International Journal for Numerical Methods in Engineering, 15,* 1747–1769, 1980.

4.4 O.C. Zienkiewicz, B.M. Irons, J. Ergatoudis, S. Ahmad, and F.C. Scott: Isoparametric and associated element families for two and three dimensional analysis, in *The Finite Element Methods in Stress Analysis* (Eds. I. Holand and K. Bell), Tapir Press, Trondheim, Norway, 1969.

4.5 P.C. Hammer and A.H. Stroud: Numerical evaluation of multiple integrals, *Mathematical Tables and Other Aids to Computation, 12,* 272–280, 1958.

4.6 L.A. Ying: Some "special" interpolation formulae for triangular and quadrilateral elements, *International Journal for Numerical Methods in Engineering, 18,* 959–966, 1982.

4.7 T. Liszka: An interpolation method for an irregular net of nodes, *International Journal for Numerical Methods in Engineering, 20,* 1599–1612, 1984.

4.8 A. El-Zafrany and R.A. Cookson: Derivation of Lagrangian and Hermitian shape functions for quadrilateral elements, *International Journal for Numerical Methods in Engineering, 23,* 1939–1958, 1986.

PROBLEMS

4.1 Consider the shape functions, $N_i(x)$, $N_j(x)$, and $N_k(x)$, corresponding to the nodes i, j, and k of the one-dimensional quadratic element described in Section 4.2.1. Show that the shape function corresponding to a particular node i (j or k) has a value of one at node i (j or k) and zero at the other two nodes j (k or i) and k (i or j).

4.2 Consider the shape functions described in Eq. (4.10) for a one-dimensional cubic element. Show that the shape function corresponding to a particular node i, $N_i(x)$, has a value of one at node i and zero at the other three nodes j, k, and l. Repeat the procedure for the shape functions $N_j(x)$, $N_k(x)$, and $N_l(x)$.

4.3 The Cartesian (global) coordinates of the corner nodes of a quadrilateral element are given by (0, −1), (−2, 3), (2, 4), and (5, 3). Find the coordinate transformation between the global and local (natural) coordinates. Using this, determine the Cartesian coordinates of the point defined by $(r, s) = (0.5, 0.5)$ in the global coordinate system.

4.4 Determine the Jacobian matrix for the quadrilateral element defined in Problem 4.3. Evaluate the Jacobian matrix at the point, $(r, s) = (0.5, 0.5)$.

4.5 The Cartesian (global) coordinates of the corners of a triangular element are given by (−2, −1), (2, 4), and (4, 1). Find expressions for the natural (triangular) coordinates L_1, L_2, and L_3. Determine the values of L_1, L_2, and L_3 at the point, $(x, y) = (0, 0)$.

4.6 Consider a triangular element with the corner nodes defined by the Cartesian coordinates (−2, −1), (2, 4), and (4, 1). Using the expressions derived in Problem 4.5 for L_1, L_2, and L_3, evaluate the following in terms of the Cartesian coordinates x and y:

(a) Shape functions N_1, N_2, and N_3 corresponding to a linear interpolation model.
(b) Shape functions $N_1, N_2, \ldots, N_6$ corresponding to a quadratic interpolation model.
(c) Shape functions $N_1, N_2, \ldots, N_{10}$ corresponding to a cubic interpolation model.

4.7 The interpolation functions corresponding to node i of a triangular element can be expressed in terms of natural coordinates L_1, L_2, and L_3 using the relationship

$$N_i = f^{(i)}(L_1) f^{(i)}(L_2) f^{(i)}(L_3) \tag{E$_1$}$$

where

$$f^{(i)}(L_j) = \begin{cases} \prod_{k=1}^{p} \frac{1}{k}(mL_j - k + 1) & \text{if } p \geq 1 \\ 1 & \text{if } p = 0 \end{cases} \tag{E$_2$}$$

with $i = 1, 2, \ldots, n$; n = total number of nodes in the element, $p = mL_j^{(i)}$, m = order of the interpolation model (2 for quadratic, 3 for cubic, etc.), and $L_j^{(i)}$ = value of the coordinate L_j at node i.

Using Eq. (E$_1$), find the interpolation function corresponding to node 1 of a quadratic triangular element.

4.8 Using Eq. (E_1) given in Problem 4.7, find the interpolation function corresponding to node 4 of a cubic triangular element.

4.9 Using Eq. (E_1) given in Problem 4.7, find the interpolation function corresponding to node 10 of a cubic triangular element.

4.10 Using Eq. (E_1) given in Problem 4.7, find the interpolation function corresponding to node 4 of a quadratic triangular element.

4.11 Using Eq. (E_1) given in Problem 4.7, find the interpolation function corresponding to node 1 of a cubic triangular element.

4.12 The interpolation functions corresponding to node i of a tetrahedron element can be expressed in terms of the natural coordinates L_1, L_2, L_3, and L_4 using the relationship

$$N_i = f^{(i)}(L_1) f^{(i)}(L_2) f^{(i)}(L_3) f^{(i)}(L_4) \tag{E$_1$}$$

where $f^{(i)}(L_j)$ is defined by Eq. (E_2) of Problem 4.7. Using this relation, find the interpolation function corresponding to node 1 of a quadratic tetrahedron element.

4.13 Using Eq. (E_1) given in Problem 4.12, find the interpolation function corresponding to node 5 of a quadratic tetrahedron element.

4.14 Using Eq. (E_1) given in Problem 4.12, find the interpolation function corresponding to node 1 of a cubic tetrahedron element.

4.15 Using Eq. (E_1) given in Problem 4.12, find the interpolation function corresponding to node 5 of a cubic tetrahedron element.

4.16 Using Eq. (E_1) given in Problem 4.12, find the interpolation function corresponding to node 17 of a cubic tetrahedron element.

4.17 The Cartesian (global) coordinates of the corners of a tetrahedron element are given by (0, 0, 0), (1, 0, 0), (0, 1, 0), and (0, 0, 1). Find expressions for the natural (tetrahedral) coordinates, L_1, L_2, L_3, and L_4. Determine the values of L_1, L_2, L_3, and L_4 at the point $(x, y, z) = (0.25, 0.25, 0.25)$.

4.18 The nodes of a quadratic one-dimensional element are located at $x = 0$, $x = l/2$, and $x = l$. Express the shape functions using Lagrange interpolation polynomials.

4.19 Derive expressions for the shape functions of the rectangular element shown in Figure 4.25 using Lagrange interpolation polynomials.

4.20 The Cartesian coordinates of the nodes of a quadratic quadrilateral isoparametric element are shown in Figure 4.26. Determine the coordinate transformation relation between the local and global coordinates. Using this relation, find the global coordinates corresponding to the point $(r, s) = (0, 0)$.

4.21 A boundary value problem, governed by the Laplace equation, is stated as

$$\frac{\partial^2 \phi}{\partial x^2} + \frac{\partial^2 \phi}{\partial y^2} = 0 \text{ in A}$$

$$\phi = \phi_0 \text{ on C}$$

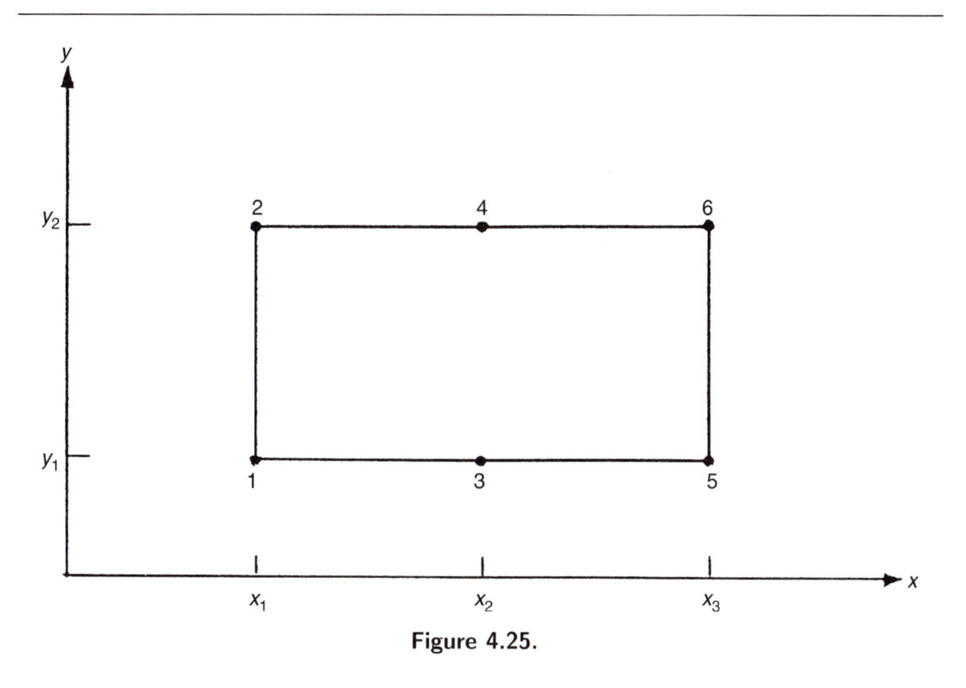

Figure 4.25.

The characteristic (stiffness) matrix of an element corresponding to this problem can be expressed as

$$[K^{(e)}] = \iint_{A^{(e)}} [B]^T [D][B] \, \mathrm{d}A$$

where

$$[D] = \begin{bmatrix} 1 & 0 \\ 0 & 1 \end{bmatrix}$$

$$[B] = \begin{bmatrix} \dfrac{\partial N_1}{\partial x} & \dfrac{\partial N_2}{\partial x} & \cdots & \dfrac{\partial N_p}{\partial x} \\ \dfrac{\partial N_1}{\partial y} & \dfrac{\partial N_2}{\partial y} & \cdots & \dfrac{\partial N_p}{\partial y} \end{bmatrix}$$

and $A^{(e)}$ is the area of the element. Derive the matrix $[B]$ for a quadratic quadrilateral isoparametric element whose nodal coordinates are shown in Figure 4.26.

4.22 Evaluate the integral

$$I = \int_{-1}^{1} (a_0 + a_1 x + a_2 x^2 + a_3 x^3 + a_4 x^4) \, \mathrm{d}x$$

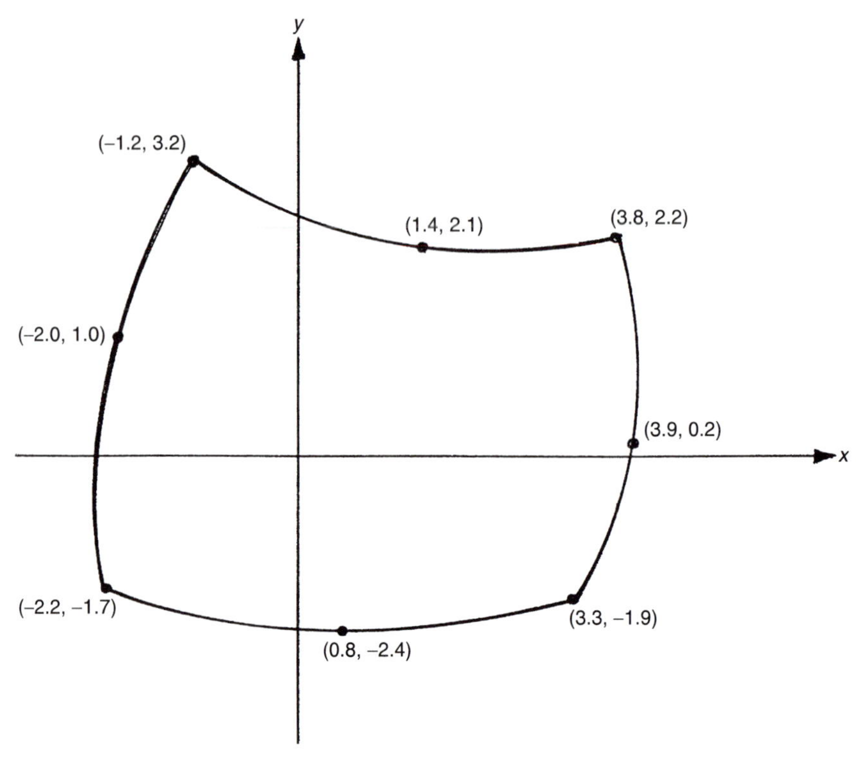

Figure 4.26.

using the following methods and compare the results:

(a) Two-point Gauss integration
(b) Analytical integration

4.23 Evaluate the integral

$$I = \int_{-1}^{1} (a_0 + a_1 x + a_2 x^2 + a_3 x^3)\, dx$$

using the following methods and compare the results:

(a) Three-point Gauss integration
(b) Analytical integration

4.24 Evaluate the integral

$$I = \int_{-1}^{1}\int_{-1}^{1} (r^2 s^3 + r s^4)\, dr\, ds$$

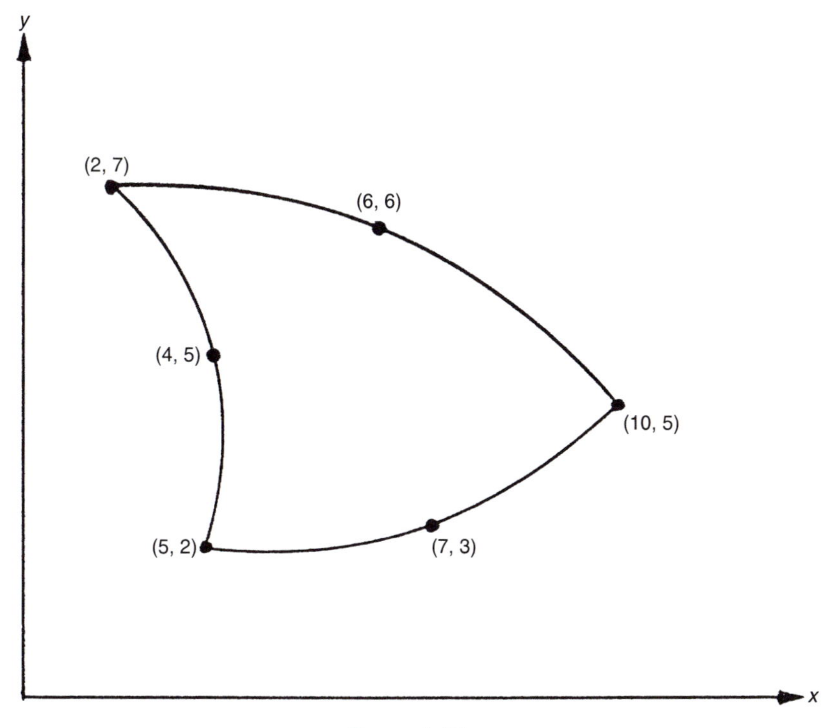

Figure 4.27.

using the following methods and compare the results:

(a) Gauss integration
(b) Analytical integration

4.25 Determine the Jacobian matrix for the quadratic isoparametric triangular element shown in Figure 4.27.

4.26 How do you generate an isoparametric quadrilateral element for C^1 continuity? (Hint: In this case we need to transform the second-order partial derivatives and the Jacobian will be a variable matrix.)

4.27 Consider a ring element with triangular cross section as shown in Figure 4.28. If the field variable ϕ does not change with respect to θ, propose linear, quadratic, and cubic interpolation models for C^0 continuity. Develop the necessary element equations for the linear case for solving the Laplace's equation

$$\frac{\partial^2 \phi}{\partial r^2} + \frac{1}{r}\frac{\partial \phi}{\partial r} + \frac{\partial^2 \phi}{\partial z^2} = 0$$

4.28 Evaluate $\partial N_4/\partial x$ and $\partial N_4/\partial y$ at the point (1.5, 2.0) for the quadratic triangular element shown in Figure 4.29. [Hint: Since the sides of the element are straight,

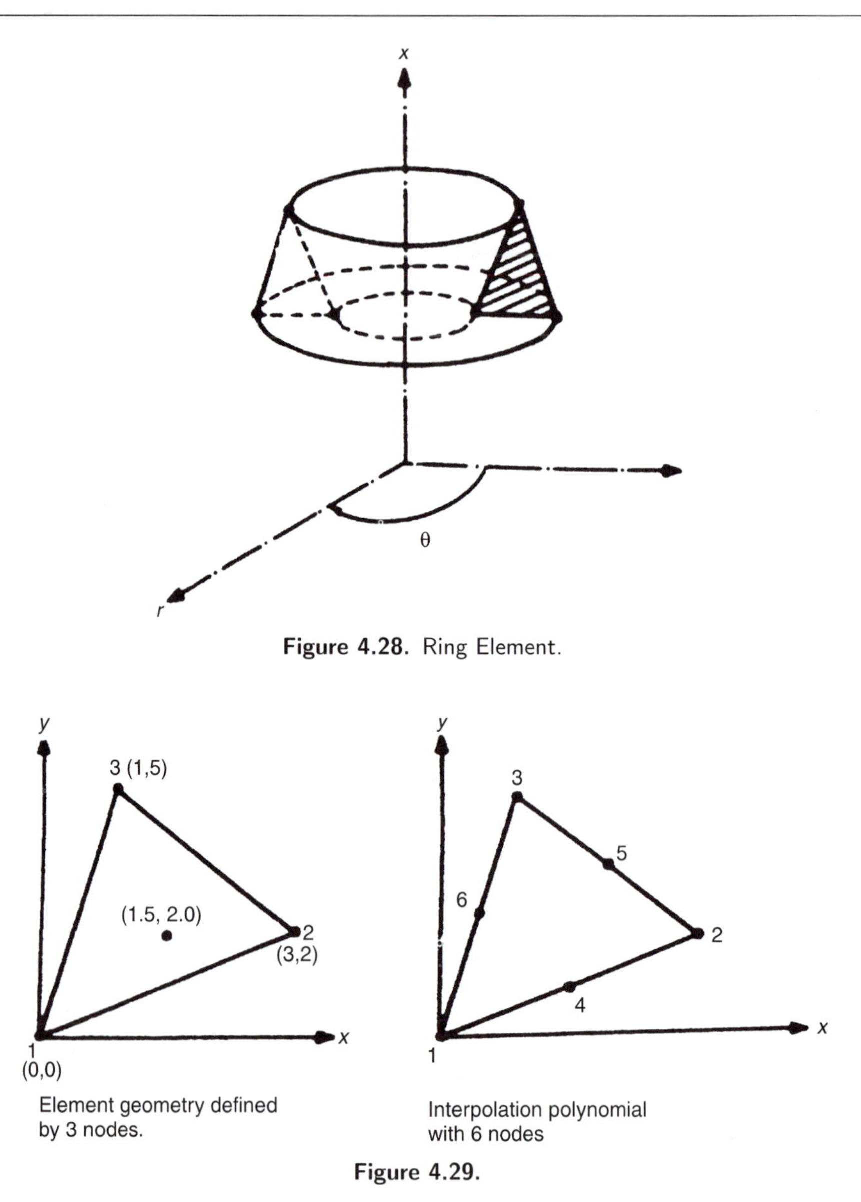

Figure 4.28. Ring Element.

Figure 4.29.

define the geometry of the element using Eq. (3.73). Evaluate the Jacobian matrix

$$[J] = \begin{bmatrix} \dfrac{\partial x}{\partial L_1} & \dfrac{\partial y}{\partial L_1} \\ \dfrac{\partial x}{\partial L_2} & \dfrac{\partial y}{\partial L_2} \end{bmatrix}$$

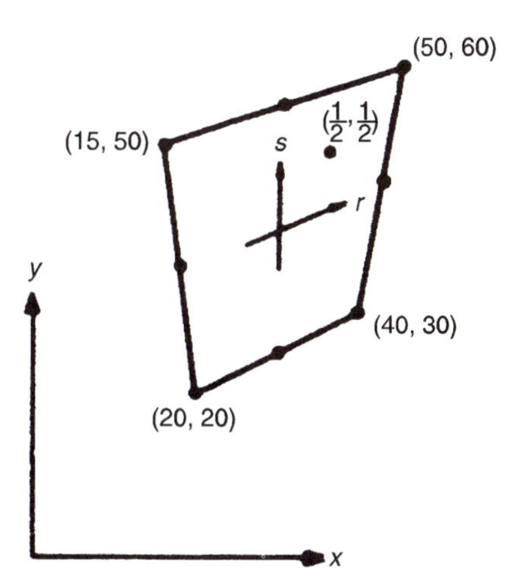

Figure 4.30. Quadrilateral Element.

Differentiate the quadratic shape function N_4 given by Eq. (4.26) with respect to x and y. Use the relation

$$\begin{Bmatrix} \dfrac{\partial N_4}{\partial x} \\ \dfrac{\partial N_4}{\partial y} \end{Bmatrix} = [J]^{-1} \begin{Bmatrix} \dfrac{\partial N_4}{\partial L_1} \\ \dfrac{\partial N_4}{\partial L_2} \end{Bmatrix}$$

and obtain the desired result.]

4.29 Evaluate the partial derivatives $(\partial N_1/\partial x)$ and $(\partial N_1/\partial y)$ of the quadrilateral element shown in Figure 4.30 at the point $(r = 1/2,\ s = 1/2)$ assuming that the scalar field variable ϕ is approximated by a quadratic interpolation model.

4.30 Derive Eqs. (4.16) and (4.17) for a one-dimensional quadratic element.

4.31 Derive Eqs. (4.20)–(4.23) for a one-dimensional cubic element.

4.32 Derive Eqs. (4.26) for a quadratic triangular element.

4.33 Derive Eqs. (4.30) for a cubic triangular element.

4.34 Derive Eq. (4.36) for a quadrilateral element.

4.35 Derive the Hermite polynomials indicated in Eqs. (4.70)–(4.72).

5

DERIVATION OF ELEMENT MATRICES AND VECTORS

5.1 INTRODUCTION

The characteristic matrices and characteristic vectors (also termed vectors of nodal actions) of finite elements can be derived by using any of the following approaches:

1. Direct approach

In this method, direct physical reasoning is used to establish the element properties (characteristic matrices and vectors) in terms of pertinent variables. Although the applicability of these methods is limited to simple types of elements, a study of these methods enhances our understanding of the physical interpretation of the finite element method.

2. Variational approach

In this method, the finite element analysis is interpreted as an approximate means for solving variational problems. Since most physical and engineering problems can be formulated in variational form, the finite element method can be readily applied for finding their approximate solutions. The variational approach has been most widely used in the literature in formulating finite element equations. A major limitation of the method is that it requires the physical or engineering problem to be stated in variational form, which may not be possible in all cases.

3. Weighted residual approach

In this method, the element matrices and vectors are derived directly from the governing differential equations of the problem without reliance on the variational statement of the problem. This method offers the most general procedure for deriving finite element equations and can be applied to almost all practical problems of science and engineering. Again, within the weighted residual approach, different procedures, such as Galerkin method and least squares method, can be used in deriving the element equations.

The details of all these approaches are considered in this chapter.

5.2 DIRECT APPROACH

Since the basic concept of discretization in the finite element method stems from the physical procedures used in structural framework analysis and network analysis, we consider a few simple examples from these areas to illustrate the direct method of deriving finite element equations.

5.2.1 Bar Element under Axial Load

Consider a stepped bar as shown in Figure 5.1(a). The different steps are assumed to have different lengths, areas of cross section, and Young's modulii. The way to discretize this system into finite elements is immediately obvious. If we define each step as an element, the system consists of three elements and four nodes [Figure 5.1(b)].

The force–displacement equations of a step constitute the required element equations. To derive these equations for a typical element e, we isolate the element as shown in Figure 5.1(c). In this figure, a force (P) and a displacement (u) are defined at each of the two nodes in the positive direction of the x axis. The field variable ϕ is the deflection u. The element equations can be expressed in matrix form as

$$[k]\vec{u} = \vec{P} \tag{5.1}$$

or

$$\begin{bmatrix} k_{11} & k_{12} \\ k_{21} & k_{22} \end{bmatrix} \begin{Bmatrix} u_1 \\ u_2 \end{Bmatrix} = \begin{Bmatrix} P_1 \\ P_2 \end{Bmatrix} \tag{5.2}$$

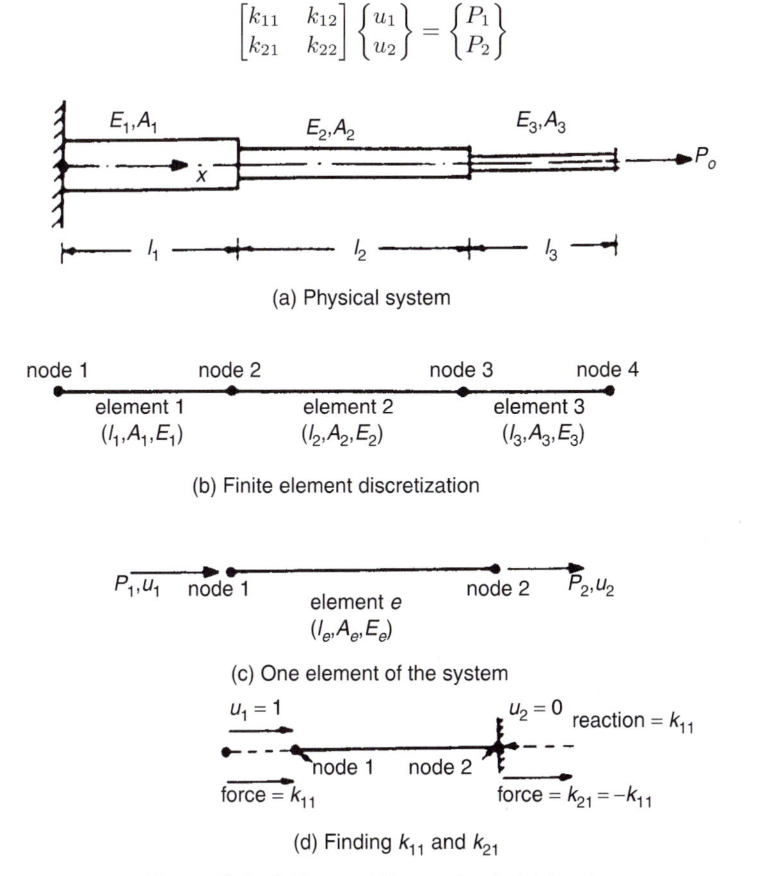

(d) Finding k_{11} and k_{21}

Figure 5.1. A Stepped Bar under Axial Load.

where $[k]$ is called the stiffness or characteristic matrix, $\vec{u}$ is the vector of nodal displacements, and $\vec{P}$ is the vector of nodal forces of the element. We shall derive the element stiffness matrix from the basic definition of the stiffness coefficient, and for this no assumed interpolation polynomials are needed. In structural mechanics [5.1], the stiffness influence coefficient k_{ij} is defined as the force needed at node i (in the direction of u_i) to produce a unit displacement at node j ($u_j = 1$) while all other nodes are restrained. This definition can be used to generate the matrix $[k]$. For example, when we apply a unit displacement to node 1 and restrain node 2 as shown in Figure 5.1(d), we induce a force (k_{11}) equal to* $A_e = (A_e E_e / l_e)$ at node 1 and a force (k_{21}) equal to $-(A_e E_e / l_e)$ at node 2. Similarly, we can obtain the values of k_{22} and k_{12} as $(A_e E_e / l_e)$ and $-(A_e E_e / l_e)$, respectively, by giving a unit displacement to node 2 and restraining node 1. Thus, the characteristic (stiffness) matrix of the element is given by

$$[k] = \begin{bmatrix} k_{11} & k_{12} \\ k_{21} & k_{22} \end{bmatrix} = \begin{bmatrix} (A_e E_e / l_e) & -(A_e E_e / l_e) \\ -(A_e E_e / l_e) & (A_e E_e / l_e) \end{bmatrix} = \frac{A_e E_e}{l_e} \begin{bmatrix} 1 & -1 \\ -1 & 1 \end{bmatrix} \tag{5.3}$$

Notes:

1. Equation (5.1) denotes the element equations regardless of the type of problem, the complexity of the element, or the way in which the element, characteristic matrix, $[k]$, is derived.
2. The stiffness matrix $[k]$ obeys the Maxwell–Betti reciprocal theorem [5.1], which states that all stiffness matrices of linear structures are symmetric.

5.2.2 Line Element for Heat Flow

Consider a composite (layered) wall through which heat flows in only the x direction [Figure 5.2(a)]. The left face is assumed to be at a uniform temperature higher than that of the right face. Each layer is assumed to be a nonhomogeneous material whose thermal conductivity is a known function of the x. Since heat flows only in the x direction, the problem can be treated as one-dimensional with each layer considered as a finite element. The nodes for any element will be the bounding planes of the layer. Thus, there are four elements and five nodes in the system. The field variable ϕ is temperature T in this problem. Thus, the nodal unknowns denote the temperatures that are uniform over the bounding planes.

We can derive the element equations by considering the basic relation between the heat flow and the temperature gradient without using any interpolation polynomials. The quantity of heat crossing a unit area per unit time in the x direction (q) is given by Eq. (1.1) as

$$q = -k(x) \cdot \frac{\mathrm{d}T}{\mathrm{d}x} \tag{5.4}$$

where $k(x)$ is the thermal conductivity of the material and $(\mathrm{d}T/\mathrm{d}x)$ is the temperature gradient.

---

* Force = stress × area of cross section = strain × Young's modulus × area of cross section = (change in length/original length) × Young's modulus × area of cross section = $(1/l_e) \cdot E_e \cdot A_e = (A_e E_e / l_e)$.

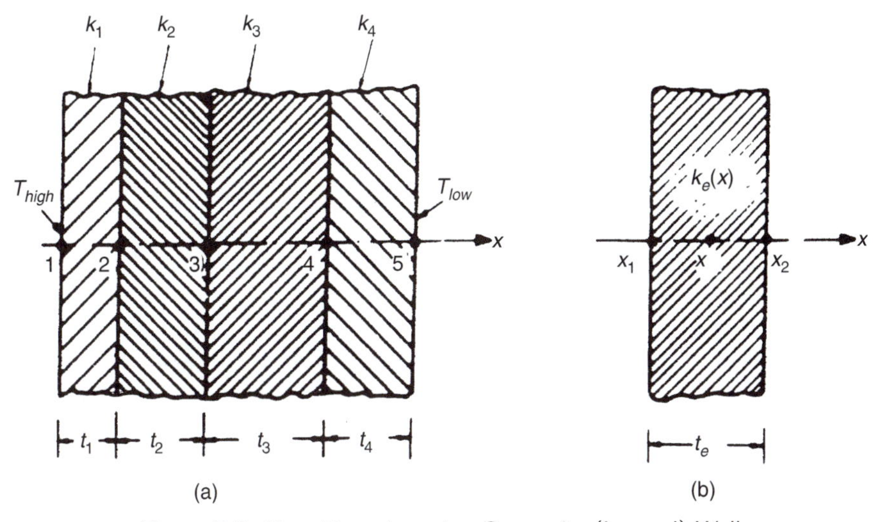

Figure 5.2. Heat Flow through a Composite (Layered) Wall.

Equation (5.4) can be integrated over the thickness of any element to obtain a relation between the nodal heat fluxes and the nodal temperatures. The integration can be avoided if we assume the thermal conductivity of a typical element e to be a constant as $k(x) = k'_e$. The temperature gradient at node 1 can be written as $(\mathrm{d}T/\mathrm{d}x)$ (at node 1) $= (T_2 - T_1)/(x_2 - x_1) = (T_2 - T_1)/t_e$ and the temperature gradient at node 2 as $(\mathrm{d}T/\mathrm{d}x)$ (at node 2) $= -$ $(\mathrm{d}T/\mathrm{d}x)$ (at node 1) $= -$ $(T_2 - T_1)/t_e$. Thus, the heat fluxes entering nodes 1 and 2 can be written as

$$\left.\begin{aligned} F_1 &= q \text{ (at node 1)} = -k'_e(T_2 - T_1)/t_e \\ F_2 &= q \text{ (at node 2)} = +k'_e(T_2 - T_1)/t_e \end{aligned}\right\} \tag{5.5}$$

By defining $k_e = k'_e/t_e$, Eq. (5.5) can be rewritten as

$$\left.\begin{aligned} F_1 &= -k_e(T_2 - T_1) \\ F_2 &= -k_e(T_1 - T_2) \end{aligned}\right\} \tag{5.6}$$

Equation (5.6) can be expressed in matrix notation as

$$[k]\vec{T} = \vec{F} \tag{5.7}$$

where

$$[k] = \begin{bmatrix} k_e & -k_e \\ -k_e & k_e \end{bmatrix} = \text{element conductivity (characteristic) matrix}$$

$$\vec{T} = \begin{Bmatrix} T_1 \\ T_2 \end{Bmatrix} = \text{vector of nodal temperatures}$$

and

$$\vec{F} = \begin{Bmatrix} F_1 \\ F_2 \end{Bmatrix} = \text{vector of nodal heat fluxes}$$

Note that Eq. (5.7) has the same form as Eq. (5.1).

5.2.3 Line Element for Fluid Flow

Consider a fluid flow network consisting of several pipeline segments as shown in Figure 5.3(a). If we assume that there is a source such as a pump at some point in the network, the problem is to find the pressures and flow rates in various paths of the network. For this we discretize the network into several finite elements, each element representing the flow path between any two connected nodes or junctions. Thus, the network shown in Figure 5.3(a) has 7 nodes and 10 finite elements. In this case, the pressure loss–flow rate relations constitute the element equations and can be derived from the basic principles of fluid mechanics without using any interpolation polynomials.

For a circular pipe of inner diameter d and length l, and for laminar flow, the pressure drop between any two nodes (sections) 1 and 2, $p_2 - p_1$, is related to the flow (F) by [5.2]

$$p_2 - p_1 = \frac{128 F l \mu}{\pi\, \mathrm{d}^4} \tag{5.8}$$

where μ is the dynamic viscosity of the fluid. For a typical element e shown in Figure 5.3(b), the flows entering at nodes 1 and 2 can be expressed using Eq. (5.8) as

$$\left.\begin{aligned} F_1 &= \frac{\pi\, \mathrm{d}_e^4}{128 l_e \mu}(p_1 - p_2) \\ F_2 &= \frac{\pi\, \mathrm{d}_e^4}{128 l_e \mu}(p_2 - p_1) \end{aligned}\right\} \tag{5.9}$$

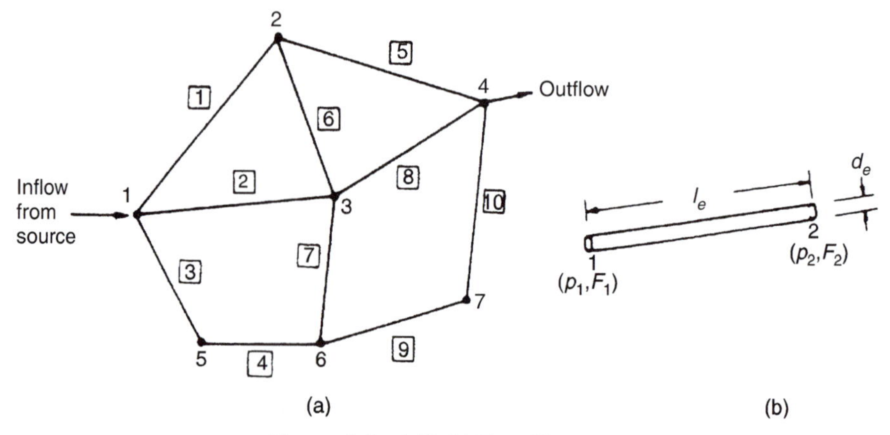

Figure 5.3. A Fluid Flow Network.

Equations (5.9) can be written in matrix form as

$$[k]\vec{p} = \vec{F} \tag{5.10}$$

where

$$[k] = \frac{\pi \, d_e^4}{128 l_e \mu} \begin{bmatrix} 1 & -1 \\ -1 & 1 \end{bmatrix} = \text{element fluidity (characteristic) matrix}$$

$\vec{p}$ = vector of nodal pressures, and $\vec{F}$ = vector of nodal flows. It can be seen that Eq. (5.10) has the same form as Eqs. (5.1) and (5.7).

5.2.4 Line Element for Current Flow

Consider a network of electrical resistances shown in Figure 5.4(a). As in the previous case, this network is discretized with 7 nodes and 10 finite elements. The current flow–voltage relations constitute the element equations in this case.

To derive the element equations for a typical element e shown in Figure 5.4(b), we use one of the basic principles of electrical engineering, namely, Ohm's law. This obviously does not need any interpolation functions. Ohm's law gives the relation between the currents entering the element at the nodes and the voltages of the nodes as

$$I_1 = \frac{1}{R_e}(V_1 - V_2)$$

$$I_2 = \frac{1}{R_e}(V_2 - V_1)$$

or

$$[k]\vec{V} = \vec{I} \tag{5.11}$$

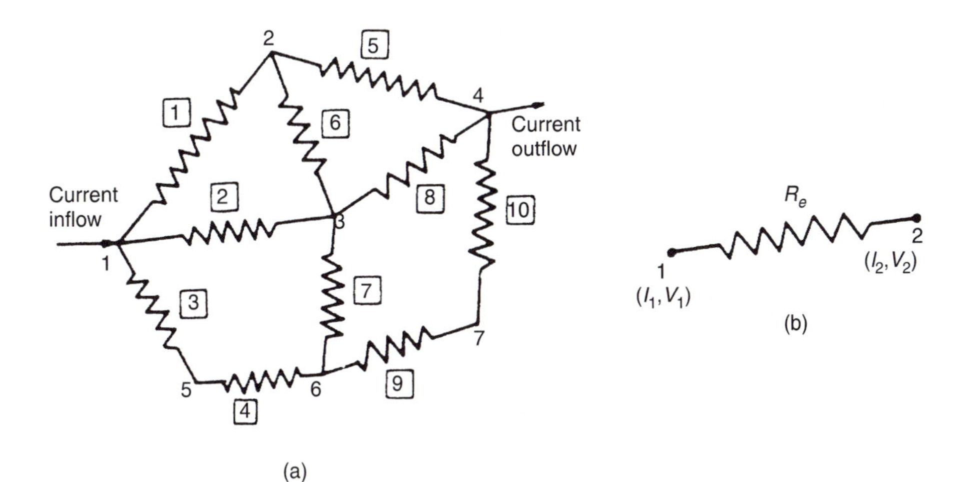

Figure 5.4. An Electrical Network.

where

$$[k] = \frac{1}{R_e}\begin{bmatrix} 1 & -1 \\ -1 & 1 \end{bmatrix} = \text{element characteristic matrix}$$

$\vec{I} = \begin{Bmatrix} I_1 \\ I_2 \end{Bmatrix}$ = vector of nodal currents, $\vec{V} = \begin{Bmatrix} V_1 \\ V_2 \end{Bmatrix}$ = vector of nodal voltages, R_e is the resistance of element e, I_i is the current entering node i $(i = 1, 2)$, and V_i is the voltage at node i $(i = 1, 2)$.

5.3 VARIATIONAL APPROACH

In the previous section the element equations were derived using the direct approach. Although the approach uses the basic principles of engineering science and hence aids in understanding the physical basis of the finite element method, insurmountable difficulties arise when we try to apply the method to complex problems (such as triangular elements). In this section we take a broader view and interpret the finite element method as an approximate method of solving variational problems. Most of the solutions reported in the literature for physical and engineering problems have been based on this approach. In this section we derive the finite element equations using the variational approach.

5.3.1 Specification of Continuum Problems

Most of the continuum problems can be specified in one of two ways. In the first, a variational principle valid over the whole domain of the problem is postulated and an integral I is defined in terms of the unknown parameters and their derivatives. The correct solution of the problem is one that minimizes the integral I. In the second, differential equations governing the behavior of a typical infinitesimal domain are given along with the boundary conditions. These two approaches are mathematically equivalent, an exact solution of one being the solution of the other. The final equations of the finite element method can be derived by proceeding either from the differential equations or from the variational principle of the problem.

Although the differential equation approach is more popular, the variational approach will be of special interest in studying the finite element method. This is due to the fact that the consideration of the finite element method as a variational approach has contributed significantly in formulating and solving problems of different branches of engineering in a unified manner. Thus, a knowledge of the basic concepts of calculus of variations is useful in understanding the general finite element method.

5.3.2 Approximate Methods of Solving Continuum Problems

If the physical problem is specified as a variational problem and the exact solution, which minimizes the integral I, cannot be found easily, we would like to find an approximate solution that approximately minimizes the integral I. Similarly, if the problem is specified in terms of differential equations and boundary conditions, and if the correct solution, which satisfies all the equations exactly, cannot be obtained easily, we would like to find an approximate solution that satisfies the boundary conditions exactly but not the governing differential equations. Of the various approximate methods available, the methods using trial functions have been more popular. Depending on the manner in which the problem is specified, two types of approximate methods, namely variational methods (e.g., Rayleigh–Ritz method) and weighted residual methods (e.g., Galerkin method), are available. The finite element can be considered as a variational (Rayleigh–Ritz) method and also as a weighted residual (Galerkin) method. The consideration of the finite element method as a variational approach (which minimizes the integral I approximately) is discussed in this section. The consideration of the finite element method as a weighted residual approach (which satisfies the governing differential equations approximately) is discussed in the next section.

5.3.3 Calculus of Variations

The calculus of variations is concerned with the determination of extrema (maxima and minima) or stationary values of functionals. A functional can be defined as a function of several other functions. The basic problem in variational calculus is to find the function $\phi(x)$ that makes the functional (integral)

$$I = \int_{x_1}^{x_2} F(x, \phi, \phi_x, \phi_{xx}) \cdot \mathrm{d}x \tag{5.12}$$

stationary. Here, x is the independent variable, $\phi_x = \mathrm{d}\phi/\mathrm{d}x$, $\phi_{xx} = \mathrm{d}^2\phi/\mathrm{d}x^2$, and I and F can be called functionals. The functional I usually possesses a clear physical meaning in most of the applications. For example, in structural and solid mechanics, the potential energy (π) plays the role of the functional (π is a function of the displacement vector $\vec{\phi}$, whose components are u, v, and w, which is a function of the coordinates x, y, and z). The integral in Eq. (5.12) is defined in the region or domain $[x_1, x_2]$. Let the value of ϕ be prescribed on the boundaries as $\phi(x_1) = \phi_1$ and $\phi(x_2) = \phi_2$. These are called the boundary conditions of the problem.

One of the procedures that can be used to solve the problem in Eq. (5.12) will be as follows:

(i) Select a series of trial or tentative solutions $\phi(x)$ for the given problem and express the functional I in terms of each of the tentative solutions.
(ii) Compare the values of I given by the different tentative solutions.
(iii) Find the correct solution to the problem as that particular tentative solution which makes the functional I assume an extreme or stationary value.

The mathematical procedure used to select the correct solution from a number of tentative solutions is called the calculus of variations.

Stationary Values of Functionals

Any tentative solution $\bar{\phi}(x)$ in the neighborhood of the exact solution $\phi(x)$ may be expressed as (Figure 5.5)

$$\underset{\text{tentative solution}}{\bar{\phi}(x)} = \underset{\text{exact solution}}{\phi(x)} + \underset{\text{variation of } \phi}{\delta\phi(x)} \tag{5.13}$$

The variation in ϕ (i.e., $\delta\phi$) is defined as an infinitesimal, arbitrary change in ϕ for a fixed value of the variable x (i.e., for $\delta x = 0$). Here, δ is called the variational operator

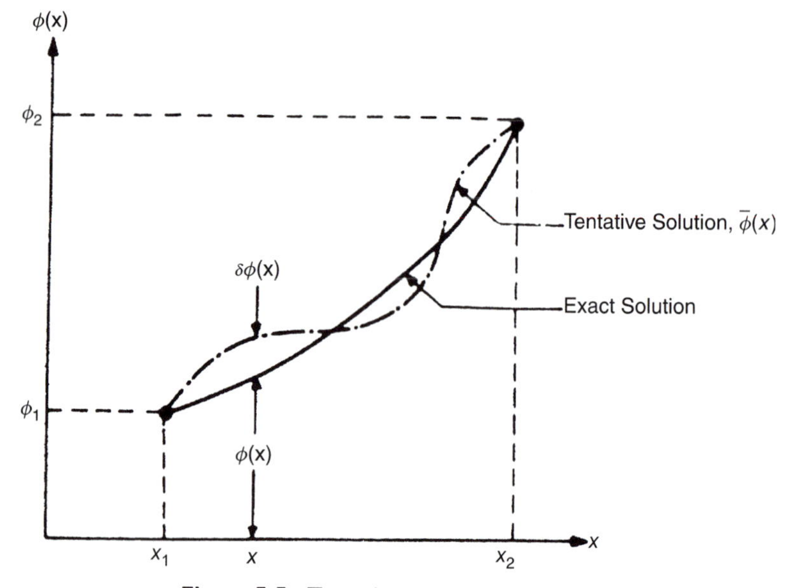

Figure 5.5. Tentative and Exact Solutions.

(similar to the differential operator d). The operation of variation is commutative with respect to both integration and differentiation:

$$\delta\left(\int F \cdot \mathrm{d}x\right) = \int (\delta F)\,\mathrm{d}x$$

and

$$\delta\left(\frac{\mathrm{d}\phi}{\mathrm{d}x}\right) = \frac{\mathrm{d}}{\mathrm{d}x}(\delta\phi)$$

Also, we define the variation of a functional or a function of several variables in a manner similar to the calculus definition of a total differential as

$$\delta F = \frac{\partial F}{\partial \phi}\delta\phi + \frac{\partial F}{\partial \phi_x}\delta\phi_x + \frac{\partial F}{\partial \phi_{xx}}\delta\phi_{xx} + \frac{\partial F}{\partial x}\cancelto{0}{\delta x} \tag{5.14}$$

(since we are finding variation of F for a fixed value of x, $\delta x = 0$).

Now, let us consider the variation in $I(\delta I)$ corresponding to variations in the solution $(\delta\phi)$. If we want the condition for the stationariness of I, we take the necessary condition as the vanishing of first derivative of I (similar to maximization or minimization of simple functions in calculus):

$$\delta I = \int_{x_1}^{x_2} \left(\frac{\partial F}{\partial \phi}\delta\phi + \frac{\partial F}{\partial \phi_x}\delta\phi_x + \frac{\partial F}{\partial \phi_{xx}}\delta\phi_{xx}\right)\mathrm{d}x = \int_{x_1}^{x_2} \delta F\,\mathrm{d}x = 0 \tag{5.15}$$

Integrate the second and third terms by parts to obtain

$$\int_{x_1}^{x_2} \frac{\partial F}{\partial \phi_x} \delta\phi_x \, \mathrm{d}x = \int_{x_1}^{x_2} \frac{\partial F}{\partial \phi_x} \delta \left(\frac{\partial \phi}{\partial x} \right) \mathrm{d}x = \int_{x_1}^{x_2} \frac{\partial F}{\partial \phi_x} \frac{\partial}{\partial x} (\delta\phi) \, \mathrm{d}x$$

$$= \frac{\partial F}{\partial \phi_x} \delta\phi \bigg|_{x_1}^{x_2} - \int_{x_1}^{x_2} \frac{\mathrm{d}}{\mathrm{d}x} \left(\frac{\partial F}{\partial \phi_x} \right) \delta\phi \, \mathrm{d}x \tag{5.16}$$

and

$$\int_{x_1}^{x_2} \frac{\partial F}{\partial \phi_{xx}} \delta\phi_{xx} \, \mathrm{d}x = \int_{x_1}^{x_2} \frac{\partial F}{\partial \phi_{xx}} \frac{\partial}{\partial x} (\delta\phi_x) \, \mathrm{d}x$$

$$= \frac{\partial F}{\partial \phi_{xx}} \delta\phi_x \bigg|_{x_1}^{x_2} - \int_{x_1}^{x_2} \frac{\mathrm{d}}{\mathrm{d}x} \left(\frac{\partial F}{\partial \phi_{xx}} \right) \delta\phi_x \, \mathrm{d}x$$

$$= \frac{\partial F}{\partial \phi_{xx}} \delta\phi_x \bigg|_{x_1}^{x_2} - \frac{\mathrm{d}}{\mathrm{d}x} \left(\frac{\partial F}{\partial \phi_{xx}} \right) \delta\phi \bigg|_{x_1}^{x_2} + \int_{x_1}^{x_2} \frac{\mathrm{d}^2}{\mathrm{d}x^2} \left(\frac{\partial F}{\partial \phi_{xx}} \right) \delta\phi \, \mathrm{d}x \tag{5.17}$$

$$\therefore \delta I = \int_{x_1}^{x_2} \left[\frac{\partial F}{\partial \phi} - \frac{\mathrm{d}}{\mathrm{d}x} \left(\frac{\partial F}{\partial \phi_x} \right) + \frac{\mathrm{d}^2}{\mathrm{d}x^2} \left(\frac{\partial F}{\partial \phi_{xx}} \right) \right] \delta\phi \, \mathrm{d}x$$

$$+ \left[\frac{\partial F}{\partial \phi_x} - \frac{\mathrm{d}}{\mathrm{d}x} \left(\frac{\partial F}{\partial \phi_{xx}} \right) \right] \delta\phi \bigg|_{x_1}^{x_2} + \left[\left(\frac{\partial F}{\partial \phi_{xx}} \right) \delta\phi_x \right] \bigg|_{x_1}^{x_2} = 0 \tag{5.18}$$

Since $\delta\phi$ is arbitrary, each term must vanish individually so that

$$\frac{\partial F}{\partial \phi} - \frac{\mathrm{d}}{\mathrm{d}x} \left(\frac{\partial F}{\partial \phi_x} \right) + \frac{\mathrm{d}^2}{\mathrm{d}x^2} \left(\frac{\partial F}{\partial \phi_{xx}} \right) = 0 \tag{5.19}$$

$$\left[\frac{\partial F}{\partial \phi_x} - \frac{d}{\mathrm{d}x} \left(\frac{\partial F}{\partial \phi_{xx}} \right) \right] \delta\phi \bigg|_{x_1}^{x_2} = 0 \tag{5.20}$$

$$\frac{\partial F}{\partial \phi_{xx}} \delta\phi_x \bigg|_{x_1}^{x_2} = 0 \tag{5.21}$$

Equation (5.19) will be the governing differential equation for the given problem and is called the Euler equation or Euler–Lagrange equation. Equations (5.20) and (5.21) give the boundary conditions. The conditions

$$\left[\frac{\partial F}{\partial \phi_x} - \frac{d}{\mathrm{d}x} \left(\frac{\partial F}{\partial \phi_{xx}} \right) \right] \bigg|_{x_1}^{x_2} = 0 \tag{5.22}$$

and

$$\left. \frac{\partial F}{\partial \phi_{xx}} \right|_{x_1}^{x_2} = 0 \tag{5.23}$$

are called natural boundary conditions (if they are satisfied, they are called free boundary conditions). If the natural boundary conditions are not satisfied, we should have

$$\delta\phi(x_1) = 0, \qquad \delta\phi(x_2) = 0$$

and

$$\delta\phi_x(x_1) = 0, \qquad \delta\phi_x(x_2) = 0 \tag{5.24}$$

in order to satisfy Eqs. (5.20) and (5.21). These are called geometric or essential or forced boundary conditions. Thus, the boundary conditions, Eqs. (5.20) and (5.21), can be satisfied by any combination of free and forced boundary conditions. If the finite element equations are derived on the basis of a variational principle, the natural boundary conditions will be automatically incorporated in the formulation; hence, only the geometric boundary conditions are to be enforced on the solution.

5.3.4 Several Dependent Variables and One Independent Variable

Although Eqs. (5.19)–(5.21) were derived for a single dependent variable, the method can be extended to the case of several dependent variables $\phi_i(x)$ to obtain the set of Euler–Lagrange equations:

$$\frac{\mathrm{d}^2}{\mathrm{d}x^2}\left\{\frac{\partial F}{\partial(\phi_i)_{xx}}\right\} - \frac{\mathrm{d}}{\mathrm{d}x}\left\{\frac{\partial F}{\partial(\phi_i)_x}\right\} + \frac{\partial F}{\partial \phi_i} = 0, \qquad i = 1, 2, \ldots, n \tag{5.25}$$

In general, the integrand F will involve derivatives of higher order than the second order so that

$$I = \int_{x_1}^{x_2} F[x, \phi_i, \phi_i^{(1)}, \phi_i^{(2)}, \ldots, \phi_i^{(j)}]\, \mathrm{d}x, \qquad i = 1, 2, \ldots, n \tag{5.26}$$

where $\phi_i^{(j)}$ indicates the jth derivative of ϕ_i with respect to x. The corresponding Euler–Lagrange equations can be expressed as [5.3]

$$\sum_{j=0}^{n} (-1)^{n-j} \frac{\mathrm{d}^{n-j}}{\mathrm{d}x^{n-j}}\left\{\frac{\partial F}{\partial \phi_i^{(n-j)}}\right\} = 0, \qquad i = 1, 2, \ldots, n \tag{5.27}$$

5.3.5 Several Independent Variables and One Dependent Variable

Consider the following functional with three independent variables:

$$I = \int_V F(x, y, z, \phi, \phi_x, \phi_y, \phi_z)\, \mathrm{d}V \tag{5.28}$$

where $\phi_x = (\partial\phi/\partial x)$, $\phi_y = (\partial\phi/\partial y)$, and $\phi_z = (\partial\phi/\partial z)$. The variation of I due to an arbitrary small change in the solution ϕ can be expressed as

$$\delta I = \int_V \left(\frac{\partial F}{\partial \phi}\delta\phi + \frac{\partial F}{\partial \phi_x}\delta\phi_x + \frac{\partial F}{\partial \phi_y}\delta\phi_y + \frac{\partial F}{\partial \phi_z}\delta\phi_z \right) \mathrm{d}V$$

$$= \int_V \left[\frac{\partial F}{\partial \phi}\delta\phi + \frac{\partial F}{\partial \phi_x}\frac{\partial}{\partial x}(\delta\phi) + \frac{\partial F}{\partial \phi_y}\frac{\partial}{\partial y}(\delta\phi) + \frac{\partial F}{\partial \phi_z}\frac{\partial}{\partial z}(\delta\phi) \right] \mathrm{d}V \tag{5.29}$$

since $\delta\phi_x = \delta(\partial\phi/\partial x) = (\partial/\partial x)(\delta\phi)$, etc. Integrating the second term in Eq. (5.29) by parts and applying the Green–Gauss theorem (given in Appendix A) gives

$$\int_V \frac{\partial F}{\partial \phi_x}\frac{\partial}{\partial x}(\delta\phi)\,\mathrm{d}V = \int_V \frac{\partial}{\partial x}\left(\frac{\partial F}{\partial \phi_x}\delta\phi\right)\mathrm{d}V - \int_V \frac{\partial}{\partial x}\left(\frac{\partial F}{\partial \phi_x}\right)\delta\phi\,\mathrm{d}V$$

$$= \int_S l_x \frac{\partial F}{\partial \phi_x}\delta\phi\,\mathrm{d}S - \int_V \frac{\partial}{\partial x}\left(\frac{\partial F}{\partial \phi_x}\right)\delta\phi\,\mathrm{d}V \tag{5.30}$$

where l_x is the direction cosine of the normal to the outer surface with respect to the x axis. Similarly, the third and fourth terms in Eq. (5.29) can be integrated and δI can be expressed as

$$\delta I = \int_V \left[\frac{\partial F}{\partial \phi} - \frac{\partial}{\partial x}\left(\frac{\partial F}{\partial \phi_x}\right) - \frac{\partial}{\partial y}\left(\frac{\partial F}{\partial \phi_y}\right) - \frac{\partial}{\partial z}\left(\frac{\partial F}{\partial \phi_z}\right) \right] \delta\phi\,\mathrm{d}V$$

$$+ \int_S \left[\frac{\partial F}{\partial \phi_x} l_x + \frac{\partial F}{\partial \phi_y}\cdot l_y + \frac{\partial F}{\partial \phi_z} l_z \right] \delta\phi \cdot \mathrm{d}S \tag{5.31}$$

The functional I assumes a stationary value only if the bracketed terms within the integrals vanish. This requirement gives the governing differential equation and the boundary conditions of the problem. Equation (5.31) is the one that is applicable to the finite element formulation of most field problems according to the variational approach.

5.3.6 Advantages of Variational Formulation

From the previous discussion it is evident that any continuum problem can be solved using a differential equation formulation or a variational formulation. The equivalence of these formulations is apparent from the previous equations, which show that the functional I is extremized or made stationary only when the corresponding Euler–Lagrange equations and boundary conditions are satisfied. These equations are precisely the governing differential equations of the given problem.

The variational formulation of a continuum problem has the following advantages over differential equation formulation:

1. The functional I usually possesses a clear physical meaning in most practical problems.

2. The functional I contains lower order derivatives of the field variable compared to the governing differential equation and hence an approximate solution can be obtained using a larger class of functions.
3. Sometimes the problem may possess a dual variational formulation, in which case the solution can be sought either by minimizing (or maximizing) the functional I or by maximizing (or minimizing) its dual functional. In such cases, one can find an upper and a lower bound to the solution and estimate the order of error in either of the approximate solutions obtained.
4. Using variational formulation, it is possible to prove the existence of solution in some cases.
5. The variational formulation permits us to treat complicated boundary conditions as natural or free boundary conditions. Thus, we need to explicitly impose only the geometric or forced boundary conditions in the finite element method, and the variational statement implicitly imposes the natural boundary conditions.

As stated in Section 1.4, the finite element method is applicable to all types of continuum problems, namely, equilibrium, eigenvalue, and propagation problems. We first present the solution of all three categories of problems using the variational approach and then derive the finite element equations using the variational approach.

5.4 SOLUTION OF EQUILIBRIUM PROBLEMS USING VARIATIONAL (RAYLEIGH–RITZ) METHOD

The differential equation formulation of a general equilibrium problem leads to the following equations:

$$A\phi = b \text{ in } V \tag{5.32}$$

$$B_j\phi = g_j, \qquad j = 1, 2, \ldots, p \text{ on } S \tag{5.33}$$

where ϕ is the unknown field variable (assumed to be a scalar for simplicity), A and B_j are differential operators, b and g_j are functions of the independent variables, p is the number of boundary conditions, V is the domain, and S is the boundary of the domain.

In variational formulation, a functional $I(\phi)$ for which the conditions of stationariness or extremization give the governing differential equation, Eq. (5.32), is identified and the problem is stated as follows:

Minimize $$I(\phi) \text{ in } V \tag{5.34}$$

subject to the essential or forced boundary conditions

$$B_j\phi = g_j, \qquad j = 1, 2, \ldots, p \tag{5.35}$$

The functional I is some integral of ϕ and its derivatives over the domain V and/or the boundary S. If the integrand of the functional is denoted by F so that

$$I = \int_V F(x,\ \phi,\ \phi_x) \cdot \mathrm{d}x \tag{5.36}$$

it can be seen from the previous discussion that F satisfies the Euler–Lagrange equation, Eq. (5.19). This Euler–Lagrange equation in ϕ is the same as the original field equation, Eq. (5.32).

In the most widely used variational method, the Rayleigh–Ritz method, an approximate solution of the following type is assumed for the field variable $\phi(x)$:

$$\underset{\sim}{\phi}(x) = \sum_{i=1}^{n} C_i f_i(x) \tag{5.37}$$

where $f_i(x)$ are known linearly independent functions (also called trial functions) defined over V and S and C_i are unknown parameters to be determined. When $\underset{\sim}{\phi}(x)$ of Eq. (5.37) is substituted, Eq. (5.36) becomes a function of the unknowns C_i. The necessary conditions for the functional to be stationary are given by

$$\frac{\partial I(\underset{\sim}{\phi})}{\partial C_i} = 0, \qquad i = 1, 2, \ldots, n \tag{5.38}$$

which yields n equations in the n unknowns C_i. If I is a quadratic function of ϕ and ϕ_x, Eq. (5.38) gives a set of n linear simultaneous equations.

It can be seen that the accuracy of the assumed solution $\underset{\sim}{\phi}(x)$ depends on the choice of the trial functions $f_i(x)$. The functions $f_j(x)$, $j = 1, 2, \ldots, n$ have to be continuous up to degree $r - 1$, where r denotes the highest degree of differentiation in the functional I [$r = 1$ in Eq. (5.36)] and have to satisfy the essential boundary conditions, Eq. (5.35). In addition, the functions $f_j(x)$ must be part of a complete set for the solution to converge to the correct solution. To assess the convergence of the process, we need to take two or more trial functions, $f_j(x)$. When the method is applied for the stationariness of a given functional I, we can study the convergence by comparing the results obtained with the following sequence of assumed solutions:

$$\begin{aligned}
\underset{\sim}{\phi}^{(1)}(x) &= C_1^{(1)} f_1(x) \\
\underset{\sim}{\phi}^{(2)}(x) &= C_1^{(2)} f_1(x) + C_2^{(2)} f_2(x) \\
&\vdots \\
\underset{\sim}{\phi}^{(i)}(x) &= C_1^{(i)} f_1(x) + C_2^{(i)} f_2(x) + \cdots + C_i^{(i)} f_i(x)
\end{aligned} \tag{5.39}$$

where the ith assumed solution includes all the functions $f_j(x)$ included in the previous solutions. Usually, the functions $f_j(x)$ are taken as polynomials or trigonometric functions. If I is a quadratic functional, the sequence of solutions given by Eq. (5.39) leads to

$$I^{(1)} \geq I^{(2)} \geq \cdots \geq I^{(i)} \tag{5.40}$$

This behavior is called monotonic convergence to the minimum of I.

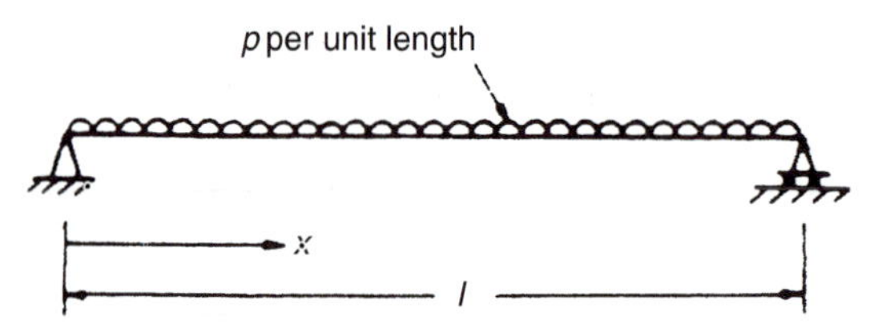

Figure 5.6. A Simply Supported Beam under Uniformly Distributed Load.

Example 5.1 Find the approximate deflection of a simply supported beam under a uniformly distributed load p (Figure 5.6) using the Rayleigh–Ritz method.

Solution Let $w(x)$ denote the deflection of the beam (field variable). The differential equation formulation leads to the following statement of the problem:

Find $w(x)$ that satisfies the governing equation [Eq. (1.24) with constant p and EI]

$$EI\frac{\mathrm{d}^4 w}{\mathrm{d}x^4} - p = 0, \qquad 0 \le x \le l \tag{E$_1$}$$

and the boundary conditions

$$\left.\begin{aligned} &w(x=0) = w(x=l) = 0 \text{ (deflection zero at ends)} \\ &EI\frac{\mathrm{d}^2 w}{\mathrm{d}x^2}(x=0) = EI\frac{\mathrm{d}^2 w}{\mathrm{d}x^2}(x=l) = 0 \text{ (bending moment zero at ends)} \end{aligned}\right\} \tag{E$_2$}$$

where E is the Young's modulus, and I is the moment of inertia of the beam. The variational formulation gives the following statement of the problem:

Find $w(x)$ that minimizes the integral*

$$A = \int_{x=0}^{l} F \cdot \mathrm{d}x = \frac{1}{2}\int_{0}^{l}\left[EI\left(\frac{\mathrm{d}^2 w}{\mathrm{d}x^2}\right)^2 - 2p \cdot w\right]\mathrm{d}x \tag{E$_3$}$$

and satisfies the boundary conditions stated in Eq. (E$_2$).

We shall approximate the solution $w(x)$ by orthogonal functions of the sinusoidal type. For example,

$$\underset{\sim}{w}(x) = C_1 \sin\left(\frac{\pi x}{l}\right) + C_2 \sin\left(\frac{3\pi x}{l}\right) \equiv C_1 f_1(x) + C_2 f_2(x) \tag{E$_4$}$$

* It can be verified that the Euler–Lagrange equation corresponding to the functional A is the same as Eq. (E$_1$). The functional A represents the potential energy of the beam.

where the functions $f_1(x) = \sin(\pi x/l)$ and $f_2(x) = \sin(3\pi x/l)$ satisfy the boundary conditions, Eq. (E_2). The substitution of Eq. (E_4) into Eq. (E_3) gives

$$A = \int_o^l \left[\frac{EI}{2} \left\{ C_1 \left(\frac{\pi}{l}\right)^2 \sin\left(\frac{\pi x}{l}\right) + C_2 \left(\frac{3\pi}{l}\right)^2 \sin\left(\frac{3\pi x}{l}\right) \right\}^2 - p \left\{ C_1 \sin\left(\frac{\pi x}{l}\right) + C_2 \sin\left(\frac{3\pi x}{l}\right) \right\} \right] dx$$

$$= \int_o^l \left[\frac{EI}{2} \left\{ C_1^2 \left(\frac{\pi}{l}\right)^4 \sin^2\left(\frac{\pi x}{l}\right) + C_2^2 \left(\frac{3\pi}{l}\right)^4 \sin^2\left(\frac{3\pi x}{l}\right) + 2C_1C_2 \left(\frac{\pi}{l}\right)^2 \left(\frac{3\pi}{l}\right)^2 \cdot \sin\left(\frac{\pi x}{l}\right) \sin\left(\frac{3\pi x}{l}\right) \right\} - p \left\{ C_1 \sin\left(\frac{\pi x}{l}\right) + C_2 \sin\left(\frac{3\pi x}{l}\right) \right\} \right] dx \quad (E_5)$$

By using the relations

$$\int_o^l \sin\left(\frac{m\pi x}{l}\right) \sin\left(\frac{n\pi x}{l}\right) dx = \begin{cases} 0 & \text{if } m \neq n \\ l/2 & \text{if } m = n \end{cases} \quad (E_6)$$

and

$$\int_o^l \sin\left(\frac{m\pi x}{l}\right) dx = \frac{2l}{m\pi} \text{ if } m \text{ is an odd integer,} \quad (E_7)$$

Eq. (E_5) can be written as

$$A = \frac{EI}{2} \left\{ C_1^2 \left(\frac{\pi}{l}\right)^4 \frac{l}{2} + C_2^2 \left(\frac{3\pi}{l}\right)^4 \frac{l}{2} \right\} - p \left\{ C_1 \frac{2l}{\pi} + C_2 \frac{2l}{3\pi} \right\} \quad (E_8)$$

where C_1 and C_2 are independent constants. For the minimum of A, we have

$$\left. \begin{aligned} \frac{\partial A}{\partial C_1} &= \frac{EI}{2} \left\{ 2C_1 \left(\frac{\pi}{l}\right)^4 \frac{l}{2} \right\} - p\frac{2l}{\pi} = 0 \\ \frac{\partial A}{\partial C_2} &= \frac{EI}{2} \left\{ 2C_2 \left(\frac{3\pi}{l}\right)^4 \frac{l}{2} \right\} - p\frac{2l}{3\pi} = 0 \end{aligned} \right\} \quad (E_9)$$

The solution of Eqs. (E_9) gives

$$C_1 = \frac{4pl^4}{\pi^5 EI} \quad \text{and} \quad C_2 = \frac{4pl^4}{243\pi^5 EI} \quad (E_{10})$$

Thus, the deflection of the beam is given by

$$\underset{\sim}{w}(x) = \frac{4pl^4}{\pi^5 EI}\left[\sin\left(\frac{\pi x}{l}\right) + \frac{1}{243}\sin\left(\frac{3\pi x}{l}\right)\right] \tag{E$_{11}$}$$

The deflection of the beam at the middle point is given by

$$\underset{\sim}{w}(x = l/2) = \frac{968}{243\pi^5}\frac{pl^4}{EI} = \frac{1}{76.5}\frac{pl^4}{EI} \tag{E$_{12}$}$$

which compares well with the exact solution

$$w(x = l/2) = \frac{5}{384}\frac{pl^4}{EI} = \frac{1}{76.8}\frac{pl^4}{EI} \tag{E$_{13}$}$$

We can find a more accurate solution by including more terms in Eq. (E_4). The n term solution converges to the exact solution as $n \to \infty$.

5.5 SOLUTION OF EIGENVALUE PROBLEMS USING VARIATIONAL (RAYLEIGH–RITZ) METHOD

An eigenvalue problem, according to the differential equation formulation, can be stated as

$$A\phi = \lambda B\phi \text{ in } V \tag{5.41}$$

subject to the boundary conditions

$$E_j\phi = \lambda F_j\phi, \qquad j = 1, 2, \ldots, p \text{ on } S \tag{5.42}$$

where A, B, E_j, and F_j are differential operators, and λ is the eigenvalue. Although the methods discussed for the solution of equilibrium problems can be extended to the case in which the boundary conditions are of the type shown in Eq. (5.42), only the special case in which $F_j = 0$ is considered here so that Eq. (5.42) becomes

$$E_j\phi = 0, \qquad j = 1, 2, \ldots, p \text{ on } S \tag{5.43}$$

It is assumed that the solution of the problem defined by Eqs. (5.41) and (5.43) gives real eigenvalues λ.

In the variational formulation corresponding to Eq. (5.41), a functional $I(\phi, \lambda)$ to be made stationary is identified as

$$I(\phi, \lambda) = I_A(\phi) - I_B(\phi) \tag{5.44}$$

where the subscripts of the functionals I_A and I_B indicate that they are derived from A and B, respectively. It can be shown [5.4] that the stationary value of a function λ_R, called the Rayleigh quotient, defined by

$$\lambda_R = \frac{I_A(\underset{\sim}{\phi})}{I_B(\underset{\sim}{\phi})} \tag{5.45}$$

gives the eigenvalue, and the corresponding $\underset{\sim}{\phi}$ gives the eigenfunction. The trial solution $\underset{\sim}{\phi}(x)$ is chosen as

$$\underset{\sim}{\phi}(x) = \sum_{i=1}^{n} C_i f_i(x) \tag{5.46}$$

where $f_i(x)$ satisfy only the essential boundary conditions. In this case, the conditions for the stationariness of λ_R can be expressed as

$$\frac{\partial \lambda_R}{\partial C_i} = \frac{\partial}{\partial C_i} \left\{ \frac{I_A\left(\sum_{i=1}^{n} C_i f_i\right)}{I_B\left(\sum_{i=1}^{n} C_i f_i\right)} \right\} = \frac{1}{I_B^2}\left(I_B \frac{\partial I_A}{\partial C_i} - I_A \frac{\partial I_B}{\partial C_i}\right) = 0, \qquad i = 1, 2, \ldots, n \tag{5.47}$$

which shows that

$$\frac{\partial I_A}{\partial C_i} = \lambda \frac{\partial I_B}{\partial C_i}, \qquad i = 1, 2, \ldots, n \tag{5.48}$$

Equation (5.48) yields a set of n simultaneous linear equations. Moreover, if the functions $f_i(x)$ also satisfy the natural (free) boundary conditions, then the Rayleigh–Ritz method gives the same equations as the Galerkin method discussed in Section 5.9. The example in Section 1.6.4 illustrates the Rayleigh–Ritz method of solution of an eigenvalue problem.

5.6 SOLUTION OF PROPAGATION PROBLEMS USING VARIATIONAL (RAYLEIGH–RITZ) METHOD

The differential equation formulation of a general propagation problem leads to the following equations:

Field equation: $A\phi = e$ in V for $t > t_0$ (5.49)

Boundary conditions: $B_i\phi = g_i$, $i = 1, 2, \ldots, k$ on S for $t \geq t_0$ (5.50)

Initial conditions: $E_j\phi = h_j$, $j = 1, 2, \ldots, l$ in V for $t = t_0$ (5.51)

where A, B_i, and E_j are differential operators; e, g_i, and h_j are functions of the independent variable; and t_0 is the initial time. In variational methods, the functionals I associated with specific types of propagation problems have been developed by several authors, such as Rosen, Biot, and Gurtin [5.5].

In the case of propagation problems, the trial solution $\underset{\sim}{\phi}(x, t)$ is taken as

$$\underset{\sim}{\phi}(x, t) = \sum_{i=1}^{n} C_i(t) f_i(x) \tag{5.52}$$

where C_i is now a function of time t. Alternatively, C_i can be taken as a constant and f_i as a function of both x and t as

$$\underset{\sim}{\phi}(x, t) = \sum_{i=1}^{n} C_i f_i(x, t) \tag{5.53}$$

As in the case of equilibrium and eigenvalue problems, the functions f_i have to satisfy the forced boundary conditions.

The solution given by Eq. (5.52) or Eq. (5.53) is substituted into the functional, and the necessary conditions for the stationariness are applied to derive a set of equations in the unknowns $C_i(t)$ or C_i. These equations can then be solved to find the functions $C_i(t)$ or the constants C_i.

5.7 EQUIVALENCE OF FINITE ELEMENT AND VARIATIONAL (RAYLEIGH–RITZ) METHODS

If we compare the basic steps of the finite element method described in Section 1.5 with the Rayleigh–Ritz method discussed in this section, we find that both are essentially equivalent. In both methods, a set of trial functions are used for obtaining an approximate solution. Both methods seek a linear combination of trial functions that extremizes (or makes stationary) a given functional. The main difference between the methods is that in the finite element method, the assumed trial functions are not defined over the entire solution domain and they need not satisfy any boundary conditions. Since the trial functions have to be defined over the entire solution domain, the Rayleigh–Ritz method can be used only for domains of simple geometric shape. Similar geometric restrictions also exist in the finite element method, but for the elements. Since elements with simple geometric shape can be assembled to approximate even complex domains, the finite element method proves to be a more versatile technique than the Rayleigh–Ritz method. The only limitation of the finite element method is that the trial functions have to satisfy the convergence (continuity and completeness) conditions stated in Section 3.6.

5.8 DERIVATION OF FINITE ELEMENT EQUATIONS USING VARIATIONAL (RAYLEIGH–RITZ) APPROACH

Let the general problem (either physical or purely mathematical), when formulated according to variational approach, require the extremization of a functional I over a domain V. Let the functional I be defined as

$$I = \iiint\limits_{V} F\left[\vec{\phi}, \frac{\partial}{\partial x}(\vec{\phi}), \cdots\right] \mathrm{d}V + \iint\limits_{S} g\left[\vec{\phi}, \frac{\partial}{\partial x}(\vec{\phi}), \cdots\right] \mathrm{d}S \tag{5.54}$$

where, in general, the field variable or the unknown function $\vec{\phi}$ is a vector. The finite element procedure for solving this problem can be stated by the following steps:

Step 1: The solution domain V is divided into E smaller parts called subdomains that we call finite elements.

Step 2: The field variable $\vec{\phi}$, which we are trying to find, is assumed to vary in each element in a suitable manner as

$$\vec{\phi} = [N]\vec{\Phi}^{(e)} \tag{5.55}$$

where $[N]$ is a matrix of shape functions ($[N]$ will be a function of the coordinates), and $\vec{\Phi}^{(e)}$ is a vector representing the nodal values of the function $\vec{\phi}$ associated with the element.

Step 3: To derive the elemental equations, we use the conditions of extremization of the functional I with respect to the nodal unknowns $\underset{\sim}{\vec{\Phi}}$ associated with the whole domain. These are

$$\frac{\partial I}{\partial \underset{\sim}{\vec{\Phi}}} = \begin{Bmatrix} \partial I/\partial \Phi_1 \\ \partial I/\partial \Phi_2 \\ \vdots \\ \partial I/\partial \Phi_M \end{Bmatrix} = 0 \tag{5.56}$$

where M denotes the total number of nodal unknowns in the problem. If the functional I can be expressed as a summation of elemental contributions as

$$I = \sum_{e=1}^{E} I^{(e)} \tag{5.57}$$

where e indicates the element number, then Eq. (5.56) can be expressed as

$$\frac{\partial I}{\partial \Phi_i} = \sum_{e=1}^{E} \frac{\partial I^{(e)}}{\partial \Phi_i} = 0, \qquad i = 1, 2, \ldots, M \tag{5.58}$$

In the special case in which I is a quadratic functional of $\vec{\phi}$ and its derivatives, we can obtain the element equations as

$$\frac{\partial I^{(e)}}{\partial \vec{\Phi}^{(e)}} = [K^{(e)}]\vec{\Phi}^{(e)} - \vec{P}^{(e)} \tag{5.59}$$

where $[K^{(e)}]$ and $\vec{P}^{(e)}$ are the element characteristic matrix and characteristic vector (or vector of nodal actions), respectively.

Step 4: To obtain the overall equations of the system, we rewrite Eq. (5.56) as

$$\frac{\partial I}{\partial \underset{\sim}{\vec{\Phi}}} = [\underset{\sim}{K}]\underset{\sim}{\vec{\Phi}} - \underset{\sim}{\vec{P}} = \vec{0} \tag{5.60}$$

where

$$[\underset{\sim}{K}] = \sum_{e=1}^{E} [K^{(e)}] \tag{5.61}$$

$$\underset{\sim}{\vec{P}} = \sum_{e=1}^{E} \vec{P}^{(e)} \tag{5.62}$$

and the summation sign indicates assembly over all finite elements. The assembly procedure is described in Chapter 6.

Step 5: The linear simultaneous equations (5.60) can be solved, after applying the boundary conditions, to find the nodal unknowns $\underset{\sim}{\vec{\Phi}}$.

Step 6: The function (or field variable) $\vec{\phi}$ in each element is found by using Eq. (5.55). If necessary, the derivatives of $\vec{\phi}$ are found by differentiating the function $\vec{\phi}$ in a suitable manner.

Convergence Requirements

As stated in Section 3.6, the following conditions have to be satisfied in order to achieve convergence of results as the subdivision is made finer:

(i) As the element size decreases, the functions F and g of Eq. (5.54) must tend to be single valued and well behaved. Thus, the shape functions $[N]$ and nodal unknowns $\vec{\Phi}^{(e)}$ chosen must be able to represent any constant value of $\vec{\phi}$ or its derivatives present in the functional I in the limit as the element size decreases to zero.

(ii) In order to make the summation $I = \Sigma_{e=1}^{E} I^{(e)}$ valid, we must ensure that terms such as F and g remain finite at interelement boundaries. This can be achieved if the highest derivatives of the field variable $\vec{\phi}$ that occur in F and g are finite. Thus, the element shape functions $[N]$ are to be selected such that at element interface, $\vec{\phi}$ and its derivatives, of one order less than that occurring in F and g, are continuous.

The step-by-step procedure outlined previously can be used to solve any problem provided that the variational principle valid over the domain is known to us. This is illustrated by the following example.

Example 5.2 Solve the differential equation

$$\frac{d^2\phi}{dx^2} + \phi + x = 0, \qquad 0 \le x \le 1 \tag{E$_1$}$$

subject to the boundary conditions $\phi(0) = \phi(1) = 0$ using the variational approach.

Solution The functional I corresponding to Eq. (E_1) is given by*

$$I = \frac{1}{2} \int_{0}^{1} \left[-\left(\frac{d\phi}{dx}\right)^2 + \phi^2 + 2\phi x \right] dx \tag{E$_2$}$$

---

* The Euler–Lagrange equation corresponding to Eq. (E_2) can be verified to be same as Eq. (E_1).

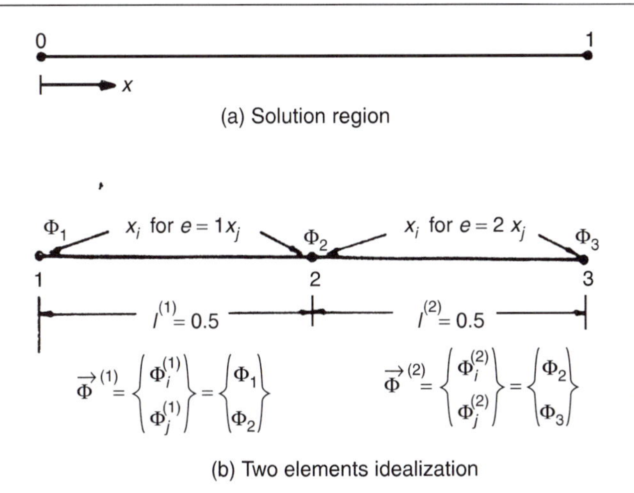

(a) Solution region

(b) Two elements idealization

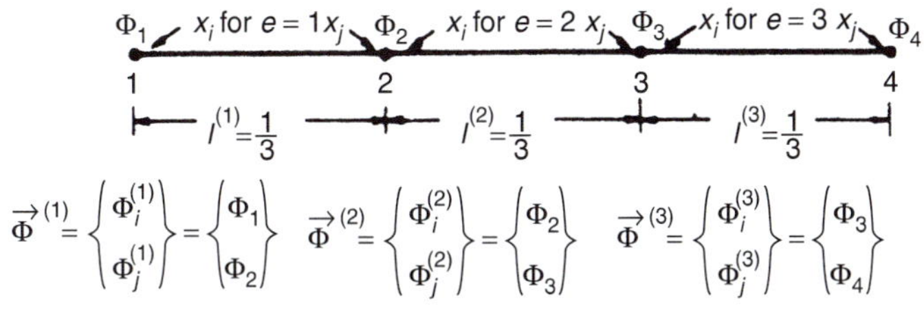

(c) Three element idealization

Figure 5.7. Discretization for Example 5.2.

Step 1: We discretize the domain ($x = 0$ to 1) using three nodes and two elements of equal length as shown in Figure 5.7.

Step 2: We assume a linear interpolation model within an element e as

$$\phi(x) = [N(x)]\vec{\Phi}^{(e)} = N_i(x) \cdot \Phi_i^{(e)} + N_j(x) \cdot \Phi_j^{(e)} \tag{E$_3$}$$

where, from Eq. (3.26),

$$N_i(x) = (x_j - x)/l^{(e)}, \tag{E$_4$}$$

$$N_j(x) = (x - x_i)/l^{(e)}, \tag{E$_5$}$$

$\vec{\Phi}^{(e)} = \begin{Bmatrix} \Phi_i^{(e)} \\ \Phi_j^{(e)} \end{Bmatrix}$ is the vector of nodal degrees of freedom; $l^{(e)}$ is the length; $\Phi_i^{(e)}$ and $\Phi_j^{(e)}$ are the values of $\phi(x)$ at nodes i ($x = x_i$) and j ($x = x_j$), respectively; and i and j are the first and the second (global) nodes of the element e.

Step 3: We express the functional I as a sum of E elemental quantities $I^{(e)}$ as

$$I = \sum_{e=1}^{E} I^{(e)} \tag{E6}$$

where

$$I^{(e)} = \frac{1}{2} \int_{x_i}^{x_j} \left[-\left(\frac{\mathrm{d}\phi}{\mathrm{d}x} \right)^2 + \phi^2 + 2x\phi \right] \cdot \mathrm{d}x \tag{E7}$$

By substituting Eq. (E3) into Eq. (E7), we obtain

$$I^{(e)} = \frac{1}{2} \int_{x_i}^{x_j} \left\{ -\vec{\Phi}^{(e)T} \left[\frac{\mathrm{d}N}{\mathrm{d}x} \right]^T \left[\frac{\mathrm{d}N}{\mathrm{d}x} \right] \vec{\Phi}^{(e)} + \vec{\Phi}^{(e)T} [N]^T [N] \vec{\Phi}^{(e)} + 2x[N]\vec{\Phi}^{(e)} \right\} \cdot \mathrm{d}x \tag{E8}$$

For the stationariness of I, we use the necessary conditions

$$\frac{\partial I}{\partial \Phi_i} = \sum_{e=1}^{E} \frac{\partial I^{(e)}}{\partial \Phi_i} = 0, \qquad i = 1, 2, \ldots, M \tag{E9}$$

where E is the number of elements, and M is the number of nodal degrees of freedom. Equation (E9) can also be expressed as

$$\sum_{e=1}^{E} \frac{\partial I^{(e)}}{\partial \vec{\Phi}^{(e)}} = \sum_{e=1}^{E} \int_{x_i}^{x_j} \left\{ -\left[\frac{\mathrm{d}N}{\mathrm{d}x} \right]^T \left[\frac{\mathrm{d}N}{\mathrm{d}x} \right] \vec{\Phi}^{(e)} + [N]^T [N] \vec{\Phi}^{(e)} + x[N]^T \right\} \mathrm{d}x = \vec{0}$$

or

$$\sum_{e=1}^{E} [K^{(e)}] \vec{\Phi}^{(e)} = \sum_{e=1}^{E} \vec{P}^{(e)} \tag{E10}$$

where

$$[K^{(e)}] = \text{element characteristic matrix} = \int_{x_i}^{x_j} \left\{ \left[\frac{\mathrm{d}N}{\mathrm{d}x} \right]^T \left[\frac{\mathrm{d}N}{\mathrm{d}x} \right] - [N]^T [N] \right\} \mathrm{d}x \tag{E11}$$

and

$$\vec{P}^{(e)} = \text{element characteristic vector} = \int_{x_i}^{x_j} x[N]^T \, \mathrm{d}x \tag{E12}$$

By substituting $[N(x)] = [N_i(x)\ N_j(x)] \equiv \left[\frac{x_j - x}{l^{(e)}}\ \frac{x - x_i}{l^{(e)}}\right]$ into Eqs. (E_{11}) and (E_{12}), we obtain

$$[K^{(e)}] = \int_{x_i}^{x_j} \left[\begin{Bmatrix} -\frac{1}{l^{(e)}} \\ \frac{1}{l^{(e)}} \end{Bmatrix} \begin{Bmatrix} -\frac{1}{l^{(e)}} & \frac{1}{l^{(e)}} \end{Bmatrix} - \begin{Bmatrix} \frac{x_j - x}{l^{(e)}} \\ \frac{x - x_i}{l^{(e)}} \end{Bmatrix} \begin{Bmatrix} \frac{x_j - x}{l^{(e)}} & \frac{x - x_i}{l^{(e)}} \end{Bmatrix} \right] \mathrm{d}x$$

$$= \frac{1}{l^{(e)}} \begin{bmatrix} 1 & -1 \\ -1 & 1 \end{bmatrix} - \frac{l^{(e)}}{6} \begin{bmatrix} 2 & 1 \\ 1 & 2 \end{bmatrix} \qquad (E_{13})$$

and

$$\vec{P}^{(e)} = \int_{x_i}^{x_j} x \begin{Bmatrix} \frac{x_j - x}{l^{(e)}} \\ \frac{x - x_i}{l^{(e)}} \end{Bmatrix} \mathrm{d}x = \frac{1}{6} \begin{Bmatrix} (x_j^2 + x_i x_j - 2x_i^2) \\ (2x_j^2 - x_i x_j - x_i^2) \end{Bmatrix} \qquad (E_{14})$$

We shall compute the results for the cases of two and three elements.

For $E = 2$

Assuming the elements to be of equal length, we have $l^{(1)} = l^{(2)} = 0.5$; $x_i = 0.0$ and $x_j = 0.5$ for $e = 1$; $x_i = 0.5$ and $x_j = 1.0$ for $e = 2$. Equations (E_{13}) and (E_{14}) yield

$$[K^{(1)}] = [K^{(2)}] = \frac{1}{0.5} \begin{bmatrix} 1 & -1 \\ -1 & 1 \end{bmatrix} - \frac{0.5}{6} \begin{bmatrix} 2 & 1 \\ 1 & 2 \end{bmatrix} = \frac{1}{12} \begin{bmatrix} 22 & -25 \\ -25 & 22 \end{bmatrix}$$

$$\vec{P}^{(1)} = \frac{1}{24} \begin{Bmatrix} 1 \\ 2 \end{Bmatrix}$$

$$\vec{P}^{(2)} = \frac{1}{24} \begin{Bmatrix} 4 \\ 5 \end{Bmatrix}$$

For $E = 3$

Assuming the elements to be of equal length, we have $l^{(1)} = l^{(2)} = l^{(3)} = 1/3$; $x_i = 0.0$ and $x_j = 1/3$ for $e = 1$; $x_i = 1/3$ and $x_j = 2/3$ for $e = 2$; $x_i = 2/3$ and $x_j = 1$ for $e = 3$. Equations (E_{13}) and (E_{14}) give

$$[K^{(1)}] = [K^{(2)}] = [K^{(3)}] = 3 \begin{bmatrix} 1 & -1 \\ -1 & 1 \end{bmatrix} - \frac{1}{18} \begin{bmatrix} 2 & 1 \\ 1 & 2 \end{bmatrix} = \frac{1}{18} \begin{bmatrix} 52 & -55 \\ -55 & 52 \end{bmatrix}$$

$$\vec{P}^{(1)} = \frac{1}{54} \begin{Bmatrix} 1 \\ 2 \end{Bmatrix}$$

$$\vec{P}^{(2)} = \frac{1}{54} \begin{Bmatrix} 4 \\ 5 \end{Bmatrix}$$

$$\vec{P}^{(3)} = \frac{1}{54} \begin{Bmatrix} 7 \\ 8 \end{Bmatrix}$$

Step 4: We assemble the element characteristic matrices and vectors and obtain the overall equations as

$$[\underset{\sim}{K}]\underset{\sim}{\vec{\Phi}} = \underset{\sim}{\vec{P}} \tag{E$_{15}$}$$

where

$$[\underset{\sim}{K}] = \frac{1}{12}\begin{bmatrix} 22 & -25 & 0 \\ -25 & (22+22) & -25 \\ 0 & -25 & 22 \end{bmatrix} = \frac{1}{12}\begin{bmatrix} 22 & -25 & 0 \\ -25 & 44 & -25 \\ 0 & -25 & 22 \end{bmatrix}$$

$$\underset{\sim}{\vec{\Phi}} = \begin{Bmatrix} \Phi_1 \\ \Phi_2 \\ \Phi_3 \end{Bmatrix}$$

and

$$\underset{\sim}{\vec{P}} = \frac{1}{24}\begin{Bmatrix} 1 \\ 2+4 \\ 5 \end{Bmatrix} = \frac{1}{24}\begin{Bmatrix} 1 \\ 6 \\ 5 \end{Bmatrix} \quad \text{for } E = 2, \text{ and}$$

$$[\underset{\sim}{K}] = \frac{1}{18}\begin{bmatrix} 52 & -55 & 0 & 0 \\ -55 & (52+52) & -55 & 0 \\ 0 & -55 & (52+52) & -55 \\ 0 & 0 & 55 & 52 \end{bmatrix}$$

$$= \frac{1}{18}\begin{bmatrix} 52 & -55 & 0 & 0 \\ -55 & 104 & -55 & 0 \\ 0 & -55 & 104 & -55 \\ 0 & 0 & -55 & 52 \end{bmatrix}$$

$$\underset{\sim}{\vec{\Phi}} = \begin{Bmatrix} \Phi_1 \\ \Phi_2 \\ \Phi_3 \\ \Phi_4 \end{Bmatrix}$$

and

$$\underset{\sim}{\vec{P}} = \frac{1}{54}\begin{Bmatrix} 1 \\ 2+4 \\ 5+7 \\ 8 \end{Bmatrix} = \frac{1}{54}\begin{Bmatrix} 1 \\ 6 \\ 12 \\ 8 \end{Bmatrix} \quad \text{for } E = 3.$$

Step 5: We can solve the system equations (E$_{15}$) after incorporating the boundary conditions.

For $E = 2$, the boundary conditions are $\Phi_1 = \Phi_3 = 0$.

The incorporation of these boundary conditions in Eq. (E_{15}) leads to (after deleting the rows and columns corresponding to Φ_1 and Φ_3 in $[\underset{\sim}{K}]$, $\underset{\sim}{\vec{\Phi}}$, and $\underset{\sim}{\vec{P}}$)

$$\frac{1}{12}(44) \quad \Phi_2 = \frac{1}{24}(6)$$

or

$$\Phi_2 = \frac{3}{44} = 0.06817 \tag{E$_{16}$}$$

For $E = 3$, the boundary conditions are $\Phi_1 = \Phi_4 = 0$.

After incorporating these boundary conditions, Eq. (E_{15}) reduces to

$$\frac{1}{18}\begin{bmatrix} 104 & -55 \\ -55 & 104 \end{bmatrix}\begin{Bmatrix} \Phi_2 \\ \Phi_3 \end{Bmatrix} = \frac{1}{54}\begin{Bmatrix} 6 \\ 12 \end{Bmatrix}$$

the solution of which is given by

$$\begin{Bmatrix} \Phi_2 \\ \Phi_3 \end{Bmatrix} = \begin{Bmatrix} 0.05493 \\ 0.06751 \end{Bmatrix} \tag{E$_{17}$}$$

There is no necessity of Step 6 in this example.

The exact solution of this problem is

$$\phi(x) = \left(\frac{\sin x}{\sin 1} - x\right)$$

which gives

$$\phi(0.5) = 0.0697 \tag{E$_{18}$}$$

and

$$\phi\left(\frac{1}{3}\right) = 0.0536, \qquad \phi\left(\frac{2}{3}\right) = 0.0649 \tag{E$_{19}$}$$

Thus, the accuracy of the two- and three-element discretizations can be seen by comparing Eqs. (E_{16}) and (E_{18}), and Eqs. (E_{17}) and (E_{19}), respectively.

5.9 WEIGHTED RESIDUAL APPROACH

The weighted residual method is a technique that can be used to obtain approximate solutions to linear and nonlinear differential equations. If we use this method the finite element equations can be derived directly from the governing differential equations of the problem without any need of knowing the "functional." We first consider the solution of equilibrium, eigenvalue, and propagation problems using the weighted residual method and then derive the finite element equations using the weighted residual approach.

5.9.1 Solution of Equilibrium Problems Using the Weighted Residual Method

A general equilibrium problem has been stated in Section 5.4 as

$$A\phi = b \text{ in } V \tag{5.63}$$

$$B_j\phi = g_j, \qquad j = 1, 2, \ldots, p \text{ on } S \tag{5.64}$$

Equation (5.63) can be expressed in a more general form as

$$F(\phi) = G(\phi) \text{ in } V \tag{5.65}$$

where F and G are functions of the field variable ϕ. In fact, Eqs. (5.63), (5.64); (5.41), (5.42); and (5.49)–(5.51) can be seen to be special cases of Eq. (5.65). In the weighted residual method, the field variable is approximated as

$$\underset{\sim}{\phi}(x) = \sum_{i=1}^{n} C_i f_i(x) \tag{5.66}$$

where C_i are constants and $f_i(x)$ are linearly independent functions chosen such that all boundary conditions are satisfied. A quantity R, known as the residual or error, is defined as

$$R = G(\underset{\sim}{\phi}) - F(\underset{\sim}{\phi}) \tag{5.67}$$

which is required to satisfy certain conditions that make this error R a minimum or maintain it small in some specified sense. More generally, a weighted function of the residual, $wf(R)$, where w is a weight or weighting function and $f(R)$ is a function of R, is taken to satisfy the smallness criterion. The function $f(R)$ is chosen so that $f(R) = 0$ when $R = 0$—that is, when $\underset{\sim}{\phi}(x)$ equals the exact solution $\phi(x)$. As stated, the trial function $\underset{\sim}{\phi}$ is chosen so as to satisfy the boundary conditions but not the governing equation in the domain V, and the smallness criterion is taken as

$$\int_V wf(R) \cdot dV = 0 \tag{5.68}$$

where the integration is taken over the domain of the problem. In the following subsections, four different methods, based on a weighted residual criterion, are given.

5.9.2 Collocation Method

In this method, the residual R is set equal to zero at n points in the domain V, thereby implying that the parameters C_i are to be selected such that the trial function $\underset{\sim}{\phi}(x)$ represents $\phi(x)$ at these n points exactly. This procedure yields n simultaneous algebraic equations in the unknowns C_i $(i = 1, 2, \ldots, n)$. The collocation points x_j at which $\underset{\sim}{\phi}(x_j) = \phi(x_j)$, $j = 1, 2, \ldots, n$ are usually chosen to cover the domain V more or less uniformly in some simple pattern. This approach is equivalent to taking, in Eq. (5.68),

$$f(R) = R \quad \text{and} \quad w = \delta(x_j - x) \tag{5.69}$$

where δ indicates the Dirac delta function, x_j denotes the position of the jth point, and x gives the position of a general point in the domain V. Thus, $w = 1$ at point $x = x_j$ and zero elsewhere in the domain V ($j = 1, 2, \ldots, n$).

5.9.3 Subdomain Collocation Method

Here, the domain V is first subdivided into n subdomains V_i, $i = 1, 2, \ldots, n$, and the integral of the residual over each subdomain is then required to be zero:

$$\int_{V_i} R \, \mathrm{d}V_i = 0, \qquad i = 1, 2, \ldots, n \tag{5.70}$$

This yields n simultaneous algebraic equations for the n unknowns C_i, $i = 1, 2, \ldots, n$. It can be seen that the method is equivalent to choosing

$$f(R) = R \quad \text{and} \quad w = \begin{cases} 1 & \text{if } x \text{ is in } V_i \\ 0 & \text{if } x \text{ is not in } V_i, \; i = 1, 2, \ldots, n \end{cases} \tag{5.71}$$

5.9.4 Galerkin Method

Here, the weights w_i are chosen to be the known functions $f_i(x)$ of the trial solution and the following n integrals of the weighted residual are set equal to zero:

$$\int_V f_i R \, \mathrm{d}V = 0, \qquad i = 1, 2, \ldots, n \tag{5.72}$$

Equations (5.72) represent n simultaneous equations in the n unknowns, $C_1, C_2, \ldots, C_n$. This method generally gives the best approximate solution.

5.9.5 Least Squares Method

In this method, the integral of the weighted square of the residual over the domain is required to be a minimum; that is,

$$\int_V w R^2 \, \mathrm{d}V = \text{minimum} \tag{5.73}$$

By using Eqs. (5.66) and (5.63), Eq. (5.73) can be written as

$$\int_V w \left[b - A \left(\sum_{i=1}^{n} C_i f_i(x) \right) \right]^2 \mathrm{d}V = \text{minimum} \tag{5.74}$$

where the unknowns in the integral are only C_i. The necessary conditions for minimizing the integral can be expressed as

$$\frac{\partial}{\partial C_i}\left[\int_V w\left\{b - A\left(\sum_{i=1}^{n} C_i f_i(x)\right)\right\}^2 \mathrm{d}V\right] = 0, \qquad i = 1, 2, \ldots, n$$

or

$$\int_V wA(f_i(x))\left[b - A\left(\sum_{i=1}^{n} C_i f_i(x)\right)\right] \mathrm{d}V = 0, \qquad i = 1, 2, \ldots, n \tag{5.75}$$

The weighting function w is usually taken as unity in this method. Equation (5.75) leads to n simultaneous linear algebraic equations in terms of the unknowns C_1, C_2, ... , C_n.

Example 5.3 Find the approximate deflection of a simply supported beam under a uniformly distributed load p (Figure 5.6) using the Galerkin method.

Solution The differential equation governing the deflection of the beam (w) is given by (see Example 5.1)

$$EI\frac{\mathrm{d}^4 w}{\mathrm{d}x^4} - p = 0, \qquad 0 \le x \le l \tag{E$_1$}$$

The boundary conditions to be satisfied are

$$\left.\begin{aligned} &w(x=0) = w(x=l) = 0 && \text{(deflection zero at ends)}\\ &EI\frac{\mathrm{d}^2 w}{\mathrm{d}x^2}(x=0) = EI\frac{\mathrm{d}^2 w}{\mathrm{d}x^2}(x=l) = 0 && \text{(bending moment zero at ends)}\end{aligned}\right\} \tag{E$_2$}$$

where E is the Young's modulus, and I is the moment of inertia of the beam.

We shall assume the trial solution as

$$\underset{\sim}{w}(x) = C_1 \sin\left(\frac{\pi x}{l}\right) + C_2 \sin\left(\frac{3\pi x}{l}\right) \equiv C_1 f_1(x) + C_2 f_2(x) \tag{E$_3$}$$

where $f_1(x)$ and $f_2(x)$ satisfy the boundary conditions, Eq. (E$_2$), and C_1 and C_2 are the unknown constants. By substituting the trial solution of Eq. (E$_3$) into Eq. (E$_1$), we obtain the residual, R, as

$$R = EIC_1\left(\frac{\pi}{l}\right)^4 \sin\left(\frac{\pi x}{l}\right) + EIC_2\left(\frac{3\pi}{l}\right)^4 \sin\left(\frac{3\pi x}{l}\right) - p \tag{E$_4$}$$

By applying the Galerkin procedure, we obtain

$$\left.\begin{aligned}\int_0^l f_1(x) R\,\mathrm{d}x &= EIC_1\left(\frac{\pi}{l}\right)^4 \frac{l}{2} - p\frac{2l}{\pi} = 0\\ \int_0^l f_2(x) R\,\mathrm{d}x &= EIC_2\left(\frac{3\pi}{l}\right)^4 \frac{l}{2} - p\frac{2l}{3\pi} = 0\end{aligned}\right\} \quad (\mathrm{E}_5)$$

The solution of Eqs. (E_5) is

$$C_1 = \frac{4pl^4}{\pi^5 EI} \quad \text{and} \quad C_2 = \frac{4pl^4}{243\pi^5 EI} \quad (\mathrm{E}_6)$$

which can be seen to be the same as the one obtained in Example 5.1.

Example 5.4 Find the approximate deflection of a simply supported beam under a uniformly distributed load p (Figure 5.6) using the least squares method.

Solution The governing differential equation and the boundary conditions are the same as those given in Eqs. (E_1) and (E_2), respectively, of Example 5.3. By assuming the trial solution as

$$\underset{\sim}{w}(x) = C_1 f_1(x) + C_2 f_2(x) \quad (\mathrm{E}_1)$$

where

$$f_1(x) = \sin\left(\frac{\pi x}{l}\right) \quad \text{and} \quad f_2(x) = \sin\left(\frac{3\pi x}{l}\right) \quad (\mathrm{E}_2)$$

the residual, R, becomes

$$R = EI\frac{\mathrm{d}^4 w}{\mathrm{d}x^4} - p = EIC_1\left(\frac{\pi}{l}\right)^4 \sin\left(\frac{\pi x}{l}\right) + EIC_2\left(\frac{3\pi}{l}\right)^4 \sin\left(\frac{3\pi x}{l}\right) - p \quad (\mathrm{E}_3)$$

The application of the least squares method gives the following equations:

$$\begin{aligned}\frac{\partial}{\partial C_1}\left(\int_o^l R^2\,\mathrm{d}x\right) = \frac{\partial}{\partial C_1}\Bigg\{\int_o^l \Bigg[&(EI)^2 C_1^2\left(\frac{\pi}{l}\right)^8 \sin^2\left(\frac{\pi x}{l}\right)\\ &+ (EI)^2 C_2^2\left(\frac{3\pi}{l}\right)^8 \sin^2\left(\frac{3\pi x}{l}\right) + p^2\\ &+ 2(EI)^2 C_1 C_2\left(\frac{\pi}{l}\right)^8 \sin\left(\frac{\pi x}{l}\right)\cdot\sin\left(\frac{3\pi x}{l}\right)\end{aligned}$$

$$- 2EIpC_1 \left(\frac{\pi}{l}\right)^4 \sin\left(\frac{\pi x}{l}\right)$$

$$-2EIpC_2 \left(\frac{3\pi}{l}\right)^4 \sin\left(\frac{3\pi x}{l}\right) \Bigg] \mathrm{d}x \Bigg\} = 0$$

or

$$(EI)^2 C_1 \left(\frac{\pi}{l}\right)^8 l - 4EIp\frac{l}{\pi}\left(\frac{\pi}{l}\right)^4 = 0 \tag{E$_4$}$$

$$\frac{\partial}{\partial C_2}\left(\int_o^l R^2 \cdot \mathrm{d}x\right) = \frac{\partial}{\partial C_2}\Bigg\{\int_o^l \Bigg[(EI)^2 C_1^2 \left(\frac{\pi}{l}\right)^8 \sin^2\left(\frac{\pi x}{l}\right)$$

$$+ (EI)^2 C_2 \left(\frac{3\pi}{l}\right)^8 \cdot \sin^2\left(\frac{3\pi x}{l}\right) + p^2$$

$$+ 2(EI)^2 C_1 C_2 \left(\frac{\pi}{l}\right)^8 \sin\left(\frac{\pi x}{l}\right) \sin\left(\frac{3\pi x}{l}\right)$$

$$-2EIp \cdot C_1 \left(\frac{\pi}{l}\right)^4 \sin\left(\frac{\pi x}{l}\right) - 2EIpC_2\left(\frac{3\pi}{l}\right)^4 \sin\left(\frac{3\pi x}{l}\right)\Bigg] \mathrm{d}x\Bigg\} = 0$$

or

$$(EI)^2 C_2 \left(\frac{3\pi}{l}\right)^8 l - 4EIp\frac{l}{3\pi}\left(\frac{3\pi}{l}\right)^4 = 0 \tag{E$_5$}$$

The solution of Eqs. (E_4) and (E_5) leads to

$$C_1 = \frac{4pl^4}{\pi^5 EI} \quad \text{and} \quad C_2 = \frac{4pl^4}{243\pi^5 EI} \tag{E$_6$}$$

which can be seen to be identical to the solutions obtained in Examples 5.1 and 5.3.*

5.10 SOLUTION OF EIGENVALUE PROBLEMS USING WEIGHTED RESIDUAL METHOD

An eigenvalue problem can be stated as

$$A\phi = \lambda B\phi \text{ in } V \tag{5.76}$$

$$E_j\phi = 0, \qquad j = 1, 2, \ldots, p \text{ on } S \tag{5.77}$$

* Although the solutions given by the Rayleigh–Ritz, Galerkin, and least squares methods happen to be the same for the example considered, in general they lead to different solutions.

where A, B, and E_j are differential operators. By using Eqs. (5.66) and (5.76), the residual R can be expressed as**

$$R = \lambda B \underset{\sim}{\phi} - A \underset{\sim}{\phi} = \sum_{i=1}^{n} C_i(\lambda B f_i - A f_i) \tag{5.78}$$

If the trial solution of Eq. (5.66) contains any true eigenfunctions, then there exists sets of C_i and values of λ for which the residual R vanishes identically over the domain V. If $\underset{\sim}{\phi}(x)$ does not contain any eigenfunctions, then only approximate solutions will be obtained.

All four residual methods discussed in the case of equilibrium problems are also applicable to eigenvalue problems. For example, if we use the Galerkin method, we set the integral of the weighted residual equal to zero as

$$\int_V f_i(x) \cdot R \, \mathrm{d}V = 0, \qquad i = 1, 2, \ldots, n \tag{5.79}$$

Equation (5.79) gives the following algebraic (matrix) eigenvalue problem:

$$[A]\vec{C} = \lambda [B]\vec{C} \tag{5.80}$$

where $[A]$ and $[B]$ denote square symmetric matrices of size $n \times n$ given by

$$[A] = [A_{ij}] = \left[\int_V f_i \, A f_j \, \mathrm{d}V \right] \tag{5.81}$$

$$[B] = [B_{ij}] = \left[\int_V f_i \, B f_j \, \mathrm{d}V \right] \tag{5.82}$$

and $\vec{C}$ denotes the vector of unknowns C_i, $i = 1, 2, \ldots, n$. Now the solution of Eq. (5.80) can be obtained by any of the methods discussed in Section 7.3.

5.11 SOLUTION OF PROPAGATION PROBLEMS USING WEIGHTED RESIDUAL METHOD

A propagation problem has been stated earlier as

$$A\phi = e \text{ in } V \text{ for } t > t_0 \tag{5.83}$$

$$B_i\phi = g_i, \qquad i = 1, 2, \ldots, k \text{ on } S \text{ for } t \geq t_0 \tag{5.84}$$

$$E_j\phi = h_j, \qquad j = 1, 2, \ldots, l \text{ in } V \text{ for } t = t_0 \tag{5.85}$$

** The trial functions $f_i(x)$ are assumed to satisfy the boundary conditions of Eq. (5.77).

The trial solution of the problem is taken as

$$\underset{\sim}{\phi}(x,t) = \sum_{i=1}^{n} C_i(t) f_i(x) \tag{5.86}$$

where $f_i(x)$ are chosen to satisfy the boundary conditions, Eq. (5.84). Since Eqs. (5.83) and (5.85) are not satisfied by $\underset{\sim}{\phi}(x,t)$, there will be two residuals, one corresponding to each of these equations. For simplicity, we will assume that Eq. (5.85) gives the initial conditions explicitly as

$$\phi(x,t) = \phi_0 \quad \text{at} \quad t = 0 \tag{5.87}$$

Thus, the residual corresponding to the initial conditions (R_1) can be formulated as

$$R_1 = \phi_0 - \underset{\sim}{\phi}(x, t = 0) \text{ for all } x \text{ in } V \tag{5.88}$$

where

$$\underset{\sim}{\phi}(x, t = 0) = \sum_{i=1}^{n} C_i(0) f_i(x) \tag{5.89}$$

Similarly, the residual corresponding to the field equation (R_2) is defined as

$$R_2 = e - A\,\underset{\sim}{\phi}(x,t) \text{ for all } x \text{ in } V \tag{5.90}$$

Now any of the four residual methods discussed in the case of equilibrium problems can be applied to find the unknown functions $C_i(t)$. For example, if we apply the Galerkin procedure to each of the residuals R_1 and R_2, we obtain

$$\int_V f_i(x) R_1 \cdot \mathrm{d}V = 0, \qquad i = 1, 2, \ldots, n \tag{5.91}$$

$$\int_V f_i(x) R_2 \cdot \mathrm{d}V = 0, \qquad i = 1, 2, \ldots, n \tag{5.92}$$

Equations (5.91) and (5.92) lead to $2n$ equations in the $2n$ unknowns $C_i(o)$ and $C_i(t)$, $i = 1, 2, \ldots, n$, which can be solved either analytically or numerically.

5.12 DERIVATION OF FINITE ELEMENT EQUATIONS USING WEIGHTED RESIDUAL (GALERKIN) APPROACH

Let the governing differential equation of the (equilibrium) problem be given by

$$A(\phi) = b \text{ in } V \tag{5.93}$$

and the boundary conditions by

$$B_j(\phi) = g_j, \qquad j = 1, 2, \ldots, p \text{ on } S \tag{5.94}$$

The Galerkin method requires that

$$\int_V [A(\underset{\sim}{\phi}) - b] f_i \, \mathrm{d}V = 0, \qquad i = 1, 2, \ldots, n \tag{5.95}$$

where the trial functions f_i in the approximate solution

$$\underset{\sim}{\phi} = \sum_{i=1}^{n} C_i f_i \tag{5.96}$$

are assumed to satisfy the boundary conditions, Eq. (5.94). Note that f_i are defined over the entire domain of the problem.

Since the field equation (5.93) holds for every point in the domain V, it also holds for any set of points lying in an arbitrary subdomain or finite element in V. This permits us to consider any one element and define a local approximation similar to Eq. (5.96). Thus, we immediately notice that the familiar interpolation model for the field variable of the finite element will be applicable here also. If Eq. (5.96) is interpreted to be valid for a typical element e, the unknowns C_i can be recognized as the nodal unknowns $\Phi_i^{(e)}$ (nodal values of the field variable or its derivatives) and the functions f_i as the shape functions $N_i^{(e)}$. Equations (5.95) can be made to be valid for element e as

$$\int_{V^{(e)}} [A(\phi^{(e)}) - b^{(e)}] N_i^{(e)} \cdot \mathrm{d}V^{(e)} = 0, \qquad i = 1, 2, \ldots, n \tag{5.97}$$

where the interpolation model is taken in the standard form as

$$\phi^{(e)} = [N^{(e)}] \vec{\Phi}^{(e)} = \sum_i N_i^{(e)} \Phi_i^{(e)} \tag{5.98}$$

Equation (5.97) gives the required finite element equations for a typical element. These element equations have to be assembled to obtain the system or overall equations as outlined in Section 6.2.

Notes:

The shape functions of individual elements $N_i^{(e)}$ need not satisfy any boundary conditions, but they have to satisfy the interelement continuity conditions necessary for the assembly of the element equations. As stated earlier, to avoid any spurious contributions in the assembly process, we have to ensure that the (assumed) field variable ϕ and its derivatives up to one order less than the highest order derivative appearing under the integral in Eq. (5.97) are continuous along element boundaries. Since the differential operator A in the integrand usually contains higher order derivatives than the ones that appear in the integrand of the functional I in the variational formulation, we notice that the Galerkin method places more restrictions on the shape functions. The boundary conditions of the problem have to be incorporated after assembling the element equations as outlined in Chapter 6.

Example 5.5 Solve the differential equation

$$\frac{\mathrm{d}^2\phi}{\mathrm{d}x^2} + \phi + x = 0, \qquad 0 \le x \le 1$$

subject to the boundary conditions $\phi(0) = \phi(1) = 0$ using the Galerkin method.

Solution In this case the residual is given by

$$R = \left(\frac{\mathrm{d}^2\phi}{\mathrm{d}x^2} + \phi + x\right) \tag{E$_1$}$$

Equation (5.95) can be expressed as

$$\int_0^1 \left[\frac{\mathrm{d}^2\phi}{\mathrm{d}x^2} + \phi + x\right] N_k(x)\,\mathrm{d}x = 0; \qquad k = i, j$$

or

$$\sum_{e=1}^{E} \int_{x_i}^{x_j} [N^{(e)}]^T \left[\frac{\mathrm{d}^2\phi^{(e)}}{\mathrm{d}x^2} + \phi^{(e)} + x\right] \mathrm{d}x = 0 \tag{E$_2$}$$

where E is the number of elements, and x_i and x_j are the values of x at the first and the second nodes of element e, respectively.

We shall assume a linear interpolation model for $\phi^{(e)}$ so that

$$\phi^{(e)}(x) = N_i(x)\Phi_i^{(e)} + N_j(x)\Phi_j^{(e)} \tag{E$_3$}$$

and hence

$$[N^{(e)}] = [N_i(x) \quad N_j(x)] \tag{E$_4$}$$

where

$$N_i(x) = \frac{x_j - x}{l^{(e)}} \quad \text{and} \quad N_j(x) = \frac{x - x_i}{l^{(e)}} \tag{E$_5$}$$

The term $\int_{x_i}^{x_j} [N^{(e)}]^T (\mathrm{d}^2\phi^{(e)}/\mathrm{d}x^2)\,\mathrm{d}x$ can be written, after substitution of Eqs. (E$_3$) and (E$_4$) and integration by parts, as

$$\int_{x_i}^{x_j} [N^{(e)}]^T \frac{\mathrm{d}^2\phi^{(e)}}{\mathrm{d}x^2}\,\mathrm{d}x = [N^{(e)}]^T \frac{d\phi^{(e)}}{\mathrm{d}x}\bigg|_{x_i}^{x_j} - \int_{x_i}^{x_j} \frac{\mathrm{d}[N^{(e)}]^T}{\mathrm{d}x}\frac{\mathrm{d}\phi^{(e)}}{\mathrm{d}x}\,\mathrm{d}x \tag{E$_6$}$$

Substitution of Eq. (E_6) into Eq. (E_2) yields, for element e,

$$[N^{(e)}]^T \frac{\mathrm{d}\phi^{(e)}}{\mathrm{d}x}\Bigg|_{x_i}^{x_j} - \int_{x_i}^{x_j} \left\{ \frac{\mathrm{d}[N^{(e)}]^T}{\mathrm{d}x} \frac{\mathrm{d}\phi^{(e)}}{\mathrm{d}x} - [N^{(e)}]^T \phi^{(e)} - [N^{(e)}]^T x \right\} \mathrm{d}x = 0 \tag{E$_7$}$$

as the governing equation.

The first two terms in the integral of Eq. (E_7) yield the element characteristic matrix $[K^{(e)}]$ and the last term in the integral produces the element characteristic vector $\vec{P}^{(e)}$ in the equation

$$[K^{(e)}]\vec{\Phi}^{(e)} = \vec{P}^{(e)} \tag{E$_8$}$$

The left-most term in Eq. (E_7) contributes to the assembled vector $\vec{P}$ provided the derivative $(\mathrm{d}\phi/\mathrm{d}x)$ is specified at either end of the element e. This term is neglected if nothing is known about the value of $(\mathrm{d}\phi/\mathrm{d}x)$ at the nodal points. The evaluation of the integrals in Eq. (E_7) proceeds as follows:

$$\frac{\mathrm{d}}{\mathrm{d}x}[N^{(e)}]^T = \frac{\mathrm{d}}{\mathrm{d}x}\begin{Bmatrix} (x_j - x)/l^{(e)} \\ (x - x_i)/l^{(e)} \end{Bmatrix} = \frac{1}{l^{(e)}}\begin{Bmatrix} -1 \\ 1 \end{Bmatrix} \tag{E$_9$}$$

$$\frac{\mathrm{d}\phi^{(e)}}{\mathrm{d}x} = \frac{\mathrm{d}}{\mathrm{d}x}[N^{(e)}\vec{\Phi}^{(e)}] = \frac{1}{l^{(e)}}\begin{bmatrix} -1 & 1 \end{bmatrix}\begin{Bmatrix} \Phi_i^{(e)} \\ \Phi_j^{(e)} \end{Bmatrix} \tag{E$_{10}$}$$

$$\int_{x_i}^{x_j} \frac{\mathrm{d}}{\mathrm{d}x}[N^{(e)}]^T \frac{\mathrm{d}\phi^{(e)}}{\mathrm{d}x}\,\mathrm{d}x = \frac{1}{l^{(e)}}\begin{bmatrix} 1 & -1 \\ -1 & 1 \end{bmatrix}\begin{Bmatrix} \Phi_i^{(e)} \\ \Phi_j^{(e)} \end{Bmatrix} \tag{E$_{11}$}$$

$$\int_{x_i}^{x_j} [N^{(e)}]^T \phi^{(e)}\,\mathrm{d}x = \frac{l^{(e)}}{6}\begin{bmatrix} 2 & 1 \\ 1 & 2 \end{bmatrix}\begin{Bmatrix} \Phi_i^{(e)} \\ \Phi_j^{(e)} \end{Bmatrix} \tag{E$_{12}$}$$

$$\int_{x_i}^{x_j} [N^{(e)}]^T x\,\mathrm{d}x = \frac{1}{6}\begin{Bmatrix} (x_j^2 + x_i x_j + 2x_i^2) \\ (2x_j^2 - x_i x_j - x_i^2) \end{Bmatrix} \tag{E$_{13}$}$$

Since the value of $(\mathrm{d}\phi/\mathrm{d}x)$ is not specified at any node, we neglect the left-most term in Eq. (E_7). Thus, we obtain from Eq. (E_7)

$$\sum_{e=1}^{E}[K^{(e)}]\vec{\Phi}^{(e)} = \sum_{e=1}^{E}\vec{P}^{(e)} \tag{E$_{14}$}$$

where

$$[K^{(e)}] = \frac{1}{l^{(e)}}\begin{bmatrix} 1 & 1 \\ -1 & 1 \end{bmatrix} - \frac{l^{(e)}}{6}\begin{bmatrix} 2 & 1 \\ 1 & 2 \end{bmatrix} \tag{E$_{15}$}$$

$$\vec{P}^{(e)} = \frac{1}{6}\begin{Bmatrix} (x_j^2 + x_i x_j - 2x_i^2) \\ (2x_j^2 - x_i x_j - x_i^2) \end{Bmatrix} \tag{E$_{16}$}$$

It can be seen that Eqs. (E_{14})–(E_{16}) are identical to those obtained in Example 5.2 and hence the solution will also be the same.

5.13 DERIVATION OF FINITE ELEMENT EQUATIONS USING WEIGHTED RESIDUAL (LEAST SQUARES) APPROACH

Let the differential equation to be solved be stated as

$$A(\phi) = f(x, y, z) \text{ in } V \tag{5.99}$$

subject to the boundary conditions

$$\phi = \phi_0 \text{ on } S_0 \tag{5.100}$$

$$B_j\left(\phi, \frac{\partial \phi}{\partial x}, \frac{\partial \phi}{\partial y}, \frac{\partial \phi}{\partial z}, \ldots\right) = g_j(x, y, z) \text{ on } S_j \tag{5.101}$$

with

$$\sum_{j=0,1,\ldots} S_j = S \tag{5.102}$$

where $A(\ )$ and $B_j(\ )$ are linear differential operators involving the (unknown) field variable and its derivatives with respect to x, y, and z; f and g_j are known functions of x, y, and z; and V is the solution domain with boundary S.

Step 1: Divide the solution domain into E finite elements each having n nodal points with m unknowns (degrees of freedom) per node. Thus, m denotes the number of parameters, such as $\phi, (\partial\phi/\partial x), (\partial\phi/\partial y), \ldots$, taken as unknowns at each node.

Step 2: Assume an interpolation model for the field variable inside an element e as

$$\phi^{(e)}(x, y, z) = \sum_i N_i(x, y, z)\Phi_i^{(e)} = [N(x, y, z)]\vec{\Phi}^{(e)} \tag{5.103}$$

where N_i is the shape function corresponding to the ith degree of freedom of element e, $\Phi_i^{(e)}$.

Step 3: Derive the element characteristic matrices and vectors. Substitution of the approximate solution of Eq. (5.103) into Eqs. (5.99) and (5.101) yields the residual errors as

$$R^{(e)}(x, y, z) = A([N]\vec{\Phi}^{(e)}) - f^{(e)} \equiv A^{(e)} - f^{(e)} \tag{5.104}$$

$$r_j^{(e)}(x, y, z) = B_j([N]\vec{\Phi}^{(e)}) - g_j^{(e)} \equiv B_j^{(e)} - g_j^{(e)} \tag{5.105}$$

where $R^{(e)}$ and $r_j^{(e)}$ represent the residual errors due to differential equation and jth boundary condition, respectively, and $A^{(e)}$ and $B_j^{(e)}$ can be expressed in terms of the vector of nodal unknowns as

$$A^{(e)} = [C^{(e)}(x, y, z)]\vec{\Phi}^{(e)} \tag{5.106}$$

$$B_j^{(e)} = [D_j^{(e)}(x, y, z)]\vec{\Phi}^{(e)} \tag{5.107}$$

In the least squares method we minimize the weighted square of the residual error over the domain; that is,

$$I = a \iiint\limits_{V} R^2 \, \mathrm{d}V + \sum_j b_j \iint\limits_{S_j} r_j^2 \, \mathrm{d}S_j = \text{minimum} \tag{5.108}$$

where a and $b_1, b_2, \ldots$ are the weighting factors, all of which can be taken to be unity for simplicity; and the errors R and r_j can be expressed as the sum of element contributions as

$$R = \sum_{e=1}^{E} R^{(e)}, \qquad r_j = \sum_{j=1}^{E} r_j^{(e)} \tag{5.109}$$

The conditions for the minimum of I are

$$\frac{\partial I}{\partial \underset{\sim}{\vec{\Phi}}} = \begin{Bmatrix} \partial I / \partial \Phi_1 \\ \partial I / \partial \Phi_2 \\ \vdots \\ \partial I / \partial \Phi_M \end{Bmatrix} = \sum_{e=1}^{E} \frac{\partial I^{(e)}}{\partial \vec{\Phi}^{(e)}} = \vec{0} \tag{5.110}$$

where M denotes the total number of nodal unknowns in the problem ($M = m \times$ total number of nodes), and $I^{(e)}$ represents the contribution of element e to the functional I:

$$I^{(e)} = \iiint\limits_{V^{(e)}} R^{(e)^2} \, \mathrm{d}V + \sum_j \iint\limits_{S_j^{(e)}} r_j^{(e)^2} \, \mathrm{d}S_j \tag{5.111}$$

The squares of the residues $R^{(e)}$ and $r_j^{(e)}$ can be expressed as

$$\begin{aligned} R^{(e)^2} &= A^{(e)^2} - 2A^{(e)} f^{(e)} + f^{(e)^2} \\ &= \vec{\Phi}^{(e)^T} [C^{(e)}]^T [C^{(e)}] \vec{\Phi}^{(e)} - 2[C^{(e)}] \vec{\Phi}^{(e)} f^{(e)} + f^{(e)^2} \end{aligned} \tag{5.112}$$

$$\begin{aligned} r_j^{(e)^2} &= B_j^{(e)^2} - 2B_j^{(e)} g_j^{(e)} + g_j^{(e)^2} \\ &= \vec{\Phi}^{(e)^T} [D_j^{(e)}]^T [D_j^{(e)}] \vec{\Phi}^{(e)} - 2[D_j^{(e)}] \vec{\Phi}^{(e)} g_j^{(e)} + g_j^{(e)^2} \end{aligned} \tag{5.113}$$

Equations (5.110) and (5.111) lead to

$$\sum_{e=1}^{E} \left[\frac{\partial}{\partial \vec{\Phi}^{(e)}} \iiint\limits_{V^{(e)}} R^{(e)^2} \, \mathrm{d}V + \frac{\partial}{\partial \vec{\Phi}^{(e)}} \left(\sum_j \iint\limits_{S_j^{(e)}} r_j^{(e)^2} \mathrm{d}S_j \right) \right] = \vec{0} \tag{5.114}$$

with

$$\frac{\partial}{\partial \vec{\Phi}^{(e)}} \left(R^{(e)^2} \right) = 2[C^{(e)}]^T [C^{(e)}] \vec{\Phi}^{(e)} - 2[C^{(e)}]^T f^{(e)} \tag{5.115}$$

and

$$\frac{\partial}{\partial \vec{\Phi}^{(e)}} \left(r_j^{(e)^2} \right) = 2[D_j^{(e)}]^T [D_j^{(e)}] \vec{\Phi}^{(e)} - 2[D_j^{(e)}] g_j^{(e)} \tag{5.116}$$

By defining

$$[K_1^{(e)}] = \iiint\limits_{V^{(e)}} [C^{(e)}]^T [C^{(e)}] \, \mathrm{d}V \tag{5.117}$$

$$[K_2^{(e)}] = \sum_j \iint\limits_{S_j^{(e)}} [D_j^{(e)}]^T [D_j^{(e)}] \, \mathrm{d}S_j \tag{5.118}$$

$$\vec{P}_1^{(e)} = \iiint\limits_{V^{(e)}} f^{(e)} [C^{(e)}]^T \, \mathrm{d}V \tag{5.119}$$

$$\vec{P}_2^{(e)} = \sum_j \iint\limits_{S_j^{(e)}} g_j^{(e)} [D_j^{(e)}]^T \, \mathrm{d}S_j \tag{5.120}$$

Equation (5.114) becomes

$$\sum_{e=1}^{E} \left([K^{(e)}] \vec{\Phi}^{(e)} - \vec{P}^{(e)} \right) = \vec{0} \tag{5.121}$$

where the summation sign indicates the familiar assembly over all the finite elements,

$$[K^{(e)}] = [K_1^{(e)}] + [K_2^{(e)}] = \text{element characteristic matrix} \tag{5.122}$$

and

$$\vec{P}^{(e)} = \vec{P}_1^{(e)} + \vec{P}_2^{(e)} = \text{element characteristic vector} \tag{5.123}$$

Step 4: Derive the overall system equations. The assembled set of equations (5.121) can be expressed in the standard form

$$[\underset{\sim}{K}] \underset{\sim}{\vec{\Phi}} = \underset{\sim}{\vec{P}} \tag{5.124}$$

where

$$[\underset{\sim}{K}] = \sum_{e=1}^{E} [K^{(e)}] \quad \text{and} \quad \underset{\sim}{\vec{P}} = \sum_{e=1}^{E} \vec{P}^{(e)} \tag{5.125}$$

Step 5: Solve for the nodal unknowns. After incorporating the boundary conditions prescribed on S_o, Eq. (5.124) can be solved to find the vector $\underset{\sim}{\vec{\Phi}}$.

Table 5.1.

Value of x	Value of ϕ		Value of $(d\phi/dx)$	
	Finite element	Exact	Finite element	Exact
0.0	0.000 000	0.000 000	0.701 837	0.701 837
0.2	0.142 641	0.142 641	0.735 988	0.735 987
0.4	0.299 034	0.299 034	0.839 809	0.839 808
0.6	0.483 481	0.483 480	1.017 47	1.017 47
0.8	0.711 411	0.711 412	1.276 09	1.276 10
1.0	1.000 000	1.000 000	1.626 07	1.626 07

Step 6: Compute the element resultants. Once the nodal unknown vector $\vec{\underset{\sim}{\Phi}}$, and hence $\vec{\Phi}^{(e)}$, is determined the element resultants can be found, if required, by using Eq. (5.103).

Example 5.6 Find the solution of the differential equation $(d^2\phi/dx^2) - \phi = x$ subject to the boundary conditions $\phi(0) = 0$ and $\phi(1) = 1$.

Solution Here, the solution domain is given by $0 \leq x \leq 1$. Five one-dimensional elements, each having two nodes with two unknowns $[\phi$ and $(d\phi/dx)]$ per node, are used for the idealization. Thus, the total number of degrees of freedom is $M = 12$. Since there are four nodal degrees of freedom per element, the first-order Hermite polynomials are used as interpolation functions (as in the case of a beam element). The exact solution of this problem is given by $\phi(x) = (2\sinh x/\sinh 1) - x$. The finite element solution obtained by Akin and Sen Gupta [5.6] is compared with the exact solution at the six nodes in Table 5.1.

REFERENCES

5.1 H.C. Martin: *Introduction to Matrix Methods of Structural Analysis*, McGraw-Hill, New York, 1966.

5.2 J.W. Daily and D.R.F. Harleman: *Fluid Dynamics*, Addison-Wesley, Reading, MA, 1966.

5.3 R.S. Schechter: *The Variational Method in Engineering*, McGraw-Hill, New York, 1967.

5.4 C. Lanczos: *The Variational Principles of Mechanics*, University of Toronto Press, Toronto, 1970.

5.5 M. Gurtin: Variational principles for linear initial-value problems, *Quarterly of Applied Mathematics, 22,* 252–256, 1964.

5.6 J.E. Akin and S.R. Sen Gupta: Finite element application of least square techniques, *Variational Methods in Engineering*, Vol. I, C.A. Brebbia and H. Tottenham (Eds.), University of Southampton, Southampton, UK, 1973.

5.7 S.S. Rao: *Mechanical Vibrations*, 3rd Ed., Addison-Wesley, Reading, MA, 1995.

PROBLEMS

5.1 Derive the Euler–Lagrange equation corresponding to the functional

$$I = \frac{1}{2} \iiint\limits_{V} \left[\left(\frac{\partial \phi}{\partial x}\right)^2 + \left(\frac{\partial \phi}{\partial y}\right)^2 + \left(\frac{\partial \phi}{\partial z}\right)^2 - 2C\phi \right] \cdot \mathrm{d}V$$

What considerations would you take while selecting the interpolation polynomial for this problem?

5.2 Show that the equilibrium equations $[K]\vec{X} = \vec{P}$, where $[K]$ is a symmetric matrix, can be interpreted as the stationary requirement for the functional

$$I = \tfrac{1}{2}\vec{X}^T[K]\vec{X} - \vec{X}^T\vec{P}$$

5.3 The deflection of a beam on an elastic foundation is governed by the equation $(\mathrm{d}^4w/\mathrm{d}x^4) + w = 1$, where x and w are dimensionless quantities. The boundary conditions for a simply supported beam are given by transverse deflection $= w = 0$ and bending moment $= (\mathrm{d}^2w/\mathrm{d}x^2) = 0$. By taking a two-term trial solution as $w(x) = C_1f_1(x) + C_2f_2(x)$ with $f_1(x) = \sin \pi x$ and $f_2(x) = \sin 3\pi x$, find the solution of the problem using the Galerkin method.

5.4 Solve Problem 5.3 using the collocation method with collocation points at $x = 1/4$ and $x = 1/2$.

5.5 Solve Problem 5.3 using the least squares method.

5.6 Find the solution of the differential equation

$$\frac{\mathrm{d}^2\phi}{\mathrm{d}x^2} + \phi + x = 0, \qquad 0 \le x \le 1$$

subject to the boundary conditions $\phi(0) = \phi(1) = 0$ using the collocation method with $x = 1/4$ and $x = 1/2$ as the collocation points.

5.7 Solve Problem 5.6 using the least squares method.

5.8 Solve Problem 5.6 using the Galerkin method.

5.9 Solve Problem 5.6 using the Rayleigh–Ritz method.

5.10 Derive the finite element equations for a simplex element in two dimensions using a variational approach for the biharmonic equation $\nabla^4\phi = C$. Discuss the continuity requirements of the interpolation model.

5.11 Derive the finite element equations for a simplex element in two dimensions using a residual method for the biharmonic equation.

5.12 The Rayleigh quotient (λ_R) for the vibrating tapered beam, shown in Figure 5.8, is given by [5.7]

$$\lambda_R = \frac{I_A(\phi)}{I_B(\phi)}$$

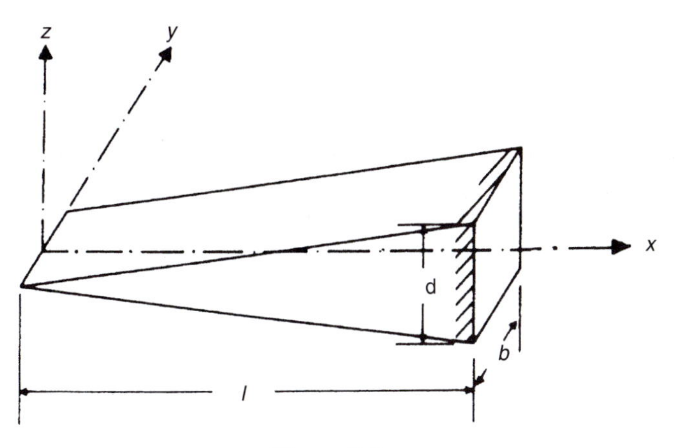

Figure 5.8.

where

$$I_A(\phi) = \frac{1}{2}\int_0^l EI\left[\frac{\mathrm{d}^2\phi(x)}{\mathrm{d}x^2}\right]^2 \mathrm{d}x$$

$$I_B(\phi) = \frac{1}{2}\int_0^l \rho A[\phi(x)]^2 \,\mathrm{d}x$$

ϕ is the assumed solution for the deflection of the beam, E is the Young's modulus, I is the area moment of inertia of the cross section $= (1/12)b[dx/l]^3$, ρ is the density, and A is the area of cross section $= b[dx/l]$. Find the eigenvalues (λ_R) of the beam using the Rayleigh–Ritz method with the assumed solution

$$\phi(x) = C_1\left[1 - \frac{x}{l}\right]^2 + C_2\left[\frac{x}{l}\right]\left[1 - \frac{x}{l}\right]^2$$

5.13 The differential equation governing the free transverse vibration of a string (Figure 5.9) is given by

$$\frac{\mathrm{d}^2\phi}{\mathrm{d}x^2} + \lambda\phi = 0, \qquad 0 \le x \le l$$

with the boundary conditions

$$\phi(x) = 0 \quad \text{at} \quad x = 0, x = l$$

where $\lambda = (\rho\omega^2 l^2/T)$ is the eigenvalue, ρ is the mass per unit length, l is the length, T is the tension in string, ω is the natural frequency of vibration, and

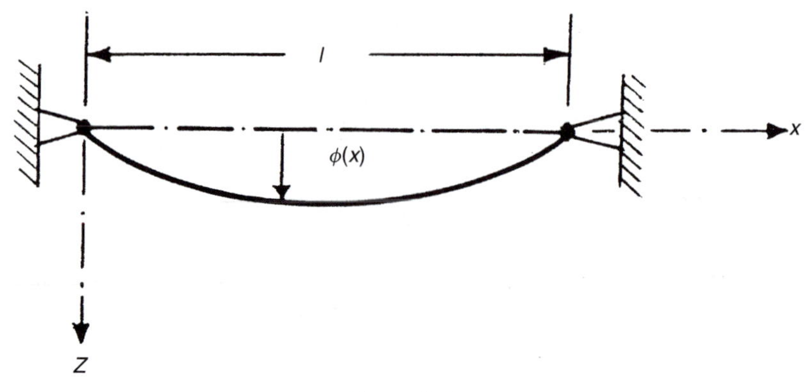

Figure 5.9.

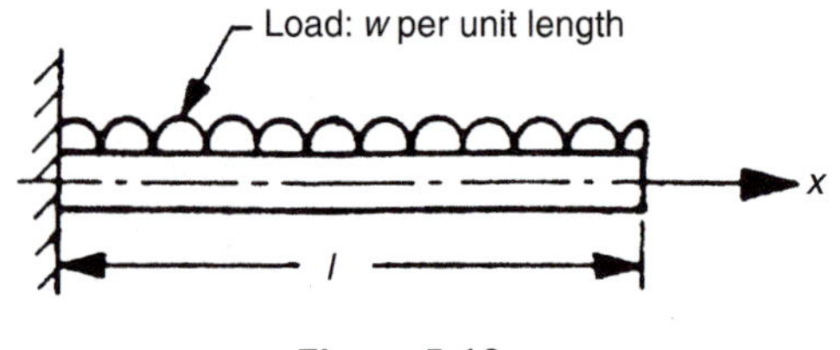

Figure 5.10.

$x = (y/l)$. Using the trial solution

$$\phi(x) = C_1 x(l - x) + C_2 x^2 (l - x)$$

where C_1 and C_2 are constants, determine the eigenvalues of the string using the Galerkin method.

5.14 Solve Problem 5.13 using the collocation method with $x = l/4$ and $x = 3l/4$ as the collocation points.

5.15 The cantilever beam shown in Figure 5.10 is subjected to a uniform load of w per unit length. Assuming the deflection as

$$\phi(x) = c_1 \sin \frac{\pi x}{2l} + c_2 \sin \frac{3\pi x}{2l}$$

determine the constants c_1 and c_2 using the Rayleigh–Ritz method.

5.16 Solve Problem 5.15 using the Galerkin method.

5.17 Solve Problem 5.15 using the least squares method.

5.18 A typical stiffness coefficient, k_{ij}, in the stiffness matrix, $[K]$, denotes the force along the degree of freedom i that results in a unit displacement along the degree of freedom j when the displacements along all other degrees of freedom are zero. Using this definition, beam deflection relations, and static equilibrium equations,

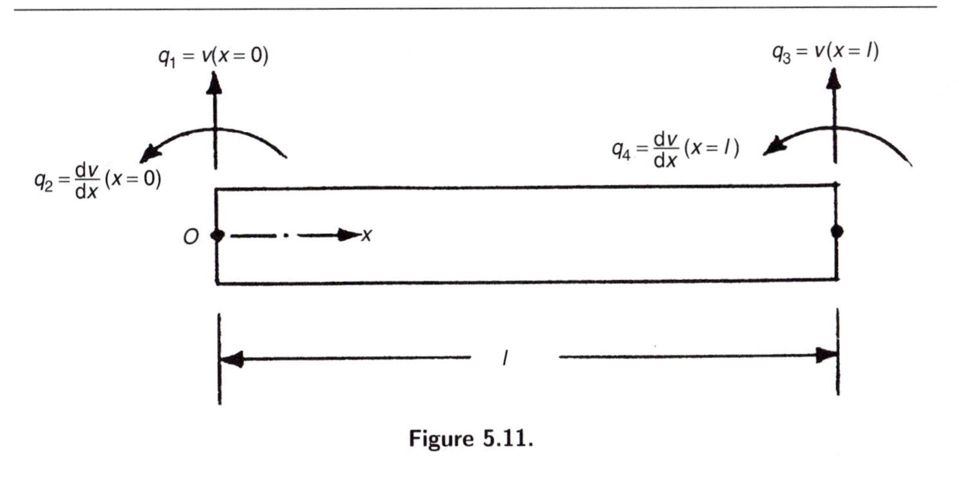

Figure 5.11.

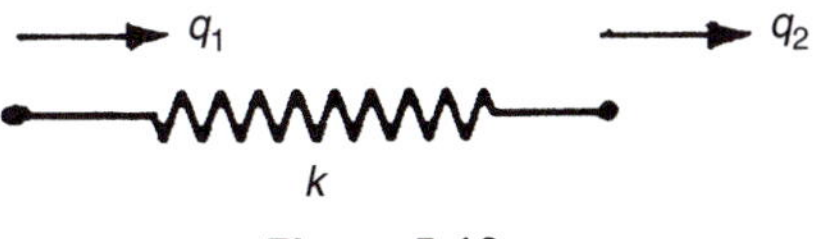

Figure 5.12.

generate expressions for k_{11}, k_{21}, k_{31}, and k_{41} for the uniform beam element shown in Figure 5.11. The Young's modulus, area moment of inertia, and length of the element are given by E, I, and l, respectively.

5.19 For the beam element considered in Problem 5.18, generate expressions for the stiffness coefficients k_{12}, k_{22}, k_{32}, and k_{42}.

5.20 For the beam element considered in Problem 5.18, generate expressions for the stiffness coefficients k_{13}, k_{23}, k_{33}, and k_{43}.

5.21 For the beam element considered in Problem 5.18, generate expressions for the stiffness coefficients k_{14}, k_{24}, k_{34}, and k_{44}.

5.22 Consider a spring with stiffness k as shown in Figure 5.12. Determine the stiffness matrix of the spring using the direct method.

5.23 Derive the stiffness matrix of a tapered bar, with linearly varying area of cross section (Figure 5.13), using a direct approach.

5.24 The heat transfer in the tapered fin shown in Figure 5.14 can be assumed to be one-dimensional due to the large value of W compared to L. Derive the element characteristic matrix of the fin using the direct approach.

5.25 Consider the differential equation

$$\frac{d^2\phi}{dx^2} + 400x^2 = 0, \qquad 0 \le x \le 1$$

with the boundary conditions

$$\phi(0) = 0, \qquad \phi(1) = 0$$

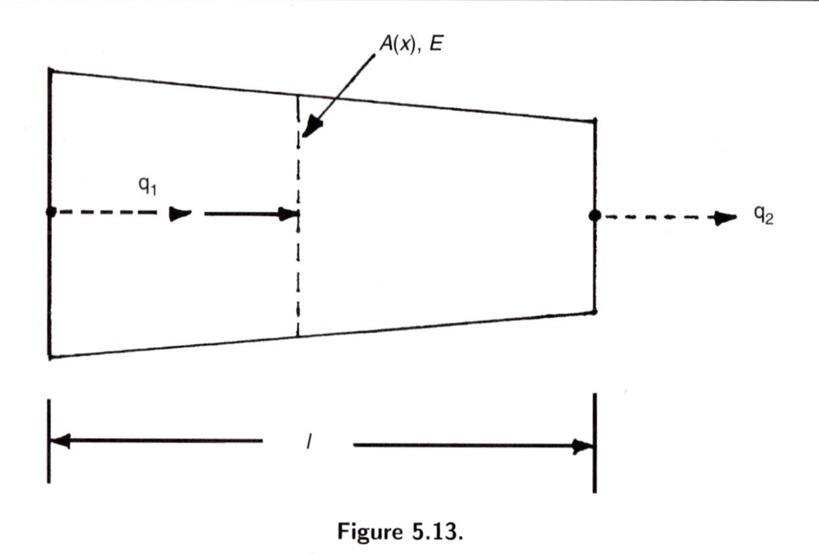

Figure 5.13.

The functional, corresponding to this problem, to be extremized is given by

$$I = \int_0^1 \left\{ -\frac{1}{2} \left[\frac{\mathrm{d}\phi}{\mathrm{d}x} \right]^2 + 400x^2\phi \right\} \mathrm{d}x$$

Find the solution of the problem using the Rayleigh–Ritz method using a one-term solution as $\phi(x) = c_1 x(1 - x)$.

5.26 Find the solution of Problem 5.25 using the Rayleigh–Ritz method using a two-term solution as $\phi(x) = c_1 x(1 - x) + c_2 x^2(1 - x)$.

5.27 Find the solution of Problem 5.25 using the Galerkin method using the solution

$$\phi(x) = c_1 x(1 - x) + c_2 x^2(1 - x)$$

5.28 Find the solution of Problem 5.25 using the two-point collocation method with the trial solution

$$\phi(x) = c_1 x(1 - x) + c_2 x^2(1 - x)$$

Assume the collocation points as $x = 1/4$ and $x = 3/4$.

5.29 Find the solution of Problem 5.25 using the least squares approach with the trial solution

$$\phi(x) = c_1 x(1 - x) + c_2 x^2(1 - x)$$

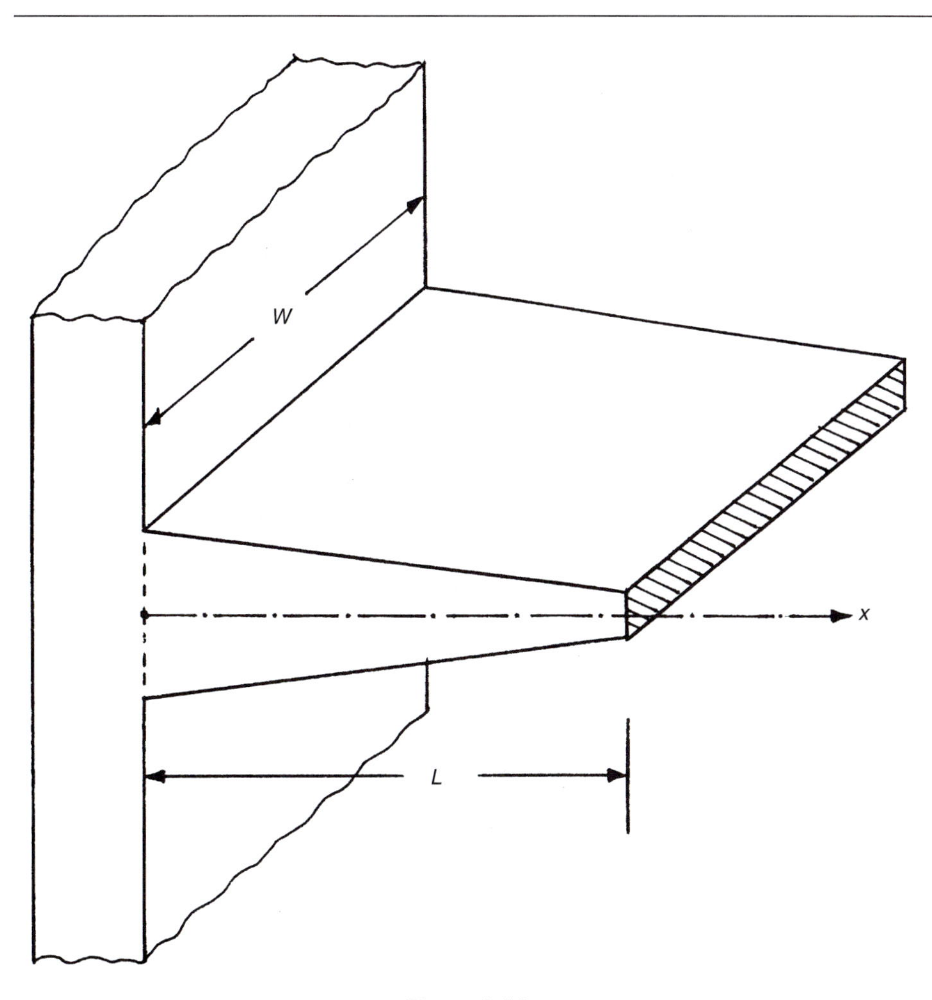

Figure 5.14.

5.30 Find the solution of Problem 5.25 using subdomain collocation with the trial solution

$$\phi(x) = c_1 x(1-x) + c_2 x^2 (1-x)$$

Assume two subdomains as $x = 0$ to $1/4$ and $x = 3/4$ to 1.

5.31 Consider the coaxial cable shown in Figure 5.15 with inside radius $r_i = 7$ mm, interface radius $r_m = 15$ mm, and outer radius $r_0 = 22$ mm. The permittivities of the inside and outside layers are $\varepsilon_1 = 1$ and $\varepsilon_2 = 2$, respectively, and the charge densities of the inside and outside layers are $\sigma_i = 50$ and $\sigma_0 = 0$, respectively. If the electric potential is specified at the inner and outer surfaces as $\phi_i = 400$ and $\phi_0 = 0$, determine the variation of $\phi(r)$ using the finite element method based on

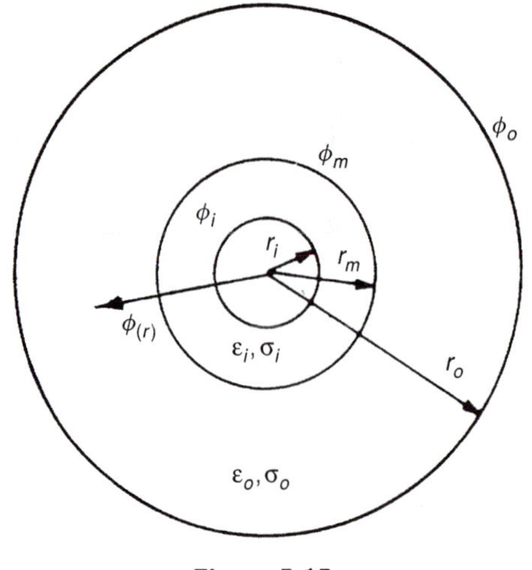

Figure 5.15.

the variational (Rayleigh–Ritz) approach. Use two linear finite elements in each layer.

Hint: The equation governing axisymmetric electrostatics is given by

$$c\frac{\mathrm{d}^2\phi}{\mathrm{d}r^2} + \frac{c}{r}\frac{\mathrm{d}\phi}{\mathrm{d}r} + p = 0 \tag{E$_1$}$$

where c is the permittivity of the material, ϕ is the electric potential, p is the charge density, and r is the radial distance. The variational function, I, corresponding to Eq. (E_1) is given by

$$I(\phi) = \int_{r_1}^{r_2} \left\{ \pi r c \left[\frac{\mathrm{d}\phi}{\mathrm{d}r}\right]^2 - 2\pi r p \phi \right\} \mathrm{d}r \tag{E$_2$}$$

where the relation, $\mathrm{d}V = 2\pi r\,\mathrm{d}r$, has been used.

5.32 Solve Problem 5.31 using the finite element method by adopting the Galerkin approach.

5.33 Solve Problem 5.31 using the finite element method by adopting the least squares approach.

6

ASSEMBLY OF ELEMENT MATRICES AND VECTORS AND DERIVATION OF SYSTEM EQUATIONS

6.1 COORDINATE TRANSFORMATION

The various methods of deriving element characteristic matrices and vectors have been discussed in Chapter 5. Before considering how these element matrices and vectors are assembled to obtain the characteristic equations of the entire system of elements, we need to discuss the aspect of coordinate transformation. The coordinate transformation is necessary when the field variable is a vector quantity such as displacement and velocity. Sometimes, the element matrices and vectors are computed in local coordinate systems suitably oriented for minimizing the computational effort. The local coordinate system may be different for different elements. When a local coordinate system is used, the directions of the nodal degrees of freedom will also be taken in a convenient manner. In such a case, before the element equations can be assembled, it is necessary to transform the element matrices and vectors derived in local coordinate systems so that all the elemental equations are referred to a common global coordinate system. The choice of the global coordinate system is arbitrary, and in practice it is generally taken to be the same as the coordinate system used in the engineering drawings, from which the coordinates of the different node points of the body can easily be found.

In general, for an equilibrium problem, the element equations in a local coordinate system can be expressed in the standard form

$$[k^{(e)}]\vec{\phi}^{(e)} = \vec{p}^{(e)} \tag{6.1}$$

where $[k^{(e)}]$ and $\vec{p}^{(e)}$ are the element characteristic matrix and vector, respectively, and $\vec{\phi}^{(e)}$ is the vector of nodal displacements of element e. We shall use lowercase and capital letters to denote the characteristics pertaining to the local and the global coordinate systems, respectively. Let a transformation matrix $[\lambda^{(e)}]$ exist between the local and the global coordinate systems such that

$$\vec{\phi}^{(e)} = [\lambda^{(e)}]\vec{\Phi}^{(e)} \tag{6.2}$$

and

$$\vec{p}^{(e)} = [\lambda^{(e)}]\vec{P}^{(e)} \tag{6.3}$$

By substituting Eqs. (6.2) and (6.3) into Eq. (6.1), we obtain

$$[k^{(e)}][\lambda^{(e)}]\vec{\Phi}^{(e)} = [\lambda^{(e)}]\vec{P}^{(e)} \tag{6.4}$$

Premultiplying this equation throughout by $[\lambda^{(e)}]^{-1}$ yields*

$$[\lambda^{(e)}]^{-1}[k^{(e)}][\lambda^{(e)}]\vec{\Phi}^{(e)} = \vec{P}^{(e)}$$

or

$$[K^{(e)}]\vec{\Phi}^{(e)} = \vec{P}^{(e)} \tag{6.5}$$

where the element characteristic matrix corresponding to the global coordinate system is given by

$$[K^{(e)}] = [\lambda^{(e)}]^{-1}[k^{(e)}][\lambda^{(e)}] \tag{6.6}$$

Notes:

1. If the vectors $\vec{\phi}^{(e)}$ and $\vec{p}^{(e)}$ are directional quantities such as nodal displacements and forces, then the transformation matrix $[\lambda^{(e)}]$ will be the matrix of direction cosines relating the two coordinate systems. In this case, the transformation matrix will be orthogonal and hence

$$[\lambda^{(e)}]^{-1} = [\lambda^{(e)}]^T \tag{6.7}$$

and

$$[K^{(e)}] = [\lambda^{(e)}]^T[k^{(e)}][\lambda^{(e)}] \tag{6.8}$$

2. In structural and solid mechanics problems, Eq. (6.8) can also be derived by equating the potential energies of the element in the two coordinate systems (see Problem 6.15).

* This assumes that $[\lambda^{(e)}]$ is a square matrix and its inverse exists.

6.2 ASSEMBLAGE OF ELEMENT EQUATIONS

Once the element characteristics, namely, the element matrices and element vectors, are found in a common global coordinate system, the next step is to construct the overall or system equations. The procedure for constructing the system equations from the element characteristics is the same regardless of the type of problem and the number and type of elements used.

The procedure of assembling the element matrices and vectors is based on the requirement of "compatibility" at the element nodes. This means that at the nodes where elements are connected, the value(s) of the unknown nodal degree(s) of freedom or variable(s) is the same for all the elements joining at that node. In solid mechanics and structural problems, the nodal variables are usually generalized displacements, which can be translations, rotations, curvatures, or other spatial derivatives of translations. When the generalized displacements are matched at a common node, the nodal stiffnesses and nodal loads of each of the elements sharing the node are added to obtain the net stiffness and the net load at that node.

Let E and M denote the total number of elements and nodal degrees of freedom (including the boundary and restrained degrees of freedom), respectively. Let $\vec{\underset{\sim}{\Phi}}$ denote the vector of M nodal degrees of freedom and $[\underset{\sim}{K}]$ the assembled system characteristic matrix of order $M \times M$. Since the element characteristic matrix $[K^{(e)}]$ and the element characteristic vector $\vec{P}^{(e)}$ are of the order $n \times n$ and $n \times 1$, respectively, they can be expanded to the order $M \times M$ and $M \times 1$, respectively, by including zeros in the remaining locations. Thus, the global characteristic matrix and the global characteristic vector can be obtained by algebraic addition as

$$[\underset{\sim}{K}] = \sum_{e=1}^{E} [\underline{K}^{(e)}] \tag{6.9}$$

and

$$\vec{\underset{\sim}{P}} = \sum_{e=1}^{E} \vec{\underline{P}}^{(e)} \tag{6.10}$$

where $[\underline{K}^{(e)}]$ is the expanded characteristic matrix of element e (of order $M \times M$), and $\vec{\underline{P}}^{(e)}$ is the expanded characteristic vector of element e (of order $M \times l$). Even if the assemblage contains many different types of elements, Eqs. (6.9) and (6.10) will be valid, although the number of element degrees of freedom, n, changes from element to element.

In actual computations, the expansion of the element matrix $[K^{(e)}]$ and the vector $\vec{P}^{(e)}$ to the sizes of the overall $[\underset{\sim}{K}]$ and $\vec{\underset{\sim}{P}}$ is not necessary. $[\underset{\sim}{K}]$ and $\vec{\underset{\sim}{P}}$ can be generated by identifying the locations of the elements of $[K^{(e)}]$ and $\vec{P}^{(e)}$ in $[\underset{\sim}{K}]$ and $\vec{\underset{\sim}{P}}$, respectively, and by adding them to the existing values as e changes from 1 to E. This procedure is illustrated with reference to the assemblage of four one-dimensional elements for the planar truss structure shown in Figure 6.1(a). Since the elements lie in the XY plane, each element has four degrees of freedom as shown in Figure 6.1(b). It is assumed that a proper coordinate transformation (Section 6.1) was used and $[K^{(e)}]$ of order 4×4

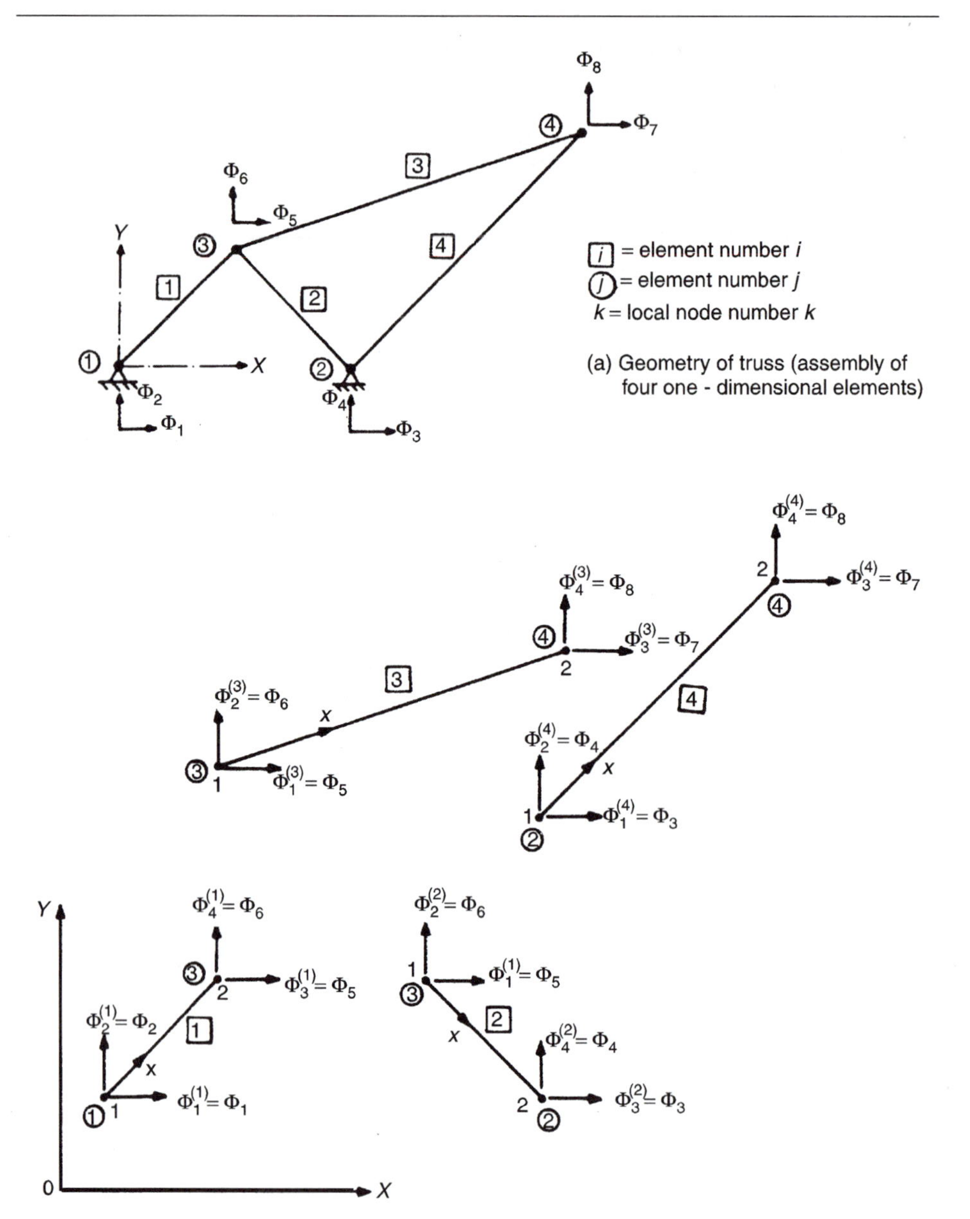

Figure 6.1. A Planar Truss as an Assembly of One-Dimensional Elements.

and $\vec{P}^{(e)}$ of order 4×1 of element e ($e = 1$–4) were obtained in the global coordinate system.

For assembling $[K^{(e)}]$ and $\vec{P}^{(e)}$, we consider the elements one after another. For $e = 1$, the element stiffness matrix $[K^{(1)}]$ and the element load vector $\vec{P}^{(1)}$ can be written as shown in Table 6.1. The location (row l and column m) of any component $K_{ij}^{(1)}$ in the global stiffness matrix $[\underset{\sim}{K}]$ is identified by the global degrees of freedom Φ_l and Φ_m corresponding to the local degrees of freedom $\Phi_i^{(1)}$ and $\Phi_j^{(1)}$, respectively, for $i = 1$–4 and $j = 1$–4. The correspondence between Φ_l and $\Phi_i^{(1)}$, and that between Φ_m and $\Phi_j^{(1)}$, is also shown in Table 6.1. Thus, the location of the components $K_{ij}^{(1)}$ in $[\underset{\sim}{K}]$ will be as shown in Table 6.2(a). Similarly, the location of the components of the element load vector $\vec{P}^{(1)}$ in $\underset{\sim}{\vec{P}}$ will also be as shown in Table 6.2(b). For $e = 2$, the element stiffness matrix $[K^{(2)}]$ and the element load vector $\vec{P}^{(2)}$ can be written as shown in Table 6.3. As in the case of $e = 1$, the locations of the elements $K_{ij}^{(2)}$ for $i = 1$–4 and $j = 1$–4 in the global stiffness matrix $[\underset{\sim}{K}]$ and $P_i^{(2)}$ for $i = 1$–4 in the global load vector $\underset{\sim}{\vec{P}}$ can be identified. Hence, these elements would be placed in $[\underset{\sim}{K}]$ and $\underset{\sim}{\vec{P}}$ at appropriate locations as shown in Table 6.4. It can be seen that if more than one element contributes to the stiffness K_{lm} of $[\underset{\sim}{K}]$, then the stiffnesses $K_{ij}^{(e)}$ for all the elements e contributing to K_{lm} are added together to obtain K_{lm}. A similar procedure is followed in obtaining P_l of $\underset{\sim}{\vec{P}}$.

For $e = 3$ and 4, the element stiffness matrices $[K^{(3)}]$ and $[K^{(4)}]$ and the element load vectors $\vec{P}^{(3)}$ and $\vec{P}^{(4)}$ are shown in Table 6.5. By proceeding with $e = 3$ and $e = 4$ as in the cases of $e = 1$ and $e = 2$, the final global stiffness matrix $[\underset{\sim}{K}]$ and load vector $\underset{\sim}{\vec{P}}$ can be obtained as shown in Table 6.6. If there is no contribution from any element to any K_{lm} in $[\underset{\sim}{K}]$, then the coefficient K_{lm} will be zero. Thus, each of the blank locations of the matrix $[\underset{\sim}{K}]$ in Table 6.6 are to be taken as zero. A similar argument applies to the blank locations, if any, of the vector $\underset{\sim}{\vec{P}}$. It is important to note that although a structure consisting of only four elements is considered in Figure 6.1 for illustration, the same procedure is applicable for any structure having any number of finite elements. In fact, the procedure is applicable equally well to all types of problems.

Table 6.1. Stiffness Matrix and Load Vector of Element 1

Local d.o.f. $(\Phi_i^{(1)})$ ↓	Corresponding global d.o.f. (Φ_l) ↓	Local $(\Phi_j^{(1)})$ → 1	2	3	4
		(Φ_m) → 1	2	5	6
1	1	$K_{11}^{(1)}$	$K_{12}^{(1)}$	$K_{13}^{(1)}$	$K_{14}^{(1)}$
2	2	$K_{21}^{(1)}$	$K_{22}^{(1)}$	$K_{23}^{(1)}$	$K_{24}^{(1)}$
3	5	$K_{31}^{(1)}$	$K_{32}^{(1)}$	$K_{33}^{(1)}$	$K_{34}^{(1)}$
4	6	$K_{41}^{(1)}$	$K_{42}^{(1)}$	$K_{43}^{(1)}$	$K_{44}^{(1)}$

$$\underset{4\times 4}{[K^{(1)}]} = \begin{bmatrix} K_{11}^{(1)} & K_{12}^{(1)} & K_{13}^{(1)} & K_{14}^{(1)} \\ K_{21}^{(1)} & K_{22}^{(1)} & K_{23}^{(1)} & K_{24}^{(1)} \\ K_{31}^{(1)} & K_{32}^{(1)} & K_{33}^{(1)} & K_{34}^{(1)} \\ K_{41}^{(1)} & K_{42}^{(1)} & K_{43}^{(1)} & K_{44}^{(1)} \end{bmatrix}$$

$\underset{4\times 1}{\vec{P}^{(1)}}$	Local d.o.f. $(\Phi_i^{(1)})$	Corresponding global d.o.f. (Φ_l)
$P_1^{(1)}$	1	1
$P_2^{(1)}$	2	2
$P_3^{(1)}$	3	5
$P_4^{(1)}$	4	6

Table 6.2. Location of the Elements of $[K^{(1)}]$ and $\vec{P}^{(1)}$ in $[\underset{\sim}{K}]$ and $\underset{\sim}{\vec{P}}$

$$
\begin{array}{cc}
 & \text{Global d.o.f.} \rightarrow \; 1 \quad 2 \quad 3 \quad 4 \quad 5 \quad 6 \quad 7 \quad 8 \\
 & \downarrow \\
[K^{(1)}] = & \begin{array}{c} 1\\2\\3\\4\\5\\6\\7\\8 \end{array}
\left[\begin{array}{cccccccc}
K_{11}^{(1)} & K_{12}^{(1)} & & & K_{13}^{(1)} & K_{14}^{(1)} & & \\
K_{21}^{(1)} & K_{22}^{(1)} & & & K_{23}^{(1)} & K_{24}^{(1)} & & \\
 & & & & & & & \\
 & & & & & & & \\
K_{31}^{(1)} & K_{32}^{(1)} & & & K_{33}^{(1)} & K_{34}^{(1)} & & \\
K_{41}^{(1)} & K_{42}^{(1)} & & & K_{43}^{(1)} & K_{44}^{(1)} & & \\
 & & & & & & & \\
 & & & & & & &
\end{array}\right]
\end{array}
$$

(a) Location of $[K^{(1)}]$ in $[\underset{\sim}{K}]$

$$
\begin{array}{cc}
 & \text{Global d.o.f.} \\
 & \downarrow \\
\vec{P}^{(1)} = & \begin{array}{c} 1\\2\\3\\4\\5\\6\\7\\8 \end{array}
\left\{\begin{array}{c}
P_{1}^{(1)} \\ P_{2}^{(1)} \\ \vdots \\ \\ P_{3}^{(1)} \\ P_{4}^{(1)} \\ \vdots \\ \\
\end{array}\right\}
\end{array}
$$

(b) Location of $\vec{P}^{(1)}$ in $\underset{\sim}{\vec{P}}$

Table 6.3. Stiffness Matrix and Load Vector of Element 2

$$
\begin{array}{cccc}
\text{Local d.o.f. } (\Phi_i^{(2)}) & (\Phi_j^{(2)}) \text{ Corresponding global d.o.f. } (\phi_l) & & \longrightarrow 1 \quad 2 \quad 3 \quad 4 \\
 & & & (\Phi_m) \rightarrow 5 \quad 6 \quad 3 \quad 4 \\
\downarrow & \downarrow & & \\
\underset{4\times 4}{[K^{(2)}]} = \begin{array}{c} 1\\2\\3\\4 \end{array} & \begin{array}{c} 5\\6\\3\\4 \end{array} & &
\left[\begin{array}{cccc}
K_{11}^{(2)} & K_{12}^{(2)} & K_{13}^{(2)} & K_{14}^{(2)} \\
K_{21}^{(2)} & K_{22}^{(2)} & K_{23}^{(2)} & K_{24}^{(2)} \\
K_{31}^{(2)} & K_{32}^{(2)} & K_{33}^{(2)} & K_{34}^{(2)} \\
K_{41}^{(2)} & K_{42}^{(2)} & K_{43}^{(2)} & K_{44}^{(2)}
\end{array}\right]
\end{array}
$$

$$
\begin{array}{ccc}
 & \text{Local d.o.f. } (\Phi_i^{(2)}) & \text{Corresponding global d.o.f. } (\Phi_l) \\
\underset{4\times 1}{\vec{P}^{(2)}} = \left\{\begin{array}{c} P_1^{(2)} \\ P_2^{(2)} \\ P_3^{(2)} \\ P_4^{(2)} \end{array}\right\} & \begin{array}{c} 1\\2\\3\\4 \end{array} & \begin{array}{c} 5\\6\\3\\4 \end{array}
\end{array}
$$

6.3 COMPUTER IMPLEMENTATION OF THE ASSEMBLY PROCEDURE

The assembly procedure outlined in the previous section is shown in Figure 6.2 in the form of a flow chart. We define the following quantities for the computer implementation of the assembly procedure:

NB = bandwidth of the overall characteristic matrix
NE = number of elements
NNE = number of nodes in each element
NN = total number of nodes in the complete body (one degree of freedom is assumed for each node)
GK = global or overall system characteristic matrix (size: NN × NN if stored as a square matrix; NN × NB if stored in band form)
EK = element characteristic matrix (size = NNE × NNE)
ID(I,J) = global node number corresponding to the Jth corner of Ith element (size of ID: NE × NNE)

Table 6.7 gives the Fortran statements that can be used for the assembly process. In a similar manner, the vectors of element nodal actions can also be assembled into a global or system action vector.

Table 6.4. Assembly of $[K^{(1)}]$, $\vec{P}^{(1)}$, $[K^{(2)}]$, and $\vec{P}^{(2)}$

Global d.o.f. → 1 2 3 4 5 6 7 8

$$[K^{(1)}]+[K^{(2)}] = \begin{array}{c} 1\\2\\3\\4\\5\\6\\7\\8 \end{array}\begin{bmatrix} K_{11}^{(1)} & K_{12}^{(1)} & & & K_{13}^{(1)} & K_{14}^{(1)} & & \\ K_{21}^{(1)} & K_{22}^{(1)} & & & K_{23}^{(1)} & K_{24}^{(1)} & & \\ & & K_{33}^{(2)} & K_{34}^{(2)} & K_{31}^{(2)} & K_{32}^{(2)} & & \\ & & K_{43}^{(2)} & K_{44}^{(2)} & K_{41}^{(2)} & K_{42}^{(2)} & & \\ K_{31}^{(1)} & K_{32}^{(1)} & K_{13}^{(2)} & K_{14}^{(2)} & K_{33}^{(1)}+K_{11}^{(2)} & K_{34}^{(1)}+K_{12}^{(2)} & & \\ K_{41}^{(1)} & K_{42}^{(1)} & K_{23}^{(2)} & K_{24}^{(2)} & K_{43}^{(1)}+K_{21}^{(2)} & K_{44}^{(1)}+K_{22}^{(2)} & & \\ & & & & & & & \\ & & & & & & & \end{bmatrix}$$

(a) Location of $[K^{(1)}]$ and $[K^{(2)}]$ in $[\underset{\sim}{K}]$

Global d.o.f. ↓

$$\vec{P}^{(1)}+\vec{P}^{(2)} = \begin{array}{c} 1\\2\\3\\4\\5\\6\\7\\8 \end{array}\begin{Bmatrix} P_1^{(1)} \\ P_2^{(1)} \\ \vec{P}_3^{(2)} \\ \vec{P}_4^{(1)} \\ P_3^{(1)}+P_1^{(2)} \\ P_4^{(1)}+P_2^{(2)} \\ \vdots \\ \end{Bmatrix}$$

(b) Location of $\vec{P}^{(1)}$ and $\vec{P}^{(2)}$ in $\underset{\sim}{P}$

6.4 INCORPORATION OF BOUNDARY CONDITIONS

After assembling the element characteristic matrices $[K^{(e)}]$ and the element characteristic vectors $\vec{P}^{(e)}$, the overall or system equations of the entire domain or body can be written (for an equilibrium problem) as

$$\underset{M \times M}{[\underset{\sim}{K}]} \; \underset{M \times 1}{\underset{\sim}{\vec{\Phi}}} = \underset{M \times 1}{\underset{\sim}{\vec{P}}} \tag{6.11}$$

These equations cannot be solved for $\underset{\sim}{\vec{\Phi}}$ since the matrix $[\underset{\sim}{K}]$ will be singular and hence its inverse does not exist. The physical significance of this, in the case of solid mechanics problems, is that the loaded body or structure is free to undergo unlimited rigid body motion unless some support constraints are imposed to keep the body or structure in equilibrium under the loads. Hence, some boundary or support conditions have to be

Table 6.5. Element Stiffness Matrices and Load Vectors for $e = 3$ and 4

Local d.o.f. $(\Phi_i^{(3)})$ ↓; $(\Phi_j^{(3)})$ → 1 2 3 4

Corresponding global d.o.f. (Φ_l) ↓; (Φ_m) → 5 6 7 8

Local d.o.f. $(\Phi_i^{(3)})$	Corresponding global d.o.f. (Φ_l)				
1	5	$K_{11}^{(3)}$	$K_{12}^{(3)}$	$K_{13}^{(3)}$	$K_{14}^{(3)}$
2	6	$K_{21}^{(3)}$	$K_{22}^{(3)}$	$K_{23}^{(3)}$	$K_{24}^{(3)}$
3	7	$K_{31}^{(3)}$	$K_{32}^{(3)}$	$K_{33}^{(3)}$	$K_{34}^{(3)}$
4	8	$K_{41}^{(3)}$	$K_{42}^{(3)}$	$K_{43}^{(3)}$	$K_{44}^{(3)}$

$$\underset{4 \times 4}{[K^{(3)}]} = \begin{bmatrix} K_{11}^{(3)} & K_{12}^{(3)} & K_{13}^{(3)} & K_{14}^{(3)} \\ K_{21}^{(3)} & K_{22}^{(3)} & K_{23}^{(3)} & K_{24}^{(3)} \\ K_{31}^{(3)} & K_{32}^{(3)} & K_{33}^{(3)} & K_{34}^{(3)} \\ K_{41}^{(3)} & K_{42}^{(3)} & K_{43}^{(3)} & K_{44}^{(3)} \end{bmatrix}$$

Local d.o.f. $(\Phi_i^{(4)})$ ↓; $(\Phi_j^{(4)})$ → 1 2 3 4

Corresponding global d.o.f. (Φ_l) ↓; (Φ_m) → 3 4 7 8

Local d.o.f. $(\Phi_i^{(4)})$	Corresponding global d.o.f. (Φ_l)				
1	3	$K_{11}^{(4)}$	$K_{12}^{(4)}$	$K_{13}^{(4)}$	$K_{14}^{(4)}$
2	4	$K_{21}^{(4)}$	$K_{22}^{(4)}$	$K_{23}^{(4)}$	$K_{24}^{(4)}$
3	7	$K_{31}^{(4)}$	$K_{32}^{(4)}$	$K_{33}^{(4)}$	$K_{34}^{(4)}$
4	8	$K_{41}^{(4)}$	$K_{42}^{(4)}$	$K_{43}^{(4)}$	$K_{44}^{(4)}$

$$\underset{4 \times 4}{[K^{(4)}]} = \begin{bmatrix} K_{11}^{(4)} & K_{12}^{(4)} & K_{13}^{(4)} & K_{14}^{(4)} \\ K_{21}^{(4)} & K_{22}^{(4)} & K_{23}^{(4)} & K_{24}^{(4)} \\ K_{31}^{(4)} & K_{32}^{(4)} & K_{33}^{(4)} & K_{34}^{(4)} \\ K_{41}^{(4)} & K_{42}^{(4)} & K_{43}^{(4)} & K_{44}^{(4)} \end{bmatrix}$$

(a) Element stiffness matrices

$\vec{P}^{(3)}$ (4 × 1)	Local d.o.f. $(\Phi_i^{(3)})$ ↓	Corresponding global d.o.f. (Φ_l) ↓	$\vec{P}^{(4)}$ (4 × 1)	Local d.o.f. $(\Phi_i^{(4)})$ ↓	Corresponding global d.o.f. (Φ_l) ↓
$P_1^{(3)}$	1	5	$P_1^{(4)}$	1	3
$P_2^{(3)}$	2	6	$P_2^{(4)}$	2	4
$P_3^{(3)}$	3	7	$P_3^{(4)}$	3	7
$P_4^{(3)}$	4	8	$P_4^{(4)}$	4	8

(b) Load vectors

applied to Eq. (6.11) before solving for $\vec{\underset{\sim}{\Phi}}$. In nonstructural problems, we have to specify the value of at least one and sometimes more than one nodal degree of freedom. The number of degrees of freedom to be specified is dictated by the physics of the problem.

As seen in Eqs. (5.20) and (5.21), there are two types of boundary conditions: forced or geometric or essential and free or natural. If we use a variational approach for deriving the system equations, we need to specify only the essential boundary conditions and the

Table 6.6. Assembled Stiffness Matrix and Load Vector

$$\underset{8\times 8}{[\underset{\sim}{K}]} = \sum_{e=1}^{4} [K^{(e)}] =$$

Global d.o.f.	1	2	3	4	5	6	7	8
1	$K_{11}^{(1)}$	$K_{12}^{(1)}$	0	0	$K_{13}^{(1)}$	$K_{14}^{(1)}$	0	0
2	$K_{21}^{(1)}$	$K_{22}^{(1)}$	0	0	$K_{23}^{(1)}$	$K_{24}^{(1)}$	0	0
3	0	0	$K_{33}^{(2)}+K_{11}^{(4)}$	$K_{34}^{(2)}+K_{12}^{(4)}$	$K_{31}^{(2)}$	$K_{32}^{(2)}$	$K_{13}^{(4)}$	$K_{14}^{(4)}$
4	0	0	$K_{43}^{(2)}+K_{21}^{(4)}$	$K_{44}^{(2)}+K_{22}^{(4)}$	$K_{41}^{(2)}$	$K_{42}^{(2)}$	$K_{23}^{(4)}$	$K_{24}^{(4)}$
5	$K_{31}^{(1)}$	$K_{32}^{(1)}$	$K_{13}^{(2)}$	$K_{14}^{(2)}$	$K_{33}^{(1)}+K_{11}^{(2)}+K_{11}^{(3)}$	$K_{34}^{(1)}+K_{12}^{(2)}+K_{12}^{(3)}$	$K_{13}^{(3)}$	$K_{14}^{(3)}$
6	$K_{41}^{(1)}$	$K_{42}^{(1)}$	$K_{23}^{(2)}$	$K_{24}^{(2)}$	$K_{43}^{(1)}+K_{21}^{(2)}+K_{21}^{(3)}$	$K_{44}^{(1)}+K_{22}^{(2)}+K_{22}^{(3)}$	$K_{23}^{(3)}$	$K_{24}^{(3)}$
7	0	0	$K_{31}^{(4)}$	$K_{32}^{(4)}$	$K_{31}^{(3)}$	$K_{32}^{(3)}$	$K_{33}^{(3)}+K_{33}^{(4)}$	$K_{34}^{(3)}+K_{34}^{(4)}$
8	0	0	$K_{41}^{(4)}$	$K_{42}^{(4)}$	$K_{41}^{(3)}$	$K_{42}^{(3)}$	$K_{43}^{(3)}+K_{43}^{(4)}$	$K_{44}^{(3)}+K_{44}^{(4)}$

(a) Global stiffness matrix

$$\underset{8\times 1}{\vec{\underset{\sim}{P}}} = \sum_{e=1}^{4} \vec{P}^{(e)} = \begin{Bmatrix} P_1^{(1)} \\ P_2^{(1)} \\ P_3^{(2)} + P_1^{(4)} \\ P_4^{(2)} + P_2^{(4)} \\ P_3^{(1)} + P_1^{(2)} + P_1^{(3)} \\ P_4^{(1)} + P_2^{(2)} + P_2^{(3)} \\ P_3^{(3)} + P_3^{(4)} \\ P_4^{(3)} + P_4^{(4)} \end{Bmatrix} \begin{matrix} \text{Global d.o.f.} \\ \downarrow \\ 1 \\ 2 \\ 3 \\ 4 \\ 5 \\ 6 \\ 7 \\ 8 \end{matrix}$$

(b) Global load vector

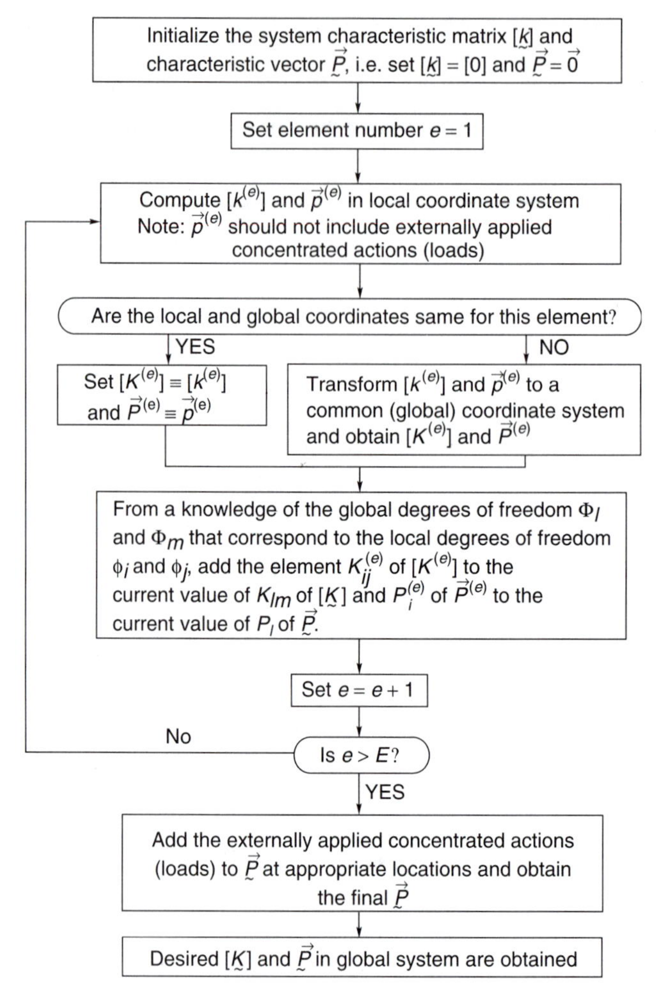

Figure 6.2. Assembly Procedure.

natural boundary conditions will be implicitly satisfied in the solution procedure. Thus, we need to apply only the geometric boundary conditions to Eq. (6.11). The geometric boundary conditions can be incorporated into Eq. (6.11) by several methods as outlined in the following paragraphs. The boundary conditions involving more than one nodal degree of freedom are known as multipoint constraints. Several methods are available to incorporate linear multipoint constraints (see Problems 6.5 and 6.6). The processing of nonlinear multipoint constraints is described by Narayanaswamy [6.1].

Method 1

To understand this method, we partition Eq. (6.11) as

$$\begin{bmatrix} [K_{11}] & [K_{12}] \\ [K_{21}] & [K_{22}] \end{bmatrix} \begin{Bmatrix} \vec{\Phi}_1 \\ \vec{\Phi}_2 \end{Bmatrix} = \begin{Bmatrix} \vec{P}_1 \\ \vec{P}_2 \end{Bmatrix} \tag{6.12}$$

Table 6.7. Assembly of Element Matrices

(i) When GK is stored as a square matrix	(ii) When GK is stored in a band form

<pre>   DO 10 I=1, NN
      DO 10 J=1, NN
 10   GK (I,J)=0.0
      DO 20 I=1, NE
      .
      .
      .
C   GENERATE ELEMENT MATRIX EK
C   FOR ELEMENT I
      DO 30 J=1, NNE
      IJ=ID(I,J)
      DO 30 K=1, NNE
      IK=ID(I,K)
      GK (IJ,IK)=GK (IJ,IK)+EK(J,K)
 30   CONTINUE
 20   CONTINUE</pre>
	<pre> DO 10 I=1, NN
	      DO 10 J=1, NB
	 10 GK (I,J)=0.0
	      DO 20 I=1, NE
	      .
	      .
	      .
	C GENERATE ELEMENT MATRIX EK
	C FOR ELEMENT I
	      DO 30 J=1, NNE
	      IJ=ID(I,J)
	      DO 30 K=1, NNE
	      IK=ID(I,K)
	      IKM=IK-IJ+1
	      IF (IKM. LT. 1) GO TO 30
	      GK (IJ,IKM)=GK (IJ,IKM)+EK(J,K)
	 30 CONTINUE
	 20 CONTINUE</pre>

where $\vec{\Phi}_2$ is assumed to be the vector of specified nodal degrees of freedom and $\vec{\Phi}_1$ as the vector of unrestricted (free) nodal degrees of freedom. Then $\vec{P}_1$ will be the vector of known nodal actions, and $\vec{P}_2$ will be the vector of unknown nodal actions.* Equation (6.12) can be written as

$$[K_{11}]\vec{\Phi}_1 + [K_{12}]\vec{\Phi}_2 = \vec{P}_1$$

or

$$[K_{11}]\vec{\Phi}_1 = \vec{P}_1 - [K_{12}]\vec{\Phi}_2 \tag{6.13}$$

and

$$[K_{12}]^T\vec{\Phi}_1 + [K_{22}]\vec{\Phi}_2 = \vec{P}_2 \tag{6.14}$$

Here, $[K_{11}]$ will not be singular and hence Eq. (6.13) can be solved to obtain

$$\vec{\Phi}_1 = [K_{11}]^{-1}(\vec{P}_1 - [K_{12}]\vec{\Phi}_2) \tag{6.15}$$

Once $\vec{\Phi}_1$ is known, the vector of unknown nodal actions $\vec{P}_2$ can be found from Eq. (6.14). In the special case in which all the prescribed nodal degrees of freedom are equal to

* In the case of solid mechanics problems, $\vec{\Phi}_2$ denotes the vector of nodal displacements that avoids the rigid body motion of the body, $\vec{P}_1$ the vector of known nodal loads, and $\vec{P}_2$ the unknown reactions at points at which the displacements $\vec{\Phi}_2$ are prescribed.

zero, we can delete the rows and columns corresponding to $\vec{\Phi}_2$ and state the equations simply as

$$[K_{11}]\vec{\Phi}_1 = \vec{P}_1 \tag{6.16}$$

Method 2

Since all the prescribed nodal degrees of freedom usually do not come at the end of the vector $\underset{\sim}{\Phi}$, the procedure of method 1 involves an awkward renumbering scheme. Even when the prescribed nodal degrees of freedom are not zero, it can be seen that the rearrangement of Eqs. (6.11) and solution of Eqs. (6.13) and (6.14) are time-consuming and require tedious bookkeeping. Hence, the following equivalent method can be used for incorporating the prescribed boundary conditions $\vec{\Phi}_2$. Equations (6.13) and (6.14) can be written together as

$$\left[\begin{array}{c|c} [k_{11}] & [0] \\ \hline [0] & [I] \end{array}\right] = \left\{\begin{array}{c} \vec{\Phi}_1 \\ \hline \vec{\Phi}_2 \end{array}\right\} = \left\{\begin{array}{c} \vec{P}_1 - [K_{12}]\vec{\Phi}_2 \\ \hline \vec{\Phi}_2 \end{array}\right\} \tag{6.17}$$

In actual computations, the process indicated in Eqs. (6.17) can be performed without reordering the equations implied by the partitioning as follows:

Step (i): If Φ_j is prescribed as Φ_j^*, the characteristic vector $\underset{\sim}{P}$ is modified as

$$P_i = P_i - K_{ij}\Phi_j^* \quad \text{for} \quad i = 1, 2, \ldots, M$$

Step (ii): The row and column of $[\underset{\sim}{K}]$ corresponding to Φ_j are made zero except the diagonal element, which is made unity; that is,

$$K_{ji} = K_{ij} = 0 \quad \text{for} \quad i = 1, 2, \ldots, M$$
$$K_{jj} = 1$$

Step (iii): The prescribed value of Φ_j is inserted in the characteristic vector as

$$P_j = \Phi_j^*$$

This procedure [steps (i)–(iii)] is repeated for all prescribed nodal degrees of freedom, Φ_j. This procedure retains the symmetry of the equations and the matrix $[\underset{\sim}{K}]$ can be stored in the band format with little extra programming effort.

Method 3

Another method of incorporating the prescribed condition $\Phi_j = \Phi_j^*$ is as follows:

Step (i): Multiply the diagonal term K_{jj} by a large number, such as 10^{10}, so that the new $K_{jj} = \text{old } K_{jj} \times 10^{10}$.

Step (ii): Take the corresponding load P_j as

$$P_j = \text{new } K_{jj} \times \Phi_j^* = \text{old } K_{jj} \times 10^{10} \times \Phi_j^*.$$

Step (iii): Keep all other elements of the characteristic matrix and the characteristic vector unaltered so that

$$\text{new } K_{ik} = \text{old } K_{ik} \text{ for all } i \text{ and } k \text{ except } i = k = j$$

and

$$\text{new } P_i = \text{old } P_i \text{ for all } i \text{ except } i = j$$

This procedure [steps (i)–(iii)] is repeated for all prescribed nodal degrees of freedom, Φ_j. This procedure will yield a solution in which Φ_j is very nearly equal to Φ_j^*. This method can also be used when the characteristic matrix is stored in banded form. We represent the equations that result from the application of the boundary conditions to Eq. (6.11) as

$$[K]\vec{\Phi} = \vec{P} \tag{6.18}$$

where $[K]$, $\vec{\Phi}$, and $\vec{P}$ denote the final (modified) characteristic matrix, vector of nodal degrees of freedom, and vector of nodal actions, respectively, of the complete body or system.

6.5 INCORPORATION OF BOUNDARY CONDITIONS IN THE COMPUTER PROGRAM

To incorporate the boundary conditions in the computer program according to method 2 of Section 6.3, a subroutine called ADJUST is written. This subroutine assumes that the global characteristic matrix GK is stored in a band form. If the degree of freedom "II" is to be set equal to a constant value "CØNST," the following Fortran statement calls the subroutine ADJUST, which modifies the matrix GK and vector P for incorporating the given boundary condition:

```
CALL ADJUST (GK,P,NN,NB,II,CØNST)
```

where NN is the total number of degrees of freedom, NB is the bandwidth of GK, P is the global vector of nodal actions (size: NN), and GK is the global characteristic matrix (size: NN × NB).

```
      SUBROUTINE ADJUST(A,B,NN,NB,II,CONST)
      DIMENSION A(NN,NB),B(NN)
      DO 10 J=2,NB
      I1=II-J+1
      I2=II+J-1
      IF(I1.GE.1) B(I1)=B(I1)-A(I1,J)*CONST
   10 IF(I2.LE.NN) B(I2)=B(I2)-A(II,J)*CONST
      B(II)=CONST
      DO 20 J=1,NB
      I1=II-J+1
      IF(I1.GE.1) A(I1,J)=0.0
   20 A(II,J)=0.0
      A(II,1)=1.0
      RETURN
      END
```

Note:

If the values of several degrees of freedom are to be prescribed, we have to incorporate these conditions one at a time by calling the subroutine ADJUST once for each prescribed degree of freedom.

To illustrate how the program works, consider the following simple example. Let the original system of equations be in the form

$$\begin{bmatrix} 1.9 & 2.1 & -5.7 & 0.0 & 0.0 \\ 2.1 & 3.4 & 1.5 & 3.3 & 0.0 \\ -5.7 & 1.5 & 2.2 & 4.5 & 2.8 \\ 0.0 & 3.3 & 4.5 & 5.6 & -1.8 \\ 0.0 & 0.0 & 2.8 & -1.8 & 4.7 \end{bmatrix} \begin{Bmatrix} \Phi_1 \\ \Phi_2 \\ \Phi_3 \\ \Phi_4 \\ \Phi_5 \end{Bmatrix} = \begin{Bmatrix} 0 \\ 0 \\ 0 \\ 0 \\ 0 \end{Bmatrix}$$

Thus, we can identify GK and P as

$$[GK] = \begin{bmatrix} 1.9 & 2.1 & -5.7 \\ 3.4 & 1.5 & 3.3 \\ 2.2 & 4.5 & 2.8 \\ 5.6 & -1.8 & 0.0 \\ 4.7 & 0.0 & 0.0 \end{bmatrix}, \qquad \vec{P} = \begin{Bmatrix} 0 \\ 0 \\ 0 \\ 0 \\ 0 \end{Bmatrix}, \qquad NN = 5, \qquad NB = 3$$

Let the boundary condition to be prescribed be $\Phi_3 = 2.0$ so that II = 3 and CØNST = 2.0. Then the calling statement

```
CALL ADJUST(GK,P,5,3,3,2.0)
```

returns the matrix GK and the vector P with the following values:

$$[GK] = \begin{bmatrix} 1.9 & 2.1 & 0.0 \\ 3.4 & 0.0 & 3.3 \\ 1.0 & 0.0 & 0.0 \\ 5.6 & -1.8 & 0.0 \\ 4.7 & 0.0 & 0.0 \end{bmatrix}, \qquad \vec{P} = \begin{Bmatrix} 11.4 \\ -3.0 \\ 2.0 \\ -9.0 \\ -5.6 \end{Bmatrix}$$

REFERENCES

6.1 O.S. Narayanaswamy: Processing nonlinear multipoint constraints in the finite element method, *International Journal for Numerical Methods in Engineering, 21,* 1283–1288, 1985.

6.2 P.E. Allaire: *Basics of the Finite Element Method—Solid Mechanics, Heat Transfer, and Fluid Mechanics*, Brown, Dubuque, IA, 1985.

PROBLEMS

6.1 Modify and solve the following system of equations using each of the methods described in Section 6.3 for the conditions $\Phi_1 = \Phi_2 = \Phi_3 = 2$, $\Phi_4 = 1$, $\Phi_8 = \Phi_9 = \Phi_{10} = 10$:

$$\begin{aligned}
&1.45\Phi_1 - 0.2\Phi_2 - 1.25\Phi_4 &= 0\\
&-0.2\Phi_1 + 2.45\Phi_2 - 1.25\Phi_5 - \Phi_6 &= 0\\
&\Phi_3 - 0.5\Phi_6 - 0.5\Phi_7 &= 0\\
&-1.25\Phi_1 + 2.90\Phi_4 - 0.4\Phi_5 - 1.25\Phi_8 &= 0\\
&-1.25\Phi_2 - 0.4\Phi_4 + 4.90\Phi_5 - \Phi_6 - 1.75\Phi_9 - 0.5\Phi_{10} &= 0\\
&-\Phi_2 - 0.5\Phi_3 - \Phi_5 + 4\Phi_6 - \Phi_7 - 0.5\Phi_{10} &= 0\\
&-0.5\Phi_3 - \Phi_6 + 2\Phi_7 - 0.5\Phi_{10} &= 0\\
&-1.25\Phi_4 + 1.45\Phi_8 - 0.2\Phi_9 &= 0\\
&-1.75\Phi_5 - 0.2\Phi_8 + 1.95\Phi_9 &= 0\\
&-0.5\Phi_5 - 0.5\Phi_6 - 0.5\Phi_7 + 1.5\Phi_{10} &= 0
\end{aligned}$$

6.2 Derive the coordinate transformation matrix for the one-dimensional element shown in Figure 6.3, where q_i and Q_i denote, respectively, the local (x, y) and the global (X, Y) nodal displacements of the element.

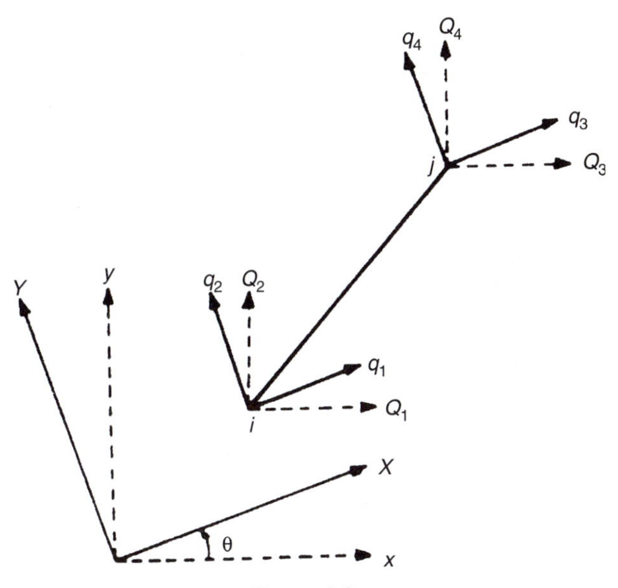

Figure 6.3.

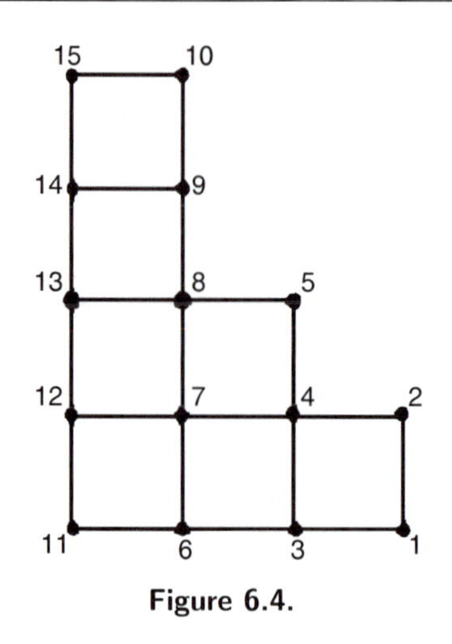

Figure 6.4.

6.3 If the element characteristic matrix of an element in the finite element grid shown in Figure 6.4 is given by

$$[K^{(e)}] = \begin{bmatrix} 3 & 1 & 1 & 1 \\ 1 & 3 & 1 & 1 \\ 1 & 1 & 3 & 1 \\ 1 & 1 & 1 & 3 \end{bmatrix}$$

find the overall or system characteristic matrix after applying the boundary conditions $\Phi_i = 0$, $i = 11$–15. Can the bandwidth be reduced by renumbering the nodes?

6.4 Incorporate the boundary conditions $\phi_1 = 3.0$ and $\phi_3 = -2.0$ using each of the methods described in Section 6.3 to the following system of equations:

$$\begin{bmatrix} 1.5 & -0.5 & 2.0 & 0.0 \\ -0.5 & 2.5 & -1.0 & -1.5 \\ 2.0 & -1.0 & 3.0 & 0.5 \\ 0.0 & -1.5 & 0.5 & 1.0 \end{bmatrix} \begin{Bmatrix} \phi_1 \\ \phi_2 \\ \phi_3 \\ \phi_4 \end{Bmatrix} = \begin{Bmatrix} 3.0 \\ -1.0 \\ 1.5 \\ -0.5 \end{Bmatrix}$$

6.5 Consider a node that is supported by rollers as indicated in Figure 6.5(a). In this case, the displacement normal to the roller surface XY must be zero:

$$Q = Q_5 \cos\alpha + Q_6 \sin\alpha = 0 \tag{E$_1$}$$

where α denotes the angle between the normal direction to the roller surface and the horizontal [Figure 6.5(b)]. Constraints, in the form of linear equations, involving multiple variables are known as multipoint constraints. Indicate two methods of incorporating the boundary condition of Eq. (E_1) in the solution of equations.

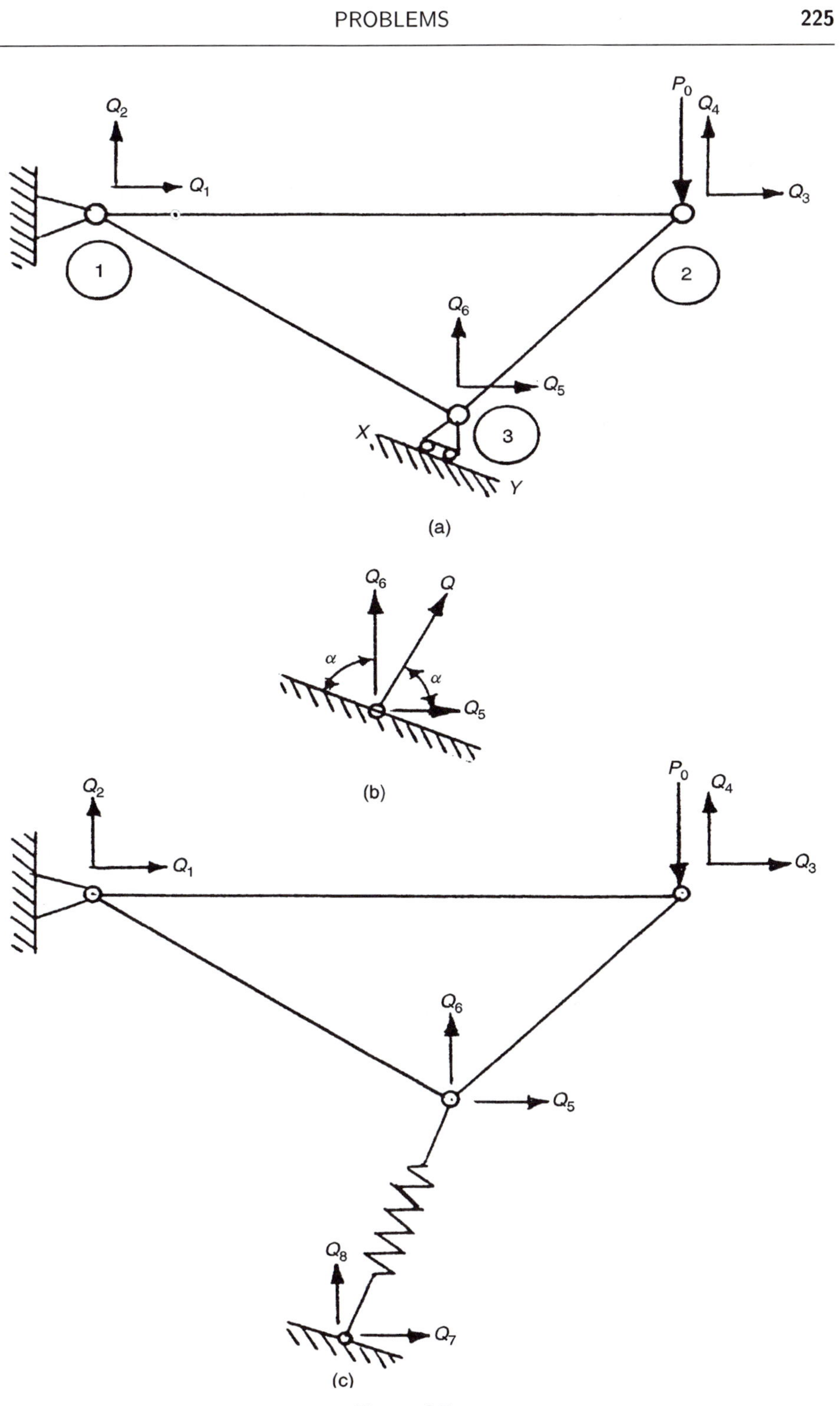

Figure 6.5.

6.6 As stated in Problem 6.5, the displacement normal to the roller support (surface) is zero. To incorporate the boundary condition, sometimes a stiff spring element is assumed perpendicular to the roller support surface as shown in Figure 6.5(c). In this case, the system will have four elements and eight degrees of freedom. The boundary conditions $Q_1 = Q_2 = Q_7 = Q_8 = 0$ are incorporated in this method. Show the structure of the assembled equations for this case and discuss the advantages and disadvantages of the approach.

6.7 The stiffness matrix of a planar frame element in the local coordinate system is given by (see Figure 6.6)

$$[k] = \begin{bmatrix} \frac{EA}{L} & 0 & 0 & \frac{-EA}{L} & 0 & 0 \\ 0 & \frac{12EI}{L^3} & \frac{6EI}{L^2} & 0 & \frac{-12EI}{L^3} & \frac{6EI}{L^2} \\ 0 & \frac{6EI}{L^2} & \frac{4EI}{L} & 0 & \frac{-6EI}{L^2} & \frac{2EI}{L} \\ \frac{-EA}{L} & 0 & 0 & \frac{EA}{L} & 0 & 0 \\ 0 & \frac{-12EI}{L^3} & \frac{-6EI}{L^2} & 0 & \frac{12EI}{L^3} & \frac{-6EI}{L^2} \\ 0 & \frac{6EI}{L^2} & \frac{2EI}{L} & 0 & \frac{-6EI}{L^2} & \frac{4EI}{L} \end{bmatrix}$$

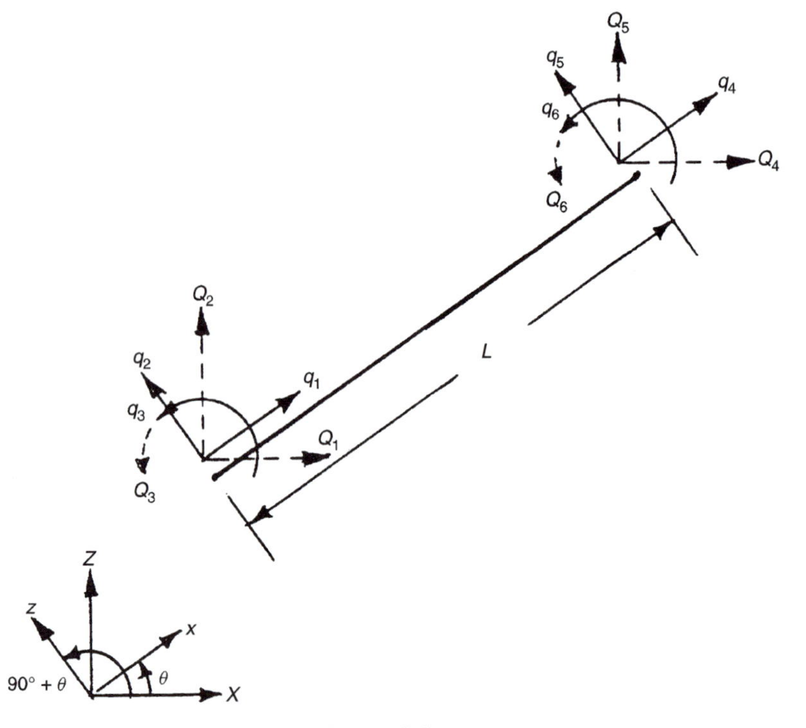

Figure 6.6.

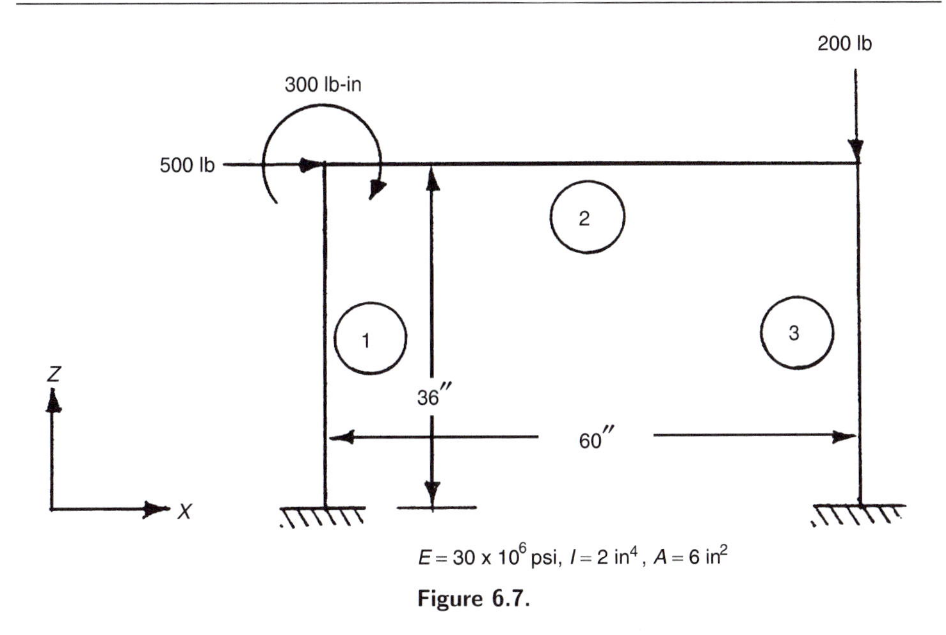

Figure 6.7.

where E is the Young's modulus, A is the area of cross section, I is the moment of inertia, and L is the length. Using this, generate the stiffness matrices of the three elements shown in Figure 6.7 in the local coordinate system and indicate the respective local degrees of freedom.

6.8 The transformation matrix between the local degrees of freedom q_i and the global degrees of freedom Q_i for the planar frame element shown in Figure 6.6 is given by

$$[\lambda] = \begin{bmatrix} l_{ox} & m_{ox} & 0 & 0 & 0 & 0 \\ l_{oz} & m_{oz} & 0 & 0 & 0 & 0 \\ 0 & 0 & 1 & 0 & 0 & 0 \\ 0 & 0 & 0 & l_{ox} & m_{ox} & 0 \\ 0 & 0 & 0 & l_{oz} & m_{oz} & 0 \\ 0 & 0 & 0 & 0 & 0 & 1 \end{bmatrix}$$

where $l_{ox} = \cos\theta$, $m_{ox} = \sin\theta$, $l_{oz} = \cos(90+\theta) = -\sin\theta$, and $m_{oz} = \sin(90+\theta) = \cos\theta$. Using this, generate the transformation matrices for the three elements shown in Figure 6.7.

6.9 Consider the coordinate transformation matrix, $[\lambda]$, of element 1 of Figure 6.7 in Problem 6.7. Show that it is orthogonal—that is, show that $[\lambda]^{-1} = [\lambda]^T$.

6.10 Consider the coordinate transformation matrix, $[\lambda]$, of element 2 of Figure 6.7 in Problem 6.7. Show that it is orthogonal—that is, show that $[\lambda]^{-1} = [\lambda]^T$.

6.11 Consider the coordinate transformation matrix, $[\lambda]$, of element 3 of Figure 6.7 in Problem 6.7. Show that it is orthogonal—that is, show that $[\lambda]^{-1} = [\lambda]^T$.

A truss
- 2 elements
- 2 pin supports

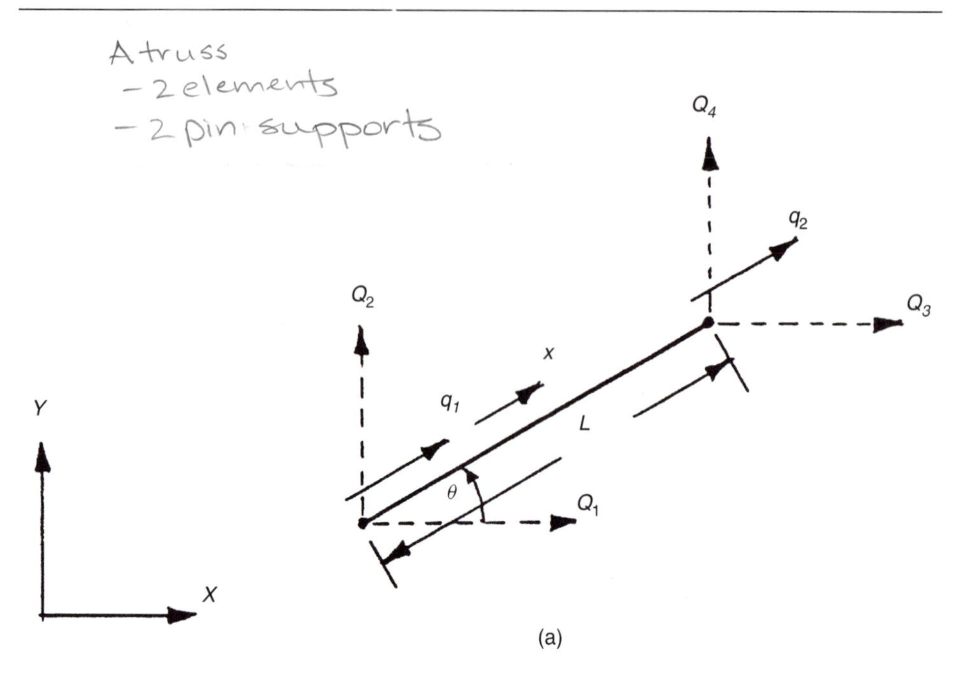

(a)

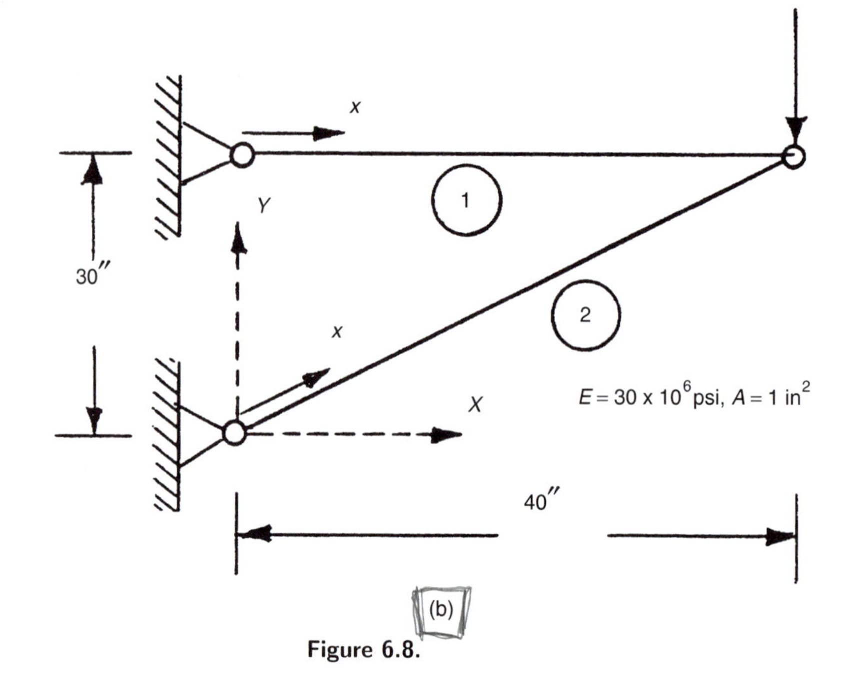

(b)

Figure 6.8.

6.12 Using the results of Problems 6.7 and 6.8, generate the stiffness matrices of the three elements shown in Figure 6.7 in the global coordinate system. Derive the assembled stiffness matrix of the system.

6.13 For the assembled stiffness matrix derived in Problem 6.12, apply the boundary conditions, derive the final equilibrium equations, and solve the resulting equations.

6.14 The local stiffness matrix, $[k]$, and the corresponding coordinate transformation matrix, $[\lambda]$, of a planar truss element [see Figure 6.8(a)] are given by

$$[k] = \frac{AE}{L}\begin{bmatrix} 1 & -1 \\ -1 & 1 \end{bmatrix}; \quad [\lambda] = \begin{bmatrix} l_{ox} & m_{ox} & 0 & 0 \\ 0 & 0 & l_{ox} & m_{ox} \end{bmatrix}$$

where A is the cross-sectional area, E is the Young's modulus, L is the length, $l_{ox} = \cos\theta$, and $m_{ox} = \sin\theta$.

(a) Generate the global stiffness matrices of the two elements shown in Figure 6.8(b).

(b) Find the assembled stiffness matrix, apply the boundary conditions, and find the displacement of node P of the two-bar truss shown in Figure 6.8(b).

6.15 Derive Eq. (6.8) using the equivalence of potential energies in the local and global coordinate systems.

7

NUMERICAL SOLUTION OF FINITE ELEMENT EQUATIONS

7.1 INTRODUCTION

Most problems in engineering mechanics can be stated either as continuous or discrete problems. Continuous problems involve infinite number of degrees of freedom, whereas discrete problems involve finite number of degrees of freedom. All discrete and continuous problems can be classified as equilibrium (static), eigenvalue, and propagation (transient) problems. The finite element method is applicable for the solution of all three categories of problems. As stated in Chapter 1, the finite element method is a numerical procedure that replaces a continuous problem by an equivalent discrete one. It will be quite convenient to use matrix notation in formulating and solving problems using the finite element procedure. When matrix notation is used in finite element analysis, the organizational properties of matrices allow for systematic compilation of the required data and the finite element analysis can then be defined as a sequence of matrix operations that can be programmed directly for a digital computer.

The governing finite element equations for various types of field problems can be expressed in matrix form as follows:

1. Equilibrium problems

$$[A]\vec{X} = \vec{b} \tag{7.1a}$$

subject to the boundary conditions

$$[B]\vec{X} = \vec{g} \tag{7.1b}$$

2. Eigenvalue problems

$$[A]\vec{X} = \lambda[B]\vec{X} \tag{7.2a}$$

subject to the boundary conditions

$$[C]\vec{X} = \vec{g} \tag{7.2b}$$

3. Propagation problems

$$[A]\frac{d^2\vec{X}}{dt^2} + [B]\frac{d\vec{X}}{dt} + [C]\vec{X} = \vec{F}(\vec{X}, t), \qquad t > 0 \tag{7.3a}$$

subject to the boundary conditions

$$[D]\vec{X} = \vec{g}, \qquad t \geq 0 \tag{7.3b}$$

and the initial conditions

$$\vec{X} = \vec{X}_0, \qquad t = 0 \tag{7.3c}$$

$$\frac{d\vec{X}}{dt} = \vec{Y}_0, \qquad t = 0 \tag{7.3d}$$

where $[A]$, $[B]$, $[C]$, and $[D]$ are square matrices whose elements are known to us; $\vec{X}$ is the vector of unknowns (or field variables) in the problem; $\vec{b}$, $\vec{g}$, $\vec{X}_0$, and $\vec{Y}_0$ are vectors of known constants; λ is the eigenvalue; t is the time parameter; and $\vec{F}$ is a vector whose elements are known functions of $\vec{X}$ and t.

In this chapter, an introduction to matrix techniques that are useful for the solution of Eqs. (7.1)–(7.3) is given along with a description of the relevant Fortran computer programs included in the disk.

7.2 SOLUTION OF EQUILIBRIUM PROBLEMS

When the finite element method is used for the solution of equilibrium or steady state or static problems, we get a set of simultaneous linear equations that can be stated in the form of Eq. (7.1). We shall consider the solution of Eq. (7.1a) in this section by assuming that the boundary conditions of Eq. (7.1b) have been incorporated already.

Equation (7.1a) can be expressed in scalar form as

$$\begin{aligned} a_{11}x_1 + a_{12}x_2 + \cdots + a_{1n}x_n &= b_1 \\ a_{21}x_1 + a_{22}x_2 + \cdots + a_{2n}x_n &= b_2 \\ &\vdots \\ a_{n1}x_1 + a_{n2}x_2 + \cdots + a_{nn}x_n &= b_n \end{aligned} \tag{7.4}$$

where the coefficients a_{ij} and the constants b_i are either given or can be generated. The problem is to find the values of x_i $(i = 1, 2, \ldots, n)$, if they exist, which satisfy Eq. (7.4). A comparison of Eqs. (7.1a) and (7.4) shows that

$$\underset{n\times n}{[A]} = \begin{bmatrix} a_{11} & a_{12} & \ldots & a_{1n} \\ a_{21} & a_{22} & \ldots & a_{2n} \\ & & \vdots & \\ a_{n1} & a_{n2} & & a_{nn} \end{bmatrix}, \quad \underset{n\times 1}{\vec{X}} = \begin{Bmatrix} x_1 \\ x_2 \\ \vdots \\ x_n \end{Bmatrix}, \quad \underset{n\times 1}{\vec{b}} = \begin{Bmatrix} b_1 \\ b_2 \\ \vdots \\ b_n \end{Bmatrix}$$

In finite element analysis, the order of the matrix $[A]$ will be very large. The solution of some of the practical problems involves matrices of order 10,000 or more.

The methods available for solving systems of linear equations can be divided into two types: direct and iterative. Direct methods are those that, in the absence of round-off and other errors, will yield the exact solution in a finite number of elementary arithmetic operations. In practice, because a computer works with a finite word length, sometimes the direct methods do not give good solutions. Indeed, the errors arising from round-off and truncation may lead to extremely poor or even useless results. Hence, many researchers working in the field of finite element method are concerned with why and how the errors arise and with the search for methods that minimize the totality of such errors. The fundamental method used for direct solutions is Gaussian elimination, but even within this class there are a variety of choices of methods that vary in computational efficiency and accuracy.

Iterative methods are those that start with an initial approximation and that by applying a suitably chosen algorithm lead to successively better approximations. When the process converges, we can expect to get a good approximate solution. The accuracy and the rate of convergence of iterative methods vary with the algorithm chosen. The main advantages of iterative methods are the simplicity and uniformity of the operations to be performed, which make them well suited for use on digital computers, and their relative insensitivity to the growth of round-off errors.

Matrices associated with linear systems are also classified as dense or sparse. Dense matrices have very few zero elements, whereas sparse matrices have very few nonzero elements. Fortunately, in most finite element applications, the matrices involved are sparse (thinly populated) and symmetric. Hence, solution techniques that take advantage of the special character of such systems of equations have been developed.

7.2.1 Gaussian Elimination Method

The basic procedure available for the solution of Eq. (7.1) is the Gaussian elimination method, in which the given system of equations is transformed into an equivalent triangular system for which the solution can be easily found.

We first consider the following system of three equations to illustrate the Gaussian elimination method:

$$x_1 - x_2 + 3x_3 = 10 \tag{E$_1$}$$

$$2x_1 + 3x_2 + x_3 = 15 \tag{E$_2$}$$

$$4x_1 + 2x_2 - x_3 = 6 \tag{E$_3$}$$

To eliminate the x_1 terms from Eqs. (E_2) and (E_3), we multiply Eq. (E_1) by -2 and -4 and add respectively to Eqs. (E_2) and (E_3) leaving the first equation unchanged. We will then have

$$x_1 - x_2 + 3x_3 = 10 \tag{E$_4$}$$

$$5x_2 - 5x_3 = -5 \tag{E$_5$}$$

$$6x_2 - 13x_3 = -34 \tag{E$_6$}$$

To eliminate the x_2 term from Eq. (E_6), multiply Eq. (E_5) by $-6/5$ and add to Eq. (E_6). We will now have the triangular system

$$x_1 - x_2 + 3x_3 = 10 \quad (E_7)$$

$$5x_2 - 5x_3 = -5 \quad (E_8)$$

$$-7x_3 = -28 \quad (E_9)$$

This triangular system can now be solved by back substitution. From Eq. (E_9) we find $x_3 = 4$. Substituting this value for x_3 into Eq. (E_8) and solving for x_2, we obtain $x_2 = 3$. Finally, knowing x_3 and x_2, we can solve Eq. (E_1) for x_1, obtaining $x_1 = 1$. This solution can also be obtained by adopting the following equivalent procedure.

Equation (E_1) can be solved for x_1 to obtain

$$x_1 = 10 + x_2 - 3x_3 \quad (E_{10})$$

Substitution of this expression for x_1 into Eqs. (E_2) and (E_3) gives

$$5x_2 - 5x_3 = -5 \quad (E_{11})$$

$$6x_2 - 13x_3 = -34 \quad (E_{12})$$

The solution of Eq. (E_{11}) for x_2 leads to

$$x_2 = -1 + x_3 \quad (E_{13})$$

By substituting Eq. (E_{13}) into Eq. (E_{12}) we obtain

$$-7x_3 = -28 \quad (E_{14})$$

It can be seen that Eqs. (E_{10}), (E_{11}), and (E_{14}) are the same as Eqs. (E_7), (E_8), and (E_9), respectively. Hence, we can obtain $x_3 = 4$ from Eq. (E_{14}), $x_2 = 3$ from Eq. (E_{13}), and $x_1 = 1$ from Eq. (E_{10}).

Generalization of the Method

Let the given system of equations be written as

$$\begin{aligned} a_{11}^{(0)} x_1 + a_{12}^{(0)} x_2 + \cdots + a_{1n}^{(0)} x_n &= b_1^{(0)} \\ a_{21}^{(0)} x_1 + a_{22}^{(0)} x_2 + \cdots + a_{2n}^{(0)} x_n &= b_2^{(0)} \\ &\vdots \\ a_{n1}^{(0)} x_1 + a_{n2}^{(0)} x_2 + \cdots + a_{nn}^{(0)} x_n &= b_n^{(0)} \end{aligned} \quad (7.5)$$

where the superscript $^{(0)}$ has been used to denote the original values. By solving the first equation of Eq. (7.5) for x_1, we obtain

$$x_1 = \frac{b_1^{(0)}}{a_{11}^{(0)}} - \frac{a_{12}^{(0)}}{a_{11}^{(0)}} x_2 - \frac{a_{13}^{(0)}}{a_{11}^{(0)}} x_3 - \cdots - \frac{a_{1n}^{(0)}}{a_{11}^{(0)}} x_n$$

Substitution of this x_1 into the remaining equations of Eq. (7.5) leads to

$$\begin{aligned} a_{22}^{(1)} x_2 + a_{23}^{(1)} x_3 + \cdots + a_{2n}^{(1)} x_n &= b_2^{(1)} \\ &\vdots \\ a_{n2}^{(1)} x_2 + a_{n3}^{(1)} x_3 + \cdots + a_{nn}^{(1)} x_n &= b_n^{(1)} \end{aligned} \tag{7.6}$$

where

$$\left. \begin{aligned} a_{ij}^{(1)} &= a_{ij}^{(0)} - \left[a_{i1}^{(0)} a_{1j}^{(0)} / a_{11}^{(0)} \right] \\ b_i^{(1)} &= b_i^{(0)} - \left[a_{i1}^{(0)} b_1^{(0)} / a_{11}^{(0)} \right] \end{aligned} \right\} \quad i, j = 2, 3, \ldots, n$$

Next, we eliminate x_2 from Eq. (7.6), and so on. In general, when x_k is eliminated we obtain

$$x_k = \frac{b_k^{(k-1)}}{a_{kk}^{(k-1)}} - \sum_{j=k+1}^{n} \frac{a_{kj}^{(k-1)}}{a_{kk}^{(k-1)}} x_j \tag{7.7}$$

where

$$\left. \begin{aligned} a_{ij}^{(k)} &= a_{ij}^{(k-1)} - \left[a_{ik}^{(k-1)} a_{kj}^{(k-1)} / a_{kk}^{(k-1)} \right] \\ b_i^{(k)} &= b_i^{(k-1)} - \left[a_{ik}^{(k-1)} b_k^{(k-1)} / a_{kk}^{(k-1)} \right] \end{aligned} \right\} \quad i, j = k+1, \ldots, n$$

After applying the previous procedure $n - 1$ times, the original system of equations reduces to the following single eqaution:

$$a_{nn}^{(n-1)} x_n = b_n^{(n-1)}$$

from which we can obtain

$$x_n = \left[b_n^{(n-1)} / a_{nn}^{(n-1)} \right]$$

The values of the remaining unknowns can be found in the reverse order $(x_{n-1}, x_{n-2}, \ldots, x_1)$ by using Eq. (7.7).

Note: In the elimination process, if at any stage one of the pivot (diagonal) elements $a_{11}^{(0)}, a_{22}^{(1)}, a_{33}^{(2)}, \ldots,$ vanishes, we attempt to rearrange the remaining rows so as to obtain a nonvanishing pivot. If this is impossible, then the matrix $[A]$ is singular and the system has no solution.

Computer Implementation

A Fortran subroutine called GAUSS is given for the solution of

$$[A]\vec{X} = \vec{b} \tag{7.1a}$$

based on the Gaussian elimination method. This subroutine can be used to find the solution of Eq. (7.1a) for several right-hand-side vectors $\vec{b}$ and/or to find the inverse of the matrix $[A]$. The arguments of the subroutine are as follows:

A = array of order $N \times N$ in which the given coefficient matrix $[A]$ is stored at the beginning. The array A returned from the subroutine GAUSS gives the inverse $[A]^{-1}$.

B = array of dimension $N \times M$. If the solution of Eq. (7.1a) is required for several right-hand-side vectors $\vec{b}_i$ ($i = 1, 2, \ldots, M$), the vectors $\vec{b}_1, \vec{b}_2, \ldots$ are stored columnwise in the array B of order $N \times M$. Upon return from the subroutine GAUSS, the ith column of B represents the solution $\vec{X}_i$ of the problem $[A]\vec{x}_i = \vec{b}_i$ ($i = 1, 2, \ldots, M$). The array B will not be used if inverse of $[A]$ only is required.

N = order of the square matrix $[A]$; same as the number of equations to be solved.

M = number of the right-hand-side vectors $\vec{b}_i$ for which solutions are required. If only the inverse of $[A]$ is required, M is set to be equal to 1.

IFLAG = 0 if only the inverse of $[A]$ is required.
= 1 if the solution of Eq. (7.1a) is required (for any value of $M \geq 1$).

LP = a dummy vector array of dimension N.

LQ = a dummy array of dimension $N \times 2$.

R = a dummy vector of dimension N.

To illustrate the use of the subroutine GAUSS, we consider the following system of equations:

$$\begin{bmatrix} 1 & 10 & 1 \\ 2 & 0 & 1 \\ 3 & 3 & 2 \end{bmatrix} \begin{Bmatrix} x_1 \\ x_2 \\ x_3 \end{Bmatrix} = \begin{Bmatrix} 7 \\ 0 \\ 14 \end{Bmatrix} \tag{E$_1$}$$

Here, the number of equations to be solved is $N = 3$, with $M = 1$ and IFLAG = 1. The main program for solving Eq. (E$_1$) using the subroutine GAUSS is given below. The result given by the program is also included at the end.

```
C======================================================================
C
C     MAIN PROGRAM TO CALL THE SUBROUTINE GAUSS
C
C======================================================================
      DIMENSION A(3,3),B(3,1),LP(3),LQ(3,2),R(3)
      DATA((A(I,J),J=1,3),I=1,3)/1.0,10.0,1.0,2.0,0.0,1.0,3.0,3.0,2.0/
      DATA(B(I,1),I=1,3)/7.0,0.0,14.0/
      DATA N,M,IFLAG/3,1,1/
      PRINT 10,((A(I,J),J=1,3),I=1,3)
      PRINT 20,(B(I,1),I=1,3)
      CALL GAUSS(A,B,N,M,IFLAG,LP,LQ,R)
      PRINT 30,((A(I,J),J=1,3),I=1,3)
      PRINT 40,(B(I,1),I=1,3)
10    FORMAT(2X,'ORIGINAL COEFFICIENT MATRIX',//,3(E13.6,1X))
20    FORMAT(/,2X,'RIGHT HAND SIDE VECTOR',//,3(E13.6,1X))
30    FORMAT(/,2X,'INVERSE OF COEFFICIENT MATRIX',//,3(E13.6,1X))
```

```
40   FORMAT(/,2X,'SOLUTION VECTOR',//,3(E13.6,1X))
     STOP
     END
```

```
 ORIGINAL COEFFICIENT MATRIX

0.100000E+01   0.100000E+02   0.100000E+01
0.200000E+01   0.000000E+00   0.100000E+01
0.300000E+01   0.300000E+01   0.200000E+01

 RIGHT HAND SIDE VECTOR

0.700000E+01   0.000000E+00   0.140000E+02

 INVERSE OF COEFFICIENT MATRIX

 0.428572E+00   0.242857E+01  -0.142857E+01
 0.142857E+00   0.142857E+00  -0.142857E+00
-0.857143E+00  -0.385714E+01   0.285714E+01

 SOLUTION VECTOR

-0.170000E+02  -0.100000E+01   0.340000E+02
```

7.2.2 Choleski Method

The Choleski method is a direct method for solving a linear system that makes use of the fact that any square matrix $[A]$ can be expressed as the product of an upper and a lower triangular matrix. The method of expressing any square matrix as a product of two triangular matrices and the subsequent solution procedure are given below.

(i) Decomposition of [A] into lower and upper triangular matrices

The given system of equations is

$$[A]\vec{X} = \vec{b} \tag{7.1a}$$

The matrix $[A]$ can be written as

$$[A] = [a_{ij}] = [L][U] \tag{7.8}$$

where $[L] = [l_{ij}]$ is a lower triangular matrix, and $[U] = [u_{ij}]$ is a unit upper triangular matrix, with

$$[A] = \begin{bmatrix} a_{11} & a_{12} & \cdots & a_{1n} \\ a_{21} & a_{22} & \cdots & a_{2n} \\ \vdots & & & \\ a_{n1} & a_{n2} & \cdots & a_{nn} \end{bmatrix} = [L][U] \tag{7.9}$$

$$[L] = \begin{bmatrix} l_{11} & 0 & 0 & \cdots & 0 \\ l_{21} & l_{22} & 0 & \cdots & 0 \\ \vdots & & & & \\ l_{n1} & l_{n2} & l_{n3} & \cdots & l_{nn} \end{bmatrix} = \text{a lower triangular matrix} \tag{7.10}$$

and

$$[U] = \begin{bmatrix} 1 & u_{12} & u_{13} & \cdots & u_{1n} \\ 0 & 1 & u_{23} & \cdots & u_{2n} \\ 0 & 0 & 1 & \cdots & u_{3n} \\ \vdots & \vdots & \vdots & & \vdots \\ 0 & 0 & 0 & \cdots & 1 \end{bmatrix} = \text{a unit upper triangular matrix} \tag{7.11}$$

The elements of $[L]$ and $[U]$ satisfying the unique factorization $[A] = [L][U]$ can be determined from the recurrence formulas

$$\begin{aligned} l_{ij} &= a_{ij} - \sum_{k=1}^{j-1} l_{ik} u_{kj}, \qquad i \geq j \\ u_{ij} &= \frac{a_{ij} - \sum_{k=1}^{i-1} l_{ik} u_{kj}}{l_{ii}}, \qquad i < j \\ u_{ii} &= 1 \end{aligned} \tag{7.12}$$

For the relevant indices i and j, these elements are computed in the order

$$l_{i1}, u_{1j}; \quad l_{i2}, u_{2j}; \quad l_{i3}, u_{3j}; \quad \ldots; \quad l_{i,n-1}, u_{n-1,j}; \quad l_{nn}$$

(ii) Solution of equations

Once the given system of equations $[A]\vec{X} = \vec{b}$ is expressed in the form $[L][U]\vec{X} = \vec{b}$, the solution can be obtained as follows:

By letting

$$[U]\vec{X} = \vec{Z} \tag{7.13}$$

the equations become $[L]\vec{Z} = \vec{b}$, which in expanded form can be written as

$$\begin{aligned} l_{11}z_1 &= b_1 \\ l_{21}z_1 + l_{22}z_2 &= b_2 \\ l_{31}z_1 + l_{32}z_2 + l_{33}z_3 &= b_3 \\ &\cdots \\ &\cdots \\ &\cdots \\ l_{n1}z_1 + l_{n2}z_2 + l_{n3}z_3 + \cdots + l_{nn}z_n &= b_n \end{aligned} \tag{7.14}$$

The first of these equations can be solved for z_1, after which the second can be solved for z_2, the third for z_3, etc. We can thus determine in succession $z_1, z_2, \ldots, z_n$, provided that none of the diagonal elements l_{ii} $(i = 1, 2, \ldots, n)$ vanishes. Once z_i are obtained the values of x_i can be found by writing Eq. (7.13) as

$$\begin{aligned} x_1 + u_{12}x_2 + u_{13}x_3 + \cdots + u_{1n}x_n &= z_1 \\ x_2 + u_{23}x_3 + \cdots + u_{2n}x_n &= z_2 \\ x_3 + \cdots + u_{3n}x_n &= z_3 \\ &\cdots \\ &\cdots \\ &\cdots \\ x_{n-1} + u_{n-1,n}x_n &= z_{n-1} \\ x_n &= z_n \end{aligned} \tag{7.15}$$

Just as in the Gaussian elimination process, this system can now be solved by back substitution for $x_n, x_{n-1}, \ldots, x_1$ in that order.

(iii) Choleski decomposition of symmetric matrices

In most applications of finite element theory, the matrices involved will be symmetric, banded, and positive definite. In such cases, the symmetric positive definite matrix $[A]$ can be decomposed uniquely as*

$$[A] = [U]^T[U] \tag{7.16}$$

where

$$[U] = \begin{bmatrix} u_{11} & u_{12} & u_{13} & \cdots & u_{1n} \\ 0 & u_{22} & u_{23} & \cdots & u_{2n} \\ 0 & 0 & u_{33} & \cdots & u_{3n} \\ \vdots & \vdots & \vdots & & \vdots \\ 0 & 0 & 0 & & u_{nn} \end{bmatrix} \tag{7.17}$$

is an upper triangular matrix including the diagonal. The elements of $[U] = [u_{ij}]$ are given by

$$\left.\begin{aligned} &u_{11} = (a_{11})^{(1/2)} \\ &u_{1j} = a_{1j}/u_{11}, \quad j = 2, 3, \ldots, n \\ &u_{ii} = \left(a_{ii} - \sum_{k=1}^{i-1} u_{ki}^2\right)^{(1/2)}, \quad i = 2, 3, \ldots, n \\ &u_{ij} = \frac{1}{u_{ii}}\left(a_{ij} - \sum_{k=1}^{i-1} u_{ki}u_{kj}\right), \quad \begin{aligned} &i = 2, 3, \ldots, n, \text{ and} \\ &j = i+1, i+2, \ldots n \end{aligned} \\ &u_{ij} = 0, \quad i > j. \end{aligned}\right\} \tag{7.18}$$

* The matrix $[A]$ can also be decomposed as $[A] = [L][L]^T$, where $[L]$ represents a lower triangular matrix. The elements of $[L]$ can be found in Ref. [7.1] and also in Problem 7.2.

(iv) Inverse of a symmetric matrix

If the inverse of the symmetric matrix $[A]$ is needed, we first decompose it as $[A] = [U]^T[U]$ using Eq. (7.18), and then find $[A]^{-1}$ as

$$[A]^{-1} = [[U]^T[U]]^{-1} = [U]^{-1}([U]^T)^{-1} \tag{7.19}$$

The elements λ_{ij} of $[U]^{-1}$ can be determined from $[U][U]^{-1} = [I]$, which leads to

$$\left.\begin{aligned} \lambda_{ii} &= \frac{1}{u_{ii}} \\ \lambda_{ij} &= \frac{-\left(\sum_{k=i+1}^{j} u_{ik}\lambda_{kj}\right)}{u_{ii}}, \quad i < j \\ \lambda_{ij} &= 0, \quad i > j \end{aligned}\right\} \tag{7.20}$$

Hence, the inverse of $[U]$ is also an upper triangular matrix. The inverse of $[U]^T$ can be obtained from the relation

$$([U]^T)^{-1} = ([U]^{-1})^T \tag{7.21}$$

Finally, the inverse of the symmetric matrix $[A]$ can be calculated as

$$[A]^{-1} = [U]^{-1}([U]^{-1})^T \tag{7.22}$$

(v) Computer implementation of the Choleski method

A FØRTRAN computer program to implement the Choleski method is given. This program requires the subroutines DECØMP and SØLVE. These subroutines can be used for solving any system of N linear equations

$$[A]\vec{X} = \vec{b} \tag{7.1a}$$

where $[A]$ is a symmetric banded matrix of order N. It is assumed that the elements of the matrix $[A]$ are stored in band form in the first N rows and NB columns of the array A, where NB denotes the semi-bandwidth of the matrix $[A]$. Thus, the diagonal terms a_{ii} of $[A]$ occupy the locations $A(I, 1)$.

The subroutine DECØMP decomposes the matrix $[A]$ (stored in the form of array A) into $[A] = [U]^T[U]$ and the elements of the upper triangular matrix $[U]$ are stored in the array A. The subroutine SØLVE solves the equations (7.1a) by using the decomposed coefficient matrix $[A]$. This subroutine has the capability of solving the equations (7.1a) for different right-hand-side vectors $\vec{b}$. If $\vec{b}_1, \vec{b}_2, \ldots, \vec{b}_M$ indicate the right-hand-side vectors* for which the corresponding solutions $\vec{X}_1, \vec{X}_2, \ldots, \vec{X}_M$ are to be found, all the vectors $\vec{b}_1, \vec{b}_2, \ldots, \vec{b}_M$ are stored columnwise in the array B. Thus, the jth element of $\vec{b}_i$ will be

* The right-hand-side vectors $\vec{b}_1, \vec{b}_2, \ldots,$ represent different load vectors (corresponding to different load conditions) in a static structural or solid mechanics problem.

stored as $B(J, I)$, $J = 1, 2, \ldots, N$. The equations to be solved for any right-hand-side vector $\vec{b}$ can be expressed as $[A]\vec{X} = [U]^T[U]\vec{X} = \vec{b}$. These equations can be solved as

$$[U]\vec{X} = ([U]^T)^{-1}\vec{b} \equiv \vec{Z} \text{ (say)}$$

and

$$\vec{X} = [U]^{-1}\vec{Z}$$

In the subroutine SØLVE, the vectors $\vec{Z}_i$ for different $\vec{b}_i$ are found in the forward pass and are stored in the array B. The solutions $\vec{X}_i$ corresponding to different $\vec{b}_i$ are found in the backward pass and are stored in the array B. Thus, the columns of the array B returned from SØLVE will give the desired solutions $\vec{X}_i$, $i = 1, 2, \ldots, M$.

As an example, consider the following system of equations:

$$\begin{bmatrix} 1 & -1 & 0 & 0 & 0 \\ -1 & 2 & -1 & 0 & 0 \\ 0 & -1 & 2 & -1 & 0 \\ 0 & 0 & -1 & 2 & -1 \\ 0 & 0 & 0 & -1 & 2 \end{bmatrix} \vec{X}_i = \vec{b}_i \tag{E$_1$}$$

where

$$\vec{X}_i = \begin{Bmatrix} x_1 \\ x_2 \\ x_3 \\ x_4 \\ x_5 \end{Bmatrix}_i, \quad \vec{b}_1 = \begin{Bmatrix} 1 \\ 0 \\ 0 \\ 0 \\ 0 \end{Bmatrix}, \quad \vec{b}_2 = \begin{Bmatrix} 0 \\ 0 \\ 0 \\ 0 \\ 1 \end{Bmatrix}, \quad \text{and} \quad \vec{b}_3 = \begin{Bmatrix} 1 \\ 1 \\ 1 \\ 1 \\ 1 \end{Bmatrix}$$

Here, the number of equations $= N = 5$, the semi-bandwidth of $[A] = NB = 2$, and the number of vectors $\vec{b}_i = M = 3$.

The main program for solving the system of equations, (E_1), along with the results, is given below.

```
C=======================================================================
C
C     MAIN PROGRAM TO CALL DECOMP AND SOLVE
C
C=======================================================================
      DIMENSION A(5,2),B(5,3)
      DOUBLE PRECISION DIFF(3)
      DATA(A(I,1),I=1,5)/1.,2.,2.,2.,2./
      DATA(A(I,2),I=1,5)/-1.,-1.,-1.,-1.,0./
      DATA(B(I,1),I=1,5)/1.,0.,0.,0.,0./
      DATA(B(I,2),I=1,5)/0.,0.,0.,0.,1./
      DATA(B(I,3),I=1,5)/1.,1.,1.,1.,1./
      DATA N,NB,M/5,2,3/
      CALL DECOMP(N,NB,A)
```

```
      CALL SOLVE(N,NB,M,A,B,DIFF)
      DO 10 J=1,M
10    PRINT 20,J,(B(I,J),I=1,N)
20    FORMAT(1X,'SOLUTION:',I5,/,(6E15.8))
      STOP
      END
```

```
SOLUTION:    1
0.50000000E+01 0.40000000E+01 0.30000000E+01 0.20000000E+01 0.10000000E+01
SOLUTION:    2
0.10000000E+01 0.10000000E+01 0.10000000E+01 0.10000000E+01 0.10000000E+01
SOLUTION:    3
0.15000000E+02 0.14000000E+02 0.12000000E+02 0.90000000E+01 0.50000000E+01
```

7.2.3 Other Methods

In the computer programs DECØMP and SØLVE given in Section 7.2.2, advantage of the properties of symmetry and bandform is taken in storing the matrix $[A]$. In fact, the obvious advantage of small bandwidth has prompted engineers involved in finite element analysis to develop schemes to model systems so as to minimize the bandwidth of resulting matrices. Despite the relative compactness of bandform storage, computer core space may be inadequate for the bandform storage of matrices of extremely large systems. In such a case, the matrix is partitioned as shown in Figure 7.1, and only a few of the triangular submatrices are stored in the computer core at a given time; the remaining ones are kept in auxiliary storage, for example, on a tape or a disk. Several other schemes, such as the frontal or wavefront solution methods, have been developed for handling large matrices [7.2–7.5].

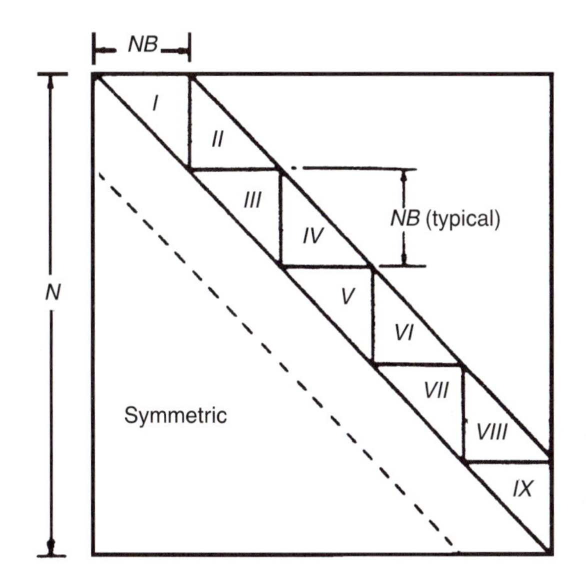

Figure 7.1. Partitioning of a Large Matrix.

The Gauss elimination and Choleski decomposition schemes fall under the category of direct methods. In the class of iterative methods, the Gauss–Seidel method is well-known [7.6]. The conjugate gradient and Newton's methods are other iterative methods based on the principle of unconstrained minimization of a function [7.7, 7.8]. Note that the indirect methods are less popular than the direct methods in solving large systems of linear equations [7.9]. Special computer programs have been developed for the solution of finite element equations on small computers [7.10].

7.3 SOLUTION OF EIGENVALUE PROBLEMS

When the finite element method is applied for the solution of eigenvalue problems, we obtain an algebraic eigenvalue problem as stated in Eq. (7.2). We consider the solution of Eq. (7.2a) in this section, assuming that the boundary conditions, Eq. (7.2b), have been incorporated already. For most engineering problems, $[A]$ and $[B]$ will be symmetric matrices of order n. λ is a scalar (called the eigenvalue), and $\vec{X}$ is a column vector with n components (called the eigenvector). If the physical problem is the free vibration analysis of a structure, $[A]$ will be the stiffness matrix, $[B]$ will be the mass matrix, λ is the square of natural frequency, and $\vec{X}$ is the mode shape of the vibrating structure.

The eigenvalue problem given by Eq. (7.2a) can be rewritten as

$$([A] - \lambda[B])\vec{X} = \vec{0} \tag{7.23}$$

which can have solutions for

$$\vec{X} = \begin{Bmatrix} x_1 \\ x_2 \\ \vdots \\ x_n \end{Bmatrix}$$

other than zero only if the determinant of the coefficients vanishes; that is,

$$\begin{vmatrix} a_{11} - \lambda b_{11} & a_{12} - \lambda b_{12} & \dots & a_{1n} - \lambda b_{1n} \\ a_{21} - \lambda b_{21} & a_{22} - \lambda b_{22} & \dots & a_{2n} - \lambda b_{2n} \\ \vdots & \vdots & & \vdots \\ a_{n1} - \lambda b_{n1} & a_{n2} - \lambda b_{n2} & \dots & a_{nn} - \lambda b_{nn} \end{vmatrix} = 0 \tag{7.24}$$

If the determinant in Eq. (7.24) is expanded, we obtain an algebraic equation of nth degree for λ. This equation is called the characteristic equation of the system. The n roots of this equation are the n eigenvalues of Eq. (7.2a). The eigenvector corresponding to any λ_j, namely $\vec{X}_j$, can be found by inserting λ_j in Eq. (7.23) and solving for the ratios of the elements in $\vec{X}_j$. A practical way to do this is to set x_n, for example, equal to unity and solve the first $n-1$ equations for $x_1, x_2, \dots, x_{n-1}$. The last equation may be used as a check.

From Eq. (7.2a), it is evident that if $\vec{X}$ is a solution, then $k\vec{X}$ will also be a solution for any nonzero value of the scalar k. Thus, the eigenvector corresponding to any eigenvalue is arbitrary to the extent of a scalar multiplier. It is convenient to choose this multiplier so that the eigenvector has some desirable numerical property, and such vectors are called normalized vectors. One method of normalization is to make the component of the vector $\vec{X}_i$ having the largest magnitude equal to unity; that is

$$\max_{j=1,2,\ldots,n} (x_{ij}) = 1 \tag{7.25}$$

where x_{ij} is the jth component of the vector $\vec{X}_i$. Another method of normalization commonly used in structural dynamics is as follows:

$$\vec{X}_i^T [B] \vec{X}_i = 1 \tag{7.26}$$

7.3.1 Standard Eigenvalue Problem

Although the procedure given previously for solving the eigenvalue problem appears to be simple, the roots of an nth degree polynomial cannot be obtained easily for matrices of high order. Hence, in most of the computer-based methods used for the solution of Eq. (7.2a), the eigenvalue problem is first converted into the form of a standard eigenvalue problem, which can be stated as

$$[H]\vec{X} = \lambda \vec{X} \text{ or } ([H] - \lambda[I])\vec{X} = \vec{0} \tag{7.27}$$

By premultiplying Eq. (7.2a) by $[B]^{-1}$, we obtain Eq. (7.27), where

$$[H] = [B]^{-1}[A] \tag{7.28}$$

However, in this form the matrix $[H]$ is in general nonsymmetric, although $[B]$ and $[A]$ are both symmetric. Since a symmetric matrix is desirable from the standpoint of storage and computer time, we adopt the following procedure to derive a standard eigenvalue problem with symmetric $[H]$ matrix.

Assuming that $[B]$ is symmetric and positive definite, we use Choleski decomposition and express $[B]$ as $[B] = [U]^T[U]$. By substituting for $[B]$ in Eq. (7.2a), we obtain

$$[A]\vec{X} = \lambda [U]^T [U] \vec{X}$$

and hence

$$([U]^T)^{-1}[A][U]^{-1}[U]\vec{X} = \lambda [U]\vec{X} \tag{7.29}$$

By defining a new vector $\vec{Y}$ as $\vec{Y} = [U]\vec{X}$, Eq. (7.29) can be written as a standard eigenvalue problem as

$$([H] - \lambda[I])\vec{Y} = \vec{0} \tag{7.30}$$

where the matrix $[H]$ is now symmetric and is given by

$$[H] = ([U]^T)^{-1}[A][U]^{-1} \tag{7.31}$$

To formulate $[H]$ according to Eq. (7.31), we decompose the symmetric matrix $[B]$ as $[B] = [U]^T[U]$, as indicated in Section 7.2.2(iii), find $[U]^{-1}$ and $([U]^T)^{-1}$ as shown in

Section 7.2.2(iv), and then carry out the matrix multiplication as stated in Eq. (7.31). The solution of the eigenvalue problem stated in Eq. (7.30) yields λ_i and $\vec{Y}_i$. Then we apply the inverse transformation to obtain the desired eigenvectors as

$$\vec{X}_i = [U]^{-1}\vec{Y}_i \tag{7.32}$$

We now discuss some of the methods of solving the special eigenvalue problem stated in Eq. (7.27).

7.3.2 Methods of Solving Eigenvalue Problems

Two general types of methods, namely transformation methods and iterative methods, are available for solving eigenvalue problems. The transformation methods, such as Jacobi, Givens, and Householder schemes, are preferable when *all* the eigenvalues and eigenvectors are required. The iterative methods, such as the power method, are preferable when *few* eigenvalues and eigenvectors are required [7.11–7.13].

7.3.3 Jacobi Method

In this section, we present the Jacobi method for solving the standard eigenvalue problem

$$[H]\vec{X} = \lambda\vec{X} \tag{7.33}$$

where $[H]$ is a symmetric matrix.

(i) Method

The method is based on a theorem in linear algebra that states that a real symmetric matrix $[H]$ has only real eigenvalues and that there exists a real orthogonal matrix $[P]$ such that $[P]^T[H][P]$ is diagonal. The diagonal elements are the eigenvalues and the columns of the matrix $[P]$ are the eigenvectors.

In the Jacobi method, the matrix $[P]$ is obtained as a product of several "rotation" matrices of the form

$$\underset{n\times n}{[P_1]} = \begin{bmatrix} 1 & 0 & & & & & \\ 0 & 1 & & & & & \\ & & \ddots & & & & \\ & & & \cos\theta & & -\sin\theta & \\ & & & & \ddots & & \\ & & & \sin\theta & & \cos\theta & \\ & & & & & & \ddots \\ & & & & & & & 1 \end{bmatrix} \begin{matrix} \\ \\ \\ i\text{th} \\ \\ j\text{th row} \\ \\ \\ \end{matrix} \tag{7.34}$$

(with $\cos\theta$ and $\sin\theta$ in the ith column and $-\sin\theta$ and $\cos\theta$ in the jth column)

where all elements other than those appearing in columns and rows i and j are identical with those of the identity matrix $[I]$. If the sine and cosine entries appear in positions (i, i), (i, j), (j, i), and (j, j), then the corresponding elements of $[P_1]^T[H]\,[P_1]$ can be computed as

$$
\begin{aligned}
\underline{h}_{ii} &= h_{ii}\cos^2\theta + 2h_{ij}\sin\theta\cos\theta + h_{jj}\sin^2\theta \\
\underline{h}_{ij} &= \underline{h}_{ji} = (h_{jj} - h_{ii})\sin\theta\cos\theta + h_{ij}(\cos^2\theta - \sin^2\theta) \\
\underline{h}_{jj} &= h_{ii}\sin^2\theta - 2h_{ij}\sin\theta\cos\theta + h_{jj}\cos^2\theta
\end{aligned}
\tag{7.35}
$$

If θ is chosen as

$$\tan 2\theta = 2h_{ij}/(h_{ii} - h_{jj}) \tag{7.36}$$

then it makes $\underline{h}_{ij} = \underline{h}_{ji} = 0$. Thus, each step of the Jacobi method reduces a pair of off-diagonal elements to zero. Unfortunately, in the next step, although the method reduces a new pair of zeros, it introduces nonzero contributions to formerly zero positions. However, successive matrices of the form

$$[P_2]^T[P_1]^T[H][P_1][P_2], \qquad [P_3]^T[P_2]^T[P_1]^T[H][P_1][P_2][P_3], \ldots$$

converge to the required diagonal form and the desired matrix $[P]$ (whose columns give the eigenvectors) would then be given by

$$[P] = [P_1][P_2][P_3]\ldots \tag{7.37}$$

(ii) Computer implementation of Jacobi method

A FØRTRAN subroutine called JACØBI is given for finding the eigenvalues of a real symmetric matrix $[H]$ using the Jacobi method. The method is assumed to have converged whenever each of the off-diagonal elements, $\underline{h}_{ij}$, is less than a small quantity *EPS*. The following arguments are used in the subroutine:

H = array of order N × N used to store the elements of the given real symmetric matrix [H]. The diagonal elements of the array H give the eigenvalues upon return to the main program.
N = order of the matrix [H].
ITMAX = maximum number of rotations permitted.
A = an array of order N × N in which the eigenvectors are stored columnwise.
EPS = a small number of order 10^{-6} used for checking the convergence of the method.

To illustrate the use of the subroutine JACØBI, consider the problem of finding the eigenvalues and eigenvectors of the matrix

$$[H] = \begin{bmatrix} 2 & -1 & 0 \\ -1 & 2 & -1 \\ 0 & -1 & 2 \end{bmatrix}$$

The main program calling the subroutine JACØBI, along with the output of the program, is given below.

```
C======================================================================
C
C    MAIN PROGRAM TO CALL THE SUBROUTINE JACOBI
C
C======================================================================

         DIMENSION H(3,3),A(3,3)
         DATA N,ITMAX,EPS/3,250,1.0E-06/
         DATA H/2.0,-1.0,0.0,-1.0,2.0,-1.0,0.0,-1.0,2.0/
         CALL JACOBI(H,N,A,EPS,ITMAX)
         PRINT 10,(H(I,I),I=1,N),((A(I,J),J=1,N),I=1,N)
  10     FORMAT(1X,17H EIGEN VALUES ARE,/,1X,3E15.6,//,1X,
        2 14H EIGEN VECTORS,/,6X,5HFIRST,10X,6HSECOND,9X,5HTHIRD,/,
        3 (1X,3E15.6))
         STOP
         END

     EIGEN VALUES ARE
       0.341421E+01   0.200000E+01   0.585786E+00

     EIGEN VECTORS
         FIRST          SECOND         THIRD
       0.500003E+00   0.707098E+00   0.500009E+00
      -0.707120E+00   0.119898E-04   0.707094E+00
       0.499979E+00  -0.707115E+00   0.500009E+00
```

7.3.4 Power Method

(i) Computing the largest eigenvalue by the power method

The power method is the simplest iterative procedure for finding the largest or principal eigenvalue (λ_1) and the corresponding eigenvector of a matrix ($\vec{X}_1$). We assume that the $n \times n$ matrix $[H]$ is symmetric and real with n independent eigenvectors $\vec{X}_1, \vec{X}_2, \ldots, \vec{X}_n$. In this method, we choose an initial vector $\vec{Z}_0$ and generate a sequence of vectors $\vec{Z}_1$, $\vec{Z}_2, \ldots$, as

$$\vec{Z}_i = [H]\vec{Z}_{i-1} \tag{7.38}$$

so that, in general, the pth vector is given by

$$\vec{Z}_p = [H]\vec{Z}_{p-1} = [H]^2\vec{Z}_{p-2} = \cdots = [H]^p\vec{Z}_0 \tag{7.39}$$

The iterative process of Eq. (7.38) is continued until the following relation is satisfied:

$$\frac{z_{p,1}}{z_{p-1},1} \simeq \frac{z_{p,2}}{z_{p-1,2}} \simeq \cdots \simeq \frac{z_{p,n}}{z_{p-1,n}} = \lambda_1 \tag{7.40}$$

where $z_{p,j}$ and $z_{p-1,j}$ are the jth components of vectors $\vec{Z}_p$ and $\vec{Z}_{p-1}$, respectively. Here, λ_1 will be the desired eigenvalue.

The convergence of the method can be explained as follows. Since the initial (any arbitrary) vector $\vec{Z}_0$ can be expressed as a linear combination of the eigenvectors, we can write

$$\vec{Z}_0 = a_1\vec{X}_1 + a_2\vec{X}_2 + \cdots + a_n\vec{X}_n \tag{7.41}$$

where $a_1, a_2, \ldots, a_n$ are constants. If λ_i is the eigenvalue of $[H]$ corresponding to $\vec{X}_i$, then

$$\begin{aligned}[H]\vec{Z}_0 &= a_1[H]\vec{X}_1 + a_2[H]\vec{X}_2 + \cdots + a_n[H]\vec{X}_n \\ &= a_1\lambda_1\vec{X}_1 + a_2\lambda_2\vec{X}_2 + \cdots + a_n\lambda_n\vec{X}_n\end{aligned} \tag{7.42}$$

and

$$\begin{aligned}[H]^p\vec{Z}_0 &= a_1\lambda_1^p\vec{X}_1 + a_2\lambda_2^p\vec{X}_2 + \cdots + a_n\lambda_n^p\vec{X}_n \\ &= \lambda_1^p\left[a_1\vec{X}_1 + \left(\frac{\lambda_2}{\lambda_1}\right)^p a_2\vec{X}_2 + \cdots + \left(\frac{\lambda_n}{\lambda_1}\right)^p a_n\vec{X}_n\right]\end{aligned} \tag{7.43}$$

If λ_1 is the largest (dominant) eigenvalue,

$$|\lambda_1| > |\lambda_2| > \cdots > |\lambda_n|, \qquad \left|\frac{\lambda_i}{\lambda_1}\right| < 1 \tag{7.44}$$

and hence $\left(\frac{\lambda_i}{\lambda_1}\right)^p \to 0$ as $p \to \infty$.

Thus, Eq. (7.43) can be written, in the limit as $p \to \infty$, as

$$[H]^{p-1}\vec{Z}_0 = \lambda_1^{p-1}a_1\vec{X}_1 \tag{7.45}$$

and

$$[H]^p\vec{Z}_0 = \lambda_1^p a_1\vec{X}_1 \tag{7.46}$$

Therefore, if we take the ratio of any corresponding components of the vectors $([H]^p\vec{Z}_0)$ and $([H]^{p-1}\vec{Z}_0)$, it should have the same limiting value, λ_1. This property can be used to stop the iterative process. Moreover,

$$\left(\frac{[H]^p\vec{Z}_0}{\lambda_1^p}\right)$$

will converge to the eigenvector $a_1\vec{X}_1$ as $p \to \infty$.

Example 7.1 Find the dominant eigenvalue and the corresponding eigenvector of the matrix

$$[H] = \begin{bmatrix} 2 & -1 & 0 \\ -1 & 2 & -1 \\ 0 & -1 & 2 \end{bmatrix}$$

Solution By choosing the initial vector as

$$\vec{Z}_0 = \begin{Bmatrix} 1 \\ 1 \\ 1 \end{Bmatrix}$$

we have

$$\vec{Z}_1 = [H]\vec{Z}_0 = \begin{Bmatrix} 1 \\ 0 \\ 1 \end{Bmatrix},$$

$$\vec{Z}_2 = [H]\vec{Z}_1 = [H]^2\vec{Z}_0 = \begin{Bmatrix} 2 \\ -2 \\ 2 \end{Bmatrix},$$

and

$$\vec{Z}_3 = [H]\vec{Z}_2 = [H]^3\vec{Z}_0 = \begin{Bmatrix} 6 \\ -8 \\ 6 \end{Bmatrix}.$$

It is convenient here to divide the components of $\vec{Z}_3$ by 8 to obtain

$$\vec{Z}_3 = k\begin{Bmatrix} 3/4 \\ -1 \\ 3/4 \end{Bmatrix}, \quad \text{where} \quad k = 8.$$

In the future, we continue to divide by some suitable factor to keep the magnitude of the numbers reasonable. Continuing the procedure, we find

$$\vec{Z}_7 = [H]^7\vec{Z}_0 = c\begin{Bmatrix} 99 \\ -140 \\ 99 \end{Bmatrix} \quad \text{and} \quad \vec{Z}_8 = [H]^8\vec{Z}_0 = c\begin{Bmatrix} 338 \\ -478 \\ 338 \end{Bmatrix}$$

where c is a constant factor. The ratios of the corresponding components of $\vec{Z}_8$ and $\vec{Z}_7$ are $338/99 = 3.41414$ and $478/140 = 3.41429$, which can be assumed to be the same for our purpose. The eigenvalue given by this method is thus $\lambda_1 \simeq 3.41414$ or 3.41429, whereas the exact solution is $\lambda_1 = 2 + \sqrt{2} = 3.41421$. By dividing the last vector $\vec{Z}_8$ by the magnitude of the largest component (478), we obtain the eigenvector as

$$\vec{X}_1 \simeq \begin{Bmatrix} 1/1.4142 \\ -1.0 \\ 1/1.4142 \end{Bmatrix}, \text{ which is very close to the correct solution}$$

$$\vec{X}_1 = \begin{Bmatrix} 1/\sqrt{2} \\ -1 \\ 1/\sqrt{2} \end{Bmatrix}$$

The eigenvalue λ_1 can also be obtained by using the Rayleigh quotient (R) defined as

$$R = \frac{\vec{X}^T[H]\vec{X}}{\vec{X}^T\vec{X}} \tag{7.47}$$

If $[H]\vec{X} = \lambda\vec{X}, R$ will be equal to λ. Thus, we can compute the Rayleigh quotient at ith iteration as

$$R_i = \frac{\vec{X}_i^T[H]\vec{X}_i}{\vec{X}_i^T\vec{X}_i}, \qquad i = 1, 2, \ldots \tag{7.48}$$

Whenever R_i is observed to be essentially the same for two consecutive iterations $i-1$ and i, we take $\lambda_1 = R_i$.

(ii) Computing the smallest eigenvalue by the power method

If it is desired to solve

$$[H]\vec{X} = \lambda\vec{X} \tag{7.49}$$

to find the smallest eigenvalue and the associated eigenvector, we premultiply Eq. (7.49) by $[H]^{-1}$ and obtain

$$[H]^{-1}\vec{X} = \left(\frac{1}{\lambda}\right)\vec{X} \tag{7.50}$$

Eq. (7.50) can be written as

$$[\underset{\sim}{H}]\vec{X} = \underset{\sim}{\lambda}\vec{X} \tag{7.51}$$

where

$$[\underset{\sim}{H}] = [H]^{-1} \quad \text{and} \quad \underset{\sim}{\lambda} = \lambda^{-1} \tag{7.52}$$

This means that the absolutely smallest eigenvalue of $[H]$ can be found by solving the problem stated in Eq. (7.51) for the largest eigenvalue according to the procedure outlined in Section 7.3.4(i). Note that $[\underset{\sim}{H}] = [H]^{-1}$ has to be found before finding $\lambda_{\text{smallest}}$. Although this involves additional computations (in finding $[H]^{-1}$), it may prove to be the best approach in some cases.

(iii) Computing intermediate eigenvalues

Let the dominant eigenvector $\vec{X}_1$ be normalized so that its first component is one.

Let
$$\vec{X}_1 = \begin{Bmatrix} 1 \\ x_2 \\ x_3 \\ \vdots \\ x_n \end{Bmatrix}$$

Let $\vec{r}^T$ denote the first row of the matrix $[H]$—that is, $\vec{r}^T = \{h_{11}\ h_{12} \dots\ h_{1n}\}$. Then form a matrix $[\underset{\sim}{H}]$ as

$$[\underset{\sim}{H}] = \vec{X}_1 \vec{r}^T = \begin{Bmatrix} 1 \\ x_2 \\ x_3 \\ \vdots \\ x_n \end{Bmatrix} \{h_{11}\ h_{12}\ \dots\ h_{1n}\} = \begin{bmatrix} h_{11} & h_{12} & \dots & h_{1n} \\ x_2 h_{11} & x_2 h_{12} & \dots & x_2 h_{1n} \\ \vdots & & & \\ x_n h_{11} & x_n h_{12} & \dots & x_n h_{1n} \end{bmatrix} \tag{7.53}$$

Let the next dominant eigenvalue be λ_2 and normalize its eigenvector $(\vec{X}_2)$ so that its first component is one.

If $\vec{X}_1$ or $\vec{X}_2$ has a zero first element, then a different element may be normalized and the corresponding row $\vec{r}^T$ of matrix $[H]$ is used. Since $[H]\vec{X}_1 = \lambda_1 \vec{X}_1$ and $[H]\vec{X}_2 = \lambda_2 \vec{X}_2$, we obtain, by considering only the row $\vec{r}^T$ of these products, that

$$\vec{r}^T \vec{X}_1 = \lambda_1 \quad \text{and} \quad \vec{r}^T \vec{X}_2 = \lambda_2$$

This is a consequence of the normalizations. We can also obtain

$$[\underset{\sim}{H}]\vec{X}_1 = (\vec{X}_1 \vec{r}^T)\vec{X}_1 = \vec{X}_1(\vec{r}^T \vec{X}_1) = \lambda_1 \vec{X}_1 \tag{7.54}$$

and

$$[\underset{\sim}{H}]\vec{X}_2 = (\vec{X}_1 \vec{r}^T)\vec{X}_2 = \vec{X}_1(\vec{r}^T \vec{X}_2) = \lambda_2 \vec{X}_1 \tag{7.55}$$

so that

$$([H] - [\underset{\sim}{H}])(\vec{X}_2 - \vec{X}_1) = \lambda_2 \vec{X}_2 - \lambda_1 \vec{X}_1 - \lambda_2 \vec{X}_1 + \lambda_1 \vec{X}_1 = \lambda_2(\vec{X}_2 - \vec{X}_1) \tag{7.56}$$

Equation (7.56) shows that λ_2 is an eigenvalue and $\vec{X}_2 - \vec{X}_1$ an eigenvector of the matrix $[H] - [\underset{\sim}{H}]$. Since $[H] - [\underset{\sim}{H}]$ has all zeros in its first row, whereas $\vec{X}_2 - \vec{X}_1$ has a zero as its first component, both the first row and first column of $[H] - [\underset{\sim}{H}]$ may be deleted to obtain the matrix $[H_2]$. We then determine the dominant eigenvalue and the corresponding eigenvector of $[H_2]$, and by attaching a zero first component, obtain a vector $\vec{Z}_1$. Finally, $\vec{X}_2 - \vec{X}_1$ must be a multiple of $\vec{Z}_1$ so that we can write

$$\vec{X}_2 = \vec{X}_1 + a\vec{Z}_1 \tag{7.57}$$

The multiplication factor a can be found by multiplying Eq. (7.57) by the row vector $\vec{r}^T$ so that

$$a = \frac{\lambda_2 - \lambda_1}{\vec{r}^T \vec{Z}_1} \tag{7.58}$$

A similar procedure can be adopted to obtain the other eigenvalues and eigenvectors. A procedure to accelerate the convergence of the power method has been suggested by Roberti [7.14].

Example 7.2 Find the second and third eigenvalues of the matrix

$$[H] = \begin{bmatrix} 2 & -1 & 0 \\ -1 & 2 & -1 \\ 0 & -1 & 2 \end{bmatrix}$$

once $\vec{X}_1 = \begin{Bmatrix} 1.0 \\ -1.4142 \\ 1.0 \end{Bmatrix}$ and $\lambda_1 = 3.4142$ are known.

Solution The first row of the matrix $[H]$ is given by $\vec{r}_1^T = \{2 \quad -1 \quad 0\}$ and hence

$$[\underset{\sim}{H}] = \vec{X}_1 \vec{r}_1^T = \begin{Bmatrix} 1.0 \\ -1.4142 \\ 1.0 \end{Bmatrix} \{2 \quad -1 \quad 0\} = \begin{bmatrix} 2.0 & -1.0 & 0.0 \\ -2.8284 & 1.4142 & 0.0 \\ 2.0 & -1.0 & 0.0 \end{bmatrix}$$

$$[H] - [\underset{\sim}{H}] = \begin{bmatrix} 2 & -1 & 0 \\ -1 & 2 & -1 \\ 0 & -1 & 2 \end{bmatrix} - \begin{bmatrix} 2 & -1 & 0 \\ -2.8284 & 1.4142 & 0 \\ 2 & -1 & 0 \end{bmatrix} = \begin{bmatrix} 0 & 0 & 0 \\ -1.8284 & 0.5858 & -1 \\ -2 & 0 & 2 \end{bmatrix}$$

and $[H_2] = \begin{bmatrix} 0.5858 & -1 \\ 0 & 2 \end{bmatrix}$

We apply the power method to obtain the dominant eigenvalue of $[H_2]$ by taking the starting vector as $\vec{X} = \begin{Bmatrix} 1 \\ 1 \end{Bmatrix}$ and compute $[H_2]^{10}\vec{X} = c\begin{Bmatrix} -0.7071 \\ 1.0000 \end{Bmatrix}$, after which there is no significant change. As usual, c is some constant of no interest to us. Thus, the eigenvector of $[H_2]$ can be taken as $\begin{Bmatrix} -0.7071 \\ 1.0000 \end{Bmatrix}$.

The Rayleigh quotient corresponding to this vector gives $R_{10} = \lambda_2 = 2.0000$. By attaching a zero first element to the present vector $\begin{Bmatrix} -0.7071 \\ 1.0000 \end{Bmatrix}$, we obtain

$$\vec{Z}_2 = \begin{Bmatrix} 0.0 \\ -0.7071 \\ 1.0 \end{Bmatrix}$$

We then compute

$$a = \frac{\lambda_2 - \lambda_1}{\vec{r}_1^T \vec{Z}_2} = \frac{2.0000 - 3.4142}{(0.0 + 0.7071 + 0.0)} = -2.00002$$

Thus, we obtain the eigenvector $\vec{X}_2$ as

$$\vec{X}_2 = \vec{X}_1 + a\vec{Z}_2 = \begin{Bmatrix} 1.0 \\ -1.4142 \\ 1.0 \end{Bmatrix} - 2.00002 \begin{Bmatrix} 0.0 \\ -0.7071 \\ 1.0 \end{Bmatrix} = \begin{Bmatrix} 1.0 \\ 0.00001 \\ -1.00002 \end{Bmatrix}$$

Next, to find $\vec{X}_3$, we take $\lambda_2 = 2$ and normalize the vector $\begin{Bmatrix} -0.7071 \\ 1.0 \end{Bmatrix}$ to obtain the vector $\vec{Y}_2 = \begin{Bmatrix} 1.0 \\ -1.4142 \end{Bmatrix}$.

The matrix $[H_2]$ is reduced as follows:

The first row of $[H_2]$ is given by $\vec{r}_2^T = \{0.5858 \quad -1.0\}$, and

$$[H_2] - [\underset{\sim}{H}] = [H_2] - \vec{Y}_2\vec{r}_2^T = \begin{bmatrix} 0.5858 & -1.0 \\ 0 & 2.0 \end{bmatrix} - \begin{Bmatrix} 1.0 \\ -1.4142 \end{Bmatrix} \{0.5858 \quad -1.0\}$$

$$= \begin{bmatrix} 0 & 0 \\ 0.8284 & 0.5858 \end{bmatrix}$$

By deleting the first row and first column, we obtain the new reduced matrix $[H_3]$ as $[H_3] = [0.5858]$. The eigenvalue of $[H_3]$ is obviously $\lambda_3 = 0.5858$, and we can choose its eigenvector as $\{1\}$. By attaching a leading zero, we obtain $\vec{U}_2 = \begin{Bmatrix} 0 \\ 1 \end{Bmatrix}$. The value of a can be computed as

$$\frac{\lambda_3 - \lambda_2}{\vec{r}_2^T \vec{U}_2} = \frac{0.5858 - 2.0000}{(0 - 1)} = 1.4142$$

and the corresponding eigenvector of $[H_2]$ can be obtained as

$$\vec{Y}_3 = \vec{Y}_2 + a\vec{U}_2 = \begin{Bmatrix} 1.0 \\ -1.4142 \end{Bmatrix} + 1.4142 \begin{Bmatrix} 0 \\ 1 \end{Bmatrix} = \begin{Bmatrix} 1.0 \\ 0.0 \end{Bmatrix}$$

The eigenvector of $[H]$ corresponding to λ_3 can be obtained by adding a leading zero to $\vec{Y}_3$ to obtain $\vec{Z}_3$ as

$$\vec{Z}_3 = \begin{Bmatrix} 0.0 \\ 1.0 \\ 0.0 \end{Bmatrix} \text{ and computing } a \text{ as } \frac{\lambda_3 - \lambda_1}{\vec{r}_1^T \vec{Z}_3} = \frac{0.5858 - 3.4142}{(0 - 1 + 0)} = 2.8284$$

Finally, the eigenvector $\vec{X}_3$ corresponding to $[H]$ can be found as

$$\vec{X}_3 = \vec{X}_1 + a\vec{Z}_3 = \begin{Bmatrix} 1.0 \\ -1.4142 \\ 1.0 \end{Bmatrix} + 2.8284 \begin{Bmatrix} 0.0 \\ 1.0 \\ 0.0 \end{Bmatrix} = \begin{Bmatrix} 1.0 \\ 1.4142 \\ 1.0 \end{Bmatrix}$$

7.3.5 Rayleigh–Ritz Subspace Iteration Method

Another iterative method that can be used to find the lowest eigenvalues and the associated eigenvectors of the general eigenvalue problem, Eq. (7.2a), is the Rayleigh–Ritz subspace iteration method [7.15, 7.16]. This method is very effective in finding the first few eigenvalues and the corresponding eigenvectors of large eigenvalue problems whose stiffness ($[A]$) and mass ($[B]$) matrices have large bandwidths. The various steps of this method are given below briefly. A detailed description of the method can be found in Ref. [7.15].

(i) Algorithm

Step 1: Start with q initial iteration vectors $\vec{X}_1, \vec{X}_2, \ldots, \vec{X}_q$, $q > p$, where p is the number of eigenvalues and eigenvectors to be calculated. Bathe and Wilson [7.15] suggested a value of $q = \min\,(2p, p+8)$ for good convergence. Define the initial modal matrix $[X_0]$ as

$$[X_0] = [\vec{X}_1\ \vec{X}_2\ \ldots\ \vec{X}_q] \tag{7.59}$$

and set the iteration number as $k = 0$. A computer algorithm for calculating efficient initial vectors for subspace iteration method was given in Ref [7.17].

Step 2: Use the following subspace iteration procedure to generate an improved modal matrix $[X_{k+1}]$:

(a) Find $[\bar{X}_{k+1}]$ from the relation

$$[A][\bar{X}_{k+1}] = [B][X_k] \tag{7.60}$$

(b) Compute

$$[A_{k+1}] = [\bar{X}_{k+1}]^T[A][\bar{X}_{k+1}] \tag{7.61}$$

$$[B_{k+1}] = [\bar{X}_{k+1}]^T[B][\bar{X}_{k+1}] \tag{7.62}$$

(c) Solve for the eigenvalues and eigenvectors of the reduced system

$$[A_{k+1}][Q_{k+1}] = [B_{k+1}][Q_{k+1}][\Lambda_{k+1}] \tag{7.63}$$

and obtain $[\Lambda_{k+1}]$ and $[Q_{k+1}]$.

(d) Find an improved approximation to the eigenvectors of the original system as

$$[X_{k+1}] = [\bar{X}_{k+1}][Q_{k+1}]. \tag{7.64}$$

Note:

(1) It is assumed that the iteration vectors converging to the exact eigenvectors $\vec{X}_1^{(\text{exact})}$, $\vec{X}_2^{(\text{exact})}$, ..., are stored as the first, second, ..., columns of the matrix $[X_{k+1}]$.

(2) It is assumed that the vectors in $[X_0]$ are not orthogonal to one of the required eigenvectors.

Step 3: If $\lambda_i^{(k)}$ and $\lambda_i^{(k+1)}$ denote the approximations to the ith eigenvalue in the iterations $k-1$ and k, respectively, we assume convergence of the process whenever the following criteria are satisfied:

$$\frac{\lambda_i^{(k+1)} - \lambda_i^{(k)}}{\lambda_i^{(k+1)}} \le \varepsilon, \qquad i = 1, 2, \ldots, p \tag{7.65}$$

where $\varepsilon \simeq 10^{-6}$. Note that although the iteration is performed with q vectors ($q > p$), the convergence is measured only on the approximations predicted for the p smallest eigenvalues.

(ii) Computer implementation of subspace iteration method

A typical FØRTRAN computer program to implement the subspace iteration method is given. This program solves the eigenvalue problem

$$[A]\vec{X} = \lambda[B]\vec{X} \tag{E$_1$}$$

where $[A]$ and $[B]$ are symmetric banded matrices. It is assumed that the elements of the matrices $[A]$ and $[B]$ are stored in band form in the first N rows and NB columns of the arrays K and GM, respectively, where N is the order and NB is the semi-bandwidth of matrices $[A]$ and $[B]$. If Eq. (E$_1$) represents a free vibration problem, $[A]$ and $[B]$ represent the stiffness and consistent mass matrices of the structure, respectively. If a lumped mass matrix is used instead of a consistent mass matrix, the matrix $[B]$ will be a diagonal matrix and in this case $[B]$ is stored as a vector in the array M (in this case, the array GM is not defined). The information regarding the type of mass matrix is given to the program through the quantity *INDEX*. If a lumped mass matrix is used, the value of *INDEX* is set equal to 1, whereas it is set equal to 2 if a consistent mass matrix is used.

The program requires the following subroutines for computing the desired number of eigenvalues and eigenvectors:

SUSPIT: To obtain the partial eigen solution by Rayleigh–Ritz subspace iteration method. It calls the subroutines DECØMP, SØLVE, GAUSS, and EIGEN for solving the generalized Ritz problem.

EIGEN: To compute all the eigenvalues and eigenvectors of the generalized Ritz problem using power method.

GAUSS: To find the inverse of a real square matrix.

DECØMP: To perform Choleski decomposition of a symmetric banded matrix [same as given in Section 7.2.2(v)].

SØLVE: To solve a system of linear algebraic equations using the upper triangular band of the decomposed matrix obtained from DECØMP [same as given in Section 7.2.2(v)].

The following input is to be given to the program:

N Number of degrees of freedom (order of matrices $[A]$ and $[B]$).

NB Semi-bandwidth of matrix $[A]$ (and of $[B]$ if $[B]$ is a consistent mass matrix).

NMØDE Number of eigenvalues and eigenvectors to be found.

INDEX $= 1$ if $[B]$ is a lumped mass (or diagonal) matrix;
$= 2$ if $[B]$ is a consistent mass (or banded) matrix.

K The elements of the banded matrix $[A]$ are to be stored in the array K(N, NB).

GM The elements of the banded matrix $[B]$ are to be stored in the array GM(N, NB) if $[B]$ is a consistent mass matrix,

or

M The diagonal elements of the diagonal matrix $[B]$ are to be stored in the array M(N) if $[B]$ is a lumped mass matrix.

X Trial eigenvectors are to be stored columnwise in the array X(N, NMØDE).

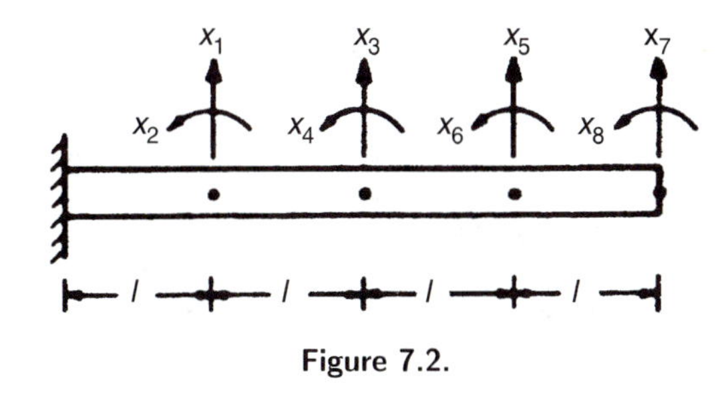

Figure 7.2.

As an example, consider an eigenvalue problem with

$$[A] = \frac{EI}{l^3}\begin{bmatrix} 24 & 0 & -12 & 6l & 0 & 0 & 0 & 0 \\ & 8l^2 & -6l & 2l^2 & 0 & 0 & 0 & 0 \\ & & 24 & 0 & -12 & 6l & 0 & 0 \\ & & & 8l^2 & -6l & 2l^2 & 0 & 0 \\ & & & & 24 & 0 & -12 & 6l \\ \text{Symmetric} & & & & & 8l^2 & -6l & 2l^2 \\ & & & & & & 12 & -6l \\ & & & & & & & 4l^2 \end{bmatrix}$$

and

$$[B] = \frac{\rho Al}{420}\begin{bmatrix} 312 & 0 & 54 & -13l & 0 & 0 & 0 & 0 \\ & 8l^2 & 13l & -3l^2 & 0 & 0 & 0 & 0 \\ & & 312 & 0 & 54 & -13l & 0 & 0 \\ & & & 8l^2 & 13l & -3l^2 & 0 & 0 \\ & & & & 312 & 0 & 54 & -13l \\ & & & & & 8l^2 & 13l & -3l^2 \\ & & & & & & 156 & -22l \\ \text{Symmetric} & & & & & & & 4l^2 \end{bmatrix}$$

This problem represents the free vibrations of a cantilever beam shown in Figure 7.2 with a four-element idealization. Here, $[B]$ is the consistent mass matrix, E is the Young's modulus, I is the moment of inertia of cross section, l is the length of an element, ρ is the mass density, and A is the area of cross section of the beam.

If the first three eigenvalues and eigenvectors are required, we will have $N = 8$, $NB = 4$, $NMØDE = 3$, and $INDEX = 2$. The trial eigenvectors are chosen as

$$\vec{X}_1 = \begin{Bmatrix} 0 \\ 0 \\ 0 \\ 0 \\ 0 \\ 0 \\ 1 \\ 0 \end{Bmatrix}, \quad \vec{X}_2 = \begin{Bmatrix} 0 \\ 0 \\ 1 \\ 0 \\ 0 \\ 0 \\ -1 \\ 0 \end{Bmatrix}, \quad \text{and} \quad \vec{X}_3 = \begin{Bmatrix} 1 \\ 0 \\ -1 \\ 0 \\ 1 \\ 0 \\ 1 \\ 0 \end{Bmatrix}$$

The main program for solving the problem of Eq. (E_1) and the results given by the program are given below.

```
C===========================================================================
C
C      COMPUTATION OF EIGENVALUES AND EIGENVECTORS
C
C===========================================================================
       DIMENSION GM(8,4),X(8,3),OMEG(3),Y(8,3),GST(3,3),GMM(3,3),
      2 VECT(3,3),ABCV(3),ABCW(3),ABCX(3),ABCY(3),ABCZ(3,3),B(3,1),
      3 LP(3),LQ(3,2),R(3)
       REAL K(8,4),M(8)
       DOUBLE PRECISION SUM(3),DIFF(3)
C      DIMENSIONS ARE: K(N,NB),M(N),GM(N,NB),X(N,NMODE),OMEG(NMODE),
C      Y(N,NMODE),B(NMODE,1),LP(NMODE),LQ(NMODE,2),R(NMODE)
C      DIMENSIONS OF MATRICES GST,GMM,VECT AND ABCZ ARE(NMODE,NMODE)
C      DIMENSION OF VECTORS ABCV,ABCW,ABCX,ABCY,SUN AND DIFF IS(NMODE)
       DATA N,NB,NMODE,INDEX/8,4,3,2/
       E=2.0E06
       AI=1.0/12.0
       AL=25.0
       AA=1.0
       RHO=0.00776
       CONK=E*AI/(AL**3)
       CONM=RHO*AA*AL/420.0
       DO 10 I=1,8
       DO 10 J=1,4
       K(I,J)=0.0
10     GM(I,J)=0.0
       K(1,1)=24.0
       K(1,3)=-12.0
       K(1,4)=6.0*AL
       K(2,1)=8.0*AL*AL
       K(2,2)=-6.0*AL
       K(2,3)=2.0*AL*AL
       K(3,1)=24.0
       K(3,3)=-12.0
       K(3,4)=6.0*AL
       K(4,1)=8.0*AL*AL
       K(4,2)=-6.0*AL
       K(4,3)=2.0*AL*AL
       K(5,1)=24.0
       K(5,3)=-12.0
       K(5,4)=6.0*AL
       K(6,1)=8.0*AL*AL
       K(6,2)=-6.0*AL
       K(6,3)=2.0*AL*AL
       K(7,1)=12.0
       K(7,2)=-6.0*AL
```

```
      K(8,1)=4.0*AL*AL
      GM(1,1)=312.0
      GM(1,3)=54.0
      GM(1,4)=-13.0*AL
      GM(2,1)=8.0*AL*AL
      GM(2,2)=13.0*AL
      GM(2,3)=-3.0*AL*AL
      GM(3,1)=312.0
      GM(3,3)=54.0
      GM(3,4)=-13.0*AL
      GM(4,1)=8.0*AL*AL
      GM(4,2)=13.0*AL
      GM(4,3)=-3.0*AL*AL
      GM(5,1)=312.0
      GM(5,3)=54.0
      GM(5,4)=-13.0*AL
      GM(6,1)=8.0*AL*AL
      GM(6,2)=13.0*AL
      GM(6,3)=-3.0*AL*AL
      GM(7,1)=156.0
      GM(7,2)=-22.0*AL
      GM(8,1)=4.0*AL*AL
      DO 20 I=1,8
      DO 20 J=1,4
      K(I,J)=CONK*K(I,J)
20    GM(I,J)=CONM*GM(I,J)
      DO 30 I=1,8
      DO 30 J=1,3
30    X(I,J)=0.0
      X(1,1)=0.1
      X(3,1)=0.3
      X(5,1)=0.6
      X(7,1)=1.0
      X(1,2)=-0.5
      X(3,2)=-1.0
      X(7.2)=1.0
      X(1,3)=1.0
      X(3,3)=0.0
      X(5,3)=1.0
      X(7,3)=1.0
      CALL SUSPIT(K,M,GM,X,OMEG,Y,GST,GMM,VECT,SUM,INDEX,N,NB,NMODE,
     2 ABCV,ABCW,ABCX,ABCY,ABCZ,DIFF,B,LP,LQ,R)
      PRINT 40
40    FORMAT(5X,'EIGENVALUES AND EIGENVECTORS',/)
      PRINT 45
45    FORMAT(6X,'J' ,3X,'OMEG(J)' ,8X,'EIGENVECTOR(J)' ,/)
      DO 50 J=1,NMODE
50    PRINT 60,J,OMEG(J),(X(I,J),I=1,N)
60    FORMAT(4X,I3,2X,E11.5,3X,4E11.4,/,23X,4E11.4)
```

```
STOP
END

EIGENVALUES AND EIGENVECTORS

J   OMEG(J)          EIGENVECTOR(J)

1   0.16295E+01    0.2209E+00    0.1653E-01    0.7709E+00    0.2641E-01
                   0.1493E+01    0.3059E-01    0.2271E+01    0.3125E-01
2   0.10224E+02   -0.9495E+00   -0.5203E-01   -0.1624E+01    0.1030E-01
                  -0.3074E+00    0.8856E-01    0.2276E+01    0.1088E+00
3   0.28863E+02    0.1726E+01    0.3402E-01   -0.6768E-02   -0.1278E+00
                  -0.1294E+01    0.6300E-01    0.2223E+01    0.1725E+00
```

7.4 SOLUTION OF PROPAGATION PROBLEMS

When the finite element method is applied for the solution of initial value problems (relating to an unsteady or transient state of phenomena), we obtain propagation problems involving a set of simultaneous linear differential equations.

Propagation problems involve time as one of the independent variables and initial conditions on the dependent variables are given in addition to the boundary conditions. A general propagation problem can be expressed (after incorporating the boundary conditions) in standard form as

$$\left.\begin{aligned} \frac{\mathrm{d}\vec{X}}{\mathrm{d}t} &= \vec{F}(\vec{X},t), & t>0 \\ \vec{X} &= \vec{X}_0, & t=0 \end{aligned}\right\} \tag{7.66}$$

where the vectors of propagation variables, forcing functions, and initial conditions are given by

$$\vec{X}=\begin{Bmatrix} x_1(t)\\ x_2(t)\\ \vdots \\ x_n(t)\end{Bmatrix},\qquad \vec{F}=\begin{Bmatrix} f_1(\vec{X},t)\\ f_2(\vec{X},t)\\ \vdots \\ f_n(\vec{X},t)\end{Bmatrix},\qquad \vec{X}_0=\begin{Bmatrix} x_1(0)\\ x_2(0)\\ \vdots \\ x_n(0)\end{Bmatrix} \tag{7.67}$$

It can be seen that Eq. (7.66) represents a system of n simultaneous ordinary differential equations with n initial conditions.

In certain propagation problems, as in the case of damped mechanical and electrical systems, the governing equations are usually stated as

$$\left.\begin{aligned} &[A]\frac{\mathrm{d}^2\vec{X}}{\mathrm{d}t^2}+[B]\frac{\mathrm{d}\vec{X}}{\mathrm{d}t}+[C]\vec{X}=\vec{F}(\vec{X},t), && t>0 \\ &\vec{X}=\vec{X}_0 \text{ and } \frac{\mathrm{d}\vec{X}}{\mathrm{d}t}=\vec{Y}_0, && t=0 \end{aligned}\right\} \tag{7.68}$$

where $[A]$, $[B]$, and $[C]$ denote known matrices of order $n\times n$. In the case of mechanical and structural systems, the matrices $[A]$, $[B]$, and $[C]$ denote mass, damping, and stiffness matrices, respectively, and the vector $\vec{F}$ represents the known spatial and time history of the external loads. It can be seen that Eqs. (7.68) denote a system of n coupled second-order ordinary differential equations with necessary initial conditions. Equation (7.66) can be used to represent any nth order differential equation (see Problem 7.21).

7.4.1 Solution of a Set of First-Order Differential Equations

Equation (7.66) can be written in scalar form as

$$\begin{aligned}
\frac{dx_1}{dt} &= f_1(t, x_1, x_2, \ldots, x_n) \\
\frac{dx_2}{dt} &= f_2(t, x_1, x_2, \ldots, x_n) \\
&\vdots \\
\frac{dx_n}{dt} &= f_n(t, x_1, x_2, \ldots, x_n)
\end{aligned} \tag{7.69}$$

with the initial conditions

$$\begin{aligned}
x_1(t=0) &= x_1(0) \\
x_2(t=0) &= x_2(0) \\
&\vdots \\
x_n(t=0) &= x_n(0)
\end{aligned} \tag{7.70}$$

These equations can be solved by any of the numerical integration methods, such as Runge–Kutta, Adams–Bashforth, Adams–Moulton, and Hamming methods [7.18]. In the fourth-order Runge–Kutta method, starting from the known initial vector $\vec{X}_0$ at $t = 0$, we compute the vector $\vec{X}$ after time Δt as

$$\vec{X}(t + \Delta t) = \vec{X}(t) + \frac{1}{6}[\vec{K}_1 + 2\vec{K}_2 + 2\vec{K}_3 + \vec{K}_4] \tag{7.71}$$

where

$$\begin{aligned}
\vec{K}_1 &= \Delta t \vec{F}(\vec{X}(t), t) \\
\vec{K}_2 &= \Delta t \vec{F}\left(\vec{X}(t) + \frac{\vec{K}_1}{2},\ t + \frac{\Delta t}{2}\right) \\
\vec{K}_3 &= \Delta t \vec{F}\left(\vec{X}(t) + \frac{\vec{K}_2}{2},\ t + \frac{\Delta t}{2}\right) \\
\vec{K}_4 &= \Delta t \vec{F}\left(\vec{X}(t) + \vec{K}_3,\ t + \Delta t\right)
\end{aligned} \tag{7.72}$$

7.4.2 Computer Implementation of Runge–Kutta Method

A computer program, in the form of the subroutine RUNGE, is given for solving a system of first-order differential equations based on the fourth-order Runge–Kutta method. The arguments of this subroutine are as follows:

T = independent variable (time). It is to be given a value of 0.0 at the beginning.
DT = desired time step for numerical integration.
NEQ = number of first-order differential equations = n.

XX = array of dimension NEQ that contains the current values of $x_1, x_2, \ldots, x_n$.

F = array of dimension NEQ $= n$ that contains the values of $f_1, f_2, \ldots, f_n$ computed at time T by a subroutine FTN (XX, F, NEQ, T) supplied by the user.

YI, YJ, YK, YL, UU = dummy arrays of dimension NEQ.

To illustrate the use of the subroutine RUNGE, we consider the solution of the following system of equations:

$$\begin{aligned}
\frac{dx_1}{dt} &= x_2 \\
\frac{dx_2}{dt} &= -G\frac{x_1}{(x_1{}^2 + x_3{}^2)^{3/2}} \\
\frac{dx_3}{dt} &= x_4 \\
\frac{dx_4}{dt} &= -G\frac{x_3}{(x_1{}^2 + x_3{}^2)^{3/2}}
\end{aligned} \tag{E$_1$}$$

with the initial conditions

$$x_1(0) = 1, \quad x_2(0) = 0, \quad x_3(0) = 0, \quad \text{and} \quad x_4(0) = 1 \tag{E$_2$}$$

Equations (E_1) represent the equations of motion of a body moving in a plane about a spherical earth that can be written in a rectangular planar coordinate system (x, y) as

$$\ddot{x} = -G\frac{x}{r^3}, \qquad \ddot{y} = -G\frac{y}{r^3}$$

where dots indicate differentiation with respect to time $t, r = (x^2 + y^2)^{1/2}$, and G is the gravitational constant. By taking $G = 1$ with the initial conditions of Eq. (E_2), the trajectory of motion described by Eqs. (E_1) will be a circle with period 2π. Now we solve Eqs. (E_1) by taking a time step of $\Delta t = 2\pi/200 = 0.031415962$ for 400 time steps (i.e., up to $t = 4\pi$). The main program that calls the subroutine RUNGE and the output of the program are given below.

```
C=========================================================================
C
C      NUMERICAL INTEGRATION OF SIMULTANEOUS DIFFERENTIAL EQUATIONS
C
C=========================================================================
       DIMENSION TIME(400),X(400,4),XX(4),F(4),YI(4),YJ(4),YK(4),YL(4),
      2 UU(4)
C      DIMENSIONS ARE: TIME(NSTEP),X(NSTEP,NEQ),XX(NEQ),F(NEQ),YI(NEQ),
C      YJ(NEQ),YK(NEQ),YL(NEQ),UU(NEQ)
C      INITIAL CONDITIONS
       XX(1)=1.0
       XX(2)=0.0
       XX(3)=0.0
```

```
      XX(4)=1.0
      NEQ=4
      NSTEP=400
      DT=6.2831853/200.0
      T=0.0
      PRINT 10
10    FORMAT(2X,'PRINTOUT OF SOLUTION',//,2X,'STEP',3X,'TIME',5X,
     2 'X(I,1)',6X,'X(I,2)',6X,'X(I,3)',6X,'X(I,4)',4X,
     3 'VALUE OF R',/)
      I=0
      R=(XX(1)**2+XX(3)**2)*SQRT(XX(1)**2+XX(3)**2)
      PRINT 30,I,T,(XX(J),J=1,NEQ),R
      DO 40 I=1,NSTEP
      CALL RUNGE(T,DT,NEQ,XX,F,YI,YJ,YK,YL,UU)
      TIME(I)=T
      DO 20 J=1,NEQ
20    X(I,J) = XX(J)
      R=SQRT(XX(1) **2+XX(3) **2)
      PRINT 30,I,TIME(I),(X(I,J),J=1,NEQ),R
30    FORMAT(2X,14,F8.4,4E12.4,E12.4)
40    CONTINUE
      STOP
      END
```

```
 PRINTOUT OF SOLUTION

 STEP   TIME    X(I,1)      X(I,2)       X(I,3)     X(I,4)    VALUE OF R

    0  0.0000 0.1000E+01  0.0000E+00  0.0000E+00 0.1000E+01 0.1000E+01
    1  0.0314 0.9995E+00 -0.3141E-01  0.3141E-01 0.9995E+00 0.1000E+01
    2  0.0628 0.9980E+00 -0.6279E-01  0.6279E-01 0.9980E+00 0.1000E+01
    3  0.0942 0.9956E+00 -0.9411E-01  0.9411E-01 0.9956E+00 0.1000E+01
    4  0.1257 0.9921E+00 -0.1253E+00  0.1253E+00 0.9921E+00 0.1000E+01
    5  0.1571 0.9877E+00 -0.1564E+00  0.1564E+00 0.9877E+00 0.1000E+01
    .
    .
    .
  396 12.4407 0.9921E+00  0.1253E+00 -0.1253E+00 0.9921E+00 0.1000E+01
  397 12.4721 0.9956E+00  0.9410E-01 -0.9410E-01 0.9956E+00 0.1000E+01
  398 12.5035 0.9980E+00  0.6279E-01 -0.6279E-01 0.9980E+00 0.1000E+01
  399 12.5350 0.9995E+00  0.3141E-01 -0.3141E-01 0.9995E+00 0.1000E+01
  400 12.5664 0.1000E+01 -0.3792E-05  0.4880E-05 0.1000E+01 0.1000E+01
```

7.4.3 Numerical Solution of Eq. (7.68)

Several methods are available for the solution of Eq. (7.68). All the methods can be divided into two classes: direct integration methods and the mode superposition method.

7.4.4 Direct Integration Methods

In these methods, Eq. (7.68), or the special case, Eq. (7.66), is integrated numerically by using a step-by-step procedure [7.19]. The term direct denotes that no transformation of the equations into a different form is used prior to numerical integration. The direct integration methods are based on the following ideas:

(a) Instead of trying to find a solution $\vec{X}(t)$ that satisfies Eq. (7.68) for any time t, we can try to satisfy Eq. (7.68) only at discrete time intervals Δt apart.

(b) Within any time interval, the nature of variation of $\vec{X}$ (displacement), $\dot{\vec{X}}$ (velocity), and $\ddot{\vec{X}}$ (acceleration) can be assumed in a suitable manner.

Here, the time interval Δt and the nature of variation of $\vec{X}$, $\dot{\vec{X}}$, and $\ddot{\vec{X}}$ within any Δt are chosen by considering factors such as accuracy, stability, and cost of solution.

The finite difference, Houbolt, Wilson, and Newmark methods fall under the category of direct methods [7.20–7.22]. The finite difference method (a direct integration method) is outlined next.

Finite Difference Method

By using central difference formulas [7.23], the velocity and acceleration at any time t can be expressed as

$$\dot{\vec{X}}_t = \frac{1}{2\Delta t}(-\vec{X}_{t-\Delta t} + \vec{X}_{t+\Delta t}) \tag{7.73}$$

$$\ddot{\vec{X}}_t = \frac{1}{(\Delta t)^2}[\vec{X}_{t-\Delta t} - 2\vec{X}_t + \vec{X}_{t+\Delta t}] \tag{7.74}$$

If Eq. (7.68) is satisfied at time t, we have

$$[A]\ddot{\vec{X}}_t + [B]\dot{\vec{X}}_t + [C]\vec{X}_t = \vec{F}_t \tag{7.75}$$

By substituting Eqs. (7.73) and (7.74) in Eq. (7.75), we obtain

$$\left(\frac{1}{(\Delta t)^2}[A] + \frac{1}{2\Delta t}[B]\right)\vec{X}_{t+\Delta t} = \vec{F}_t - \left([C] - \frac{2}{(\Delta t)^2}[A]\right)\vec{X}_t - \left(\frac{1}{(\Delta t)^2}[A] - \frac{1}{2\Delta t}[B]\right)\vec{X}_{t-\Delta x} \tag{7.76}$$

Equation (7.76) can now be solved for $\vec{X}_{t+\Delta t}$. Thus, the solution $\vec{X}_{t+\Delta t}$ is based on the equilibrium conditions at time t.

Since the solution of $\vec{X}_{t+\Delta t}$ involves $\vec{X}_t$ and $\vec{X}_{t-\Delta t}$, we need to know $\vec{X}_{-\Delta t}$ for finding $\vec{X}_{\Delta t}$. For this we first use the initial conditions $\vec{X}_0$ and $\dot{\vec{X}}_0$ to find $\ddot{\vec{X}}_0$ using Eq. (7.75) for $t = 0$. Then we compute $\vec{X}_{-\Delta t}$ using Eqs. (7.73)–(7.75) as

$$\vec{X}_{-\Delta t} = \vec{X}_o - \Delta t\dot{\vec{X}}_o + \frac{(\Delta t)^2}{2}\ddot{\vec{X}}_o \tag{7.77}$$

A disadvantage of the finite difference method is that it is conditionally stable—that is, the time step Δt has to be smaller than a critical time step $(\Delta t)_{cri}$. If the time step Δt is larger than $(\Delta t)_{cri}$, the integration is unstable in the sense that any errors resulting from the numerical integration or round-off in the computations grow and makes the calculation of $\vec{X}$ meaningless in most cases.

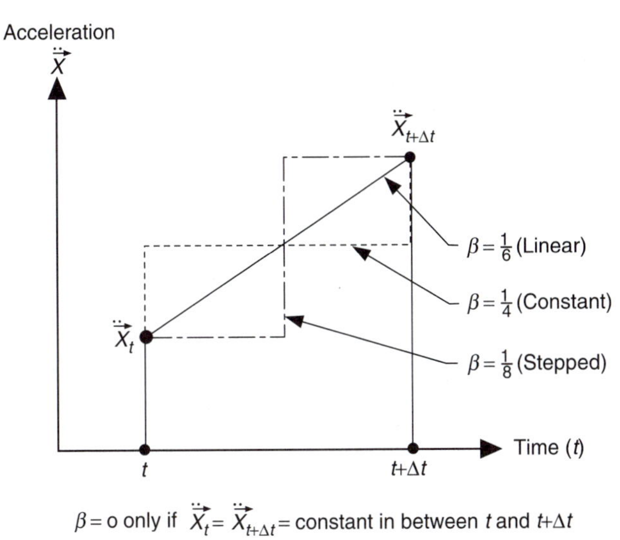

Figure 7.3. Values of β for Different Types of Variation of $\ddot{\vec{X}}$.

7.4.5 Newmark Method

The basic equations of the Newmark method (or Newmark's β method) are given by [7.20]

$$\dot{\vec{X}}_{t+\Delta t} = \dot{\vec{X}}_t + (1-\gamma)\Delta t \ddot{\vec{X}}_t + \Delta t \gamma \ddot{\vec{X}}_{t+\Delta t} \tag{7.78}$$

$$\vec{X}_{t+\Delta t} = \vec{X}_t + \Delta t \dot{\vec{X}}_t + (\frac{1}{2} - \beta)(\Delta t)^2 \ddot{\vec{X}}_t + \beta(\Delta t)^2 \ddot{\vec{X}}_{t+\Delta t} \tag{7.79}$$

where γ and β are parameters that can be determined depending on the desired accuracy and stability. Newmark suggested a value of $\gamma = 1/2$ for avoiding artificial damping. The value of β depends on the way in which the acceleration, $\ddot{\vec{X}}$, is assumed to vary during the time interval t and $t + \Delta t$. The values of β to be taken for different types of variation of $\ddot{\vec{X}}$ are shown in Figure 7.3.

In addition to Eqs. (7.78) and (7.79), Eqs. (7.68) are also assumed to be satisfied at time $t + \Delta t$ so that

$$[A]\ddot{\vec{X}}_{t+\Delta t} + [B]\dot{\vec{X}}_{t+\Delta t} + [C]\vec{X}_{t+\Delta t} = [\vec{F}]_{t+\Delta t} \tag{7.80}$$

To find the solution at the $t + \Delta t$, we solve Eq. (7.79) to obtain $\ddot{\vec{X}}_{t+\Delta t}$ in terms of $\vec{X}_{t+\Delta t}$, substitute this $\ddot{\vec{X}}_{t+\Delta t}$ into Eq. (7.78) to obtain $\dot{\vec{X}}_{t+\Delta t}$ in terms of $\vec{X}_{t+\Delta t}$, and then use Eq. (7.80) to find $\vec{X}_{t+\Delta t}$. Once $\vec{X}_{t+\Delta t}$ is known, $\dot{\vec{X}}_{t+\Delta t}$ and $\ddot{\vec{X}}_{t+\Delta t}$ can be calculated from Eqs. (7.78) and (7.79).

7.4.6 Mode Superposition Method

It can be seen that the computational work involved in the direct integration methods is proportional to the number of time steps used in the analysis. Hence, in general, the use of direct integration methods is expected to be effective when the response over only a relatively short duration (involving few time steps) is required. On the other hand, if the integration has to be carried for many time steps, it may be more effective to transform Eqs. (7.68) into a form in which the step-by-step solution is less costly. The mode superposition or normal mode method is a technique wherein Eq. (7.68) is first transformed into a convenient form before integration is carried. Thus, the vector $\vec{X}$ is transformed as

$$\underset{n \times 1}{\vec{X}(t)} = \underset{n \times r}{[T]} \underset{r \times 1}{\vec{Y}(t)} \tag{7.81}$$

where $[T]$ is a rectangular matrix of order $n \times r$, and $\vec{Y}(t)$ is a time-dependent vector of order $r (r \leq n)$. The transformation matrix $[T]$ is still unknown and will have to be determined. Although the components of $\vec{X}$ have physical meaning (like displacements), the components of $\vec{Y}$ need not have any physical meaning and hence are called generalized displacements. By substituting Eq. (7.81) into Eq. (7.68), and premultiplying throughout by $[T]^T$, we obtain

$$[\underset{\sim}{A}]\ddot{\vec{Y}} + [\underset{\sim}{B}]\dot{\vec{Y}} + [\underset{\sim}{C}]\vec{Y} = \underset{\sim}{\vec{F}} \tag{7.82}$$

where

$$[\underset{\sim}{A}] = [T]^T [A][T], \tag{7.83}$$

$$[\underset{\sim}{B}] = [T]^T [B][T], \tag{7.84}$$

$$[\underset{\sim}{C}] = [T]^T [C][T], \tag{7.85}$$

and

$$\underset{\sim}{\vec{F}} = [T]^T \vec{F} \tag{7.86}$$

The basic idea behind using the transformation of Eq. (7.81) is to obtain the new system of equations (7.82) in which the matrices $[\underset{\sim}{A}], [\underset{\sim}{B}]$, and $[\underset{\sim}{C}]$ will be of much smaller order than the original matrices $[A], [B]$, and $[C]$. Furthermore, the matrix $[T]$ can be chosen so as to obtain the matrices $[\underset{\sim}{A}]$, $[\underset{\sim}{B}]$, and $[\underset{\sim}{C}]$ in diagonal form, in which case Eq. (7.82) represents a system of r uncoupled second-order differential equations. The solution of these independent equations can be found by standard techniques, and the solution of the original problem can be found with the help of Eq. (7.81).

In the case of structural mechanics problems, the matrix $[T]$ denotes the modal matrix and Eqs. (7.82) can be expressed in scalar form as (see Section 12.6)

$$\ddot{Y}_i(t) + 2\zeta_i \omega_i \dot{Y}_i(t) + \omega_i^2 Y_i(t) = N_i(t), \qquad i = 1, 2, \ldots, r \tag{7.87}$$

where the matrices $[\underset{\sim}{A}]$, $[\underset{\sim}{B}]$, and $[\underset{\sim}{C}]$ have been expressed in diagonal form as

$$[\underset{\sim}{A}] = [I], \qquad [\underset{\sim}{B}] = \left[\diagdown 2\zeta_i \omega_i \diagdown \right], \qquad [\underset{\sim}{C}] = \left[\diagdown \omega_i^2 \diagdown \right] \tag{7.88}$$

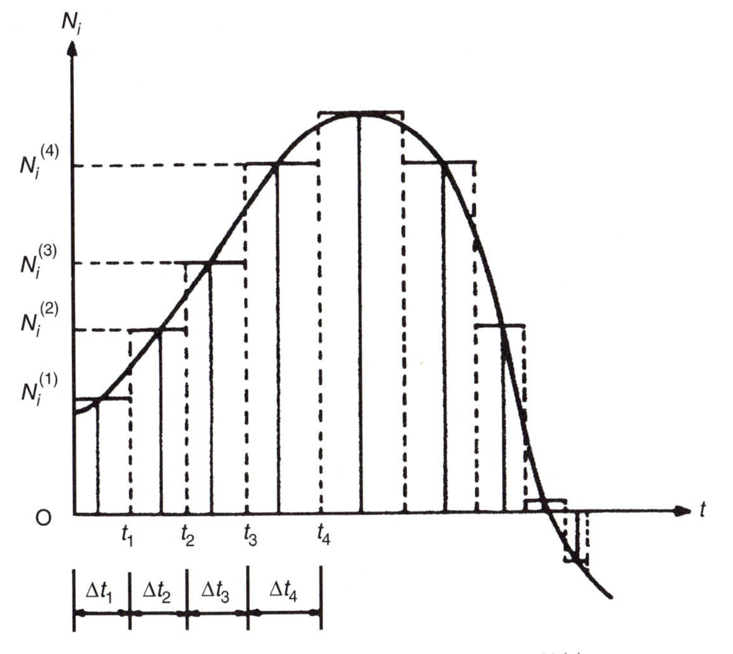

Figure 7.4. Arbitrary Forcing Function $N_i(t)$.

and the vector $\underset{\sim}{\vec{F}}$ as

$$\underset{\sim}{\vec{F}} = \begin{Bmatrix} N_1(t) \\ \vdots \\ N_r(t) \end{Bmatrix} \tag{7.89}$$

Here, ω_i is the rotational frequency (square root of the eigenvalue) corresponding to the ith natural mode (eigenvector), and ζ_i is the modal damping ratio in the ith natural mode.

7.4.7 Solution of a General Second-Order Differential Equation

We consider the solution of Eq. (7.87) in this section. In many practical problems the forcing functions $f_1(t), f_2(t), \ldots, f_n(t)$ (components of $\vec{F}$) are not analytical expressions but are represented by a series of points on a diagram or a list of numbers in a table. Furthermore, the forcing functions $N_1(t), N_2(t), \ldots, N_r(t)$ of Eq. (7.87) are given by premultiplying $\vec{F}$ by $[T]^T$ as indicated in Eq. (7.86). Hence, in many cases, the solution of Eq. (7.87) can only be obtained numerically by using a repetitive series of calculations. Let the function $N_i(t)$ vary with time in some general manner, such as that represented by the curve in Figure 7.4.

This forcing function may be approximated by a series of rectangular impulses of various magnitudes and durations as indicated in Figure 7.4. For good accuracy the magnitude

$N_i^{(j)}$ of a typical impulse should be chosen as the ordinate of the curve at the middle of the time interval Δt_j as shown in Figure 7.4. In any time interval $t_{j-1} \le t \le t_j$, the solution of Eq. (7.87) can be computed as the sum of the effects of the initial conditions at time t_{j-1} and the effect of the impulse within the interval Δt_j as follows [7.24]:

$$\begin{aligned} Y_i(t) = e^{-\zeta_i \omega_i (t-t_{j-1})} & \left[Y_i^{(j-1)} \cos \omega_{\mathrm{d}i}(t-t_{j-1}) \right. \\ & \left. + \frac{\dot{Y}_i^{(j-1)} + \zeta_i \omega_i Y_i^{(j-1)}}{\omega_{\mathrm{d}i}} \sin \omega_{\mathrm{d}i}(t-t_{j-1}) \right] \\ & + \frac{N_i^{(j)}}{\omega_i^2} \left\{ 1 - e^{-\zeta_i \omega_i (t-t_{j-1})} \left[\cos \omega_{\mathrm{d}i}(t-t_{j-1}) \right. \right. \\ & \left. \left. + \frac{\zeta_i \omega_i}{\omega_{\mathrm{d}i}} \sin \omega_{\mathrm{d}i}(t-t_{j-1}) \right] \right\} \end{aligned} \tag{7.90}$$

where
$$\omega_{\mathrm{d}i} = \omega_i (1 - \zeta_i^2)^{1/2} \tag{7.91}$$

At the end of the interval, Eq. (7.90) becomes

$$\begin{aligned} Y_i^{(j)} = Y_i(t = t_j) = e^{-\zeta_i \omega_i \Delta t_j} & \left[Y_i^{(j-1)} \cos \omega_{\mathrm{d}i} \Delta t_j \right. \\ & \left. + \frac{\dot{Y}_i^{(j-1)} + \zeta_i \omega_i Y_i^{(j-1)}}{\omega_{\mathrm{d}i}} \sin \omega_{\mathrm{d}i} \Delta t_j \right] \\ & + \frac{N_i^{(j)}}{\omega_i^2} \left[1 - e^{-\zeta_i \omega_i \Delta t_j} \left(\cos \omega_{\mathrm{d}i} \Delta t_j + \frac{\zeta_i \omega_i}{\omega_{\mathrm{d}i}} \sin \omega_{\mathrm{d}i} \Delta t_j \right) \right] \end{aligned} \tag{7.92}$$

By differentiating Eq. (7.90) with respect to time, we obtain

$$\begin{aligned} \dot{Y}_i^{(j)} = \dot{Y}_i(t = t_j) = \omega_{\mathrm{d}i} e^{-\zeta_i \omega_i \Delta t_j} & \left[-Y_i^{(j-1)} \sin \omega_{\mathrm{d}i} \Delta t_j + \frac{\dot{Y}_i^{(j-1)} + \zeta_i \omega_i Y_i^{(j-1)}}{\omega_{\mathrm{d}i}} \right. \\ & \left. \times \cos \omega_{\mathrm{d}i} \Delta t_j - \frac{\zeta_i \omega_i}{\omega_{\mathrm{d}i}} \left(Y_i^{(j-1)} \cos \omega_{\mathrm{d}i} \Delta t_j + \frac{\dot{Y}_i^{(j-1)} + \zeta_i \omega_i Y_i^{(j-1)}}{\omega_{\mathrm{d}i}} \sin \omega_{\mathrm{d}i} \Delta t_j \right) \right] \\ & + \frac{N_i^{(j)} \omega_{\mathrm{d}i}}{\omega_i^2} e^{-\zeta_i \omega_i \Delta t_j} \left(1 + \frac{\zeta_i^2 \omega_i^2}{\omega_{\mathrm{d}i}^2} \right) \sin \omega_{\mathrm{d}i} \Delta t_j \end{aligned} \tag{7.93}$$

Thus, Eqs. (7.92) and (7.93) represent recurrence formulas for calculating the solution at the end of the jth time step. They also provide the initial conditions of Y_j and $\dot{Y}_j$ at the beginning of step $j+1$. These formulas may be applied repetitively to obtain the time history of response for each of the normal modes i. Then the results for each time station can be transformed back, using Eq. (7.81), to obtain the solution of the original problem.

7.4.8 Computer Implementation of Mode Superposition Method

A subroutine called MØDAL is given to implement the mode superposition or normal mode method. This subroutine calls a matrix multiplication subroutine called MATMUL. The arguments of the subroutine MØDAL are as follows:

NMØDE = number of modes to be considered in the analysis = r of Eq. (7.81) = input.
N = number of degrees of freedom = order of the square matrices [A], [B], and [C] = input.
GM = array of N × N in which the (mass) matrix [A] is stored = input.
ØMEG = array of size NMØDE in which the natural frequencies (square root of eigenvalues) are stored = input.
T = array of size N × NMØDE in which the eigenvectors (modes) are stored columnwise = matrix [T] = input.
ZETA = array of size NMØDE in which the modal damping ratios of various modes are stored = input.
NSTEP = number of integration points = input.
XØ = array of size N in which the initial conditions $x_1(0), x_2(0), \ldots, x_n(0)$ are stored = input.
XDØ = array of size N in which the initial conditions

$$\frac{dx_1}{dt}(0) = Y_1(0), \quad \frac{dx_2}{dt}(0) = Y_2(0), \ldots, \frac{dx_n}{dt}(0) = Y_n(0)$$

are stored = input.

XX = array of size NMØDE × NSTEP.
TT = array of size NMØDE × N = transpose of the matrix [T].
F = array of size NSTEP in which the magnitudes of the force applied at coordinate M at times $t_1, t_2, \ldots, t_{\text{NSTEP}}$ are stored = input.
YØ, YDØ = arrays of size NMØDE.
U, V = arrays of size NMODE × NSTEP in which the values of $Y_i^{(j)}$ and $\dot{Y}_i^{(j)}$ are stored.
X = array of size N × NSTEP in which the solution of the original problem, $x_i^{(j)}$, is stored.
TIME = array of size NSTEP in which the times $t_1, t_2, \ldots, t_{\text{NSTEP}}$ are stored = input.
DT = array of size NSTEP in which the time intervals $\Delta t_1, \Delta t_2, \ldots, \Delta t_{\text{NSTEP}}$ are stored.
M = coordinate number at which the force is applied = input.
TGM = array of size NMØDE × N used to store the product $[T]^T[A]$.
TGMT = array of size NMØDE × NMØDE used to store the matrix $[\underset{\sim}{A}] = [T]^T[A][T]$.

To demonstrate the use of the subroutine MØDAL, the solution of the following problem is considered:

$$[A]\ddot{\vec{X}} + [B]\dot{\vec{X}} + [C]\vec{X} = \vec{F}$$

$$\text{with } \vec{X}(0) = \vec{X}_o \quad \text{and} \quad \dot{\vec{X}}(0) = \vec{Y}_o \tag{E$_1$}$$

Known data:

$$n = 3, \qquad r = 3; \qquad \zeta_i = 0.05 \quad \text{for} \quad i = 1, 2, 3;$$

$$[A] = \begin{bmatrix} 1 & 0 & 0 \\ 0 & 1 & 0 \\ 0 & 0 & 1 \end{bmatrix}; \qquad [B] = [0]; \qquad [C] = \begin{bmatrix} 2 & -1 & 0 \\ -1 & 2 & -1 \\ 0 & -1 & 1 \end{bmatrix};$$

$$[T] = \begin{bmatrix} 1.000 & 1.000 & 1.000 \\ 1.802 & 0.445 & -1.247 \\ 2.247 & -0.802 & 0.555 \end{bmatrix};$$

$$\omega_1 = 0.445042, \qquad \omega_2 = 1.246978, \qquad \omega_3 = 1.801941;$$

$$\vec{X}_o = \vec{0}; \qquad \vec{Y}_o = \vec{0};$$

$$\vec{F} = \begin{Bmatrix} f_1 \\ f_2 \\ f_3 \end{Bmatrix} = \begin{Bmatrix} 0 \\ 0 \\ f \end{Bmatrix}$$

Value of time t	1	2	3	4	5	6	7	8	9	10
Magnitude of f	1	1	1	1	1	1	1	1	1	1

Thus, in this case, $NMODE = 3$, $N = 3$, $NSTEP = 10$, $M = 3$, $TIME(I) = I$ for $I = 1$–10, and $F(I) = 1$ for $I = 1$–10. The main program for this problem and the results given by the program are given below.

```
C===============================================================================
C
C     RESPONSE OF MULTI-DEGREE-OF-FREEDOM SYSTEM BY MODAL ANALYSIS
C
C===============================================================================
      DIMENSION GM(3,3),OMEG(3),T(3,3),ZETA(3),TT(3,3),TGMT(3,3),
     2 XO(3),XDO(3),YO(3),YDO(3),WN(3,10),F(10),U(3,10),TGM(3,3),
     3 V(3,10),X(3,10),TIME(10),DT(10)
      DATA NMODE,N,NSTEP,M/3,3,10,3/
      DATA GM/1.0,0.0,0.0,0.0,1.0,0.0,0.0,0.0,1.0/
      DATA OMEG/0.445042,1.246978,1.801941/
      DATA ZETA/0.05,0.05,0.05/
      DATA(T(I,1),I=1,3)/0.445042,0.8019375,1.0/
      DATA(T(I,2),I=1,3)/-1.246984, -0.5549535,1.0/
      DATA(T(I,3),I=1,3)/1.801909,-2.246983,1.0/
      DATA XO/0.0,0.0,0.0/
      DATA XDO/0.0,0.0,0.0/
      DATA TIME/1.0,2.0,3.0,4.0,5.0,6.0,7.0,8.0,9.0,10.0/
      DATA F/1.0,1.0,1.0,1.0,1.0,1.0,1.0,1.0,1.0,1.0/
      DO 10 I=1,NMODE
      DO 10 J=1,NMODE
10    TT(I,J) = T(J,I)
      CALL-MODAL(GM,OMEG,T,ZETA,XO,XDO,YO,YDO,WN,F,U,V,X,
      TIME,DT,TT,M,
     2 NSTEP,N,NMODE,TGMT,TGM)
      PRINT 20,M
20    FORMAT(/,69H RESPONSE OF THE SYSTEM TO A TIME VARYING FORCE APPL
     2IED AT COORDINATE,I2,/)
```

```
      DO 30 I=1,N
30    PRINT 40,I,(X(I,J),J=1,NSTEP)
40    FORMAT(/,11H COORDINATE,I5,/,1X,5E14.8,/,1X,5E14.8)
      STOP
      END
```

```
RESPONSE OF THE SYSTEM TO A TIME VARYING FORCE APPLIED AT COORDINATE 3

COORDINATE    1
0.30586943E-020.75574949E-010.44995812E+000.11954379E+010.18568037E+01
0.20145943E+010.18787326E+010.18112489E+010.17405561E+010.14387581E+01

COORDINATE    2
0.42850435E-010.45130622E+000.13615156E+010.23700531E+010.31636245E+01
0.36927490E+010.38592181E+010.36473811E+010.32258503E+010.26351447E+01

COORDINATE    3
0.44888544E+000.14222714E+010.24165692E+010.33399298E+010.42453833E+01
0.50193300E+010.54301090E+010.52711205E+010.45440755E+010.35589526E+01
```

7.5 PARALLEL PROCESSING IN FINITE ELEMENT ANALYSIS

Parallel processing is defined as the exploitation of parallel or concurrent events in the computing process [7.25]. Parallel processing techniques are being investigated because of the high degree of sophistication of the computational models required for future aerospace, transportation, nuclear, and microelectronic systems. Most of the present-day supercomputers, such as CRAY X-MP, CRAY-2, CYBER-205, and ETA-10, achieve high performance through vectorization/parallelism. Efforts have been devoted to the development of vectorized numerical algorithms for performing the matrix operations, solution of algebraic equations, and extraction of eigenvalues [7.26, 7.27]. However, the progress has been slow, and no effective computational strategy exists that performs the entire finite element solution in the parallel processing mode.

The various phases of the finite element analysis can be identified as (a) input of problem characteristics, element and nodal data, and geometry of the system; (b) data preprocessing; (c) evaluation of element characteristics; (d) assembly of elemental contributions; (e) incorporation of boundary conditions; (f) solution of system equations; and (g) postprocessing of the solution and evaluation of secondary fields.

The input and preprocessing phases can be parallelized. Since the element characteristics require only information pertaining to the elements in question, they can be evaluated in parallel. The assembly cannot utilize the parallel operation efficiently since the element and global variables are related through a Boolean transformation. The incorporation of boundary conditions, although usually not time-consuming, can be done in parallel.

The solution of system equations is the most critical phase. For static linear problems, the numerical algorithm should be selected to take advantage of the symmetric banded structure of the equations and the type of hardware used. A variety of efficient direct iterative and noniterative solution techniques have been developed for different computers by exploiting the parallelism, pipeline (or vector), and chaining capabilities [7.28]. For nonlinear steady-state problems, the data structure is essentially the same as for linear problems. The major difference lies in the algorithms for evaluating the nonlinear terms and solving the nonlinear algebraic equations. For transient problems, several parallel integration techniques have been proposed [7.29].

The parallel processing techniques are still evolving and are expected to be the dominant methodologies in the computing industry in the near future. Hence, it can be hoped that the full potentialities of parallel processing in finite element analysis will be realized in the next decade.

REFERENCES

7.1 S.S. Rao: *Applied Numerical Methods for Engineers and Scientists*, Prentice Hall, Upper Saddle River, NJ, 2002.

7.2 G. Cantin: An equation solver of very large capacity, *International Journal for Numerical Methods in Engineering, 3*, 379–388, 1971.

7.3 B.M. Irons: A frontal solution problem for finite element analysis, *International Journal for Numerical Methods in Engineering, 2*, 5–32, 1970.

7.4 A. Razzaque: Automatic reduction of frontwidth for finite element analysis, *International Journal for Numerical Methods in Engineering, 15*, 1315–1324, 1980.

7.5 G. Beer and W. Haas: A partitioned frontal solver for finite element analysis, *International Journal for Numerical Methods in Engineering, 18*, 1623–1654, 1982.

7.6 R.S. Varga: *Matrix Iterative Analysis*, Prentice Hall, Englewood Cliffs, NJ, 1962.

7.7 I. Fried: A gradient computational procedure for the solution of large problems arising from the finite element discretization method, *International Journal for Numerical Methods in Engineering, 2*, 477–494, 1970.

7.8 G. Gambolati: Fast solution to finite element flow equations by Newton iteration and modified conjugate gradient method, *International Journal for Numerical Methods in Engineering, 15*, 661–675, 1980.

7.9 O.C. Zienkiewicz and R. Lohner: Accelerated "relaxation" or direct solution? Future prospects for finite element method, *International Journal for Numerical Methods in Engineering, 21*, 1–11, 1985.

7.10 N. Ida and W. Lord: Solution of linear equations for small computer systems, *International Journal for Numerical Methods in Engineering, 20*, 625–641, 1984.

7.11 J.H. Wilkinson: *The Algebraic Eigenvalue Problem*, Clarendon, Oxford, UK, 1965.

7.12 A.R. Gourlay and G.A. Watson: *Computational Methods for Matrix Eigen Problems*, Wiley, London, 1973.

7.13 M. Papadrakakis: Solution of the partial eigenproblem by iterative methods, *International Journal for Numerical Methods in Engineering, 20*, 2283–2301, 1984.

7.14 P. Roberti: The accelerated power method, *International Journal for Numerical Methods in Engineering, 20*, 1179–1191, 1984.

7.15 K.J. Bathe and E.L. Wilson: Large eigenvalue problems in dynamic analysis, *Journal of Engineering Mechanics Division, Proc. of ASCE, 98*(EM6), 1471–1485, 1972.

7.16 F.A. Akl, W.H. Dilger, and B.M. Irons: Acceleration of subspace iteration, *International Journal for Numerical Methods in Engineering, 18*, 583–589, 1982.

7.17 T.C. Cheu, C.P. Johnson, and R.R. Craig, Jr.: Computer algorithms for calculating efficient initial vectors for subspace iteration method, *International Journal for Numerical Methods in Engineering, 24*, 1841–1848, 1987.

7.18 A. Ralston: *A First Course in Numerical Analysis*, McGraw-Hill, New York, 1965.

7.19 L. Brusa and L. Nigro: A one-step method for direct integration of structural dynamic equations, *International Journal for Numerical Methods in Engineering, 15*, 685–699, 1980.

7.20 S.S. Rao: *Mechanical Vibrations*, Addison-Wesley, Reading, MA, 1986.
7.21 W.L. Wood, M. Bossak, and O.C. Zienkiewicz: An alpha modification of Newmark's method, *International Journal for Numerical Methods in Engineering, 15*, 1562–1566, 1980.
7.22 W.L. Wood: A further look at Newmark, Houbolt, etc., time-stepping formulae, *International Journal for Numerical Methods in Engineering, 20*, 1009–1017, 1984.
7.23 S.H. Crandall: *Engineering Analysis: A Survey of Numerical Procedures*, McGraw-Hill, New York, 1956.
7.24 S. Timoshenko, D.H. Young, and W. Weaver: *Vibration Problems in Engineering* (4th Ed.), Wiley, New York, 1974.
7.25 A.K. Noor: Parallel processing in finite element structural analysis, *Engineering with Computers, 3*, 225–241, 1988.
7.26 C. Farhat and E. Wilson: Concurrent iterative solution of large finite element systems, *Communications in Applied Numerical Methods, 3*, 319–326, 1987.
7.27 S.W. Bostic and R.E. Fulton: Implementation of the Lanczos method for structural vibration analysis on a parallel computer, *Computers and Structures, 25*, 395–403, 1987.
7.28 L. Adams: Reordering computations for parallel execution, *Communications in Applied Numerical Methods, 2*, 263–271, 1986.
7.29 M. Ortiz and B. Nour-Omid: Unconditionally stable concurrent procedures for transient finite element analysis, *Computer Methods in Applied Mechanics and Engineering, 58*, 151–174, 1986.

PROBLEMS

7.1 Find the inverse of the following matrix using the decomposition $[A] = [U]^T[U]$:

$$[A] = \begin{bmatrix} 5 & -1 & 1 \\ -1 & 6 & -4 \\ 1 & -4 & 3 \end{bmatrix}$$

7.2 Find the inverse of the matrix $[A]$ given in Problem 7.1 using the decomposition $[A] = [L][L]^T$, where $[L]$ is a lower triangular matrix.

Hint: If a symmetric matrix $[A]$ of order n is decomposed as $[A] = [L][L]^T$, the elements of $[L]$ are given by

$$l_{ii} = \left[a_{ii} - \sum_{k=1}^{i-1} l_{ik}^2 \right]^{(1/2)}, \qquad i = 1, 2, \ldots, n$$

$$l_{mi} = \frac{1}{l_{ii}} \left[a_{mi} - \sum_{k=1}^{i-1} l_{ik} l_{mk} \right], \qquad m = i+1, \ldots, n \quad \text{and} \quad i = 1, 2, \ldots, n$$

$$l_{ij} = 0, \qquad i < j$$

The elements of $[L]^{-1} = [\lambda_{ij}]$ can be obtained from the relation $[L][L]^{-1} = [l_{ij}][\lambda_{ij}] = [I]$ as

$$\lambda_{ii} = \frac{1}{l_{ii}}, \qquad i = 1, 2, \ldots, n$$

$$\lambda_{ij} = -\left(\sum_{k=j}^{i-1} l_{ik} \lambda_{kj} \right) \Bigg/ l_{ii}, \qquad i > j$$

$$\lambda_{ij} = 0, \qquad i < j$$

7.3 Express the following functions in matrix form as $f = (1/2)\vec{X}^T[A]\vec{X}$ and identify the matrix $[A]$:

(i) $f = 6x_1^2 + 49x_2^2 + 51x_3^2 - 82x_2x_3 + 20x_1x_3 - 4x_1x_2$

(ii) $f = 6x_1^2 + 3x_2^2 + 3x_3^2 - 4x_1x_2 - 2x_2x_3 + 4x_1x_3$

7.4 Find the eigenvalues and eigenvectors of the following problem by solving the characteristic polynomial equation:

$$\begin{bmatrix} 2 & -1 & 0 \\ -1 & 2 & -1 \\ 0 & -1 & 3 \end{bmatrix} \begin{Bmatrix} x_1 \\ x_2 \\ x_3 \end{Bmatrix} = \lambda \begin{bmatrix} 1 & 0 & 0 \\ 0 & 1 & 0 \\ 0 & 0 & 2 \end{bmatrix} \begin{Bmatrix} x_1 \\ x_2 \\ x_3 \end{Bmatrix}$$

7.5 Find the eigenvalues and eigenvectors of the following matrix by solving the characteristic equation:

$$[A] = \begin{bmatrix} 1 & 2 & 0 \\ 2 & 2 & 0 \\ 0 & 0 & -1 \end{bmatrix}$$

7.6 Find the eigenvalues and eigenvectors of the following matrix using the Jacobi method:

$$[A] = \begin{bmatrix} 3 & 2 & 1 \\ 2 & 2 & 1 \\ 1 & 1 & 1 \end{bmatrix}$$

7.7 Find the eigenvalues and eigenvectors of the matrix $[A]$ given in Problem 7.6 using the power method.

7.8 Solve the following system of equations using the finite difference method:

$$[A]\ddot{\vec{X}} + [C]\vec{X} = \vec{F}$$

where $[A] = \begin{bmatrix} 2 & 0 \\ 0 & 1 \end{bmatrix}$, $[C] = \begin{bmatrix} 6 & -2 \\ -2 & 4 \end{bmatrix}$, and $\vec{F} = \begin{Bmatrix} 0 \\ 10 \end{Bmatrix}$

with the initial conditions $\vec{X}(t=0) = \dot{\vec{X}}(t=0) = \begin{Bmatrix} 0 \\ 0 \end{Bmatrix}$.

Take the time step Δt as 0.28 and find the solution at $t = 4.2$.

7.9 Use the subroutine GAUSS of Section 7.2.1 to find the solution of the following equations:

$$\begin{bmatrix} 4 & 2 & 4 & 5 \\ 3 & 9 & 12 & 15 \\ 2 & 4 & 11 & 10 \\ 1 & 2 & 4 & 10 \end{bmatrix} \begin{Bmatrix} x_1 \\ x_2 \\ x_3 \\ x_4 \end{Bmatrix} = \begin{Bmatrix} 1 \\ 1 \\ 1 \\ 1 \end{Bmatrix}$$

7.10 Use the subroutine GAUSS of Section 7.2.1 to find the solution of the following equations:

$$\begin{bmatrix} 5 & -4 & 1 & 0 \\ -4 & 6 & -4 & 1 \\ 1 & -4 & 6 & -4 \\ 0 & 1 & -4 & 5 \end{bmatrix} \begin{Bmatrix} x_1 \\ x_2 \\ x_3 \\ x_4 \end{Bmatrix} = \begin{Bmatrix} 0 \\ 1 \\ 1 \\ 0 \end{Bmatrix}$$

7.11 Use the subroutines DECØMP and SØLVE of Section 7.2.2(v) to find the inverse of the following matrix with $n = 20$:

$$\begin{bmatrix} \frac{n+2}{2n+2} & -\frac{1}{2} & 0 & 0 & \cdots & 0 & 0 & \frac{1}{2n+2} \\ -\frac{1}{2} & 1 & -\frac{1}{2} & 0 & \cdots & 0 & 0 & 0 \\ 0 & -\frac{1}{2} & 1 & -\frac{1}{2} & \cdots & 0 & 0 & 0 \\ \vdots & & & & & & & \\ 0 & 0 & 0 & 0 & \cdots & -\frac{1}{2} & 1 & -\frac{1}{2} \\ \frac{1}{2n+2} & 0 & 0 & 0 & \cdots & 0 & -\frac{1}{2} & \frac{n+2}{2n+2} \end{bmatrix}$$

Hint: The first, second, $\ldots, n$th columns of $[A]^{-1}$ are nothing but the solutions $\vec{X}_1, \vec{X}_2, \ldots, \vec{X}_n$ corresponding to the right-hand-side vectors $\vec{b}_1, \vec{b}_2, \ldots, \vec{b}_n$, respectively,

$$\text{where } \vec{b}_1 = \begin{Bmatrix} 1 \\ 0 \\ 0 \\ \vdots \\ 0 \end{Bmatrix}, \vec{b}_2 = \begin{Bmatrix} 0 \\ 1 \\ 0 \\ \vdots \\ 0 \end{Bmatrix}, \ldots, \vec{b}_n = \begin{Bmatrix} 0 \\ 0 \\ 0 \\ \vdots \\ 1 \end{Bmatrix}$$

7.12 Use the subroutine GAUSS of Section 7.2.1 to find the inverse of the following matrix with $n = 10$:

$$\underset{n \times n}{[A]} = \begin{bmatrix} n & n-1 & n-2 & \cdots & 2 & 1 \\ n-1 & n-1 & n-2 & & 2 & 1 \\ n-2 & n-2 & n-2 & & 2 & 1 \\ \vdots & & & & & \\ 2 & 2 & 2 & & 2 & 1 \\ 1 & 1 & 1 & & 1 & 1 \end{bmatrix}$$

7.13 Using the subroutines SUSPIT and EIGEN of Section 7.3.5(ii), find the first two eigenvalues and the corresponding eigenvectors of the following problem:

$$\begin{bmatrix} 4 & -6 & 2 & 0 \\ -6 & 24 & 0 & 6 \\ 2 & 0 & 8 & 2 \\ 0 & 6 & 2 & 4 \end{bmatrix} \vec{X} = \frac{\lambda}{420} \begin{bmatrix} 4 & 13 & -3 & 0 \\ 13 & 312 & 0 & -13 \\ -3 & 0 & 8 & -3 \\ 0 & -13 & -3 & 4 \end{bmatrix} \vec{X}$$

Assume the trial eigenvectors as

$$\vec{X}_1 = \begin{Bmatrix} 0 \\ 1 \\ 0 \\ 0 \end{Bmatrix} \quad \text{and} \quad \vec{X}_2 = \begin{Bmatrix} 1 \\ 0 \\ 0 \\ 1 \end{Bmatrix}$$

7.14 Find the eigenvalues and eigenvectors of the matrix $[A]$ given in Problem 7.12 (with $n = 10$) using the subroutine JACØBI of Section 7.3.3(ii).

7.15 Solve the following system of equations using the Cholesky decomposition method using (i) $[L][L]^T$ decomposition and (ii) $[U]^T[U]$ decomposition:

$$5x_1 + 3x_2 + x_3 = 14$$
$$3x_1 + 6x_2 + 2x_3 = 21$$
$$x_1 + 2x_2 + 3x_3 = 14$$

7.16 Express the following set of equations as a system of first-order equations:

$$\frac{d^2x}{dt^2} = x^2 - y + e^t$$
$$\frac{d^2y}{dt^2} = x - y^2 - e^t$$
$$t = 0; \quad x(0) = \frac{dx}{dt}(0) = 0; \quad y(0) = 1, \quad \frac{dy}{dt}(0) = -2$$

Obtain the solution of these equations using the subroutine RUNGE of Section 7.4.2.

7.17 Solve the following equations using the Gauss elimination method:

$$2x_1 + 3x_2 + x_3 = 9$$
$$x_1 + 2x_2 + 3x_3 = 6$$
$$3x_1 + x_2 + 2x_3 = 8$$

7.18 The finite element analysis of certain systems leads to a tridiagonal system of equations, $[A]\vec{x} = \vec{b}$, where

$$[A] = \begin{bmatrix} a_{11} & a_{12} & 0 & 0 & \cdots & 0 & 0 & 0 \\ a_{21} & a_{22} & a_{23} & 0 & \cdots & 0 & 0 & 0 \\ 0 & a_{32} & a_{33} & a_{34} & \cdots & 0 & 0 & 0 \\ \cdot & \cdot & \cdot & \cdot & \cdots & \cdot & \cdot & \cdot \\ \cdot & \cdot & \cdot & \cdot & \cdots & \cdot & \cdot & \cdot \\ \cdot & \cdot & \cdot & \cdot & \cdots & \cdot & \cdot & \cdot \\ 0 & 0 & 0 & 0 & \cdots & a_{n-1,n-2} & a_{n-1,n-1} & a_{n-1,n} \\ 0 & 0 & 0 & 0 & \cdots & 0 & a_{n,n-1} & a_{n,n} \end{bmatrix}$$

$$\vec{x} = \begin{Bmatrix} x_1 \\ x_2 \\ \cdot \\ \cdot \\ \cdot \\ x_n \end{Bmatrix}; \quad \vec{b} = \begin{Bmatrix} b_1 \\ b_2 \\ \cdot \\ \cdot \\ \cdot \\ b_n \end{Bmatrix}$$

Indicate a method of solving these equations.

7.19 Solve the following system of equations using a suitable procedure:

$$[A]\vec{x} = \vec{b}$$

with

$$[A] = \begin{bmatrix} 5 & -5 & 0 & 0 & 0 \\ -5 & 10 & -5 & 0 & 0 \\ 0 & -5 & 10 & -5 & 0 \\ 0 & 0 & -5 & 10 & -5 \\ 0 & 0 & 0 & -5 & 10 \end{bmatrix}$$

$$\vec{x} = \begin{Bmatrix} x_1 \\ x_2 \\ \cdot \\ \cdot \\ \cdot \\ x_5 \end{Bmatrix}; \quad \text{and} \quad \vec{b} = \begin{Bmatrix} b_1 \\ b_2 \\ \cdot \\ \cdot \\ \cdot \\ b_5 \end{Bmatrix}$$

7.20 The elements of the Hillbert matrix, $[A] = [a_{ij}]$, are given by

$$a_{ij} = \frac{1}{i+j-1}; \qquad i,j = 1,2,\ldots,n$$

Find the inverse of the matrix, $[A]^{-1} = [b_{ij}]$, with $n = 10$ using the subroutine GAUSS, and compare the result with the exact solution given by

$$b_{ij} = \frac{(-1)^{i+j}(n+i-1)!(n+j-1)!}{(i+j-1)\left\{(i-1)!(j-1)!\right\}^2(n-i)!(n-j)!}; \qquad i,j = 1,2,\ldots,n$$

7.21 Express the nth-order differential equation

$$\frac{\mathrm{d}^n x}{\mathrm{d}t^n} = f\left(t, x, \frac{\mathrm{d}x}{\mathrm{d}t}, \frac{\mathrm{d}^2 x}{\mathrm{d}t^2}, \ldots \frac{\mathrm{d}^{n-1}x}{\mathrm{d}t^{n-1}}\right)$$

as a set of n first-order differential equations.

APPLICATION TO SOLID MECHANICS PROBLEMS

8

BASIC EQUATIONS AND SOLUTION PROCEDURE

8.1 INTRODUCTION

As stated in Chapter 1, the finite element method has been most extensively used in the field of solid and structural mechanics. The various types of problems solved by the finite element method in this field include the elastic, elastoplastic, and viscoelastic analysis of trusses, frames, plates, shells, and solid bodies. Both static and dynamic analysis have been conducted using the finite element method. We consider the finite element elastic analysis of one-, two-, and three-dimensional problems as well as axisymmetric problems in this book.

In this chapter, the general equations of solid and structural mechanics are presented. The displacement method (or equivalently the principle of minimum potential energy) is used in deriving the finite element equations. The application of these equations to several specific cases is considered in subsequent chapters.

8.2 BASIC EQUATIONS OF SOLID MECHANICS

8.2.1 Introduction

The primary aim of any stress analysis or solid mechanics problem is to find the distribution of displacements and stresses under the stated loading and boundary conditions. If an analytical solution of the problem is to be found, one has to satisfy the following basic equations of solid mechanics:

	Number of equations		
Type of equations	In 3-dimensional problems	In 2-dimensional problems	In 1-dimensional problems
Equilibrium equations	3	2	1
Stress–strain relations	6	3	1
Strain–displacement relations	6	3	1
Total number of equations	15	8	3

The unknown quantities, whose number is equal to the number of equations available, in various problems are given below:

Unknowns	In 3-dimensional problems	In 2-dimensional problems	In 1-dimensional problems
Displacements	u, v, w	u, v	u
Stresses	σ_{xx}, σ_{yy}, σ_{zz}, σ_{xy}, σ_{yz}, σ_{zx}	σ_{xx}, σ_{yy}, σ_{xy}	σ_{xx}
Strains	ε_{xx}, ε_{yy}, ε_{zz}, ε_{xy} ε_{yz}, ε_{zx}	ε_{xx}, ε_{yy}, ε_{xy}	ε_{xx}
Total number of unknowns	15	8	3

Thus, we have as many equations as there are unknowns to find the solution of any stress analysis problem. In practice, we will also have to satisfy some additional equations, such as external equilibrium equations (which pertain to the overall equilibrium of the body under external loads), compatibility equations (which pertain to the continuity of strains and displacements), and boundary conditions (which pertain to the prescribed conditions on displacements and/or forces at the boundary of the body).

Although any analytical (exact) solution has to satisfy all the equations stated previously, the numerical (approximate) solutions, like the ones obtained by using the finite element method, generally do not satisfy all the equations. However, a sound understanding of all the basic equations of solid mechanics is essential in deriving the finite element relations and also in estimating the order of error involved in the finite element solution by knowing the extent to which the approximate solution violates the basic equations, including the compatibility and boundary conditions. Hence, the basic equations of solid mechanics are summarized in the following section for ready reference in the formulation of finite element equations.

8.2.2 Equations

(i) External equilibrium equations

If a body is in equilibrium under specified static loads, the reactive forces and moments developed at the support points must balance the externally applied forces and moments. In other words, the force and moment equilibrium equations for the overall body (overall or external equilibrium equations) have to be satisfied. If ϕ_x, ϕ_y, and ϕ_z are the body forces, Φ_x, Φ_y, and Φ_z are the surface (distributed) forces, P_x, P_y, and P_z are the external concentrated loads (including reactions at support points such as B, C, and D in Figure 8.1), and Q_x, Q_y, and Q_z are the external concentrated moments (including reactions at support points such as B, C, and D in Figure 8.1), the external equilibrium equations can be stated as follows [8.1]:

$$\left.\begin{aligned} \int_S \Phi_x \, \mathrm{d}s + \int_V \phi_x \mathrm{d}v + \sum P_x = 0 \\ \int_S \Phi_y \, \mathrm{d}s + \int_V \phi_y \mathrm{d}v + \sum P_y = 0 \\ \int_S \Phi_z \, \mathrm{d}s + \int_V \phi_z \mathrm{d}v + \sum P_x = 0 \end{aligned}\right\} \tag{8.1}$$

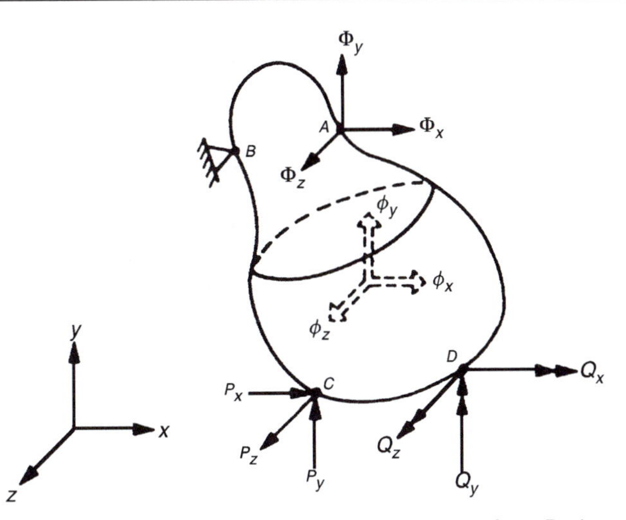

Figure 8.1. Force System for Macroequilibrium for a Body.

For moment equilibrium:

$$\left.\begin{aligned}
&\int_S (\Phi_z y - \Phi_y z)\,\mathrm{d}s + \int_V (\phi_z y - \phi_y z)\,\mathrm{d}v + \sum Q_x = 0\\
&\int_S (\Phi_x z - \Phi_z x)\,\mathrm{d}s + \int_V (\phi_x z - \phi_z x)\,\mathrm{d}v + \sum Q_y = 0\\
&\int_S (\Phi_y x - \Phi_x y)\,\mathrm{d}s + \int_V (\phi_y x - \phi_x y)\,\mathrm{d}v + \sum Q_z = 0
\end{aligned}\right\} \tag{8.2}$$

where S is the surface and V is the volume of the solid body.

(ii) Equations of internal equilibrium:
Due to the application of loads, stresses will be developed inside the body. If we consider an element of material inside the body, it must be in equilibrium due to the internal stresses developed. This leads to equations known as internal equilibrium equations.

Theoretically, the state of stress at any point in a loaded body is completely defined in terms of the nine components of stress σ_{xx}, σ_{yy}, σ_{zz}, σ_{xy}, σ_{yx}, σ_{yz}, σ_{zy}, σ_{zx}, and σ_{xz}, where the first three are the normal components and the latter six are the components of shear stress. The equations of internal equilibrium relating the nine components of stress can be derived by considering the equilibrium of moments and forces acting on the elemental volume shown in Figure 8.2. The equilibrium of moments about the x, y, and z axes, assuming that there are no body moments, leads to the relations

$$\sigma_{yx} = \sigma_{xy}, \qquad \sigma_{zy} = \sigma_{yz}, \qquad \sigma_{xz} = \sigma_{zx} \tag{8.3}$$

These equations show that the state of stress at any point can be completely defined by the six components σ_{xx}, σ_{yy}, σ_{zz}, σ_{xy}, σ_{yz}, and σ_{zx}. The equilibrium of forces in x, y,

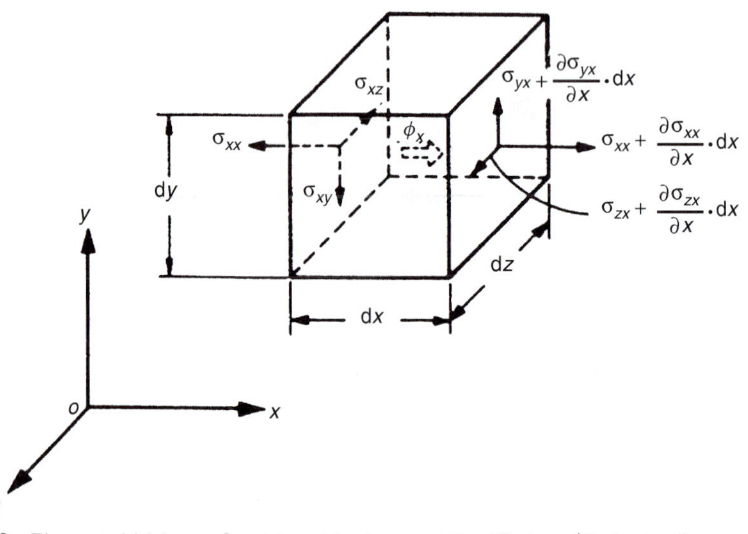

Figure 8.2. Elemental Volume Considered for Internal Equilibrium (Only the Components of Stress Acting on a Typical Pair of Faces Are Shown for Simplicity).

and z directions gives the following differential equilibrium equations:

$$\left.\begin{aligned}\frac{\partial \sigma_{xx}}{\partial x}+\frac{\partial \sigma_{xy}}{\partial y}+\frac{\partial \sigma_{zx}}{\partial z}+\phi_x=0\\\frac{\partial \sigma_{xy}}{\partial x}+\frac{\partial \sigma_{yy}}{\partial y}+\frac{\partial \sigma_{yz}}{\partial z}+\phi_y=0\\\frac{\partial \sigma_{zx}}{\partial x}+\frac{\partial \sigma_{yz}}{\partial y}+\frac{\partial \sigma_{zz}}{\partial z}+\phi_z=0\end{aligned}\right\}\tag{8.4}$$

where ϕ_x, ϕ_y, and ϕ_z are the body forces per unit volume acting along the directions x, y, and z, respectively.

For a two-dimensional problem, there will be only three independent stress components (σ_{xx}, σ_{yy}, σ_{xy}) and the equilibrium equations, Eqs. (8.4), reduce to

$$\left.\begin{aligned}\frac{\partial \sigma_{xx}}{\partial x}+\frac{\partial \sigma_{xy}}{\partial y}+\phi_x=0\\\frac{\partial \sigma_{xy}}{\partial x}+\frac{\partial \sigma_{yy}}{\partial y}+\phi_y=0\end{aligned}\right\}\tag{8.5}$$

In one-dimensional problems, only one component of stress, namely σ_{xx}, will be there and hence Eqs. (8.4) reduce to

$$\frac{\partial \sigma_{xx}}{\partial x}+\phi_x=0\tag{8.6}$$

(iii) Stress–strain relations (constitutive relations) for isotropic materials

Three-dimensional case In the case of linearity elastic isotropic three-dimensional solid, the stress–strain relations are given by Hooke's law as

$$\vec{\varepsilon} = \begin{Bmatrix} \varepsilon_{xx} \\ \varepsilon_{yy} \\ \varepsilon_{zz} \\ \varepsilon_{xy} \\ \varepsilon_{yz} \\ \varepsilon_{zx} \end{Bmatrix} = [C]\vec{\sigma} + \vec{\varepsilon}_0 \equiv [C] \begin{Bmatrix} \sigma_{xx} \\ \sigma_{yy} \\ \sigma_{zz} \\ \sigma_{xy} \\ \sigma_{yz} \\ \sigma_{zx} \end{Bmatrix} + \begin{Bmatrix} \varepsilon_{xx0} \\ \varepsilon_{yy0} \\ \varepsilon_{zz0} \\ \varepsilon_{xy0} \\ \varepsilon_{yz0} \\ \varepsilon_{zx0} \end{Bmatrix} \tag{8.7}$$

where $[C]$ is a matrix of elastic coefficients given by

$$[C] = \frac{1}{E} \begin{bmatrix} 1 & -v & -v & 0 & 0 & 0 \\ -v & 1 & -v & 0 & 0 & 0 \\ -v & -v & 1 & 0 & 0 & 0 \\ 0 & 0 & 0 & 2(1+v) & 0 & 0 \\ 0 & 0 & 0 & 0 & 2(1+v) & 0 \\ 0 & 0 & 0 & 0 & 0 & 2(1+v) \end{bmatrix} \tag{8.8}$$

$\vec{\varepsilon}_0$ is the vector of initial strains, E is Young's modules, and v is Poisson's ratio of the material. In the case of heating of an isotropic material, the initial strain vector is given by

$$\vec{\varepsilon}_0 = \begin{Bmatrix} \varepsilon_{xx0} \\ \varepsilon_{yy0} \\ \varepsilon_{zz0} \\ \varepsilon_{xy0} \\ \varepsilon_{yz0} \\ \varepsilon_{zx0} \end{Bmatrix} = \alpha T \begin{Bmatrix} 1 \\ 1 \\ 1 \\ 0 \\ 0 \\ 0 \end{Bmatrix} \tag{8.9}$$

where α is the coefficient of thermal expansion, and T is the temperature charge.

Sometimes, the expressions for stresses in terms of strains will be needed. By including thermal strains, Eqs. (8.7) can be inverted to obtain

$$\vec{\sigma} = \begin{Bmatrix} \sigma_{xx} \\ \sigma_{yy} \\ \sigma_{zz} \\ \sigma_{xy} \\ \sigma_{yz} \\ \sigma_{zx} \end{Bmatrix} = [D](\vec{\varepsilon} - \vec{\varepsilon}_0) \equiv [D] \begin{Bmatrix} \varepsilon_{xx} \\ \varepsilon_{yy} \\ \varepsilon_{zz} \\ \varepsilon_{xy} \\ \varepsilon_{yz} \\ \varepsilon_{zx} \end{Bmatrix} - \frac{E\alpha T}{1-2v} \begin{Bmatrix} 1 \\ 1 \\ 1 \\ 0 \\ 0 \\ 0 \end{Bmatrix} \tag{8.10}$$

where the matrix $[D]$ is given by

$$[D] = \frac{E}{(1+v)(1-2v)} \begin{bmatrix} 1-v & v & v & 0 & 0 & 0 \\ v & 1-v & v & 0 & 0 & 0 \\ v & v & 1-v & 0 & 0 & 0 \\ 0 & 0 & 0 & \frac{1-2v}{2} & 0 & 0 \\ 0 & 0 & 0 & 0 & \frac{1-2v}{2} & 0 \\ 0 & 0 & 0 & 0 & 0 & \frac{1-2v}{2} \end{bmatrix} \tag{8.11}$$

In the case of two-dimensional problems, two types of stress distributions, namely plane stress and plane strain, are possible.

Two-dimensional case (plane stress) The assumption of plane stress is applicable for bodies whose dimension is very small in one of the coordinate directions. Thus, the analysis of thin plates loaded in the plane of the plate can be made using the assumption of plane stress. In plane stress distribution, it is assumed that

$$\sigma_{zz} = \sigma_{zx} = \sigma_{yz} = 0 \tag{8.12}$$

where z represents the direction perpendicular to the plane of the plate as shown in Figure 8.3, and the stress components do not vary through the thickness of the plate (i.e., in z direction). Although these assumptions violate some of the compatibility conditions, they are sufficiently accurate for all practical purposes provided the plate is thin. In this case, the stress–strain relations, Eqs. (8.7) and (8.10), reduce to

$$\vec{\varepsilon} = [C]\vec{\sigma} + \vec{\varepsilon}_0 \tag{8.13}$$

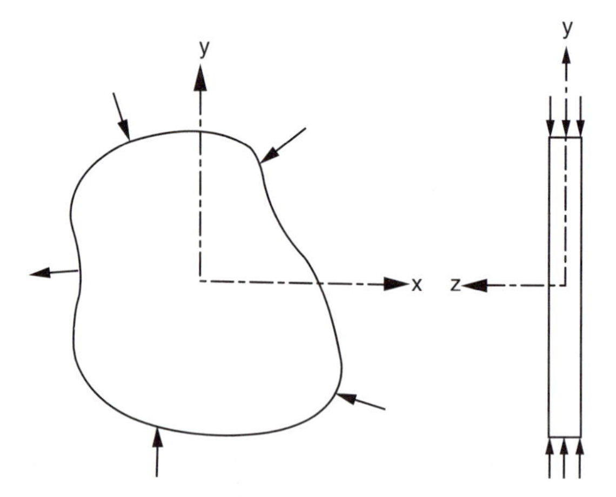

Figure 8.3. Example of a Plane Stress Problem; A Thin Plate under Inplane Loading.

where

$$\vec{\varepsilon} = \begin{Bmatrix} \varepsilon_{xx} \\ \varepsilon_{yy} \\ \varepsilon_{xy} \end{Bmatrix}, \qquad \vec{\sigma} = \begin{Bmatrix} \sigma_{xx} \\ \sigma_{yy} \\ \sigma_{xy} \end{Bmatrix}$$

$$[C] = \frac{1}{E}\begin{bmatrix} 1 & -v & 0 \\ -v & 1 & 0 \\ 0 & 0 & 2(1+v) \end{bmatrix} \tag{8.14}$$

$$\vec{\varepsilon}_0 = \begin{Bmatrix} \varepsilon_{xx0} \\ \varepsilon_{yy0} \\ \varepsilon_{xy0} \end{Bmatrix} = \alpha T \begin{Bmatrix} 1 \\ 1 \\ 0 \end{Bmatrix} \quad \text{in the case of thermal strains} \tag{8.15}$$

and

$$\vec{\sigma} = [D](\vec{\varepsilon} - \vec{\varepsilon}_0) = [D]\vec{\varepsilon} - \frac{E\alpha T}{1-v}\begin{Bmatrix} 1 \\ 1 \\ 0 \end{Bmatrix} \tag{8.16}$$

with

$$[D] = \frac{E}{1-v^2}\begin{bmatrix} 1 & v & 0 \\ v & 1 & 0 \\ 0 & 0 & \dfrac{1-v}{2} \end{bmatrix} \tag{8.17}$$

In the case of plane stress, the component of strain in the z direction will be nonzero and is given by (from Eq. 8.7)

$$\varepsilon_{zz} = -\frac{v}{E}(\sigma_{xx} + \sigma_{yy}) + \alpha T = \frac{-v}{1-v}(\varepsilon_{xx} + \varepsilon_{yy}) + \frac{1+v}{1-v}\alpha T \tag{8.18}$$

while

$$\varepsilon_{yz} = \varepsilon_{zx} = 0 \tag{8.19}$$

Two-dimensional case (plane strain) The assumption of plane strain is applicable for bodies that are long and whose geometry and loading do not vary significantly in the longitudinal direction. Thus, the analysis of dams, cylinders, and retaining walls shown in Figure 8.4 can be made using the assumption of plane strain. In plane strain distribution, it is assumed that $w = 0$ and $(\partial w/\partial z) = 0$ at every cross section. Here, the dependent variables are assumed to be functions of only the x and y coordinates provided we consider a cross section of the body away from the ends. In this case, the three-dimensional stress–strain relations given by Eqs. (8.7) and (8.10) reduce to

$$\vec{\varepsilon} = [C]\vec{\sigma} + \vec{\varepsilon}_0 \tag{8.20}$$

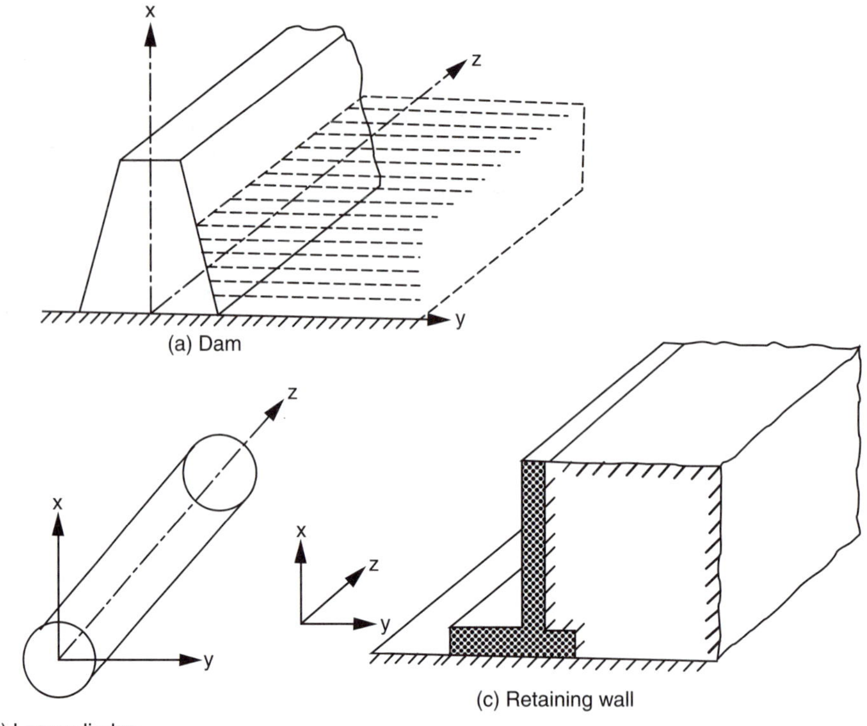

Figure 8.4. Examples of Plane Strain Problems.

where

$$\vec{\varepsilon} = \begin{Bmatrix} \varepsilon_{xx} \\ \varepsilon_{yy} \\ \varepsilon_{xy} \end{Bmatrix}, \qquad \vec{\sigma} = \begin{Bmatrix} \sigma_{xx} \\ \sigma_{yy} \\ \sigma_{xy} \end{Bmatrix},$$

$$[C] = \frac{1+y}{E} \begin{bmatrix} 1-v & -v & 0 \\ -v & 1-v & 0 \\ 0 & 0 & 2 \end{bmatrix}, \tag{8.21}$$

$$\vec{\varepsilon}_0 = \begin{Bmatrix} \varepsilon_{xx0} \\ \varepsilon_{yy0} \\ \varepsilon_{xy0} \end{Bmatrix} = (1+v)\alpha T \begin{Bmatrix} 1 \\ 1 \\ 0 \end{Bmatrix} \quad \text{in the case of thermal strains} \tag{8.22}$$

and

$$\vec{\sigma} = [D](\vec{\varepsilon} - \vec{\varepsilon}_0) = [D]\vec{\varepsilon} - \frac{E\alpha T}{1-2v} \begin{Bmatrix} 1 \\ 1 \\ 0 \end{Bmatrix} \tag{8.23}$$

with

$$[D] = \frac{E}{(1+v)(1-2v)} \begin{bmatrix} 1-v & v & 0 \\ v & 1-v & 0 \\ 0 & 0 & \frac{1-2v}{2} \end{bmatrix} \tag{8.24}$$

The component of stress in the z direction will be nonzero and is given by

$$\sigma_{zz} = v(\sigma_{xx} + \sigma_{yy}) - E\alpha T \tag{8.25}$$

and

$$\sigma_{yz} = \sigma_{zx} = 0 \tag{8.26}$$

One-dimensional case In the case of one-dimensional problems, all stress components except for one normal stress are zero and the stress–strain relations degenerate to

$$\vec{\varepsilon} = [C]\vec{\sigma} + \vec{\varepsilon}_0 \tag{8.27}$$

where

$$\vec{\varepsilon} = \{\varepsilon_{xx}\}, \qquad \vec{\sigma} = \{\sigma_{xx}\},$$

$$[C] = \left[\frac{1}{E}\right], \tag{8.28}$$

$$\vec{\varepsilon}_0 = \{\varepsilon_{xx_0}\} = \alpha T \text{ in the case of thermal strain} \tag{8.29}$$

and

$$\vec{\sigma} = [D](\vec{\varepsilon} - \vec{\varepsilon}_0) = [D]\vec{\varepsilon} - E\alpha T\{1\} \tag{8.30}$$

with

$$[D] = [E] \tag{8.31}$$

Axisymmetric case In the case of solids of revolution (axisymmetric solids), the stress–strain relations are given by

$$\vec{\varepsilon} = [C]\vec{\sigma} + \vec{\varepsilon}_0 \tag{8.32}$$

where

$$\vec{\varepsilon} = \begin{Bmatrix} \varepsilon_{rr} \\ \varepsilon_{\theta\theta} \\ \varepsilon_{zz} \\ \varepsilon_{rz} \end{Bmatrix}, \qquad \vec{\sigma} = \begin{Bmatrix} \sigma_{rr} \\ \sigma_{\theta\theta} \\ \sigma_{zz} \\ \sigma_{rz} \end{Bmatrix},$$

$$[C] = \frac{1}{E}\begin{bmatrix} 1 & -v & -v & 0 \\ -v & 1 & -v & 0 \\ -v & -v & 1 & 0 \\ 0 & 0 & 0 & 2(1+v) \end{bmatrix}, \tag{8.33}$$

$$\vec{\varepsilon}_0 = \begin{Bmatrix} \varepsilon_{rr0} \\ \varepsilon_{\theta\theta0} \\ \varepsilon_{zz0} \\ \varepsilon_{rz0} \end{Bmatrix} = \alpha T \begin{Bmatrix} 1 \\ 1 \\ 1 \\ 0 \end{Bmatrix} \text{ in the case of theremal strains} \tag{8.34}$$

and

$$\vec{\sigma} = [D](\vec{\varepsilon} - \vec{\varepsilon}_0) = [D]\vec{\varepsilon} - \frac{E\alpha T}{1-2v}\begin{Bmatrix} 1 \\ 1 \\ 1 \\ 0 \end{Bmatrix} \tag{8.35}$$

with

$$[D] = \frac{E}{(1+v)(1-2v)}\begin{bmatrix} 1-v & v & v & 0 \\ v & 1-v & v & 0 \\ v & v & 1-v & 0 \\ 0 & 0 & 0 & \left(\frac{1-2v}{2}\right) \end{bmatrix} \tag{8.36}$$

In these equations, the subscripts r, θ, and z denote the radial, tangential, and axial directions, respectively.

(iv) Stress–strain relations for anisotropic materials

The stress–strain relations given earlier are valid for isotropic elastic materials. The term "isotropic" indicates that the material properties at a point in the body are not a function of orientation. In other words, the material properties are constant in any plane passing through a point in the material. There are certain materials (e.g., reinforced concrete, fiber-reinforced composites, brick, and wood) for which the material properties at any point depend on the orientation also. In general, such materials are called anisotropic materials. The generalized Hooke's law valid for anisotropic materials is given in this section. The special cases of the Hooke's law for orthotropic and isotropic materials will also be indicated.

For a linearly elastic anisotropic material, the strain–stress relations are given by the generalized Hooke's law as [8.7, 8.8]

$$\begin{Bmatrix} \varepsilon_1 \\ \varepsilon_2 \\ \varepsilon_3 \\ \varepsilon_{23} \\ \varepsilon_{13} \\ \varepsilon_{12} \end{Bmatrix} = \begin{bmatrix} C_{11} & C_{12} & \cdots & C_{16} \\ C_{12} & C_{22} & \cdots & C_{26} \\ \vdots & & & \\ C_{16} & C_{26} & \cdots & C_{66} \end{bmatrix} \begin{Bmatrix} \sigma_1 \\ \sigma_2 \\ \sigma_3 \\ \sigma_{23} \\ \sigma_{13} \\ \sigma_{12} \end{Bmatrix} \tag{8.37}$$

where the matrix $[C]$ is symmetric and is called the compliance matrix. Thus, 21 independent elastic constants (equal to the number of independent components of $[C]$) are needed to describe an anisotropic material. Note that subscripts 1, 2, and 3 are used instead of x, y, and z in Eq. (8.37) for convenience.

Certain materials exhibit symmetry with respect to certain planes within the body. In such cases, the number of elastic constants will be reduced from 21. For an orthotropic material, which has three planes of material property symmetry, Eq. (8.37) reduces to

$$\begin{Bmatrix} \varepsilon_1 \\ \varepsilon_2 \\ \varepsilon_3 \\ \varepsilon_{23} \\ \varepsilon_{13} \\ \varepsilon_{12} \end{Bmatrix} = \begin{bmatrix} C_{11} & C_{12} & C_{13} & 0 & 0 & 0 \\ C_{12} & C_{22} & C_{23} & 0 & 0 & 0 \\ C_{13} & C_{23} & C_{33} & 0 & 0 & 0 \\ 0 & 0 & 0 & C_{44} & 0 & 0 \\ 0 & 0 & 0 & 0 & C_{55} & 0 \\ 0 & 0 & 0 & 0 & 0 & C_{66} \end{bmatrix} \begin{Bmatrix} \sigma_1 \\ \sigma_2 \\ \sigma_3 \\ \sigma_{23} \\ \sigma_{13} \\ \sigma_{12} \end{Bmatrix} \tag{8.38}$$

where the elements C_{ij} are given by

$$\left.\begin{aligned} C_{11} &= \frac{1}{E_{11}}, & C_{12} &= -\frac{v_{21}}{E_{22}}, & C_{13} &= -\frac{v_{31}}{E_{33}}, \\ C_{22} &= \frac{1}{E_{22}}, & C_{23} &= -\frac{v_{32}}{E_{33}}, & C_{33} &= \frac{1}{E_{33}}, \\ C_{44} &= \frac{1}{G_{23}}, & C_{55} &= \frac{1}{G_{13}}, & C_{66} &= \frac{1}{G_{12}} \end{aligned}\right\} \tag{8.39}$$

Here, E_{11}, E_{22}, and E_{33} denote the Young's modulus in the planes defined by axes 1, 2, and 3, respectively; G_{12}, G_{23}, and G_{13} represent the shear modulus in the planes 12, 23, and 13, respectively; and v_{12}, v_{13}, and v_{23} indicate the major Poisson's ratios. Thus, nine independent elastic constants are needed to describe an orthotropic material under three-dimensional state of stress. For the specially orthotropic material that is in a state of plane stress, $\sigma_3 = \sigma_{23} = \sigma_{13} = 0$ and Eq. (8.38) reduces to

$$\begin{Bmatrix} \varepsilon_1 \\ \varepsilon_2 \\ \varepsilon_{12} \end{Bmatrix} = \begin{bmatrix} C_{11} & C_{12} & 0 \\ C_{12} & C_{22} & 0 \\ 0 & 0 & C_{66} \end{bmatrix} \begin{Bmatrix} \sigma_1 \\ \sigma_2 \\ \sigma_{12} \end{Bmatrix} \tag{8.40}$$

which involves four independent elastic constants. The elements of the compliance matrix, in this case, can be expressed as

$$\begin{aligned} C_{11} &= \frac{1}{E_{11}} \\ C_{22} &= \frac{1}{E_{22}} \\ C_{12} &= -\frac{v_{12}}{E_{11}} = -\frac{v_{21}}{E_{22}} \\ C_{66} &= \frac{1}{G_{12}} \end{aligned} \tag{8.41}$$

The stress–strain relations can be obtained by inverting the relations given by Eqs. (8.37), (8.38), and (8.40). Specifically, the stress–strain relations for a specially orthotropic material (under plane stress) can be expressed as

$$\begin{Bmatrix} \sigma_1 \\ \sigma_2 \\ \sigma_{12} \end{Bmatrix} = \begin{bmatrix} Q_{11} & Q_{12} & 0 \\ Q_{12} & Q_{22} & 0 \\ 0 & 0 & Q_{66} \end{bmatrix} \begin{Bmatrix} \varepsilon_1 \\ \varepsilon_2 \\ \varepsilon_{12} \end{Bmatrix} \equiv [Q]\vec{\varepsilon} \tag{8.42}$$

where the elements of the matrix $[Q]$ are given by

$$\left.\begin{aligned} Q_{11} &= \frac{E_{11}}{1 - v_{12}v_{21}}, \quad Q_{22} = \frac{E_{22}}{1 - v_{12}v_{21}} \\ Q_{12} &= \frac{v_{21}E_{11}}{1 - v_{12}v_{21}} = \frac{v_{12}E_{22}}{1 - v_{12}v_{21}} \\ Q_{66} &= 2G_{12} \end{aligned}\right\} \tag{8.43}$$

If the material is linearly elastic and isotropic, only two elastic constants are needed to describe the behavior and the stress–strain relations are given by Eq. (8.7) or Eq. (8.10).

(v) Strain–displacement relations

The deformed shape of an elastic body under any given system of loads and temperature distribution conditions can be completely described by the three components of displacement u, v, and w parallel to the directions x, y, and z, respectively. In general, each of these components u, v, and w is a function of the coordinates x, y, and z. The strains induced in the body can be expressed in terms of the displacements u, v, and w. In this section, we assume the deformations to be small so that the strain–displacement relations remain linear.

To derive expressions for the normal strain components ε_{xx} and ε_{yy} and the shear strain component ε_{xy}, consider a small rectangular element $OACB$ whose sides (of lengths $\mathrm{d}x$ and $\mathrm{d}y$) lie parallel to the coordinate axes before deformation. When the body undergoes deformation under the action of external load and temperature distributions, the element $OACB$ also deforms to the shape $O'A'C'B'$ as shown in Figure 8.5. We can observe that the element $OACB$ has two basic types of deformation, one of change in size and the other of angular distortion.

Since the normal strain is defined as change in length divided by original length, the strain components ε_{xx} and ε_{yy} can be found as

$$\begin{aligned} \varepsilon_{xx} &= \frac{\begin{array}{c}\text{change in length of the fiber } OA \text{ that lies} \\ \text{in the } x \text{ directon before deformation}\end{array}}{\text{original length of the fiber } OA} \\ &= \frac{\left[\mathrm{d}x + \left(u + \dfrac{\partial u}{\partial x} \cdot \mathrm{d}x\right) - u\right] - \mathrm{d}x}{\mathrm{d}x} = \frac{\partial u}{\partial x} \end{aligned} \tag{8.44}$$

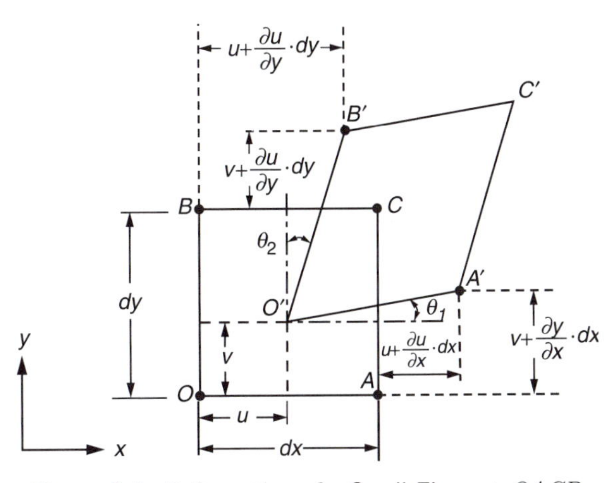

Figure 8.5. Deformation of a Small Element $OACB$.

and

$$\varepsilon_{yy} = \frac{\text{change in length of the fiber } OB \text{ that lies in the } y \text{ directon before deformation}}{\text{original length of the fiber } OB}$$

$$= \frac{\left[dy + \left(v + \frac{\partial v}{\partial y} \cdot dy\right) - v\right] - dy}{dy} = \frac{\partial v}{\partial y} \tag{8.45}$$

The shear strain is defined as the decrease in the right angle between the fibers OA and OB, which were at right angles to each other before deformation. Thus, the expression for the shear strain ε_{xy} can be obtained as

$$\varepsilon_{xy} = \theta_1 + \theta_2 \simeq \tan\theta_1 + \tan\theta_2 \simeq \frac{\left(v + \frac{\partial v}{\partial x}\cdot dx\right) - v}{\left[dx + \left(u + \frac{\partial u}{\partial x}\cdot dx\right) - u\right]} + \frac{\left(u + \frac{\partial u}{\partial y}\cdot dy\right) - u}{\left[dy + \left(v + \frac{\partial v}{\partial y}\cdot dy\right) - v\right]}$$

If the displacements are assumed to be small, ε_{xy} can be expressed as

$$\varepsilon_{xy} = \frac{\partial u}{\partial y} + \frac{\partial v}{\partial x} \tag{8.46}$$

The expressions for the remaining normal strain component ε_{zz} and shear strain components ε_{yz} and ε_{zx} can be derived in a similar manner as

$$\varepsilon_{zz} = \frac{\partial w}{\partial z}, \tag{8.47}$$

$$\varepsilon_{yz} = \frac{\partial w}{\partial y} + \frac{\partial v}{\partial z}, \tag{8.48}$$

and

$$\varepsilon_{zx} = \frac{\partial u}{\partial z} + \frac{\partial w}{\partial x} \tag{8.49}$$

In the case of two-dimensional problems, Eqs. (8.44)–(8.46) are applicable, whereas Eq. (8.44) is applicable in the case of one-dimensional problems.

In the case of an axisymmetric solid, the strain–displacement relations can be derived as

$$\begin{aligned} \varepsilon_{rr} &= \frac{\partial u}{\partial r} \\ \varepsilon_{\theta\theta} &= \frac{u}{r} \\ \varepsilon_{zz} &= \frac{\partial w}{\partial z} \\ \varepsilon_{rz} &= \frac{\partial u}{\partial z} + \frac{\partial w}{\partial r} \end{aligned} \tag{8.50}$$

where u and w are the radial and the axial displacements, respectively.

(vi) Boundary conditions

Boundary conditions can be either on displacements or on stresses. The boundary conditions on displacements require certain displacements to prevail at certain points on the boundary of the body, whereas the boundary conditions on stresses require that the stresses induced must be in equilibrium with the external forces applied at certain points on the boundary of the body. As an example, consider the flat plate under inplane loading shown in Figure 8.6.

In this case, the boundary conditions can be expressed as

$$u = v = 0 \text{ along the edge } AB$$

(displacement boundary conditions)

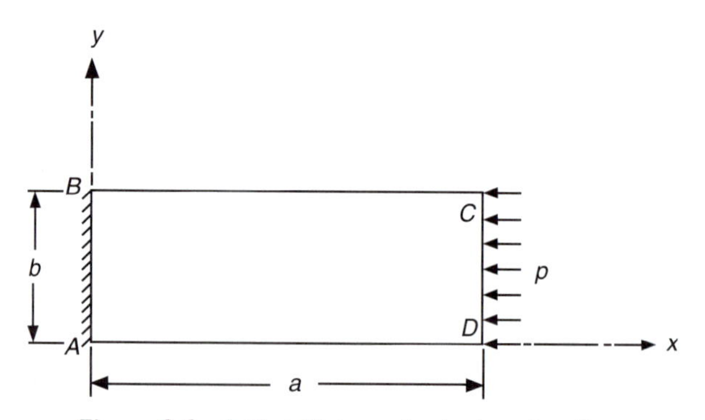

Figure 8.6. A Flat Plate under Inplane Loading.

and

$$\sigma_{yy} = \sigma_{xy} = 0 \text{ along the edges } BC \text{ and } AD$$
$$\sigma_{xx} = -p, \quad \sigma_{yy} = \sigma_{xy} = 0 \text{ along the edge } CD$$
(stress boundary conditions)

It can be observed that the displacements are unknown and are free to assume any values dictated by the solution wherever stresses are prescribed and vice versa. This is true of all solid mechanics problems.

For the equilibrium of induced stresses and applied surface forces at point A of Figure 8.7, the following equations must be satisfied:

$$\begin{aligned} \ell_x\sigma_{xx} + \ell_y \quad \sigma_{xy} + \ell_z \quad \sigma_{xz} &= \Phi_x \\ \ell_x\sigma_{xy} + \ell_y \quad \sigma_{yy} + \ell_z \quad \sigma_{yz} &= \Phi_y \\ \ell_x\sigma_{xz} + \ell_y \quad \sigma_{yz} + \ell_z \quad \sigma_{zz} &= \Phi_z \end{aligned} \tag{8.51}$$

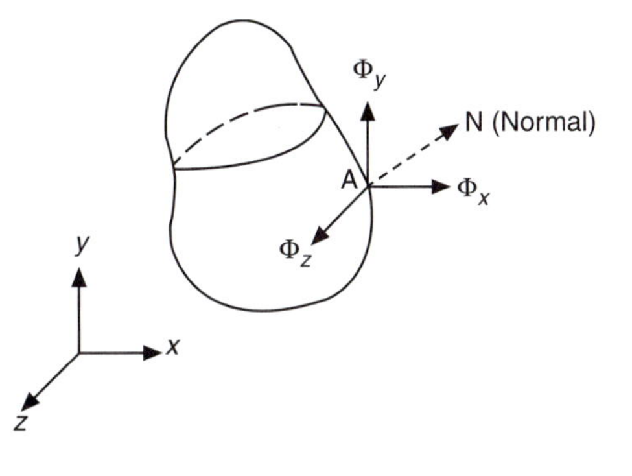

(a) Components of the surface force

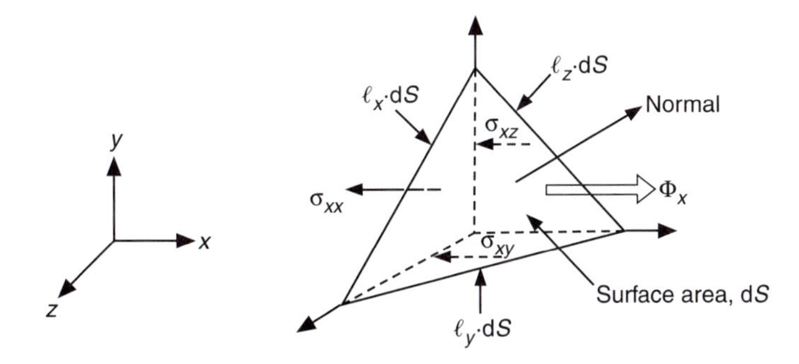

(b) Equillibrium of internal stresses and surface forces around point A

Figure 8.7. Forces Acting at the Surface of a Body.

where ℓ_x, ℓ_y, and ℓ_z are the direction cosines of the outward drawn normal (AN) at point A; and Φ_x, Φ_y, and Φ_z are the components of surface forces (tractions) acting at point A in the directions x, y, and z, respectively. The surface (distributed) forces Φ_x, Φ_y, and Φ_z have dimensions of force per unit area. Equation (8.51) can be specialized to two- and one-dimensional problems without much difficulty.

(vii) Compatibility equations

When a body is continuous before deformation, it should remain continuous after deformation. In other words, no cracks or gaps should appear in the body and no part should overlap another due to deformation. Thus, the displacement field must be continuous as well as single-valued. This is known as the "condition of compatibility." The condition of compatibility can also be seen from another point of view. For example, we can see from Eqs. (8.44)–(8.49) that the three strains ε_{xx}, ε_{yy}, and ε_{xy} can be derived from only two displacements u and v. This implies that a definite relation must exist between ε_{xx}, ε_{yy}, and ε_{xy} if these strains correspond to a compatible deformation. This definite relation is called the "compatibility equation." Thus, in three-dimensional elasticity problems, there are six compatibility equations [8.2]:

$$\frac{\partial^2 \varepsilon_{xx}}{\partial y^2} + \frac{\partial^2 \varepsilon_{yy}}{\partial x^2} = \frac{\partial^2 \varepsilon_{xy}}{\partial x \partial y} \tag{8.52}$$

$$\frac{\partial^2 \varepsilon_{yy}}{\partial z^2} + \frac{\partial^2 \varepsilon_{zz}}{\partial y^2} = \frac{\partial^2 \varepsilon_{yz}}{\partial y \partial z} \tag{8.53}$$

$$\frac{\partial^2 \varepsilon_{zz}}{\partial x^2} + \frac{\partial^2 \varepsilon_{xx}}{\partial z^2} = \frac{\partial^2 \varepsilon_{zx}}{\partial x \partial z} \tag{8.54}$$

$$\frac{1}{2}\frac{\partial}{\partial x}\left(\frac{\partial \varepsilon_{xy}}{\partial z} - \frac{\partial \varepsilon_{yz}}{\partial x} + \frac{\partial \varepsilon_{zx}}{\partial y}\right) = \frac{\partial^2 \varepsilon_{xx}}{\partial y \partial z} \tag{8.55}$$

$$\frac{1}{2}\frac{\partial}{\partial y}\left(\frac{\partial \varepsilon_{xy}}{\partial z} + \frac{\partial \varepsilon_{yz}}{\partial x} - \frac{\partial \varepsilon_{zx}}{\partial y}\right) = \frac{\partial^2 \varepsilon_{yy}}{\partial z \partial x} \tag{8.56}$$

$$\frac{1}{2}\frac{\partial}{\partial x}\left(-\frac{\partial \varepsilon_{xy}}{\partial z} + \frac{\partial \varepsilon_{yz}}{\partial x} + \frac{\partial \varepsilon_{zx}}{\partial y}\right) = \frac{\partial^2 \varepsilon_{zz}}{\partial x \partial y} \tag{8.57}$$

In the case of two-dimensional plane strain problems, Eqs. (8.52)–(8.57) reduce to a single equation as

$$\frac{\partial^2 \varepsilon_{xx}}{\partial y^2} + \frac{\partial^2 \varepsilon_{yy}}{\partial x^2} = \frac{\partial^2 \varepsilon_{xy}}{\partial x \partial y} \tag{8.58}$$

For plane stress problems, Eqs. (8.52)–(8.57) reduce to the following equations:

$$\frac{\partial^2 \varepsilon_{xx}}{\partial y^2} + \frac{\partial^2 \varepsilon_{yy}}{\partial x^2} = \frac{\partial^2 \varepsilon_{xy}}{\partial x \partial y}, \qquad \frac{\partial^2 \varepsilon_{zz}}{\partial y^2} = \frac{\partial^2 \varepsilon_{zz}}{\partial x^2} = \frac{\partial^2 \varepsilon_{zz}}{\partial x \partial y} = 0 \tag{8.59}$$

In the case of one-dimensional problems the conditions of compatibility will be automatically satisfied.

8.3 FORMULATIONS OF SOLID AND STRUCTURAL MECHANICS

As stated in Section 5.3, most continuum problems, including solid and structural mechanics problems, can be formulated according to one of the two methods: differential equation method and variational method. Hence, the finite element equations can also be derived by using either a differential equation formulation method (e.g., Galerkin approach) or variational formulation method (e.g., Rayleigh–Ritz approach). In the case of solid and structural mechanics problems, each of the differential equation and variational formulation methods can be classified into three categories as shown in Table 8.1.

The displacement, force, and displacement–force methods of differential equation formulation are closely related to the principles of minimum potential energy, minimum complementary energy, and stationary Reissner energy formulations, respectively. We use the displacement method or the principle of minimum potential energy for presenting the various concepts of the finite element method because they have been extensively used in the literature.

8.3.1 Differential Equation Formulation Methods

(i) Displacement method

As stated in Section 8.2.1, for a three-dimensional continuum or elasticity problem, there are six stress–strain relations [Eq. (8.10)], six strain–displacement relations [Eqs. (8.44)–(8.49)], and three equilibrium equations [Eqs. (8.4)], and the unknowns are six stresses (σ_{ij}), six strains (ε_{ij}), and three displacements (u, v, and w). By substituting Eqs. (8.44)–(8.49) into Eqs. (8.10), we obtain the stresses in terms of the displacements. By substituting these stress–displacement relations into Eqs. (8.4), we obtain three equilibrium equations in terms of the three unknown displacement components u, v, and w. Now these equilibrium equations can be solved for u, v, and w. Of course, the additional requirements such as boundary and compatibility conditions also have to be satisfied while finding the solution for u, v, and w. Since the displacements u, v, and w are made the final unknowns, the method is known as the displacement method.

(ii) Force method

For a three-dimensional elasticity problem, there are three equilibrium equations, Eqs. (8.4), in terms of six unknown stresses σ_{ij}. At the same time, there are six compatibility equations, Eqs. (8.52)–(8.57), in terms of the six strain components ε_{ij}. Now we take any three strain components, for example, ε_{xy}, ε_{yz}, and ε_{zx}, as independent strains and

Table 8.1. Methods of Formulating Solid and Structural Mechanics Problems

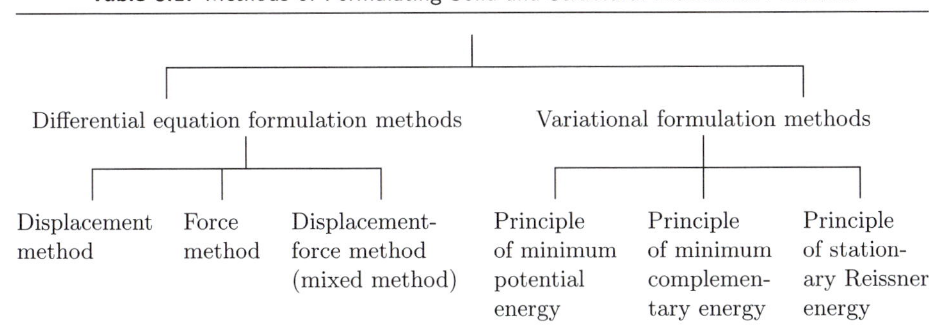

write the compatibility equations in terms of ε_{xy}, ε_{yz}, and ε_{zx} only. By substituting the known stress–strain relations, Eq. (8.10), we express the three independent compatibility equations in terms of the stresses σ_{ij}. By using these three equations, three of the stresses out of σ_{xx}, σ_{yy}, σ_{zz}, σ_{xy}, σ_{yz}, and σ_{zx} can be eliminated from the original equilibrium equations. Thus, we get three equilibrium equations in terms of three stress components only, and hence the problem can be solved. Since the final equations are in terms of stresses (or forces), the method is known as the force method.

(iii) Displacement–force method

In this method, we use the strain–displacement relations to eliminate strains from the stress–strain relations. These six equations, in addition to the three equilibrium equations, will give us nine equations in the nine unknowns σ_{xx}, σ_{yy}, σ_{zz}, σ_{xy}, σ_{yz}, σ_{zx}, u, v, and w. Thus, the solution of the problem can be found by using the additional conditions such as compatibility and boundary conditions. Since both the displacements and the stresses (or forces) are taken as the final unknowns, the method is known as the displacement–force method.

8.3.2 Variational Formulation Methods

(i) Principle of minimum potential energy

The potential energy of an elastic body π_p is defined as

$$\pi_p = \pi - W_p \tag{8.60}$$

where π is the strain energy, and W_p is the work done on the body by the external forces. The principle of minimum potential energy can be stated as follows: Of all possible displacement states (u, v, and w) a body can assume that satisfy compatibility and given kinematic or displacement boundary conditions, the state that satisfies the equilibrium equations makes the potential energy assume a minimum value. If the potential energy, π_p, is expressed in terms of the displacements u, v, and w, the principle of minimum potential energy gives, at the equilibrium state,

$$\delta\pi_p(u, v, w) = \delta\pi(u, v, w) - \delta W_p(u, v, w) = 0 \tag{8.61}$$

It is important to note that the variation is taken with respect to the displacements in Eq. (8.61), whereas the forces and stresses are assumed constant. The strain energy of a linear elastic body is defined as

$$\pi = \frac{1}{2}\iiint_V \vec{\varepsilon}^{\,T}\vec{\sigma}\,\mathrm{d}V \tag{8.62}$$

where V is the volume of the body. By using the stress–strain relations of Eq. (8.10), the strain energy, in the presence of initial strains $\vec{\varepsilon}_0$, can be expressed as

$$\pi = \frac{1}{2}\iiint_V \vec{\varepsilon}^{\,T}[D]\vec{\varepsilon}\,\mathrm{d}V - \iiint_V \vec{\varepsilon}^{\,T}[D]\vec{\varepsilon}_0\,\mathrm{d}V \tag{8.63}$$

The work done by the external forces can be expressed as

$$W_p = \iiint_V \vec{\bar{\phi}}^{\,T}\vec{U}\cdot\mathrm{d}V + \iint_{S_1} \vec{\bar{\Phi}}^{\,T}\vec{U}\cdot\mathrm{d}S_1 \tag{8.64}$$

where $\vec{\bar{\phi}} = \begin{Bmatrix} \bar{\phi}_x \\ \bar{\phi}_y \\ \bar{\phi}_z \end{Bmatrix}$ = known body force vector, $\vec{\bar{\Phi}} = \begin{Bmatrix} \bar{\Phi}_x \\ \bar{\Phi}_y \\ \bar{\Phi}_z \end{Bmatrix}$ = vector of prescribed surface forces (tractions), $\vec{U} = \begin{Bmatrix} u \\ v \\ w \end{Bmatrix}$ = vector of displacements, and S_1 is the surface of the body on which surface forces are prescribed. Using Eqs. (8.63) and (8.64), the potential energy of the body can be expressed as

$$\pi_p(u,v,w) = \frac{1}{2}\iiint_V \vec{\varepsilon}^{\,T}[D](\vec{\varepsilon} - 2\vec{\varepsilon}_0)\,\mathrm{d}V - \iiint_V \vec{\bar{\phi}}^{\,T}\vec{U}\cdot \mathrm{d}V - \iint_{S_1} \vec{\bar{\Phi}}^{\,T}\vec{U}\cdot \mathrm{d}S_1 \tag{8.65}$$

If we use the principle of minimum potential energy to derive the finite element equations, we assume a simple form of variation for the displacement field within each element and derive conditions that will minimize the functional I (same as π_p in this case). The resulting equations are the approximate equilibrium equations, whereas the compatibility conditions are identically satisfied. This approach is called the "displacement" or "stiffness" method of finite element analysis.

(ii) Principle of minimum complementary energy

The complementary energy of an elastic body (π_c) is defined as

π_c = complementary strain energy in terms of stresses ($\tilde{\pi}$) − work done by the applied loads during stress changes ($\tilde{W}_p$)

The principle of minimum complementary energy can be stated as follows: Of all possible stress states that satisfy the equilibrium equations and the stress boundary conditions, the state that satisfies the compatibility conditions will make the complementary energy assume a minimum value.

If the complementary energy π_c is expressed in terms of the stresses σ_{ij}, the principle of minimum complementary energy gives, for compatibility,

$$\delta\pi_c(\sigma_{xx},\sigma_{yy},\ldots,\sigma_{zx}) = \delta\tilde{\pi}(\sigma_{xx},\sigma_{yy},\ldots,\sigma_{zx}) - \delta\tilde{W}_p(\sigma_{xx},\sigma_{yy},\ldots,\sigma_{zx}) = 0 \tag{8.66}$$

It is important to note that the variation is taken with respect to the stress components in Eq. (8.66), whereas the displacements are assumed constant. The complementary strain energy of a linear elastic body is defined as

$$\tilde{\pi} = \frac{1}{2}\iiint_V \vec{\sigma}^{\,T}\vec{\varepsilon}\,\mathrm{d}V \tag{8.67}$$

By using the strain–stress relations of Eqs. (8.7), the complementary strain energy, in the presence of known initial strain $\vec{\varepsilon}_0$, can be expressed as*

$$\tilde{\pi} = \frac{1}{2}\iiint_V \vec{\sigma}^{\,T}([C]\vec{\sigma} + 2\vec{\varepsilon}_0)\,\mathrm{d}V \tag{8.68}$$

* The correctness of this expression can be verified from the fact that the partial derivative of $\tilde{\pi}$ with respect to the stresses should yield the strain–stress relations of Eq. (8.7).

The work done by applied loads during stress change (also known as complementary work) is given by

$$\tilde{W}_p = \iint_{S_2} (\Phi_x \bar{u} + \Phi_y \bar{v} + \Phi_z \bar{w})\, \mathrm{d}S_2 = \iint_{S_2} \vec{\Phi}^T \vec{\bar{U}}\, \mathrm{d}S_2 \tag{8.69}$$

where S_2 is the part of the surface of the body on which the values of the displacements are prescribed as $\vec{\bar{U}} = \begin{Bmatrix} \bar{u} \\ \bar{v} \\ \bar{w} \end{Bmatrix}$. Equations (8.68) and (8.69) can be used to express the complementary energy of the body as

$$\pi_c(\sigma_{xx}, \sigma_{yy}, \ldots, \sigma_{zx}) = \frac{1}{2} \iiint_V \vec{\sigma}^T([C]\vec{\sigma} + 2\vec{\varepsilon}_0) \cdot \mathrm{d}V - \iint_{S_2} \vec{\Phi}^T \vec{\bar{U}} \cdot \mathrm{d}S_2 \tag{8.70}$$

If we use the principle of minimum complementary energy in the finite element analysis, we assume a simple form of variation for the stress field within each element and derive conditions that will minimize the functional I (same as π_c in this case). The resulting equations are the approximate compatibility equations, whereas the equilibrium equations are identically satisfied. This approach is called the "force" or "flexibility" method of finite element analysis.

(iii) Principle of stationary Reissner energy

In the case of the principle of minimum potential energy, we expressed π_p in terms of displacements and permitted variations of u, v, and w. Similarly, in the case of the principle of minimum complementary energy, we expressed π_c in terms of stresses and permitted variations of $\sigma_{xx}, \ldots, \sigma_{zx}$. In the present case, the Reissner energy (π_R) is expressed in terms of both displacements and stresses and variations are permitted in $\vec{U}$ and $\vec{\sigma}$. The Reissner energy for a linearly elastic material is defined as

$$\begin{aligned}
\pi_R = {} & \iiint_V [(\text{internal stresses}) \times (\text{strains expressed in terms of} \\
& \quad \text{displacements}) - \text{complementary energy in terms of stresses}] \cdot \mathrm{d}V \\
& \quad - \text{work done by applied forces} \\
= {} & \iiint_V \left[\left\{\sigma_{xx} \cdot \frac{\partial u}{\partial x} + \sigma_{yy} \cdot \frac{\partial v}{\partial y} + \cdots + \sigma_{zx}\left(\frac{\partial w}{\partial x} + \frac{\partial u}{\partial z}\right)\right\} - \tilde{\pi}\right] \cdot \mathrm{d}V \\
& - \iiint_V (\bar{\phi}_x \cdot u + \bar{\phi}_y \cdot v + \bar{\phi}_z \cdot w) \cdot \mathrm{d}V - \iint_{S_1} (\bar{\Phi}_x \cdot u + \bar{\Phi}_y \cdot v + \bar{\Phi}_z \cdot w) \cdot \mathrm{d}S_1 \\
& - \iint_{S_2} \{(u - \bar{u})\Phi_x + (v - \bar{v})\Phi_y + (w - \bar{w})\Phi_z\} \cdot \mathrm{d}S_2 \\
= {} & \iiint_V \left[\vec{\sigma}^T \vec{\varepsilon} - \frac{1}{2}\vec{\sigma}^T [C]\vec{\sigma} - \vec{\bar{\phi}}^T \vec{U}\right] \cdot \mathrm{d}V \\
& - \iint_{S_1} \vec{U}^T \vec{\bar{\Phi}}\, \mathrm{d}S_1 - \iint_{S_2} (\vec{U} - \vec{\bar{U}})^T \vec{\Phi} \cdot \mathrm{d}S_2
\end{aligned} \tag{8.71}$$

The variation of π_R is set equal to zero by considering variations in both displacements and stresses:

$$\delta\pi_R = \underbrace{\Sigma \frac{\partial \pi_R}{\partial \sigma_{ij}} \delta\sigma_{ij}}_{\text{gives stress–displacement equations}} + \underbrace{\left(\frac{\partial \pi_R}{\partial u} \delta u + \frac{\partial \pi_R}{\partial v} \delta v + \frac{\partial \pi_R}{\partial w} \delta w \right)}_{\text{gives equilibrium equations and boundary conditions}} = 0 \tag{8.72}$$

The principle of stationary Reissner energy can be stated as follows: Of all possible stress and displacement states the body can have, the particular set that makes the Reissner energy stationary gives the correct stress–displacement and equilibrium equations along with the boundary conditions. To derive the finite element equations using the principle of stationary Reisssner energy, we must assume the form of variation for both displacement and stress fields within an element.

(iv) Hamilton's principle

The variational principle that can be used for dynamic problems is called the Hamilton's principle. In this principle, the variation of the functional is taken with respect to time. The functional (similar to π_p, π_c, and π_R) for this principle is the Lagrangian (L) defined as

$$L = T - \pi_p = \text{kinetic energy} - \text{potential energy} \tag{8.73}$$

The kinetic energy (T) of a body is given by

$$T = \frac{1}{2} \iiint\limits_V \rho \dot{\vec{U}}^T \dot{\vec{U}} \, \mathrm{d}V \tag{8.74}$$

where ρ is the density of the material, and $\dot{\vec{U}} = \begin{Bmatrix} \dot{u} \\ \dot{v} \\ \dot{w} \end{Bmatrix}$ is the vector of velocity components at any point inside the body. Thus, the Lagrangian can be expressed as

$$L = \frac{1}{2} \iiint\limits_V \left[\rho \dot{\vec{U}}^T \dot{\vec{U}} - \vec{\varepsilon}^T [D] \vec{\varepsilon} + 2\vec{U}^T \vec{\bar{\phi}} \right] \mathrm{d}V + \iint\limits_{S_1} \vec{U}^T \vec{\bar{\Phi}} \, \mathrm{d}S_1 \tag{8.75}$$

Hamilton's principle can be stated as follows: Of all possible time histories of displacement states that satisfy the compatibility equations and the constraints or the kinematic boundary conditions and that also satisfy the conditions at initial and final times (t_1 and t_2), the history corresponding to the actual solution makes the Lagrangian functional a minimum.

Thus, Hamilton's principle can be stated as

$$\delta \int_{t_1}^{t_2} L \, \mathrm{d}t = 0 \tag{8.76}$$

8.4 FORMULATION OF FINITE ELEMENT EQUATIONS (STATIC ANALYSIS)

We use the principle of minimum potential energy for deriving the equilibrium equations for a three-dimensional problem in this section. Since the nodal degrees of freedom are treated as unknowns in the present (displacement) formulation, the potential energy π_p has to be first expressed in terms of nodal degrees of freedom. Then the necessary equilibrium equations can be obtained by setting the first partial derivatives of π_p with respect to each of the nodal degrees of freedom equal to zero. The various steps involved in the derivation of equilibrium equations are given below.

Step 1: The solid body is divided into E finite elements.

Step 2: The displacement model within an element "e" is assumed as

$$\vec{U} = \begin{Bmatrix} u(x, y, z) \\ v(x, y, z) \\ w(x, y, z) \end{Bmatrix} = [N]\vec{Q}^{(e)} \tag{8.77}$$

where $\vec{Q}^{(e)}$ is the vector of nodal displacement degrees of freedom of the element, and $[N]$ is the matrix of shape functions.

Step 3: The element characteristic (stiffness) matrices and characteristic (load) vectors are to be derived from the principle of minimum potential energy. For this, the potential energy functional of the body π_p is written as (by considering only the body and surface forces)

$$\pi_p = \sum_{e=1}^{E} \pi_p^{(e)}$$

where $\pi_p^{(e)}$ is the potential energy of element e given by (see Eq. 8.65)

$$\pi_p^{(e)} = \frac{1}{2} \iiint\limits_{V^{(e)}} \vec{\varepsilon}^T [D](\vec{\varepsilon} - 2\vec{\varepsilon}_0) \, \mathrm{d}V - \iint\limits_{S_1^{(e)}} \vec{U}^T \vec{\Phi} \, \mathrm{d}S_1 - \iiint\limits_{V^{(e)}} \vec{U}^T \vec{\phi} \, \mathrm{d}V \tag{8.78}$$

where $V^{(e)}$ is the volume of the element, $S_1^{(e)}$ is the portion of the surface of the element over which distributed surface forces or tractions, $\vec{\Phi}$, are prescribed, and $\vec{\phi}$ is the vector of body forces per unit volume.

The strain vector $\vec{\varepsilon}$ appearing in Eq. (8.78) can be expressed in terms of the nodal displacement vector $\vec{Q}^{(e)}$ by differentiating Eq. (8.77) suitably as

$$\vec{\varepsilon} = \begin{Bmatrix} \varepsilon_{xx} \\ \varepsilon_{yy} \\ \varepsilon_{zz} \\ \varepsilon_{xy} \\ \varepsilon_{yz} \\ \varepsilon_{zx} \end{Bmatrix} = \begin{Bmatrix} \frac{\partial u}{\partial x} \\ \frac{\partial v}{\partial y} \\ \frac{\partial w}{\partial z} \\ \frac{\partial u}{\partial y} + \frac{\partial v}{\partial x} \\ \frac{\partial v}{\partial z} + \frac{\partial w}{\partial y} \\ \frac{\partial w}{\partial x} + \frac{\partial u}{\partial z} \end{Bmatrix} = \begin{bmatrix} \frac{\partial}{\partial x} & 0 & 0 \\ 0 & \frac{\partial}{\partial y} & 0 \\ 0 & 0 & \frac{\partial}{\partial z} \\ \frac{\partial}{\partial y} & \frac{\partial}{\partial x} & 0 \\ 0 & \frac{\partial}{\partial z} & \frac{\partial}{\partial y} \\ \frac{\partial}{\partial z} & 0 & \frac{\partial}{\partial x} \end{bmatrix} \begin{Bmatrix} u \\ v \\ w \end{Bmatrix} = [B]\vec{Q}^{(e)} \tag{8.79}$$

where

$$[B] = \begin{bmatrix} \frac{\partial}{\partial x} & 0 & 0 \\ 0 & \frac{\partial}{\partial y} & 0 \\ 0 & 0 & \frac{\partial}{\partial z} \\ \frac{\partial}{\partial y} & \frac{\partial}{\partial x} & 0 \\ 0 & \frac{\partial}{\partial z} & \frac{\partial}{\partial y} \\ \frac{\partial}{\partial z} & 0 & \frac{\partial}{\partial x} \end{bmatrix} [N] \tag{8.80}$$

The stresses $\vec{\sigma}$ can be obtained from the strains $\vec{\varepsilon}$ using Eq. (8.10) as

$$\vec{\sigma} = [D](\vec{\varepsilon} - \vec{\varepsilon}_0) = [D][B]\vec{Q}^{(e)} - [D]\vec{\varepsilon}_0 \tag{8.81}$$

Substitution of Eqs. (8.77) and (8.79) into Eq. (8.78) yields the potential energy of the element as

$$\begin{aligned} \pi_p^{(e)} = \frac{1}{2} \iiint\limits_{V^{(e)}} \vec{Q}^{(e)^T} [B]^T [D][B]\vec{Q}^{(e)} \,\mathrm{d}V - \iiint\limits_{V^{(e)}} \vec{Q}^{(e)^T} [B]^T [D]\vec{\varepsilon}_0 \,\mathrm{d}V \\ - \iint\limits_{S_1^{(e)}} \vec{Q}^{(e)^T} [N]^T \vec{\Phi} \,\mathrm{d}S_1 - \iiint\limits_{V^{(e)}} \vec{Q}^{(e)^T} [N]^T \vec{\phi} \,\mathrm{d}V \end{aligned} \tag{8.82}$$

In Eqs. (8.78) and (8.82), only the body and surface forces are considered. However, generally some external concentrated forces will also be acting at various nodes. If $\vec{\underset{\sim}{P}}_c$ denotes the vector of nodal forces (acting in the directions of the nodal displacement vector $\underset{\sim}{\vec{Q}}$ of the total structure or body), the total potential energy of the structure or body can be expressed as

$$\pi_p = \sum_{e=1}^{E} \pi_p^{(e)} - \underset{\sim}{\vec{Q}}^T \underset{\sim}{\vec{P}}_c \tag{8.83}$$

where $\underset{\sim}{\vec{Q}} = \begin{Bmatrix} Q_1 \\ Q_2 \\ \vdots \\ Q_M \end{Bmatrix}$ is the vector of nodal displacements of the entire structure or body, and M is the total number of nodal displacements or degrees of freedom.

Note that each component of the vector $\vec{Q}^{(e)}$, $e = 1, 2, \ldots, E$, appears in the global nodal displacement vector of the structure or body, $\underset{\sim}{\vec{Q}}$. Accordingly, $\vec{Q}^{(e)}$ for each element may be replaced by $\underset{\sim}{\vec{Q}}$ if the remaining element matrices and vectors (e.g., $[B]$, $[N]$, $\vec{\bar{\Phi}}$, and $\vec{\bar{\phi}}$) in the expression for $\pi_p^{(e)}$ are enlarged by adding the required number of zero elements and, where necessary, by rearranging their elements. In other words, the summation of Eq. (8.83) implies the expansion of element matrices to "structure" or "body" size followed by summation of overlapping terms. Thus, Eqs. (8.82) and (8.83) give

$$\pi_p = \frac{1}{2}\underset{\sim}{\vec{Q}}^T \left[\sum_{e=1}^{E} \iiint_{V^{(e)}} [B]^T[D][B]\,\mathrm{d}V \right] \underset{\sim}{\vec{Q}} - \underset{\sim}{\vec{Q}}^T \sum_{e=1}^{E} \left(\iiint_{V^{(e)}} [B]^T[D]\vec{\varepsilon}_0\,\mathrm{d}V \right.$$

$$\left. + \iint_{S_1^{(e)}} [N]^T \vec{\bar{\Phi}}\,\mathrm{d}S_1 + \iiint_{V^{(e)}} [N]^T \vec{\bar{\phi}}\,\mathrm{d}V \right) - \underset{\sim}{\vec{Q}}^T \underset{\sim}{\vec{P}}_c \tag{8.84}$$

Equation (8.84) expresses the total potential energy of the structure or body in terms of the nodal degrees of freedom, $\underset{\sim}{\vec{Q}}$. The static equilibrium configuration of the structures can be found by solving the following necessary conditions (for the minimization of potential energy):

$$\frac{\partial \pi_p}{\partial \underset{\sim}{\vec{Q}}} = \vec{0} \text{ or } \frac{\partial \pi_p}{\partial Q_1} = \frac{\partial \pi_p}{\partial Q_2} = \cdots = \frac{\partial \pi_p}{\partial Q_M} = 0 \tag{8.85}$$

With the help of Eq. (8.84), Eqs. (8.85) can be expressed as

$$\underbrace{\left(\sum_{e=1}^{E} \underbrace{\iiint_{V^{(e)}} [B]^T[D][B]\,\mathrm{d}V}_{\substack{\text{element stiffness} \\ \text{matrix, } [K^{(e)}]}} \right)}_{\substack{\text{global or overall stiffness} \\ \text{matrix of the structure or} \\ \text{body, } [\underset{\sim}{K}]}} \quad \underbrace{\underset{\sim}{\vec{Q}}}_{\substack{\text{global vector of} \\ \text{nodal displacements}}} \quad =$$

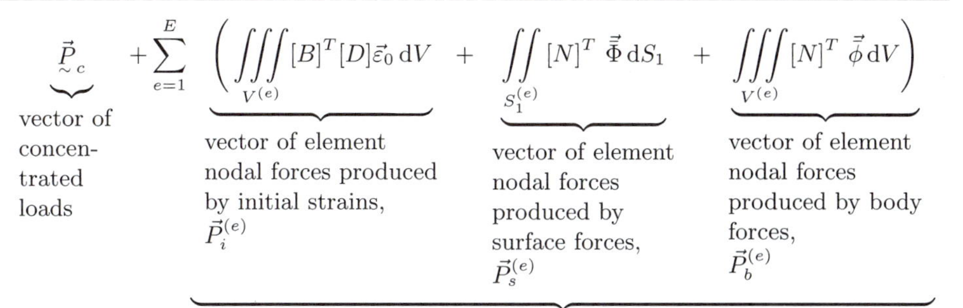

$$\underset{\sim}{\vec{P}}_c + \sum_{e=1}^{E} \left(\iiint_{V^{(e)}} [B]^T [D] \vec{\varepsilon}_0 \, dV + \iint_{S_1^{(e)}} [N]^T \, \vec{\Phi} \, dS_1 + \iiint_{V^{(e)}} [N]^T \, \vec{\phi} \, dV \right)$$

That is,

$$\left(\sum_{e=1}^{E} [K^{(e)}] \right) \underset{\sim}{\vec{Q}} = \underset{\sim}{\vec{P}}_c + \sum_{e=1}^{E} \left(\vec{P}_i^{(e)} + \vec{P}_s^{(e)} + \vec{P}_b^{(e)} \right) = \underset{\sim}{\vec{P}} \tag{8.86}$$

where

$$[K^{(e)}] = \iiint_{V^{(e)}} [B]^T [D][B] \, dV = \text{element stiffness matrix} \tag{8.87}$$

$$\vec{P}_i^{(e)} = \iiint_{V^{(e)}} [B]^T [D] \vec{\varepsilon}_0 \, dV = \text{element load vector due to initial strains} \tag{8.88}$$

$$\vec{P}_s^{(e)} = \iint_{S_1^{(e)}} [N]^T \, \vec{\Phi} \, dS_1 = \text{element load vector due to surface forces} \tag{8.89}$$

$$\vec{P}_b^{(e)} = \iiint_{V^{(e)}} [N]^T \, \vec{\phi} \, dV = \text{element load vector due to body forces} \tag{8.90}$$

Some of the contributions to the load vector $\underset{\sim}{\vec{P}}$ may be zero in a particular problem. In particular, the contribution of surface forces will be nonzero only for those element boundaries that are also part of the boundary of the structure or body that is subjected to externally applied distributed loading.

The load vectors $\vec{P}_i^{(e)}$, $\vec{P}_s^{(e)}$, and $\vec{P}_b^{(e)}$ given in Eqs. (8.88)–(8.90) are called kinematically consistent nodal load vectors [8.3]. Some of the components of $\vec{P}_i^{(e)}$, $\vec{P}_s^{(e)}$, and $\vec{P}_b^{(e)}$ may be moments or even higher order quantities if the corresponding nodal displacements represent strains or curvatures. These load vectors are called "kinematically consistent" because they satisfy the virtual work (or energy) equation. That is, the virtual work done by a particular generalized load P_j when the corresponding displacement δQ_j is permitted (while all other nodal displacements are prohibited) is equal to the work done by the distributed (nonnodal) loads in moving through the displacements dictated by δQ_j and the assumed displacement field.

Step 4: The desired equilibrium equations of the overall structure or body can now be expressed, using Eq. (8.86), as

$$[\underset{\sim}{K}]\underset{\sim}{\vec{Q}} = \underset{\sim}{\vec{P}} \tag{8.91}$$

where

$$[\underset{\sim}{K}] = \sum_{e=1}^{E} [K^{(e)}] = \text{assembled (global) stiffness matrix} \tag{8.92}$$

and

$$\underset{\sim}{\vec{P}} = \underset{\sim}{\vec{P}_c} + \sum_{e=1}^{E} \vec{P}_i^{(e)} + \sum_{e=1}^{E} \vec{P}_s^{(e)} + \sum_{e=1}^{E} \vec{P}_b^{(e)} = \text{assembled (global) nodal load vector} \tag{8.93}$$

Steps 5 and 6: The required solution for the nodal displacements and element stresses can be obtained after solving Eq. (8.91).

The following observations can be made from the previous derivation:

1. The formulation of element stiffness matrices, $[K^{(e)}]$, and element load vectors, $\vec{P}_i^{(e)}$, $\vec{P}_s^{(e)}$, and $\vec{P}_b^{(e)}$, which is basic to the development of finite element equations [Eq. (8.91), requires integrations as indicated in Eqs. (8.87)–(8.90). For some elements, the evaluation of these integrals is simple. However, in certain cases, it is often convenient to perform the integrations numerically [8.4].
2. The formulae for the element stiffness and load vector in Eqs. (8.87)–(8.90) remain the same irrespective of the type of element. However, the orders of the stiffness matrix and load vector will change for different types of elements. For example, in the case of a triangular element under plane stress, the order of $[K^{(e)}]$ is 6×6 and of $\vec{Q}^{(e)}$ is 6×1. For a rectangular element under plane stress, the orders of $[K^{(e)}]$ and $\vec{Q}^{(e)}$ are 8×8 and 8×1, respectively. It is assumed that the displacement model is linear in both these cases.
3. The element stiffness matrix given by Eq. (8.87) and the assembled stiffness matrix given by Eq. (8.92) are always symmetric. In fact, the matrix $[D]$ and the product $[B]^T[D][B]$ appearing in Eq. (8.87) are also symmetric.
4. In the analysis of certain problems, it is generally more convenient to compute the element stiffness matrices $[k^{(e)}]$ and element load vectors $\vec{p}_i^{(e)}$, $\vec{p}_s^{(e)}$, and $\vec{p}_b^{(e)}$ in local coordinate systems* suitably set up (differently for different elements) for minimizing the computational effort. In such cases, the matrices $[k^{(e)}]$ and vectors $\vec{p}_i^{(e)}$, $\vec{p}_s^{(e)}$, and $\vec{p}_b^{(e)}$ have to be transformed to a common global coordinate system before using them in Eqs. (8.92) and (8.93).
5. The equilibrium equations given by Eq. (8.91) cannot be solved since the stiffness matrices $[K^{(e)}]$ and $[\underset{\sim}{K}]$ are singular, and hence their inverses do not exist.

* When a local coordinate system is used, the resulting quantities are denoted by lowercase letters as $[k^{(e)}]$, $\vec{p}_i^{(e)}$, $\vec{p}_s^{(e)}$, and $\vec{p}_b^{(e)}$ instead of $[K^{(e)}]$, $\vec{P}_i^{(e)}$, $\vec{P}_s^{(e)}$, and $\vec{P}_b^{(e)}$.

The physical significance of this is that a loaded structure or body is free to undergo unlimited rigid body motion (translation and/or rotation) unless some support or boundary constraints are imposed on the structure or body to suppress the rigid body motion. These constraints are called boundary conditions. The method of incorporating boundary conditions was considered in Chapter 6.

6. To obtain the (displacement) solution of the problem, we have to solve Eq. (8.91) after incorporating the prescribed boundary conditions. The methods of solving the resulting equations were discussed in Chapter 7.

REFERENCES

8.1 J.S. Przemieniecki: *Theory of Matrix Structural Analysis*, McGraw-Hill, New York, 1968.
8.2 S. Timoshenko and J.N. Goodier: *Theory of Elasticity*, 2nd Ed., McGraw-Hill, New York, 1951.
8.3 L.R. Calcote: *The Analysis of Laminated Composite Structures*, Van Nostrand Reinhold, New York, 1969.
8.4 A.K. Gupta: Efficient numerical integration of element stiffness matrices, *International Journal for Numerical Methods in Engineering, 19*, 1410–1413, 1983.
8.5 R.J. Roark and W.C. Young: *Formulas for Stress and Strain*, 6th Ed., McGraw-Hill, New York, 1989.
8.6 G. Sines: *Elasticity and Strength*, Allyn & Bacon, Boston, 1969.
8.7 R.F.S. Hearman: *An Introduction to Applied Anisotropic Elasticity*, Oxford University Press, London, 1961.
8.8 S.G. Lekhnitskii: *Anisotropic Plates* (translation from Russian, 2nd Ed., by S.W. Tsai and T. Cheron), Gordon & Breach, New York, 1968.

PROBLEMS

8.1 Consider an infinitesimal element of a solid body in the form of a rectangular parallelopiped as shown in Figure 8.2. In this figure, the components of stress acting on one pair of faces only are shown for simplicity. Apply the moment equilibrium equations about the x, y, and z axes and show that the shear stresses are symmetric; that is, $\sigma_{yx} = \sigma_{xy}$, $\sigma_{zy} = \sigma_{yz}$, and $\sigma_{xz} = \sigma_{zx}$.

8.2 Determine whether the following state of strain is physically realizable:

$$\varepsilon_{xx} = c(x^2 + y^2), \quad \varepsilon_{yy} = cy^2, \quad \varepsilon_{xy} = 2cxy, \quad \varepsilon_{zz} = \varepsilon_{yz} = \varepsilon_{zx} = 0$$

where c is a constant.

8.3 When a body is heated nonuniformly and each element of the body is allowed to expand nonuniformly, the strains are given by

$$\varepsilon_{xx} = \varepsilon_{yy} = \varepsilon_{zz} = \alpha T, \quad \varepsilon_{xy} = \varepsilon_{yz} = \varepsilon_{zx} = 0 \tag{E$_1$}$$

where α is the coefficient of thermal expansion (constant), and $T = T(x, y, z)$ is the temperature. Determine the nature of variation of $T(x, y, z)$ for which Eqs. (E$_1$) are valid.

8.4 Consider the following state of stress and strain:

$$\sigma_{xx} = x^2, \quad \sigma_{yy} = y^2, \quad \varepsilon_{xy} = -2xy, \quad \sigma_{zz} = \varepsilon_{xz} = \varepsilon_{yz} = 0$$

Determine whether the equilibrium equations are satisfied.

8.5 Consider the following condition:

$$\sigma_{xx} = x^2, \quad \sigma_{yy} = y^2, \quad \varepsilon_{xy} = -2xy, \quad \sigma_{zz} = \varepsilon_{xz} = \varepsilon_{yz} = 0$$

Determine whether the compatibility equations are satisfied.

8.6 Consider the following state of strain:

$$\varepsilon_{xx} = c_1 x, \quad \varepsilon_{yy} = c_2, \quad \varepsilon_{zz} = c_3 x + c_4 y + c_5, \quad \varepsilon_{yz} = c_6 y, \quad \varepsilon_{xy} = \varepsilon_{zx} = 0$$

where c_i, $i = 1, 2, \ldots, 6$ are constants. Determine whether the compatibility equations are satisfied.

8.7 The internal equilibrium equations of a two-dimensional solid body, in polar coordinates, are given by

$$\frac{\partial \sigma_{rr}}{\partial r} + \frac{1}{r}\frac{\partial \sigma_{r\theta}}{\partial \theta} + \frac{\sigma_{rr} - \sigma_{\theta\theta}}{r} + \phi_r = 0$$

$$\frac{1}{r}\frac{\partial \sigma_{\theta\theta}}{\partial \theta} + \frac{\partial \sigma_{r\theta}}{\partial r} + 2\frac{\sigma_{r\theta}}{r} + \phi_\theta = 0$$

where ϕ_r and ϕ_θ are the body forces per unit volume in the radial (r) and circumferential (θ) directions, respectively.

If the state of stress in a loaded, thick-walled cylinder is given by

$$\sigma_{rr} = \frac{a^2 p}{b^2 - a^2}\left[1 - \frac{b^2}{r^2}\right],$$

$$\sigma_{\theta\theta} = \frac{a^2 p}{b^2 - a^2}\left[1 + \frac{b^2}{r^2}\right],$$

$$\sigma_{r\theta} = 0$$

where a, b, and p denote the inner radius, outer radius, and internal pressure, respectively, determine whether this state of stress satisfies the equilibrium equations.

8.8 Determine whether the following displacement represents a feasible deformation state in a solid body:

$$u = ax, \quad v = ay, \quad w = az$$

where a is a constant.

8.9 Consider a plate with a hole of radius a subjected to an axial stress p. The state of stress around the hole is given by [8.5]

$$\sigma_{rr} = \frac{1}{2}p\left[1 - \frac{a^2}{r^2}\right] + \frac{1}{2}p\left[1 + 3\frac{a^4}{r^4} - 4\frac{a^2}{r^2}\right]\cos 2\theta$$

$$\sigma_{\theta\theta} = \frac{1}{2}p\left[1 + \frac{a^2}{r^2}\right] - \frac{1}{2}p\left[1 + 3\frac{a^4}{r^4}\right]\cos 2\theta$$

$$\sigma_{r\theta} = -\frac{1}{2}p\left[1 - 3\frac{a^4}{r^4} + 2\frac{a^2}{r^2}\right]\sin 2\theta$$

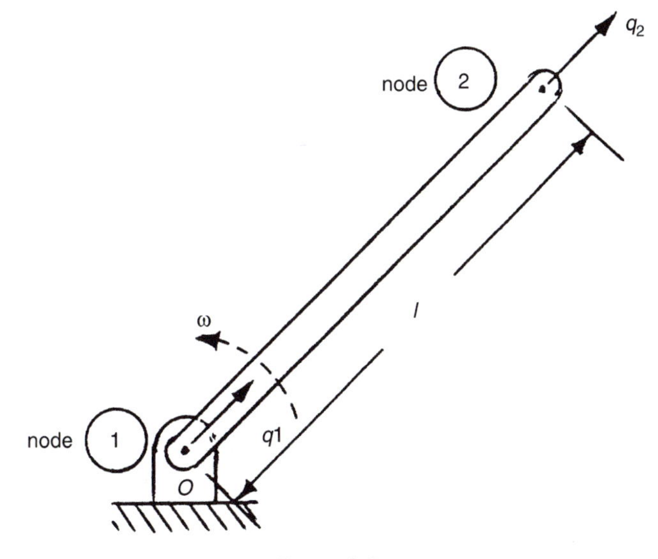

Figure 8.8.

Determine whether these stresses satisfy the equilibrium equations stated in Problem 8.7.

8.10 Consider a uniform bar of length l and cross-sectional area A rotating about a pivot point O as shown in Figure 8.8. Using the centrifugal force as the body force, determine the stiffness matrix and load vector of the element using a linear displacement model:

$$u(x) = N_1(x)q_1 + N_2(x)q_2$$

where $N_1(x) = 1-(x/l)$, $N_2(x) = (x/l)$ and the stress–strain relation, $\sigma_{xx} = E\varepsilon_{xx}$, where E is the Young's modulus.

9

ANALYSIS OF TRUSSES, BEAMS, AND FRAMES

9.1 INTRODUCTION

The derivation of element equations for one-dimensional structural elements is considered in this chapter. These elements can be used for the analysis of skeletal-type systems such as planar trusses, space trusses, beams, continuous beams, planar frames, grid systems, and space frames. Pin-jointed bar elements are used in the analysis of trusses. A truss element is a bar that can resist only axial forces (compressive or tensile) and can deform only in the axial direction. It will not be able to carry transverse loads or bending moments. In planar truss analysis, each of the two nodes can have components of displacement parallel to X and Y axes. In three-dimensional truss analysis, each node can have displacement components in X, Y, and Z directions. Rigidly jointed bar (beam) elements are used in the analysis of frames. Thus, a frame or a beam element is a bar that can resist not only axial forces but also transverse loads and bending moments. In the analysis of planar frames, each of the two nodes of an element will have two translational displacement components (parallel to X and Y axes) and a rotational displacement (in the plane XY). For a space frame element, each of the two ends is assumed to have three translational displacement components (parallel to X, Y, and Z axes) and three rotational displacement components (one in each of the three planes XY, YZ, and ZX). In the present development, we assume the members to be uniform and linearly elastic.

9.2 SPACE TRUSS ELEMENT

Consider the pin-jointed bar element shown in Figure 9.1, in which the local x axis is taken in the axial direction of the element with origin at corner (or local node) 1. A linear displacement model is assumed as

$$u(x) = q_1 + (q_2 - q_1)\frac{x}{l}$$

or

$$\underset{1\times 1}{\{u(x)\}} = \underset{1\times 2}{[N]}\ \underset{2\times 1}{\vec{q}^{(e)}} \tag{9.1}$$

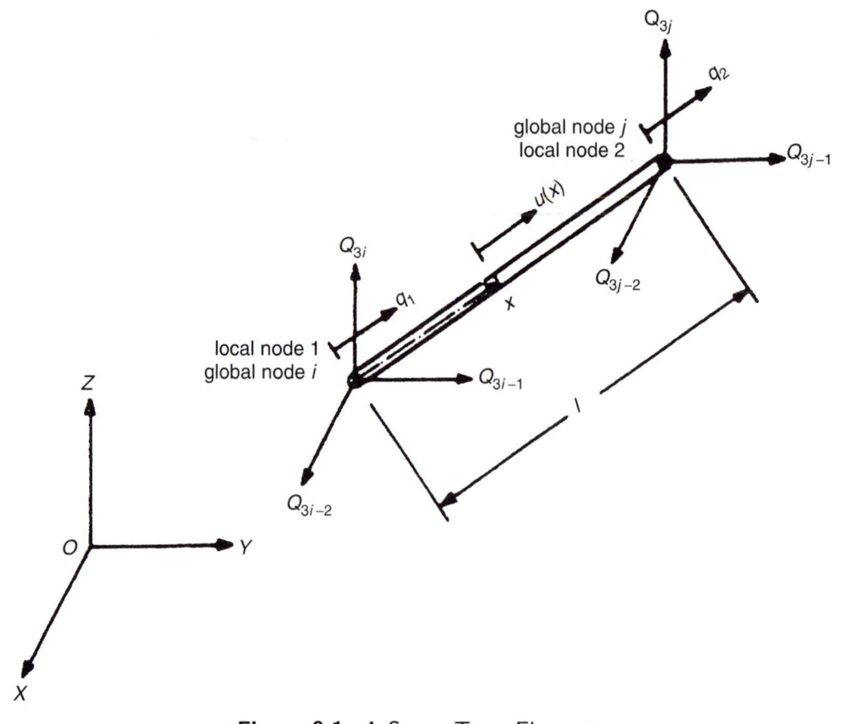

Figure 9.1. A Space Truss Element.

where

$$[N] = \left[\left(1 - \frac{x}{l}\right) \ \frac{x}{l}\right] \tag{9.2}$$

$$\vec{q}^{\,(e)} = \begin{Bmatrix} q_1 \\ q_2 \end{Bmatrix} \tag{9.3}$$

where q_1 and q_2 represent the nodal degrees of freedom in the local coordinate system (unknowns), l denotes the length of the element, and the superscript e denotes the element number. The axial strain can be expressed as

$$\varepsilon_{xx} = \frac{\partial u(x)}{\partial x} = \frac{q_2 - q_1}{l}$$

or

$$\underset{1\times 1}{\{\varepsilon_{xx}\}} = \underset{1\times 2}{[B]}\ \underset{2\times 1}{\vec{q}^{\,(e)}} \tag{9.4}$$

where

$$[B] = \left[-\frac{1}{l} \ \ \frac{1}{l}\right] \tag{9.5}$$

The stress–stain relation is given by

$$\begin{aligned} \sigma_{xx} &= E\varepsilon_{xx} \\ \underset{1\times 1}{\{\sigma_{xx}\}} &= \underset{1\times 1}{[D]}\ \underset{1\times 1}{\{\varepsilon_{xx}\}} \end{aligned} \tag{9.6}$$

where $[D] = [E]$, and E is the Young's modulus of the material. The stiffness matrix of the element (in the local coordinate system) can be obtained, from Eq. (8.87), as

$$\begin{aligned} \underset{2\times 2}{[k^{(e)}]} &= \iiint\limits_{V^{(e)}} [B]^T[D][B]\,\mathrm{d}V = A\int_{x=0}^{l} \begin{Bmatrix} -\frac{1}{l} \\ \frac{1}{l} \end{Bmatrix} E \begin{Bmatrix} -\frac{1}{l} & \frac{1}{l} \end{Bmatrix} \mathrm{d}x \\ &= \frac{AE}{l}\begin{bmatrix} 1 & -1 \\ -1 & 1 \end{bmatrix} \end{aligned} \tag{9.7}$$

where A is the area of cross section of the bar. To find the stiffness matrix of the bar in the global coordinate system, we need to find the transformation matrix. In general, the element under consideration will be one of the elements of a space truss. Let the (local) nodes 1 and 2 of the element correspond to nodes i and j, respectively, of the global system as shown in Figure 9.1. The local displacements q_1 and q_2 can be resolved into components Q_{3i-2}, Q_{3i-1}, Q_{3i} and Q_{3j-2}, Q_{3j-1}, Q_{3j} parallel to the global X, Y, Z, axes, respectively. Then the two sets of displacements are related as

$$\vec{q}^{\,(e)} = [\lambda]\vec{Q}^{(e)} \tag{9.8}$$

where the transformation matrix $[\lambda]$ and the vector of nodal displacements of element e in the global coordinate system, $\vec{Q}^{(e)}$, are given by

$$[\lambda] = \begin{bmatrix} l_{ij} & m_{ij} & n_{ij} & 0 & 0 & 0 \\ 0 & 0 & 0 & l_{ij} & m_{ij} & n_{ij} \end{bmatrix} \tag{9.9}$$

$$\vec{Q}^{(e)} = \begin{Bmatrix} Q_{3i-2} \\ Q_{3i-1} \\ Q_{3i} \\ Q_{3j-2} \\ Q_{3j-1} \\ Q_{3j} \end{Bmatrix} \tag{9.10}$$

and l_{ij}, m_{ij}, and n_{ij} denote the direction cosines of angles between the line ij and the directions OX, OY, and OZ, respectively. The direction cosines can be computed in terms of the global coordinates of nodes i and j as

$$l_{ij} = \frac{X_j - X_i}{l}, \qquad m_{ij} = \frac{Y_j - Y_i}{l}, \qquad n_{ij} = \frac{Z_j - Z_i}{l} \tag{9.11}$$

where (X_i, Y_i, Z_i) and (X_j, Y_j, Z_j) are the global coordinates of nodes i and j, respectively, and l is the length of the element ij given by

$$l = \left\{(X_j - X_i)^2 + (Y_j - Y_i)^2 + (Z_j - Z_i)^2\right\}^{1/2} \tag{9.12}$$

Thus, the stiffness matrix of the element in the global coordinate system can be obtained, using Eq. (6.8), as

$$\underset{6\times 6}{[K^{(e)}]} = \underset{6\times 2}{[\lambda]^T}\ \underset{2\times 2}{[k^{(e)}]}\ \underset{2\times 6}{[\lambda]}$$

$$= \frac{AE}{l}\begin{bmatrix} l_{ij}^2 & l_{ij}m_{ij} & l_{ij}n_{ij} & -l_{ij}^2 & -l_{ij}m_{ij} & -l_{ij}n_{ij} \\ l_{ij}m_{ij} & m_{ij}^2 & m_{ij}n_{ij} & -l_{ij}m_{ij} & -m_{ij}^2 & -m_{ij}n_{ij} \\ l_{ij}n_{ij} & m_{ij}n_{ij} & n_{ij}^2 & -l_{ij}n_{ij} & -m_{ij}n_{ij} & -n_{ij}^2 \\ -l_{ij}^2 & -l_{ij}m_{ij} & -l_{ij}n_{ij} & l_{ij}^2 & l_{ij}m_{ij} & l_{ij}n_{ij} \\ -l_{ij}m_{ij} & -m_{ij}^2 & -m_{ij}n_{ij} & l_{ij}m_{ij} & m_{ij}^2 & m_{ij}n_{ij} \\ -l_{ij}n_{ij} & -m_{ij}n_{ij} & -n_{ij}^2 & l_{ij}n_{ij} & m_{ij}n_{ij} & n_{ij}^2 \end{bmatrix} \tag{9.13}$$

Consistent Load Vector

The consistent load vectors can be computed using Eqs. (8.88)–(8.90):

$$\vec{p}_i^{(e)} = \text{load vector due to initial (thermal) strains} = \iiint_{V^{(e)}} [B]^T [D] \vec{\varepsilon}_0 \, dV$$

$$= A \int_0^l \begin{Bmatrix} -1/l \\ 1/l \end{Bmatrix} [E]\{\alpha T\} \cdot dx = AE\alpha T \begin{Bmatrix} -1 \\ 1 \end{Bmatrix} \tag{9.14}$$

$$\vec{p}_b^{(e)} = \text{load vector due to constant body force } (\phi_0) = \iiint_{V^{(e)}} [N]^T \vec{\phi} \, dV$$

$$= A \int_0^l \begin{Bmatrix} \left(1 - \dfrac{x}{l}\right) \\ (x/l) \end{Bmatrix} \{\phi_0\} \, dx = \frac{\phi_0 A l}{2} \begin{Bmatrix} 1 \\ 1 \end{Bmatrix} \tag{9.15}$$

The only surface stress that can exist is p_x and this must be applied at one of the nodal points. Assuming that p_x is applied at node 1, the load vector becomes

$$\vec{p}_{S_1}^{(e)} = \iint_{S_1^{(e)}} [N]^T \{p_x\} \, dS_1 = p_0 \begin{Bmatrix} 1 \\ 0 \end{Bmatrix} \iint_{S_1} dS_1 = p_0 A_1 \begin{Bmatrix} 1 \\ 0 \end{Bmatrix} \tag{9.16}$$

where $p_x = p_0$ is assumed to be a constant and the subscript 1 is used to denote the node. The matrix of shape functions $[N]$ reduces to $\begin{Bmatrix} 1 \\ 0 \end{Bmatrix}$ since the stress is located at node 1. Similarly, when $p_x = p_0$ is applied at node 2, the load vector becomes

$$\vec{p}_{S_2}^{(e)} = \iint_{S_2^{(e)}} [N]^T \{p_x\} \, dS_1 = p_0 \begin{Bmatrix} 0 \\ 1 \end{Bmatrix} \iint_{S_2} dS_1 = p_0 \cdot A_2 \begin{Bmatrix} 0 \\ 1 \end{Bmatrix} \tag{9.17}$$

The total consistent load vector in the local coordinate system is given by

$$\vec{p}^{(e)} = \vec{p}_i^{(e)} + \vec{p}_b^{(e)} + \vec{p}_{S_1}^{(e)} + \vec{p}_{S_2}^{(e)} \tag{9.18}$$

This load vector, when referred to the global coordinate system, will be

$$\vec{P}^{(e)} = [\lambda]^T \vec{p}^{(e)} \tag{9.19}$$

where $[\lambda]$ is given by Eq. (9.9).

9.2.1 Computation of Stresses

After finding the displacement solution of the system, the nodal displacement vector $\vec{Q}^{(e)}$ of element e can be identified. The stress induced in element e can be determined, using Eqs. (9.6), (9.4), and (9.8), as

$$\sigma_{xx} = E[B][\lambda]\vec{Q}^{(e)} \tag{9.20}$$

where $[B]$ and $[\lambda]$ are given by Eqs. (9.5) and (9.9), respectively.

Example 9.1 Find the nodal displacements developed in the planar truss shown in Figure 9.2 when a vertically downward load of 1000 N is applied at node 4. The pertinent data are given in Table 9.1.

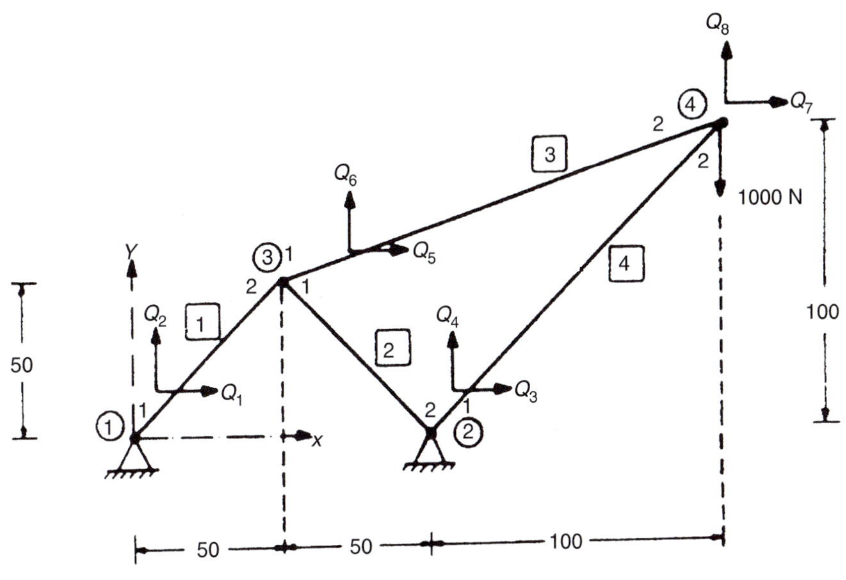

Figure 9.2. Geometry of the Planar Truss of Example 9.1.

Solution The numbering system for the nodes, members, and global displacements is indicated in Figure 9.2. The nodes 1 and 2 in the local system and the local x direction assumed for each of the four elements are shown in Figure 9.3. For convenience, the global node numbers i and j corresponding to the local nodes 1 and 2 for each element and the direction cosines of the line ij (x axis) with respect to the global X and Y axes are given in Table 9.2.

Table 9.1.

Member number "e"	Cross-sectional area $A^{(e)}$ cm^2	Length $l^{(e)}$ cm	Young's modulus $E^{(e)}$ N/cm^2
1	2.0	$\sqrt{2}\ 50$	2×10^6
2	2.0	$\sqrt{2}\ 50$	2×10^6
3	1.0	$\sqrt{2.5}\ 100$	2×10^6
4	1.0	$\sqrt{2}\ 100$	2×10^6

Table 9.2.

Member number "e"	Global node corresponding to local node 1 (i)	Global node corresponding to local node 2 (j)	Coordinates of local nodes 1 (i) and 2 (j) in global system X_i	Y_i	X_j	Y_j	Direction cosines of line ij: l_{ij}	m_{ij}
1	1	3	0.0	0.0	50.0	50.0	$1/\sqrt{2}$	$1/\sqrt{2}$
2	3	2	50.0	50.0	100.0	0.0	$1/\sqrt{2}$	$-1/\sqrt{2}$
3	3	4	50.0	50.0	200.0	100.0	$1.5/\sqrt{2.5}$	$0.5/\sqrt{2.5}$
4	2	4	100.0	0.0	200.0	100.0	$1/\sqrt{2}$	$1/\sqrt{2}$

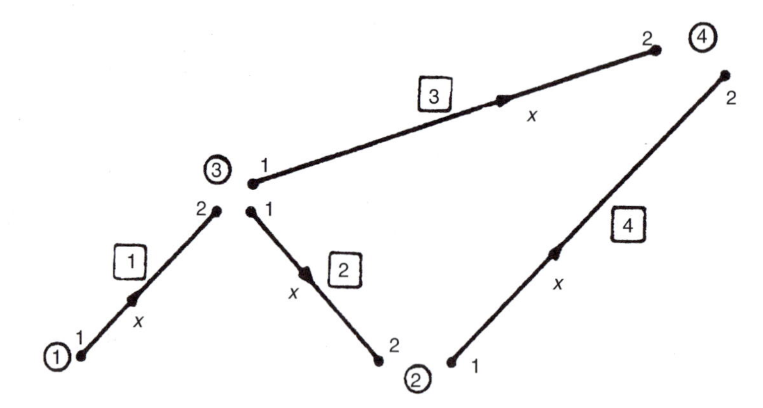

Figure 9.3. Finite Element Idealization.

The stiffness matrix of element e in the global coordinate system can be computed from [obtained by deleting the rows and columns corresponding to the Z degrees of freedom from Eq. (9.13)]

$$[K^{(e)}] = \frac{A^{(e)}E^{(e)}}{l^{(e)}} \begin{bmatrix} l_{ij}^2 & l_{ij}m_{ij} & -l_{ij}^2 & -l_{ij}m_{ij} \\ l_{ij}m_{ij} & m_{ij}^2 & -l_{ij}m_{ij} & -m_{ij}^2 \\ -l_{ij}^2 & -l_{ij}m_{ij} & l_{ij}^2 & l_{ij}m_{ij} \\ -l_{ij}m_{ij} & -m_{ij}^2 & l_{ij}m_{ij} & m_{ij}^2 \end{bmatrix} \begin{matrix} Q_{2i-1} \\ Q_{2i} \\ Q_{2j-1} \\ Q_{2j} \end{matrix}$$

Rows ($Q_{2i-1}, Q_{2i}, Q_{2j-1}, Q_{2j}$): Global degrees of freedom corresponding to different rows

Columns ($Q_{2i-1}, Q_{2i}, Q_{2j-1}, Q_{2j}$): Global degrees of freedom corresponding to different columns

Hence,

$$[K^{(1)}] = \frac{(2.0)(2\times 10^6)}{\sqrt{2}\,50} \begin{bmatrix} 1/2 & 1/2 & -1/2 & -1/2 \\ 1/2 & 1/2 & -1/2 & -1/2 \\ -1/2 & -1/2 & 1/2 & 1/2 \\ -1/2 & -1/2 & 1/2 & 1/2 \end{bmatrix} \begin{matrix} Q_1 \\ Q_2 \\ Q_5 \\ Q_6 \end{matrix} \quad (\text{columns } Q_1\; Q_2\; Q_5\; Q_6)$$

$$= \begin{bmatrix} 1 & 1 & -1 & -1 \\ 1 & 1 & -1 & -1 \\ -1 & -1 & 1 & 1 \\ -1 & -1 & 1 & 1 \end{bmatrix} (2\sqrt{2}) \times 10^4 \text{ N/cm}$$

$$[K^{(2)}] = \frac{(2.0)(2\times 10^6)}{\sqrt{2}\,50} \begin{bmatrix} 1/2 & -1/2 & -1/2 & 1/2 \\ -1/2 & 1/2 & 1/2 & -1/2 \\ -1/2 & 1/2 & 1/2 & -1/2 \\ 1/2 & -1/2 & -1/2 & 1/2 \end{bmatrix} \begin{matrix} Q_5 \\ Q_6 \\ Q_3 \\ Q_4 \end{matrix} \quad (\text{columns } Q_5\; Q_6\; Q_3\; Q_4)$$

$$= \begin{bmatrix} 1 & -1 & -1 & 1 \\ -1 & 1 & 1 & -1 \\ -1 & 1 & 1 & -1 \\ 1 & -1 & -1 & 1 \end{bmatrix} (2\sqrt{2}) \times 10^4 \text{ N/cm}$$

$$[K^{(3)}] = \frac{(1.0)(2\times 10^6)}{\sqrt{2.5}\,100} \begin{bmatrix} \frac{2.25}{2.50} & \frac{0.75}{2.50} & \frac{-2.25}{2.50} & \frac{-0.75}{2.50} \\ \frac{0.75}{2.50} & \frac{0.25}{2.50} & \frac{-0.75}{2.50} & \frac{-0.25}{2.50} \\ \frac{-2.25}{2.50} & \frac{-0.75}{2.50} & \frac{2.25}{2.50} & \frac{0.75}{2.50} \\ \frac{-0.75}{2.50} & \frac{-0.25}{2.50} & \frac{0.75}{2.50} & \frac{0.25}{2.50} \end{bmatrix} \begin{matrix} Q_5 \\ Q_6 \\ Q_7 \\ Q_8 \end{matrix} \quad (\text{columns } Q_5\; Q_6\; Q_7\; Q_8)$$

$$= \begin{bmatrix} 9 & 3 & -9 & -3 \\ 3 & 1 & -3 & -1 \\ -9 & -3 & 9 & 3 \\ -3 & -1 & 3 & 1 \end{bmatrix} (8\sqrt{2.5}) \times 10^2 \text{ N/cm}$$

$$[K^{(4)}] = \frac{(1.0)(2 \times 10^6)}{\sqrt{2}\ 100} \begin{array}{c} \begin{array}{cccc} Q_3 & Q_4 & Q_7 & Q_8 \end{array} \\ \begin{bmatrix} \frac{1}{2} & \frac{1}{2} & -\frac{1}{2} & -\frac{1}{2} \\ \frac{1}{2} & \frac{1}{2} & -\frac{1}{2} & -\frac{1}{2} \\ -\frac{1}{2} & -\frac{1}{2} & \frac{1}{2} & \frac{1}{2} \\ -\frac{1}{2} & -\frac{1}{2} & \frac{1}{2} & \frac{1}{2} \end{bmatrix} \end{array} \begin{array}{c} Q_3 \\ Q_4 \\ Q_7 \\ Q_8 \end{array}$$

$$= \begin{bmatrix} 1 & 1 & -1 & -1 \\ 1 & 1 & -1 & -1 \\ -1 & -1 & 1 & 1 \\ -1 & -1 & 1 & 1 \end{bmatrix} (5\sqrt{2}) \times 10^3 \text{ N/cm}$$

These element matrices can be assembled to obtain the global stiffness matrix, $[\underset{\sim}{K}]$, as

$$[\underset{\sim}{K}] = 10^3 \begin{array}{c} \begin{array}{cccccccc} Q_1 & Q_2 & Q_3 & Q_4 & Q_5 & Q_6 & Q_7 & Q_8 \end{array} \\ \left[\begin{array}{cccccccc}
20\sqrt{2} & 20\sqrt{2} & 0 & 0 & -20\sqrt{2} & -20\sqrt{2} & 0 & 0 \\
20\sqrt{2} & 20\sqrt{2} & 0 & 0 & -20\sqrt{2} & -20\sqrt{2} & 0 & 0 \\
0 & 0 & \begin{matrix} 20\sqrt{2} \\ +5\sqrt{2} \end{matrix} & \begin{matrix} -20\sqrt{2} \\ +5\sqrt{2} \end{matrix} & -20\sqrt{2} & 20\sqrt{2} & -5\sqrt{2} & -5\sqrt{2} \\
0 & 0 & \begin{matrix} -20\sqrt{2} \\ +5\sqrt{2} \end{matrix} & \begin{matrix} 20\sqrt{2} \\ +5\sqrt{2} \end{matrix} & 20\sqrt{2} & -20\sqrt{2} & -5\sqrt{2} & -5\sqrt{2} \\
-20\sqrt{2} & -20\sqrt{2} & -20\sqrt{2} & 20\sqrt{2} & \begin{matrix} 20\sqrt{2} \\ +20\sqrt{2} \\ +7.2\sqrt{2.5} \end{matrix} & \begin{matrix} 20\sqrt{2} \\ -20\sqrt{2} \\ +2.4\sqrt{2.5} \end{matrix} & -7.2\sqrt{2.5} & -2.4\sqrt{2.5} \\
-20\sqrt{2} & -20\sqrt{2} & 20\sqrt{2} & -20\sqrt{2} & \begin{matrix} 20\sqrt{2} \\ -20\sqrt{2} \\ +2.4\sqrt{2.5} \end{matrix} & \begin{matrix} 20\sqrt{2} \\ +20\sqrt{2} \\ +0.8\sqrt{2.5} \end{matrix} & -2.4\sqrt{2.5} & -0.8\sqrt{2.5} \\
0 & 0 & -5\sqrt{2} & -5\sqrt{2} & -7.2\sqrt{2.5} & -2.4\sqrt{2.5} & \begin{matrix} 7.2\sqrt{2.5} \\ +5\sqrt{2} \end{matrix} & \begin{matrix} 2.4\sqrt{2.5} \\ +5\sqrt{2} \end{matrix} \\
0 & 0 & -5\sqrt{2} & -5\sqrt{2} & -2.4\sqrt{2.5} & -0.8\sqrt{2.5} & \begin{matrix} 2.4\sqrt{2.5} \\ +5\sqrt{2} \end{matrix} & \begin{matrix} 0.8\sqrt{2.5} \\ +5\sqrt{2} \end{matrix}
\end{array} \right] \end{array} \begin{array}{c} Q_1 \\ Q_2 \\ Q_3 \\ Q_4 \\ Q_5 \\ Q_6 \\ Q_7 \\ Q_8 \end{array} \text{ N/cm}$$

Thus, the global equations of equilibrium can be expressed as

$$[\underset{\sim}{K}]\underset{\sim}{\vec{Q}} = \underset{\sim}{\vec{p}} \tag{E$_1$}$$

where

$$\underset{\sim}{\vec{Q}} = \begin{Bmatrix} Q_1 \\ Q_2 \\ \vdots \\ Q_8 \end{Bmatrix} \quad \text{and} \quad \underset{\sim}{\vec{P}} = \begin{Bmatrix} P_1 \\ P_2 \\ \vdots \\ P_8 \end{Bmatrix}$$

By deleting the rows and columns corresponding to the restrained degrees of freedom ($Q_1 = Q_2 = Q_3 = Q_4 = 0$), Eq. ($E_1$) can be written as

$$[K]\vec{Q} = \vec{P} \tag{E$_2$}$$

where

$$[K] = 10^3 \begin{bmatrix} 40\sqrt{2} + 7.2\sqrt{2.5} & 2.4\sqrt{2.5} & -7.2\sqrt{2.5} & -2.4\sqrt{2.5} \\ 2.4\sqrt{2.5} & 40\sqrt{2} + 0.8\sqrt{2.5} & -2.4\sqrt{2.5} & -0.8\sqrt{2.5} \\ -7.2\sqrt{2.5} & -2.4\sqrt{2.5} & 5\sqrt{2} + 7.2\sqrt{2.5} & 5\sqrt{2} + 2.4\sqrt{2.5} \\ -2.4\sqrt{2.5} & -0.8\sqrt{2.5} & 5\sqrt{2} + 2.4\sqrt{2.5} & 5\sqrt{2} + 0.8\sqrt{2.5} \end{bmatrix} \text{N/cm}$$

$$\vec{Q} = \begin{Bmatrix} Q_5 \\ Q_6 \\ Q_7 \\ Q_8 \end{Bmatrix} \quad \text{and} \quad \vec{P} = \begin{Bmatrix} P_5 \\ P_6 \\ P_7 \\ P_8 \end{Bmatrix} \equiv \begin{Bmatrix} 0 \\ 0 \\ 0 \\ -1000 \end{Bmatrix} \text{N}$$

The solution of Eqs. (E_2) gives the displacements as

$$Q_5 = 0.026517 \text{ cm}$$
$$Q_6 = 0.008839 \text{ cm}$$
$$Q_7 = 0.347903 \text{ cm}$$
$$Q_8 = -0.560035 \text{ cm}$$

Example 9.2 Find the stresses developed in the various elements of the truss considered in Example 9.1.

Solution The nodal displacements of the truss in the global coordinate system (including the fixed degrees of freedom) are given by

$$\underset{\sim}{\vec{Q}} = \begin{Bmatrix} Q_1 \\ Q_2 \\ Q_3 \\ Q_4 \\ Q_5 \\ Q_6 \\ Q_7 \\ Q_8 \end{Bmatrix} = \begin{Bmatrix} 0 \\ 0 \\ 0 \\ 0 \\ 0.026517 \\ 0.008839 \\ 0.347903 \\ -0.560035 \end{Bmatrix} \text{cm} \tag{E$_1$}$$

The nodal degrees of freedom of various elements, in the global coordinate system, can be identified from Figures 9.2 and 9.3 and Eq. (E_1) as

$$\vec{Q}^{(1)} = \begin{Bmatrix} Q_1^{(1)} \\ Q_2^{(1)} \\ Q_3^{(1)} \\ Q_4^{(1)} \end{Bmatrix} \equiv \begin{Bmatrix} Q_1 \\ Q_2 \\ Q_5 \\ Q_6 \end{Bmatrix} = \begin{Bmatrix} 0 \\ 0 \\ 0.026517 \\ 0.008839 \end{Bmatrix} \text{cm} \tag{E$_2$}$$

$$\vec{Q}^{(2)} = \begin{Bmatrix} Q_1^{(2)} \\ Q_2^{(2)} \\ Q_3^{(2)} \\ Q_4^{(2)} \end{Bmatrix} \equiv \begin{Bmatrix} Q_5 \\ Q_6 \\ Q_3 \\ Q_4 \end{Bmatrix} = \begin{Bmatrix} 0.026517 \\ 0.008839 \\ 0 \\ 0 \end{Bmatrix} \text{ cm} \tag{E$_3$}$$

$$\vec{Q}^{(3)} = \begin{Bmatrix} Q_1^{(3)} \\ Q_2^{(3)} \\ Q_3^{(3)} \\ Q_4^{(3)} \end{Bmatrix} \equiv \begin{Bmatrix} Q_5 \\ Q_6 \\ Q_7 \\ Q_8 \end{Bmatrix} = \begin{Bmatrix} 0.026517 \\ 0.008839 \\ 0.347903 \\ -0.560035 \end{Bmatrix} \text{ cm} \tag{E$_4$}$$

$$\vec{Q}^{(4)} = \begin{Bmatrix} Q_1^{(4)} \\ Q_2^{(4)} \\ Q_3^{(4)} \\ Q_4^{(4)} \end{Bmatrix} \equiv \begin{Bmatrix} Q_3 \\ Q_4 \\ Q_7 \\ Q_8 \end{Bmatrix} = \begin{Bmatrix} 0 \\ 0 \\ 0.347903 \\ -0.560035 \end{Bmatrix} \text{ cm} \tag{E$_5$}$$

As indicated in Eq. (9.20), the axial stress developed in element e is given by

$$\sigma_{xx}^{(e)} = E^{(e)}[B^{(e)}][\lambda^{(e)}]\vec{Q}^{(e)}$$

$$= E^{(e)} \begin{bmatrix} -\frac{1}{l^{(e)}} & \frac{1}{l^{(e)}} \end{bmatrix} \begin{bmatrix} l_{12}{}^{(e)} & m_{12}{}^{(e)} & 0 & 0 \\ 0 & 0 & l_{12}{}^{(e)} & m_{12}{}^{(e)} \end{bmatrix} \begin{Bmatrix} Q_1^{(e)} \\ Q_2^{(e)} \\ Q_3^{(e)} \\ Q_4^{(e)} \end{Bmatrix} \tag{E$_6$}$$

Equation (E$_2$) can be simplified as

$$\sigma_{xx}^{(e)} = E^{(e)} \left\{ -\frac{1}{l^{(e)}} \left(l_{12}^{(e)} Q_1^{(e)} + m_{12}^{(e)} Q_2^{(e)} \right) + \frac{1}{l^{(e)}} \left(l_{12}^{(e)} Q_3^{(e)} + m_{12}^{(e)} Q_4^{(e)} \right) \right\} \tag{E$_7$}$$

which yields the following results:

Element 1: $E^{(1)} = 2 \times 10^6$ N/cm^2, $l^{(1)} = 70.7107$ cm, $l_{12}^{(1)} = m_{12}^{(1)} = 0.707107$, $Q_i^{(1)}$, $i = 1, 2, 3, 4$, are given by Eq. (E$_2$) so that $\sigma_{xx}^{(1)} = 707.1200$ N/cm^2.

Element 2: $E^{(2)} = 2 \times 10^6$ N/cm^2, $l^{(2)} = 70.7107$ cm, $l_{12}^{(2)} = 0.707107$, $m_{12}^{(1)} = -0.707107$, $Q_i^{(2)}$, $i = 1, 2, 3, 4$, are given by Eq. (E$_3$) so that $\sigma_{xx}^{(2)} = -353.560$ N/cm^2.

Element 3: $E^{(3)} = 2 \times 10^6$ N/cm^2, $l^{(3)} = 158.114$ cm, $l_{12}^{(3)} = 0.948683$, $m_{12}^{(2)} = 0.316228$, $Q_i^{(3)}$, $i = 1, 2, 3, 4$, are given by Eq. (E$_4$) so that $\sigma_{xx}^{(3)} = 1581.132$ N/cm^2.

Element 4: $E^{(4)} = 2 \times 10^6$ N/cm^2, $l^{(4)} = 141.421$ cm, $l_{12}^{(4)} = m_{12}^{(4)} = 0.707107$, $Q_i^{(4)}$, $i = 1, 2, 3, 4$, are given by Eq. (E$_5$) so that $\sigma_{xx}^{(4)} = -2121.326$ N/cm^2.

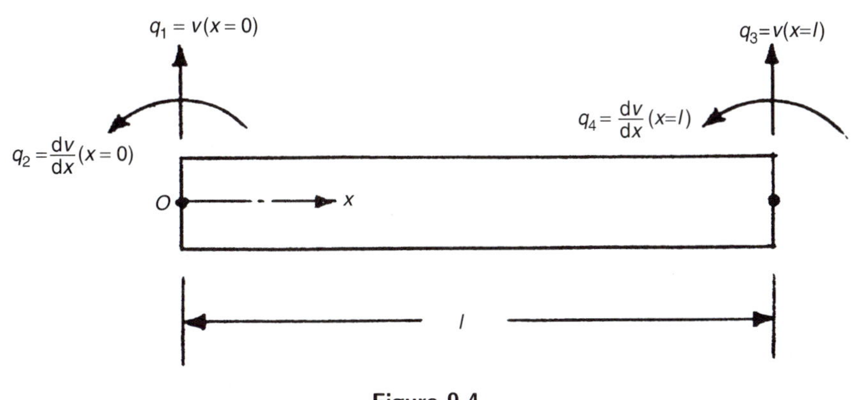

Figure 9.4.

9.3 BEAM ELEMENT

A beam is a straight bar element that is primarily subjected to transverse loads. The deformed shape of a beam is described by the transverse displacement and slope (rotation) of the beam. Hence, the transverse displacement and rotation at each end of the beam element are treated as the unknown degrees of freedom. Consider a beam element of length l in the xy plane as shown in Figure 9.4. The four degrees of freedom in the local (xy) coordinate system are indicated as q_1, q_2, q_3, and q_4. Because there are four nodal displacements, we assume a cubic displacement model for $v(x)$ as (Figure 9.4)

$$v(x) = \alpha_1 + \alpha_2 x + \alpha_3 x^2 + \alpha_4 x^3 \tag{9.21}$$

where the constants α_1–α_4 can be found by using the conditions

$$v(x) = q_1 \quad \text{and} \quad \frac{dv}{dx}(x) = q_2 \quad \text{at} \quad x = 0$$

and

$$v(x) = q_3 \quad \text{and} \quad \frac{dv}{dx}(x) = q_4 \quad \text{at} \quad x = l$$

Equation (9.21) can thus be expressed as

$$\underset{1\times 1}{v(x)} = \underset{1\times 4}{[N]} \; \underset{4\times 1}{\vec{q}} \tag{9.22}$$

where $[N]$ is given by

$$[N] = \begin{bmatrix} N_1 & N_2 & N_3 & N_4 \end{bmatrix}$$

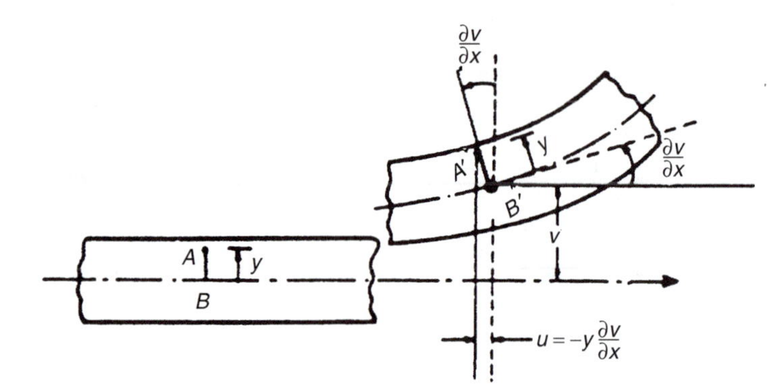

Figure 9.5. Deformation of an Element of Beam in *xy* Plane.

with

$$\begin{aligned} N_1(x) &= \left(2x^3 - 3lx^2 + l^3\right)/l^3 \\ N_2(x) &= \left(x^3 - 2lx^2 + l^2x\right)/l^2 \\ N_3(x) &= -\left(2x^3 - 3lx^2\right)/l^3 \\ N_4(x) &= \left(x^3 - lx^2\right)/l^2 \end{aligned} \tag{9.23}$$

and

$$\vec{q} = \begin{Bmatrix} q_1 \\ q_2 \\ q_3 \\ q_4 \end{Bmatrix} \tag{9.24}$$

According to simple beam theory, plane sections of the beam remain plane after deformation and hence the axial displacement u due to the transverse displacement v can be expressed as (from Figure 9.5)

$$u = -y\frac{\partial v}{\partial x}$$

where y is the distance from the neutral axis. The axial strain is given by[†]

$$\varepsilon_{xx} = \frac{\partial u}{\partial x} = -y\frac{\partial^2 v}{\partial x^2} = [B]\vec{q} \tag{9.25}$$

where

$$[B] = -\frac{y}{l^3}\left\{(12x - 6l) \quad l(6x - 4l) - (12x - 6l) \quad l(6x - 2l)\right\} \tag{9.26}$$

[†] If the nodal displacements of the element, q_1, q_2, q_3 and q_4, are known the stress distribution in the element, σ_{xx}, can be found as

$$\sigma_{xx} = \sigma_{xx}(x, y) = E\varepsilon_{xx} = [E][B]\vec{q}$$

where $\sigma_{xx}(x, y)$ denotes the stress in the element at a point located at distance x from the origin (in horizontal direction from the left node) and y from the neutral axis (in the vertical direction).

Using Eqs. (9.26) and (8.87) with $[D] = [E]$, the stiffness matrix can be found as

$$[k^{(e)}] = \iiint_{V^{(e)}} [B]^T \, [D] \, [B] \, \mathrm{d}V = E \int_o^l \mathrm{d}x \iint_A [B]^T \, [B] \, \mathrm{d}A$$

$$= \frac{EI_{zz}}{l^3} \begin{bmatrix} 12 & 6l & -12 & 6l \\ 6l & 4l^2 & -6l & 2l^2 \\ -12 & -6l & 12 & -6l \\ 6l & 2l^2 & -6l & 4l^2 \end{bmatrix} \begin{matrix} q_1 \\ q_2 \\ q_3 \\ q_4 \end{matrix} \qquad (9.27)$$

(columns: $q_1 \quad q_2 \quad q_3 \quad q_4$)

where $I_{zz} = \iint_A y^2 \cdot \mathrm{d}A$ is the area moment of inertia of the cross section about the z axis. Notice that the nodal interpolation functions $N_i(x)$ of Eq. (9.23) are the same as the first-order Hermite polynomials defined in Section 4.4.5.

9.4 SPACE FRAME ELEMENT

A space frame element is a straight bar of uniform cross section that is capable of resisting axial forces, bending moments about the two principal axes in the plane of its cross section, and twisting moment about its centroidal axis. The corresponding displacement degrees of freedom are shown in Figure 9.6(a). It can be seen that the stiffness matrix of a frame element will be of order 12×12. If the local axes (xyz system) are chosen to coincide with the principal axes of the cross section, it is possible to construct the 12×12 stiffness matrix from 2×2 and 4×4 submatrices. According to the engineering theory of bending and torsion of beams, the axial displacements q_1 and q_7 depend only on the axial forces, and the torsional displacements q_4 and q_{10} depend only on the torsional moments. However, for arbitrary choice of xyz coordinate system, the bending displacements in xy plane, namely q_2, q_6, q_8, and q_{12}, depend not only on the bending forces acting in that plane (i.e., shear forces acting in the y direction and the bending moments acting in the xy plane) but also on the bending forces acting in the plane xz. On the other hand, if the xy and xz planes coincide with the principle axes of the cross section, the bending displacements and forces in the two planes can be considered to be independent of each other.

In this section, we choose the local xyz coordinate system to coincide with the principal axes of the cross section with the x axis representing the centroidal axis of the frame element. Thus, the displacements can be separated into four groups, each of which can be considered independently of the others. We first consider the stiffness matrices corresponding to different independent sets of displacements and then obtain the total stiffness matrix of the element by superposition.

9.4.1 Axial Displacements

The nodal displacements are q_1 and q_7 (Figure 9.6b) and a linear displacement model leads to the stiffness matrix (corresponding to the axial displacement) as

$$[k_a^{(e)}] = \iiint_{V^{(e)}} [B]^T [D] [B] \, \mathrm{d}V = \frac{AE}{l} \begin{bmatrix} 1 & -1 \\ -1 & 1 \end{bmatrix} \begin{matrix} q_1 \\ q_7 \end{matrix} \qquad (9.28)$$

(columns: $q_1 \quad q_7$)

where A, E, and l are the area of cross section, Young's modulus, and length of the element, respectively. Notice that the elements of the matrix $[k_a^{(e)}]$ are identified by the degrees of freedom indicated at the top and right-hand side of the matrix in Eq. (9.28).

9.4.2 Torsional Displacements

Here, the degrees of freedom (torsional displacements) are given by q_4 and q_{10}. By assuming a linear variation of the torsional displacement or twist angle, the displacement model can be expressed as

$$\theta(x) = [N]\vec{q}_t \tag{9.29}$$

where

$$[N] = \left[\left(1 - \frac{x}{l}\right) \left(\frac{x}{l}\right)\right] \tag{9.30}$$

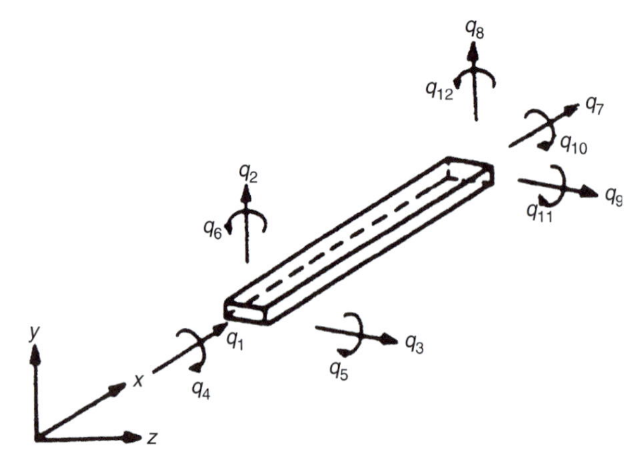

(a) Element with 12 degrees of freedom

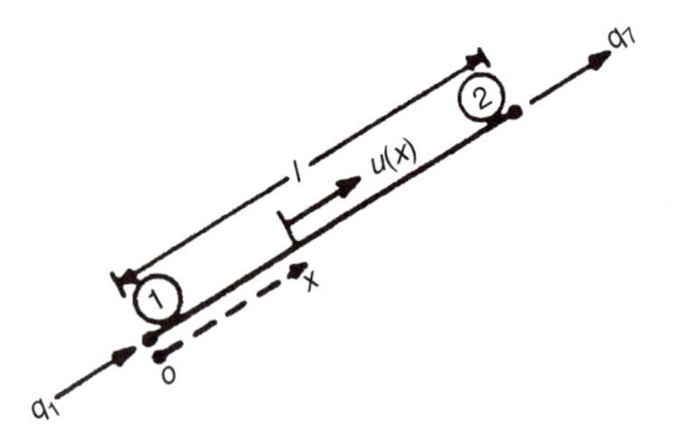

(b) Axial degrees of freedom

Figure 9.6. A Space Frame Element.

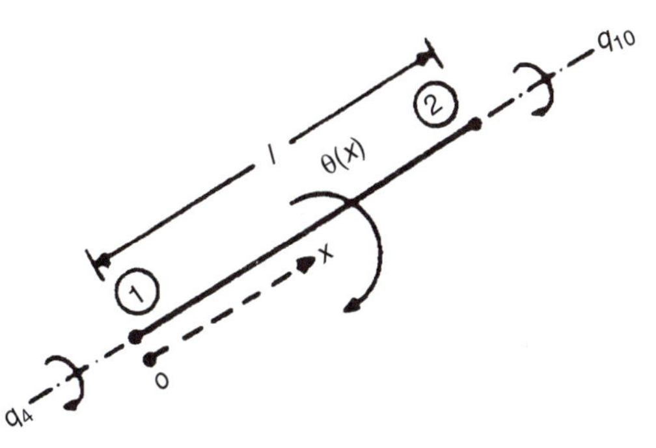

(c) Torsional degrees of freedom

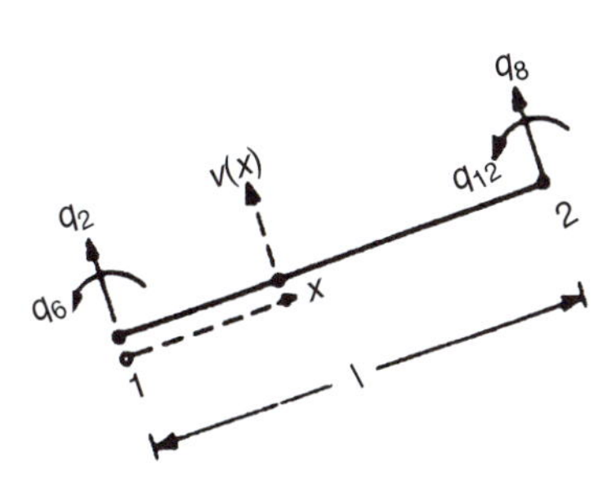

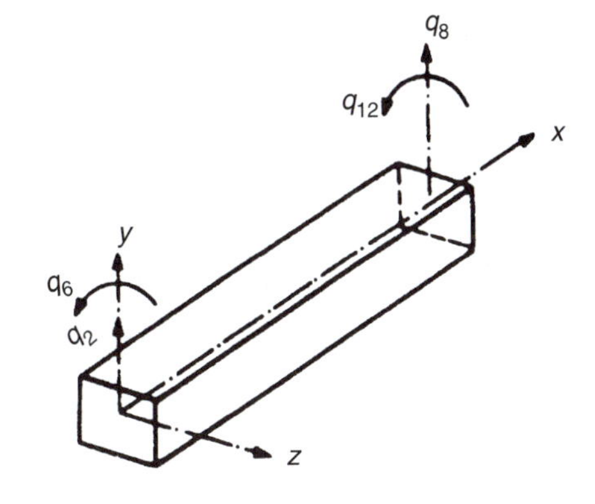

(d) Bending degrees of freedom in *xy* plane

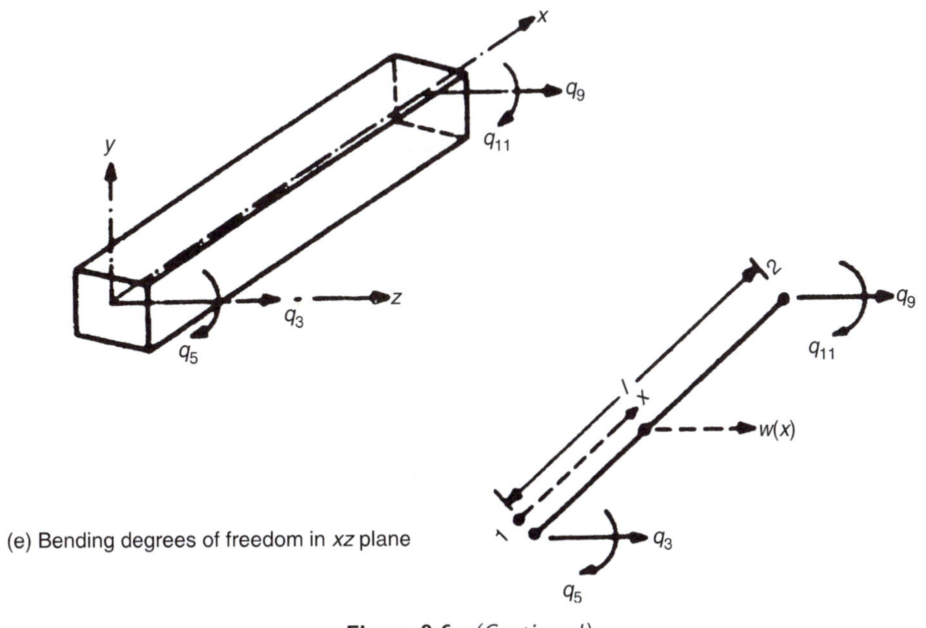

(e) Bending degrees of freedom in *xz* plane

Figure 9.6. *(Continued)*

and

$$\vec{q}_t = \begin{Bmatrix} q_4 \\ q_{10} \end{Bmatrix} \tag{9.31}$$

Assuming the cross section of the frame element to be circular, the shear strain induced in the element can be expressed as [9.1]

$$\varepsilon_{\theta x} = r \frac{\mathrm{d}\theta}{\mathrm{d}x} \tag{9.32}$$

where r is the distance of the fiber from the centroidal axis of the element. Thus, the strain–displacement relation can be expressed as

$$\vec{\varepsilon} = [B]\vec{q}_t \tag{9.33}$$

where $\vec{\varepsilon} = \{\varepsilon_{\theta x}\}$ and $[B] = \begin{bmatrix} -\frac{r}{l} & \frac{r}{l} \end{bmatrix}$

From Hooke's law, the stress–strain relation can be expressed as

$$\vec{\sigma} = [D]\vec{\varepsilon} \tag{9.34}$$

where $\vec{\sigma} = \{\sigma_{\theta x}\}, \qquad [D] = [G],$

and G is the shear modulus of the material. The stiffness matrix of the element corresponding to torsional displacement degrees of freedom can be derived as

$$[k_t^{(e)}] = \iiint\limits_{V^{(e)}} [B]^T [D][B]\,\mathrm{d}V$$

$$= G \int\limits_{x=0}^{l} \mathrm{d}x \iint\limits_{A} r^2\,\mathrm{d}A \begin{Bmatrix} -\frac{1}{l} \\ \frac{1}{l} \end{Bmatrix} \begin{Bmatrix} -\frac{1}{l} & \frac{1}{l} \end{Bmatrix} \tag{9.35}$$

Since $\iint_A r^2\,\mathrm{d}A = J$ = polar moment of inertia of the cross section, Eq. (9.35) can be rewritten as

$$[k_t^{(e)}] = \frac{GJ}{l} \begin{bmatrix} 1 & -1 \\ -1 & 1 \end{bmatrix} \begin{matrix} q_4 \\ q_{10} \end{matrix} \quad (\text{columns } q_4,\ q_{10}) \tag{9.36}$$

Note that the quantity GJ/l is called the torsional stiffness of the frame element [9.1]. If the cross section of the frame element is rectangular as shown in Figure 9.7, the torsional stiffness is given by $(GJ/l) = cG(ab^3/l)$, where the value of the constant c is given below:

Value of a/b	1.0	1.5	2.0	3.0	5.0	10.0
Value of c	0.141	0.196	0.229	0.263	0.291	0.312

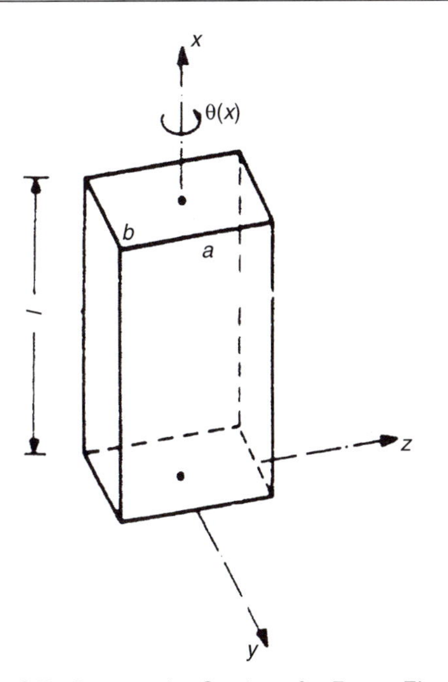

Figure 9.7. Rectangular Section of a Frame Element.

9.4.3 Bending Displacements in the Plane *xy*

The four bending degrees of freedom are q_2, q_6, q_8, and q_{12} [Figure 9.6(d)] and the corresponding stiffness matrix can be derived as (see Section 9.3)

$$[k_{xy}^{(e)}] = \frac{EI_{zz}}{l^3} \begin{bmatrix} 12 & 6l & -12 & 6l \\ 6l & 4l^2 & -6l & 2l^2 \\ -12 & -6l & 12 & -6l \\ 6l & 2l^2 & -6l & 4l^2 \end{bmatrix} \begin{matrix} q_2 \\ q_6 \\ q_8 \\ q_{12} \end{matrix} \quad \text{(columns: } q_2, q_6, q_8, q_{12}) \tag{9.37}$$

where $I_{zz} = \iint_A y^2 \, dA$ is the area moment of inertia of the cross section about the z axis.

9.4.4 Bending Displacements in the Plane *xz*

Here, bending of the element takes place in the xz plane instead of the xy plane. Thus, we have the degrees of freedom q_3, q_5, q_9, and q_{11} [Figure 9.6(e)] in place of q_2, q_6, q_8, and q_{12} [Figure 9.6(d)], respectively. By proceeding as in the case of bending in the plane xy, we can derive the stiffness matrix as

$$[k_{xz}^{(e)}] = \frac{EI_{yy}}{l^3} \begin{bmatrix} 12 & 6l & -12 & 6l \\ 6l & 4l^2 & -6l & 2l^2 \\ -12 & -6l & 12 & -6l \\ 6l & 2l^2 & -6l & 4l^2 \end{bmatrix} \begin{matrix} q_3 \\ q_5 \\ q_9 \\ q_{11} \end{matrix} \quad \text{(columns: } q_3, q_5, q_9, q_{11}) \tag{9.38}$$

where I_{yy} denotes the area moment of inertia of the cross section of the element about the y axis.

9.4.5 Total Element Stiffness Matrix

The stiffness matrices derived for different sets of independent displacements can now be compiled (superposed) to obtain the overall stiffness matrix of the frame element as

$$\underset{12\times 12}{[k^{(e)}]} = \begin{array}{c} \begin{matrix} q_1 & q_2 & q_3 & q_4 & q_5 & q_6 & q_7 & q_8 & q_9 & q_{10} & q_{11} & q_{12} \end{matrix} \\ \left[\begin{matrix}
\frac{EA}{l} & & & & & & & & & & & \\
0 & \frac{12EI_{zz}}{l^3} & & & & & \text{Symmetric} & & & & & \\
0 & 0 & \frac{12EI_{yy}}{l^3} & & & & & & & & & \\
0 & 0 & 0 & \frac{GJ}{l} & & & & & & & & \\
0 & 0 & \frac{-6EI_{yy}}{l^2} & 0 & \frac{4EI_{yy}}{l} & & & & & & & \\
0 & \frac{6EI_{zz}}{l^2} & 0 & 0 & 0 & \frac{4EI_{zz}}{l} & & & & & & \\
-\frac{EA}{l} & 0 & 0 & 0 & 0 & 0 & \frac{EA}{l} & & & & & \\
0 & \frac{-12EI_{zz}}{l^3} & 0 & 0 & 0 & \frac{-6EI_{zz}}{l^2} & 0 & \frac{12EI_{zz}}{l^3} & & & & \\
0 & 0 & \frac{-12EI_{yy}}{l^3} & 0 & \frac{6EI_{yy}}{l^2} & 0 & 0 & 0 & \frac{12EI_{yy}}{l^3} & & & \\
0 & 0 & 0 & \frac{-GJ}{l} & 0 & 0 & 0 & 0 & 0 & \frac{GJ}{l} & & \\
0 & 0 & \frac{-6EI_{yy}}{l^2} & 0 & \frac{2EI_{yy}}{l} & 0 & 0 & 0 & \frac{6EI_{yy}}{l^2} & 0 & \frac{4EI_{yy}}{l} & \\
0 & \frac{6EI_{zz}}{l^2} & 0 & 0 & 0 & \frac{2EI_{zz}}{l} & 0 & \frac{-6EI_{zz}}{l^2} & 0 & 0 & 0 & \frac{4EI_{zz}}{l}
\end{matrix}\right] \end{array} \begin{matrix} q_1 \\ q_2 \\ q_3 \\ q_4 \\ q_5 \\ q_6 \\ q_7 \\ q_8 \\ q_9 \\ q_{10} \\ q_{11} \\ q_{12} \end{matrix} \tag{9.39}$$

9.4.6 Global Stiffness Matrix

It can be seen that the 12×12 stiffness matrix given in Eq. (9.39) is with respect to the local xyz coordinate system. Since the nodal displacements in the local and global coordinate systems are related by the relation (from Figure 9.8)

$$\begin{Bmatrix} q_1 \\ q_2 \\ q_3 \\ q_4 \\ q_5 \\ q_6 \\ q_7 \\ q_8 \\ q_9 \\ q_{10} \\ q_{11} \\ q_{12} \end{Bmatrix} = \begin{bmatrix}
l_{ox} & m_{ox} & n_{ox} & 0 & 0 & 0 & 0 & 0 & 0 & 0 & 0 & 0 \\
l_{oy} & m_{oy} & n_{oy} & 0 & 0 & 0 & 0 & 0 & 0 & 0 & 0 & 0 \\
l_{oz} & m_{oz} & n_{oz} & 0 & 0 & 0 & 0 & 0 & 0 & 0 & 0 & 0 \\
0 & 0 & 0 & l_{ox} & m_{ox} & n_{ox} & 0 & 0 & 0 & 0 & 0 & 0 \\
0 & 0 & 0 & l_{oy} & m_{oy} & n_{oy} & 0 & 0 & 0 & 0 & 0 & 0 \\
0 & 0 & 0 & l_{oz} & m_{oz} & n_{oz} & 0 & 0 & 0 & 0 & 0 & 0 \\
0 & 0 & 0 & 0 & 0 & 0 & l_{ox} & m_{ox} & n_{ox} & 0 & 0 & 0 \\
0 & 0 & 0 & 0 & 0 & 0 & l_{oy} & m_{oy} & n_{oy} & 0 & 0 & 0 \\
0 & 0 & 0 & 0 & 0 & 0 & l_{oz} & m_{oz} & n_{oz} & 0 & 0 & 0 \\
0 & 0 & 0 & 0 & 0 & 0 & 0 & 0 & 0 & l_{ox} & m_{ox} & n_{ox} \\
0 & 0 & 0 & 0 & 0 & 0 & 0 & 0 & 0 & l_{oy} & m_{oy} & n_{oy} \\
0 & 0 & 0 & 0 & 0 & 0 & 0 & 0 & 0 & l_{oz} & m_{oz} & n_{oz}
\end{bmatrix} \begin{Bmatrix} Q_{6i-5} \\ Q_{6i-4} \\ Q_{6i-3} \\ Q_{6i-2} \\ Q_{6i-1} \\ Q_{6i} \\ Q_{6j-5} \\ Q_{6j-4} \\ Q_{6j-3} \\ Q_{6j-2} \\ Q_{6j-1} \\ Q_{6j} \end{Bmatrix} \tag{9.40}$$

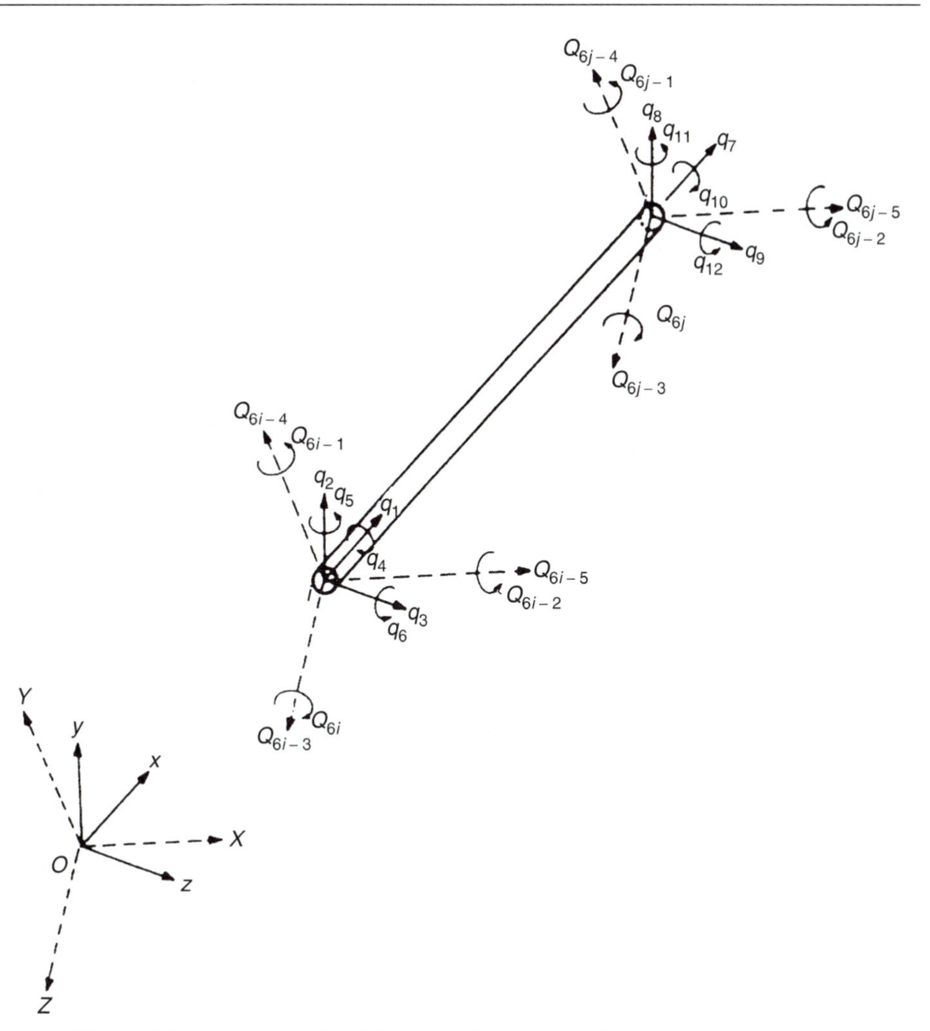

Figure 9.8. Local and Global Degrees of Freedom of a Space Frame Element.

the transformation matrix, $[\lambda]$, can be identified as

$$\underset{12\times 12}{[\lambda]} = \begin{bmatrix} [\underline{\lambda}] & [0] & [0] & [0] \\ [0] & [\underline{\lambda}] & [0] & [0] \\ [0] & [0] & [\underline{\lambda}] & [0] \\ [0] & [0] & [0] & [\underline{\lambda}] \end{bmatrix} \tag{9.41}$$

where

$$\underset{3\times 3}{[\underline{\lambda}]} = \begin{bmatrix} l_{ox} & m_{ox} & n_{ox} \\ l_{oy} & m_{oy} & n_{oy} \\ l_{oz} & m_{oz} & n_{oz} \end{bmatrix} \tag{9.42}$$

and

$$\underset{3\times 3}{[0]} = \begin{bmatrix} 0 & 0 & 0 \\ 0 & 0 & 0 \\ 0 & 0 & 0 \end{bmatrix} \tag{9.43}$$

Here, l_{ox}, m_{ox}, and n_{ox} denote the direction cosines of the x axis (line ij in Figure 9.8); l_{oy}, m_{oy}, and n_{oy} represent the direction cosines of the y axis; and l_{oz}, m_{oz}, and n_{oz} indicate the direction cosines of the z axis with respect to the global X, Y, Z axes. It can be seen that finding the direction cosines of the x axis is a straightforward computation since

$$l_{ox} = \frac{X_j - X_i}{l}, \qquad m_{ox} = \frac{Y_j - Y_i}{l}, \qquad n_{ox} = \frac{Z_j - Z_i}{l} \tag{9.44}$$

where X_k, Y_k, Z_k indicate the coordinates of node k $(k = i, j)$ in the global system. However, the computation of the direction cosines of the y and z axes requires some special effort. Finally, the stiffness matrix of the element with reference to the global coordinate system can be obtained as

$$[K^{(e)}] = [\lambda]^T [k^{(e)}][\lambda] \tag{9.45}$$

Transformation Matrix

We shall derive the transformation matrix $[\underline{\lambda}]$ between the local and global coordinate systems in two stages. In the first stage, we derive a transformation matrix $[\lambda_1]$ between the global coordinates XYZ and the coordinates $\bar{x}\ \bar{y}\ \bar{z}$ by assuming the $\bar{z}$ axis to be parallel to the XZ plane [Figure 9.9(a)]:

$$\begin{Bmatrix} \bar{x} \\ \bar{y} \\ \bar{z} \end{Bmatrix} = [\lambda_1] \begin{Bmatrix} X \\ Y \\ Z \end{Bmatrix} \tag{9.46}$$

In the second stage, we drive a transformation matrix $[\lambda_2]$ between the local coordinates xyz (principal member axes) and the coordinates $\bar{x}\ \bar{y}\ \bar{z}$ as

$$\begin{Bmatrix} x \\ y \\ z \end{Bmatrix} = [\lambda_2] \begin{Bmatrix} \bar{x} \\ \bar{y} \\ \bar{z} \end{Bmatrix} \tag{9.47}$$

by assuming that the local coordinate system (xyz) can be obtained by rotating the $\bar{x}\ \bar{y}\ \bar{z}$ system about the $\bar{x}$ axis by an angle α as shown in Figure 9.9(b). Thus, the desired transformation between the xyz system and the XYZ system can be obtained as

$$[\underline{\lambda}] = [\lambda_2][\lambda_1] \tag{9.48}$$

where

$$\begin{Bmatrix} x \\ y \\ z \end{Bmatrix} = [\underline{\lambda}] \begin{Bmatrix} X \\ Y \\ Z \end{Bmatrix} \tag{9.49}$$

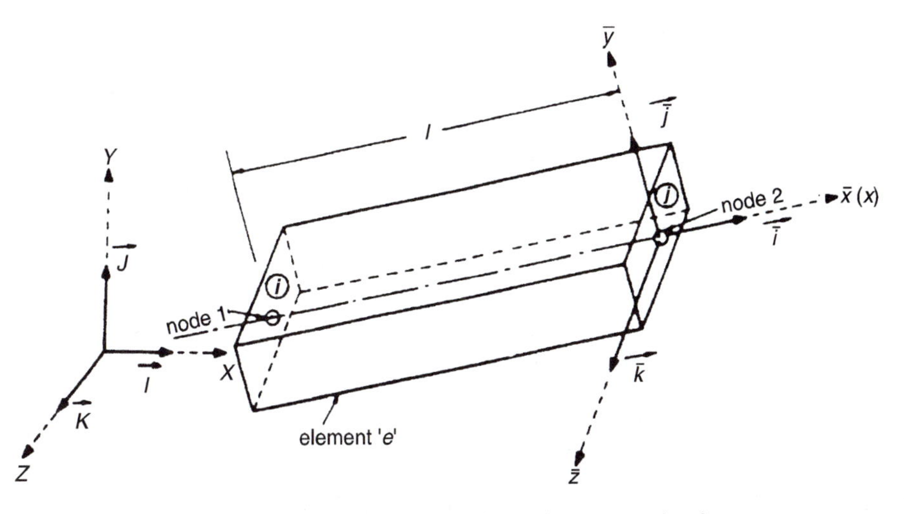

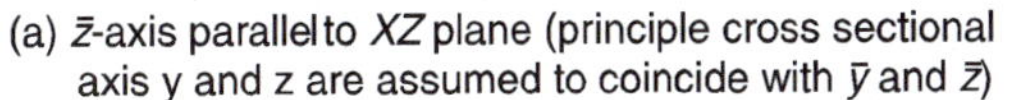
(a) $\bar{z}$-axis parallel to XZ plane (principle cross sectional axis y and z are assumed to coincide with $\bar{y}$ and $\bar{z}$)

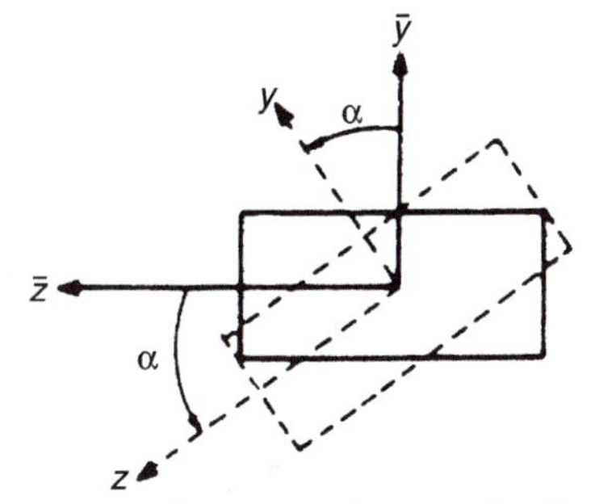

view as seen from +x direction

(b) General case (y and z do not coincide with $\bar{y}$ and $\bar{z}$)

Figure 9.9. Local and Global Coordinate Systems.

Expression for $[\lambda_1]$

From Figure 9.9(a), the direction cosines of the longitudinal axis of the frame element ($\bar{x}$ or x or first local axis) can be obtained as

$$
\begin{aligned}
l_{o\bar{x}} &= l_{ox} = \frac{X_j - X_i}{l} \\
m_{o\bar{x}} &= m_{ox} = \frac{Y_j - Y_i}{l} \\
n_{o\bar{x}} &= n_{ox} = \frac{Z_j - Z_i}{l}
\end{aligned}
\tag{9.50}
$$

where i and j denote the first and second nodes of the element e in the global system, and l represents the length of the element e:

$$l = \{(X_j - X_i)^2 + (Y_j - Y_i)^2 + (Z_j - Z_i)^2\}^{1/2} \tag{9.51}$$

Since the unit vector $\vec{\bar{k}}$ (which is parallel to the $\bar{z}$ axis) is normal to both the unit vectors $\vec{J}$ (parallel to Y axis) and $\vec{\bar{i}}$ (parallel to $\bar{x}$ axis), we have, from vector analysis [9.2],

$$\vec{\bar{k}} = \frac{\vec{\bar{i}} \times \vec{J}}{\|\vec{\bar{i}} \times \vec{J}\|} = \frac{1}{d}\begin{bmatrix} \vec{I} & \vec{J} & \vec{K} \\ l_{ox} & m_{ox} & n_{ox} \\ 0 & 1 & 0 \end{bmatrix} = \frac{1}{d}(-\vec{I}n_{ox} + \vec{K}l_{ox}) \tag{9.52}$$

where

$$d = (l_{ox}^2 + n_{ox}^2)^{1/2} \tag{9.53}$$

Thus, the direction cosines of the $\bar{z}$ axis with respect to the global XYZ system are given by

$$l_{o\bar{z}} = -\frac{n_{ox}}{d}, \qquad m_{o\bar{z}} = 0, \qquad n_{o\bar{z}} = \frac{l_{ox}}{d} \tag{9.54}$$

To find the direction cosines of the $\bar{y}$ axis, we use the condition that the $\bar{y}$ axis (unit vector $\vec{\bar{j}}$) is normal to the $\bar{x}$ axis ($\vec{\bar{i}}$) and $\bar{z}$ axis ($\vec{\bar{k}}$). Hence, we can express $\vec{\bar{j}}$ as

$$\vec{\bar{j}} = \vec{\bar{k}} \times \vec{\bar{i}} = \begin{vmatrix} \vec{I} & \vec{J} & \vec{K} \\ l_{o\bar{z}} & -m_{o\bar{z}} & n_{o\bar{z}} \\ l_{ox} & m_{ox} & n_{ox} \end{vmatrix}$$

$$= \frac{1}{d}\left[\vec{I}(-l_{ox}m_{ox}) - \vec{J}(-n_{ox}^2 - l_{ox}^2) + \vec{K}(-m_{ox}n_{ox})\right] \tag{9.55}$$

Thus, the direction cosines of the $\bar{y}$ axis are given by

$$\left.\begin{aligned} l_{o\bar{y}} &= -\frac{l_{ox}m_{ox}}{d} \\ m_{o\bar{y}} &= \frac{n_{ox}^2 + l_{ox}^2}{d} \\ n_{o\bar{y}} &= -\frac{m_{ox}n_{ox}}{d} \end{aligned}\right\} \tag{9.56}$$

Thus, the $[\lambda_1]$ matrix is given by

$$[\lambda_1] = \begin{bmatrix} l_{o\bar{x}} & m_{o\bar{x}} & n_{o\bar{x}} \\ l_{o\bar{y}} & m_{o\bar{y}} & n_{o\bar{y}} \\ l_{o\bar{z}} & m_{o\bar{z}} & n_{o\bar{z}} \end{bmatrix}$$

$$= \begin{bmatrix} l_{ox} & m_{ox} & n_{ox} \\ -(l_{ox}m_{ox})/d & (l_{ox}^2 + n_{ox}^2)/d & -(m_{ox}n_{ox})/d \\ -n_{ox}/d & 0 & l_{ox}/d \end{bmatrix} \tag{9.57}$$

where l_{ox}, m_{ox}, n_{ox} are given by Eq. (9.50) and d by Eq. (9.53).

Expression for $[\lambda_2]$

When the principal cross-sectional axes of the frame element (xyz axes) are arbitrary, making an angle α with the $\bar{x}\ \bar{y}\ \bar{z}$ axes (x axis is same as $\bar{x}$ axis), the transformation between the two systems can be expressed as

$$\begin{Bmatrix} x \\ y \\ z \end{Bmatrix} = \begin{bmatrix} 1 & 0 & 0 \\ 0 & \cos\alpha & \sin\alpha \\ 0 & -\sin\alpha & \cos\alpha \end{bmatrix} \begin{Bmatrix} \bar{x} \\ \bar{y} \\ \bar{z} \end{Bmatrix} \equiv [\lambda_2] \begin{Bmatrix} \bar{x} \\ \bar{y} \\ \bar{z} \end{Bmatrix} \tag{9.58}$$

so that

$$[\lambda_2] = \begin{bmatrix} 1 & 0 & 0 \\ 0 & \cos\alpha & \sin\alpha \\ 0 & -\sin\alpha & \cos\alpha \end{bmatrix} \tag{9.59}$$

Thus, the transformation between the coordinate axes XYZ and xyz can be found using Eq. (9.48).

Notes:

(i) When $\alpha = 0$, the matrix $[\lambda_2]$ degenerates to a unit matrix.
(ii) When the element e lies vertical [i.e., when the x (or $\bar{x}$) axis coincides with the Y axis], $l_{ox} = n_{ox} = 0$ and hence d in Eq. (9.53) becomes zero. This makes some of the terms in the $[\lambda_2]$ matrix indeterminate. Thus, the previous procedure breaks down.

In this case, we can redefine the angle α as the angle in the horizontal (XZ) plane between the axes Z and z, positive when turning from Z to the X axis as shown in Figure 9.10. In this case, the $[\underset{\sim}{\lambda}]$ matrix can be derived, by going through the same procedure as before, as

$$[\underset{\sim}{\lambda}] = \begin{bmatrix} 0 & m_{ox} & 0 \\ -m_{ox}\cos\alpha & 0 & m_{ox}\sin\alpha \\ \sin\alpha & 0 & \cos\alpha \end{bmatrix} \tag{9.60}$$

where $m_{ox} = 1$ for this case.

$[\lambda]$ Matrix

Finally, the transformation matrix, $[\lambda]$, relating the degrees of freedom in the local and global coordinate systems is given by Eq. (9.41).

9.5 PLANAR FRAME ELEMENT

In the case of two-dimensional (planar) frame analysis, we need to use an element having six degrees of freedom as shown in Figure 9.11. This element is assumed to lie in the XZ plane and has two axial and four bending degrees of freedom. By using a linear interpolation model for axial displacement and a cubic model for the transverse displacement, and superimposing the resulting two stiffness matrices, the following stiffness matrix can be

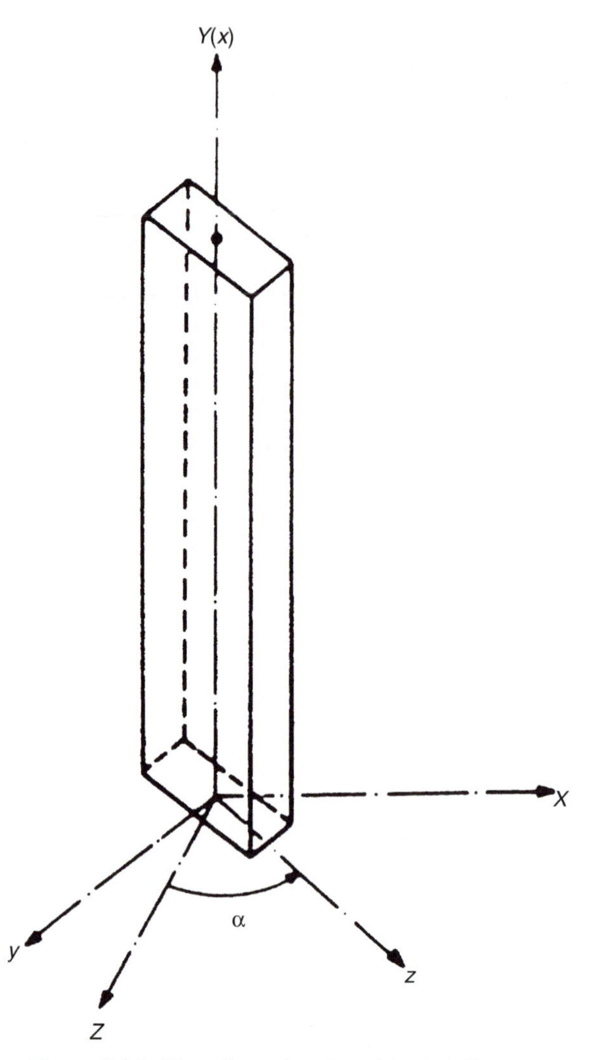

Figure 9.10. Transformation for a Vertical Element.

obtained (the vector $\vec{q}^{\,(e)}$ is taken as $\vec{q}^{\,(e)^T} = \{q_1\ q_2 \ldots q_6\}$):

$$[k^{(e)}] = \frac{EI_{yy}}{l^3} \begin{bmatrix} \dfrac{Al^2}{I_{yy}} & & & & \text{Symmetric} & \\ 0 & 12 & & & & \\ 0 & 6l & 4l^2 & & & \\ -\dfrac{Al^2}{I_{yy}} & 0 & 0 & \dfrac{Al^2}{I_{yy}} & & \\ 0 & -12 & -6l & 0 & 12 & \\ 0 & 6l & 2l^2 & 0 & -6l & 4l^2 \end{bmatrix} \tag{9.61}$$

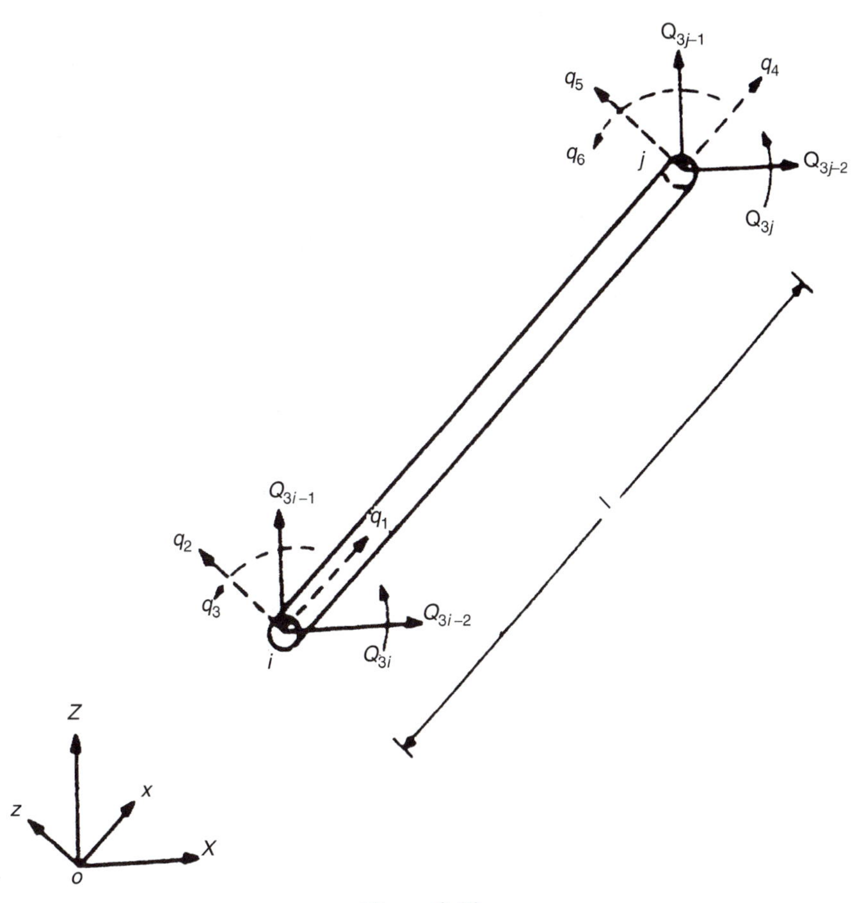

Figure 9.11.

Note that the bending and axial deformation effects are uncoupled while deriving Eq. (9.61). Equation (9.61) can also be obtained as a special case of Eq. (9.39) by deleting rows and columns 2, 4, 5, 8, 10, and 11. In this case the stiffness matrix of the element in the global XZ coordinate system can be found as

$$[K^{(e)}] = [\lambda]^T [k^{(e)}][\lambda] \tag{9.62}$$

where

$$[\lambda] = \begin{bmatrix} l_{ox} & m_{ox} & 0 & 0 & 0 & 0 \\ l_{oz} & m_{oz} & 0 & 0 & 0 & 0 \\ 0 & 0 & 1 & 0 & 0 & 0 \\ 0 & 0 & 0 & l_{ox} & m_{ox} & 0 \\ 0 & 0 & 0 & l_{oz} & m_{oz} & 0 \\ 0 & 0 & 0 & 0 & 0 & 1 \end{bmatrix} \tag{9.63}$$

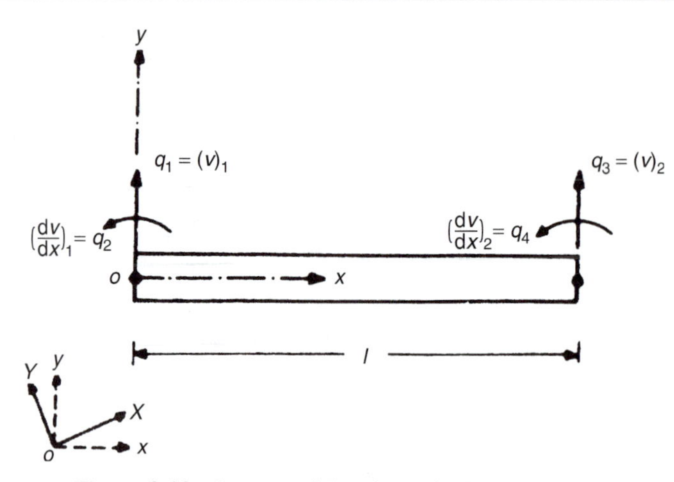

Figure 9.12. Degrees of Freedom of a Beam Element.

with (l_{ox}, m_{ox}) denoting the direction cosines of the x axis and (l_{oz}, m_{oz}) indicating the direction cosines of the z axis with respect to the global XZ system.

9.5.1 Beam Element

For a beam element lying in the local xy plane, the stiffness matrix is given by

$$[k^{(e)}] = \frac{EI_{zz}}{l^3} \begin{bmatrix} 12 & 6l & -12 & 6l \\ 6l & 4l^2 & -6l & 2l^2 \\ -12 & -6l & 12 & -6l \\ 6l & 2l^2 & -6l & 4l^2 \end{bmatrix} \tag{9.64}$$

The stiffness matrix in the global XY plane (Figure 9.12) can be found as

$$\underset{6\times 6}{[K^{(e)}]} = \underset{6\times 4}{[\lambda]^T} \; \underset{4\times 4}{[k^{(e)}]} \; \underset{4\times 6}{[\lambda]} \tag{9.65}$$

where the transformation matrix $[\lambda]$ is given by

$$[\lambda] = \begin{bmatrix} l_{oy} & n_{oy} & 0 & 0 & 0 & 0 \\ 0 & 0 & 1 & 0 & 0 & 0 \\ 0 & 0 & 0 & l_{oy} & n_{oy} & 0 \\ 0 & 0 & 0 & 0 & 0 & 1 \end{bmatrix} \tag{9.66}$$

where (l_{oy}, n_{oy}) are the direction cosines of the y axis with respect to the XY system.

9.6 COMPUTER PROGRAM FOR FRAME ANALYSIS

A Fortran subroutine called FRAME is written for the displacement analysis of three-dimensional frame structures. The load vector has to be either generated or given as data in the main program that calls the subroutine FRAME. The subroutine FRAME requires the following quantities as input:

NN = total number of nodes in the frame structure (including the fixed nodes).

NE = number of finite elements.

ND = total number of degrees of freedom (including the fixed degrees of freedom). Six degrees of freedom are considered at each node as shown in Figure 9.8.

NB = bandwidth of the overall stiffness matrix.

M = number of load conditions.

LØC = an array of dimension NE × 2. LØC (I,J) denotes the global node number corresponding to Jth corner of element I. It is assumed that for vertical elements the bottom node represents the first corner and the top node the second corner.

CX,CY,CZ = vector arrays of dimension NN each. CX (I), CY (I), CZ (I) denote the global X, Y, Z coordinates of node I. The global (X, Y, Z) coordinate system must be a right-handed system and must be set up such that the XZ plane represents the horizontal plane and the Y direction denotes the vertical axis.

E = Young's modulus of the material.

G = shear modulus of the material.

A = a vector array of dimension NE. A (I) denotes the cross-sectional area of element I.

YI,ZI = vector arrays of dimension NE each. YI (I) and ZI (I) denote the area moment of inertia of the cross section about the local y and z axis, respectively, of element I. It is assumed that the local xyz system is also right-handed, with the line going from corner 1 toward corner 2 representing the x axis of the element. For convenience, we take the principal cross-sectional axis running parallel to the longer dimension of the cross section (in the case of rectangular, I and H sections) as the z axis. If this is not possible (as in the case of circular section), we take any of the principal cross-sectional axes as the z axis.

TJ = a vector array of dimension NE. TJ (I) denotes the torsional constant (J) of element I.

ALPHA = a vector array of dimension NE. ALPHA (I) represents the value of α for element I (radians). If the element I is vertical, α denotes the angle that the z axis makes with the Z axis as shown in Figure 9.10. For nonvertical elements, α denotes the angle that the z axis makes with the horizontal ($\bar{z}$ axis) as shown in Figure 9.9(b).

IVERT = a vector array of dimension NE denoting the orientation of the element. IVERT (I) is set equal to 1 if element I is vertical and 0 otherwise.

NFIX = number of fixed degrees of freedom (zero displacements).

IFIX = a vector array of dimension NFIX. IFIX (I) denotes the Ith fixed degree of freedom number.

P = an array of size ND × M to store the global load vectors. Upon return from the subroutine FRAME, it gives the global displacement vectors. P (I,J) denotes the Ith component of global load (or displacement) vector in Jth load condition.

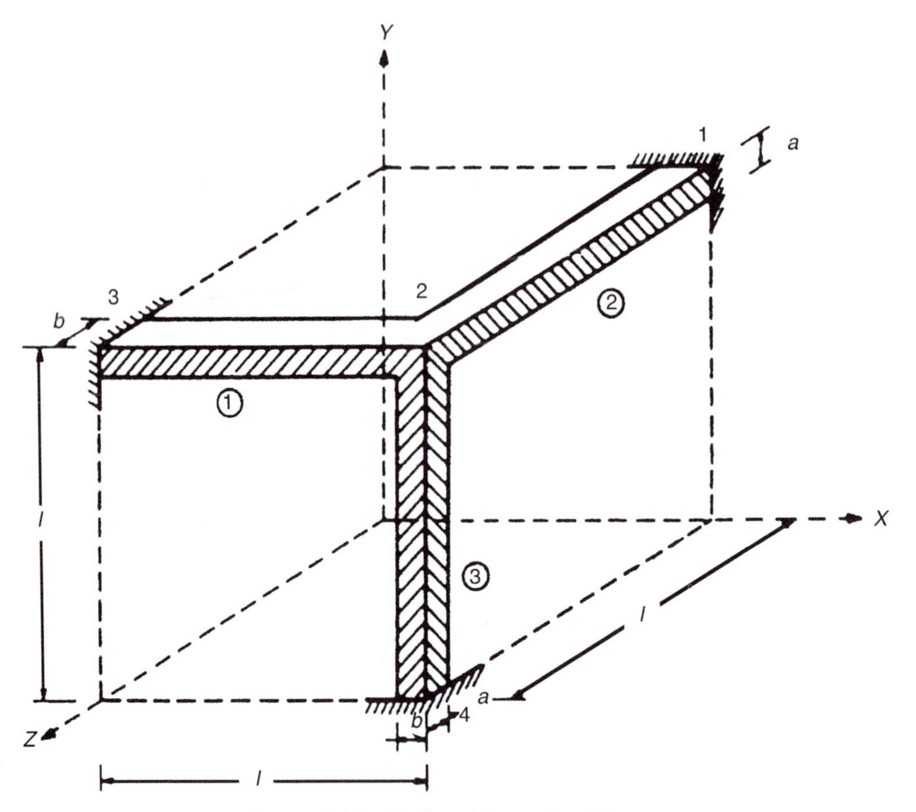

Figure 9.13. A Three-Dimensional Frame.

In addition to the input data, the array GS of size ND × NB and the vector array DIFF of size M are included as arguments for the subroutine FRAME. The array GS represents the global stiffness matrix of the structure in band form. The array DIFF is a dummy array defined in double precision. The subroutine FRAME calls the following subroutines:

MATMUL : for multiplication of two matrices
DECØMP and SØLVE : for the solution of load deflection equations
[Given in Section 7.2.2(v)]

Example 9.2 To illustrate the use of the subroutine FRAME, the three-dimensional rigid frame shown in Figure 9.13 is analyzed for the following load conditions:

(i) When a vertically downward load of 10 N acts at node 2.
(ii) When a uniformly distributed downward load of intensity 1 N/cm acts on members 1 and 2.

Nodes 1, 3, and 4 are fixed. The data of the elements are

$$a = 1 \text{ cm}, \qquad b = 2 \text{ cm}, \qquad l = 20 \text{ cm},$$

$$E = 2 \times 10^6 \text{ N/cm}^2, \qquad G = 0.8 \times 10^6 \text{ N/cm}^2$$

Data for the Subroutine

The global (X, Y, Z) coordinate system is selected as shown in Figure 9.13. The corner numbers and the local (x, y, z) coordinate axes of the various elements and the global degrees of freedom chosen are indicated in Figure 9.14. The cross-sectional properties of element I can be computed as follows:

$$\left.\begin{aligned}
&\text{A (I)} = \text{cross-sectional area} = ab = 2\ \text{cm}^2\\
&\text{ZI (I)} = \text{area moment of inertia about the } z \text{ axis} = \frac{1}{12}ba^3 = \frac{1}{6}\ \text{cm}^4\\
&\text{YI (I)} = \text{area moment of inertia about the } y \text{ axis} = \frac{1}{12}ab^3 = \frac{2}{3}\ \text{cm}^4\\
&\text{TJ (I)} = cba^3 (\text{where } c = 0.229 \text{ for } b/a = 2) = 0.458\ \text{cm}^4
\end{aligned}\right\}; \quad I = 1, 2, 3$$

The other data of the problem are as follows:

NN = 4, NE = 3, ND = 24, NB = (2 + 1) × 6 = 18

M = 2, NFIX = 18

CX (1) = 20, CY (1) = 20, CZ (1) = 0

CX (2) = 20, CY (2) = 20, CZ (2) = 20

CX (3) = 0, CY (3) = 20, CZ (3) = 20

CX (4) = 20, CY (4) = 0, CZ (4) = 20

LØC(1, 1) = 3, LØC(1, 2) = 2; LØC(2, 1) = 1, LØC(2, 2) = 2;

LØC(3, 1) = 4, LØC(3, 2) = 2

IVERT (1) = IVERT (2) = 0 (since elements 1 and 2 are not vertical)

IVERT (3) = 1 (since element 3 is vertical)

ALPHA (1) = ALPHA (2) = 0

ALPHA (3) = angle made by the z axis of element 3 with the Z axis = 90°

IFIX (1), IFIX (2), ..., IFIX (18) = fixed degrees of freedom numbers = 1, 2, 3, 4, 5, 6, 13, 14, ..., 24

First load condition:

$$P(I, 1) = 0 \quad \text{for all I except} \quad I = 8; \qquad P(8, 1) = -10$$

Second load condition:

The distributed load acting on elements 1 and 2 has to be converted into equivalent nodal forces. For this we use Eq. (8.89):

$$\vec{P}_s^{(e)} = \iint\limits_{S_1^{(e)}} [N]^T \ \vec{\Phi}\, \mathrm{d}S_1 \equiv \int_o^l [N]^T p_o \cdot \mathrm{d}x \tag{E$_1$}$$

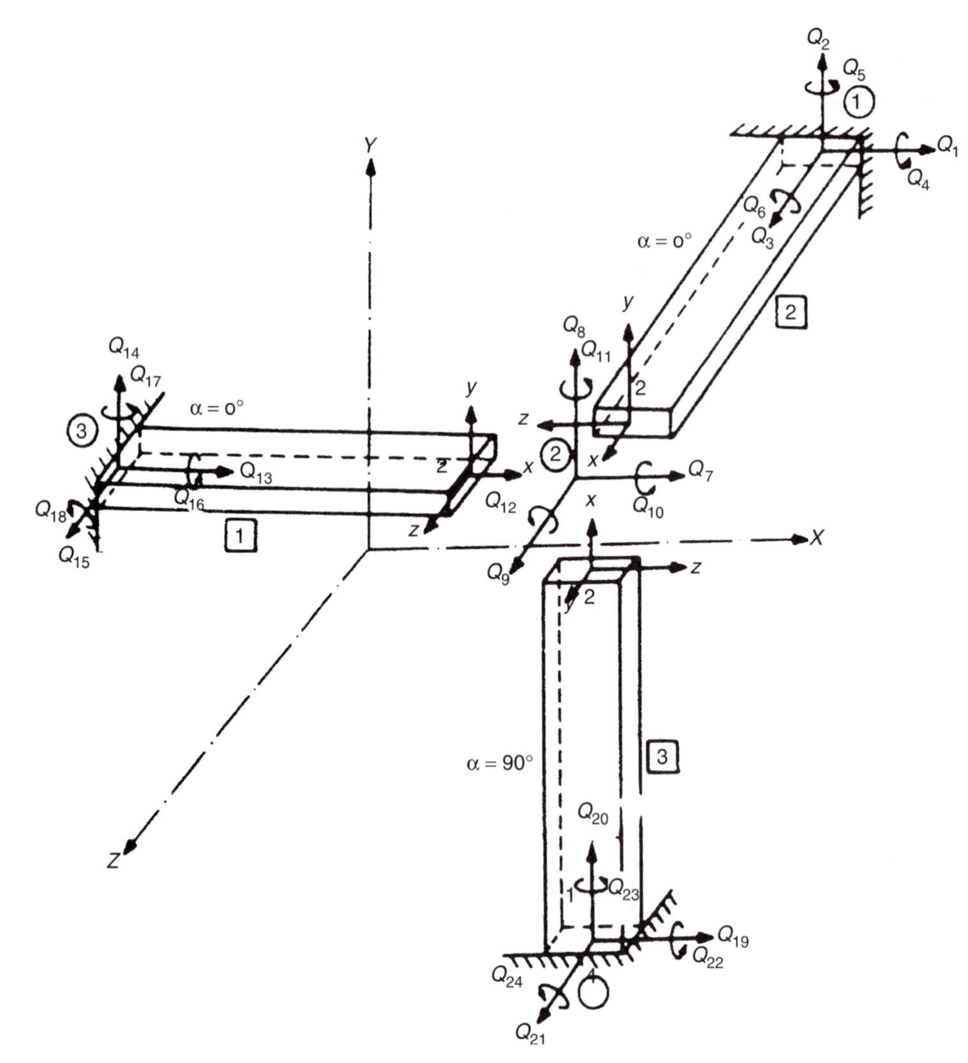

Figure 9.14. Global Degrees of Freedom of the Structure.

where $\vec{\vec{\Phi}} = -p_o$ is the intensity of load per unit length. By substituting the matrix of interpolation functions $[N]$ defined in Eq. (9.21), we obtain

$$\vec{P}_s^{(e)} = -p_o \int_o^l \begin{Bmatrix} (2x^3 - 3lx^2 + l^3)/l^3 \\ (x^3 - 2lx^2 + l^2x)/l^2 \\ -(2x^3 - 3lx^2)/l^3 \\ (x^3 - lx^2)/l^2 \end{Bmatrix} dx = \begin{Bmatrix} -p_ol/2 \\ -p_ol^2/12 \\ -p_ol/2 \\ -p_ol^2/12 \end{Bmatrix} = \begin{Bmatrix} -10 \\ -100/3 \\ -10 \\ -100/3 \end{Bmatrix} \tag{E$_2$}$$

since $p_o = 1$ and $l = 20$. Note that the components of the load vector given by Eq. (E_2) act in the directions of q_1, q_2, q_3, and q_4 shown in Figure 9.15(a). In the present case, for

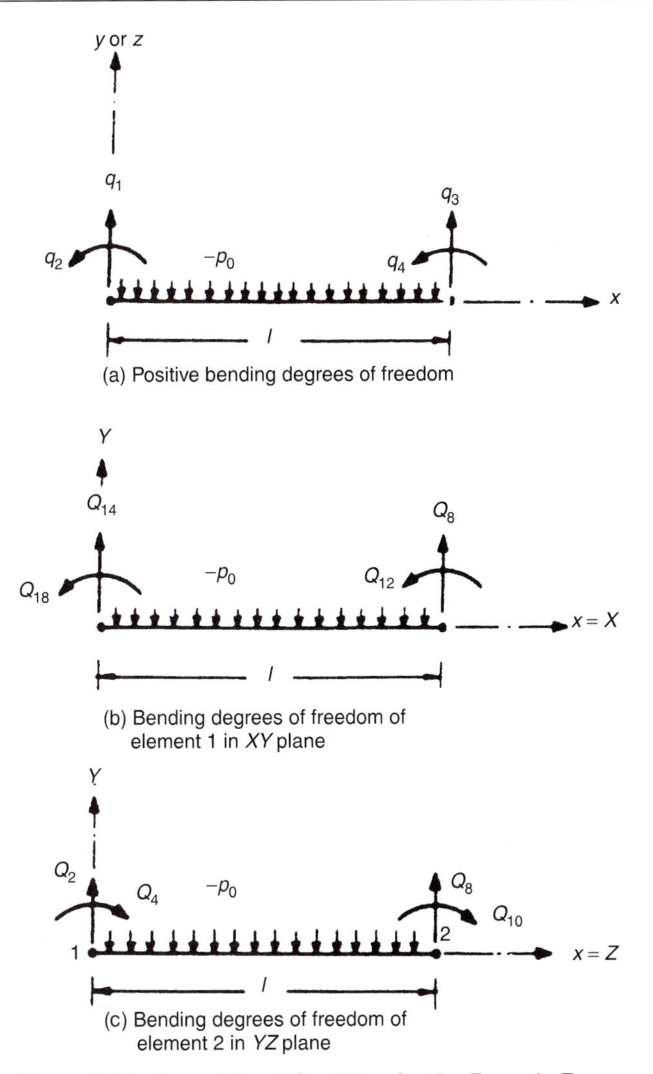

Figure 9.15. Second Load Condition for the Example Frame.

$e = 1$, we have the situation shown in Figure 9.15(b). Here, the degrees of freedom Q_{14}, Q_{18}, Q_8, and Q_{12} correspond to q_1, q_2, q_3, and q_4 of Figure 9.15(a) and hence the load vector becomes

$$\vec{P}_s^{(1)} = \begin{Bmatrix} P_{14} \\ P_{18} \\ P_8 \\ P_{12} \end{Bmatrix} = \begin{Bmatrix} -10 \\ -100/3 \\ -10 \\ -100/3 \end{Bmatrix} \tag{E$_3$}$$

However, for $e = 2$, the situation is as shown in Figure 9.15(c). Here, the degrees of freedom Q_2, Q_4, Q_8, and Q_{10} correspond to q_1, $-q_2$, q_3, and $-q_4$ of Figure 9.15(a) and hence the element load vector becomes

$$\vec{P}_s^{(2)} = \begin{Bmatrix} P_2 \\ P_4 \\ P_8 \\ P_{10} \end{Bmatrix} = \begin{Bmatrix} -10 \\ 100/3 \\ -10 \\ 100/3 \end{Bmatrix} \tag{E$_4$}$$

By superposing the two load vectors $\vec{P}_s^{(1)}$ and $\vec{P}_s^{(2)}$ and neglecting the components corresponding to the fixed degrees of freedom (Q_{14}, Q_{18}, Q_2, and Q_4), we obtain the nonzero components of the load vector of the second load condition as

$$P(8,2) = -20$$
$$P(10,2) = +100/3$$
$$P(12,2) = -100/3$$

Main Program

The main program for solving the example problem (which calls the subroutine FRAME) and the results obtained from the program are given below.

```
C=========================================================================
C
C     THREE DIMENSIONAL FRAME ANALYSIS
C
C=========================================================================
      DIMENSION LOC(3,2),CX(4),CY(4),CZ(4),A(3),YI(3),ZI(3),
     2 TJ(3),ALPHA(3),IVERT(3),IFIX(18),P(24,2),GS(24,18)
      DOUBLE PRECISION DIFF(2)
      DATA NN,NE,ND,NB,NFIX,M,E,G/4,3,24,18,18,2,2.0E6,0.8E6/
      DATA LOC/3,1,4,2,2,2/
      DATA CX/20.0,20.0,0.0,20.0/
      DATA CY/20.0,20.0,20.0,0.0/
      DATA CZ/0.0,20.0,20.0,20.0/
      DATA A/2.0,2.0,2.0/
      DATA YI/0.6667,0.6667,0.6667/
      DATA ZI/0.1667,0.1667,0.1667/
      DATA TJ/0.458,0.458,0.458/
      DATA ALPHA/0.0,0.0,4.7124/
      DATA IVERT/0,0,1/
      DATA IFIX/1,2,3,4,5,6,13,14,15,16,17,18,19,20,21,22,23,24/
      DO 10 I=1,ND
      DO 10 J=1,M
 10   P(I,J)=0.0
      P(8,1)=-10.0
      P(8,2)=-20.0
      P(12,2)=-100.0/3.0
      P(10,2)=100.0/3.0
```

```
      CALL FRAME(NN,NE,ND,NB,M,LOC,CX,CY,CZ,E,G,A,YI,ZI,TJ,
     2 ALPHA,IVERT,NFIX,IFIX,P,GS,DIFF)
      PRINT 20
20    FORMAT(1X,'DISPLACEMENT OF FRAME STRUCTURE')
      DO 30 J=1,M
30    PRINT 40,J,(P(I,J),I=1,ND)
40    FORMAT(/,1X,'INLOAD CONDITION',I4/(1X,6E12.4))
      STOP
      END
```

```
DISPLACEMENT OF FRAME STRUCTURE

INLOAD CONDITION 1
  0.5926E-13 -0.3337E-10  0.4049E-13  0.2914E-11  0.4712E-14 -0.7123E-12
  0.6994E-07 -0.4981E-04  0.4049E-07  0.1644E-05  0.1068E-08 -0.7123E-06
  0.6994E-13 -0.4269E-10  0.5116E-13  0.1644E-11 -0.3570E-14 -0.3380E-11
 -0.7053E-11 -0.4981E-10 -0.1640E-10 -0.8188E-12  0.1068E-14  0.3509E-12

INLOAD CONDITION 2
  0.8058E-11  0.2127E-08  0.5505E-11 -0.1037E-09  0.6406E-12 -0.9685E-10
  0.9510E-05 -0.1075E-03  0.5505E-05  0.2235E-03  0.1452E-06 -0.9685E-04
  0.9510E-11  0.8611E-09  0.6957E-11  0.2235E-09 -0.4855E-12  0.4037E-10
 -0.9590E-09 -0.1075E-09 -0.2229E-08 -0.1113E-09  0.1452E-12  0.4771E-10
```

REFERENCES

9.1 W. Weaver, Jr.: *Computer Programs for Structural Analysis*, Van Nostrand, Princeton, NJ, 1967.

9.2 M.R. Spiegel: *Schaum's Outline of Theory and Problems of Vector Analysis and an Introduction to Tensor Analysis*, Schaum, New York, 1959.

PROBLEMS

9.1 Derive the transformation matrices for the members of the frame shown in Figure 9.16. Indicate clearly the local and global degrees of freedom for each member separately.

9.2 Find the deflections of nodes 2 and 3 of the frame shown in Figure 9.16 under the following load conditions:

(i) When a load of 100 N acts in the direction of $-Y$ at node 2.

(ii) When a load of 100 N acts at node 3 in the direction of Z.

(iii) When a distributed load of magnitude 1 N per unit length acts on member 2 in the direction of $-Y$.

Assume the material properties as $E = 2 \times 10^7$ N/cm^2 and $G = 0.8 \times 10^7$ N/cm^2. Use the subroutine FRAME for the purpose.

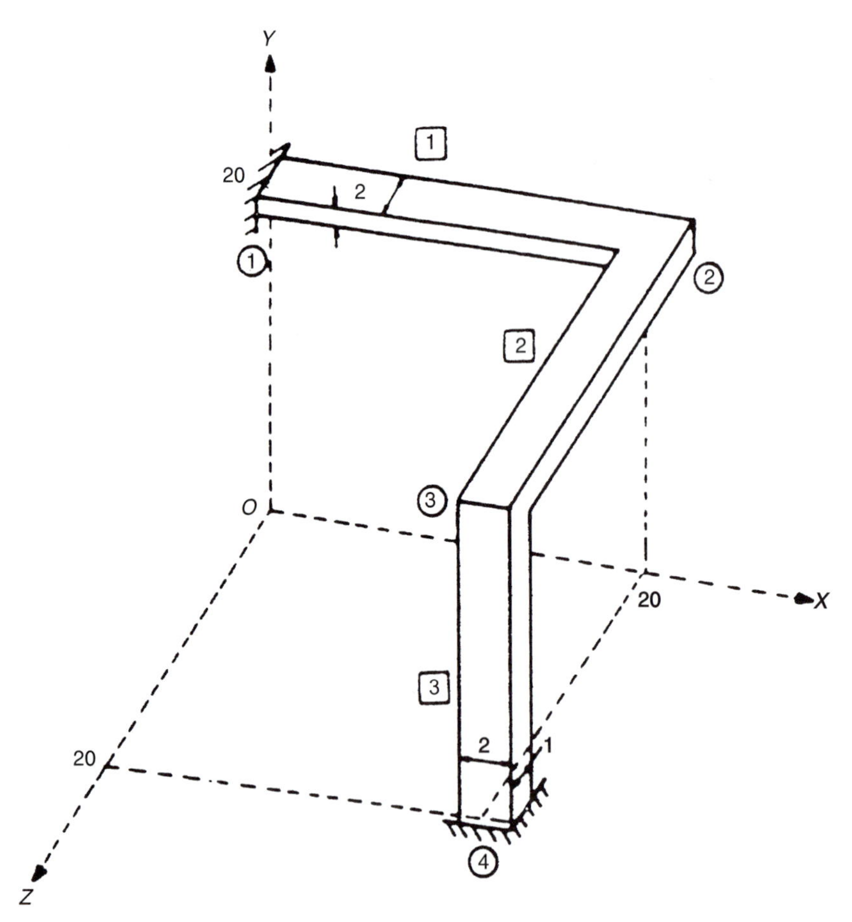

Figure 9.16.

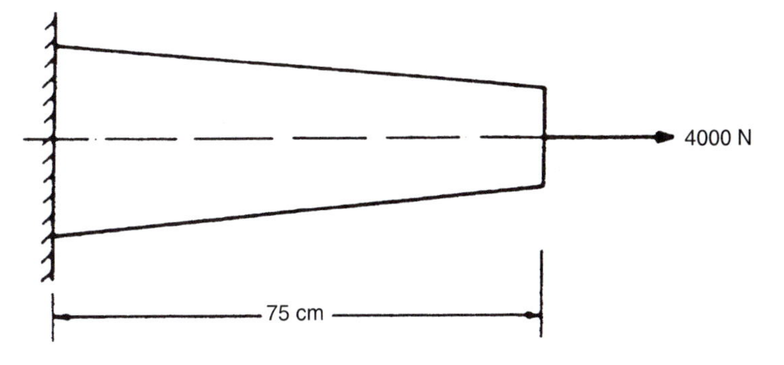

Figure 9.17.

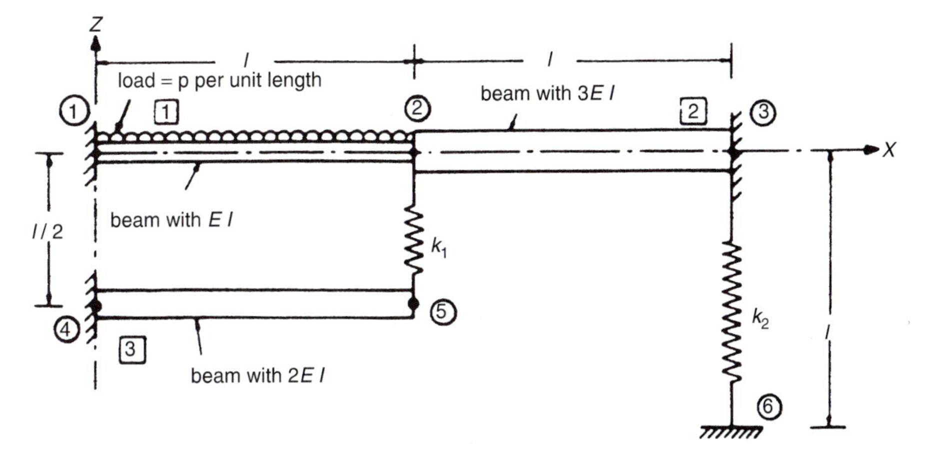

Figure 9.18.

9.3 Derive the stiffness matrix and load vector of a three-dimensional truss element whose area of cross section varies linearly along its length.

9.4 Derive the transformation relation $[K^{(e)}] = [\lambda]^T[k^{(e)}][\lambda]$ from the equivalence of potential energy in the local and global coordinate systems.

9.5 Find the nodal displacements in the tapered one-dimensional member subjected to an end load of 4000 N (shown in Figure 9.17). The cross-sectional area decreases linearly from 10 cm^2 at the left end to 5 cm^2 at the right end. Furthermore, the member experiences a temperature increase of 25 °C. Use three 25-cm elements to idealize the member. Assume $E = 2 \times 10^7$ N/cm^2, $\nu = 0.3$, and $\alpha = 6 \times 10^{-6}$ cm/cm-°C.

9.6 Derive the equilibrium equations for the beam-spring system shown in Figure 9.18. Start from the principle of minimum potential energy and indicate briefly the various steps involved in your finite element derivation.

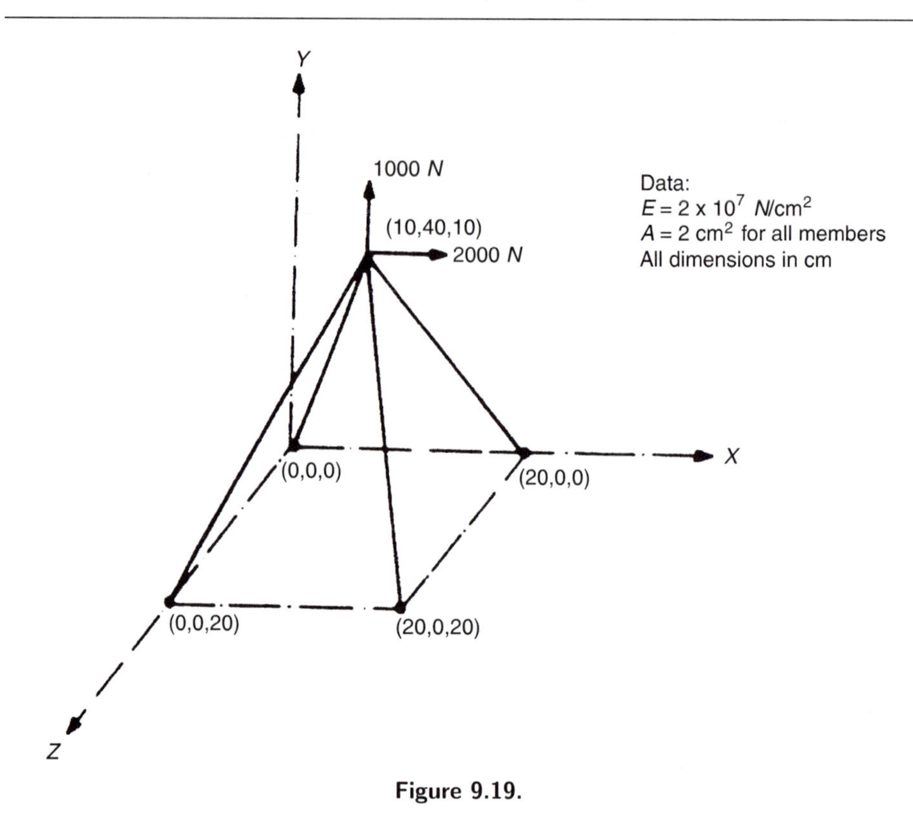

Figure 9.19.

9.7 Write a subroutine called TRUSS for the displacement and stress analysis of three-dimensional truss structures. Using this subroutine, find the stresses developed in the members of the truss shown in Figure 9.19.

9.8 Find the stresses developed in the members of the truss shown in Figure 9.2.

9.9 The stiffness matrix of a spring, of stiffness c, in the local (x) coordinate system is given by (see Figure 9.20):

$$[k] = c\begin{bmatrix} 1 & -1 \\ -1 & 1 \end{bmatrix}$$

Derive the stiffness matrix of the element in the global (XY) coordinate system.

9.10 Explain why the stiffness matrix given by Eq. (9.7) or Eq. (9.13) is symmetric.

9.11 Explain why the stiffness matrix given by Eq. (9.7) or Eq. (9.13) is singular.

9.12 Explain why the sum of elements in any row of the stiffness matrix given by Eq. (9.7) or Eq. (9.13) is zero.

9.13 The members 1 and 2 of Figure 9.21 are circular with diameters of 1 and 2 in., respectively. Determine the displacement of node P by assuming the joints to be pin connected.

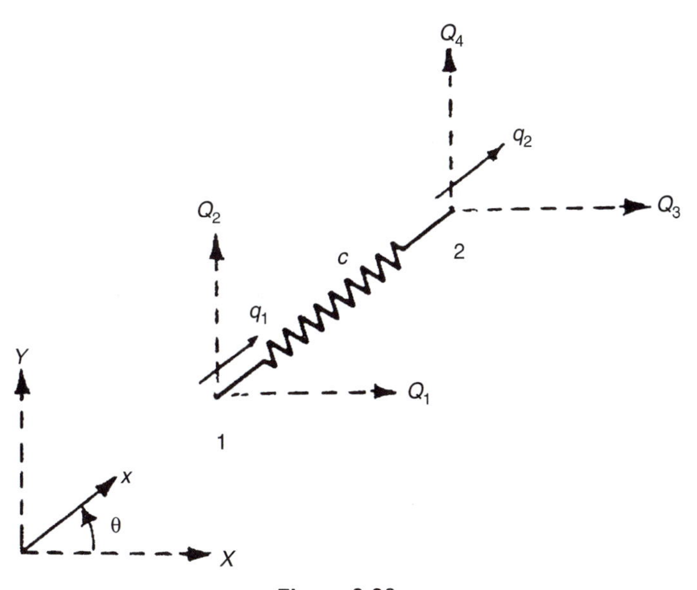

Figure 9.20.

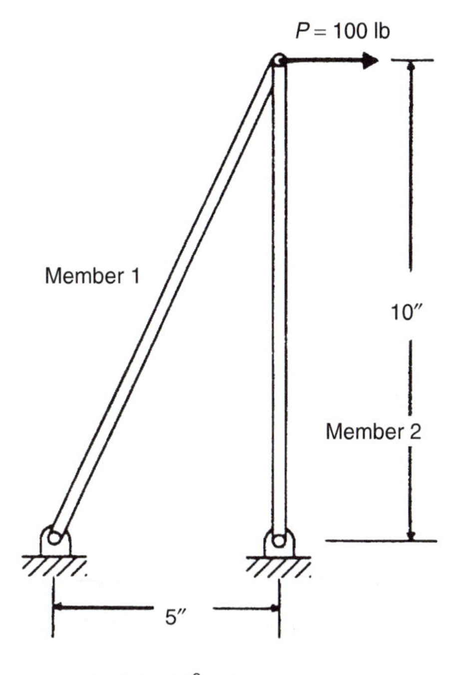

Figure 9.21.

9.14 The members 1 and 2 of Figure 9.21 are circular with diameters of 1 and 2 in., respectively. Determine the displacement of node P by assuming the joints to be welded.

9.15 The stepped bar shown in Figure 9.22 is subjected to an axial load of 100 lb at node 2. The Young's moduli of elements 1, 2, and 3 are given by 30×10^6, 20×10^6, and 10×10^6 psi, respectively. If the cross-sectional areas of elements 1, 2, and 3 are given by 3×3, 2×2, and 1×1 in., respectively, determine the following:

(a) Displacements of nodes 2 and 3.

(b) Stresses in elements 1, 2, and 3.

(c) Reactions at nodes 1 and 4.

9.16 Loads of magnitude 100 and 200 lb are applied at points C and D of a rigid bar AB that is supported by two cables as shown in Figure 9.23. If cables 1 and 2 have cross-sectional areas of 1 and 2 in.^{2} and Young's moduli of 30×10^6 and

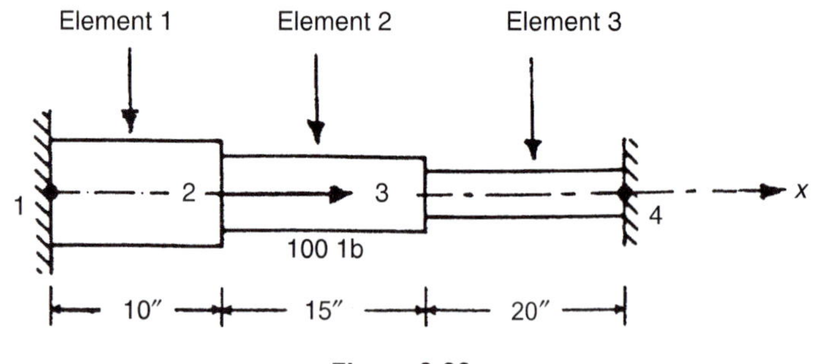

Figure 9.22.

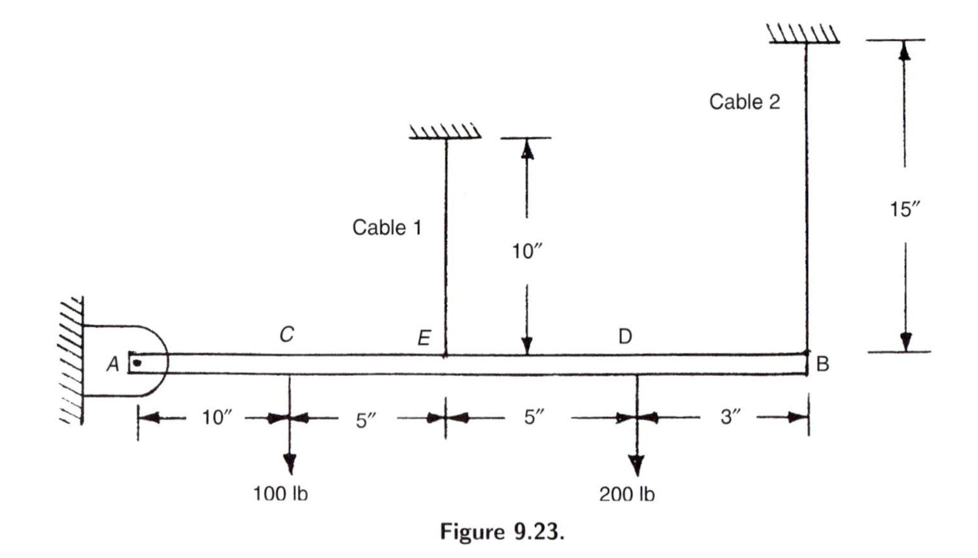

Figure 9.23.

20×10^6 psi, respectively, determine the following:

(a) The finite element equilibrium equations of the system by modeling each cable as a bar element.

(b) The boundary conditions of the system.

(c) The nodal displacements of the system.

Hint: A boundary condition involving the degrees of freedom Q_i and Q_j in the form of a linear equation:

$$a_i Q_i + a_j Q_j = a_0$$

where a_i, a_j, and a_0 are known constants (also known as multipoint boundary condition), can be incorporated as follows.

Add the quantities ca_i^2, $ca_i a_j$, $ca_i a_j$, and ca_j^2 to the elements located at (i,i), (i,j), (j,i), and (j,j), respectively, in the assembled stiffness matrix and add the quantities $ca_0 a_i$ and $ca_0 a_j$ to the elements in rows i and j of the load vector. Here, c is a large number compared to the magnitude of the elements of the stiffness matrix and the load vector.

9.17 The stepped bar shown in Figure 9.24 is heated by 100 °F. The cross-sectional areas of elements 1 and 2 are given by 2 and 1 in.^{2} and the Young's moduli by 30×10^6 and 20×10^6 psi, respectively.

(a) Derive the stiffness matrices and the load vectors of the two elements.

(b) Derive the assembled equilibrium equations of the system and find the displacement of node C.

(c) Find the stresses induced in elements 1 and 2.

Assume the value of α for elements 1 and 2 to be 15×10^{-6} and 10×10^{-6} per °F, respectively.

9.18 Consider the two-bar truss shown in Figure 9.25. The element properties are given below:

Element 1: $E_1 = 30 \times 10^6$ psi, $A_1 = 1$ in.^{2}

Element 2: $E_2 = 20 \times 10^6$ psi, $A_2 = 0.5$ in.^{2}

The loads acting at node A are given by $P_1 = 100$ and $P_2 = 200$ lb.

(a) Derive the assembled equilibrium equations of the truss.

(b) Find the displacement of node A.

(c) Find the stresses in elements 1 and 2.

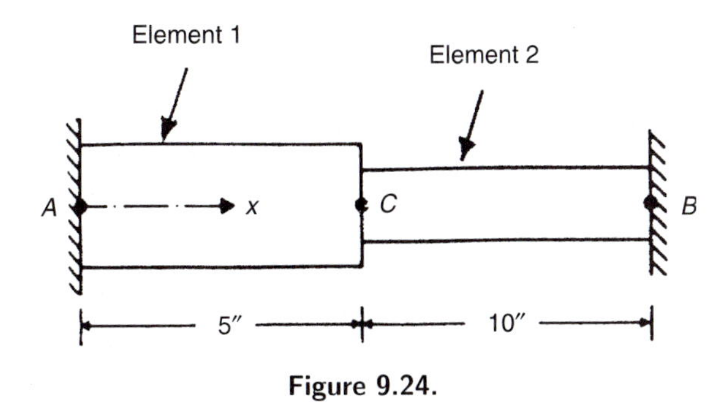

Figure 9.24.

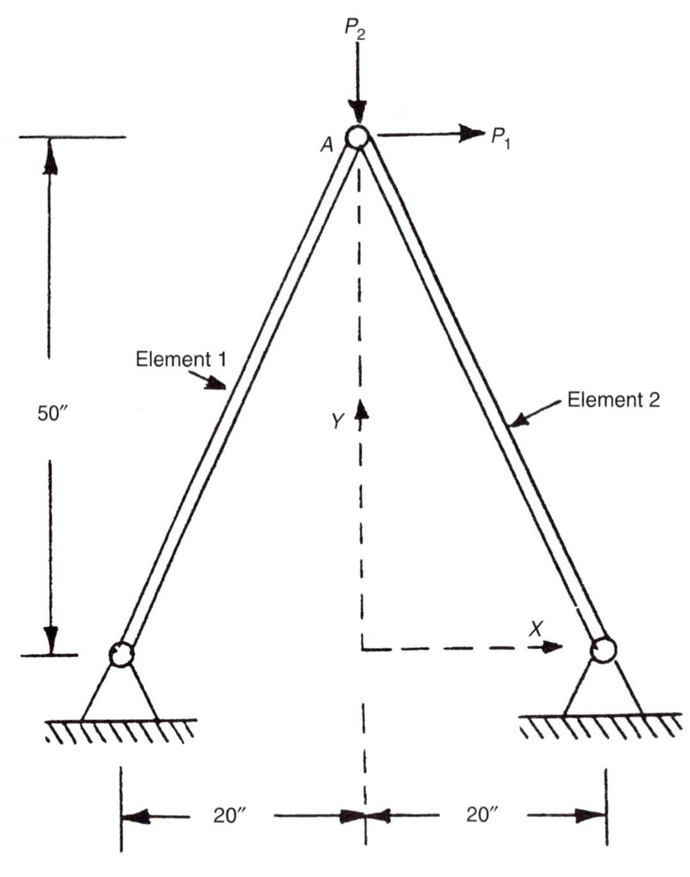

Figure 9.25.

9.19 A beam is fixed at one end, supported by a cable at the other end, and subjected to a uniformly distributed load of 50 lb/in. as shown in Figure 9.26.

(a) Derive the finite element equilibrium equations of the system by using one finite element for the beam and one finite element for the cable.

(b) Find the displacement of node 2.

(c) Find the stress distribution in the beam.

(d) Find the stress distribution in the cable.

9.20 A beam is fixed at one end and is subjected to three forces and three moments at the other end as shown in Figure 9.27. Find the stress distribution in the beam using a one-beam element idealization.

9.21 Determine the stress distribution in the two members of the frame shown in Figure 9.28. Use one finite element for each member of the frame.

9.22 Find the displacement of node 3 and the stresses in the two members of the truss shown in Figure 9.29. Assume the Young's modulus and the cross-sectional areas of the two members are the same, with $E = 30 \times 10^6$ psi and $A = 1$ in.^{2}.

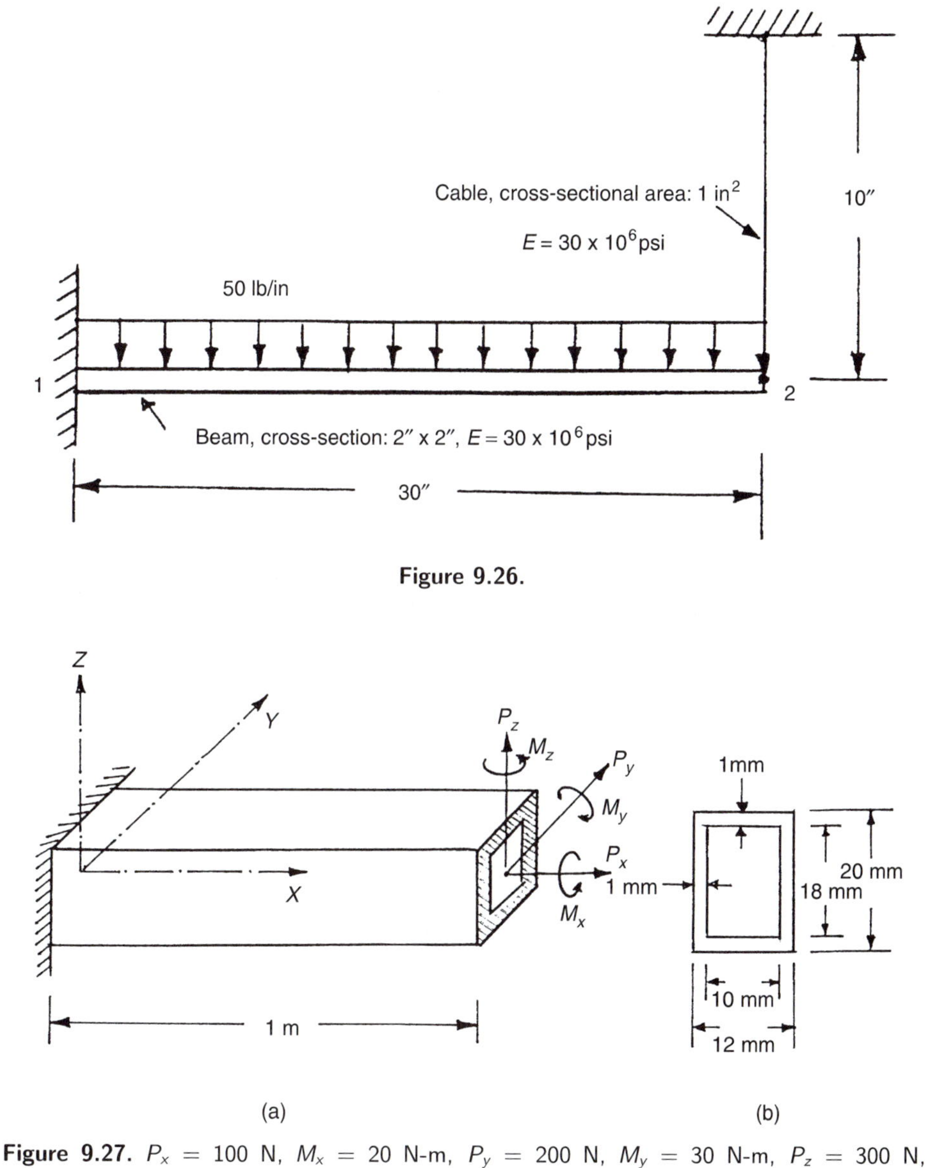

Figure 9.26.

Figure 9.27. $P_x = 100$ N, $M_x = 20$ N-m, $P_y = 200$ N, $M_y = 30$ N-m, $P_z = 300$ N, $M_z = 40$ N-m, $E = 205$ GPa.

9.23 A simple model of a radial drilling machine structure is shown in Figure 9.30. Using two beam elements for the column and one beam element for the arm, derive the stiffness matrix of the system. Assume the material of the structure is steel and the foundation is a rigid block. The cross section of the column is tubular with inside diameter 350 mm and outside diameter 400 mm. The cross section of the

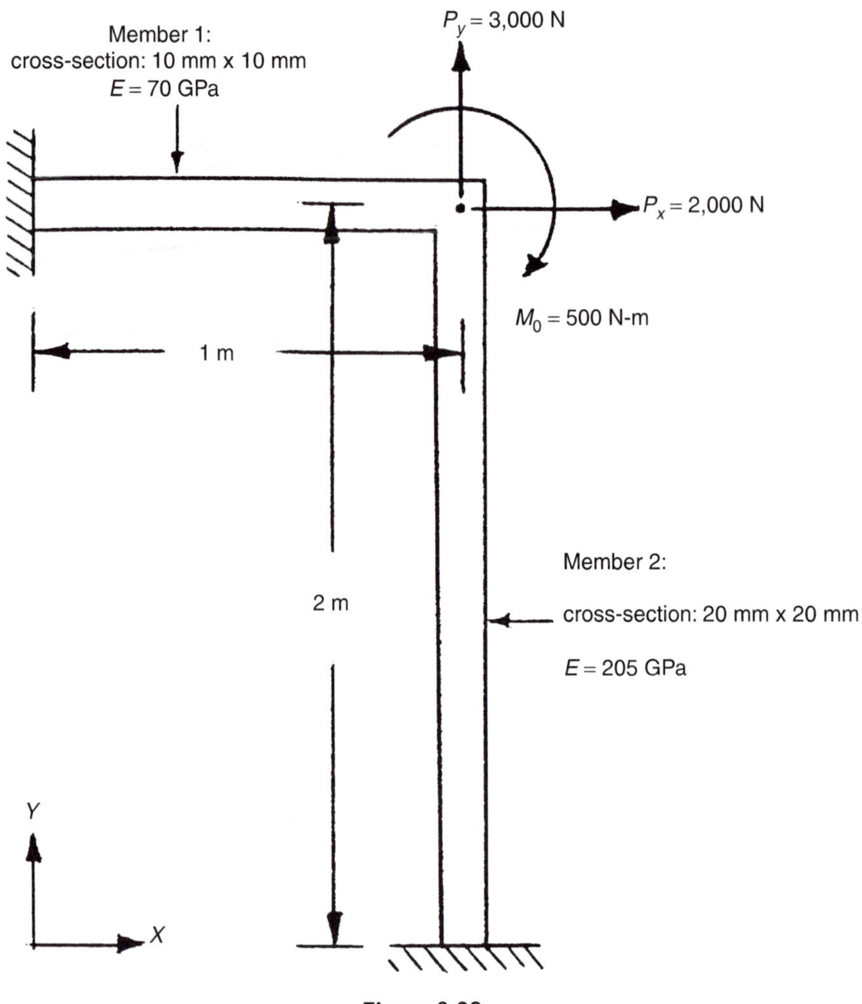

Figure 9.28.

arm is hollow rectangular with an overall depth of 400 mm and overall width of 300 mm, with a wall thickness of 10 mm.

9.24 If a vertical force of 4000 N along the z direction and a bending moment of 1000 N-m in the xz plane are developed at point A during a metal-cutting operation, find the stresses developed in the machine tool structure shown in Figure 9.30.

9.25 The crank in the slider–crank mechanism shown in Figure 9.31 rotates at a constant angular speed of 1500 rpm. Find the stresses in the connecting rod and the crank when the pressure acting on the piston is 100 psi and $\theta = 30°$. The diameter of the piston is 10 in. and the material of the mechanism is steel. Model the connecting rod and the crank by one beam element each. The lengths of the crank and the connecting rod are 10 and 45 in., respectively.

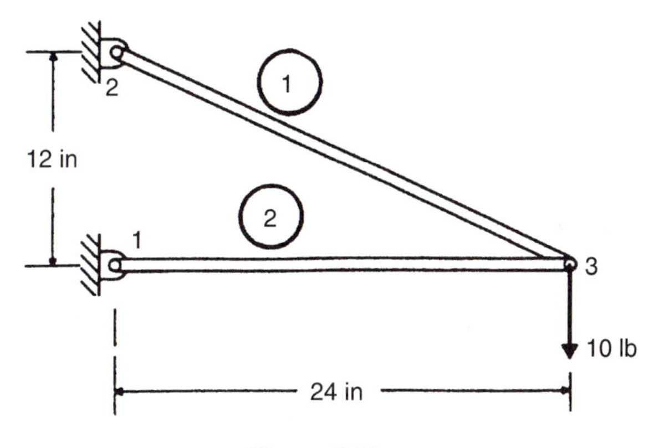

Figure 9.29.

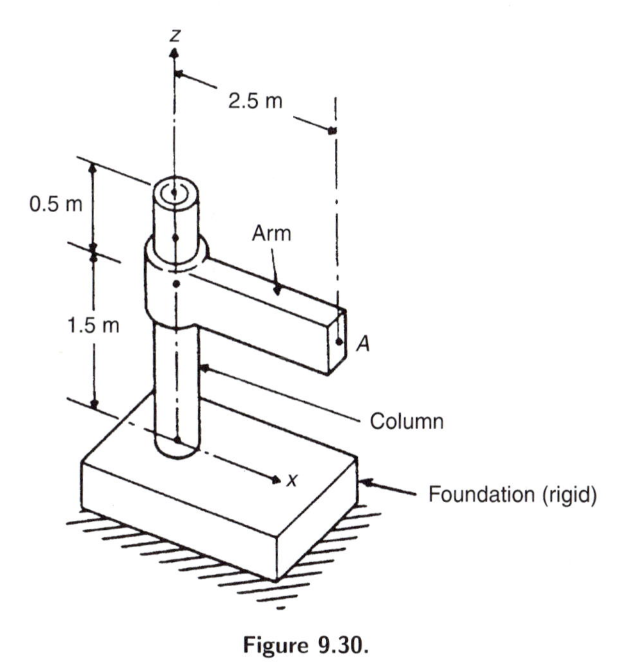

Figure 9.30.

9.26 A water tank of weight W is supported by a hollow circular steel column of inner diameter d, wall thickness t, and height h. The wind pressure acting on the column can be assumed to vary linearly from 0 to $p_{\max}$, as shown in Figure 9.32. Find the bending stress induced in the column under the loads using a one-beam element idealization with the following data:

$$W = 15{,}000 \text{ lb}, \qquad h = 30 \text{ ft}, \qquad d = 2 \text{ ft}, \qquad t = 2 \text{ in.}, \qquad p_{\max} = 200 \text{ psi}$$

9.27 Find the nodal displacements and stresses in elements 1, 2, and 3 of the system shown in Figure 9.33. Use three bar elements and one spring element for modeling.

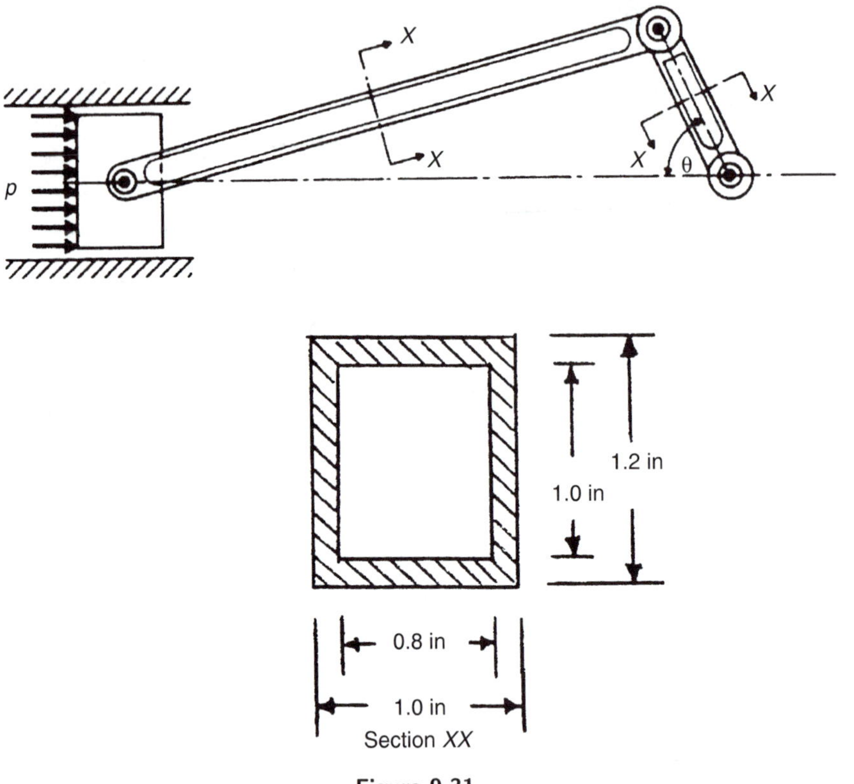

Figure 9.31.

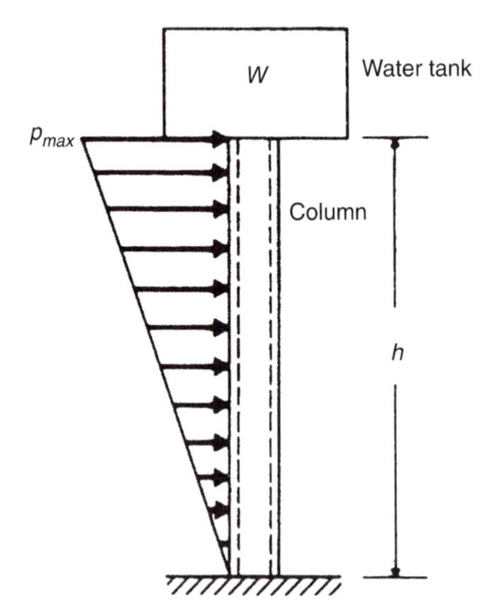

Figure 9.32.

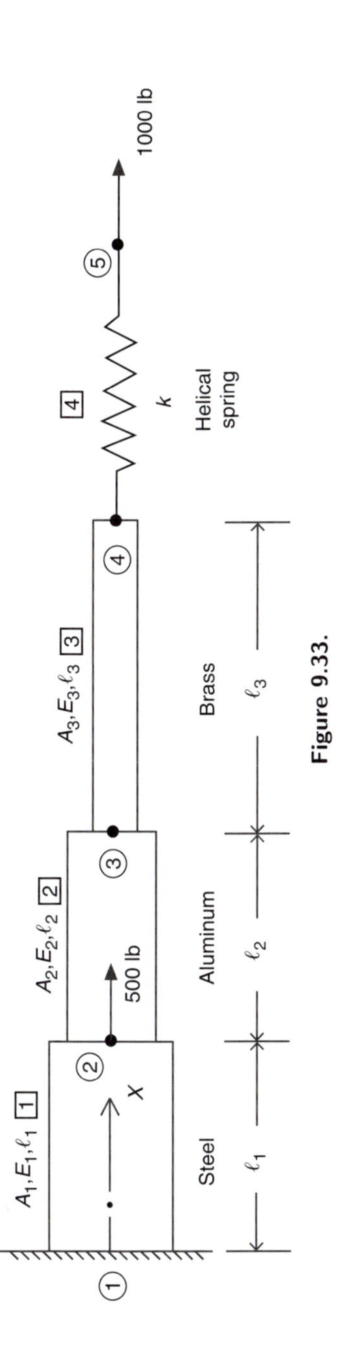

1
2
3
4
5
A_1, E_1, ℓ_1 1
A_2, E_2, ℓ_2 2
A_3, E_3, ℓ_3 3
4
x
500 lb
1000 lb
k
Helical spring
Steel
Aluminum
Brass
ℓ_1
ℓ_2
ℓ_3

Figure 9.33.

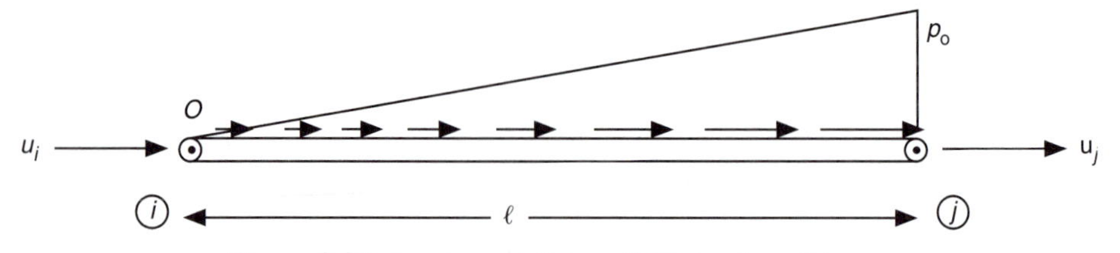

Figure 9.34. Load per Unit Length Varies from 0 to p_0.

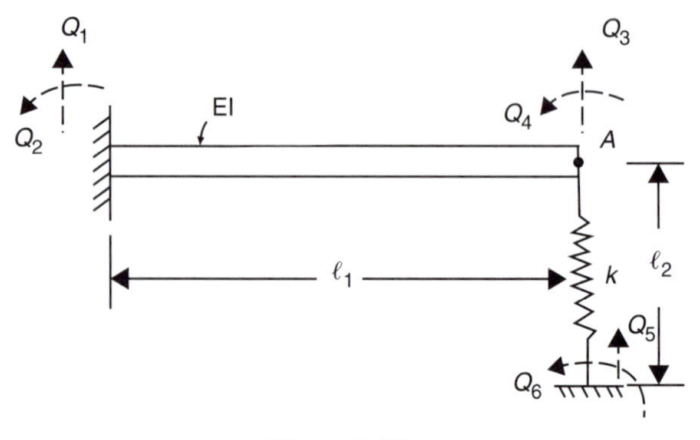

Figure 9.35.

Data: $A_1 = 3$ in.^{2}, $A_2 = 2$ in.^{2}, $A_3 = 1$ in.^{2}, $E_1 = 30 \times 10^6$ psi, $E_2 = 10 \times 10^6$ psi, $E_3 = 15 \times 10^6$ psi, $l_1 = 10$ in., $l_2 = 20$ in., $l_3 = 30$ in., $k = 10^5$ lb/in.

9.28 A truss element of length l and cross-sectional area A is subjected to a linearly varying load acting on the surface in the axial direction as shown in Figure 9.34. Derive the consistent load vector of the element using a linear interpolation model. Also indicate the lumped load vector of the element.

9.29 A beam of flexural rigidity EI, fixed at the left end, is supported on a spring of stiffness k at the right end as shown in Figure 9.35, where Q_i, $i = 1, 2, \ldots, 6$ denote the global degrees of freedom.

(a) Derive the stiffness matrix of the system before applying the boundary conditions.

(b) Find the stiffness matrix of the system after applying the boundary conditions.

(c) Find the displacement and slope of the beam at point A for the following data: $EI = 25 \times 10^6$ lb-in.^{2}, $l_1 = 20$ in., $l_2 = 10$ in., $k = 10^4$ lb/in., load at A (acting in a vertically downward direction) $= 100$ lb.

9.30 Find the stress in the bar shown in Figure 9.36 using the finite element method with one bar and one spring element.

Data: Cross-sectional area of bar $(A) = 2$ in.^{2}, Young's modulus of the bar $(E) = 30 \times 10^6$ psi, spring constant of the spring $(k) = 10^5$ lb/in.

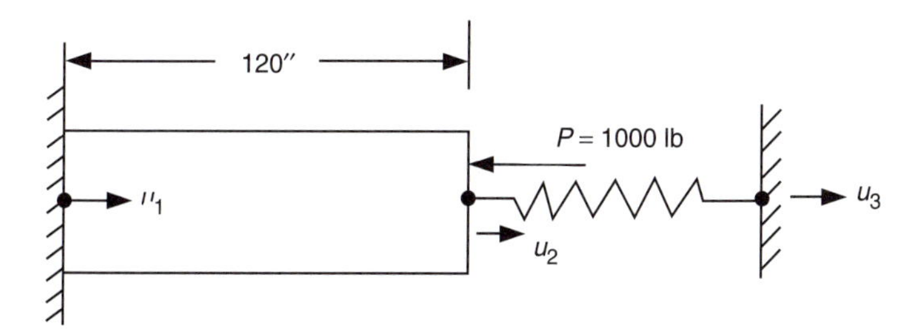

Figure 9.36.

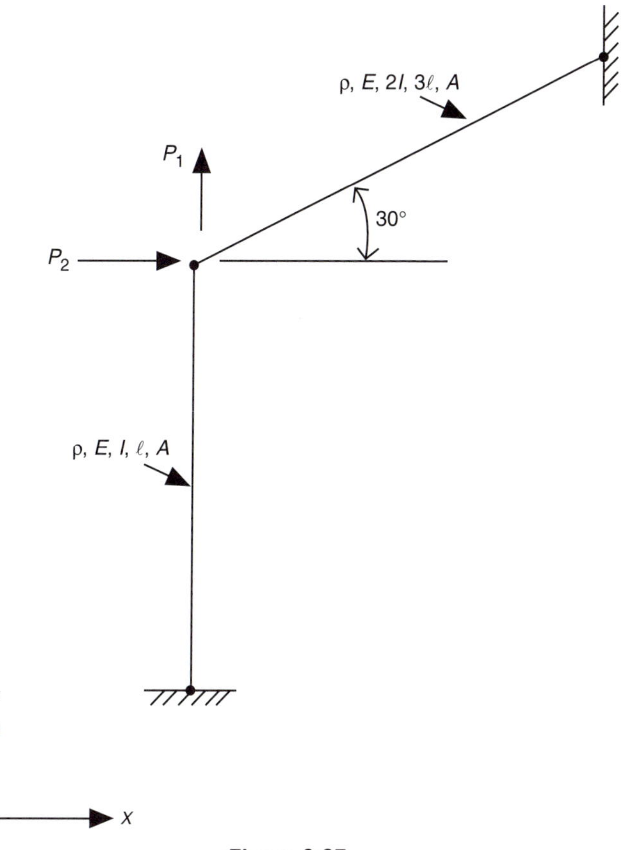

Figure 9.37.

9.31 A two-dimensional frame is shown in Figure 9.37. Using three degrees of freedom per node, derive the following:

(a) Global stiffness and mass matrices of order 9×9 before applying the boundary conditions.

(b) Global stiffness and mass matrices of order 3×3 after applying the boundary conditions.

(c) Nodal displacement vector under the given load.

(d) Natural frequencies and mode shapes of the frame.

Data: $E = 30 \times 10^6$ psi, $I = 2$ in.^{4}, $A = 1$ in.^{2}, $l = 30$ in., $\rho = 0.283$ lb/in.^{3} (weight density), $g = 384$ in./sec^2 (gravitational constant), $P_1 = 1000$ lb, $P_2 = 500$ lb.

9.32 Derive the stiffness matrix of a beam element in bending using trignometric functions (instead of a cubic equation) for the interpolation model. Discuss the convergence of the resulting element.

10

ANALYSIS OF PLATES

10.1 INTRODUCTION

When a flat plate is subjected to both inplane and transverse or normal loads as shown in Figure 10.1 any point inside the plate can have displacement components u, v, and w parallel to x, y, and z axes, respectively. In the small deflection (or linear) theory of thin plates, the transverse deflection w is uncoupled from the inplane deflections u and v. Consequently, the stiffness matrices for the inplane and transverse deflections are also uncoupled and they can be calculated independently. Thus, if a plate is subjected to inplane loads only, it will undergo deformation in its plane only. In this case, the plate is said to be under the action of "membrane" forces. Similarly, if the plate is subjected to transverse loads (and/or bending moments), any point inside the plate experiences essentially a lateral displacement w (inplane displacements u and v are also experienced because of the rotation of the plate element). In this case, the plate is said to be under the action of bending forces. The inplane and bending analysis of plates is considered in this chapter. If the plate elements are used for the analysis of three-dimensional structures, such as folded plate structures, both inplane and bending actions have to be considered in the development of element properties. This aspect of coupling the membrane and bending actions of a plate element is also considered in this chapter.

10.2 TRIANGULAR MEMBRANE ELEMENT

The triangular membrane element is considered to lie in the xy plane of a local xy coordinate system as shown in Figure 10.2. By assuming a linear displacement variation inside the element, the displacement model can be expressed as

$$\begin{aligned} u(x,y) &= \alpha_1 + \alpha_2 x + \alpha_3 y \\ v(x,y) &= \alpha_4 + \alpha_5 x + \alpha_6 y \end{aligned} \tag{10.1}$$

By considering the displacements u_i and v_i as the local degrees of freedom of node $i(i = 1, 2, 3)$, the constants $\alpha_1, \ldots, \alpha_6$ can be evaluated. Thus, by using the conditions

$$\left.\begin{aligned} u(x,y) = u_1 = q_1 \text{ and } v(x,y) = v_1 = q_2 \text{ at } (x_1, y_1) \\ u(x,y) = u_2 = q_3 \text{ and } v(x,y) = v_2 = q_4 \text{ at } (x_2, y_2) \\ u(x,y) = u_3 = q_5 \text{ and } v(x,y) = v_3 = q_6 \text{ at } (x_3, y_3) \end{aligned}\right\} \tag{10.2}$$

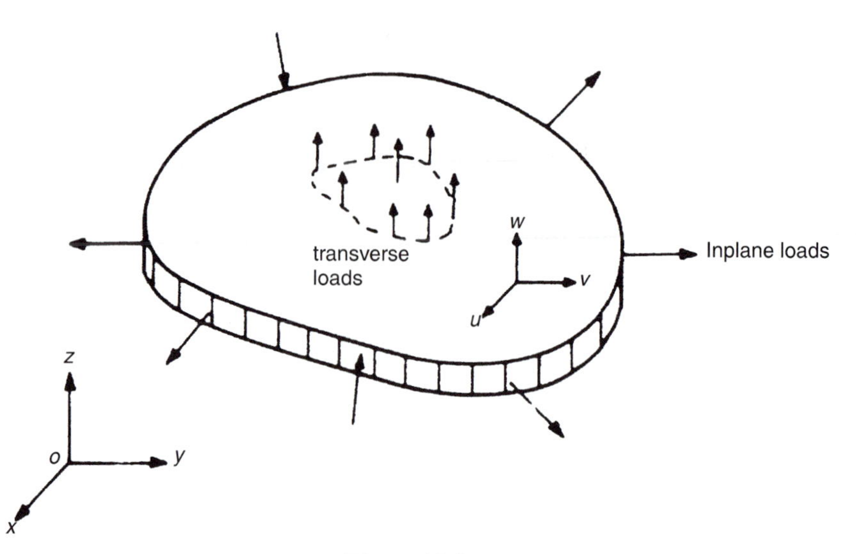

Figure 10.1.

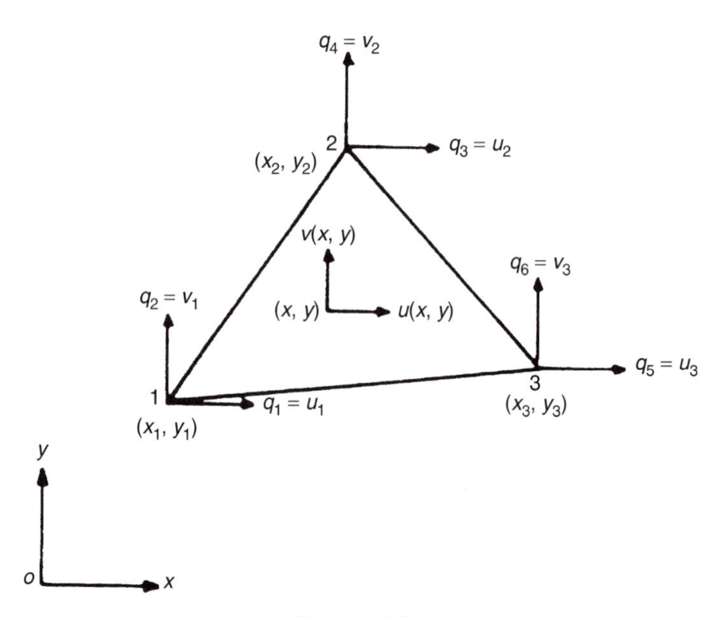

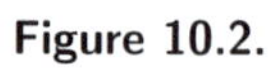

Figure 10.2.

we can express the constants $\alpha_1, \ldots, \alpha_6$ in terms of the nodal degrees of freedom as outlined in Section 3.4. This leads to the displacement model:

$$\vec{U} = \begin{Bmatrix} u(x,y) \\ v(x,y) \end{Bmatrix} = [N]\vec{q}^{\,(e)} \tag{10.3}$$

where

$$[N(x,y)] = \begin{bmatrix} N_1(x,y) & 0 & N_2(x,y) & 0 & N_3(x,y) & 0 \\ 0 & N_1(x,y) & 0 & N_2(x,y) & 0 & N_3(x,y) \end{bmatrix} \tag{10.4}$$

$$\left.\begin{aligned} N_1(x,y) &= \frac{1}{2A}[y_{32}(x-x_2) - x_{32}(y-y_2)] \\ N_2(x,y) &= \frac{1}{2A}[-y_{31}(x-x_3) + x_{31}(y-y_3)] \\ N_3(x,y) &= \frac{1}{2A}[y_{21}(x-x_1) - x_{21}(y-y_1)] \end{aligned}\right\} \tag{10.5}$$

$$A = \frac{1}{2}(x_{32}y_{21} - x_{21}y_{32}) = \text{area of the triangle 1 2 3} \tag{10.6}$$

$$\left.\begin{aligned} x_{ij} &= x_i - x_j \\ y_{ij} &= y_i - y_j \end{aligned}\right\} \tag{10.7}$$

$$\vec{U} = \begin{Bmatrix} u(x,y) \\ v(x,y) \end{Bmatrix} \tag{10.8}$$

$$\vec{q}^{\,(e)} = \begin{Bmatrix} q_1 \\ q_2 \\ q_3 \\ q_4 \\ q_5 \\ q_6 \end{Bmatrix}^{(e)} = \begin{Bmatrix} u_1 \\ v_1 \\ u_2 \\ v_2 \\ u_3 \\ v_3 \end{Bmatrix}^{(e)} \tag{10.9}$$

By using the relations

$$\vec{\varepsilon} = \begin{Bmatrix} \varepsilon_{xx} \\ \varepsilon_{yy} \\ \varepsilon_{xy} \end{Bmatrix} = \begin{Bmatrix} \partial u/\partial x \\ \partial v/\partial y \\ \dfrac{\partial u}{\partial y} + \dfrac{\partial v}{\partial x} \end{Bmatrix} \tag{10.10}$$

and Eq. (10.3), the components of strain can be expressed in terms of nodal displacements as

$$\vec{\varepsilon} = [B]\vec{q}^{\,(e)} \tag{10.11}$$

where

$$[B] = \frac{1}{2A}\begin{bmatrix} y_{32} & 0 & -y_{31} & 0 & y_{21} & 0 \\ 0 & -x_{32} & 0 & x_{31} & 0 & -x_{21} \\ -x_{32} & y_{32} & x_{31} & -y_{31} & -x_{21} & y_{21} \end{bmatrix} \tag{10.12}$$

If the element is in a state of plane stress, the stress–strain relations are given by (Eq. 8.35)

$$\vec{\sigma} = [D]\vec{\varepsilon} \tag{10.13}$$

where

$$\vec{\sigma} = \begin{Bmatrix} \sigma_{xx} \\ \sigma_{yy} \\ \sigma_{xy} \end{Bmatrix} \tag{10.14}$$

and

$$[D] = \frac{E}{1-\nu^2} \begin{bmatrix} 1 & \nu & 0 \\ \nu & 1 & 0 \\ 0 & 0 & \dfrac{1-\nu}{2} \end{bmatrix} \tag{10.15}$$

The stiffness matrix of the element $[k^{(e)}]$ can be found by using Eq. (8.87):

$$[k^{(e)}] = \iiint\limits_{V^{(e)}} [B]^T [D][B] \, \mathrm{d}V \tag{10.16}$$

where $V^{(e)}$ denotes the volume of the element. If the plate thickness is taken as a constant (t), the evaluation of the integral in Eq. (10.16) presents no difficulty since the elements of the matrices $[B]$ and $[D]$ are all constants (not functions of x and y). Hence, Eq. (10.16) can be rewritten as

$$[k^{(e)}] = [B]^T [D][B] t \iint\limits_{A} \mathrm{d}A = tA[B]^T [D][B] \tag{10.17}$$

Although the matrix products involved in Eq. (10.17) can be performed conveniently on a computer, the explicit form of the stiffness matrix is given below for convenience:

$$[k^{(e)}] = [k_n^{(e)}] + [k_s^{(e)}] \tag{10.18}$$

where the matrix $[k^{(e)}]$ is separated into two parts; one due to normal stresses, $[k_n^{(e)}]$, and the other due to shear stresses, $[k_s^{(e)}]$. The components of the matrices $[k_n^{(e)}]$ and $[k_s^{(e)}]$ are given by

$$[k_n^{(e)}] = \frac{Et}{4A(1-\nu^2)} \begin{bmatrix} y_{32}^2 & & & \text{Symmetric} & & \\ -\nu y_{32} x_{32} & x_{32}^2 & & & & \\ -y_{32} y_{31} & \nu x_{32} y_{31} & y_{31}^2 & & & \\ \nu y_{32} x_{31} & -x_{32} x_{31} & -\nu y_{31} x_{31} & x_{31}^2 & & \\ y_{32} y_{21} & -\nu x_{32} y_{21} & -y_{31} y_{21} & \nu x_{31} y_{21} & y_{21}^2 & \\ -\nu y_{32} x_{21} & x_{32} x_{21} & \nu y_{31} x_{21} & -x_{31} x_{21} & -\nu y_{21} x_{21} & x_{21}^2 \end{bmatrix} \tag{10.19}$$

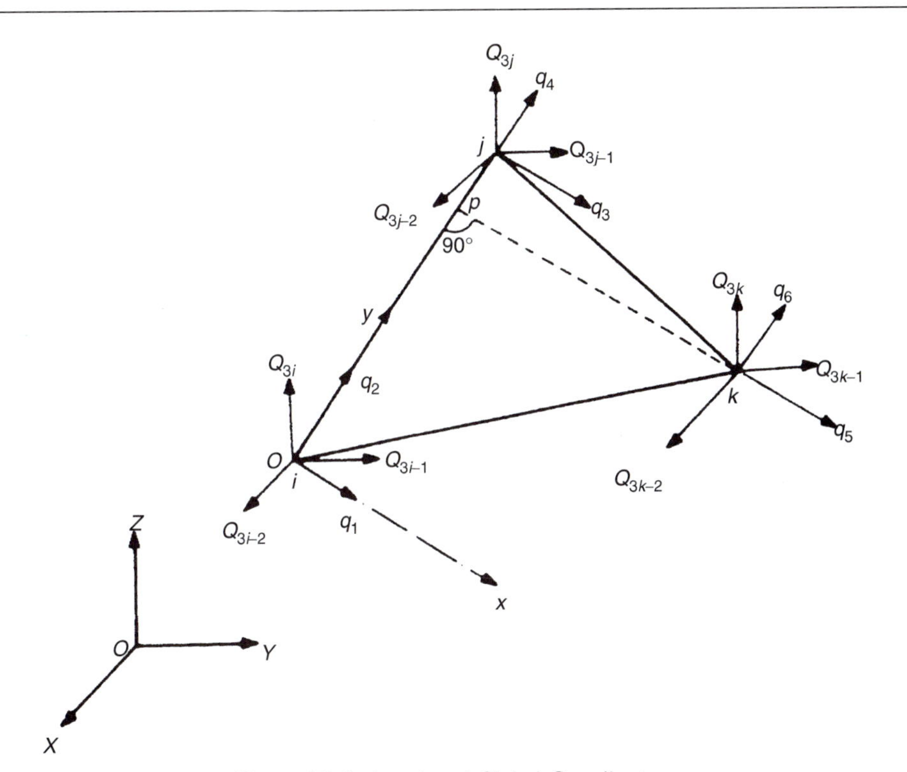

Figure 10.3. Local and Global Coordinates.

and

$$[k_s^{(e)}] = \frac{Et}{8A(1+\nu)} \begin{bmatrix} x_{32}^2 & & & & & \\ -x_{32}y_{32} & y_{32}^2 & & \text{Symmetric} & & \\ -x_{32}x_{31} & y_{32}x_{31} & x_{31}^2 & & & \\ x_{32}y_{31} & -y_{32}y_{31} & -x_{31}y_{31} & y_{31}^2 & & \\ x_{32}x_{21} & -y_{32}x_{21} & -x_{31}x_{21} & y_{31}x_{21} & x_{21}^2 & \\ -x_{32}y_{21} & y_{32}y_{21} & x_{31}y_{21} & -y_{31}y_{21} & -x_{21}y_{21} & y_{21}^2 \end{bmatrix} \tag{10.20}$$

Transformation Matrix

In actual computations, it will be convenient, from the standpoint of calculating the transformation matrix $[\lambda]$, to select the local xy coordinate system as follows. Assuming that the triangular element under consideration is an interior element of a large structure, let the node numbers 1, 2, and 3 of the element correspond to the node numbers i, j, and k, respectively, of the global system. Then place the origin of the local xy system at node 1 (node i), and take the y axis along the edge 1 2 (edge ij) and the x axis perpendicular to the y axis directed toward node 3 (node k) as shown in Figure 10.3.

To generate the transformation matrix $[\lambda]$, the direction cosines of lines ox and oy with respect to the global X, Y, and Z axes are required. Since the direction cosines of the line oy are the same as those of line ij, we obtain

$$l_{ij} = \frac{X_j - X_i}{d_{ij}}, \qquad m_{ij} = \frac{Y_j - Y_i}{d_{ij}}, \qquad n_{ij} = \frac{Z_j - Z_i}{d_{ij}} \tag{10.21}$$

where the distance between the points i and j (d_{ij}) is given by

$$d_{ij} = [(X_j - X_i)^2 + (Y_j - Y_i)^2 + (Z_j - Z_i)^2]^{1/2} \tag{10.22}$$

and (X_i, Y_i, Z_i) and (X_j, Y_j, Z_j) denote the (X, Y, Z) coordinates of points i and j, respectively. Since the direction cosines of the line ox cannot be computed unless we know the coordinates of a second point on the line ox (in addition to those of point i), we draw a perpendicular line kp from node k onto the line ij as shown in Figure 10.3. Then the direction cosines of the line ox will be the same as those of the line pk:

$$l_{pk} = \frac{X_k - X_p}{d_{pk}}, \qquad m_{pk} = \frac{Y_k - Y_p}{d_{pk}}, \qquad n_{pk} = \frac{Z_k - Z_p}{d_{pk}} \tag{10.23}$$

where d_{pk} is the distance between the points p and k. The coordinates (X_p, Y_p, Z_p) of the point p in the global coordinate system can be computed as

$$\begin{aligned} X_p &= X_i + l_{ij}\, d_{ip} \\ Y_p &= Y_i + m_{ij}\, d_{ip} \\ Z_p &= Z_i + n_{ij}\, d_{ip} \end{aligned} \tag{10.24}$$

where d_{ip} is the distance between the points i and p. To find the distance d_{ip}, we use the condition that the lines ij and pk are perpendicular to each other:

$$l_{ij} l_{pk} + m_{ij} m_{pk} + n_{ij} n_{pk} = 0 \tag{10.25}$$

Using Eqs. (10.23) and (10.24), Eq. (10.25) can be rewritten as

$$\frac{1}{d_{pk}}[l_{ij}(X_k - X_i - l_{ij}\, d_{ip}) + m_{ij}(Y_k - Y_i - m_{ij}\, d_{ip}) + n_{ij}(Z_k - Z_i - n_{ij}\, d_{ip})] = 0 \tag{10.26}$$

Equation (10.26) can be solved for d_{ip} as

$$d_{ip} = l_{ij}(X_k - X_i) + m_{ij}(Y_k - Y_i) + n_{ij}(Z_k - Z_i) \tag{10.27}$$

where the condition $l_{ij}^2 + m_{ij}^2 + n_{ij}^2 = 1$ has been used. Finally, the distance d_{pk} can be found by considering the right-angle triangle ikp as

$$d_{pk} = (d_{ik}^2 - d_{ip}^2)^{1/2} = [(X_k - X_i)^2 + (Y_k - Y_i)^2 + (Z_k - Z_i)^2 - d_{ip}^2]^{1/2} \tag{10.28}$$

The transformation matrix $[\lambda]$ can now be constructed by using the direction cosines of lines ij and pk as

$$[\lambda] = \begin{bmatrix} \vec{\lambda}_{pk} & \vec{0} & \vec{0} \\ \vec{\lambda}_{ij} & \vec{0} & \vec{0} \\ \vec{0} & \vec{\lambda}_{pk} & \vec{0} \\ \vec{0} & \vec{\lambda}_{ij} & \vec{0} \\ \vec{0} & \vec{0} & \vec{\lambda}_{pk} \\ \vec{0} & \vec{0} & \vec{\lambda}_{ij} \end{bmatrix} \tag{10.29}$$

where

$$\underset{1\times 3}{\vec{\lambda}_{pk}} = \begin{pmatrix} l_{pk} & m_{pk} & n_{pk} \end{pmatrix} \tag{10.30}$$

$$\underset{1\times 3}{\vec{\lambda}_{ij}} = \begin{pmatrix} l_{ij} & m_{ij} & n_{ij} \end{pmatrix} \tag{10.31}$$

$$\underset{1\times 3}{\vec{0}} = \begin{pmatrix} 0 & 0 & 0 \end{pmatrix} \tag{10.32}$$

Finally, the stiffness matrix of the element in the global XYZ coordinate system can be computed as

$$[K^{(e)}] = [\lambda]^T [k^{(e)}][\lambda] \tag{10.33}$$

Consistent Load Vector

The consistent load vectors can be evaluated using Eqs. (8.88)–(8.90):

$$\begin{aligned} p_i^{(e)} &= \text{load vector due to initial strains} \\ &= \iiint\limits_{V^{(e)}} [B]^T [D] \vec{\varepsilon}_0 \, \mathrm{d}V \end{aligned} \tag{10.34}$$

In the case of thermal loading, Eq. (10.34) becomes

$$\vec{p}_i^{\,(e)} = [B]^T [D] \vec{\varepsilon}_0 t A = \frac{E\alpha t T}{2(1-\nu)} \begin{Bmatrix} y_{32} \\ -x_{32} \\ -y_{31} \\ x_{31} \\ y_{21} \\ -x_{21} \end{Bmatrix} \tag{10.35}$$

$$\begin{aligned} \vec{p}_b^{\,(e)} &= \text{load vector due to constant body forces } \phi_{xo} \text{ and } \phi_{yo} \\ &= \iiint\limits_{V^{(e)}} [N]^T \vec{\phi}_0 \, \mathrm{d}V \end{aligned} \tag{10.36}$$

By using Eq. (10.4), Eq. (10.36) can be rewritten as

$$\vec{p}_b^{(e)} = \iiint_{V^{(e)}} \begin{Bmatrix} N_1\phi_{xo} \\ N_1\phi_{yo} \\ N_2\phi_{xo} \\ N_2\phi_{yo} \\ N_3\phi_{xo} \\ N_3\phi_{yo} \end{Bmatrix} \mathrm{d}V \tag{10.37}$$

Substituting the expressions for N_1, N_2, and N_3 from Eq. (10.5) into Eq. (10.37) and carrying out the integration yields

$$\vec{p}_b^{(e)} = \frac{At}{3} \begin{Bmatrix} \phi_{xo} \\ \phi_{yo} \\ \phi_{xo} \\ \phi_{yo} \\ \phi_{xo} \\ \phi_{yo} \end{Bmatrix} \tag{10.38}$$

The following relations have been used in deriving Eq. (10.38):

$$\iint_A x \cdot dA = x_c A \quad \text{and} \quad \iint_A y \cdot \mathrm{d}A = y_c A \tag{10.39}$$

where x_c and y_c are the coordinates of the centroid of the triangle 1 2 3 given by

$$x_c = (x_1 + x_2 + x_3)/3 \quad \text{and} \quad y_c = (y_1 + y_2 + y_3)/3 \tag{10.40}$$

The load vector due to the surface stresses $\vec{\vec{\phi}} = \begin{Bmatrix} p_{xo} \\ p_{yo} \end{Bmatrix}$, where p_{xo} and p_{yo} are constants, can be evaluated as

$$\vec{p}_s^{(e)} = \iint_{S_1^{(e)}} [N]^T \begin{Bmatrix} p_{xo} \\ p_{yo} \end{Bmatrix} \mathrm{d}S_1 \tag{10.41}$$

There are three different vectors $\vec{p}_s^{(e)}$ corresponding to the three sides of the element. Let the side between the nodes 1 and 2 be subjected to surface stresses of magnitude p_{xo} and p_{yo}. Then

$$\vec{p}_s^{(e)} = \iint_{S_1^{(e)}} \begin{bmatrix} N_1 & 0 \\ 0 & N_1 \\ N_2 & 0 \\ 0 & N_2 \\ N_3 & 0 \\ 0 & N_3 \end{bmatrix} \begin{Bmatrix} p_{xo} \\ p_{yo} \end{Bmatrix} \mathrm{d}S_1 = \frac{S_{12}}{2} \begin{Bmatrix} p_{xo} \\ p_{yo} \\ p_{xo} \\ p_{yo} \\ 0 \\ 0 \end{Bmatrix} \tag{10.42}$$

where S_{12} is the surface area between nodes 1 and 2 given by

$$S_{12} = t \cdot \mathrm{d}_{12} \tag{10.43}$$

with d_{12} denoting the length of side 12. Since the stress components p_{xo} and p_{yo} are parallel to the x and y coordinate directions, Eq. (10.42) shows that the total force in either coordinate direction is $(p_{xo} \cdot S_{12})$ and $(p_{yo} \cdot S_{12})$, respectively. Thus, one-half of the total force in each direction is allotted to each node on the side under consideration. The total load vector in the local coordinate system is thus given by

$$\vec{p}^{\,(e)} = \vec{p}_i^{\,(e)} + \vec{p}_b^{\,(e)} + \vec{p}_s^{\,(e)} \tag{10.44}$$

This load vector, when referred to the global system, becomes

$$\vec{P}^{(e)} = [\lambda]^T \vec{p}^{\,(e)} \tag{10.45}$$

Characteristics of the Element

1. The displacement model chosen (Eq. 10.1) guarantees continuity of displacements with adjacent elements because the displacements vary linearly along any side of the triangle (due to linear model).
2. From Eqs. (10.11) and (10.12), we find that the $[B]$ matrix is independent of the position within the element and hence the strains are constant throughout it. This is the reason why this element is often referred to as CST element (constant strain triangular element). Obviously, the criterion of constant strain mentioned in the convergence requirements in Section 3.6 is satisfied by the displacement model.
3. From Eq. (10.13), we can notice that the stresses are also constant inside an element. Since the stresses are independent of x and y, the equilibrium equations (Eqs. 8.1) are identically satisfied inside the element since there are no body forces.
4. If the complete plate structure being analyzed lies in a single (e.g., XY) plane as in the case of Figure 10.4, the vector $\vec{Q}^{(e)}$ will also contain six components. In such a case, the matrices $[\lambda]$ and $[K^{(e)}]$ will each be of order 6×6.

10.3 NUMERICAL RESULTS WITH MEMBRANE ELEMENT

The following examples are considered to illustrate the application of the membrane element in the solution of some problems of linear elasticity.

10.3.1 A Plate under Tension

The uniform plate under tension, shown in Figure 10.4(a), is analyzed by using the CST elements. Due to symmetry of geometry and loading, only a quadrant is considered for analysis. The finite element modeling is done with eight triangular elements as shown in Figure 10.4(b). The total number of nodes is nine and the displacement unknowns are 18. However, the x components of displacement of nodes 3, 4, and 5 (namely Q_5, Q_7, and Q_9) and the y components of displacement of nodes 5, 6, and 7 (namely Q_{10}, Q_{12}, and Q_{14})

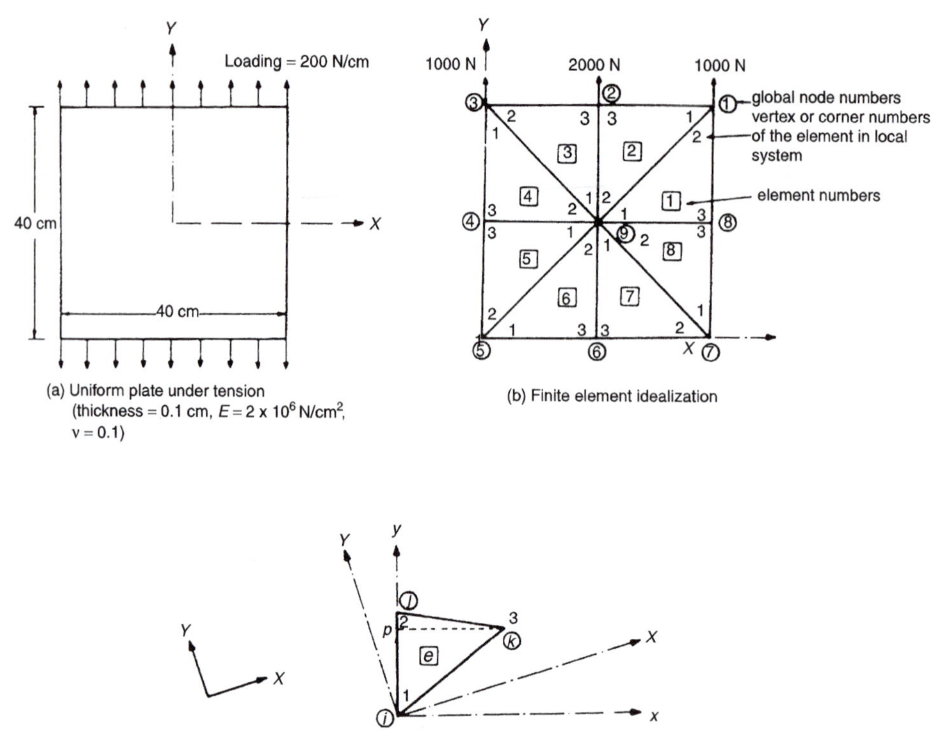

Figure 10.4. A Uniform Plate under Tensile Load.

are set equal to zero for maintaining symmetry conditions. After solving the equilibrium equations, the global displacement components can be obtained as

$$Q_i = \begin{cases} 0.020, & i = 2, 4, 6 \\ 0.010, & i = 8, 16, 18 \\ -0.002, & i = 1, 13, 15 \\ -0.001, & i = 3, 11, 17 \\ 0.000, & i = 5, 7, 9, 10, 12, 14 \end{cases}$$

Computation of Stresses

For finding the stresses inside any element "e," shown in Figure 10.4(c), the following procedure can be adopted:

Step 1: Convert the global displacements of the nodes of element e into local displacements as

$$\underset{6\times 1}{\vec{q}^{(e)}} = \underset{6\times 6}{[\lambda]}\ \underset{6\times 1}{\vec{Q}^{(e)}}$$

where

$$\vec{q}^{\,(e)} = \begin{Bmatrix} u_1 \\ v_1 \\ u_2 \\ v_2 \\ u_3 \\ v_3 \end{Bmatrix}, \qquad \vec{Q}^{(e)} = \begin{Bmatrix} Q_{2i-1} \\ Q_{2i} \\ Q_{2j-1} \\ Q_{2j} \\ Q_{2k-1} \\ Q_{2k} \end{Bmatrix}$$

and $[\lambda]$ is the transformation matrix of the element given by [two-dimensional specialization of Eq. (10.29)]

$$[\lambda] = \begin{bmatrix} l_{pk} & m_{pk} & 0 & 0 & 0 & 0 \\ l_{ij} & m_{ij} & 0 & 0 & 0 & 0 \\ 0 & 0 & l_{pk} & m_{pk} & 0 & 0 \\ 0 & 0 & l_{ij} & m_{ij} & 0 & 0 \\ 0 & 0 & 0 & 0 & l_{pk} & m_{pk} \\ 0 & 0 & 0 & 0 & l_{ij} & m_{ij} \end{bmatrix} \tag{10.46}$$

Here, (l_{pk}, m_{pk}) and (l_{ij}, m_{ij}) denote the direction cosines of lines pk (x axis) and ij (y axis) with respect to the global (X, Y) system.

Step 2: Using the local displacement vector $\vec{q}^{\,(e)}$ of element e, find the stresses inside the element in the local system by using Eqs. (10.13) and (10.11) as

$$\vec{\sigma} = \begin{Bmatrix} \sigma_{xx} \\ \sigma_{yy} \\ \sigma_{xy} \end{Bmatrix} = [D][B]\vec{q}^{\,(e)} \tag{10.47}$$

where $[D]$ and $[B]$ are given by Eqs. (10.15) and (10.12), respectively.

Step 3: Convert the local stresses σ_{xx}, σ_{yy}, and σ_{xy} of the element into global stresses σ_{XX}, σ_{YY}, and σ_{XY} by using the stress transformation relations [10.1]:

$$\begin{aligned} \sigma_{XX} &= \sigma_{xx}\, l_{pk}^2 + \sigma_{yy}\, l_{ij}^2 + 2\sigma_{xy}\, l_{pk}\, l_{ij} \\ \sigma_{YY} &= \sigma_{xx}\, m_{pk}^2 + \sigma_{yy}\, m_{ij}^2 + 2\sigma_{xy}\, m_{pk}\, m_{ij} \\ \sigma_{XY} &= \sigma_{xx}\, l_{pk}\, m_{pk} + \sigma_{yy}\, l_{ij}\, m_{ij} + \sigma_{xy}\, (l_{pk}\, m_{ij} + m_{pk}\, l_{ij}) \end{aligned}$$

The results of computation are shown in Table 10.1. It can be noticed that the stresses in the global system exactly match the correct solution given by

$$\sigma_{YY} = \frac{\text{total tensile load}}{\text{area of cross section}} = \frac{(200 \times 40)}{(40 \times 0.1)} = 2000\ \text{N/cm}^2$$

Table 10.1. Computation of Stresses inside the Elements

Element e	Displacements (cm): In global system $\vec{Q}^{(e)}$	Displacements (cm): In local system $\vec{q}^{\,(e)}$	Stress vector in local system $\begin{Bmatrix} \sigma_{xx} \\ \sigma_{yy} \\ \sigma_{xy} \end{Bmatrix}$ N/cm^2	Stress vector in global system $\begin{Bmatrix} \sigma_{XX} \\ \sigma_{YY} \\ \sigma_{XY} \end{Bmatrix}$ N/cm^2
1	−0.001	0.01556	1000	0
	0.010	−0.01273	1000	2000
	−0.002	0.00778	−1000	0
	0.020	−0.00636		
	−0.002	0.01485		
	0.010	−0.01344		
2	−0.002	0.00636	1000	0
	0.020	0.00778	1000	2000
	−0.001	0.01414	1000	0
	0.010	0.01414		
	−0.001	0.01344		
	0.020	0.01485		
3	−0.001	−0.01414	1000	0
	0.010	−0.01414	1000	2000
	0.000	−0.00636	1000	0
	0.020	−0.00778		
	−0.001	−0.00707		
	0.020	−0.00707		
4	0.000	0.00778	1000	0
	0.020	−0.00636	1000	2000
	−0.001	0.0	−1000	0
	0.010	0.0		
	0.000	0.00707		
	0.010	−0.00707		
5	−0.001	0.0	1000	0
	0.010	0.0	1000	2000
	0.000	−0.00778	−1000	0
	0.000	0.00636		
	0.000	−0.00071		
	0.010	−0.00071		
6	0.000	−0.00636	1000	0
	0.000	−0.00778	1000	2000
	−0.001	0.00141	1000	0
	0.010	−0.00141		
	−0.001	0.00071		
	0.000	−0.00071		
7	−0.001	−0.00141	1000	0
	0.010	0.00141	1000	2000
	−0.002	0.00636	1000	0
	0.000	0.00778		
	−0.001	0.00566		
	0.000	0.00849		
8	−0.002	−0.00778	1000	0
	0.000	0.00636	1000	2000
	−0.001	−0.01556	−1000	0
	0.010	0.01273		
	−0.002	−0.00849		
	0.010	0.00566		

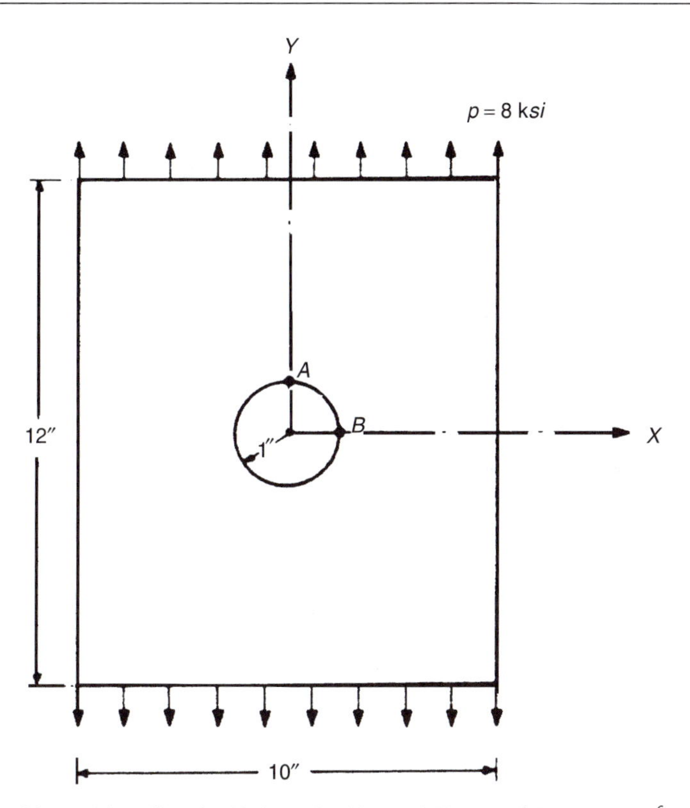

Figure 10.5. Plate with a Circular Hole under Uniaxial Tension ($E = 30 \times 10^6$ psi, $v = 0.25$, t = Plate Thickness = 1″).

10.3.2 A Plate with a Circular Hole [10.2]

The performance of the membrane elements for problems of stress concentration due to geometry is studied by considering a tension plate with a circular hole (Figure 10.5). Due to the symmetry of geometry and loading, only a quadrant was analyzed using four different finite element idealizations as shown in Figure 10.6. The results are shown in Table 10.2. The results indicate that the stress concentration is predicted to be smaller than the exact value consistently.

10.3.3 A Cantilevered Box Beam

The cantilevered box beam shown in Figure 10.7 is analyzed by using CST elements. The finite element idealization consists of 24 nodes, 72 degrees of freedom (in global XYZ system), and 40 elements as shown in Figure 10.8. The displacement results obtained for two different load conditions are compared with those given by simple beam theory in Table 10.3. It can be seen that the finite element results compare well with those of simple beam theory.

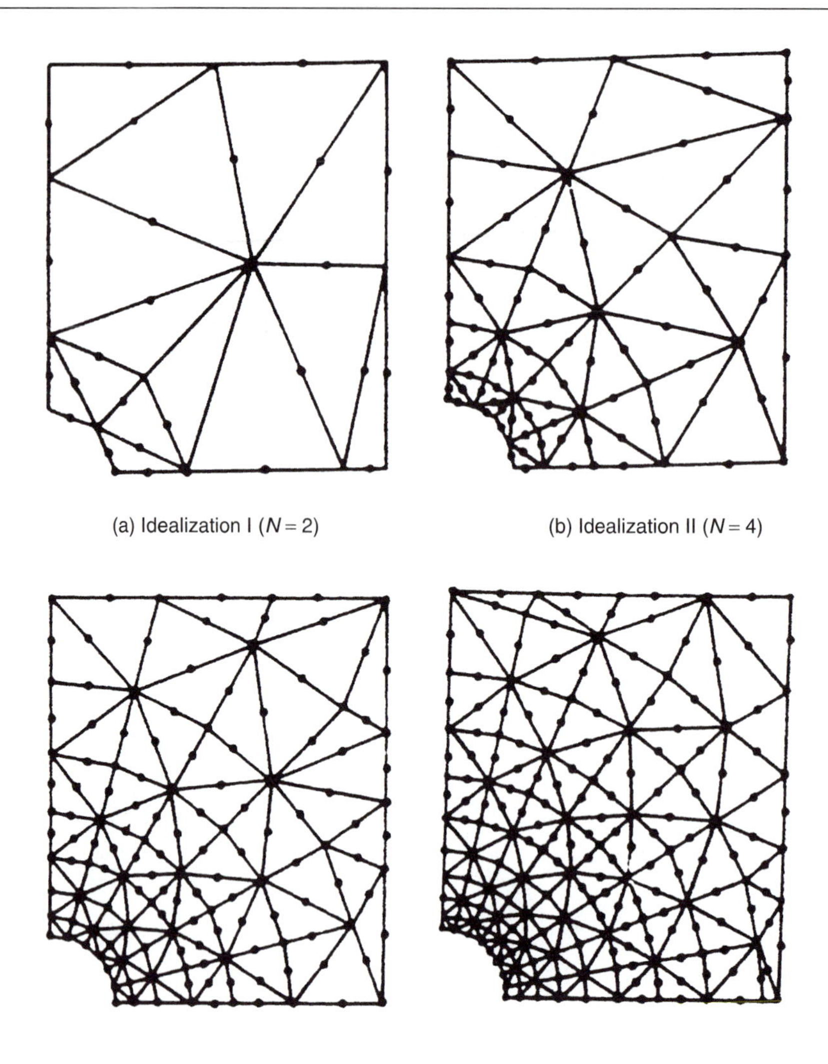

(a) Idealization I ($N = 2$)

(b) Idealization II ($N = 4$)

(c) Idealization III ($N = 6$)

(d) Idealization IV ($N = 8$)

Figure 10.6. Finite Element Idealization of the Plate with a Circular Hole [10.2] (N = number of subdivisions of $\frac{1}{4}$ hole).

Table 10.2. Stress Concentration Factors Given by Finite Element Method

Idealization (Figure 10.6)	Value of (σ_{xx}/p) at A	Value of (σ_{yy}/p) at B
I	−0.229	1.902
II	−0.610	2.585
III	−0.892	2.903
IV	−1.050	3.049
Exact (theory)	−1.250	3.181

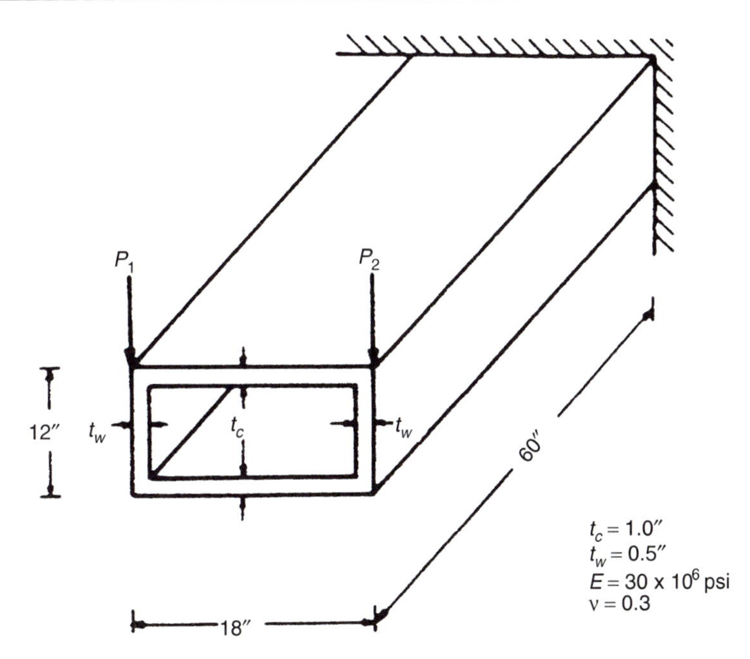

Figure 10.7. A Cantilevered Box Beam.

10.4 COMPUTER PROGRAM FOR PLATES UNDER INPLANE LOADS

A Fortran subroutine called CST is given for the stress analysis of plates under inplane loads using CST elements. It is assumed that the plate structure lies in the XY plane. The subroutine CST requires the following quantities as input:

NN = total number of nodes (including the fixed nodes).
NE = number of triangular elements.
ND = total number of degrees of freedom (including the fixed degrees of freedom). Two degrees of freedom (one parallel to X axis and the other parallel to Y axis) are considered at each node.
NB = bandwidth of the overall stiffness matrix.
M = number of load conditions.
LØC = an array of size NE × 3. LØC(I, J) denotes the global node number corresponding to Jth corner of element I.
CX,CY = vector arrays of size NN each. CX (I) and CY (I) denote the global X and Y coordinates of node I.
E = Young's modulus of the material.
ANU = Poisson's ratio of the material.
T = thickness of the plate.
NFIX = number of fixed degrees of freedom (zero displacements).
IFIX = a vector array of size NFIX. IFIX (I) denotes the Ith fixed degree of freedom number.
P = an array of size ND × M representing the global load vectors.
The array P returned from CST to the main program represents the global displacement vectors. P(I, J) denotes the Ith component of global load (or displacement) vector in Jth load condition.

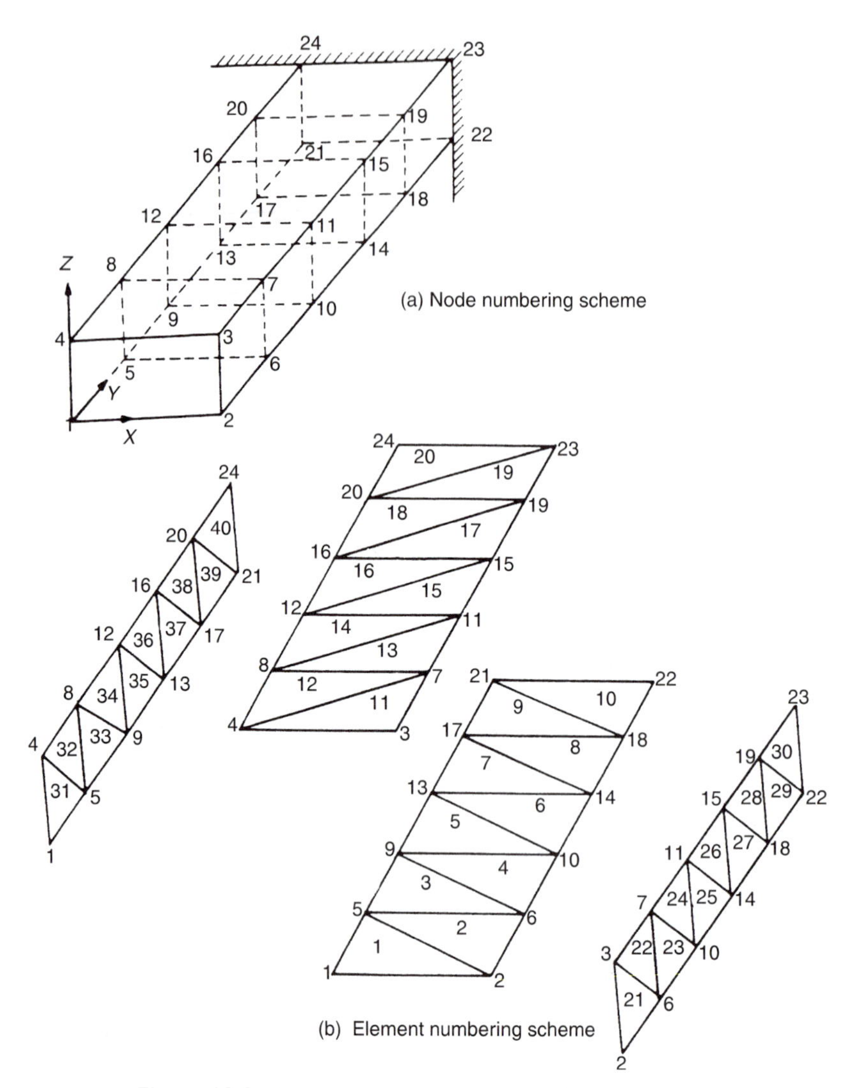

Figure 10.8. Finite Element Idealization of the Box Beam.

Table 10.3. Tip Deflection of Box Beam in the Direction of Load

Load condition	Finite element method	Simple beam theory
$P_1 = P_2 = 5000$ lb	0.0195 in.	0.0204 in.
$P_1 = -P_2 = 5000$ lb	0.0175 in.	—

In addition to this input, the arrays GS of size ND × NB and STRES of size NE × 3 and the double precision vector array DIFF of size M are included as arguments for the subroutine CST. The array GS represents the global stiffness matrix, whereas the array DIFF denotes a dummy array. The array STRES represents the output of the subroutine CST. STRES (I, 1), STRES (I, 2), and STRES (I, 3) denote the stresses σ_{XX}, σ_{YY}, and σ_{XY} in the global coordinate system of element I.

Example 10.1 To illustrate the use of the subroutine CST, the plate shown in Figure 10.4(a) is analyzed for the stresses. Due to the double symmetry, only a quadrant of the plate is used for idealization. The finite element idealization and the corner numbers used are indicated in Figure 10.4(b). The boundary (symmetry) conditions are

$Q_5 = Q_7 = Q_9 = 0$ (X component of displacement of nodes 3, 4, and 5 is zero).

$Q_{10} = Q_{12} = Q_{14} = 0$ (Y component of displacement of nodes 5, 6, and 7 is zero).

The only load condition is

$$P(6,1) = 1000.0 \text{ N}$$
$$P(4,1) = 2000.0 \text{ N}$$
$$P(2,1) = 1000.0 \text{ N}$$

Note

The node numbering scheme used in Figure 10.4(b) leads to a high bandwidth (NB = 18). We can reduce NB to 10 by relabeling the nodes 8, 9, 4, 7, 6, and 5 of Figure 10.4(b) as 4, 5, 6, 7, 8, and 9, respectively. The main program for solving this example and the output of the program are given below.

```
C=====================================================================
C
C      ANALYSIS OF PLATES UNDER INPLANE LOADS
C
C=====================================================================
       DIMENSION LOC(8,3),CX(9),CY(9),IFIX(6),P(18,1),GS(18,18),
      2STRES(8,3)
       DOUBLE PRECISION DIFF(1)
       DATA NN,NE,ND,NB,NFIX,M,E,ANU/9,8,18,18,6,1,2.0E6,0.1/
       DATA LOC/9,1,9,3,9,5,9,7,1,9,3,9,5,9,7,9,8,2,2,4,4,6,6,8/
       DATA CX/20.0,10.0,0.0,0.0,0.0,10.0,20.0,20.0,10.0/
       DATA CY/20.0,20.0,20.0,10.0,0.0,0.0,0.0,10.0,10.0/
       DATA T/0.1/
       DATA IFIX/5,7,9,10,12,14/
       DO 10 I=1,ND
10     P(I,1) = 0.0
       P(6,1) = 1000.0
       P(4,1) = 2000.0
       P(2,1) = 1000.0
       CALL CST(NN,NE,ND,NB,M,LOC,CX,CY,E,ANU,T,NFIX,IFIX,P,GS,DIFF,
```

```
     2STRES)
      PRINT 20
20    FORMAT(1X, 'DISPLACEMENT OF NODES',/)
      PRINT 30,(P(I,1),I=1,ND)
30    FORMAT(6E12.4)
      PRINT 40
40    FORMAT(/,1X,'STRESSES IN ELEMENTS',/)
      DO 50 I=1,NE
50    PRINT 60,I,(STRES(I,J),J=1,3)
60    FORMAT(1X,I3,5X,3E12.4)
      STOP
      END
```

```
DISPLACEMENT OF NODES

-0.2000E-02  0.2000E-01  -0.1000E-02  0.2000E-01  -0.1502E-14  0.2000E-01
-0.1924E-15  0.1000E-01   0.2543E-15  0.6828E-08  -0.1000E-02  0.6828E-08
-0.2000E-02  0.6828E-08  -0.2000E-02  0.1000E-01  -0.1000E-02  0.1000E-01

STRESSES IN ELEMENTS

  1       -0.2441E-03  0.2000E+04  -0.9155E-04
  2       -0.3052E-03  0.2000E+04  -0.3052E-04
  3       -0.4883E-03  0.2000E+04   0.9155E-04
  4       -0.1221E-03  0.2000E+04   0.1221E-03
  5       -0.6104E-04  0.2000E+04   0.2136E-03
  6        0.0000E+00  0.2000E+04   0.0000E+00
  7        0.6104E-04  0.2000E+04   0.0000E+00
  8       -0.1831E-03  0.2000E+04  -0.9155E-04
```

10.5 BENDING BEHAVIOR OF PLATES

The following assumptions are made in the classical theory of thin plates [10.3]:

1. The thickness of the plate is small compared to its other dimensions.
2. The deflections are small.
3. The middle plane of the plate does not undergo inplane deformation.
4. The transverse shear deformation is zero.

The stresses induced in an element of a flat plate subjected to bending forces (transverse load and bending moments) are shown in Figure 10.9(a). These stresses are shear stresses σ_{yz}, σ_{xz}, and σ_{xy} and normal stresses σ_{xx} and σ_{yy}. It can be noticed that in beams, which can be considered as one-dimensional analogs of plates, the shear stress σ_{xy} will not be present. As in beam theory, the stresses σ_{xx} (and σ_{yy}) and σ_{xz} (and σ_{yz}) are assumed to vary linearly and parabolically, respectively, over the thickness of the plate. The shear stress σ_{xy} is assumed to vary linearly. The stresses σ_{xx}, σ_{yy}, σ_{xy}, σ_{xz}, and σ_{yz} lead to the following force and moment resultants per unit length:

$$\left.\begin{aligned} M_x &= \int_{-t/2}^{t/2} \sigma_{xx} z \, dz, \qquad M_y = \int_{-t/2}^{t/2} \sigma_{yy} z \, dz, \\ M_{xy} &= \int_{-t/2}^{t/2} \sigma_{xy} z \, dz, \qquad Q_x = \int_{-t/2}^{t/2} \sigma_{xz} \, dz, \qquad Q_y = \int_{-t/2}^{t/2} \sigma_{yz} \, dz \end{aligned}\right\} \tag{10.48}$$

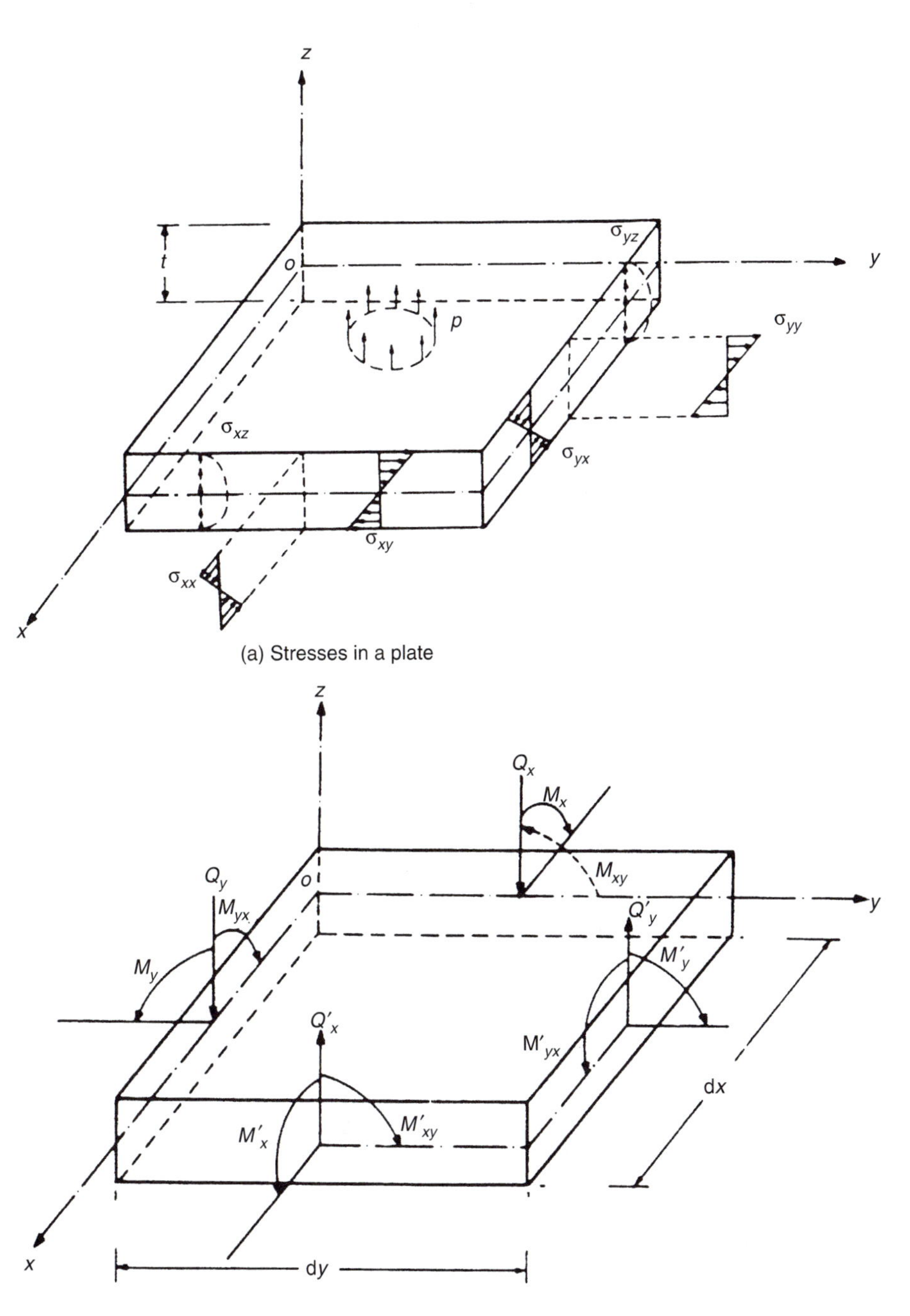

(a) Stresses in a plate

(b) Forces and moments in a plate

Figure 10.9.

These forces and moments are indicated in Figure 10.9(b). By considering an element of the plate, the differential equations of equilibrium in terms of force resultants can be derived. For this, we consider the bending moments and shear forces to be functions of x and y so that, if M_x acts on one side of the element, $M'_x = M_x + \frac{\partial M_x}{\partial x} \cdot dx$ acts on the opposite side. The resulting equations can be written as

$$\left.\begin{aligned} &\frac{\partial Q_x}{\partial x} + \frac{\partial Q_y}{\partial y} + p = 0 \\ &\frac{\partial M_x}{\partial x} + \frac{\partial M_{xy}}{\partial y} = Q_x \\ &\frac{\partial M_{xy}}{\partial x} + \frac{\partial M_y}{\partial y} = Q_y \end{aligned}\right\} \tag{10.49}$$

where p is the distributed surface load. Because the plate is thin in comparison to its length and width, any body force may be converted to an equivalent load p and hence no body force is considered separately in Eqs. (10.49).

To derive the strain–displacement relations for a plate, consider the bending deformation of a small element (by neglecting shear deformation). Any point A in this element experiences both transverse (w) and inplane (u and v) displacements. The strains can be expressed as

$$\left.\begin{aligned} \varepsilon_{xx} &= \frac{\partial u}{\partial x} = -z\frac{\partial^2 w}{\partial x^2} \\ \varepsilon_{yy} &= \frac{\partial v}{\partial y} = -z\frac{\partial^2 w}{\partial y^2} \\ \varepsilon_{xy} &= \frac{\partial u}{\partial y} + \frac{\partial v}{\partial x} = -2z\frac{\partial^2 w}{\partial x \partial y} \end{aligned}\right\} \tag{10.50}$$

Equations (10.50) show that the transverse displacement w, which is a function of x and y only, completely describes the deformation state.

The moment–displacement relations can also be derived for plates. For this, we assume the plate to be in a state of plane stress by considering the transverse stress σ_{zz} to be negligible in comparison to σ_{xx} and σ_{yy}. Thus, the stress–strain relations are given by (Eq. 8.35):

$$\vec{\sigma} = \begin{Bmatrix} \sigma_{xx} \\ \sigma_{yy} \\ \sigma_{xy} \end{Bmatrix} = [D]\vec{\varepsilon} = [D]\begin{Bmatrix} \varepsilon_{xx} \\ \varepsilon_{yy} \\ \varepsilon_{xy} \end{Bmatrix} \tag{10.51}$$

where

$$[D] = \frac{E}{(1-\nu^2)}\begin{bmatrix} 1 & \nu & 0 \\ \nu & 1 & 0 \\ 0 & 0 & \frac{1-\nu}{2} \end{bmatrix} \tag{10.52}$$

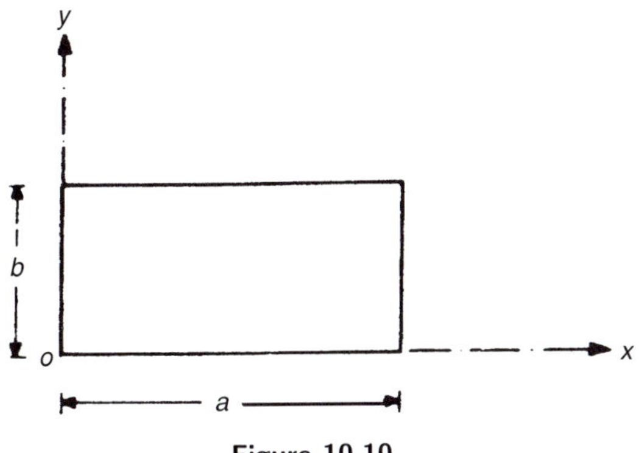

Figure 10.10.

By substituting Eqs. (10.50) into Eqs. (10.51) and the resulting stresses into Eqs. (10.48), we obtain after integration,

$$\left.\begin{aligned} M_x &= -D\left(\frac{\partial^2 w}{\partial x^2} + \nu\frac{\partial^2 w}{\partial y^2}\right) \\ M_y &= -D\left(\frac{\partial^2 w}{\partial y^2} + \nu\frac{\partial^2 w}{\partial x^2}\right) \\ M_{yx} = M_{xy} &= -(1-\nu)\,D\frac{\partial^2 w}{\partial x\partial y} = -\frac{Gt^3}{6}\cdot\frac{\partial^2 w}{\partial x\partial y} \end{aligned}\right\} \tag{10.53}$$

where D is called the flexural rigidity of the plate and is given by

$$D = \frac{Et^3}{12(1-\nu^2)} \tag{10.54}$$

The flexural rigidity D corresponds to the bending stiffness of a beam (EI). In fact, $D = EI$ for a plate of unit width when ν is taken as zero. Equations (10.49) and (10.53) give

$$\left.\begin{aligned} Q_x &= -D\cdot\frac{\partial}{\partial x}\left(\frac{\partial^2 w}{\partial x^2} + \frac{\partial^2 w}{\partial y^2}\right) \\ Q_y &= -D\cdot\frac{\partial}{\partial y}\left(\frac{\partial^2 w}{\partial x^2} + \frac{\partial^2 w}{\partial y^2}\right) \end{aligned}\right\} \tag{10.55}$$

The following boundary conditions have to be satisfied for plates (Figure 10.10):

1. Simply supported edge (along $y =$ constant):

$$\left.\begin{aligned} &w(x,y) = 0 \\ &M_y = -D\left(\frac{\partial^2 w}{\partial y^2} + \nu\frac{\partial^2 w}{\partial x^2}\right) = 0 \end{aligned}\right\} \text{for } y = \text{constant, and } 0 \le x \le a \tag{10.56}$$

2. Clamped edge (along $y =$ constant):

$$\left.\begin{aligned} &w(x,y) = 0 \\ &\frac{\partial w}{\partial y}(x,y) = 0 \end{aligned}\right\} \text{ for } y = \text{constant, and } 0 \le x \le a \tag{10.57}$$

3. Free edge (along $y =$ constant):

$$\left.\begin{aligned} M_y &= -D\left(\frac{\partial^2 w}{\partial y^2} + \nu\frac{\partial^2 w}{\partial x^2}\right) = 0 \\ Q_y + \frac{\partial M_{yx}}{\partial x} &= \text{vertical shear} \\ &= -(2-\nu)D\frac{\partial^3 w}{\partial x^2 \partial y} - D\frac{\partial^3 w}{\partial y^3} = 0 \end{aligned}\right\} \begin{aligned} &\text{for } y = \text{constant, and} \\ &0 \le x \le a \end{aligned} \tag{10.58}$$

In the classical theory of plates, first the displacement $w(x, y)$ is found by solving the equilibrium equations (10.49) under the prescribed loading condition $p(x, y)$. By substituting Eqs. (10.55) into Eqs. (10.49), we notice that the second and third equilibrium equations are automatically satisfied and the first one gives

$$\frac{\partial^4 w}{\partial x^4} + 2\frac{\partial^4 w}{\partial x^2 \partial y^2} + \frac{\partial^4 w}{\partial y^4} = \frac{p}{D} \tag{10.59}$$

Thus, the problem is to solve the fourth-order partial differential equation (10.59) by using appropriate boundary conditions. Once $w(x, y)$ is found, the strains, stresses, and moments developed in the plate can be determined by using Eqs. (10.50), (10.51), and (10.53), respectively. It can be noticed that the closed-form solution of Eq. (10.59) cannot be obtained except for plates having simple configuration (e.g., rectangular and circular plates) and simple loading and boundary conditions. However, the finite element method can be used for analyzing problems involving plates of arbitrary planform and loading conditions that may sometimes have cutouts or cracks.

10.6 FINITE ELEMENT ANALYSIS OF PLATE BENDING

A large number of plate bending elements have been developed and reported in the literature [10.4, 10.5]. In the classical theory of thin plates discussed in this section, certain simplifying approximations are made. One of the important assumptions made is that shear deformation is negligible. Some elements have been developed by including the effect of transverse shear deformation also.

According to thin plate theory, the deformation is completely described by the transverse deflection of the middle surface of the plate (w) only. Thus, if a displacement model is assumed for w, the continuity of not only w but also its derivatives has to be maintained between adjacent elements. According to the convergence requirements stated in Section 3.6, the polynomial for w must be able to represent constant strain states. This means, from Eqs. (10.50), that the assumed displacement model must contain constant curvature states $(\partial^2 w/\partial x^2)$ and $(\partial^2 w/\partial y^2)$ and constant twist $(\partial^2 w/\partial x \partial y)$. Also, the polynomial for w should have geometric isotropy.

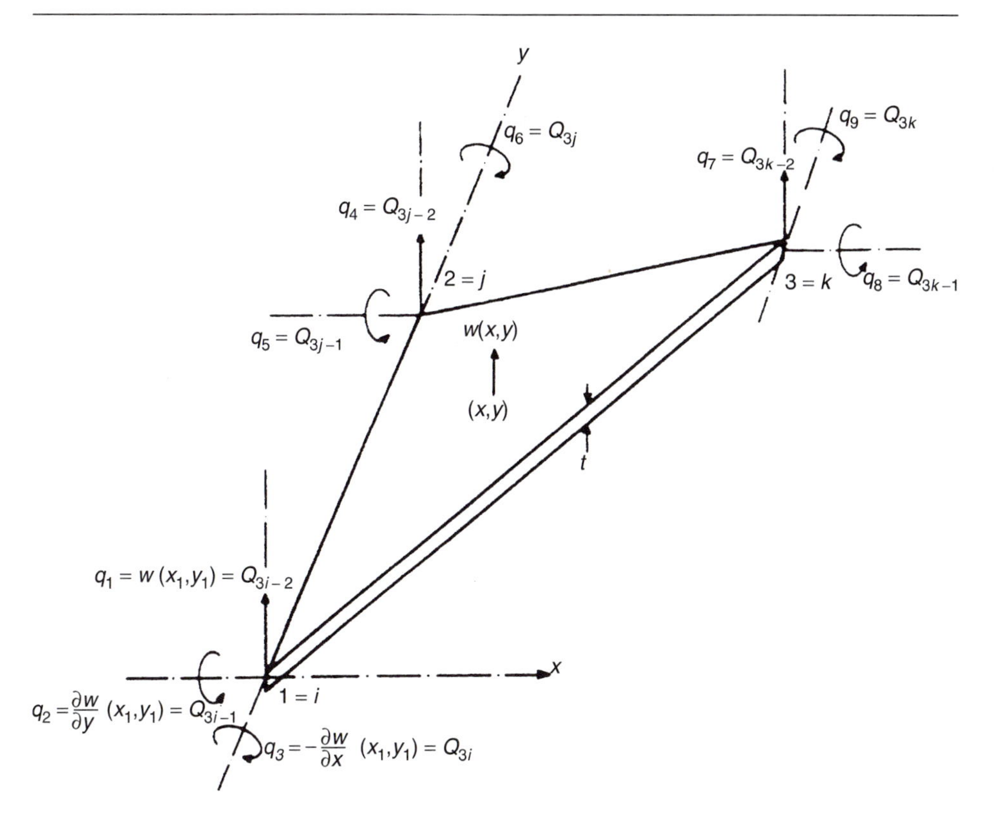

Figure 10.11. Nodal Degrees of Freedom of a Triangular Plate in Bending.

Thus, it becomes evident that it is much more difficult to choose a displacement model satisfying all these requirements. In surmounting these difficulties, especially for triangular and general quadrilateral elements, different investigators have developed different elements, some of them quite complicated. In the following section, a simple triangular plate bending element is described along with its characteristics.

10.7 TRIANGULAR PLATE BENDING ELEMENT

At each node of the triangular plate element shown in Figure 10.11, the transverse displacement w and slopes (rotations) about the x and y axes $[(\partial w/\partial y)$ and $-(\partial w/\partial x)]$ are taken as the degrees of freedom. The minus sign for the third degree of freedom indicates that if we take a positive displacement dw at a distance dx from node 1, the rotation (dw/dx) about the y axis at node 1 will be opposite to the direction of the degree of freedom q_3 indicated in Figure 10.11. Since there are nine displacement degrees of freedom in the element, the assumed polynomial for $w\ (x, y)$ must also contain nine constant terms. To maintain geometric isotropy, the displacement model is taken as

$$\begin{aligned} w(x,y) &= \alpha_1 + \alpha_2 x + \alpha_3 y + \alpha_4 x^2 + \alpha_5 xy + \alpha_6 y^2 + \alpha_7 x^3 + \alpha_8 (x^2 y + xy^2) + \alpha_9 y^3 \\ &= [\eta]\vec{\alpha} \end{aligned} \tag{10.60}$$

where

$$[\eta] = [1 \ x \ y \ x^2 \ xy \ y^2 \ x^3 \ (x^2y + xy^2) \ y^3] \tag{10.61}$$

and

$$\vec{\alpha} = \begin{Bmatrix} \alpha_1 \\ \alpha_2 \\ \vdots \\ \alpha_9 \end{Bmatrix} \tag{10.62}$$

The constants $\alpha_1, \alpha_2, \ldots, \alpha_9$ have to be determined from the nodal conditions

$$\left.\begin{aligned} &w(x,y) = q_1, \quad \frac{\partial w}{\partial y}(x,y) = q_2, \quad -\frac{\partial w}{\partial x}(x,y) = q_3 \quad \text{at} \quad (x_1, y_1) = (0,0) \\ &w(x,y) = q_4, \quad \frac{\partial w}{\partial y}(x,y) = q_5, \quad -\frac{\partial w}{\partial x}(x,y) = q_6 \quad \text{at} \quad (x_2, y_2) = (0, y_2) \\ &w(x,y) = q_7, \quad \frac{\partial w}{\partial y}(x,y) = q_8, \quad -\frac{\partial w}{\partial x}(x,y) = q_9 \quad \text{at} \quad (x_3, y_3) \end{aligned}\right\} \tag{10.63}$$

Note that the local y axis is taken to be the same as the line connecting the nodes 1 and 2 with the origin placed at node 1. The local x axis is taken toward node 3 as shown in Figure 10.11. The local node numbers 1, 2, and 3 are assumed to correspond to nodes i, j, and k, respectively, in the global system. By using Eq. (10.60), Eqs. (10.63) can be stated in matrix form as

$$\vec{q}^{\,(e)} = \begin{Bmatrix} q_1 \\ q_2 \\ \vdots \\ q_9 \end{Bmatrix}^{(e)} = [\underset{\sim}{\eta}]\vec{\alpha} \tag{10.64}$$

where

$$[\underset{\sim}{\eta}] = \begin{bmatrix} 1 & 0 & 0 & 0 & 0 & 0 & 0 & 0 & 0 \\ 0 & 0 & 1 & 0 & 0 & 0 & 0 & 0 & 0 \\ 0 & -1 & 0 & 0 & 0 & 0 & 0 & 0 & 0 \\ 1 & 0 & y_2 & 0 & 0 & y_2^2 & 0 & 0 & y_2^3 \\ 0 & 0 & 1 & 0 & 0 & 2y_2 & 0 & 0 & 3y_2^2 \\ 0 & -1 & 0 & 0 & -y_2 & 0 & 0 & -y_2^2 & 0 \\ 1 & x_3 & y_3 & x_3^2 & x_3y_3 & y_3^2 & x_3^3 & (x_3^2y_3 + x_3y_3^2) & y_3^3 \\ 0 & 0 & 1 & 0 & x_3 & 2y_3 & 0 & (2x_3y_3 + x_3^2) & 3y_3^2 \\ 0 & -1 & 0 & -2x_3 & -y_3 & 0 & -3x_3^2 & (-y_3^2 + 2x_3y_3) & 0 \end{bmatrix} \tag{10.65}$$

By using Eqs. (10.60) and (10.64), Eqs. (10.50) can be expressed as

$$\vec{\varepsilon} = [\underset{\sim}{B}]\vec{\alpha} = [B]\vec{q}^{\,(e)} \tag{10.66}$$

where

$$[\underset{\sim}{B}] = -z \begin{bmatrix} 0 & 0 & 0 & 2 & 0 & 0 & 6x & 2y & 0 \\ 0 & 0 & 0 & 0 & 0 & 2 & 0 & 2x & 6y \\ 0 & 0 & 0 & 0 & 2 & 0 & 0 & 4(x+y) & 0 \end{bmatrix} \tag{10.67}$$

and

$$[B] = [\underset{\sim}{B}][\underset{\sim}{\eta}]^{-1} \tag{10.68}$$

Finally, the element stiffness matrix in the local (xy) coordinate system can be derived as

$$[k^{(e)}] = \iiint\limits_{V^{(e)}} [B]^T[D][B]\,\mathrm{d}V \tag{10.69}$$

where $V^{(e)}$ indicates the volume of the element, and the matrix $[D]$ is given by Eq. (10.52). By substituting for $[B]$ from Eq. (10.68), Eq. (10.69) can be expressed as

$$[k^{(e)}] = ([\underset{\sim}{\eta}]^{-1})^T \left\{ \iint\limits_{\text{area}} \mathrm{d}A \left(\int\limits_{-t/2}^{t/2} [\underset{\sim}{B}]^T[D][\underset{\sim}{B}]\,\mathrm{d}z \right) \right\} [\underset{\sim}{\eta}]^{-1} \tag{10.70}$$

where t denotes the thickness of the plate. The integrals within the curved brackets of Eq. (10.70) can be rewritten as

$$\iint\limits_{\text{area}} \mathrm{d}A \int\limits_{-t/2}^{t/2} [\underset{\sim}{B}]^T[D][\underset{\sim}{B}]\,\mathrm{d}z = \frac{Et^3}{(12(1-\nu^2))} \iint\limits_{\text{area}} \mathrm{d}x\,\mathrm{d}y$$

$$\times \begin{bmatrix} 0 & & & & & & & & \\ 0 & 0 & & & & & & & \\ 0 & 0 & 0 & & & \text{Symmetric} & & & \\ 0 & 0 & 0 & 4 & & & & & \\ 0 & 0 & 0 & 0 & 2(1-\nu) & & & & \\ 0 & 0 & 0 & 4\nu & 0 & 4 & & & \\ 0 & 0 & 0 & 12x & 0 & 12\nu x & 36x^2 & & \\ 0 & 0 & 0 & 4(\nu x+y) & 4(1-\nu)(x+y) & 4(x+\nu y) & 12x(\nu x+y) & \begin{matrix}\{(12-8\nu)(x+y)^2 \\ -8(1-\nu)xy\}\end{matrix} & \\ 0 & 0 & 0 & 12\nu y & 0 & 12y & 36\nu xy & 12(x+\nu y)y & 36y^2 \end{bmatrix} \tag{10.71}$$

The area integrals appearing on the right-hand side of Eq. (10.71) can be evaluated in the general XY coordinate system as well as in the particular local xy system chosen in

Figure 10.11 using the following relations:

$$\iint_{\text{area}} \mathrm{d}x\,\mathrm{d}y = A = \frac{1}{2}x_3y_2 \tag{10.72}$$

$$\iint_{\text{area}} x\,\mathrm{d}x\,\mathrm{d}y = X_cA = \frac{1}{6}x_3^2y_2 \tag{10.73}$$

$$\iint_{\text{area}} y\,\mathrm{d}x\,\mathrm{d}y = Y_cA = \frac{1}{6}x_3y_2(y_2 + y_3) \tag{10.74}$$

$$\begin{aligned}\iint_{\text{area}} x^2\,\mathrm{d}x\,\mathrm{d}y &= X_c^2A + \frac{A}{12}[(X_i - X_c)^2 + (X_j - X_c)^2 + (X_k - X_c)^2] \\ &= \frac{1}{12}x_3^3y_2\end{aligned} \tag{10.75}$$

$$\begin{aligned}\iint_{\text{area}} xy\,\mathrm{d}x\cdot\mathrm{d}y &= X_cY_cA + \frac{A}{12}[(X_i - X_c)(Y_i - Y_c) \\ &\quad + (X_j - X_c)(Y_j - Y_c) + (X_k - X_c)(Y_k - Y_c)] \\ &= \frac{1}{24}x_3^2y_2(y_2 + 2y_3)\end{aligned} \tag{10.76}$$

$$\begin{aligned}\iint_{\text{area}} y^2\,\mathrm{d}x\cdot\mathrm{d}y &= Y_c^2A + \frac{A}{12}[(Y_i - Y_c)^2 + (Y_j - Y_c)^2 + (Y_k - Y_c)^2] \\ &= \frac{1}{12}x_3y_2(y_2^2 + y_2y_3 + y_3^2)\end{aligned} \tag{10.77}$$

where

$$X_c = (X_i + X_j + X_k)/3 \tag{10.78}$$

and

$$Y_c = (Y_i + Y_j + Y_k)/3 \tag{10.79}$$

It can be seen that the evaluation of the element stiffness matrix from Eqs. (10.70) and (10.71) involves the numerical determination of the inverse of the 9×9 matrix, $[\underset{\sim}{\eta}]$, for each element separately. Finally, the element stiffness matrix in the global coordinate system (whose XY plane is assumed to be the same as the local xy plane) can be obtained from

$$[K^{(e)}] = [\lambda]^T[k^{(e)}][\lambda] \tag{10.80}$$

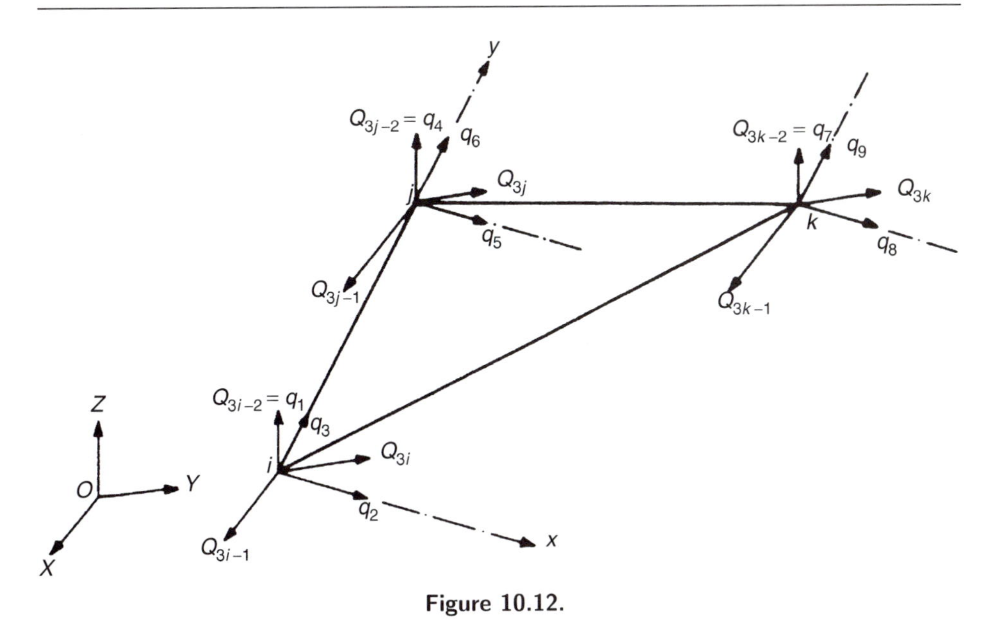

Figure 10.12.

where the transformation matrix $[\lambda]$ is given by

$$\underset{9\times 9}{[\lambda]} = \begin{bmatrix} 1 & 0 & 0 & 0 & 0 & 0 & 0 & 0 & 0 \\ 0 & l_{ox} & m_{ox} & 0 & 0 & 0 & 0 & 0 & 0 \\ 0 & l_{oy} & m_{oy} & 0 & 0 & 0 & 0 & 0 & 0 \\ 0 & 0 & 0 & 1 & 0 & 0 & 0 & 0 & 0 \\ 0 & 0 & 0 & 0 & l_{ox} & m_{ox} & 0 & 0 & 0 \\ 0 & 0 & 0 & 0 & l_{oy} & m_{oy} & 0 & 0 & 0 \\ 0 & 0 & 0 & 0 & 0 & 0 & 1 & 0 & 0 \\ 0 & 0 & 0 & 0 & 0 & 0 & 0 & l_{ox} & m_{ox} \\ 0 & 0 & 0 & 0 & 0 & 0 & 0 & l_{oy} & m_{oy} \end{bmatrix} \tag{10.81}$$

where (l_{ox}, m_{ox}) and (l_{oy}, m_{oy}) represent the direction cosines of the lines ox and oy, respectively (Figure 10.12).

10.8 NUMERICAL RESULTS WITH BENDING ELEMENTS

The triangular plate bending element considered in Section 10.7 is one of the simplest elements. Several other elements were developed for the analysis of plates. Since the strains developed in a plate under bending involve second derivatives of the transverse displacement w, the expression for w must contain a complete second-degree polynomial in x and y. Furthermore, the interelement compatibility requires the continuity of w as well as of the normal derivative $(\partial w/\partial n)$ across the boundaries of two elements.

For a rectangular element (Figure 10.10), the simplest thing to do is to take the values of w, $(\partial w/\partial x)$, and $(\partial w/\partial y)$ at each of the four corners as nodal degrees of freedom. This gives a total of 12 degrees of freedom for the element. Thus, the polynomial for w

must also contain 12 constants α_i. Since a complete polynomial of degree three in x and y contains 10 terms, we need to include 2 additional terms. These terms can be selected arbitrarily, but we should preserve the symmetry of the expansion to ensure geometric isotropy. Thus, we have three possibilities, namely to take x^3y and xy^3, x^3y^2 and x^2y^3, or x^2y^2 and x^3y^3 in the expression of w. All these choices satisfy the condition that along any edge of the element w varies as a cubic. This can be verified by setting $x = 0$ or a (or $y = 0$ or b) in the expression of w. Since there are four nodal unknowns for any edge [e.g., along the edge $x = 0$, we have w and $(\partial w/\partial y)$ at the two corners as degrees of freedom], w is uniquely specified along that edge. This satisfies the continuity condition of w across the boundaries. For the continuity of $(\partial w/\partial n)$, we need to have $(\partial w/\partial n)$ vary linearly on a side since it is specified only at the node points. Irrespective of what combination of 12 polynomial terms we choose for w, we cannot avoid ending up with a cubic variation for $(\partial w/\partial n)$ ($n = x$ for the sides defined by $x = 0$ and a and $n = y$ for the edges defined by $y = 0$ and b). Therefore, it is not possible to satisfy the interelement compatibility conditions [continuity of both w and $(\partial w/\partial n)$] with 12 degrees of freedom only. A similar reasoning will reveal that the triangular element considered in Section 10.7 is also nonconforming.

The displacement models of some of the plate bending elements available in the literature are given next.

10.8.1 Rectangular Elements

1. Nonconforming element due to Adini–Clough–Melosh (ACM):

$$\begin{aligned} w(x,y) &= \alpha_1 + \alpha_2 x + \alpha_3 y + \alpha_4 x^2 + \alpha_5 y^2 + \alpha_6 xy + \alpha_7 x^3 + \alpha_8 y^3 \\ &\quad + \alpha_9 x^2 y + \alpha_{10} xy^2 + \alpha_{11} x^3 y + \alpha_{12} xy^3 \end{aligned} \tag{10.82}$$

   Degrees of freedom at each node: w, $(\partial w/\partial x)$, $(\partial w/\partial y)$. Ref. [10.6].

2. Conforming element due to Bogner–Fox–Schmit (BFS-16):

$$\begin{aligned} w(x,y) = \sum_{i=1}^{2}\sum_{j=1}^{2} \Bigg[& H_{0i}^{(1)}(x)H_{0j}^{(1)}(y)w_{ij} + H_{1i}^{(1)}(x)H_{0j}^{(1)}(y)\left(\frac{\partial w}{\partial x}\right)_{ij} \\ & + H_{0i}^{(1)}(x)H_{1j}^{(1)}(y)\left(\frac{\partial w}{\partial y}\right)_{ij} + H_{1i}^{(1)}(x)H_{1j}^{(1)}(y)\left(\frac{\partial^2 w}{\partial x \partial y}\right)_{ij}\Bigg] \end{aligned} \tag{10.83}$$

   Degrees of freedom at each node: w_{ij}, $(\partial w/\partial x)_{ij}$, $(\partial w/\partial y)_{ij}$, $(\partial^2 w/\partial x\partial y)_{ij}$ (node numbering scheme shown in Figure 4.16). Ref. [10.7].

3. More accurate conforming element due to Bogner–Fox–Schmit (BFS-24):

$$\begin{aligned} w(x,y) = \sum_{i=1}^{2}\sum_{j=1}^{2} \Bigg[& H_{0i}^{(2)}(x)H_{0j}^{(2)}(y)w_{ij} + H_{1i}^{(2)}(x)H_{0j}^{(2)}(y)\left(\frac{\partial w}{\partial x}\right)_{ij} \\ & + H_{0i}^{(2)}(x)H_{1j}^{(2)}(y)\left(\frac{\partial w}{\partial y}\right)_{ij} + H_{2i}^{(2)}(x)H_{0j}^{(2)}(y)\left(\frac{\partial^2 w}{\partial x^2}\right)_{ij} \\ & + H_{0i}^{(2)}(x)H_{2j}^{(2)}(y)\left(\frac{\partial^2 w}{\partial y^2}\right)_{ij} + H_{1i}^{(2)}(x)H_{1j}^{(2)}(y)\left(\frac{\partial^2 w}{\partial x \partial y}\right)_{ij}\Bigg] \end{aligned} \tag{10.84}$$

Degrees of freedom at each node: w_{ij}, $(\partial w/\partial x)_{ij}$, $(\partial w/\partial y)_{ij}$,

$$\left(\frac{\partial^2 w}{\partial x^2}\right)_{ij}, \quad \left(\frac{\partial^2 w}{\partial y^2}\right)_{ij}, \quad \left(\frac{\partial^2 w}{\partial x \partial y}\right)_{ij}$$

(node numbering scheme shown in Figure 4.16). Ref. [10.7].

10.8.2 Triangular Elements

1. Nonconforming element due to Tocher (T-9):

$$w(x, y) = \text{same as Eq. (10.60)}$$

Degrees of freedom at each node: w, $\partial w/\partial x$, $\partial w/\partial y$. Ref. [10.8].

2. Nonconforming element due to Tocher (T-10):

$$w(y, x) = \alpha_1 + \alpha_2 x + \alpha_3 y + \alpha_4 x^2 + \alpha_5 y^2 + \alpha_6 xy + \alpha_7 x^3 + \alpha_8 y^3 + \alpha_9 x^2 y + \alpha_{10} xy^2 \quad (10.85)$$

Degrees of freedom at each node: w, $\partial w/\partial x$, $\partial w/\partial y$
(The 10th constant was suppressed using the Ritz method). Ref. [10.8].

3. Nonconforming element due to Adini (A):

$$w(x, y) = \alpha_1 + \alpha_2 x + \alpha_3 y + \alpha_4 x^2 + \alpha_5 y^2 + \alpha_6 x^3 + \alpha_7 y^3 + \alpha_8 x^2 y + \alpha_9 xy^2 \quad (10.86)$$

(The uniform twist term xy was neglected.)
Degrees of freedom at each node: w, $\partial w/\partial x$, $\partial w/\partial y$. Ref. [10.9].

4. Conforming element due to Cowper *et al.* (C):

$$w(x, y) = \alpha_1 + \alpha_2 x + \alpha_3 y + \alpha_4 x^2 + \alpha_5 y^2 + \alpha_6 xy + \alpha_7 x^3 + \alpha_8 y^3 + \alpha_9 x^2 y + \alpha_{10} xy^2 + \alpha_{11} x^4 + \alpha_{12} y^4 + \alpha_{13} x^3 y + \alpha_{14} xy^3 + \alpha_{15} x^2 y^2 + \alpha_{16} x^5 + \alpha_{17} y^5 + \alpha_{18} x^4 y + \alpha_{19} xy^4 + \alpha_{20} x^3 y^2 + \alpha_{21} x^2 y^3 \quad (10.87)$$

(Three constraints are imposed to reduce the number of unknowns from 21 to 18. These are that the normal slope $\partial w/\partial n$ along any edge must have a cubic variation.)
Degrees of freedom at each node: w, $\partial w/\partial x$, $\partial w/\partial y$, $\partial^2 w/\partial x^2$, $\partial^2 w/\partial y^2$, $\partial^2 w/\partial x \partial y$. Ref. [10.10].

10.8.3 Numerical Results

Typical numerical results obtained for a clamped square plate subjected to uniformly distributed load with nonconforming and conforming bending elements are shown in Figure 10.13 and Table 10.4, respectively. The finite element idealizations considered are shown in Figure 10.14. Due to symmetry of geometry and load condition, only a quarter

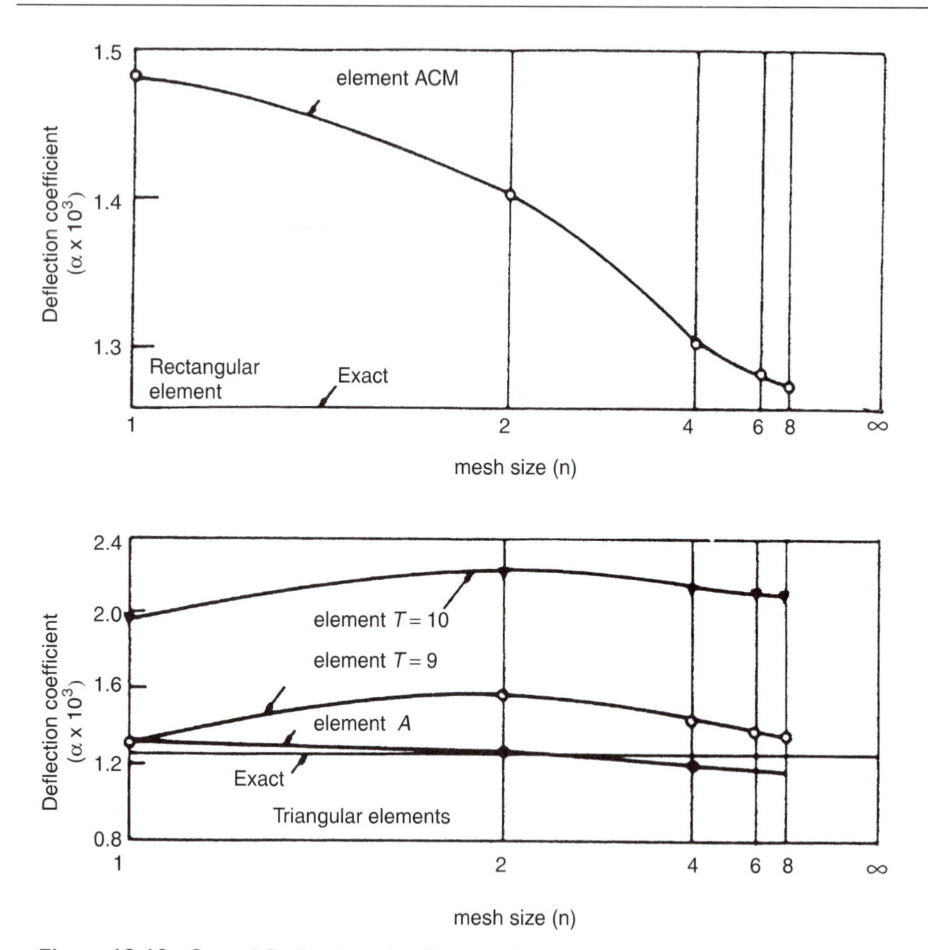

Figure 10.13. Central Deflection of a Clamped Plate under Uniformly Distributed Load.

of the plate is considered for analysis. Of course, the symmetry conditions have to be imposed before solving the problem. For example, if the quarter plate 1, 2, 3, 4 shown in Figure 10.14 is to be analyzed, $\partial w/\partial x$ has to be set equal to zero along line 2, 4, and $\partial w/\partial y$ has to be equated to zero along line 3, 4. The deflection of the center of the clamped plate (w_c) is taken as the measure of the quality of the approximation and the deflection coefficient α of Figure 10.13 is defined by

$$w_c = \frac{\alpha q a^4}{D}$$

where q denotes the intensity of the uniformly distributed load, a is the side of the plate, and D is the flexural rigidity. An important conclusion that can be drawn from the results of Figure 10.13 is that monotonic convergence of deflection cannot be expected always from any of the nonconforming elements considered.

Table 10.4. Central Deflection of a Square Clamped Plate under Uniformly Distributed Load ($\nu = 0.3$)
(a) Results given by the triangular element due to Cowper *et al.*:

Idealization (Figure 10.14)	Number of d.o.f. for one-quarter plate	Value of $w_c(10^3 D/qa^4)$
$n = 1$	5	1.14850
$n = 2$	21	1.26431
$n = 3$ (not shown in Figure 10.14)	49	1.26530
Exact [10.3]	—	1.26

(b) Results given by the rectangular elements due to Bogner *et al.*:

	16 d.o.f. element (BFS-16)		24 d.o.f. element (BFS-24)	
Number of elements in a quadrant	Number of degrees of freedom	Value of w_c^*	Number of degrees of freedom	Value of w_c^*
1	1	0.042393″	5	0.0405″
4 (2 × 2 grid)	9	0.040475″	21	0.0402″
9 (3 × 3 grid)	25	0.040482″	—	—
16 (4 × 4 grid)	49	0.040487″	—	—
Exact [10.3]	—	0.0403″	—	0.0403″

*For $a = 20''$, $q = 0.2$ psi, $E = 10.92 \times 10^6$ psi, $t = 0.1''$.

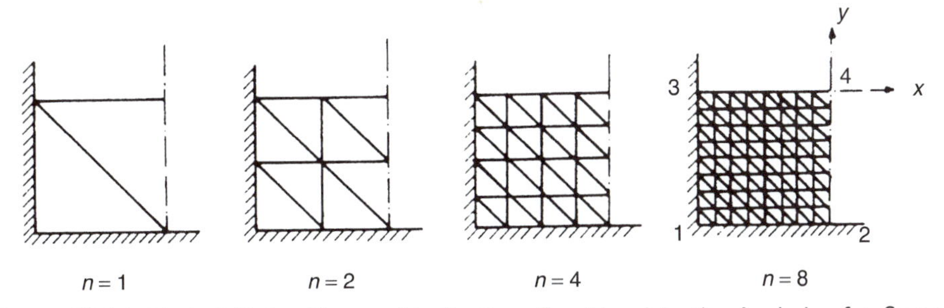

$n=1$ $n=2$ $n=4$ $n=8$

Figure 10.14. Typical Finite Element Idealizations Considered in the Analysis of a Square Plate.

10.9 ANALYSIS OF THREE-DIMENSIONAL STRUCTURES USING PLATE ELEMENTS

If three-dimensional structures under arbitrary load conditions are to be analyzed using plate elements, we have to provide both inplane and bending load-carrying capacity for the elements. The procedure to be adopted will be illustrated with reference to a triangular element. If a linear displacement field is assumed under inplane loads (as in Eq. 10.1), the resulting 6×6 inplane stiffness matrix (in local coordinate system) can be expressed as

$$\underset{6\times 6}{[k^{(e)}]_m} = \begin{bmatrix} \underset{2\times 2}{[k_{11}]_m} & \underset{2\times 2}{[k_{12}]_m} & \underset{2\times 2}{[k_{13}]_m} \\ \underset{2\times 2}{[k_{21}]_m} & \underset{2\times 2}{[k_{22}]_m} & \underset{2\times 2}{[k_{23}]_m} \\ \underset{2\times 2}{[k_{31}]_m} & \underset{2\times 2}{[k_{32}]_m} & \underset{2\times 2}{[k_{33}]_m} \end{bmatrix} \tag{10.88}$$

where the submatrices $[k_{ij}]_m$ correspond to the stiffness coefficients associated with nodes i and j, and the subscript m is used to indicate membrane action. In this case, the relationship between the nodal displacements and nodal forces can be written as

$$\begin{Bmatrix} P_{x1} \\ P_{y1} \\ P_{x2} \\ P_{y2} \\ P_{x3} \\ P_{y3} \end{Bmatrix} = [k^{(e)}]_m \begin{Bmatrix} u_1 \\ v_1 \\ u_2 \\ v_2 \\ u_3 \\ v_3 \end{Bmatrix} \tag{10.89}$$

where u_i and v_i denote the components of displacement of node $i(i = 1, 2, 3)$ parallel to the local x and y axes, respectively. Similarly, P_{x_i} and P_{y_i} indicate the components of force at node $i(i = 1, 2, 3)$ parallel to the x and y axes, respectively.

Similarly, the relation between the forces and displacements corresponding to the bending of the plate (obtained from Eq. 10.60) can be written as

$$\begin{Bmatrix} P_{z1} \\ M_{y1} \\ -M_{x1} \\ P_{z2} \\ M_{y2} \\ -M_{x2} \\ P_{z3} \\ M_{y3} \\ -M_{x3} \end{Bmatrix} = [k^{(e)}]_b \begin{Bmatrix} w_1 \\ w_{y1} \\ -w_{x1} \\ w_2 \\ w_{y2} \\ -w_{x2} \\ w_3 \\ w_{y3} \\ -w_{x3} \end{Bmatrix} \tag{10.90}$$

where w_i and P_{zi} indicate the components of displacement and force parallel to the z axis at node i, M_{yi} and M_{xi} represent the generalized forces corresponding to the rotations (generalized displacements) $w_{yi}(\theta_{xi})$ and w_{xi} (θ_{yi}) at node i $(i = 1, 2, 3)$, respectively, and the subscript b has been used to denote the bending stiffness matrix. The 9×9 bending stiffness matrix (in local coordinate system) can be written as

$$[k^{(e)}]_b = \begin{bmatrix} \underset{3\times 3}{[k_{11}]_b} & \underset{3\times 3}{[k_{12}]_b} & \underset{3\times 3}{[k_{13}]_b} \\ \underset{3\times 3}{[k_{21}]_b} & \underset{3\times 3}{[k_{22}]_b} & \underset{3\times 3}{[k_{23}]_b} \\ \underset{3\times 3}{[k_{31}]_b} & \underset{3\times 3}{[k_{32}]_b} & \underset{3\times 3}{[k_{33}]_b} \end{bmatrix} \tag{10.91}$$

In the analysis of three-dimensional structures the inplane and bending stiffnessses have to be combined in accordance with the following observations:

(i) For small displacements, the inplane (membrane) and bending stiffnesses are uncoupled.

(ii) The inplane rotation θ_z (rotation about the local z axis) is not necessary for a single element. However, θ_z and its conjugate force M_z have to be considered in the analysis by including the appropriate number of zeroes to obtain the element stiffness matrix for the purpose of assembling several elements.

Therefore, to obtain the total element stiffness matrix $[k^{(e)}]$, the inplane and bending stiffnesses are combined as shown below.

$$
\underset{18\times18}{[k^{(e)}]} = \left[\begin{array}{c:c:c:c:c:c:c:c:c}
\begin{matrix}[k_{11}]_m\\2\times2\end{matrix} & \begin{matrix}0&0&0\\0&0&0\end{matrix} & \begin{matrix}0\\0\end{matrix} & \begin{matrix}[k_{12}]_m\\2\times2\end{matrix} & \begin{matrix}0&0&0\\0&0&0\end{matrix} & \begin{matrix}0\\0\end{matrix} & \begin{matrix}[k_{13}]_m\\2\times2\end{matrix} & \begin{matrix}0&0&0\\0&0&0\end{matrix} & \begin{matrix}0\\0\end{matrix} \\ \hdashline
\begin{matrix}0&0\\0&0\\0&0\end{matrix} & \begin{matrix}[k_{11}]_b\\3\times3\end{matrix} & \begin{matrix}0\\0\\0\end{matrix} & \begin{matrix}0&0\\0&0\\0&0\end{matrix} & \begin{matrix}[k_{12}]_b\\3\times3\end{matrix} & \begin{matrix}0\\0\\0\end{matrix} & \begin{matrix}0&0\\0&0\\0&0\end{matrix} & \begin{matrix}[k_{13}]_b\\3\times3\end{matrix} & \begin{matrix}0\\0\\0\end{matrix} \\ \hdashline
0\;\;0 & 0\;\;0\;\;0 & 0 & 0\;\;0 & 0\;\;0\;\;0 & 0 & 0\;\;0 & 0\;\;0\;\;0 & 0 \\ \hdashline
\begin{matrix}[k_{21}]_m\\2\times2\end{matrix} & \begin{matrix}0&0&0\\0&0&0\end{matrix} & \begin{matrix}0\\0\end{matrix} & \begin{matrix}[k_{22}]_m\\2\times2\end{matrix} & \begin{matrix}0&0&0\\0&0&0\end{matrix} & \begin{matrix}0\\0\end{matrix} & \begin{matrix}[k_{23}]_m\\2\times2\end{matrix} & \begin{matrix}0&0&0\\0&0&0\end{matrix} & \begin{matrix}0\\0\end{matrix} \\ \hdashline
\begin{matrix}0&0\\0&0\\0&0\end{matrix} & \begin{matrix}[k_{21}]_b\\3\times3\end{matrix} & \begin{matrix}0\\0\\0\end{matrix} & \begin{matrix}0&0\\0&0\\0&0\end{matrix} & \begin{matrix}[k_{22}]_b\\3\times3\end{matrix} & \begin{matrix}0\\0\\0\end{matrix} & \begin{matrix}0&0\\0&0\\0&0\end{matrix} & \begin{matrix}[k_{23}]_b\\3\times3\end{matrix} & \begin{matrix}0\\0\\0\end{matrix} \\ \hdashline
0\;\;0 & 0\;\;0\;\;0 & 0 & 0\;\;0 & 0\;\;0\;\;0 & 0 & 0\;\;0 & 0\;\;0\;\;0 & 0 \\ \hdashline
\begin{matrix}[k_{31}]_m\\2\times2\end{matrix} & \begin{matrix}0&0&0\\0&0&0\end{matrix} & \begin{matrix}0\\0\end{matrix} & \begin{matrix}[k_{32}]_m\\2\times2\end{matrix} & \begin{matrix}0&0&0\\0&0&0\end{matrix} & \begin{matrix}0\\0\end{matrix} & \begin{matrix}[k_{33}]_m\\2\times2\end{matrix} & \begin{matrix}0&0&0\\0&0&0\end{matrix} & \begin{matrix}0\\0\end{matrix} \\ \hdashline
\begin{matrix}0&0\\0&0\\0&0\end{matrix} & \begin{matrix}[k_{31}]_b\\3\times3\end{matrix} & \begin{matrix}0\\0\\0\end{matrix} & \begin{matrix}0&0\\0&0\\0&0\end{matrix} & \begin{matrix}[k_{32}]_b\\3\times3\end{matrix} & \begin{matrix}0\\0\\0\end{matrix} & \begin{matrix}0&0\\0&0\\0&0\end{matrix} & [k_{33}]_b & \begin{matrix}0\\0\\0\end{matrix} \\ \hdashline
0\;\;0 & 0\;\;0\;\;0 & 0 & 0\;\;0 & 0\;\;0\;\;0 & 0 & 0\;\;0 & 0\;\;0\;\;0 & 0
\end{array}\right] \tag{10.92}
$$

The stiffness matrix given by Eq. (10.92) is with reference to the local xyz coordinate system shown in Figure 10.15. In the analysis of three-dimensional structures in which different finite elements have different orientations, it is necessary to transform the local stiffness matrices to a common set of global coordinates. In this case, the global stiffness matrix of the element can be obtained as

$$
[K^{(e)}] = [\lambda]^T [k^{(e)}][\lambda] \tag{10.93}
$$

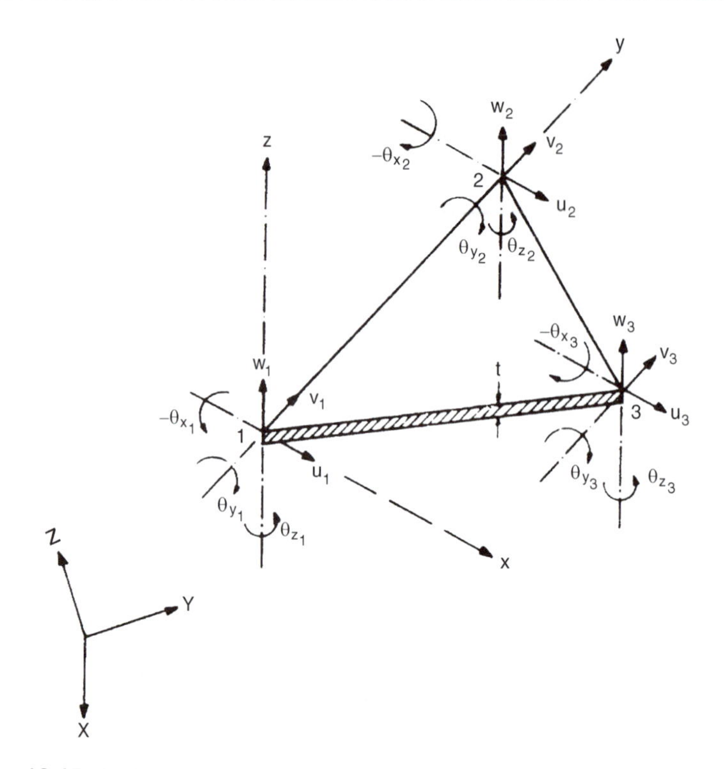

Figure 10.15. Inplane and Bending Displacements in a Local *xyz* Coordinate System.

where the transformation matrix, $[\lambda]$, is given by

$$\underset{18\times 18}{[\lambda]} = \begin{bmatrix} [\underline{\lambda}] & [0] & [0] \\ [0] & [\underline{\lambda}] & [0] \\ [0] & [0] & [\underline{\lambda}] \end{bmatrix} \tag{10.94}$$

and

$$\underset{6\times 6}{[\underline{\lambda}]} = \begin{bmatrix} l_{ox} & m_{ox} & n_{ox} & 0 & 0 & 0 \\ l_{oy} & m_{oy} & n_{oy} & 0 & 0 & 0 \\ l_{oz} & m_{oz} & n_{oz} & 0 & 0 & 0 \\ 0 & 0 & 0 & l_{ox} & m_{ox} & n_{ox} \\ 0 & 0 & 0 & l_{oy} & m_{oy} & n_{oy} \\ 0 & 0 & 0 & l_{oz} & m_{oz} & n_{oz} \end{bmatrix} \tag{10.95}$$

Here, (l_{ox}, m_{ox}, n_{ox}), for example, denotes the set of direction cosines of the x axis, and $[0]$ represents a null square matrix of order six.

10.10 COMPUTER PROGRAM FOR THREE-DIMENSIONAL STRUCTURES USING PLATE ELEMENTS

A Fortran subroutine called PLATE is written for the equilibrium and eigenvalue analysis of three-dimensional structures using triangular plate elements. The description and listing of the program are given in Chapter 12. A numerical example is also considered to illustrate the use of the program.

REFERENCES

10.1 I.H. Shames: *Mechanics of Deformable Solids*, Prentice Hall of India, New Delhi, 1965.

10.2 C.A. Felippa: Refined finite element analysis of linear and nonlinear two dimensional structures, *Ph.D. dissertation*, Department of Civil Engineering, University of California, Berkeley, 1966.

10.3 S. Timoshenko and S. Woinowsky-Krieger: *Theory of Plates and Shells*, 2nd Ed., McGraw-Hill, New York, 1959.

10.4 J.L. Batoz, K.J. Bathe, and L.W. Ho: A study of three-node triangular plate bending elements, *International Journal for Numerical Methods in Engineering, 15*, 1771–1812, 1980.

10.5 J.L. Batoz: An explicit formulation for an efficient triangular plate bending element, *International Journal for Numerical Methods in Engineering, 18*, 1077–1089, 1982.

10.6 A. Adini and R.W. Clough: *Analysis of Plate Bending by the Finite Element Method*, Report submitted to the National Science Foundation, Grant G7337, 1960.

10.7 F.K. Bogner, R.L. Fox, and L.A. Schmit, Jr.: The generation of interelement compatible stiffness and mass matrices by the use of interpolation formulas, *Proceedings of the First Conference on Matrix Methods in Structural Mechanics AFFDL-TR-66-80*, pp. 397–443, November 1966.

10.8 J.L. Tocher: Analysis of plate bending using triangular elements, *Ph.D. dissertation*, University of California, Berkeley, 1962.

10.9 A. Adini: Analysis of shell structures by the finite element method, *Ph.D. dissertation*, Department of Civil Engineering, University of California, Berkeley, 1961.

10.10 G.R. Cowper, E. Kosko, G.M. Lindberg, and M.D. Olson: Static and dynamic applications of a high-precision triangular plate element, *AIAA Journal, 7*, 1957–1965, 1969.

PROBLEMS

10.1 Find the stresses in the plate shown in Figure 10.16 using one triangular membrane element.

10.2 Find the stresses in the plate shown in Figure 10.17 using two triangular membrane elements.

10.3 Find the coordinate transformation matrix for the triangular membrane element shown in Figure 10.18.

10.4 The plate shown in Figure 10.16 is heated by 50°C. Determine the load vector. Assume the coefficient of expansion of the material as $\alpha = 12 \times 10^{-6}$ per °C.

10.5 The nodal coordinates and the nodal displacements of a triangular element, under a specific load condition, are given below:

$X_i = 0$, $Y_i = 0$, $X_j = 1$ in., $Y_j = 3$ in., $X_k = 4$ in., $Y_k = 1$ in.

$Q_{2i-1} = 0.001$ in., $Q_{2i} = 0.0005$ in., $Q_{2j-1} = -0.0005$ in., $Q_{2j} = 0.0015$ in., $Q_{2k-1} = 0.002$ in., $Q_{2k} = -0.001$ in.

If $E = 30 \times 10^6$ psi and $v = 0.3$, find the stresses in the element.

10.6 For a triangular element in a state of plane stress, it is proposed to consider three corner and three midside nodes. Suggest a suitable displacement model and discuss its convergence and other properties.

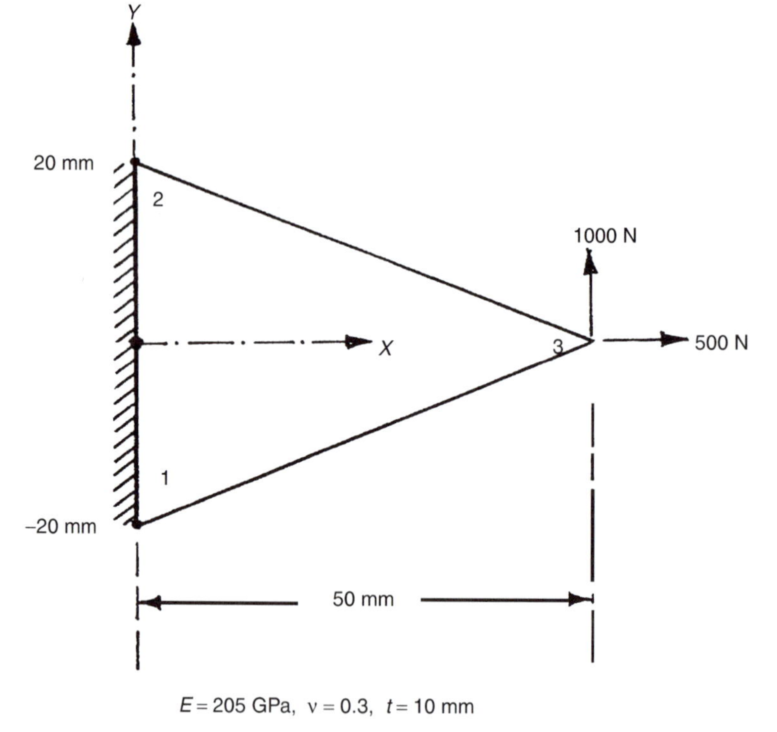

Figure 10.16.

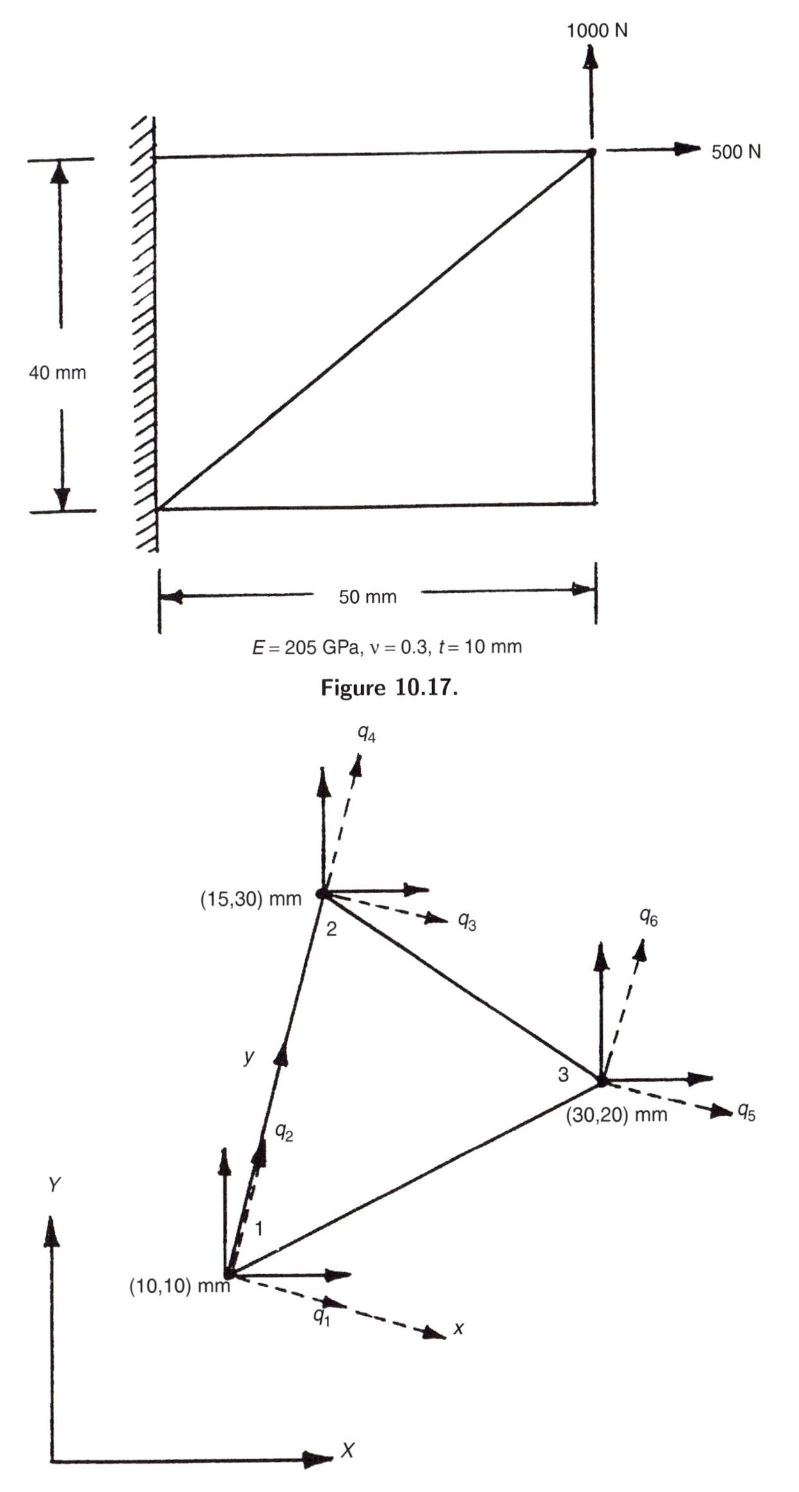

Figure 10.17.

Figure 10.18.

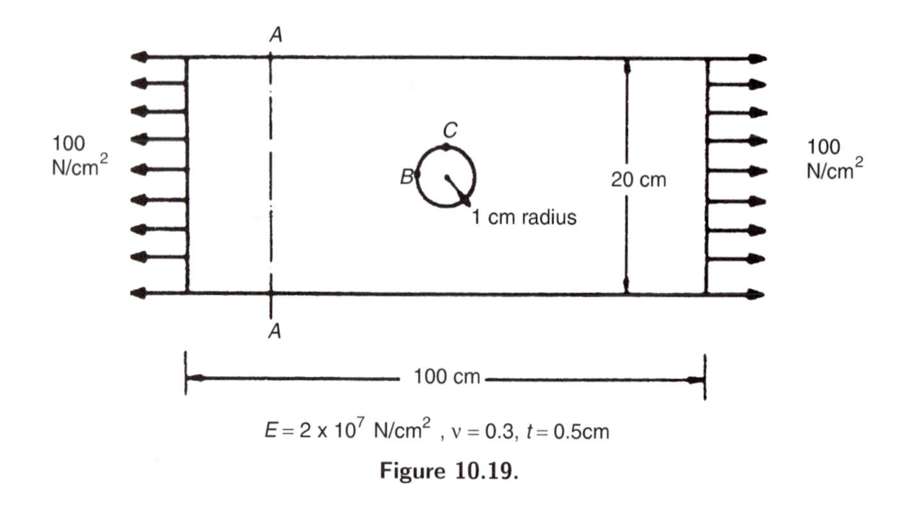

Figure 10.19.

10.7 Modify the subroutine CST so as to make it applicable for the stress analysis of three-dimensional structures using constant strain triangles. Using this subroutine, find the deflections and stresses in the box beam of Section 10.3.3.

10.8 Find the stress concentration factors at points B and C of the plate with a hole shown in Figure 10.19 using the subroutine CST. Definition: Stress concentration factor at B or C

$$= \left(\frac{\text{stress at point B or C}}{\text{stress along section AA}} \right)$$

10.9 Explain why the sum of coefficients of the stiffness matrix in any row for triangular plates with only inplane loads is equal to zero; that is, $\Sigma_j k_{ij} = 0$ for any row i.

10.10 Consider two rectangular plate elements joined as shown in Figure 10.20. If both inplane and bending actions are considered, what conditions do you impose on the nodal displacements of the two elements if the edge AB is (i) hinged and (ii) welded?

10.11 A triangular plate is subjected to a transverse load of 1000 N as shown in Figure 10.21. Find the transverse displacement and the stresses induced in the plate using a one-element idealization. Assume $E = 205$ GPa, $v = 0.33$, and $t = 10$ mm.

10.12 Consider a rectangular element in plane stress (Figure 10.22) with a bilinear displacement model:

$$u(\xi, \eta) = \sum_{i=1}^{4} N_i(\xi, \eta) u_i$$

$$v(\xi, \eta) = \sum_{i=1}^{4} N_i(\xi, \eta) v_i$$

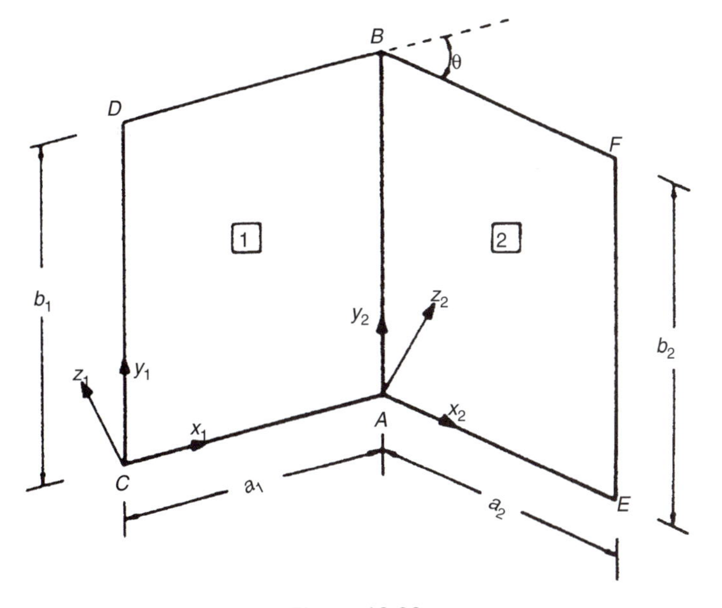

Figure 10.20.

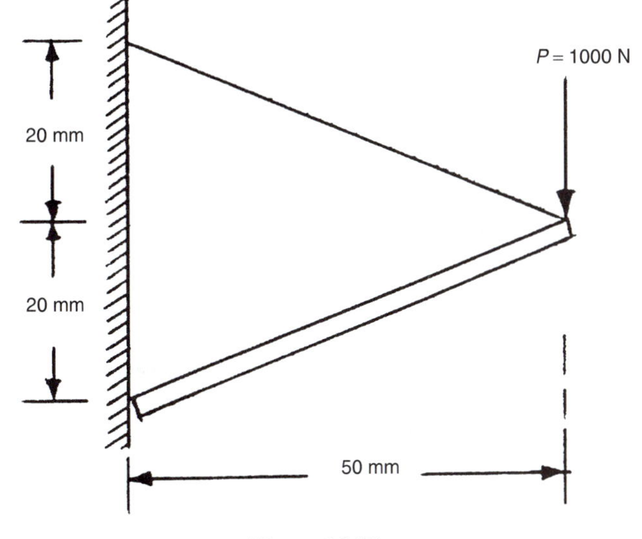

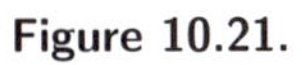

Figure 10.21.

where

$N_1(\xi,\eta) = (1-\xi)(1-\eta)$, $N_2(\xi,\eta) = \xi(1-\eta)$,

$N_3(\xi,\eta) = \xi\eta$, $N_4(\xi,\eta) = (1-\xi)\eta$

If the strains are given by $\varepsilon_x = (\partial u/\partial x)$, $\varepsilon_y = (\partial v/\partial y)$, and $\varepsilon_{xy} = (\partial u/\partial y) + (\partial v/\partial x)$, derive the element stiffness matrix.

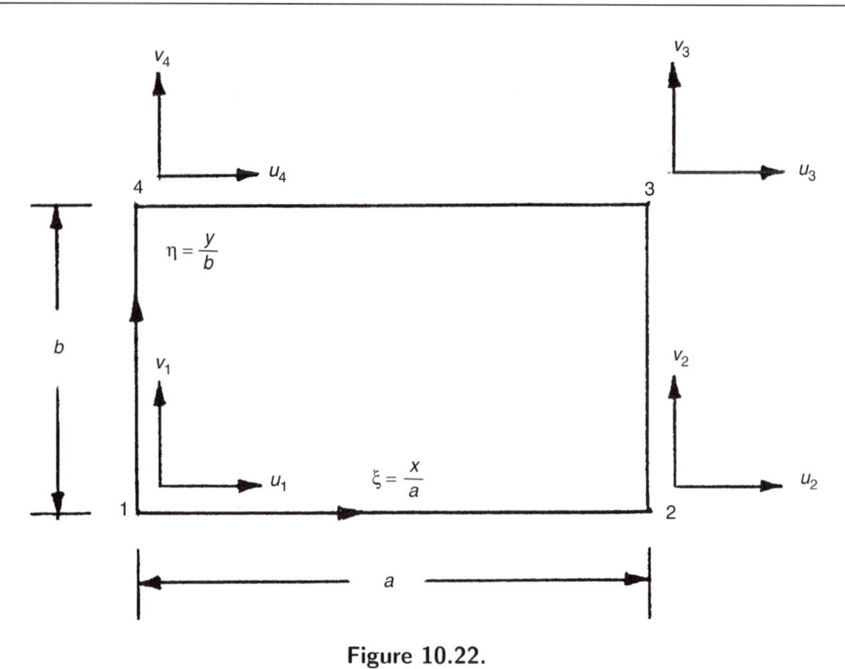

Figure 10.22.

10.13 A rectangular plate, simply supported on all the edges, is subjected to a distributed transverse load of

$$p(x, y) = p_0 \sin \frac{\pi x}{a} \sin \frac{\pi y}{b}$$

where a and b are the dimensions of the plate (Figure 10.23).

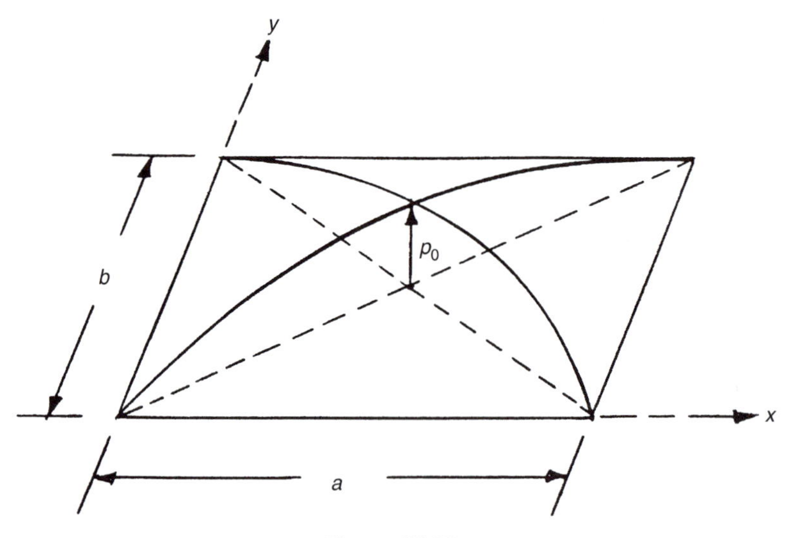

Figure 10.23.

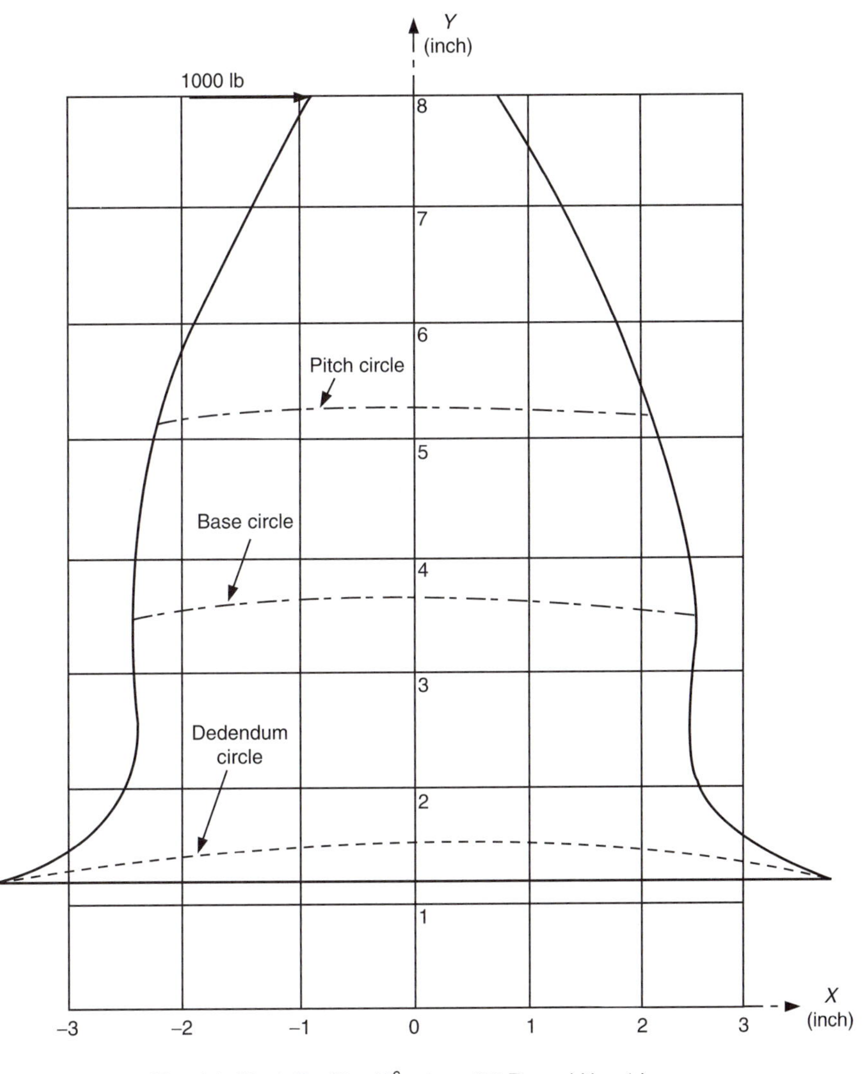

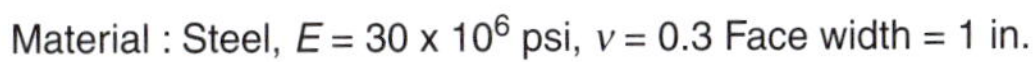
Material : Steel, $E = 30 \times 10^6$ psi, $\nu = 0.3$ Face width = 1 in.

Figure 10.24.

(a) Verify that the displacement solution

$$w(x, y) = c \sin \frac{\pi x}{a} \sin \frac{\pi y}{b} \tag{E$_1$}$$

where

$$c = \frac{p_0}{\pi^4 D \left[\dfrac{1}{a^2} + \dfrac{1}{b^2} \right]^2}$$

satisfies the equilibrium equation and the boundary conditions.

(b) Using the solution of Eq. (E_1), find exprressions for the moments and reactions in the plate.

10.14 Using the subroutine CST, find the nodal displacements and element stresses of the gear tooth shown in Figure 10.24. Compare the finite element solution with the approximate solution used in the machine design literature (Lewis solution). Use at least 50 finite elements for modeling the gear tooth.

11

ANALYSIS OF THREE-DIMENSIONAL PROBLEMS

11.1 INTRODUCTION

For the realistic analysis of certain problems such as thick short beams, thick pressure vessels, elastic half space acted on by a concentrated load, and machine foundations, we have to use three-dimensional finite elements. Just like a triangular element is a basic element for analyzing two-dimensional problems, the tetrahedron element, with four corner nodes, is the basic element for modeling three-dimensional problems. One of the major difficulties associated with the use of three-dimensional elements (e.g., tetrahedra, hexahedra, and rectangular parallelepiped elements) is that a large number of elements have to be used for obtaining reasonably accurate results. This will result in a very large number of simultaneous equations to be solved in static analyses. Despite this difficulty, we may not have any other choice except to use three-dimensional elements in certain situations. Hence, the tetrahedron and hexahedron elements are considered in this chapter [11.1–11.3].

11.2 TETRAHEDRON ELEMENT

The tetrahedron element, with three translational degrees of freedom per node, is shown in the global xyz coordinate system in Figure 11.1 (the global coordinates are denoted as x, y, z instead of X, Y, Z, for simplicity). For this element, there will be no advantage in setting up a local coordinate system, and hence we shall derive all the elemental equations in the global system. Since there are 12 nodal degrees of freedom $Q_{3i-2}, Q_{3i-1}, Q_{3i}, Q_{3j-2} \dots, Q_{3l}$ and three displacement components u, v, and w, we choose the displacement variation to be linear as

$$\left.\begin{aligned} u(x,y,z) &= \alpha_1 + \alpha_2 x + \alpha_3 y + \alpha_4 z \\ v(x,y,z) &= \alpha_5 + \alpha_6 x + \alpha_7 y + \alpha_8 z \\ w(x,y,z) &= \alpha_9 + \alpha_{10} x + \alpha_{11} y + \alpha_{12} z \end{aligned}\right\} \tag{11.1}$$

where $\alpha_1, \alpha_2, \dots, \alpha_{12}$ are constants. By using the nodal conditions

$$\begin{aligned} &u = Q_{3i-2}, \quad && v = Q_{3i-1}, \quad && w = Q_{3i} \quad \text{at} \quad (x_i, y_i, z_i) \\ &u = Q_{3j-2}, \quad && v = Q_{3j-1}, \quad && w = Q_{3j} \quad \text{at} \quad (x_j, y_j, z_j) \end{aligned}$$

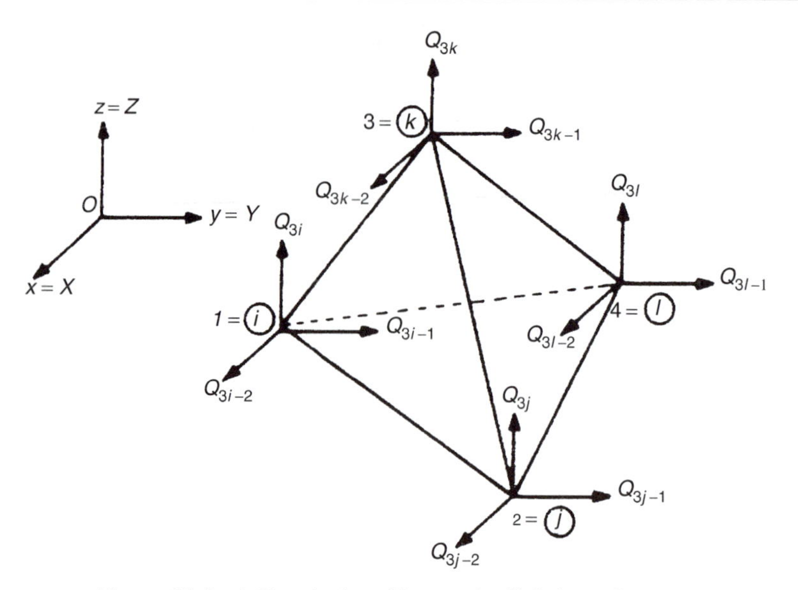

Figure 11.1. A Tetrahedron Element in Global xyz System.

$$
\begin{aligned}
u &= Q_{3k-2}, \quad v = Q_{3k-1}, \quad w = Q_{3k} \quad \text{at} \quad (x_k, y_k, z_k) \\
u &= Q_{3l-2}, \quad v = Q_{3l-1}, \quad w = Q_{3l} \quad \text{at} \quad (x_l, y_l, z_l)
\end{aligned} \tag{11.2}
$$

we can obtain

$$
\begin{aligned}
u(x, y, z) = {} & N_i(x, y, z) Q_{3i-2} + N_j(x, y, z) Q_{3j-2} + N_k(x, y, z) Q_{3k-2} \\
& + N_l(x, y, z) Q_{3l-2}
\end{aligned} \tag{11.3}
$$

where N_i, N_j, N_k, and N_l are given by Eq. (3.48), and similar expressions for $v(x, y, z)$ and $w(x, y, z)$. Thus, the displacement field can be expressed in matrix form as

$$
\underset{3 \times 1}{\vec{U}} = \begin{Bmatrix} u(x, y, z) \\ v(x, y, z) \\ w(x, y, z) \end{Bmatrix} = \underset{3 \times 12}{[N]} \; \underset{12 \times 1}{\vec{Q}^{(e)}} \tag{11.4}
$$

where

$$
[N] = \begin{bmatrix} N_i & 0 & 0 & N_j & 0 & 0 & N_k & 0 & 0 & N_l & 0 & 0 \\ 0 & N_i & 0 & 0 & N_j & 0 & 0 & N_k & 0 & 0 & N_l & 0 \\ 0 & 0 & N_i & 0 & 0 & N_j & 0 & 0 & N_k & 0 & 0 & N_l \end{bmatrix} \tag{11.5}
$$

and

$$
\vec{Q}^{(e)} = \begin{Bmatrix} Q_{3i-2} \\ Q_{3i-1} \\ Q_{3i} \\ \vdots \\ Q_{3l} \end{Bmatrix} \tag{11.6}
$$

Noting that all six strain components are relevant in three-dimensional analysis, the strain–displacement relations can be expressed, using Eq. (11.4), as

$$\underset{6\times 1}{\vec{\varepsilon}} = \begin{Bmatrix} \varepsilon_{xx} \\ \varepsilon_{yy} \\ \varepsilon_{zz} \\ \varepsilon_{xy} \\ \varepsilon_{yz} \\ \varepsilon_{zx} \end{Bmatrix} = \begin{Bmatrix} \partial u/\partial x \\ \partial v/\partial y \\ \partial w/\partial z \\ \dfrac{\partial u}{\partial y} + \dfrac{\partial v}{\partial x} \\ \dfrac{\partial v}{\partial z} + \dfrac{\partial w}{\partial y} \\ \dfrac{\partial w}{\partial x} + \dfrac{\partial u}{\partial z} \end{Bmatrix} = \underset{6\times 12}{[B]} \; \underset{12\times 1}{\vec{Q}^{(e)}} \tag{11.7}$$

where

$$[B] = \frac{1}{6V}\begin{bmatrix} b_i & 0 & 0 & b_j & 0 & 0 & b_k & 0 & 0 & b_l & 0 & 0 \\ 0 & c_i & 0 & 0 & c_j & 0 & 0 & c_k & 0 & 0 & c_l & 0 \\ 0 & 0 & d_i & 0 & 0 & d_j & 0 & 0 & d_k & 0 & 0 & d_l \\ c_i & b_i & 0 & c_j & b_j & 0 & c_k & b_k & 0 & c_l & b_l & 0 \\ 0 & d_i & c_i & 0 & d_j & c_j & 0 & d_k & c_k & 0 & d_l & c_l \\ d_i & 0 & b_i & d_j & 0 & b_j & d_k & 0 & b_k & d_l & 0 & b_l \end{bmatrix} \tag{11.8}$$

The stress–strain relations, in the case of three-dimensional analysis, are given by Eq. (8.10) as

$$\vec{\sigma} = [D]\vec{\varepsilon} \tag{11.9}$$

where

$$\vec{\sigma}^T = \{\sigma_{xx} \quad \sigma_{yy} \quad \sigma_{zz} \quad \sigma_{xy} \quad \sigma_{yz} \quad \sigma_{zx}\}$$

and

$$[D] = \frac{E}{(1+v)(1-2v)}\begin{bmatrix} (1-v) & v & v & 0 & 0 & 0 \\ v & (1-v) & v & 0 & 0 & 0 \\ v & v & (1-v) & & 0 & 0 \\ 0 & 0 & 0 & \left(\dfrac{1-2v}{2}\right) & 0 & 0 \\ 0 & 0 & 0 & 0 & \left(\dfrac{1-2v}{2}\right) & 0 \\ 0 & 0 & 0 & 0 & 0 & \left(\dfrac{1-2v}{2}\right) \end{bmatrix} \tag{11.10}$$

The stiffness matrix of the element (in the global system) can be obtained as

$$[K^{(e)}] = \iiint\limits_{V^{(e)}} [B]^T[D][B]\,\mathrm{d}V \tag{11.11}$$

Since the matrices $[B]$ and $[D]$ are independent of x, y, and z, the stiffness matrix can be obtained by carrying out matrix multiplications as

$$[K^{(e)}] = V^{(e)}[B]^T[D][B] \tag{11.12}$$

In this case, since the assumed displacement model is linear, the continuity of displacement along the interface between neighboring elements will be satisfied automatically.

11.2.1 Consistent Load Vector

The total load vector due to initial (thermal) strains, body forces $\vec{\phi} = \begin{Bmatrix} \phi_x \\ \phi_y \\ \phi_z \end{Bmatrix}$, and surface (distributed) forces $\vec{\bar{\Phi}} = \begin{Bmatrix} p_{x0} \\ p_{y0} \\ p_{z0} \end{Bmatrix}$ can be computed using Eqs. (8.88), (8.90), and (8.89) as

$$\vec{P}^{(e)} = \iiint\limits_{V^{(e)}} [B]^T[D]\cdot \alpha T \begin{Bmatrix} 1\\1\\1\\0\\0\\0 \end{Bmatrix} \mathrm{d}V + \iiint\limits_{V^{(e)}} [N]^T \begin{Bmatrix} \phi_x \\ \phi_y \\ \phi_z \end{Bmatrix} \cdot \mathrm{d}V + \iint\limits_{S_1^{(e)}} [N]^T \begin{Bmatrix} p_{x0} \\ p_{y0} \\ p_{z0} \end{Bmatrix} \cdot \mathrm{d}S_1$$

$$= \frac{E\cdot\alpha\cdot T\cdot V^{(e)}}{(1-2\nu)}[B]^T \begin{Bmatrix} 1\\1\\1\\0\\0\\0 \end{Bmatrix} + \frac{V^{(e)}}{4} \begin{Bmatrix} \phi_x\\ \phi_y\\ \phi_z\\ \phi_x\\ \phi_y\\ \phi_z\\ \phi_x\\ \phi_y\\ \phi_z\\ \phi_x\\ \phi_y\\ \phi_z \end{Bmatrix} + \frac{S_{ijk}^{(e)}}{3} \begin{Bmatrix} p_{x0}\\ p_{y0}\\ p_{z0}\\ p_{x0}\\ p_{y0}\\ p_{z0}\\ p_{x0}\\ p_{y0}\\ p_{z0}\\ 0\\ 0\\ 0 \end{Bmatrix} \tag{11.13}$$

Equation (11.13) shows that the body force is distributed equally between the four nodes of the element. It is assumed in deriving Eq. (11.13) that the surface forces are distributed only on the face $i\ j\ k$ of the element e. These surface forces can be seen to be equally distributed between the three nodes i, j, and k, which define the loaded face. $S_{ijk}^{(e)}$ denotes the area of the face $i\ j\ k$ of element e. The last three components of the surface load vector are zero since they are related to the term $\iint N_l \cdot \mathrm{d}S_1$ and N_l is zero on the face $i\ j\ k$. Note that the location of the zero terms changes in the last column of Eq. (11.13), and their location depends on which face the surface forces are acting. If more than one face of the element e is subjected to the surface forces, then there will be additional surface load vectors in Eq. (11.13).

11.3 HEXAHEDRON ELEMENT

In this section, we consider the simplest hexahedron element having eight corner nodes with three degrees of freedom per node. For convenience, we derive the element matrices by treating it as an isoparametric element. This element is also known as Zienkiewicz–Irons–Brick with eight nodes (ZIB 8) and is shown in Figure 11.2(a).

11.3.1 Natural Coordinate System

As shown in Figure 11.2(a), the natural coordinates are r, s, and t with the origin of the system taken at the centroid of the element. It can be seen that each of the coordinate axes r, s, and t is associated with a pair of opposite faces, which are given by the coordinate values ± 1. Thus, in the local (natural) coordinates, the element is a cube as shown in Figure 11.2(b), although in the global Cartesian coordinate system it may be an arbitrarily warped and distorted six-sided solid as shown in Figure 11.2(a). The relationship between the local and global coordinates can be expressed as

$$\begin{Bmatrix} x \\ y \\ z \end{Bmatrix} = [N] \begin{Bmatrix} x_1 \\ y_1 \\ z_1 \\ x_2 \\ \vdots \\ z_8 \end{Bmatrix} \tag{11.14}$$

where

$$[N] = \begin{bmatrix} N_1 & 0 & 0 & N_2 & \dots & 0 \\ 0 & N_1 & 0 & 0 & \dots & 0 \\ 0 & 0 & N_1 & 0 & \dots & N_8 \end{bmatrix} \tag{11.15}$$

and

$$N_i(r,s,t) = \frac{1}{8}(1 + rr_i)(1 + ss_i)(1 + tt_i); \qquad i = 1, 2, \dots, 8 \tag{11.16}$$

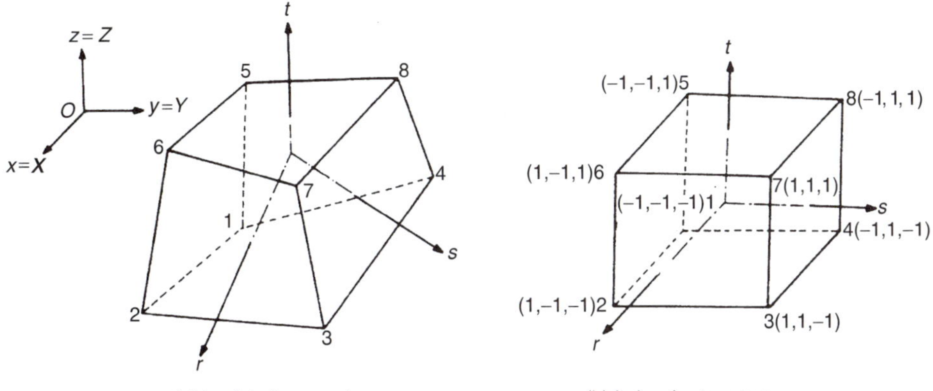

(a) In global *xyz* system (b) In local *rst* system

Figure 11.2. A Hexahedron Element with Eight Nodes.

or

$$\begin{Bmatrix} x \\ y \\ z \end{Bmatrix} = \begin{Bmatrix} \sum_{i=1}^{8} N_i x_i \\ \sum_{i=1}^{8} N_i y_i \\ \sum_{i=1}^{8} N_i z_i \end{Bmatrix} \tag{11.17}$$

11.3.2 Displacement Model

By assuming the variations of the displacements in between the nodes to be linear, the displacements can be expressed by the same interpolation functions used to describe the geometry as (analogous to Eq. 11.14)

$$\begin{Bmatrix} u \\ v \\ w \end{Bmatrix} = [N] \begin{Bmatrix} u_1 \\ v_1 \\ w_1 \\ u_2 \\ \vdots \\ w_8 \end{Bmatrix} = [N]\vec{Q}^{(e)} \tag{11.18}$$

where $\vec{Q}^{(e)}$ is the vector of nodal displacement degrees of freedom, and (u_i, v_i, w_i) denote the displacements of node i, $i = 1$–8.

11.3.3 Strain–Displacement and Stress–Strain Relations

Using Eq. (11.18), the three-dimensional strain–displacement relations can be expressed as

$$\vec{\varepsilon} = \begin{Bmatrix} \varepsilon_{xx} \\ \varepsilon_{yy} \\ \varepsilon_{zz} \\ \varepsilon_{xy} \\ \varepsilon_{yz} \\ \varepsilon_{zx} \end{Bmatrix} = \begin{Bmatrix} \frac{\partial u}{\partial x} \\ \frac{\partial v}{\partial y} \\ \frac{\partial w}{\partial z} \\ \frac{\partial u}{\partial y} + \frac{\partial v}{\partial x} \\ \frac{\partial v}{\partial z} + \frac{\partial w}{\partial y} \\ \frac{\partial w}{\partial x} + \frac{\partial u}{\partial z} \end{Bmatrix} = \underset{6 \times 24}{[B]} \; \underset{24 \times 1}{\vec{Q}^{(e)}} \tag{11.19}$$

where

$$\underset{6 \times 24}{[B]} = [[B_1][B_2] \dots [B_8]] \tag{11.20}$$

and

$$\underset{6\times 3}{[B_i]} = \begin{bmatrix} \dfrac{\partial N_i}{\partial x} & 0 & 0 \\ 0 & \dfrac{\partial N_i}{\partial y} & 0 \\ 0 & 0 & \dfrac{\partial N_i}{\partial z} \\ \dfrac{\partial N_i}{\partial y} & \dfrac{\partial N_i}{\partial x} & 0 \\ 0 & \dfrac{\partial N_i}{\partial z} & \dfrac{\partial N_i}{\partial y} \\ \dfrac{\partial N_i}{\partial z} & 0 & \dfrac{\partial N_i}{\partial x} \end{bmatrix}, \quad i = 1\text{–}8 \tag{11.21}$$

The derivatives in the matrix $[B_i]$ may be evaluated by applying the chain rule of differentiation as follows:

$$\begin{Bmatrix} \dfrac{\partial N_i}{\partial r} \\ \dfrac{\partial N_i}{\partial s} \\ \dfrac{\partial N_i}{\partial t} \end{Bmatrix} = \begin{Bmatrix} \dfrac{\partial N_i}{\partial x}\dfrac{\partial x}{\partial r} + \dfrac{\partial N_i}{\partial y}\dfrac{\partial y}{\partial r} + \dfrac{\partial N_i}{\partial z}\dfrac{\partial z}{\partial r} \\ \dfrac{\partial N_i}{\partial x}\dfrac{\partial x}{\partial s} + \dfrac{\partial N_i}{\partial y}\dfrac{\partial y}{\partial s} + \dfrac{\partial N_i}{\partial z}\dfrac{\partial z}{\partial s} \\ \dfrac{\partial N_i}{\partial x}\dfrac{\partial x}{\partial t} + \dfrac{\partial N_i}{\partial y}\dfrac{\partial y}{\partial t} + \dfrac{\partial N_i}{\partial z}\dfrac{\partial z}{\partial t} \end{Bmatrix}$$

$$= \begin{bmatrix} \dfrac{\partial x}{\partial r} & \dfrac{\partial y}{\partial r} & \dfrac{\partial z}{\partial r} \\ \dfrac{\partial x}{\partial s} & \dfrac{\partial y}{\partial s} & \dfrac{\partial z}{\partial s} \\ \dfrac{\partial x}{\partial t} & \dfrac{\partial y}{\partial t} & \dfrac{\partial z}{\partial t} \end{bmatrix} \begin{Bmatrix} \dfrac{\partial N_i}{\partial x} \\ \dfrac{\partial N_i}{\partial y} \\ \dfrac{\partial N_i}{\partial z} \end{Bmatrix} = [J] \begin{Bmatrix} \dfrac{\partial N_i}{\partial x} \\ \dfrac{\partial N_i}{\partial y} \\ \dfrac{\partial N_i}{\partial z} \end{Bmatrix} \tag{11.22}$$

where $[J]$ is the Jacobian matrix, which can be expressed, using Eq. (11.17), as

$$\underset{3\times 3}{[J]} = \begin{bmatrix} \dfrac{\partial x}{\partial r} & \dfrac{\partial y}{\partial r} & \dfrac{\partial z}{\partial r} \\ \dfrac{\partial x}{\partial s} & \dfrac{\partial y}{\partial s} & \dfrac{\partial z}{\partial s} \\ \dfrac{\partial x}{\partial t} & \dfrac{\partial y}{\partial t} & \dfrac{\partial z}{\partial t} \end{bmatrix} = \begin{bmatrix} \sum_{i=1}^{8}\left(\dfrac{\partial N_i}{\partial r}x_i\right) & \sum_{i=1}^{8}\left(\dfrac{\partial N_i}{\partial r}y_i\right) & \sum_{i=1}^{8}\left(\dfrac{\partial N_i}{\partial r}z_i\right) \\ \sum_{i=1}^{8}\left(\dfrac{\partial N_i}{\partial s}x_i\right) & \sum_{i=1}^{8}\left(\dfrac{\partial N_i}{\partial s}y_i\right) & \sum_{i=1}^{8}\left(\dfrac{\partial N_i}{\partial s}z_i\right) \\ \sum_{i=1}^{8}\left(\dfrac{\partial N_i}{\partial t}x_i\right) & \sum_{i=1}^{8}\left(\dfrac{\partial N_i}{\partial t}y_i\right) & \sum_{i=1}^{8}\left(\dfrac{\partial N_i}{\partial t}z_i\right) \end{bmatrix} \tag{11.23}$$

The derivatives of the interpolation functions can be obtained from Eq. (11.16) as

$$\left.\begin{aligned}\frac{\partial N_i}{\partial r} &= \frac{1}{8} r_i (1 + s s_i)(1 + t t_i)\\ \frac{\partial N_i}{\partial s} &= \frac{1}{8} s_i (1 + r r_i)(1 + t t_i)\\ \frac{\partial N_i}{\partial t} &= \frac{1}{8} t_i (1 + r r_i)(1 + s s_i)\end{aligned}\right\}, \quad i = 1\text{–}8 \tag{11.24}$$

and the coordinates of the nodes in the local system (r_i, s_i, t_i) are shown in Figure 11.2. By inverting Eq. (11.22), we obtain

$$\begin{Bmatrix} \dfrac{\partial N_i}{\partial x} \\ \dfrac{\partial N_i}{\partial y} \\ \dfrac{\partial N_i}{\partial z} \end{Bmatrix} = [J]^{-1} \begin{Bmatrix} \dfrac{\partial N_i}{\partial r} \\ \dfrac{\partial N_i}{\partial s} \\ \dfrac{\partial N_i}{\partial t} \end{Bmatrix} \tag{11.25}$$

from which the matrix $[B_i]$ can be evaluated. The stress–strain relations are the same as those given in Eqs. (11.9) and (11.10).

11.3.4 Element Stiffness Matrix

The element stiffness matrix is given by

$$[K^{(e)}] = \iiint\limits_{V^{(e)}} [B]^T [D][B] \, \mathrm{d}V \tag{11.26}$$

Since the matrix $[B]$ is expressed in natural coordinates [evident from Eqs. (11.20), (11.21), and (11.25)], it is necessary to carry out the integration in Eq. (11.26) in natural coordinates too, using the relationship

$$\mathrm{d}V = \mathrm{d}x \, \mathrm{d}y \, \mathrm{d}z = \det[J] \cdot \mathrm{d}r \, \mathrm{d}s \, \mathrm{d}t \tag{11.27}$$

Thus, Eq. (11.26) can be rewritten as

$$[K^{(e)}] = \int_{-1}^{1} \int_{-1}^{1} \int_{-1}^{1} [B]^T [D][B] \det[J] \, \mathrm{d}r \, \mathrm{d}s \, \mathrm{d}t \tag{11.28}$$

11.3.5 Numerical Computation

Since the matrix $[B]$ is an implicit (not explicit!) function of r, s, and t, a numerical method has to be used to evaluate the multiple integral of Eq. (11.28). The Gaussian quadrature has been proven to be the most efficient method of numerical integration for this class of problems. By using the two-point Gaussian quadrature, which yields sufficiently accurate results, Eq. (11.28) can be evaluated as [11.4]

$$[K^{(e)}] = \sum_{r=R_i=R_1}^{R_2} \sum_{s=S_j=S_1}^{S_2} \sum_{t=T_k=T_1}^{T_2} [([B]^T[D][B] \cdot \det[J])|_{(R_i,S_j,T_k)}] \tag{11.29}$$

where

$$[([B]^T[D][B] \cdot \det[J])|_{(R_i,S_j,T_k)}] \tag{11.30}$$

indicates the value of

$$([B]^T[D][B] \det[J])$$

evaluated at $r = R_i$, $s = S_j$, and $t = T_k$, and $R_1 = S_1 = T_1 = -0.57735$ and $R_2 = S_2 = T_2 = +0.57735$.

11.3.6 Numerical Results

The performance of the three-dimensional elements considered in Sections 11.2 and 11.3, namely the tetrahedron and hexahedron elements, is studied by taking the short cantilever beam shown in Figure 11.3 as the test case. This cantilever is modeled as an assemblage of 42 identical hexahedra, each $2 \times 2 \times 3$ in. In the case of the tetrahedron element, each of the 42 hexahedra is considered to be composed of 5 tetrahedron elements. The cantilever

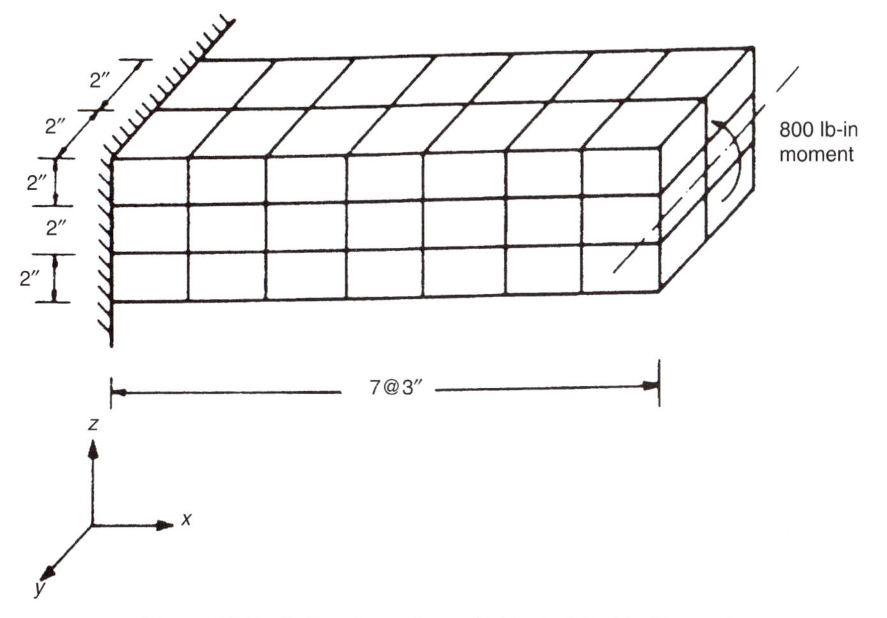

Figure 11.3. A Cantilever Beam Subjected to Tip Moment.

beam is subjected to a tip moment of 800 lb-in. as indicated in Figure 11.3. The numerical results obtained are indicated below [11.4]:

Element type	Maximum stress		Maximum deflection at the c.g. of tip
	σ_{xx}	σ_{zz}	
Tetrahedron	—	—	0.606×10^{-4} in.
ZIB 8	31.6 psi	1.4 psi	0.734×10^{-4} in.
Beam theory	33.3 psi	0.0 psi	0.817×10^{-4} in.

It can be seen that ZIB 8 is superior to the tetrahedron element.

11.4 ANALYSIS OF SOLIDS OF REVOLUTION

11.4.1 Introduction

The problem of stress analysis of solids of revolution (axisymmetric solids) under axisymmetric loads is of considerable practical interest. This problem is similar to those of plane stress and plane strain since the displacements are confined to only two directions (radial and axial) [11.5, 11.6]. The basic element that can be used for modeling solids of revolution is the axisymmetric ring element having triangular cross section. This element was originally developed by Wilson [11.7]. This element is useful for analyzing thick axisymmetric shells, solid bodies of revolution, turbine disks (Figure 11.4), and circular footings on a soil mass. In this section, the derivation of the element stiffness matrix and load vectors for the axisymmetric ring element is presented.

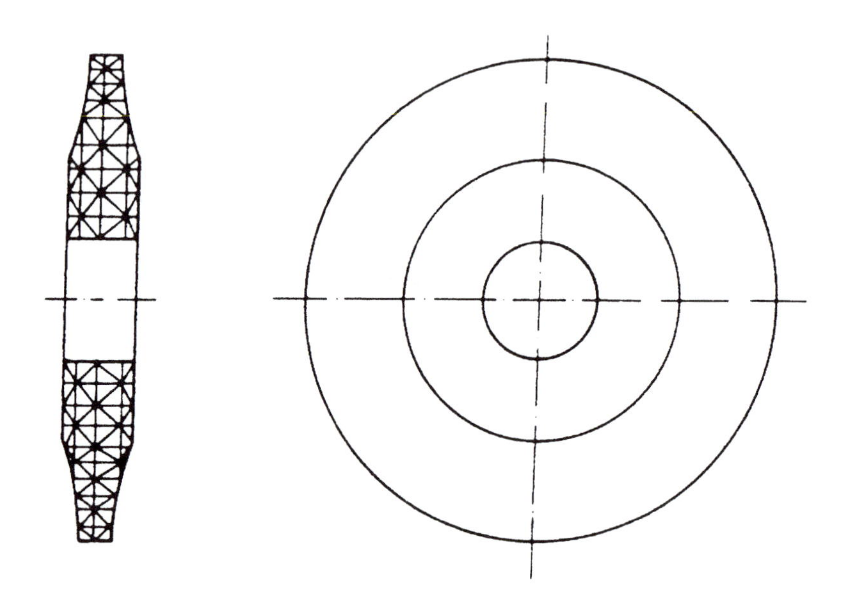

Figure 11.4. Turbine Disk Modeled by Triangular Ring Elements.

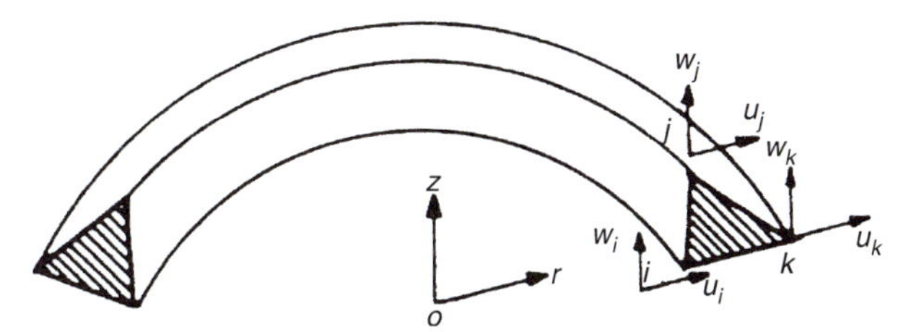

Figure 11.5. An Axisymmetric Ring Element with Triangular Cross Section.

11.4.2 Formulation of Elemental Equations for an Axisymmetric Ring Element

An axisymmetric ring element with a triangular cross section is shown in cylindrical coordinates in Figure 11.5. For axisymmetric deformation, since the displacement v along θ direction is zero (due to symmetry), the relevant displacement components are only u and w in the r and z directions, respectively. By taking the nodal values of u and w as the degrees of freedom, a linear displacement model can be assumed in terms of triangular coordinates L_i, L_j, and L_k as

$$\vec{U} = \begin{Bmatrix} u(r,z) \\ w(r,z) \end{Bmatrix} = [N]\vec{Q}^{(e)} \tag{11.31}$$

where

$$[N] = \begin{bmatrix} N_i & O & N_j & O & N_k & O \\ O & N_i & O & N_j & O & N_k \end{bmatrix} \tag{11.32}$$

$$\vec{Q}^{(e)} = \begin{Bmatrix} Q_{2i-1} \\ Q_{2i} \\ Q_{2j-1} \\ Q_{2j} \\ Q_{2k-1} \\ Q_{2k} \end{Bmatrix}^{(e)} \equiv \begin{Bmatrix} u_i \\ w_i \\ u_j \\ w_j \\ u_k \\ w_k \end{Bmatrix}^{(e)} \tag{11.33}$$

$$\begin{Bmatrix} N_i \\ N_j \\ N_k \end{Bmatrix} = \begin{Bmatrix} L_i \\ L_j \\ L_k \end{Bmatrix} = \frac{1}{2A} \begin{Bmatrix} a_i + b_i r + c_i z \\ a_j + b_j r + c_j z \\ a_k + b_k r + c_k z \end{Bmatrix} \tag{11.34}$$

$$A = \frac{1}{2}(r_i z_j + r_j z_k + r_k z_i - r_i z_k - r_j z_i - r_k z_j) \tag{11.35}$$

(r_i, z_i) are the (r, z) coordinates of node i, and $a_i, a_j, a_k, \ldots, c_k$ can be obtained from Eq. (3.32) by substituting r and z in place of x and y, respectively. In this case, there are four relevant strains, namely ε_{rr}, $\varepsilon_{\theta\theta}$, ε_{zz}, and ε_{rz}, for the axisymmetric case.

The strain–displacement relations can be expressed as

$$\vec{\varepsilon} = \begin{Bmatrix} \varepsilon_{rr} \\ \varepsilon_{\theta\theta} \\ \varepsilon_{zz} \\ \varepsilon_{rz} \end{Bmatrix} = \begin{Bmatrix} \dfrac{\partial u}{\partial r} \\ \dfrac{u}{r} \\ \dfrac{\partial w}{\partial z} \\ \dfrac{\partial u}{\partial z} + \dfrac{\partial w}{\partial r} \end{Bmatrix} = [B]\vec{Q}^{(e)} \tag{11.36}$$

where

$$[B] = \frac{1}{2A} \begin{bmatrix} b_i & 0 & b_j & 0 & b_k & 0 \\ (N_i/r) & 0 & (N_j/r) & 0 & (N_k/r) & 0 \\ 0 & c_i & 0 & c_j & 0 & c_k \\ c_i & b_i & c_j & b_j & c_k & b_k \end{bmatrix} \tag{11.37}$$

The stress–strain relations are given by

$$\vec{\sigma} = [D]\vec{\varepsilon} \tag{11.38}$$

where $\vec{\sigma} = \{\sigma_{rr}\ \sigma_{\theta\theta}\ \sigma_{zz}\ \sigma_{rz}\}^T$ and

$$[D] = \frac{E}{(1+\nu)(1-2\nu)} \begin{bmatrix} 1-\nu & \nu & \nu & 0 \\ \nu & 1-\nu & \nu & 0 \\ \nu & \nu & 1-\nu & 0 \\ 0 & 0 & 0 & \left(\dfrac{1-2\nu}{2}\right) \end{bmatrix} \tag{11.39}$$

Since the matrix $[B]$ contains terms that are functions of the coordinates r and z the product $[B]^T$ $[D]$ $[B]$ cannot be removed from under the integral sign in the expression of the element stiffness matrix $[K^{(e)}]$, Eq. (8.87). However, we can adopt an approximate procedure for evaluating the integral involved in the expression of $[K^{(e)}]$. If we evaluate the matrix $[B]$ using the r and z values at the centroid of the element, the product $[B]^T$ $[D]$ $[B]$ can be removed from under the integral sign as

$$[K^{(e)}] \simeq [\underline{B}]^T [D] [\underline{B}] \iiint\limits_{V^{(e)}} \mathrm{d}V \tag{11.40}$$

where the bar below $[B]$ denotes that the matrix $[B]$ is evaluated at the point $(\underline{r}, \underline{z})$ with

$$\underline{r} = (r_i + r_j + r_k)/3 \quad \text{and} \quad \underline{z} = (z_i + z_j + z_k)/3 \tag{11.41}$$

By using the relation

$$\iiint\limits_{V^{(e)}} \mathrm{d}V = V^{(e)} = 2\pi \underline{r} A \tag{11.42}$$

Eq. (11.40) can be expressed as

$$[K^{(e)}] = [\underline{B}]^T [D] [\underline{B}] 2\pi \underline{r} A \tag{11.43}$$

Although Eq. (11.43) is approximate, it yields reasonably accurate results. The components of the load vector of the element are given by Eqs. (8.88)–(8.90). The load vector due to initial strains (caused by the temperature change T) can be handled as in the case of $[K^{(e)}]$ since $[B]$ occurs in the integral. Thus,

$$\vec{P}_i^{(e)} = \iiint\limits_{V^{(e)}} [B]^T [D] \vec{\varepsilon}_0 \, \mathrm{d}V = \iiint\limits_{V^{(e)}} [B]^T [D] E\alpha T \begin{Bmatrix} 1 \\ 1 \\ 1 \\ 0 \end{Bmatrix} \mathrm{d}V$$

$$\simeq \frac{E\alpha T}{(1-2\nu)} [\underline{B}] \begin{Bmatrix} 1 \\ 1 \\ 1 \\ 0 \end{Bmatrix} 2\pi \underline{r} A \tag{11.44}$$

If $\bar{\phi}_r$ and $\bar{\phi}_z$ denote the components of the body force in the directions of r and z, respectively, the load vector $\vec{P}_b^{(e)}$ can be evaluated either exactly using the area coordinates or approximately using the procedure adopted earlier. If we use the area coordinates, Eq. (8.90) can be expressed as

$$\vec{P}_b^{(e)} = \iiint\limits_{V^{(e)}} [N]^T \vec{\bar{\phi}} \, \mathrm{d}V = \iint\limits_{A} \begin{bmatrix} L_i & 0 \\ 0 & L_i \\ L_j & 0 \\ 0 & L_j \\ L_k & 0 \\ 0 & L_k \end{bmatrix} \begin{Bmatrix} \bar{\phi}_r \\ \bar{\phi}_z \end{Bmatrix} 2\pi r \, \mathrm{d}A \tag{11.45}$$

The radial distance r can be written in terms of the area coordinates as

$$r = r_i L_i + r_j L_j + r_k L_k \tag{11.46}$$

By substituting Eq. (11.46) into Eq. (11.45) and evaluating the resulting area integrals using Eq. (3.78), we obtain

$$\vec{P}_b^{(e)} = \frac{2\pi A}{12} \begin{Bmatrix} (2r_i + r_j + r_k) & \bar{\phi}_r \\ (2r_i + r_j + r_k) & \bar{\phi}_z \\ (r_i + 2r_j + r_k) & \bar{\phi}_r \\ (r_i + 2r_j + r_k) & \bar{\phi}_z \\ (r_i + r_j + 2r_k) & \bar{\phi}_r \\ (r_i + r_j + 2r_k) & \bar{\phi}_z \end{Bmatrix} \tag{11.47}$$

It can be seen from Eq. (11.47) that the body forces are not distributed equally between the three nodes i, j, and k.

If $\bar{\Phi}_r$ and $\bar{\Phi}_z$ denote the applied stresses in the r and z directions, the load vector $\vec{P}_s^{(e)}$ can be evaluated using the area coordinates as in the case of $\vec{P}_b^{(e)}$. If we assume that only the edge $i\,j$ lies on the surface $S_1^{(e)}$ on which the stresses $\bar{\phi}_r$ and $\bar{\Phi}_z$ are acting (this implies that $L_k = 0$), we can write

$$\vec{P}_s^{(e)} = \iint\limits_{S_1^{(e)}} [N]^T \begin{Bmatrix} \bar{\Phi}_r \\ \bar{\Phi}_z \end{Bmatrix} \mathrm{d}S_1 = \int\limits_{s_{ij}} \begin{bmatrix} L_i & 0 \\ 0 & L_i \\ L_j & 0 \\ 0 & L_j \\ L_k & 0 \\ 0 & L_k \end{bmatrix} \begin{Bmatrix} \bar{\Phi}_r \\ \bar{\Phi}_z \end{Bmatrix} 2\pi r \cdot \mathrm{d}s \tag{11.48}$$

where $\mathrm{d}S_1 = 2\pi r\,\mathrm{d}s$, and s_{ij} denotes the length of the edge $i\,j$. By substituting Eq. (11.46) into Eq. (11.48), Eq. (3.77) can be used to evaluate the line integral of Eq. (11.48). This results in

$$\vec{P}_s^{(e)} = \frac{\pi s_{ij}}{3} \begin{Bmatrix} (2r_i + r_j) & \bar{\Phi}_r \\ (2r_i + r_j) & \bar{\Phi}_z \\ (r_i + 2r_j) & \bar{\Phi}_r \\ (r_i + 2r_j) & \bar{\Phi}_z \\ 0 \\ 0 \end{Bmatrix} \tag{11.49}$$

Note:
If the edge, for example, $i\,j$, is vertical, we have $r = r_i = r_j$ along this edge and hence Eq. (11.48) leads to

$$\vec{P}_s^{(e)} = \pi r_i s_{ij} \begin{Bmatrix} \bar{\Phi}_r \\ \bar{\Phi}_z \\ \bar{\Phi}_r \\ \bar{\Phi}_z \\ 0 \\ 0 \end{Bmatrix} \tag{11.50}$$

11.4.3 Numerical Results

An infinite cylinder subjected to an internal pressure, for which an exact solution is known, is selected as a means of demonstrating the accuracy of the finite element considered. In Figure 11.6(a), three finite element meshes are shown [11.7]. The resulting radial and hoop stresses are plotted in Figure 11.6(b). Except for the very coarse mesh, agreement with the exact solution is excellent. In this figure, stresses are plotted at the center of the quadrilaterals and are obtained by averaging the stresses in the four connecting triangles. In general, good boundary stresses are estimated by plotting the interior stresses and extrapolating to the boundary. This type of engineering judgment is always necessary in evaluating results from a finite element analysis.

11.4.4 Computer Program

A Fortran subroutine called STRESS is given for the thermal stress analysis of axisymmetric solids. It requires the following quantities as input:

NN = total number of nodes.
NE = number of elements.

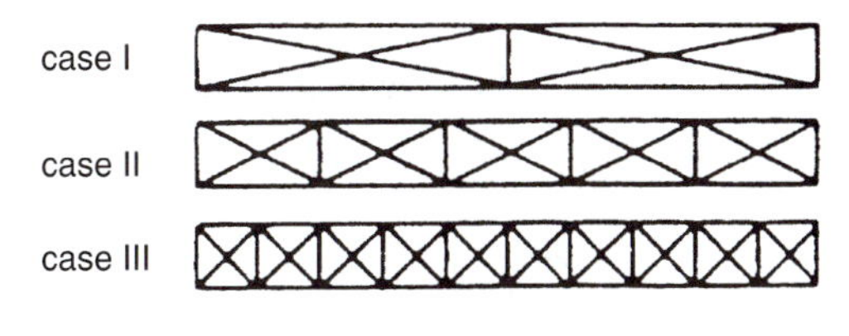

(a) Finite element idealization

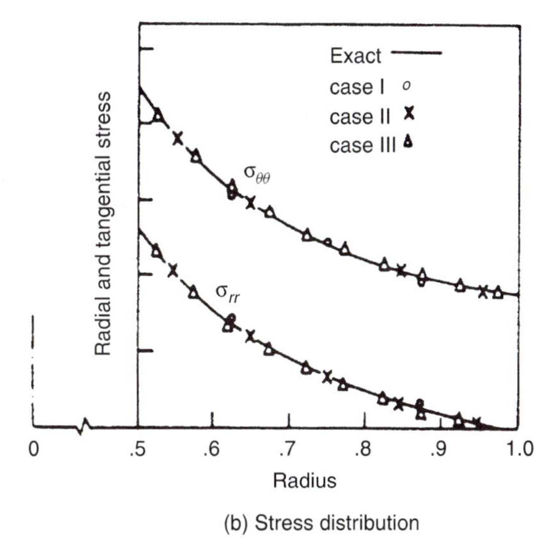

(b) Stress distribution

Figure 11.6.

NB = bandwidth of the overall stiffness matrix.
ND = total number of degrees of freedom (2NN).
ANU = Poisson's ratio.
E = Young's modulus.
EXPAN = coefficient of expansion.
TINF = ambient temperature.
R = array of size NN; R(I) = r coordinate of node I.
Z = array of size NN; Z(I) = z coordinate of node I.
LØC = array of size NE × 3; LØC(I,J) = global node number corresponding to Jth corner of element I.
T = array of size NN; T(I) = specified temperature at node I.
NØM = array of size NN; NØM(I) = number of elements connected to node I.
QS = array of size ND; QS(I) = prescribed value of displacement of Ith degree of freedom. If its value is not known, QS(I) is to be set equal to −1.0 E6. It is assumed that the radial displacement degrees of freedom are numbered first at every node.

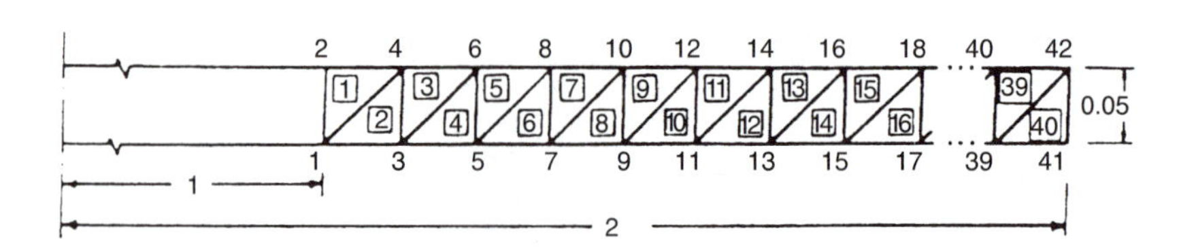

Figure 11.7. Analysis of an Axisymmetric Cylinder.

The stresses computed at the various nodes are given by the array SIGMA of size 4 × NN (output). SIGMA(1,I), SIGMA(2,I), SIGMA(3,I), and SIGMA(4,I) denote the radial, hoop, axial, and shear stresses, respectively, at node I.

Example 11.1 To illustrate the use of the subroutine STRESS, the thermal stresses developed in an infinitely long hollow cylinder with inner radius 1 and outer radius 2 are considered. Since the stress distribution does not vary along the axial length, a disk with an axial thickness of 0.05 is considered for the analysis. The finite element idealization is shown in Figure 11.7. The values of NN, NE, ND, and NB can be seen to be 42, 40, 84, and 8, respectively. The values of ANU, E, EXPAN, and TINF are taken as 0.3, 1.0, 1.0, and 0.0, respectively. The axial displacements of all the nodes are restrained to be zero. The radial temperature distribution is taken as

$$\text{Temperature } (r) = -\frac{(T_i - T_o)\, ln\, r}{ln(R_o/R_i)} + \frac{T_i ln\, R_o - T_o\, ln\, R_i}{ln(R_o/R_i)} \tag{E$_1$}$$

where r is the radial distance, T_i is the temperature at the inner surface, T_o is the temperature at the outer surface, and R_i is the inner radius and R_o is the outer radius of the cylinder.

Thus, the nodal temperatures $T(I)$ are computed from Eq. (E_1) by substituting the appropriate value of r. The main program that calls the subroutine STRESS and the output of the program are given below.

```
C=========================================================================
C
C      STRESS ANALYSIS OF AXISYMMETRIC SOLIDS
C
C=========================================================================
       DIMENSION LOC(40,3),R(42),Z(42),QS(84),T(42),NOM(42),SIGMA(4,42)
       COMMON /AREA1/R,Z,LOC
       COMMON /AREA2/NOM
       COMMON /AREA4/QS, SIGMA
       COMMON /AREA7/T
       DATA NE,NN,NB,ND,ANU,TINF/40,42,8,84,0.3,0.0/
       DATA E,EXPAN/1.0,1.0/
       DATA T (1),T (2),T(41),T(42)/1000.0,1000.0,0.0,0.0/
       LOC(1,1)=1
       LOC(1,2)=4
       LOC(1,3)=2
       LOC(2,1)=4
       LOC(2,2)=1
```

```
      LOC(2,3)=3
      DO 10 J=1,3
      DO 10 I=3, NE
      JJ=I-2
      LOC(I,J)=LOC(JJ,J)+2
 10   CONTINUE
      R(1)=1.0
      R(2)=1.0
      DO 20 I=3,NN,2
      JJ=I-2
      JK=I+1
      R(I)=R(JJ)+0.05
      R(JK)=R(I)
 20   CONTINUE
      DO 30 I=1, NN, 2
      Z (I)=0.0
      KK=I+1
      Z (KK)=0.05
 30   CONTINUE
      CA=-(T(1)-T(42))/ALOG(R(42)/R(1))
      DA=(T(1)*ALOG(R(42))-T(42)*ALOG(R(1)))/(ALOG(R(42)/R(1)))
      DO 40 I=1,NN
      RR=R(I)
      T(I)=CA*ALOG(RR)+DA
 40   CONTINUE
      DO 50 I=1,ND
 50   QS(I)=-1.0E+6
      DO 60 I=1,NN
      NFIX=2*I
 60   QS(NFIX)=0.0
C     NOM(I)=NUMBER OF ELEMENTS CONNECTED TO NODE I
      DO 70 I=1,NN
 70   NOM(I)=0
      DO 80 I=1,NE
      DO 80 J=1,3
 80   NOM(LOC(I,J))=NOM(LOC(I,J))+1
      CALL STRESS (NN,NE,NB,ND,E,ANU,EXPAN,TINF)
      PRINT 90
 90   FORMAT (/,1X,'NODE',1X,'RADIAL',1X,'AXIAL',1X,'TEMPERATURE',
     2 3X,'RADIAL',5X,'HOOP',6X,'AXIAL',7X,'SHEAR'/2X,'NO.',1X,'COORD.'
     3 ,1X,'COORD.',14X,'STRESS',4X,'STRESS',5X,'STRESS',6X,'STRESS'/
     4 73 (1H-)/)
      DO 100 I=1,NN
      PRINT 110, I,R(I),Z(I),T(I),SIGMA(1,I),SIGMA(2,I),SIGMA(3,I),
     2 SIGMA(4,I)
100   CONTINUE
110   FORMAT (I4,2(2X,F5.2),5(F11.4),3X,F8.5)
      STOP
      END
```

NODE NO.	RADIAL COORD.	AXIAL COORD.	TEMPERATURE	RADIAL STRESS	HOOP STRESS	AXIAL STRESS	SHEAR STRESS
1	1.00	0.00	1000.0001	-19.9552	-803.7952	-1211.9308	-11.9011
2	1.00	0.05	1000.0001	-31.9503	-835.0752	-1236.6448	-18.2606
3	1.05	0.00	929.6107	-38.2937	-705.7072	-1138.2609	-4.1018
4	1.05	0.05	929.6107	-39.1378	-771.0350	-1188.6688	-9.7813
5	1.10	0.00	862.4966	-62.0330	-582.1711	-1041.8383	-0.7541
6	1.10	0.05	862.4966	-70.7238	-643.3868	-1091.9757	-2.6620
.							
.							
.							
39	1.95	0.00	36.5259	-7.4387	525.9069	127.0259	3.1661
40	1.95	0.05	36.5259	-18.4688	506.3225	101.3971	1.6569
41	2.00	0.00	0.0001	0.8151	544.0840	151.2943	5.3831
42	2.00	0.05	0.0001	-6.8447	535.4181	140.3090	3.7203

REFERENCES

11.1 K.S. Surana: Transition finite elements for three-dimensional stress analysis, *International Journal for Numerical Methods in Engineering, 15*, 991–1020, 1980.

11.2 P.P. Silvester: Universal finite element matrices for tetrahedra, *International Journal for Numerical Methods in Engineering, 18*, 1055–1061, 1982.

11.3 M.J. Loikkanen and B.M. Irons: An 8-node brick finite element, *International Journal for Numerical Methods in Engineering, 20*, 523–528, 1984.

11.4 R.W. Clough: Comparison of three dimensional finite elements, *Proceedings of the Symposium on Application of Finite Element Methods in Civil Engineering*, Vanderbilt University, Nashville, pp. 1–26, 1969.

11.5 K.S. Surana: Transition finite elements for axisymmetric stress analysis, *International Journal for Numerical Methods in Engineering, 15*, 809–832, 1980.

11.6 J.A. Palacios and M. Henriksen: An analysis of alternatives for computing axisymmetric element stiffness matrices, *International Journal for Numerical Methods in Engineering, 18*, 161–164, 1982.

11.7 E.L. Wilson: Structural analysis of axisymmetric solids, *AIAA Journal, 3*, 2269–2274, 1965.

11.8 T.R. Chandrupatla and A.D. Belegundu: *Introduction to Finite Elements in Engineering*, 2nd Ed., Prentice Hall, Upper Saddle River, NJ, 1997.

PROBLEMS

11.1 The X, Y, Z coordinates of the nodes of a tetrahedron element, in inches, are shown in Figure 11.8.

(a) Derive the matrix $[B]$.

(b) Derive the stiffness matrix of the element assuming that $E = 30 \times 10^6$ psi and $\nu = 0.32$.

11.2 Find the nodal displacements and the stress distribution in the element shown in Figure 11.8 by fixing the face 123. Assume the loads applied at node 4 as $P_X = 50$ lb, $P_Y = 100$ lb, and $P_Z = -150$ lb.

11.3 A uniform pressure of 100 psi is applied on the face 234 of the tetrahedron element shown in Figure 11.8. Determine the corresponding load vector of the element.

11.4 If the temperature of the element shown in Figure 11.8 is increased by 50°F while all the nodes are constrained, determine the corresponding load vector. Assume the coefficient of expansion as $\alpha = 6.5 \times 10^{-6}$ per °F.

11.5 The X, Y, Z coordinates of a hexahedron element are shown in Figure 11.9. Derive the matrix $[J]$.

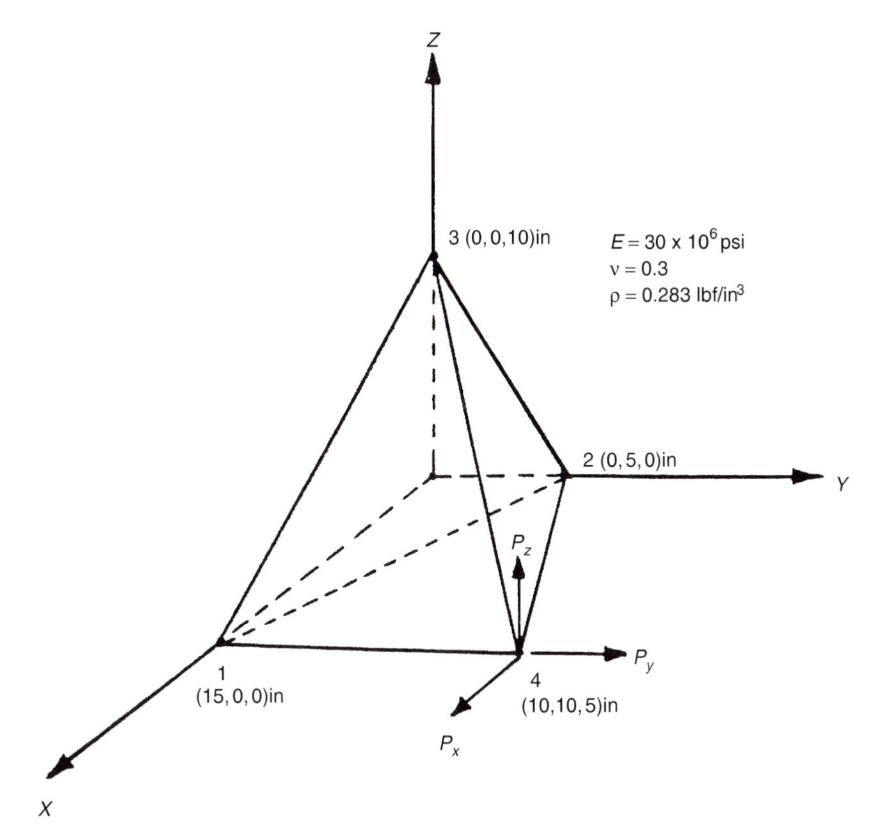

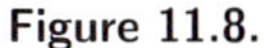
Figure 11.8.

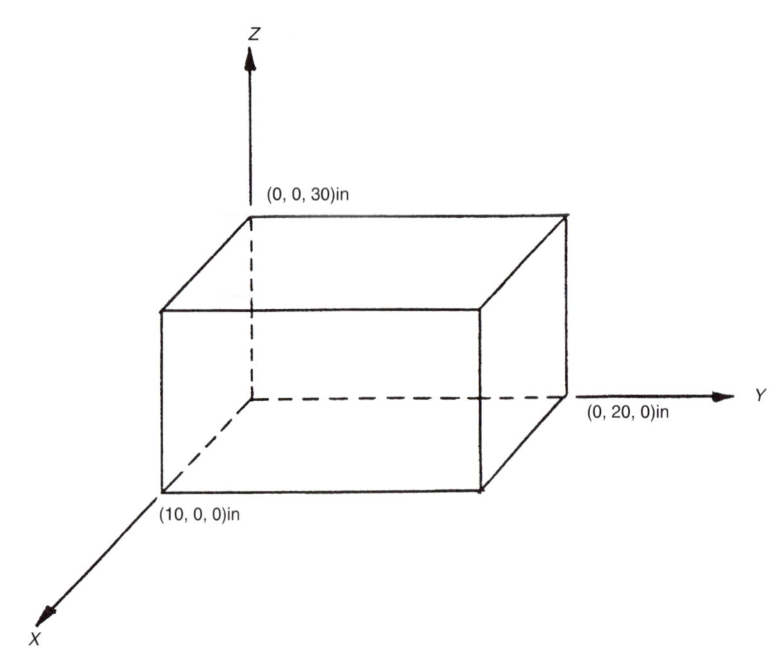

Figure 11.9.

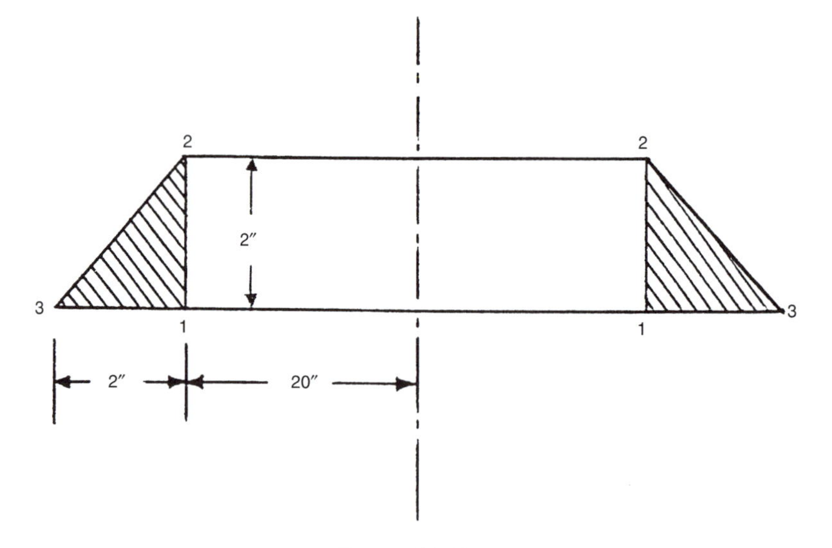

Figure 11.10.

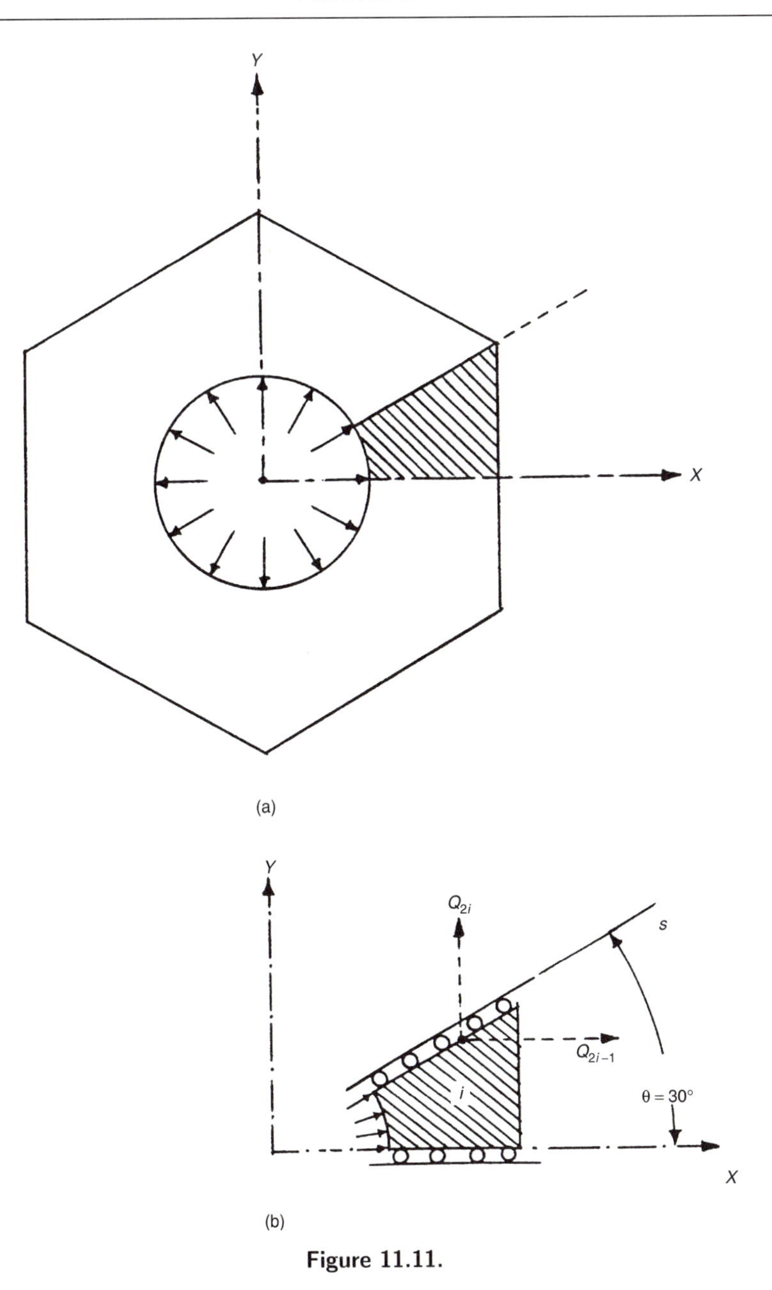

Figure 11.11.

11.6 An axisymmetric ring element is shown in Figure 11.10.

(a) Derive the matrix $[B]$.

(b) Derive the matrix $[D]$, for steel with $E = 30 \times 10^6$ psi and $\nu = 0.33$.

(c) Derive the element stiffness matrix, $[K]$.

11.7 If the element shown in Figure 11.10 is subjected to an initial strain, due to an increase in temperature of 50°F, determine the corresponding load vector. Assume a value of $\alpha = 6.5 \times 10^{-6}$ per °F.

11.8 If the face 23 of the element shown in Figure 11.10 is subjected to a uniform pressure of 200 psi, determine the corresponding load vector.

11.9 A hexagonal plate with a circular hole is subjected to a uniform pressure on the inside surface as shown in Figure 11.11(a). Due to the symmetry of the geometry and the load, only a 30° segment of the plate can be considered for the finite element analysis [Figure 11.11(b)]. Indicate a procedure for incorporating the boundary conditions along the X and s axes.

Hint: The symmetry conditions require that the nodes along the X and s axes should have zero displacement in a direction normal to the X and s axes, respectively. If the global degrees of freedom at node are denoted as Q_{2i-1} and Q_{2i}, then the boundary condition becomes a multipoint constraint that can be expressed as [11.8]

$$-Q_{2i-1} \sin\theta + Q_{2i} \cos\theta = 0$$

A method of incorporating this type of constraint was indicated in Problem 9.16.

11.10 Write a subroutine called SØLID for the analysis of three-dimensional solid bodies using tetrahedron elements. Find the tip deflection of the short cantilever beam discussed in Section 11.3.6 using this subroutine SØLID.

12

DYNAMIC ANALYSIS

12.1 DYNAMIC EQUATIONS OF MOTION

In dynamic problems the displacements, velocities, strains, stresses, and loads are all time dependent. The procedure involved in deriving the finite element equations of a dynamic problem can be stated by the following steps:

Step 1: Idealize the body into E finite elements.

Step 2: Assume the displacement model of element e as

$$\vec{U}(x, y, z, t) = \begin{Bmatrix} u(x, y, z, t) \\ v(x, y, z, t) \\ w(x, y, z, t) \end{Bmatrix} = [N(x, y, z)]\vec{Q}^{(e)}(t) \tag{12.1}$$

where $\vec{U}$ is the vector of displacements, $[N]$ is the matrix of shape functions, and $\vec{Q}^{(e)}$ is the vector of nodal displacements that is assumed to be a function of time t.

Step 3: Derive the element characteristic (stiffness and mass) matrices and characteristic (load) vector.

From Eq. (12.1), the strains can be expressed as

$$\vec{\varepsilon} = [B]\vec{q}^{(e)} \tag{12.2}$$

and the stresses as

$$\vec{\sigma} = [D]\vec{\varepsilon} = [D][B]\vec{Q}^{(e)} \tag{12.3}$$

By differentiating Eq. (12.1) with respect to time, the velocity field can be obtained as

$$\dot{\vec{U}}(x, y, z, t) = [N(x, y, z)]\dot{\vec{Q}}^{(e)}(t) \tag{12.4}$$

where $\dot{\vec{Q}}^{(e)}$ is the vector of nodal velocities. To derive the dynamic equations of motion of a structure, we can use either Lagrange equations [12.1] or Hamilton's principle stated in Section 8.3.2. The Lagrange equations are given by

$$\frac{\mathrm{d}}{\mathrm{d}t}\left\{\frac{\partial L}{\partial \dot{Q}}\right\} - \left\{\frac{\partial L}{\partial Q}\right\} + \left\{\frac{\partial R}{\partial \dot{Q}}\right\} = \{0\} \tag{12.5}$$

where

$$L = T - \pi_p \tag{12.6}$$

is called the Lagrangian function, T is the kinetic energy, π_p is the potential energy, R is the dissipation function, Q is the nodal displacement, and $\dot{Q}$ is the nodal velocity. The kinetic and potential energies of an element "e" can be expressed as

$$T^{(e)} = \frac{1}{2} \iiint\limits_{V^{(e)}} \rho \dot{\vec{U}}^T \dot{\vec{U}} \, \mathrm{d}V \tag{12.7}$$

and

$$\pi_P^{(e)} = \frac{1}{2} \iiint\limits_{V^{(e)}} \vec{\varepsilon}^T \vec{\sigma} \, \mathrm{d}V - \iint\limits_{S_1^{(e)}} \vec{U}^T \vec{\Phi} \, \mathrm{d}S_1 - \iiint\limits_{V^{(e)}} \vec{U}^T \vec{\phi} \, \mathrm{d}V \tag{12.8}$$

where $V^{(e)}$ is the volume, ρ is the density, and $\dot{\vec{U}}$ is the vector of velocities of element e. By assuming the existence of dissipative forces proportional to the relative velocities, the dissipation function of the element e can be expressed as

$$R^{(e)} = \frac{1}{2} \iiint\limits_{V^{(e)}} \mu \dot{\vec{U}}^T \dot{\vec{U}} \, \mathrm{d}V \tag{12.9}$$

where μ can be called the damping coefficient. In Eqs. (12.7)–(12.9), the volume integral has to be taken over the volume of the element, and in Eq. (12.8) the surface integral has to be taken over that portion of the surface of the element on which distributed surface forces are prescribed.

By using Eqs. (12.1)–(12.3), the expressions for T, π_p, and R can be written as

$$T = \sum_{e=1}^{E} T^{(e)} = \frac{1}{2} \dot{\vec{\underset{\sim}{Q}}}^T \left[\sum_{e=1}^{E} \iiint\limits_{V^{(e)}} \rho [N]^T [N] \, \mathrm{d}V \right] \dot{\vec{\underset{\sim}{Q}}} \tag{12.10}$$

$$\begin{aligned} \pi_p = \sum_{e=1}^{E} \pi_p^{(e)} &= \frac{1}{2} \vec{\underset{\sim}{Q}}^T \left[\sum_{e=1}^{E} \iiint\limits_{V^{(e)}} [B]^T [D] [B] \, \mathrm{d}V \right] \vec{\underset{\sim}{Q}} \\ &- \vec{\underset{\sim}{Q}}^T \left(\sum_{e=1}^{E} \iint\limits_{S_1^{(e)}} [N]^T \vec{\Phi}(t) \, \mathrm{d}S_1 + \iiint\limits_{V^{(e)}} [N]^T \vec{\phi}(t) \, \mathrm{d}V \right) - \vec{\underset{\sim}{Q}}^T \vec{\underset{\sim}{P}}_c(t) \end{aligned} \tag{12.11}$$

$$R = \sum_{e=1}^{E} R^{(e)} = \frac{1}{2} \dot{\vec{\underset{\sim}{Q}}}^T \left[\sum_{e=1}^{E} \iiint\limits_{V^{(e)}} \mu [N]^T [N] \, \mathrm{d}V \right] \dot{\vec{\underset{\sim}{Q}}} \tag{12.12}$$

where $\underset{\sim}{\vec{Q}}$ is the global nodal displacement vector, $\underset{\sim}{\dot{\vec{Q}}}$ is the global nodal velocity vector, and $\underset{\sim}{\vec{P}}_c$ is the vector of concentrated nodal forces of the structure or body. By defining the matrices involving the integrals as

$$[M^{(e)}] = \text{element mass matrix} = \iiint\limits_{V^{(e)}} \rho [N]^T [N] \, \mathrm{d}V \tag{12.13}$$

$$[K^{(e)}] = \text{element stiffness matrix} = \iiint\limits_{V^{(e)}} [B]^T [D][B] \, \mathrm{d}V \tag{12.14}$$

$$[C^{(e)}] = \text{element damping matrix} = \iiint\limits_{V^{(e)}} \mu [N]^T [N] \, \mathrm{d}V \tag{12.15}$$

$$\begin{aligned} \vec{P}_s^{(e)} &= \text{vector of element nodal forces produced by surface forces} \\ &= \iint\limits_{S_1^{(e)}} [N]^T \vec{\Phi} \cdot \mathrm{d}S_1 \end{aligned} \tag{12.16}$$

$$\begin{aligned} \vec{P}_b^{(e)} &= \text{vector of element nodal forces produced by body forces} \\ &= \iiint\limits_{V^{(e)}} [N]^T \vec{\phi} \cdot \mathrm{d}V \end{aligned} \tag{12.17}$$

Step 4: Assemble the element matrices and vectors and derive the overall system equations of motion.

Equations (12.10)–(12.12) can be written as

$$T = \tfrac{1}{2} \underset{\sim}{\dot{\vec{Q}}}^T [\underset{\sim}{M}] \underset{\sim}{\dot{\vec{Q}}} \tag{12.18}$$

$$\pi_p = \tfrac{1}{2} \underset{\sim}{\vec{Q}}^T [\underset{\sim}{K}] \underset{\sim}{\vec{Q}} - \underset{\sim}{\vec{Q}}^T \underset{\sim}{\vec{P}} \tag{12.19}$$

$$R = \tfrac{1}{2} \underset{\sim}{\dot{\vec{Q}}}^T [\underset{\sim}{C}] \underset{\sim}{\dot{\vec{Q}}} \tag{12.20}$$

where

$$[\underset{\sim}{M}] = \text{master mass matrix of the structure} = \sum_{e=1}^{E} [M^{(e)}]$$

$$[\underset{\sim}{K}] = \text{master stiffness matrix of the structure} = \sum_{e=1}^{E} [K^{(e)}]$$

$$[\underset{\sim}{C}] = \text{master damping matrix of the structure} = \sum_{e=1}^{E} [C^{(e)}]$$

$$\underset{\sim}{\vec{P}}(t) = \text{total load vector} = \sum_{e=1}^{E} \left(P_s^{(e)}(t) + P_b^{(e)}(t) \right) + \underset{\sim}{\vec{P}}_c(t)$$

By substituting Eqs. (12.18)–(12.20) into Eq. (12.5), we obtain the desired dynamic equations of motion of the structure or body as

$$[\underset{\sim}{M}]\ddot{\vec{\underset{\sim}{Q}}}(t) + [\underset{\sim}{C}]\dot{\vec{\underset{\sim}{Q}}}(t) + [\underset{\sim}{K}]\vec{\underset{\sim}{Q}}(t) = \vec{\underset{\sim}{P}}(t) \tag{12.21}$$

where $\ddot{\vec{\underset{\sim}{Q}}}$ is the vector of nodal accelerations in the global system. If damping is neglected, the equations of motion can be written as

$$[\underset{\sim}{M}]\ddot{\vec{\underset{\sim}{Q}}} + [\underset{\sim}{K}]\vec{\underset{\sim}{Q}} = \vec{\underset{\sim}{P}} \tag{12.22}$$

Steps 5 and 6: Solve the equations of motion by applying the boundary and initial conditions. Equations (12.21) or (12.22) can be solved by using any of the techniques discussed in Section 7.4 for propagation problems. Once the time history of nodal displacements, $\vec{\underset{\sim}{Q}}(t)$, is known, the time histories of stresses and strains in the elements can be found as in the case of static problems. Special space–time finite elements have also been developed for the solution of dynamic solid and structural mechanics problems [12.2, 12.3].

12.2 CONSISTENT AND LUMPED MASS MATRICES

Equation (12.13) for the mass matrix was first derived by Archer [12.4] and is called the "consistent" mass matrix of the element. It is called consistent because the same displacement model that is used for deriving the element stiffness matrix is used for the derivation of mass matrix. It is of interest to note that several dynamic problems have been and are being solved with simpler forms of mass matrices. The simplest form of mass matrix that can be used is that obtained by placing point (concentrated) masses m_i at node points i in the directions of the assumed displacement degrees of freedom. The concentrated masses refer to translational and rotational inertia of the element and are calculated by assuming that the material within the mean locations on either side of the particular displacement behaves like a rigid body while the remainder of the element does not participate in the motion. Thus, this assumption excludes the dynamic coupling that exists between the element displacements, and hence the resulting element mass matrix is purely diagonal and is called the "lumped" mass matrix.

As an example, consider the pin-jointed bar element that can deform only in the local x direction as shown in Figure 9.1. For a linear displacement model, we have

$$u(x) = [N]\vec{q}^{\,(e)} \tag{12.23}$$

where

$$[N] = \left[\left(1 - \frac{x}{l}\right)\ \left(\frac{x}{l}\right)\right] \tag{12.24}$$

$$\vec{q}^{\,(e)} = \begin{Bmatrix} q_1 \\ q_2 \end{Bmatrix}^{(e)} = \begin{Bmatrix} u(x=0) \\ u(x=l) \end{Bmatrix}^{(e)} \tag{12.25}$$

and u is the axial displacement parallel to the x axis. The consistent mass matrix of the element is given by

$$[m^{(e)}] = \iiint_{V^{(e)}} \rho [N]^T [N] \, \mathrm{d}V = \frac{\rho A l}{6} \begin{bmatrix} 2 & 1 \\ 1 & 2 \end{bmatrix} \tag{12.26}$$

where A is the uniform cross-sectional area, and l is the length of the element. Thus, the consistent mass matrices, in general, are fully populated. On the other hand, the lumped mass matrix of the element can be obtained (by dividing the total mass of the element equally between the two nodes) as

$$[m^{(e)}] = \frac{\rho A l}{2} \begin{bmatrix} 1 & 0 \\ 0 & 1 \end{bmatrix} \tag{12.27}$$

The lumped mass matrices will lead to nearly exact results if small but massive objects are placed at the nodes of a lightweight structure. The consistent mass matrices will be exact if the actual deformed shape (under dynamic conditions) is contained in the displacement shape functions $[N]$. Since the deformed shape under dynamic conditions is not known, frequently the static displacement distribution is used for $[N]$. Hence, the resulting mass distribution will only be approximate; however, the accuracy is generally adequate for most practical purposes. Since lumped element matrices are diagonal, the assembled or overall mass matrix of the structure requires less storage space than the consistent mass matrix. Moreover, the diagonal lumped mass matrices greatly facilitate the desired computations.

12.3 CONSISTENT MASS MATRICES IN GLOBAL COORDINATE SYSTEM

To reduce the computational effort, generally the consistent mass matrices of unassembled elements are derived in suitable local coordinate systems and then transformed into the global system selected for the assembled structure. If $[m^{(e)}]$, $\vec{q}^{\,(e)}$, and $\dot{\vec{q}}^{\,(e)}$ denote the mass matrix, nodal displacement vector, and nodal velocity vector in the local coordinate system, the kinetic energy associated with the motion of the element can be expressed as

$$T = \tfrac{1}{2} \dot{\vec{q}}^{\,(e)T} [m^{(e)}] \dot{\vec{q}}^{\,(e)} \tag{12.28}$$

If the element nodal displacements and nodal velocities are denoted as $\vec{Q}^{\,(e)}$ and $\dot{\vec{Q}}^{(e)}$ in the global system, we have the transformation relations

$$\vec{q}^{\,(e)} = [\lambda] \vec{Q}^{\,(e)} \tag{12.29}$$

and

$$\dot{\vec{q}}^{\,(e)} = [\lambda] \dot{\vec{Q}}^{(e)} \tag{12.30}$$

By substituting Eq. (12.30) into Eq. (12.28), we obtain

$$T = \tfrac{1}{2} \dot{\vec{Q}}^{(e)T} [\lambda]^T [m^{(e)}] [\lambda] \dot{\vec{Q}}^{(e)} \tag{12.31}$$

By denoting the mass matrix of the element in the global coordinate system as $[M^{(e)}]$, the kinetic energy associated with the motion of the element can be expressed as

$$\underset{\sim}{T} = \frac{1}{2}\dot{\vec{Q}}^{(e)^T}[M^{(e)}]\dot{\vec{Q}}^{(e)} \tag{12.32}$$

Since kinetic energy is a scalar quantity, it must be independent of the coordinate system. By equating Eqs. (12.31) and (12.32), we obtain the consistent mass matrix of the element in the global system as

$$[M^{(e)}] = [\lambda]^T[m^{(e)}][\lambda] \tag{12.33}$$

Notice that this transformation relation is similar to the one used in the case of the element stiffness matrix.

Notes:

(i) In deriving the element mass matrix from the relation

$$[m^{(e)}] = \iiint\limits_{V^{(e)}} \rho[N]^T[N] \cdot \mathrm{d}V \tag{12.34}$$

the matrix $[N]$ must refer to all nodal displacements even in the local coordinate system. Thus, for thin plates subjected to inplane forces only (membrane elements), the transverse deflection must also be considered (in addition to the inplane displacements considered in the derivation of element stiffness matrices) in formulating the matrix $[N]$.

(ii) For elements whose nodal degrees of freedom correspond to translational displacements only, the consistent mass matrix is invariant with respect to the orientation and position of the coordinate axes. Thus, the matrices $[m^{(e)}]$ and $[M^{(e)}]$ will be the same for pin-jointed bars, membrane elements, and three-dimensional elements such as solid tetrahedra having only translational degrees of freedom. On the other hand, for elements such as frame elements and plate bending elements, which have bending stiffness, the consistent mass matrices $[m^{(e)}]$ and $[M^{(e)}]$ will be different.

12.3.1 Consistent Mass Matrix of a Pin-Jointed (Space Truss) Element

As in the case of the derivation of stiffness matrix, a linear displacement model is assumed as (Figure 12.1)

$$\underset{3\times 1}{\vec{U}}(x) = \begin{Bmatrix} u(x) \\ v(x) \\ w(x) \end{Bmatrix} = \underset{3\times 6}{[N]}\ \underset{6\times 1}{\vec{Q}^{(e)}} \tag{12.35}$$

where

$$[N] = \begin{bmatrix} \left(1-\frac{x}{l}\right) & 0 & 0 & \frac{x}{l} & 0 & 0 \\ 0 & \left(1-\frac{x}{l}\right) & 0 & 0 & \frac{x}{l} & 0 \\ 0 & 0 & \left(1-\frac{x}{l}\right) & 0 & 0 & \frac{x}{l} \end{bmatrix} \tag{12.36}$$

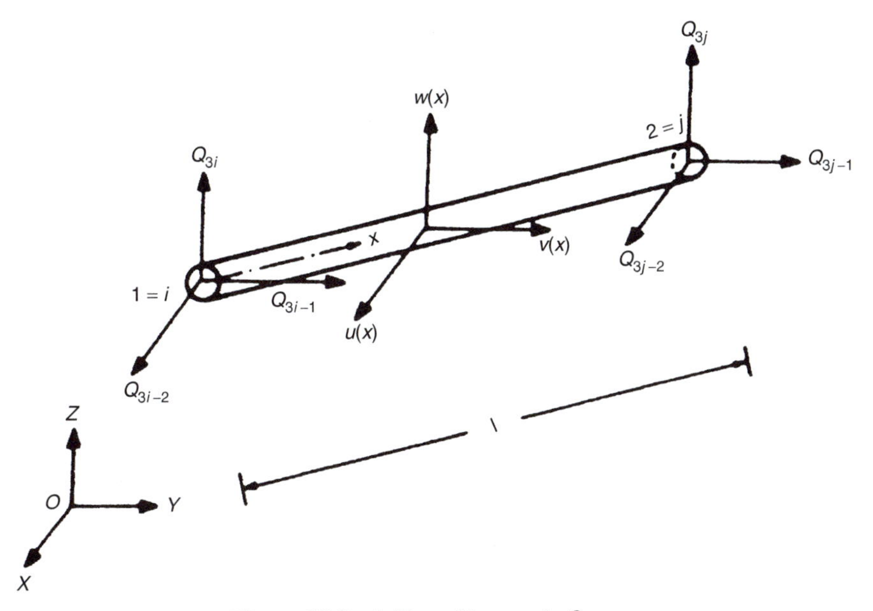

Figure 12.1. A Truss Element in Space.

and

$$\vec{Q}^{(e)} = \begin{Bmatrix} Q_{3i-2} \\ Q_{3i-1} \\ Q_{3i} \\ Q_{3j-2} \\ Q_{3j-1} \\ Q_{3j} \end{Bmatrix} \tag{12.37}$$

where Q_{3i-2}, Q_{3i-1}, and Q_{3i} are the components of displacement of node i (local node 1), and Q_{3j-2}, Q_{3j-1}, and Q_{3j} are the components of displacement of node j (local node 2) in the global XYZ system. If the density (ρ) and cross-sectional area (A) of the bar are constant, the consistent mass matrix of the element can be obtained as

$$[m^{(e)}] = [M^{(e)}] = \iiint\limits_{V^{(e)}} \rho [N]^T [N] \cdot \mathrm{d}V$$

$$= \frac{\rho A l}{6} \begin{bmatrix} 2 & 0 & 0 & 1 & 0 & 0 \\ 0 & 2 & 0 & 0 & 1 & 0 \\ 0 & 0 & 2 & 0 & 0 & 1 \\ 1 & 0 & 0 & 2 & 0 & 0 \\ 0 & 1 & 0 & 0 & 2 & 0 \\ 0 & 0 & 1 & 0 & 0 & 2 \end{bmatrix} \tag{12.38}$$

12.3.2 Consistent Mass Matrix of a Space Frame Element

A space frame element will have 12 degrees of freedom, six deflections, and six rotations, as shown in Figure 9.6(a). By taking the origin of the local coordinate system at node 1, the x axis along the length of the element, and the y and z axes along the principal axes of the element cross section, the displacement model can be expressed as

$$\vec{U}(x) = \begin{Bmatrix} u(x) \\ v(x) \\ w(x) \end{Bmatrix} = [N(x)]\vec{q}^{\,(e)} \tag{12.39}$$

where

$$[N(x)] = \left[\begin{array}{cccccccccccc} 1-\frac{x}{l} & 0 & 0 & 0 & 0 & 0 & \frac{x}{l} & 0 & 0 & 0 & 0 & 0 \\ 0 & \frac{1}{l^3}(2x^3-3lx^2+l^3) & 0 & 0 & 0 & \frac{1}{l^2}(x^3-2lx^2+l^2x) & 0 & -\frac{1}{l^3}(2x^3-3lx^2) & 0 & 0 & 0 & \frac{1}{l^2}(x^3-lx^2) \\ 0 & 0 & \frac{1}{l^3}(2x^3-3lx^2+l^3) & 0 & -\frac{1}{l^2}(x^3-2lx^2+l^2x) & 0 & 0 & 0 & -\frac{1}{l^3}(2x^3-3lx^2) & 0 & \frac{1}{l^2}(lx^2-x^3) & 0 \end{array}\right] \tag{12.40}$$

and

$$\vec{q}^{\,(e)} = \begin{Bmatrix} q_1 \\ q_2 \\ \vdots \\ q_{12} \end{Bmatrix}^{(e)} \tag{12.41}$$

The consistent mass matrix of the element in the local xyz system can be derived as

$$[m^{(e)}] = \iiint\limits_{V^{(e)}} \rho [N]^T [N] \, \mathrm{d}V$$

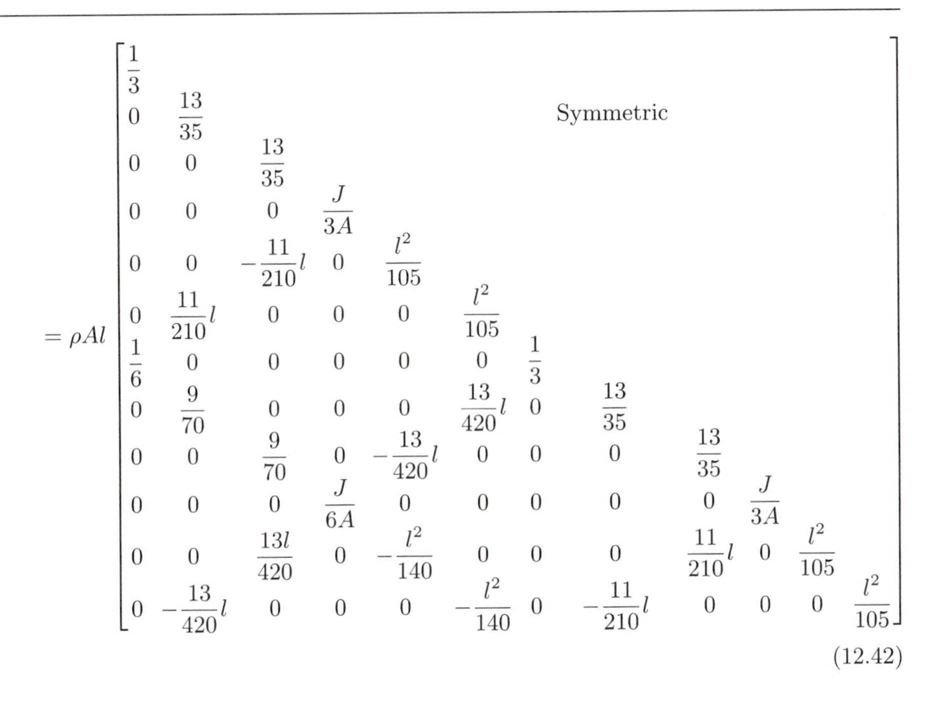

$$
= \rho A l \begin{bmatrix}
\frac{1}{3} & & & & & & & & & & & \\
0 & \frac{13}{35} & & & & & & & \text{Symmetric} & & & \\
0 & 0 & \frac{13}{35} & & & & & & & & & \\
0 & 0 & 0 & \frac{J}{3A} & & & & & & & & \\
0 & 0 & -\frac{11}{210}l & 0 & \frac{l^2}{105} & & & & & & & \\
0 & \frac{11}{210}l & 0 & 0 & 0 & \frac{l^2}{105} & & & & & & \\
\frac{1}{6} & 0 & 0 & 0 & 0 & 0 & \frac{1}{3} & & & & & \\
0 & \frac{9}{70} & 0 & 0 & 0 & \frac{13}{420}l & 0 & \frac{13}{35} & & & & \\
0 & 0 & \frac{9}{70} & 0 & -\frac{13}{420}l & 0 & 0 & 0 & \frac{13}{35} & & & \\
0 & 0 & 0 & \frac{J}{6A} & 0 & 0 & 0 & 0 & 0 & \frac{J}{3A} & & \\
0 & 0 & \frac{13l}{420} & 0 & -\frac{l^2}{140} & 0 & 0 & 0 & \frac{11}{210}l & 0 & \frac{l^2}{105} & \\
0 & -\frac{13}{420}l & 0 & 0 & 0 & -\frac{l^2}{140} & 0 & -\frac{11}{210}l & 0 & 0 & 0 & \frac{l^2}{105}
\end{bmatrix} \tag{12.42}
$$

where ρ is the density, A is the cross-sectional area, l is the length, and j is the polar moment of inertia of the element.

12.3.3 Consistent Mass Matrix of a Planar Frame Element

For the planar frame element shown in Figure 9.11, only axial and inplane bending degrees of freedom will be there and the consistent mass matrix will be

$$
[m^{(e)}] = \rho A l \begin{bmatrix}
1/3 & & & \text{Symmetric} & & \\
0 & 13/35 & & & & \\
0 & 11l/210 & l^2/105 & & & \\
1/6 & 0 & 0 & 1/3 & & \\
0 & 9/70 & 13l/420 & 0 & 13/35 & \\
0 & -13l/420 & -l^2/140 & 0 & -11l/210 & l^2/105
\end{bmatrix} \tag{12.43}
$$

12.3.4 Consistent Mass Matrix of a Beam Element

For a beam bending element, the axial displacement degrees of freedom need not be considered (Figure 9.12) and the consistent mass matrix becomes

$$
[m^{(e)}] = \frac{\rho A l}{420} \begin{bmatrix}
156 & 22l & 54 & -13l \\
22l & 4l^2 & 13l & -3l^2 \\
54 & 13l & 156 & -22l \\
-13l & -3l^2 & -22l & 4l^2
\end{bmatrix} \tag{12.44}
$$

The transformation matrices needed for the derivation of element mass matrices in the global coordinate system from those given by Eqs. (12.42), (12.43), and (12.44) are given by Eqs. (9.41), (9.63), and (9.66), respectively.

If the cross section of the frame (or beam) element is not small, the effects of rotatory inertia and shear deformation become important in the dynamic analysis. The derivation of stiffness and mass matrices of beam elements, including the effects of rotatory inertia and shear deformation, can be found in Refs. [12.5] and [12.6].

12.3.5 Consistent Mass Matrix of a Triangular Membrane Element

By considering all the nine degrees of freedom of the element (shown in Figure 10.3), linear shape functions in terms of the local coordinates x and y can be used to express the displacement field as

$$\vec{U} = \begin{Bmatrix} u(x,y) \\ v(x,y) \\ w(x,y) \end{Bmatrix} = [N]\vec{Q}^{(e)} \tag{12.45}$$

where

$$[N(x,y)] = \begin{bmatrix} N_1 & 0 & 0 & N_2 & 0 & 0 & N_3 & 0 & 0 \\ 0 & N_1 & 0 & 0 & N_2 & 0 & 0 & N_3 & 0 \\ 0 & 0 & N_1 & 0 & 0 & N_2 & 0 & 0 & N_3 \end{bmatrix} \tag{12.46}$$

with $N_1(x,y)$, $N_2(x,y)$, and $N_3(x,y)$ given by Eq. (10.5), and

$$\vec{Q}^{(e)} = \{Q_{3i-2}\ Q_{3i-1}\ Q_{3i}\ Q_{3j-2}\ Q_{3j-1}\ Q_{3j}\ Q_{3k-2}\ Q_{3k-1}\ Q_{3k}\}^T \tag{12.47}$$

The consistent mass matrix of the element (applicable in any coordinate system) can be obtained as

$$[m^{(e)}] = \iiint\limits_{V^{(e)}} \rho [N]^T [N]\, \mathrm{d}V \tag{12.48}$$

By carrying out the necessary integration (in the local xy coordinate system, for simplicity), the mass matrix can be derived as

$$[M^{(e)}] = [m^{(e)}] = \frac{\rho A t}{12} \begin{bmatrix} 2 & 0 & 0 & 1 & 0 & 0 & 1 & 0 & 0 \\ 0 & 2 & 0 & 0 & 1 & 0 & 0 & 1 & 0 \\ 0 & 0 & 2 & 0 & 0 & 1 & 0 & 0 & 1 \\ 1 & 0 & 0 & 2 & 0 & 0 & 1 & 0 & 0 \\ 0 & 1 & 0 & 0 & 2 & 0 & 0 & 1 & 0 \\ 0 & 0 & 1 & 0 & 0 & 2 & 0 & 0 & 1 \\ 1 & 0 & 0 & 1 & 0 & 0 & 2 & 0 & 0 \\ 0 & 1 & 0 & 0 & 1 & 0 & 0 & 2 & 0 \\ 0 & 0 & 1 & 0 & 0 & 1 & 0 & 0 & 2 \end{bmatrix} \tag{12.49}$$

where t is the thickness of the element.

12.3.6 Consistent Mass Matrix of a Triangular Bending Element

For the triangular plate bending element shown in Figure 10.11, the stiffness matrix has been derived in Section 10.7 by assuming the displacement model

$$w(x,y) = [\eta]\vec{\alpha} \tag{12.50}$$

where $[\eta]$ and $\vec{\alpha}$ are given by Eqs. (10.61) and (10.62), respectively. By using Eqs. (12.50) and (10.64), the transverse displacement w can be expressed as

$$w(x,y) = ([\eta][\underset{\sim}{\eta}]^{-1})\vec{q}^{\,(e)} \tag{12.51}$$

where $[\underset{\sim}{\eta}]$ is given by Eq. (10.65). Due to rotation of normals to the middle plane about the x and y axes, any point located at a distance of z from the middle plane will have inplane displacement components given by

$$\left.\begin{aligned} u &= -z\cdot\frac{\partial w}{\partial x} \\ v &= -z\cdot\frac{\partial w}{\partial y} \end{aligned}\right\} \tag{12.52}$$

Thus, the three translational displacements can be expressed, using Eqs. (12.51) and (12.52), as

$$\begin{aligned} \underset{3\times 1}{\vec{U}}(x,y) = \begin{Bmatrix} u(x,y) \\ v(x,y) \\ w(x,t) \end{Bmatrix} &= \begin{bmatrix} -z\dfrac{\partial[\eta]}{\partial x} \\ -z\dfrac{\partial[\eta]}{\partial y} \\ [\eta] \end{bmatrix} [\underset{\sim}{\eta}]^{-1}\vec{q}^{\,(e)} \\ &= \underset{3\times 9}{[N_1]}\ \underset{9\times 9}{[\underset{\sim}{\eta}]^{-1}}\ \underset{9\times 1}{\vec{q}^{\,(e)}} \equiv [N]\vec{q}^{\,(e)} \end{aligned} \tag{12.53}$$

where

$$[N_1] = \begin{bmatrix} 0 & -z & 0 & -2xz & -yz & 0 & -3x^2z & -z(y^2+2xy) & 0 \\ 0 & 0 & -z & 0 & -xz & -2yz & 0 & -z(2xy+x^2) & -3y^2z \\ 1 & x & y & x^2 & xy & y^2 & x^3 & (x^2y+xy^2) & y^3 \end{bmatrix} \tag{12.54}$$

and

$$[N] = [N_1][\underset{\sim}{\eta}]^{-1} \tag{12.55}$$

The consistent mass matrix of the element can now be evaluated as

$$
\begin{aligned}
[m^{(e)}] &= \iiint\limits_{V(e)} \rho [N]^T [N] \, \mathrm{d}V \\
&= \iiint\limits_{V(e)} \rho ([\underset{\sim}{\eta}]^{-1})^T [N_1]^T [N_1] [\underset{\sim}{\eta}]^{-1} \, \mathrm{d}V
\end{aligned} \tag{12.56}
$$

Equation (12.56) denotes the mass matrix obtained by considering both translational (due to w) and rotatory (due to u and v) inertia of the element. If rotatory inertia is neglected, as is done in most of the practical computations, the consistent mass matrix can be obtained by setting simply $[N_1] \equiv [\eta]$ in Eq. (12.56). In this case we have

$$
\begin{aligned}
[m^{(e)}] &= \iiint\limits_{V^{(e)}} \rho ([\underset{\sim}{\eta}]^{-1})^T [\eta]^T [\eta] [\underset{\sim}{\eta}]^{-1} \, \mathrm{d}V \\
&= \rho t ([\underset{\sim}{\eta}]^{-1})^T \left(\iint\limits_{A} [\eta]^T [\eta] \, \mathrm{d}x \, \mathrm{d}y \right) [\underset{\sim}{\eta}]^{-1} \\
&= \rho t ([\eta]^{-1})^T \iint\limits_{\text{area}} \\
&\times \begin{bmatrix}
1 & & & & & & & & \\
x & x^2 & & & & & & \text{Symmetric} & \\
y & xy & y^2 & & & & & & \\
x^2 & x^3 & x^2y & x^4 & & & & & \\
xy & x^2y & xy^2 & x^3y & x^2y^2 & & & & \\
y^2 & xy^2 & y^3 & x^2y^2 & xy^3 & y^4 & & & \\
x^3 & x^4 & x^3y & x^5 & x^4y & x^3y^2 & x^6 & & \\
(x^2y+xy^2) & (x^2y^2+x^3y) & (xy^3+x^2y^2) & (x^3y^2+x^4y) & (x^2y^3+x^3y^2) & (xy^4+x^2y^3) & (x^4y^2+x^5y) & (xy^2+x^2y)^2 & \\
y^3 & xy^3 & y^4 & x^2y^3 & xy^4 & y^5 & x^2y^3 & (xy^5+x^3y^3) & y^6
\end{bmatrix} \mathrm{d}x \, \mathrm{d}y [\underset{\sim}{\eta}]^{-1}
\end{aligned} \tag{12.57}
$$

Thus, the determination of the mass matrix $[m^{(e)}]$ involves the evaluation of integrals of the form

$$
\iint\limits_{\text{area}} x^i y^j \, \mathrm{d}x \, \mathrm{d}y, \qquad i = 0\text{–}6 \quad \text{and} \quad j = 0\text{–}6 \tag{12.58}
$$

Notice that the highest powers of x and y appearing in the integrand of Eq. (12.58) are larger than the highest powers involved in the derivation of the stiffness matrix of the same element [see Eq. (10.71)]. This characteristic is true for all finite elements.

12.3.7 Consistent Mass Matrix of a Tetrahedron Element

For the solid tetrahedron element shown in Figure 11.1, the displacement field is given by Eq. (11.4). The element mass matrix in the global coordinate system can be found from the relation

$$[M^{(e)}] = \iiint\limits_{V^{(e)}} \rho [N]^T [N] \,\mathrm{d}V \tag{12.59}$$

After carrying out the lengthy volume integrations (using tetrahedral coordinates, for simplicity), the mass matrix can be obtained as

$$[M^{(e)}] = \frac{\rho V^{(e)}}{20} \begin{bmatrix} 2 & 0 & 0 & 1 & 0 & 0 & 1 & 0 & 0 & 1 & 0 & 0 \\ 0 & 2 & 0 & 0 & 1 & 0 & 0 & 1 & 0 & 0 & 1 & 0 \\ 0 & 0 & 2 & 0 & 0 & 1 & 0 & 0 & 1 & 0 & 0 & 1 \\ 1 & 0 & 0 & 2 & 0 & 0 & 1 & 0 & 0 & 1 & 0 & 0 \\ 0 & 1 & 0 & 0 & 2 & 0 & 0 & 1 & 0 & 0 & 1 & 0 \\ 0 & 0 & 1 & 0 & 0 & 2 & 0 & 0 & 1 & 0 & 0 & 1 \\ 1 & 0 & 0 & 1 & 0 & 0 & 2 & 0 & 0 & 1 & 0 & 0 \\ 0 & 1 & 0 & 0 & 1 & 0 & 0 & 2 & 0 & 0 & 1 & 0 \\ 0 & 0 & 1 & 0 & 0 & 1 & 0 & 0 & 2 & 0 & 0 & 1 \\ 1 & 0 & 0 & 1 & 0 & 0 & 1 & 0 & 0 & 2 & 0 & 0 \\ 0 & 1 & 0 & 0 & 1 & 0 & 0 & 1 & 0 & 0 & 2 & 0 \\ 0 & 0 & 1 & 0 & 0 & 1 & 0 & 0 & 1 & 0 & 0 & 2 \end{bmatrix} \tag{12.60}$$

12.4 FREE VIBRATION ANALYSIS

If we disturb any elastic structure in an appropriate manner initially at time $t = 0$ (i.e., by imposing properly selected initial displacements and then releasing these constraints), the structure can be made to oscillate harmonically. This oscillatory motion is a characteristic property of the structure and it depends on the distribution of mass and stiffness in the structure. If damping is present, the amplitudes of oscillations will decay progressively and if the magnitude of damping exceeds a certain critical value, the oscillatory character of the motion will cease altogether. On the other hand, if damping is absent, the oscillatory motion will continue indefinitely, with the amplitudes of oscillations depending on the initially imposed disturbance or displacement. The oscillatory motion occurs at certain frequencies known as natural frequencies or characteristic values, and it follows well-defined deformation patterns known as mode shapes or characteristic modes. The study of such free vibrations (free because the structure vibrates with no external forces after $t = 0$) is very important in finding the dynamic response of the elastic structure.

By assuming the external force vector $\vec{P}$ to be zero and the displacements to be harmonic as

$$\vec{Q} = \underline{\vec{Q}} \cdot e^{i\omega t} \tag{12.61}$$

Eq. (12.22) gives the following free vibration equation:

$$[[K] - \omega^2 [M]]\underline{\vec{Q}} = \vec{O} \tag{12.62}$$

where $\underset{\sim}{\vec{Q}}$ represents the amplitudes of the displacements $\vec{Q}$ (called the mode shape or eigenvector), and ω denotes the natural frequency of vibration. Equation (12.62) is called a "linear" algebraic eigenvalue problem since neither $[K]$ nor $[M]$ is a function of the circular frequency ω, and it will have a nonzero solution for $\underset{\sim}{\vec{Q}}$ provided that the determinant of the coefficient matrix $([K]-\omega^2[M])$ is zero—that is,

$$\left|[K]-\omega^2[M]\right| = 0 \tag{12.63}$$

The various methods of finding the natural frequencies and mode shapes were discussed in Section 7.3. In general, all the eigenvalues of Eq. (12.63) will be different, and hence the structure will have n different natural frequencies. Only for these natural frequencies, a nonzero solution can be obtained for $\underset{\sim}{\vec{Q}}$ from Eq. (12.62). We designate the eigenvector (mode shape) corresponding to the jth natural frequency (ω_j) as $\underset{\sim}{\vec{Q}}_j$.

It was assumed that the rigid body degrees of freedom were eliminated in deriving Eq. (12.62). If rigid body degrees of freedom are not eliminated in deriving the matrices $[K]$ and $[M]$, some of the natural frequencies ω would be zero. In such a case, for a general three-dimensional structure, there will be six rigid body degrees of freedom and hence six zero frequencies. It can be easily seen why $\omega = 0$ is a solution of Eq. (12.62). For $\omega = 0$, $\vec{Q} = \underset{\sim}{\vec{Q}} =$ constant vector in Eq. (12.61) and Eq. (12.62) gives

$$[K]\underset{\sim}{\vec{Q}}_{\text{rigid body}} = \vec{0} \tag{12.64}$$

which is obviously satisfied due to the fact that rigid body displacements alone do not produce any elastic restoring forces in the structure. The rigid body degrees of freedom in dynamic analysis can be eliminated by deleting the rows and columns corresponding to these degrees of freedom from the matrices $[K]$ and $[M]$ and by deleting the corresponding elements from displacement ($\vec{Q}$) and load ($\vec{P}$) vectors.

Example 12.1 (*Longitudinal Vibrations of a Stepped Bar*) Find the natural frequencies of longitudinal vibration of the unconstrained stepped bar shown in Figure 12.2.

Solution We shall idealize the bar with two elements as shown in Figure 12.2(a). The stiffness and mass matrices of the two elements are given by

$$[K^{(1)}] = \frac{A^{(1)}E^{(1)}}{l^{(1)}}\begin{bmatrix} 1 & -1 \\ -1 & 1 \end{bmatrix} = \frac{4AE}{L}\begin{bmatrix} 1 & -1 \\ -1 & 1 \end{bmatrix}$$

$$[K^{(2)}] = \frac{A^{(2)}E^{(2)}}{l^{(2)}}\begin{bmatrix} 1 & -1 \\ -1 & 1 \end{bmatrix} = \frac{2AE}{L}\begin{bmatrix} 1 & -1 \\ -1 & 1 \end{bmatrix}$$

$$[M^{(1)}] = \frac{\rho^{(1)}A^{(1)}l^{(1)}}{6}\begin{bmatrix} 2 & 1 \\ 1 & 2 \end{bmatrix} = \frac{\rho AL}{6}\begin{bmatrix} 2 & 1 \\ 1 & 2 \end{bmatrix}$$

$$[M^{(2)}] = \frac{\rho^{(2)}A^{(2)}l^{(2)}}{6}\begin{bmatrix} 2 & 1 \\ 1 & 2 \end{bmatrix} = \frac{\rho AL}{12}\begin{bmatrix} 2 & 1 \\ 1 & 2 \end{bmatrix}$$

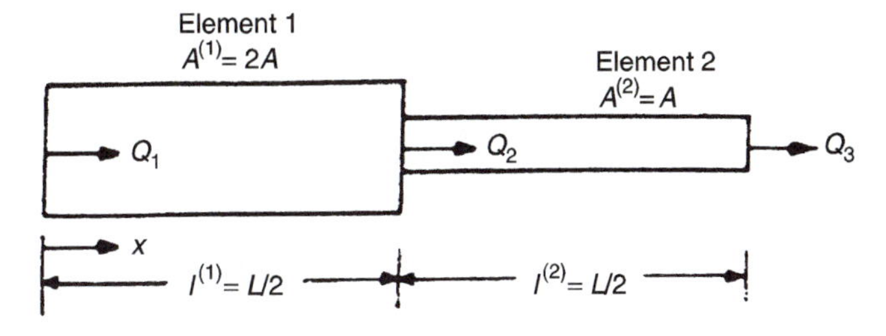

(a) A stepped bar with axial degrees of freedom

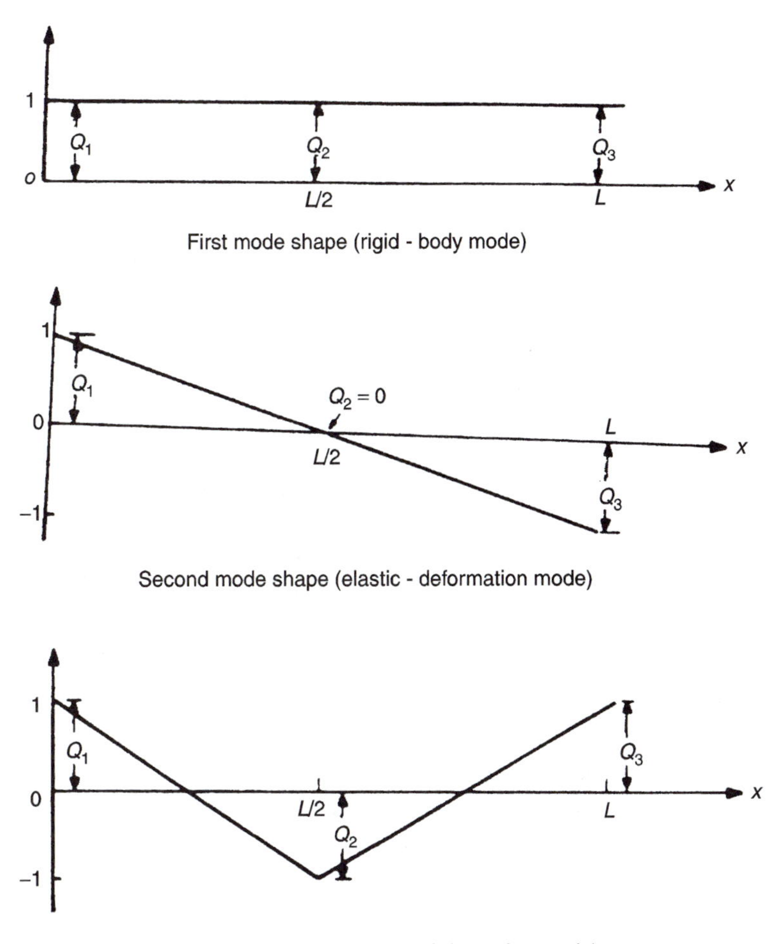

(b) Longitudinal vibration modes

Figure 12.2. An Unconstrained Stepped Bar and Its Mode Shapes.

The assembled stiffness and mass matrices are given by

$$[\underset{\sim}{K}] = \frac{2AE}{L} \begin{bmatrix} 2 & -2 & 0 \\ -2 & 3 & -1 \\ 0 & -1 & 1 \end{bmatrix} \tag{E$_1$}$$

$$[\underset{\sim}{M}] = \frac{\rho AL}{12} \begin{bmatrix} 4 & 2 & 0 \\ 2 & 6 & 1 \\ 0 & 1 & 2 \end{bmatrix} \tag{E$_2$}$$

Since the bar is unconstrained (no degree of freedom is fixed), the frequency equation (12.63) becomes

$$\left| \frac{2AE}{L} \begin{bmatrix} 2 & -2 & 0 \\ -2 & 3 & -1 \\ 0 & -1 & 1 \end{bmatrix} - \omega^2 \frac{\rho AL}{12} \begin{bmatrix} 4 & 2 & 0 \\ 2 & 6 & 1 \\ 0 & 1 & 2 \end{bmatrix} \right| = 0 \tag{E$_3$}$$

By defining

$$\beta^2 = \frac{\rho L^2 \omega^2}{24E} \tag{E$_4$}$$

Equation (E$_3$) can be rewritten as

$$\begin{vmatrix} 2(1-2\beta^2) & -2(1+\beta^2) & 0 \\ -2(1+\beta^2) & 3(1-2\beta^2) & -(1+\beta^2) \\ 0 & -(1+\beta^2) & (1-2\beta^2) \end{vmatrix} = 0 \tag{E$_5$}$$

The expansion of this determinantal equation leads to

$$18\beta^2(1-2\beta^2)(\beta^2-2) = 0 \tag{E$_6$}$$

The roots of Eq. (E$_6$) give the natural frequencies of the bar as

$$\left.\begin{array}{lll} \text{When} & \beta^2 = 0: \ \omega_1^2 = 0 & \text{or } \omega_1 = 0 \\ \text{When} & \beta^2 = \dfrac{1}{2}: \ \omega_2^2 = \dfrac{12E}{\rho L^2} & \text{or } \omega_2 = 3.46[E/(\rho L^2)]^{1/2} \\ \text{When} & \beta^2 = 2: \ \omega_3^2 = \dfrac{48E}{\rho L^2} & \text{or } \omega_3 = 6.92[E/(\rho L^2)]^{1/2} \end{array}\right\} \tag{E$_7$}$$

It is to be observed that the first frequency, $\omega_1 = 0$, corresponds to the rigid-body mode, whereas the second and third frequencies correspond to elastic-deformation modes. To find the mode shape $\vec{\underline{Q}}_i$ corresponding to the natural frequency ω_i, we solve Eq. (12.62). Since Eq. (12.62) represents a system of homogeneous equations, we will be able to find only the relative magnitudes of the components of $\vec{\underline{Q}}_i$.

For $\omega_1^2 = 0$, Eq. (12.62) gives $\underline{\vec{Q}}_1 = \begin{Bmatrix} 1 \\ 1 \\ 1 \end{Bmatrix}$, whereas for $\omega_2^2 = 12E/(\rho L^2)$ and $\omega_3^2 = 48E/(\rho L^2)$, it gives $\underline{\vec{Q}}_2 = \begin{Bmatrix} 1 \\ 0 \\ -1 \end{Bmatrix}$ and $\underline{\vec{Q}}_3 = \begin{Bmatrix} 1 \\ -1 \\ 1 \end{Bmatrix}$, respectively. These mode shapes are plotted in Figure 12.2(b), where the variation of displacement between the nodes has been assumed to be linear in accordance with the assumed displacement distribution of Eqs. (9.1) and (12.23).

[*M*]—Orthogonalization of Modes

Since only the relative magnitudes of the components of the mode shapes $\underline{\vec{Q}}_i$, $i = 1, 2, 3$, are known, the mode shapes can also be written as $a_i \underline{\vec{Q}}_i$, where a_i is an arbitrary nonzero constant. In most of the dynamic response calculations, it is usual to choose the values of a_i so as to make the mode shapes orthogonal with respect to the mass matrix $[M]$ used in obtaining the modes $\underline{\vec{Q}}_i$. This requires that

$$a_i \underline{\vec{Q}}_i^T [M] a_j \underline{\vec{Q}}_j = \begin{cases} 1 & \text{if } i = j \\ 0 & \text{if } i \neq j \end{cases} \tag{E$_8$}$$

for all i and j. In the current example, the mass matrix is given by Eq. (E$_2$) and it can be verified that the condition $a_i \underline{\vec{Q}}_i^T [M] a_j \underline{\vec{Q}}_j = 0$ for $i \neq j$ is automatically satisfied for any a_i and a_j. To satisfy the condition $a_i \underline{\vec{Q}}_i^T [M] a_j \underline{\vec{Q}}_j = 1$ for $i = j$, we impose the conditions

$$a_i^2 \underline{\vec{Q}}_i^T [M] \underline{\vec{Q}}_i = \frac{\rho A L a_i^2}{12} \underline{\vec{Q}}_i^T \begin{bmatrix} 4 & 2 & 0 \\ 2 & 6 & 1 \\ 0 & 1 & 2 \end{bmatrix} \underline{\vec{Q}}_i = 1$$

for $i = 1, 2, 3$ and obtain

$$a_i^2 = \frac{12}{\rho A L} \frac{1}{\underline{\vec{Q}}_i^T \begin{bmatrix} 4 & 2 & 0 \\ 2 & 6 & 1 \\ 0 & 1 & 2 \end{bmatrix} \underline{\vec{Q}}_i}, \qquad i = 1, 2, 3 \tag{E$_9$}$$

Equation (E$_9$) gives

$$\left.\begin{aligned} a_1 &= \left(\frac{2}{3\rho A L}\right)^{1/2} \\ a_2 &= \left(\frac{2}{\rho A L}\right)^{1/2} \\ a_3 &= \left(\frac{2}{\rho A L}\right)^{1/2} \end{aligned}\right\} \tag{E$_{10}$}$$

Thus, the $[M]$ orthogonal mode shapes of the stepped bar corresponding to the natural frequencies ω_1, ω_2, and ω_3 are given by

$$\left(\frac{2}{3\rho AL}\right)^{1/2}\begin{Bmatrix}1\\1\\1\end{Bmatrix},\quad \left(\frac{2}{\rho AL}\right)^{1/2}\begin{Bmatrix}1\\0\\-1\end{Bmatrix}$$

and

$$\left(\frac{2}{\rho AL}\right)^{1/2}\begin{Bmatrix}1\\-1\\1\end{Bmatrix},\ \text{respectively.}$$

Example 12.2 Find the natural frequencies of longitudinal vibration of the constrained stepped bar shown in Figure 12.3.

Solution Since the left end of the bar is fixed, $Q_1 = 0$ and this degree of freedom has to be eliminated from the stiffness and mass matrices of Eqs. (E_1) and (E_2) of Example 12.1 to find the natural frequencies. This amounts to eliminating the rigid-body mode of the structure. For this, we delete the row and column corresponding to Q_1 from Eqs. (E_1) and (E_2) of Example 12.1 and write the frequency equation as

$$\left|\frac{2AE}{L}\begin{bmatrix}3 & -1\\-1 & 1\end{bmatrix} - \frac{\rho AL\omega^2}{12}\begin{bmatrix}6 & 1\\1 & 2\end{bmatrix}\right| = 0 \tag{E$_1$}$$

Equation (E_1) can be rewritten as

$$\begin{vmatrix}3(1-2\beta^2) & -(1+\beta^2)\\-(1+\beta^2) & (1-2\beta^2)\end{vmatrix} = 0 \tag{E$_2$}$$

The solution of Eq. (E_2) is given by

$$\beta_1^2 = \frac{7-3\sqrt{3}}{11} = 0.1640 \quad \text{and} \quad \beta_2^2 = \frac{7+3\sqrt{3}}{11} = 1.1087$$

or

$$\omega_1 = 1.985\sqrt{\frac{E}{\rho L^2}} \quad \text{and} \quad \omega_2 = 5.159\sqrt{\frac{E}{\rho L^2}} \tag{E$_3$}$$

The mode shapes corresponding to these natural frequencies can be found by solving the equation

$$\left[\frac{2AE}{L}\begin{bmatrix}3 & -1\\-1 & 1\end{bmatrix} - \frac{\rho AL\omega_i^2}{12}\begin{bmatrix}6 & 1\\1 & 2\end{bmatrix}\right]\vec{\underset{\sim}{Q}}_i = \vec{0}, \qquad i = 1, 2 \tag{E$_4$}$$

as

$$\vec{\underset{\sim}{Q}}_1 = \begin{Bmatrix}0.5775\\1.0\end{Bmatrix} \quad \text{and} \quad \vec{\underset{\sim}{Q}}_2 = \begin{Bmatrix}-0.5775\\1.0\end{Bmatrix} \tag{E$_5$}$$

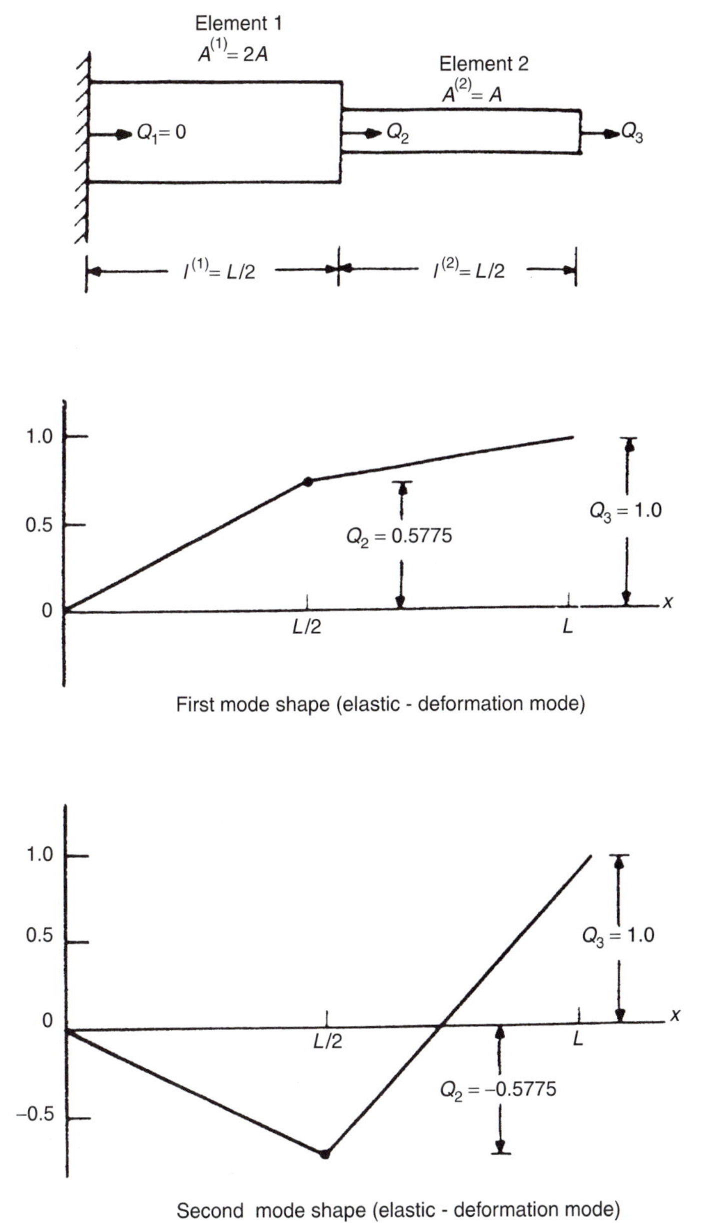

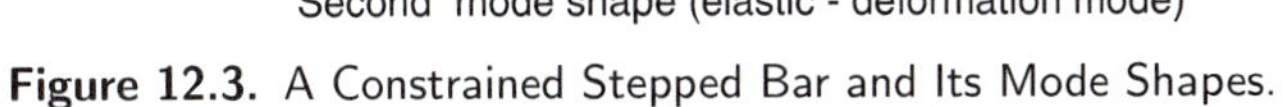

Figure 12.3. A Constrained Stepped Bar and Its Mode Shapes.

These mode shapes are plotted in Figure 12.3(b). These mode shapes, when orthogonalized with respect to the matrix $[M]$, give

$$a_1^2 \vec{\underline{Q}}_1^T [M] \vec{\underline{Q}}_1 = 1$$

or
$$a_1^2 = \frac{1}{(0.5775\ 1.0)\dfrac{\rho AL}{12}\begin{bmatrix}6 & 1\\1 & 2\end{bmatrix}\begin{Bmatrix}0.5775\\1.0\end{Bmatrix}}$$

or
$$a_1 = 1.526/(\rho AL)^{1/2} \tag{E$_6$}$$

$$a_2^2 \vec{\underline{Q}}_2^T [M] \vec{\underline{Q}}_2 = 1$$

or
$$a_2^2 = \frac{1}{(-0.5775\ 1.0)\dfrac{\rho AL}{12}\begin{bmatrix}6 & 1\\1 & 2\end{bmatrix}\begin{Bmatrix}-0.5775\\1.0\end{Bmatrix}}$$

or
$$a_2 = 2.053/(\rho AL)^{1/2} \tag{E$_7$}$$

Thus, the $[M]$-orthogonal mode shapes of the stepped bar corresponding to ω_1 and ω_2 are respectively given by

$$\frac{1.526}{(\rho AL)^{1/2}}\begin{Bmatrix}0.5775\\1.0\end{Bmatrix} = \frac{1}{(\rho AL)^{1/2}}\begin{Bmatrix}0.8812\\1.5260\end{Bmatrix} \tag{E$_8$}$$

and
$$\frac{2.053}{(\rho AL)^{1/2}}\begin{Bmatrix}-0.5775\\1.0\end{Bmatrix} = \frac{1}{(\rho AL)^{1/2}}\begin{Bmatrix}-1.186\\2.053\end{Bmatrix} \tag{E$_9$}$$

12.5 COMPUTER PROGRAM FOR EIGENVALUE ANALYSIS OF THREE-DIMENSIONAL STRUCTURES

A Fortran subroutine called PLATE is written for the deflection and eigenvalue analysis of three-dimensional structures using triangular plate elements. Both membrane and bending stiffnesses are considered in deriving the element stiffness matrices [i.e., Eqs. (10.92) and (10.93) are used]. The consistent mass matrices are used in generating the overall mass matrix. The subroutine PLATE requires the following input data:

NN = total number of nodes (including the fixed nodes).
NE = number of triangular elements.
ND = total number of degrees of freedom (including the fixed degrees of freedom). Six degrees of freedom are considered at each node as shown in Figure 10.15.
NB = bandwidth of the overall stiffness matrix.
MM = number of load conditions.
LØC = an array of size NE × 3. LØC (I,J) denotes the global node number corresponding to Jth corner of element I.
N = number of degrees of freedom considered in the analysis (taken same as ND in this program).
INDEX = 1 if lumped mass matrices are used and 2 if consistent mass matrices are used.
NMØDE = number of eigenvalues required.
X = array of size N × NMØDE representing trial eigenvectors; X(I,J) = Ith component of Jth eigenvector.
CX, CY, CZ = vector arrays of size NN each. CX(I), CY(I), CZ(I) denote the global X, Y, Z coordinates of node I.

E = Young's modulus.
ANU = Poisson's ratio.
RHØ = mass density.
T = a vector array of size NE. T(I) denotes the thickness of element I.
NFIX = number of fixed degrees of freedom (zero displacements).
IFIX = a vector array of size NFIX. IFIX(I) denotes the Ith fixed degree of freedom number.
P = an array of size ND $\times$ MM representing the global load vectors. The array P returned from the subroutine PLATE to the main program represents the global displacement vectors. P(I,J) denotes the Ith component of global load (or displacement) vector in Jth load condition.

In addition to this input, the following arrays are included as arguments for the subroutine PLATE:

K, GS, GM : each of size N $\times$ NB
AREA : size NE
M : size N
GMM, GST, ABCZ, VECT : each of size NMØDE $\times$ NMØDE
ABCV, ABCW, ABCX, ABCY, ØMEG, SUM, DIFF : each of size NMØDE
Y : size N $\times$ NMØDE

The arrays K and M are to be declared as real, whereas SUM and DIFF are to be declared as double precision quantities. The subroutine PLATE requires the subroutines DECØMP, SØLVE, and SUSPIT given in Chapter 7.

To illustrate the use of the subroutine PLATE, the box beam shown in Figures 10.7 and 10.8 is considered. The deflections under the two load conditions stated in Table 10.3 and the first three natural frequencies are computed. The main program that gives the data of the problem and calls the subroutine PLATE and the computer output are shown below.

```
C=======================================================================
C
C     STATIC AND DYNAMIC ANALYSIS USING TRIANGULAR PLATE ELEMENTS
C
C=======================================================================
      DIMENSION CX(24),CY(24),CZ(24),P(144,2),LOC(40,3),AREA(40),T(40)
     2 ,IFIX(24),GM(144,36),X(144,3),OMEG(3),Y(144,3),GST(3,3),
     3 GMM(3,3),VECT(3,3),ABCV(3),ABCW(3),ABCX(3),ABCY(3),ABCZ(3,3)
      DIMENSION B1(3,1),LP1(3),LQ1(3,2),R1(3)
      REAL K(144,36),M(144)
      DOUBLE PRECISION SUM(3),DIFF(3)
      E=30.0E6
      ANU=0.3
      DATA ND,N,NE,NB,NN,MM/144,144,40,36,24,2/
      NFIX=24
      DATA IFIX/121,122,123,124,125,126,127,128,129,130,131,132,133,
     2 134,135,136,137,138,139,140,141,142,143,144/
```

```
      DATA CX/0.,18.,18.,0.,0.,18.,18.,0.,0.,18.,18.,0.,0.,
     2 18.,18.,0.,0.,18.,18.,0.,0.,18.,18.,0./
      DATA CY/0.,0.,0.,0.,12.,12.,12.,12.,24.,24.,24.,24.,
     2 36.,36.,36.,36.,48.,48.,48.,48.,60.,60.,60.,60./
      DATA CZ/0.,0.,12.,12.,0.,0.,12.,12.,0.,0.,12.,12.,0.,0.,
     2 12.,12.,0.,0.,12.,12.,0.,0.,12.,12./
      DATA LOC/6,6,10,10,14,14,18,18,22,22,7,7,11,11,15,15,19,19,23,23,
     2 6,6,10,10,14,14,18,18,22,22,5,5,9,9,13,13,17,17,21,21,2,1,6,5,
     3 10,9,14,13,18,17,3,4,7,8,11,12,15,16,19,20,2,3,6,7,10,11,14,15,
     4 18,19,1,4,5,8,9,12,13,16,17,20,1,5,5,9,9,13,13,17,17,21,4,8,8,
     5 12,12,16,16,20,20,24,3,7,7,11,11,15,15,19,19,23,4,8,8,12,12,16,
     6 16,20,20,24/
      DO 11 LM=1,20
11    T(LM)=1.0
      DO 12 LM=21,40
12    T(LM)=0.5
      RHO=0.28/384.
      INDEX=2
      NMODE=3
      NMODE2=6
      DO 20 I=1,ND
      DO 20 J=1,MM
20    P(I,J)=0.0
      P(15,1)=-5000.0
      P(21,1)=-5000.0
      P(15,2)=5000.0
      P(21,2)=-5000.0
      DO 30 I=1,ND
      DO 30 J=1,NMODE
30    X(I,J)=0.0
      DO 31 I=3,21,6
31    X(I,1)=1.0
      DO 32 I=27,45,6
32    X(I,1)=0.75
      DO 33 I=51,69,6
33    X(I,1)=0.5
      DO 34 I=75,93,6
34    X(I,1)=0.25
      DO 36 I=99,117,6
36    X(I,1)=0.1
      DO 37 I=3,21,6
37    X(I,2)=1.0
      DO 38 I=27,45,6
38    X(I,2)=0.6
      DO 39 I=51,69,6
39    X(I,2)=0.0
      DO 41 I=75,93,6
41    X(I,2)=-0.6
      DO 42 I=99,117,6
```

```
42    X(1,2) =-0.2
      DO 43 I=3,9,6
43    X(I,3)=1.0
      DO 44 I=15,21,6
44    X(I,3)=-1.0
      DO 47 I=51,57,6
47    X(I,3)=0.8
      DO 49 I=63,69,6
49    X(I,3)=-0.8
      CALL PLATE(CX,CY,CZ,LOC,ND,N,NE,NB,NN,MM,T,ANU,E,RHO,NMODE,AREA
     2 ,INDEX,NFIX,IFIX,K,M,GM,X,OMEG,Y,GST,GMM,VECT,SUM,ABCV,ABCW,
     3 ABCX,ABCY,ABCZ,DIFF,P,B1,LP1,LQ1,R1)
      DO 50 J=1,MM
50    PRINT 51,J,(P(I,J),I=1,N)
51    FORMAT(/,'NODAL DISPLACEMENTS(INCH) IN LOAD CONDITION',I3,/
     2(6E12.4))
      PRINT 52
52    FORMAT(/,'EIGENVALUES:',/)
      DO 54 J=1,NMODE
54    PRINT 55,J,OMEG(J)
55    FORMAT(/,'EIGENVALUE(RAD/SEC)',I3,'=',E12.4,/)
      STOP
      END
```

```
NODAL DISPLACEMENTS(INCH) IN LOAD CONDITION  1

 0.6224E-04  0.1820E-02 -0.1611E-01  0.3604E-03  0.9518E-05  0.6902E-05
 0.9354E-04  0.1814E-02 -0.1632E-01  0.3603E-03  0.6629E-05  0.5110E-05
-0.8835E-04 -0.1805E-02 -0.1655E-01  0.3864E-03  0.5457E-05 -0.7891E-05
-0.5107E-04 -0.1811E-02 -0.1633E-01  0.3903E-03  0.9135E-05 -0.3322E-05
 0.4545E-04  0.1764E-02 -0.1180E-01  0.3561E-03  0.9450E-05  0.1427E-05
 0.1114E-03  0.1769E-02 -0.1202E-01  0.3550E-03  0.4191E-05 -0.1613E-05
-0.1051E-03 -0.1744E-02 -0.1199E-01  0.3624E-03  0.2790E-05  0.3171E-05
-0.3730E-04 -0.1742E-02 -0.1176E-01  0.3615E-03  0.1109E-04 -0.6870E-05
 0.2941E-05  0.1530E-02 -0.7662E-02  0.3256E-03  0.1047E-04  0.3722E-05
 0.1367E-03  0.1554E-02 -0.7864E-02  0.3222E-03 -0.2309E-05 -0.9664E-06
-0.1302E-03 -0.1535E-02 -0.7816E-02  0.3207E-03 -0.1302E-05 -0.8311E-05
 0.8558E-06 -0.1514E-02 -0.7613E-02  0.3244E-03  0.1213E-04 -0.2094E-05
-0.4960E-04  0.1146E-02 -0.4130E-02  0.2436E-03  0.1298E-04  0.5303E-05
 0.1472E-03  0.1189E-02 -0.4274E-02  0.2530E-03 -0.8034E-05  0.1201E-05
-0.1429E-03 -0.1177E-02 -0.4225E-02  0.2547E-03 -0.8257E-05  0.8508E-05
 0.5050E-04 -0.1138E-02 -0.4084E-02  0.2435E-03  0.1263E-04 -0.8578E-05
-0.9712E-04  0.6181E-03 -0.1489E-02  0.1913E-03  0.5438E-05 -0.6437E-05
 0.1314E-03  0.6753E-03 -0.1554E-02  0.1856E-03 -0.1562E-04  0.6436E-05
-0.1296E-03 -0.6712E-03 -0.1505E-02  0.1810E-03 -0.1484E-04 -0.1650E-04
 0.9669E-04 -0.6160E-03 -0.1459E-02  0.1877E-03  0.3409E-05 -0.1098E-04
-0.1981E-12  0.6316E-12  0.6512E-13  0.4807E-13  0.5447E-14 -0.5515E-14
 0.1965E-12  0.6245E-12  0.9185E-13  0.8325E-13 -0.2293E-14  0.1101E-13
-0.1955E-12 -0.6245E-12 -0.5310E-12  0.7956E-13 -0.2458E-14 -0.6196E-14
 0.1972E-12 -0.6316E-12 -0.5026E-12  0.4698E-13  0.5082E-14  0.4654E-14
```

```
NODAL DISPLACEMENTS(INCH) IN LOAD CONDITION  2
-0.4201E-02  0.1033E-02 -0.1009E-01  0.2580E-03 -0.7130E-03 -0.5733E-03
-0.4238E-02 -0.1089E-02 -0.1033E-01 -0.2937E-03 -0.8261E-03 -0.6046E-03
 0.4213E-02  0.1094E-02  0.1056E-01 -0.3123E-03 -0.8947E-03  0.4511E-03
 0.4166E-02 -0.1039E-02 -0.1034E-01  0.2798E-03 -0.7818E-03  0.4136E-03
-0.2818E-02  0.1014E-02 -0.7393E-02  0.2345E-03 -0.5519E-03 -0.7400E-06
-0.2875E-02 -0.1055E-02  0.7508E-02 -0.1734E-03 -0.6655E-03 -0.4519E-05
 0.2848E-02  0.1043E-02  0.7495E-02 -0.1770E-03 -0.7108E-03  0.2025E-03
 0.2797E-02 -0.1002E-02 -0.7380E-02  0.2419E-03 -0.6078E-03  0.1987E-03
-0.1616E-02  0.8692E-03 -0.4820E-02  0.2133E-03 -0.3771E-03 -0.1029E-03
-0.1658E-02 -0.8945E-03  0.4877E-02 -0.1829E-03 -0.4264E-03 -0.1006E-03
 0.1646E-02  0.8869E-03  0.4854E-02 -0.1833E-03 -0.4466E-03  0.1826E-03
 0.1606E-02 -0.8610E-03 -0.4796E-02  0.2090E-03 -0.3973E-03  0.1724E-03
-0.6872E-03  0.6470E-03 -0.2648E-02  0.1548E-03 -0.2050E-03 -0.3911E-04
-0.7217E-03 -0.6607E-03  0.2667E-02 -0.1331E-03 -0.2431E-03 -0.4226E-04
 0.7178E-03  0.6566E-03  0.2645E-02 -0.1320E-03 -0.2575E-03  0.7476E-04
 0.6832E-03 -0.6423E-03 -0.2626E-02  0.1563E-03 -0.2239E-03  0.7223E-04
-0.1180E-03  0.3577E-03 -0.1001E-02  0.1275E-03 -0.7815E-04 -0.1258E-04
-0.1469E-03 -0.3619E-03  0.1005E-02 -0.1033E-03 -0.9323E-04 -0.1426E-04
 0.1471E-03  0.3607E-03  0.9855E-03 -0.1023E-03 -0.9545E-04  0.1596E-04
 0.1175E-03 -0.3562E-03 -0.9801E-03  0.1253E-03 -0.8230E-04  0.1414E-04
-0.1419E-12  0.3841E-12  0.5679E-15  0.3738E-13 -0.5725E-14  0.2393E-13
 0.2363E-12 -0.3845E-12 -0.2801E-14 -0.4438E-13 -0.1954E-13  0.2588E-13
-0.2350E-12  0.3846E-12  0.3335E-12 -0.4343E-13 -0.2208E-13 -0.1001E-13
 0.1408E-12 -0.3841E-12 -0.3268E-12  0.3631E-13 -0.9895E-14 -0.1053E-13

EIGENVALUES

EIGENVALUE(RAD/SEC) 1= 0.7070E+03

EIGENVALUE(RAD/SEC) 2= 0.1173E+04

EIGENVALUE(RAD/SEC) 3= 0.2023E+04
```

12.6 DYNAMIC RESPONSE USING FINITE ELEMENT METHOD

When a structure is subjected to dynamic (time-dependent) loads, the displacements, strains, and stresses induced will also vary with time. The dynamic loads arise for a variety of reasons, such as gust loads due to atmospheric turbulence and impact forces due to landing on airplanes, wind and earthquake loads on buildings, etc. The dynamic response calculations include the determination of displacements and stresses as functions of time at each point of the body or structure. The dynamic equations of motion for a damped elastic body have already been derived in Section 12.1 using the finite element procedure. These equations of motion can be solved by using any of the methods presented in Section 7.4 for solving propagation problems.

The direct integration approach considered in Section 7.4.4 involves the numerical integration of the equations of motion by marching in a series of time steps Δt evaluating accelerations, velocities, and displacements at each step. The basis of the mode superposition method discussed in Section 7.4.5 is that the modal matrix (i.e., the matrix formed by using the modes of the system) can be used to diagonalize the mass, damping and stiffness matrices and thus uncouple the equations of motion. The solution of these independent equations, one corresponding to each degree of freedom, can be found by standard techniques and, finally, the solution of the original problem can be found by the superposition of the individual solutions. In this section, we consider the normal mode (or mode superposition or modal analysis) method of finding the dynamic response of an elastic body in some detail.

12.6.1 Uncoupling the Equations of Motion of an Undamped System

The equations of motion of an undamped elastic system are given by (derived in Section 12.1)

$$[M]\ddot{\vec{Q}} + [K]\vec{Q} = \vec{P} \tag{12.65}$$

where $\vec{Q}$ and $\vec{P}$ are the time-dependent displacement and load vectors, respectively. Equation (12.65) represents a system of n coupled second-order differential equations, where n is the number of degrees of freedom of the structure. We now present a method of uncoupling these equations.

Let the natural frequencies of the undamped eigenvalue problem

$$-\omega^2[M]\underline{\vec{Q}} + [K]\underline{\vec{Q}} = \vec{0} \tag{12.66}$$

be given by $\omega_1, \omega_2, \ldots, \omega_n$ with the corresponding eigenvectors given by $\underline{\vec{Q}}_1, \underline{\vec{Q}}_2, \ldots, \underline{\vec{Q}}_n$, respectively. By arranging the eigenvectors (normal modes) as columns, a matrix $[\underline{Q}]$, known as modal matrix, can be defined as

$$[\underline{Q}] = [\underline{\vec{Q}}_1 \quad \underline{\vec{Q}}_2 \quad \cdots \quad \underline{\vec{Q}}_n] \tag{12.67}$$

Since the eigenvectors are $[M]$-orthogonal, we have

$$\underline{\vec{Q}}_i^T [M] \underline{\vec{Q}}_j = \begin{cases} 0 & \text{for} \quad i \neq j \\ 1 & \text{for} \quad i = j \end{cases} \tag{12.68}$$

Equations (12.67) and (12.68) lead to

$$[\underline{Q}]^T [M] [\underline{Q}] = [I] \tag{12.69}$$

where $[I]$ is the identity matrix of order n, and the eigenvalue problem, Eq. (12.66), can be restated as

$$[\grave{}\omega^2_{\grave{}}][M][\underline{Q}] = [K][\underline{Q}] \tag{12.70}$$

where

$$[\grave{}\omega^2_{\grave{}}] = \begin{bmatrix} \omega_1^2 & & & \bigcirc \\ & \omega_2^2 & & \\ & & \ddots & \\ \bigcirc & & & \omega_n^2 \end{bmatrix} \tag{12.71}$$

By premultiplying Eq. (12.70) by $[\underline{Q}]^T$, we obtain

$$[{}^{\backprime}\omega^2_{\diagdown}][\underline{Q}]^T[M][\underline{Q}] = [\underline{Q}]^T[K][\underline{Q}] \tag{12.72}$$

which, in view of Eq. (12.69), becomes

$$[{}^{\backprime}\omega^2_{\diagdown}] = [\underline{Q}]^T[K][\underline{Q}] \tag{12.73}$$

Since any n-dimensional vector can be expressed by superposing the eigenvectors,* one can express $\vec{Q}(t)$ as

$$\vec{Q}(t) = [\underline{Q}]\vec{\eta}(t) \tag{12.74}$$

where $\vec{\eta}(t)$ is a column vector consisting of a set of time-dependent generalized coordinates $\eta_1(t), \eta_2(t), \ldots, \eta_n(t)$. By substituting Eq. (12.74) in Eq. (12.65), we obtain

$$[M][\underline{Q}]\ddot{\vec{\eta}} + [K][\underline{Q}]\vec{\eta} = \vec{P} \tag{12.75}$$

Premultiply both sides of Eq. (12.75) by $[\underline{Q}]^T$ and write

$$[\underline{Q}]^T[M][\underline{Q}]\ddot{\vec{\eta}} + [\underline{Q}]^T[K][\underline{Q}]\vec{\eta} = [\underline{Q}]^T\vec{P} \tag{12.76}$$

However, the normal modes satisfy Eqs. (12.69) and (12.73), and hence Eq. (12.76) reduces to

$$\ddot{\vec{\eta}} + [{}^{\backprime}\omega^2_{\diagdown}]\vec{\eta} = \vec{N} \tag{12.77}$$

where

$$\vec{N} = [\underline{Q}]^T\vec{P}(t) \tag{12.78}$$

Equation (12.77) represents a set of n uncoupled second-order differential equations of the type

$$\ddot{\eta}_i(t) + \omega_i^2\eta_i(t) = N_i(t), \qquad i = 1,\ 2, \ldots, n \tag{12.79}$$

The reason for uncoupling the original equations of motion, Eq. (12.65), into the form of Eqs. (12.79) is that the solution of n uncoupled differential equations is considerably easier than the solution of n coupled differential equations.

---

* Because the eigenvectors are orthogonal, they will form an independent set of vectors and hence they can be used as a basis for the decomposition of any arbitrary n-dimensional vector $\vec{Q}$. A proof of this statement, also known as the expansion theorem, can be found in Ref. [12.7].

12.6.2 Uncoupling the Equations of Motion of a Damped System

The equations of motion of a damped elastic system are given by

$$[M]\ddot{\vec{Q}} + [C]\dot{\vec{Q}} + [K]\vec{Q} = \vec{P} \tag{12.80}$$

Generally little is known about the evaluation of the damping coefficients that are the elements of the damping matrix $[C]$. However, since the effect of damping is small compared to those of inertia and stiffness, the damping matrix $[C]$ is represented by simplified expressions. One simple way of expressing the damping matrix involves the representation of $[C]$ as a linear combination of mass and stiffness matrices as

$$[C] = a[M] + b[K] \tag{12.81}$$

where the constants a and b must be chosen to suit the problem at hand. In this case, the equations of motion, Eq. (12.80), will be uncoupled by the same transformation Eq. (12.74) as that for the undamped system. Thus, the use of Eqs. (12.74) and (12.81) in Eq. (12.80) leads to

$$[\underline{Q}]^T[M][\underline{Q}]\ddot{\vec{\eta}} + (a[\underline{Q}]^T[M][\underline{Q}] + b[\underline{Q}]^T[K][\underline{Q}])\dot{\vec{\eta}} + [\underline{Q}]^T[K][\underline{Q}]\vec{\eta} = [\underline{Q}]^T\vec{P} \tag{12.82}$$

In view of Eqs. (12.69) and (12.73), Eq. (12.82) can be expressed as

$$\ddot{\vec{\eta}} + (a[I] + b[\grave{}\omega_{\grave{}}^2])\dot{\vec{\eta}} + [\grave{}\omega_{\grave{}}^2]\vec{\eta} = \vec{N} \tag{12.83}$$

where $\vec{N}$ is given by Eq. (12.78). Equation (12.83) can be written in scalar form as

$$\ddot{\eta}_i(t) + (a + b\omega_i^2)\dot{\eta}_i(t) + \omega_i^2\eta_i(t) = N_i(t), \qquad i = 1, 2, \ldots, n \tag{12.84}$$

The quantity $(a + b\omega_i^2)$ is known as the modal damping constant in ith normal mode, and it is common to define a quantity ζ_i known as modal damping ratio in ith normal mode as

$$\zeta_i = \frac{a + b\omega_i^2}{2\omega_i} \tag{12.85}$$

so that the equations of motion in terms of generalized coordinates become

$$\ddot{\eta}_i(t) + 2\zeta_i\omega_i\dot{\eta}_i(t) + \omega_i^2\eta_i(t) = N_i(t), \qquad i = 1, 2, \ldots, n \tag{12.86}$$

Thus, Eq. (12.86) denotes a set of n uncoupled second-order differential equations for the damped elastic system.

12.6.3 Solution of a General Second-Order Differential Equation

A general second-order differential equation (or one of the uncoupled equations of motion of a damped elastic system) can be expressed as Eq. (12.86). The solution of Eq. (12.86) consists of two parts: one called the homogeneous solution and the other known as the particular integral.

Homogeneous Solution

The homogeneous solution can be obtained by solving the equation

$$\ddot{\eta}_i(t) + 2\zeta_i\omega_i\dot{\eta}_i(t) + \omega_i^2\eta_i(t) = 0 \tag{12.87}$$

By assuming a solution of the type

$$\eta_i(t) = A \cdot e^{\alpha t} \tag{12.88}$$

where A is a constant, Eq. (12.87) gives the following characteristic equation:

$$\alpha^2 + 2\zeta_i\omega_i\alpha + \omega_i^2 = 0 \tag{12.89}$$

The roots of Eq. (12.89) are given by

$$\alpha_{1,2} = -\zeta_i\omega_i \pm \omega_i\sqrt{\zeta_i^2 - 1} \tag{12.90}$$

Thus, the homogeneous solution of Eq. (12.86) can be expressed as

$$\eta_i(t) = A_1 e^{\alpha_1 t} + A_2 e^{\alpha_2 t} \tag{12.91}$$

where A_1 and A_2 are constants to be determined from the known initial displacement and velocity. Depending on the magnitude of ζ_i, the system is classified as underdamped, critically damped, and overdamped as follows:

1. Underdamped case (when $\zeta_i < 1$): If $\zeta_i < 1$, the solution given in Eq. (12.91) can be rewritten as

$$\begin{aligned}\eta_i(t) &= e^{-\zeta_i\omega_i t}(A_1 e^{i\omega_{id}t} + A_2 e^{-i\omega_{id}t}) \\ &= e^{-\zeta_i\omega_i t}(B_1 \cos\omega_{id}t + B_2 \sin\omega_{id}t) \\ &= C_1 e^{-\zeta_i\omega_i t}\cos(\omega_{id}t - \phi)\end{aligned} \tag{12.92}$$

where $\omega_{id} = \omega_i\sqrt{1 - \zeta_i^2}$, and the constants B_1 and B_2 or C_1 and ϕ (ϕ is also known as phase angle) can be found from the initial conditions. Here, ω_{id} can be regarded as a natural frequency associated with the damped system.

2. Critically damped case ($\zeta_i = 1$): In this case, the roots α_1 and α_2 given by Eq. (12.90) will be equal:

$$\alpha_1 = \alpha_2 = -\omega_i \tag{12.93}$$

The solution of Eq. (12.87) is given by

$$\eta_i(t) = e^{-\omega_i t}(A_1 + A_2 t) \tag{12.94}$$

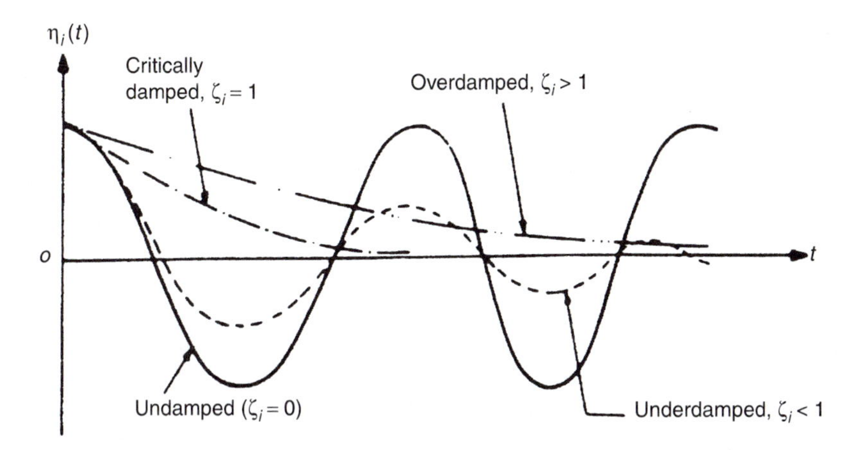

Figure 12.4. Response with Different Degrees of Damping.

where A_1 and A_2 are constants of integration to be determined from the known initial conditions.

3. Overdamped case ($\zeta_i > 1$): When $\zeta_i > 1$, both the roots given by Eq. (12.90) will be negative and the solution given by Eq. (12.91) can be rewritten as

$$\begin{aligned}\eta_i(t) &= e^{-\zeta_i \omega_i t}(A_1 e^{\sqrt{\zeta_i^2 - 1}\omega_i t} + A_2 e^{-\sqrt{\zeta_i^2 - 1}\omega_i t}) \\ &= e^{-\zeta_i \omega_i t}(B_1 \cosh \sqrt{\zeta_i^2 - 1}\omega_i t + B_2 \sinh \sqrt{\zeta_i^2 - 1}\omega_i t)\end{aligned} \tag{12.95}$$

The solutions given by Eqs. (12.92), (12.94), and (12.95) are shown graphically in Figure 12.4. It can be noticed that in the case of an underdamped system, the response oscillates within an envelope defined by $\eta_i(t) = \pm C_1 e^{\zeta_i \omega_i t}$ and the response dies out as time (t) increases. In the case of critical damping, the response is not periodic but dies out with time. Finally, in the case of overdamping, the response decreases monotonically with increasing time.

Particular Integral

By solving Eq. (12.86), the particular integral in the case of an underdamped system can be obtained as [12.7]

$$\eta_i(t) = \frac{1}{\omega_{id}} \int_0^t N_i(\tau) e^{-\zeta_i \omega_i (t-\tau)} \sin \omega_{id}(t-\tau)\, \mathrm{d}\tau \tag{12.96}$$

Total Solution

The total solution is given by the sum of homogeneous solution and the particular integral. If $\eta_i(o)$ and $\dot{\eta}_i(o)$ denote the initial conditions [i.e., values of $\eta_i(t)$ and $(\mathrm{d}\eta_i/\mathrm{d}t)(t)$ at $t = 0$], the total solution can be expressed as

$$\eta_i(t) = \frac{1}{\omega_{id}} \int_0^t N_i(\tau) \cdot e^{-\zeta_i \omega_i (t-\tau)} \sin \omega_{id}(t-\tau)\, d\tau + e^{-\zeta_i \omega_i t}$$
$$\times \left[\omega_{id} t + \frac{\zeta_i}{(1-\zeta_i^2)^{1/2}} \sin \omega_{id} t \right] \eta_i(o)$$
$$+ \left[\frac{1}{\omega_{id}} e^{-\zeta_i \omega_i t} \sin \omega_{id} t \right] \dot{\eta}_i(o) \tag{12.97}$$

Solution When the Forcing Function Is an Arbitrary Function of Time

The numerical solution of Eq. (12.96) when the forcing function $N_i(t)$ is an arbitrary function of time was given in Section 7.4.6. The recurrence formulas useful for computing the solution of Eq. (12.96) were given by Eqs. (7.90) and (7.93). Thus, by using the uncoupling procedure outlined in Sections 12.6.1 and 12.6.2, the response of any multi-degree of freedom system under any arbitrary loading conditions can be found with the help of Eqs. (7.90) and (7.93).

Example 12.3 Find the dynamic response of the stepped bar shown in Figure 12.5(a) when an axial load of magnitude P_o is applied at node 3 for a duration of time t_o as shown in Figure 12.5(b).

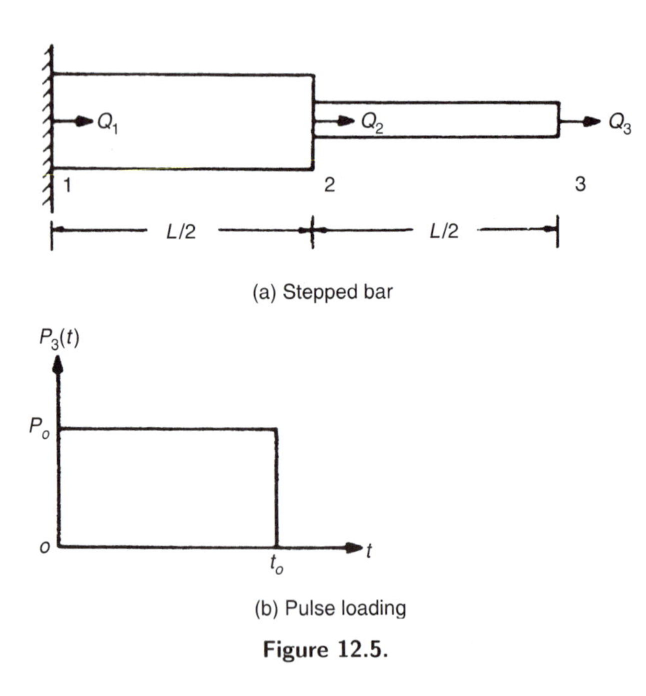

(a) Stepped bar

(b) Pulse loading

Figure 12.5.

Solution The free vibration characteristics of this bar have already been determined in Example 12.2 as

$$\omega_1 = 1.985[E/(\rho L^2)]^{1/2} \tag{E$_1$}$$

$$\omega_2 = 5.159[E/(\rho L^2)]^{1/2} \tag{E$_2$}$$

$$\vec{\underline{Q}}_1 = \frac{1}{(\rho AL)^{1/2}} \begin{Bmatrix} 0.8812 \\ 1.5260 \end{Bmatrix} \tag{E$_3$}$$

$$\vec{\underline{Q}}_2 = \frac{1}{(\rho AL)^{1/2}} \begin{Bmatrix} -1.186 \\ 2.053 \end{Bmatrix} \tag{E$_4$}$$

$$[\underline{Q}] = \frac{1}{(\rho AL)^{1/2}} \begin{bmatrix} 0.8812 & -1.1860 \\ 1.5260 & 2.0530 \end{bmatrix} \tag{E$_5$}$$

$$[M] = \frac{\rho AL}{12} \begin{bmatrix} 6 & 1 \\ 1 & 2 \end{bmatrix} \tag{E$_6$}$$

$$[K] = \frac{2AE}{L} \begin{bmatrix} 3 & -1 \\ -1 & 1 \end{bmatrix} \tag{E$_7$}$$

Thus, it can be verified that

$$[\underline{Q}]^T[M][\underline{Q}] = \begin{bmatrix} 1 & 0 \\ 0 & 1 \end{bmatrix} \tag{E$_8$}$$

and

$$[\underline{Q}]^T[K][\underline{Q}] = \left(\frac{E}{\rho L^2}\right) \begin{bmatrix} (1.985)^2 & 0 \\ 0 & (5.159)^2 \end{bmatrix} \tag{E$_9$}$$

The generalized load vector is given by

$$\begin{aligned} \vec{N} = [\underline{Q}]^T \vec{P}(t) &= \frac{1}{(\rho AL)^{1/2}} \begin{bmatrix} 0.8812 & 1.5260 \\ -1.1860 & 2.0530 \end{bmatrix} \begin{Bmatrix} 0 \\ P_3 \end{Bmatrix} \\ &= \frac{1}{(\rho AL)^{1/2}} \begin{Bmatrix} 1.526 \\ 2.053 \end{Bmatrix} \cdot P_3 \end{aligned} \tag{E$_{10}$}$$

The undamped equations of motion, Eq. (12.77), are given by

$$\ddot{\vec{\eta}} + \frac{E}{\rho L^2} \begin{bmatrix} (1.985)^2 & 0 \\ 0 & (5.159)^2 \end{bmatrix} \vec{\eta} = \frac{1}{(\rho AL)^{1/2}} \begin{Bmatrix} 1.526 \\ 2.053 \end{Bmatrix} P_3 \tag{E$_{11}$}$$

which, in scalar form, represent

$$\ddot{\eta}_1 + \frac{3.941E}{\rho L^2} \eta_1 = \frac{1.526 P_3}{(\rho AL)^{1/2}} \tag{E$_{12}$}$$

and

$$\ddot{\eta}_2 + \frac{26.62E}{\rho L^2}\eta_2 = \frac{2.063P_3}{(\rho AL)^{1/2}} \tag{E$_{13}$}$$

By assuming that all the initial displacements and velocities are zero, we obtain

$$\vec{Q}(t=0) = \vec{0} = [Q]\vec{\eta}(0)$$

$$\dot{\vec{Q}}(t=0) = \vec{0} = [Q]\dot{\vec{\eta}}(0) \tag{E$_{14}$}$$

so that

$$\vec{\eta}(0) = \vec{0} \quad \text{and} \quad \dot{\vec{\eta}}(0) = \dot{\vec{0}} \tag{E$_{15}$}$$

Thus, the solutions of Eqs. (E$_{12}$) and (E$_{13}$) can be expressed as [from Eq. (12.97)]

$$\eta_1(t) = \frac{1}{\omega_1}\int_0^t N_1(\tau)\sin\omega_1(t-\tau)\,\mathrm{d}\tau$$

that is,

$$\eta_1(t) = \left(\frac{\rho L^2}{E}\right)^{1/2}\left(\frac{1}{1.985}\right)\int_0^t \left\{\frac{1.526P_3}{(\rho AL)^{1/2}}\right\}\sin\left\{\left(\frac{E}{\rho L^2}\right)^{1/2}(1.985)(t-\tau)\right\}\mathrm{d}\tau$$

$$= \left(\frac{L}{AE}\right)^{1/2}(0.7686)\int_0^t P_3(\tau)\sin\left\{1.985\left(\frac{E}{\rho L^2}\right)^{1/2}(t-\tau)\right\}\mathrm{d}\tau \tag{E$_{16}$}$$

and

$$\eta_2(t) = \frac{1}{\omega_2}\int_0^t N_2(\tau)\sin\omega_2(t-\tau)\,\mathrm{d}\tau$$

that is,

$$\eta_2(t) = \left(\frac{\rho L^2}{E}\right)^{1/2}\left(\frac{1}{5.159}\right)\int_0^t \left\{\frac{2.053P_3}{(\rho AL)^{1/2}}\right\}\sin\left\{\left(\frac{E}{\rho L^2}\right)^{1/2}(5.159)(t-\tau)\right\}\mathrm{d}\tau$$

$$= \left(\frac{L}{AE}\right)^{1/2}(0.3979)\int_0^t P_3(\tau)\sin\left\{0.159\left(\frac{E}{\rho L^2}\right)^{1/2}(t-\tau)\right\}\mathrm{d}\tau \tag{E$_{17}$}$$

The solutions of Eqs. (E$_{16}$) and (E$_{17}$) for the given loading can be expressed as follows:

For $t < t_0$:

$$\eta_1(t) = 0.38720\,P_0\left(\frac{\rho L^3}{AE^2}\right)^{1/2}\left[1-\cos\left\{1.985\left(\frac{E}{\rho L^2}\right)^{1/2}t\right\}\right] \tag{E$_{18}$}$$

and

$$\eta_2(t) = 0.07713\ P_0 \left(\frac{\rho L^3}{AE^2}\right)^{1/2} \left[1 - \cos\left\{5.159\left(\frac{E}{\rho L^2}\right)^{1/2} t\right\}\right] \tag{E$_{19}$}$$

For $t > t_0$:

$$\eta_1(t) = 0.38720\ P_0 \left(\frac{\rho L^3}{AE^2}\right)^{1/2} \left[\cos\left\{1.985\left(\frac{E}{\rho L^2}\right)^{1/2} (t - t_0)\right\} - \cos\left\{1.985\left(\frac{E}{\rho L^2}\right)^{1/2} t\right\}\right] \tag{E$_{20}$}$$

and

$$\eta_2(t) = 0.07713\ P_0 \left(\frac{\rho L^3}{AE^2}\right)^{1/2} \left[\cos\left\{5.159\left(\frac{E}{\rho L^2}\right)^{1/2} (t - t_0)\right\} - \cos\left\{5.159\left(\frac{E}{\rho L^2}\right)^{1/2} t\right\}\right] \tag{E$_{21}$}$$

The physical displacements are given by

$$\begin{aligned}\vec{Q}(t) &= \begin{Bmatrix} Q_2(t) \\ Q_3(t) \end{Bmatrix} = [\underset{\sim}{Q}]\vec{\eta}(t) = \frac{1}{(\rho A L)^{1/2}} \begin{bmatrix} 0.8812 & -1.186 \\ 1.526 & 2.053 \end{bmatrix} \begin{Bmatrix} \eta_1(t) \\ \eta_2(t) \end{Bmatrix} \\ &= \frac{1}{(\rho A L)^{1/2}} \begin{Bmatrix} 0.8812 \quad \eta_1(t) - 1.186 \quad \eta_2(t) \\ 1.526 \quad \eta_1(t) + 2.053 \quad \eta_2(t) \end{Bmatrix}\end{aligned} \tag{E$_{22}$}$$

Thus, for $t < t_0$:

$$Q_2(t) = \frac{P_0 L}{AE}\left[0.24991 - 0.34140\cos\left\{1.985\left(\frac{E}{\rho L^2}\right)^{1/2} t\right\} + 0.09149\cos\left\{5.159\left(\frac{E}{\rho L^2}\right)^{1/2} t\right\}\right] \tag{E$_{23}$}$$

$$Q_3(t) = \frac{P_0 L}{AE}\left[0.4324 - 0.5907\cos\left\{1.985\left(\frac{E}{\rho L^2}\right)^{1/2} t\right\} + 0.1583\cos\left\{5.159\left(\frac{E}{\rho L^2}\right)^{1/2} t\right\}\right] \tag{E$_{24}$}$$

and for $t > t_0$:

$$\begin{aligned}Q_2(t) = \frac{P_0 L}{AE}\Bigg[&0.34140\cos\left\{1.985\left(\frac{E}{\rho L^2}\right)^{1/2} (t - t_0)\right\} - 0.34140\cos\left\{1.985\left(\frac{E}{\rho L^2}\right)^{1/2} t\right\} \\ &- 0.09149\cos\left\{5.159\left(\frac{E}{\rho L^2}\right)^{1/2} (t - t_0)\right\} + 0.09149\cos\left\{5.159\left(\frac{E}{\rho L^2}\right)^{1/2} t\right\}\Bigg]\end{aligned} \tag{E$_{25}$}$$

$$Q_3(t) = \frac{P_0 L}{AE}\left[0.5907\cos\left\{1.985\left(\frac{E}{\rho L^2}\right)^{1/2}(t-t_0)\right\} - 0.5907\cos\left\{1.985\left(\frac{E}{\rho L^2}\right)^{1/2} t\right\}\right.$$
$$\left. + 0.1583\cos\left\{5.159\left(\frac{E}{\rho L^2}\right)^{1/2}(t-t_0)\right\} - 0.1583\cos\left\{5.159\left(\frac{E}{\rho L^2}\right)^{1/2} t\right\}\right] \tag{E$_{26}$}$$

To determine the average dynamic stresses in the two elements, we use

$$\sigma^{(1)}(t) = 2E\left(\frac{Q_2 - Q_1}{L}\right) = \frac{2P_0}{A}\left[0.24991 - 0.34140\cos\left\{1.985\left(\frac{E}{\rho L^2}\right)^{1/2} t\right\}\right.$$
$$\left. + 0.09149\cos\left\{5.159\left(\frac{E}{\rho L^2}\right)^{1/2} t\right\}\right] \quad \text{for } t < t_0$$
$$= \frac{2P_0}{A}\left[0.34140\cos\left\{1.985\left(\frac{E}{\rho L^2}\right)^{1/2}(t-t_0)\right\}\right.$$
$$- 0.34140\cos\left\{1.985\left(\frac{E}{\rho L^2}\right)^{1/2} t\right\}$$
$$- 0.09149\cos\left\{5.159\left(\frac{E}{\rho L^2}\right)^{1/2}(t-t_0)\right\}$$
$$\left. + 0.09149\cos\left\{5.159\left(\frac{E}{\rho L^2}\right)^{1/2} t\right\}\right] \quad \text{for } t > t_0 \tag{E$_{27}$}$$

and

$$\sigma^{(2)}(t) = 2E\left(\frac{Q_3 - Q_2}{L}\right) = \frac{2P_0}{A}\left[0.18249 - 0.24930\cos\left\{1.985\left(\frac{E}{\rho L^2}\right)^{1/2} t\right\}\right.$$
$$\left. + 0.06681\cos\left\{5.159\left(\frac{E}{\rho L^2}\right)^{1/2} t\right\}\right] \quad \text{for } t < t_0$$
$$= \frac{2P_0}{A}\left[0.24930\cos\left\{1.985\left(\frac{E}{\rho L^2}\right)^{1/2}(t-t_0)\right\}\right.$$
$$- 0.24930\cos\left\{1.985\left(\frac{E}{\rho L^2}\right)^{1/2} t\right\}$$
$$+ 0.24979\cos\left\{5.159\left(\frac{E}{\rho L^2}\right)^{1/2}(t-t_0)\right\}$$
$$\left. - 0.24979\cos\left\{5.159\left(\frac{E}{\rho L^2}\right)^{1/2} t\right\}\right] \quad \text{for } t > t_0 \tag{E$_{28}$}$$

12.7 NONCONSERVATIVE STABILITY AND FLUTTER PROBLEMS

The stability of nonconservative systems was considered by the finite element method in Refs. [12.8] and [12.9]. The problem of panel flutter was treated by Olson [12.10] and Kariappa and Somashekar [12.11]. The flutter analysis of three-dimensional structures (e.g., supersonic aircraft wing structures) that involve modeling by different types of finite elements was presented by Rao [12.12, 12.13]. Flutter analysis involves the solution of a double eigenvalue problem that can be expressed as

$$[[K] - \omega^2[M] + [Q]]\vec{\xi} = \vec{0} \tag{12.98}$$

where $[K]$ and $[M]$ are the usual stiffness and mass matrices, respectively, ω is the flutter frequency, $[Q]$ is the aerodynamic matrix, and $\vec{\xi}$ is the vector of generalized coordinates. The matrix $[Q]$ is a function of flutter frequency ω and flutter velocity V, which are both unknown. For a nontrivial solution of $\vec{\xi}$, the determinant of the coefficient matrix of $\vec{\xi}$ must vanish. Thus, the flutter equation becomes

$$\left|[K] - \omega^2[M] + [Q]\right| = 0 \tag{12.99}$$

Since two unknowns, namely ω and V, are in Eq. (12.99), the problem is called a double eigenvalue problem. The details of the generation of aerodynamic matrix $[Q]$ and the solution of Eq. (12.99) are given in Refs. [12.10] and [12.12].

12.8 SUBSTRUCTURES METHOD

In the finite element analysis of large systems, the number of equations to be solved for an accurate solution will be quite large. In such cases, the method of substructures can be used to reduce the number of equations to manageable size. The system (or structure) is divided into a number of parts or segments, each called a substructure (see Figure 12.6). Each substructure, in turn, is divided into several finite elements. The element matrix equations of each substructure are assembled to generate the substructure equations. By treating each substructure as a large element with many interior and exterior (boundary) nodes, and using a procedure known as static condensation [12.14], the equations of the substructure are reduced to a form involving only the exterior nodes of that particular substructure. The reduced substructure equations can then be assembled to obtain the overall system equations involving only the boundary unknowns of the various substructures. The number of these system equations is much less compared to the total number of unknowns. The solution of the system equations gives values of the boundary unknowns of each substructure. The known boundary nodal values can then be used as prescribed boundary conditions for each substructure to solve for the respective interior nodal unknowns. The concept of substructuring has been used for the analysis of static, dynamic, as well as nonlinear analyses [12.15, 12.16].

REFERENCES

12.1 D.T. Greenwood: *Principles of Dynamics*, Prentice-Hall, Englewood Cliffs, NJ, 1965.

12.2 C.I. Bajer: Triangular and tetrahedral space–time finite elements in vibration analysis, *International Journal for Numerical Methods in Engineering, 23*, 2031–2048, 1986.

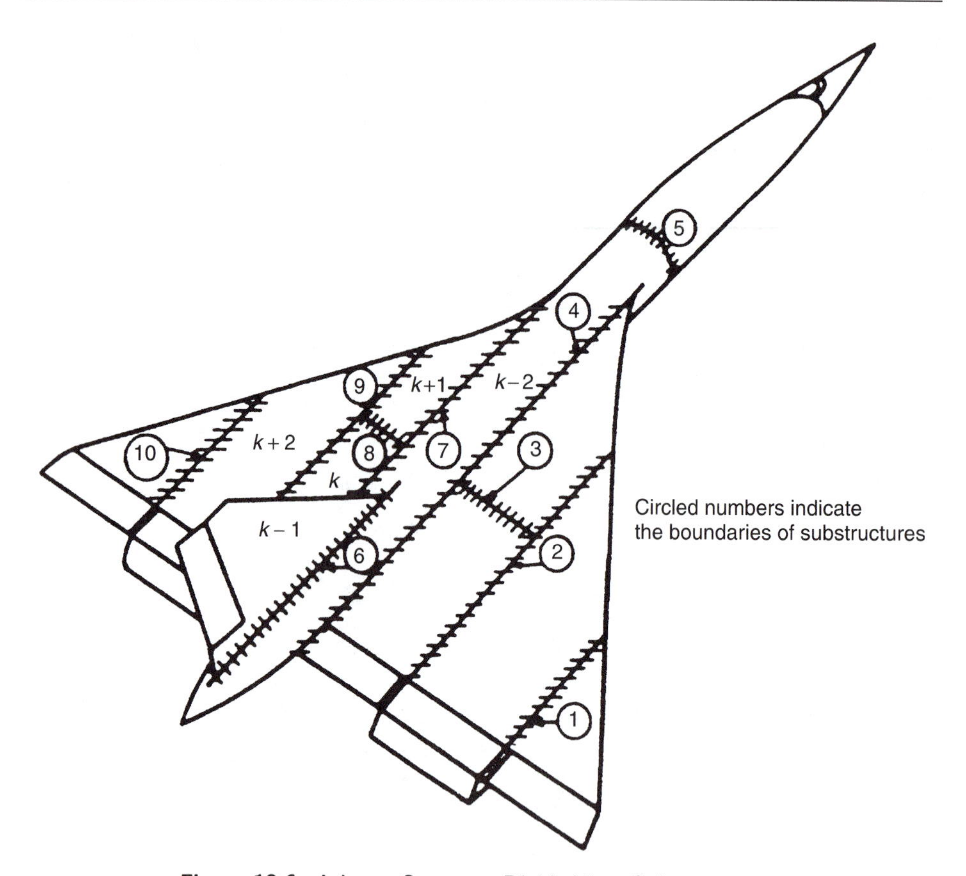

Figure 12.6. A Large Structure Divided into Substructures.

12.3 C.I. Bajer: Notes on the stability of non-rectangular space–time finite elements, *International Journal for Numerical Methods in Engineering, 24*, 1721–1739, 1987.

12.4 J.S. Archer: Consistent mass matrix for distributed mass systems, *Journal of Structural Division, Proc. ASCE, 89*, No. ST4, 161–178, 1963.

12.5 A.K. Gupta: Effect of rotary inertia on vibration of tapered beams, *International Journal for Numerical Methods in Engineering, 23*, 871–882, 1986.

12.6 R.S. Gupta and S.S. Rao: Finite element eigenvalue analysis of tapered and twisted Timoshenko beams, *Journal of Sound and Vibration, 56*, 187–200, 1978.

12.7 L. Meirovitch: *Analytical Methods in Vibrations*, Macmillan, New York, 1967.

12.8 R.S. Barsoum: Finite element method applied to the problem of stability of a nonconservative system, *International Journal for Numerical Methods in Engineering, 3*, 63–87, 1971.

12.9 C.D. Mote and G.Y. Matsumoto: Coupled, nonconservative stability–finite element, *Journal of Engineering Mechanics Division, 98*, No. EM3, 595–608, 1972.

12.10 M.D. Olson: Finite elements applied to panel flutter, *AIAA Journal, 5*, 2267–2270, 1967.

12.11 V. Kariappa and B.R. Somashekar: Application of matrix displacement methods in the study of panel flutter, *AIAA Journal, 7*, 50–53, 1969.

12.12 S.S. Rao: Finite element flutter analysis of multiweb wing structures, *Journal of Sound and Vibration, 38*, 233–244, 1975.

12.13 S.S. Rao: A finite element approach to the aeroelastic analysis of lifting surface type structures, *International Symposium on Discrete Methods in Engineering, Proceedings*, 512–525, Milan, September 1974.

12.14 J.S. Przemieniecki: *Theory of Matrix Structural Analysis*, McGraw-Hill, New York, 1968.

12.15 R.H. Dodds, Jr., and L.A. Lopez: Substructuring in linear and nonlinear analysis, *International Journal for Numerical Methods in Engineering, 15*, 583–597, 1980.

12.16 M. Kondo and G.B. Sinclair: A simple substructuring procedure for finite element analysis of stress concentrations, *Communications in Applied Numerical Methods, 1*, 215–218, 1985.

PROBLEMS

12.1 Find the solution of Example 12.1 using the lumped mass matrix.

12.2 Find the solution of Example 12.2 using the lumped mass matrix.

12.3–12.5 Find the natural frequencies and modes of vibration for the following cases:

12.3 one-element cantilever beam

12.4 one-element simply supported beam

12.5 two-element simply supported beam by taking advantage of the symmetry about the midpoint.

12.6 Find the natural frequencies and mode shapes of the rod shown in Figure 12.7 in axial vibration.

12.7 Sometimes it is desirable to suppress less important or unwanted degrees of freedom from the original system of equations

$$\underset{n \times n}{[K]} \underset{n \times 1}{\vec{X}} = \underset{n \times 1}{\vec{P}} \tag{E$_1$}$$

to reduce the size of the problem to be solved. This procedure, known as *static condensation or condensation of unwanted d.o.f.*, consists of partitioning Eq. (E$_1$) as

$$\begin{bmatrix} \underset{p \times p}{K_{11}} & \vdots & \underset{p \times q}{K_{12}} \\ \cdots & & \cdots \\ \underset{q \times p}{K_{21}} & \vdots & \underset{q \times q}{K_{22}} \end{bmatrix} \begin{Bmatrix} \underset{p \times 1}{\vec{X}_1} \\ \cdots \\ \underset{q \times 1}{\vec{X}_2} \end{Bmatrix} = \begin{Bmatrix} \underset{p \times 1}{\vec{P}_1} \\ \cdots \\ \underset{q \times 1}{\vec{P}_2} \end{Bmatrix}; \quad p + q = n \tag{E$_2$}$$

where $\vec{X}_2$ is the vector of unwanted degrees of freedom. Equation (E$_2$) gives

$$[K_{11}]\vec{X}_1 + [K_{12}]\vec{X}_2 = \vec{P}_1 \tag{E$_3$}$$

$$[K_{21}]\vec{X}_1 + [K_{22}]\vec{X}_2 = \vec{P}_2 \tag{E$_4$}$$

Solving Eq. (E$_4$) for $\vec{X}_2$ and substituting the result in Eq. (E$_3$) lead to the desired condensed set of equations

$$\underset{p \times p}{[\underline{K}]} \underset{p \times 1}{\vec{X}_1} = \underset{p \times 1}{\vec{\underline{P}}} \tag{E$_5$}$$

Derive the expressions of $[\underline{K}]$ and $\vec{\underline{P}}$.

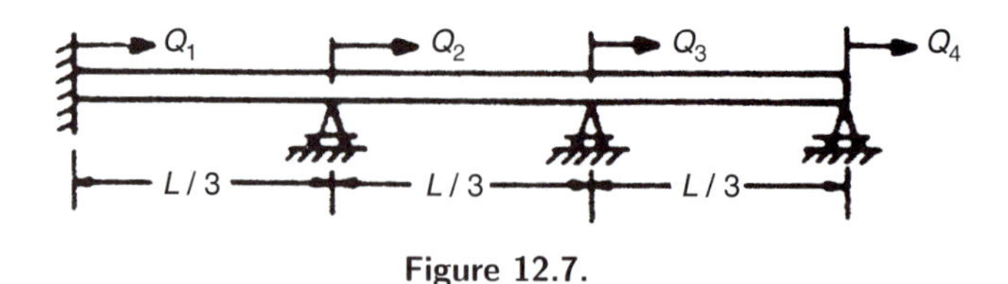

Figure 12.7.

12.8 Using the subroutine PLATE, find the displacements and the first two natural frequencies of a box beam (similar to the one shown in Figure 10.7) with the following data:

Length = 100 in., width = 20 in., depth = 10 in., $t_c = 0.5$ in., $t_\omega = 1.0$ in., $E = 30 \times 10^6$ psi, $\nu = 0.3$, $P_1 = P_2 = 1000$ lb

12.9 Find the natural frequencies of longitudinal vibration of the stepped bar shown in Figure 12.8 using consistent mass matrices.

12.10 Solve Problem 12.9 using lumped mass matrices.

12.11 Find the natural frequencies of longitudinal vibration of the stepped bar shown in Figure 12.9 using consistent mass matrices.

12.12 Solve Problem 12.11 using lumped mass matrices.

12.13 Find the mode shapes of the stepped bar shown in Figure 12.8 corresponding to the natural frequencies found in Problem 12.9.

12.14 Find the mode shapes of the stepped bar shown in Figure 12.8 corresponding to the natural frequencies found in Problem 12.10.

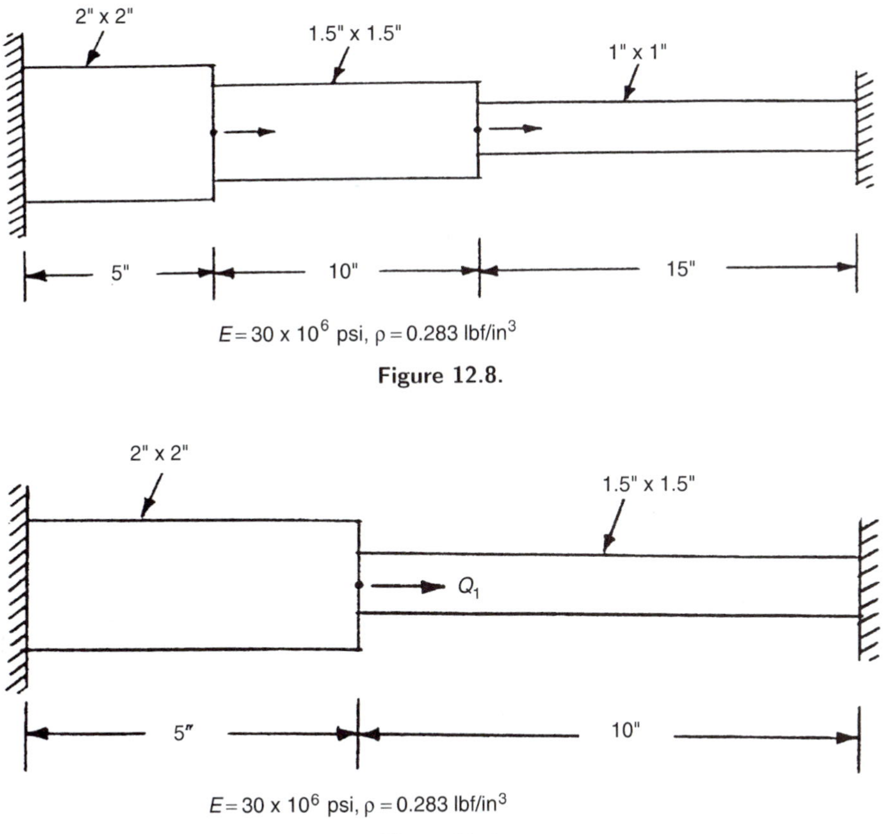

Figure 12.8.

Figure 12.9.

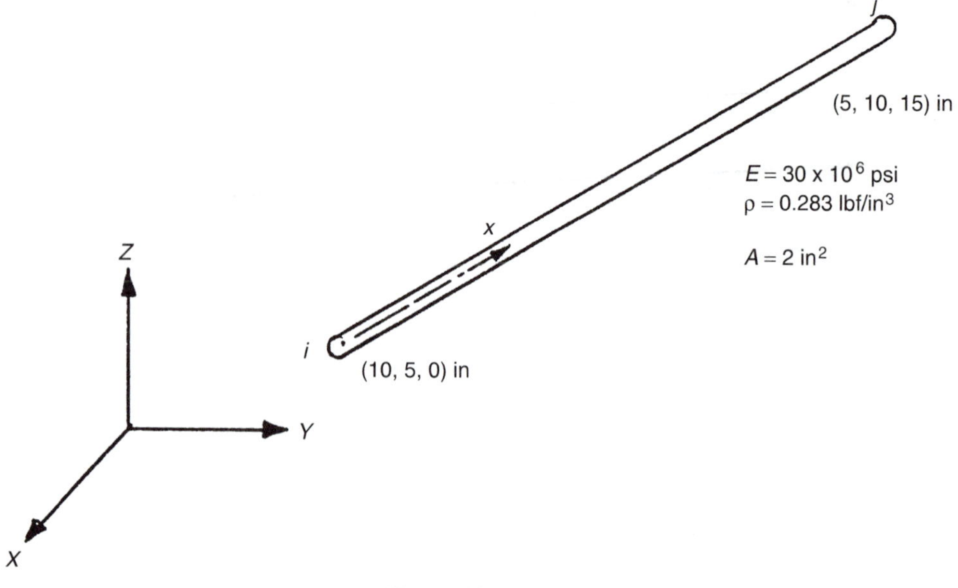

Figure 12.10.

12.15 Orthogonalize the mode shapes found in Problem 12.13 with respect to the corresponding mass matrix.

12.16 Orthogonalize the mode shapes found in Problem 12.14 with respect to the corresponding mass matrix.

12.17 Find the consistent and lumped mass matrices of the bar element shown in Figure 12.10 in the XYZ coordinate system.

12.18 (a) Derive the stiffness and consistent mass matrices of the two-bar truss shown in Figure 12.11.

(b) Determine the natural frequencies of the truss (using the consistent mass matrix).

12.19 (a) Derive the lumped mass matrix of the two-bar truss shown in Figure 12.11.

(b) Determine the natural frequencies of the truss (using the lumped mass matrix).

12.20 The properties of the two elements in the stepped beam shown in Figure 12.12 are given below:

Element 1: E $= 30 \times 10^6$ psi, $\rho = 0.283$ lbf/in.^{3}, cross section = circular, 2-in. diameter

Element 2: E $= 11 \times 10^6$ psi, $\rho = 0.1$ lbf/in.^{3}, cross section = circular, 1-in. diameter

Find the natural frequencies of the stepped beam.

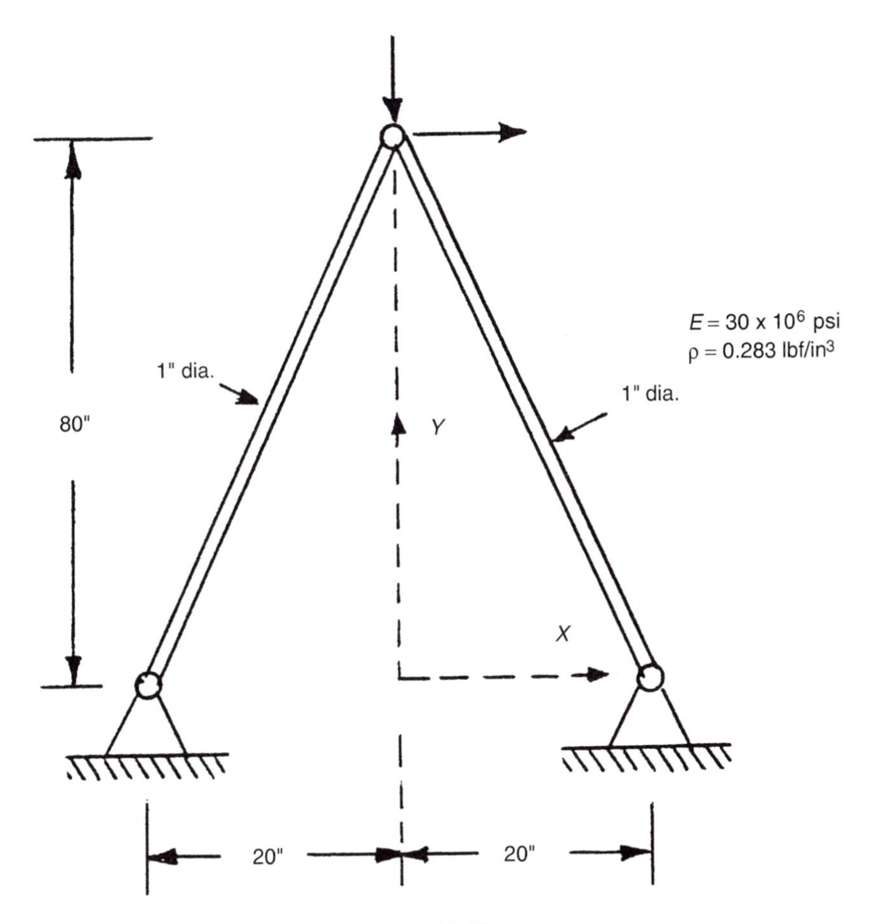

Figure 12.11.

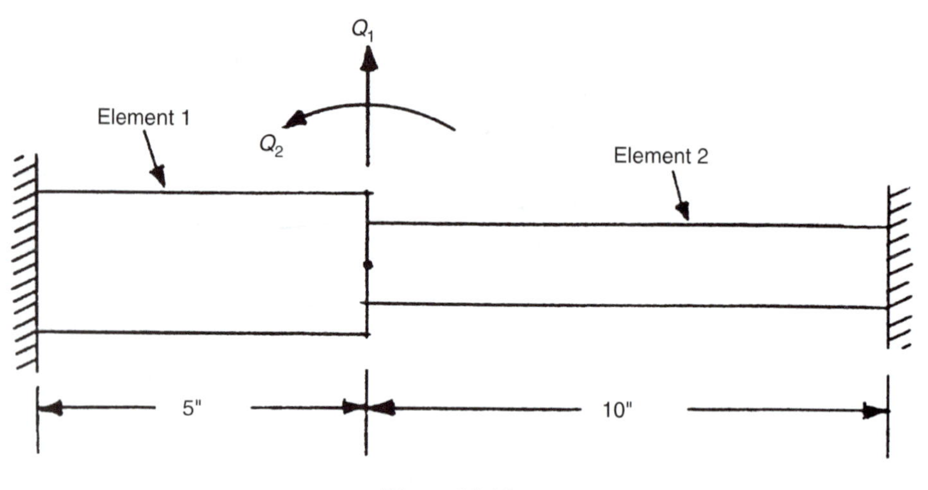

Figure 12.12.

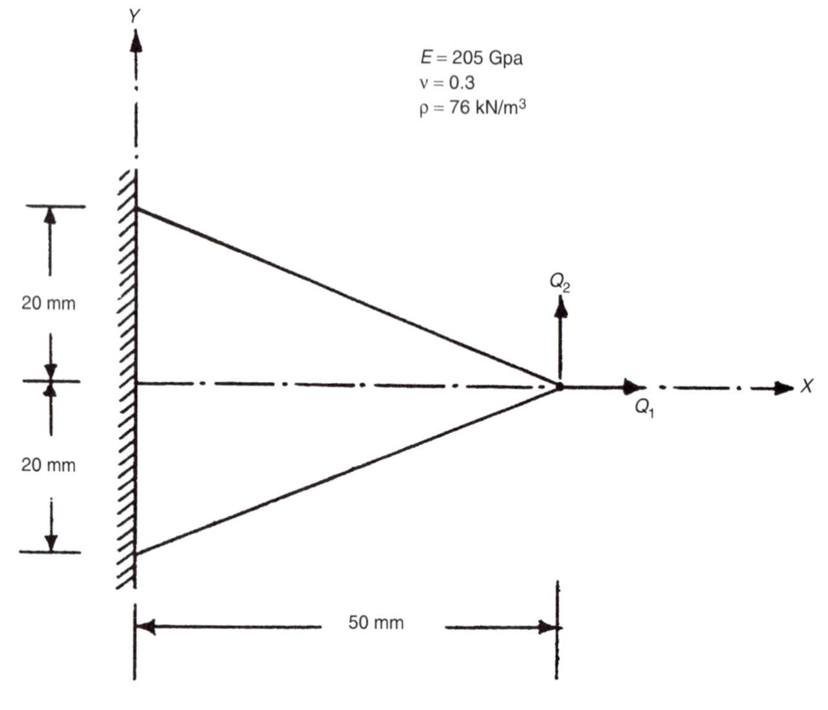

Figure 12.13.

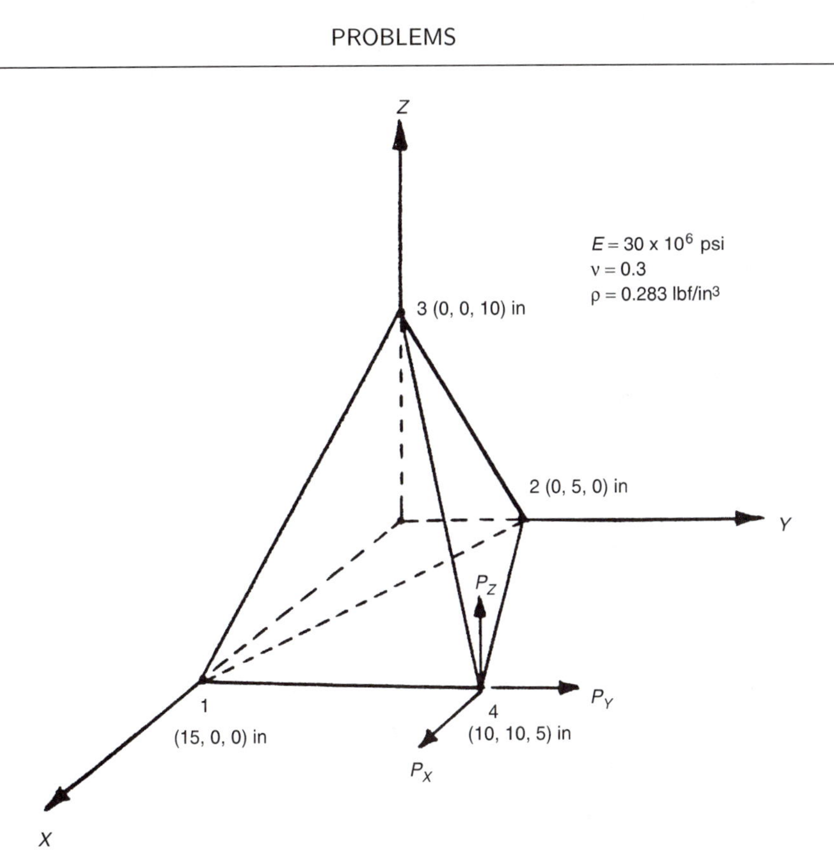

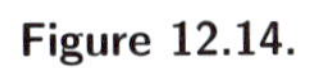
Figure 12.14.

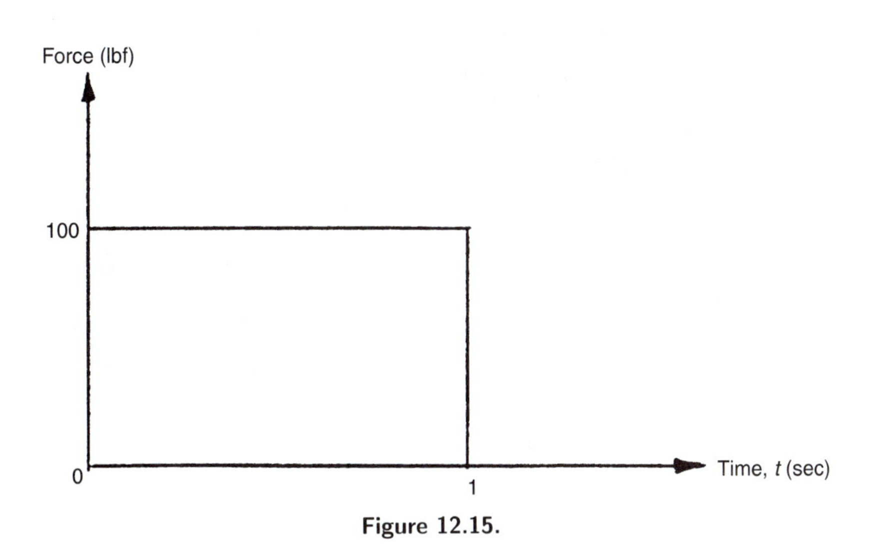

Figure 12.15.

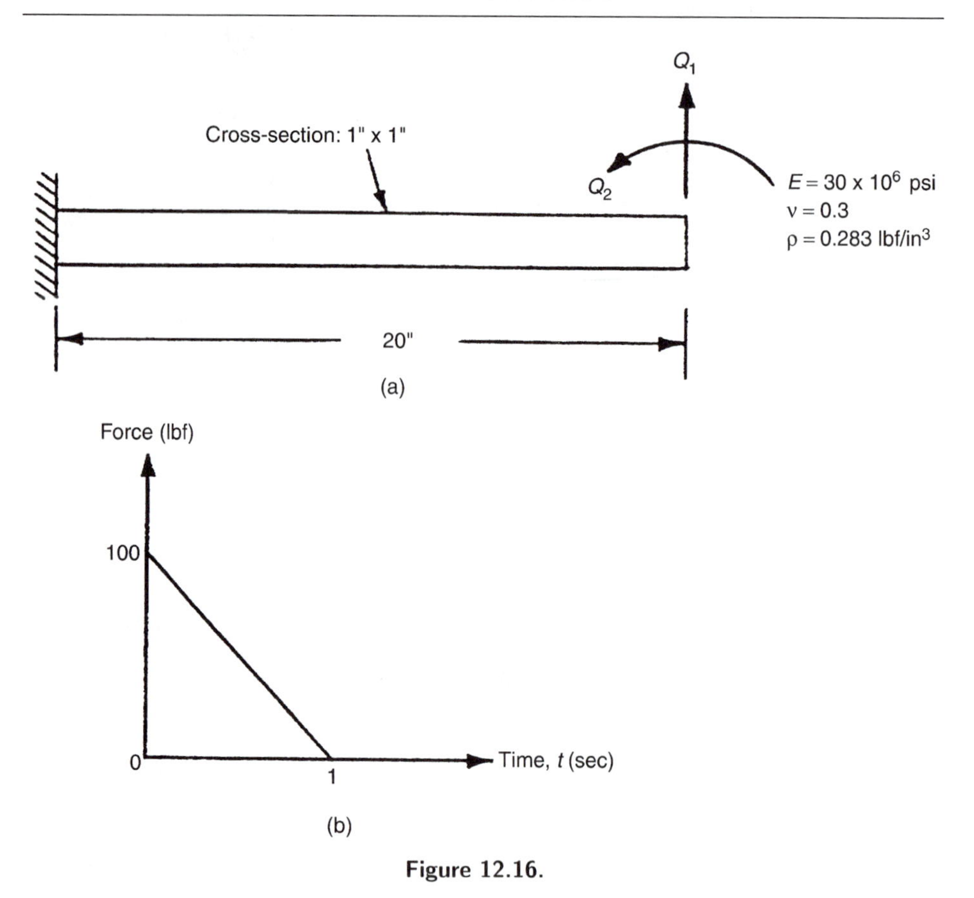

Figure 12.16.

12.21 Find the mode shapes of the stepped beam considered in Problem 12.20.

12.22 Find the natural frequencies of the triangular plate shown in Figure 12.13 using the consistent mass matrix. Use one triangular membrane element for modeling.

12.23 Solve Problem 12.22 using the lumped mass matrix.

12.24 Consider the tetrahedron element shown in Figure 12.14. Find the natural frequencies of the element by fixing the face 123.

12.25 Consider the stepped bar shown in Figure 12.9. If the force shown in Figure 12.15 is applied along Q_1, determine the dynamic response, $Q_1(t)$.

12.26 The cantilever beam shown in Figure 12.16(a) is subjected to the force indicated in Figure 12.16(b) along the direction of Q_1. Determine the responses $Q_1(t)$ and $Q_2(t)$.

APPLICATION TO HEAT TRANSFER PROBLEMS

13

FORMULATION AND SOLUTION PROCEDURE

13.1 INTRODUCTION

A knowledge of the temperature distribution within a body is important in many engineering problems. This information will be useful in computing the heat added to or removed from a body. Furthermore, if a heated body is not permitted to expand freely in all the directions, some stresses will be developed inside the body. The magnitude of these thermal stresses will influence the design of devices such as boilers, steam turbines, and jet engines. The first step in calculating the thermal stresses is to determine the temperature distribution within the body.

The objective of this chapter is to derive the finite element equations for the determination of temperature distribution within a conducting body. The basic unknown in heat transfer problems is temperature, similar to displacement in stress analysis problems. As indicated in Chapter 5, the finite element equations can be derived either by minimizing a suitable functional using a variational (Rayleigh–Ritz) approach or from the governing differential equation using a weighted residual (Galerkin) approach.

13.2 BASIC EQUATIONS OF HEAT TRANSFER

The basic equations of heat transfer, namely the energy balance and rate equations, are summarized in this section.

13.2.1 Energy Balance Equation

In the heat transfer analysis of any system, the following energy balance equation has to be satisfied because of conservation of energy:

$$\dot{E}_{in} + \dot{E}_g = \dot{E}_{out} + \dot{E}_{ie} \tag{13.1}$$

where the dot above a symbol signifies a time rate, E_{in} is the energy inflow into the system, E_g is the energy generated inside the system, E_{out} is the energy outflow from the system, and E_{ie} is the change in internal energy of the system.

13.2.2 Rate Equations

The rate equations, which describe the rates of energy flow, are given by the following equations.

(i) For conduction

Definition Conduction is the transfer of heat through materials without any net motion of the mass of the material.

The rate of heat flow in x direction by conduction (q) is given by

$$q = kA\frac{\partial T}{\partial x} \tag{13.2}$$

where k is the thermal conductivity of the material, A is the area normal to x direction through which heat flows, T is the temperature, and x is the length parameter.

(ii) For convection

Definition Convection is the process by which thermal energy is transferred between a solid and a fluid surrounding it.

The rate of heat flow by convection (q) can be expressed as

$$q = hA(T - T_\infty) \tag{13.3}$$

where h is the heat transfer coefficient, A is the surface area of the body through which heat flows, T is the temperature of the surface of the body, and T_∞ is the temperature of the surrounding medium.

(iii) For radiation

Definition Radiation heat transfer is the process by which the thermal energy is exchanged between two surfaces obeying the laws of electromagnetics.

The rate of heat flow by radiation (q) is governed by the relation

$$q = \sigma\varepsilon A(T^4 - T_\infty^4) \tag{13.4}$$

where σ is the Stefan–Boltzmann constant, ε is the emissivity of the surface, A is the surface area of the body through which heat flows, T is the absolute surface temperature of the body, and T_∞ is the absolute surrounding temperature.

(iv) Energy generated in a solid

Energy will be generated in a solid body whenever other forms of energy, such as chemical, nuclear, or electrical energy, are converted into thermal energy. The rate of heat generated (E_g) is governed by he equation

$$\dot{E}_g = \dot{q}V \tag{13.5}$$

where $\dot{q}$ is the strength of the heat source (rate of heat generated per unit volume per unit time), and V is the volume of the body.

(v) Energy stored in a solid

Whenever the temperature of a solid body increases, thermal energy will be stored in it. The equation describing this phenomenon is given by

$$\dot{E}_S = \rho c V \frac{\partial T}{\partial t} \tag{13.6}$$

where $\dot{E}_S$ is the rate of energy storage in the body, ρ is the density of the material, c is the specific heat of the material, V is the volume of the body, T is the temperature of the body, and t is the time parameter.

13.3 GOVERNING EQUATION FOR THREE-DIMENSIONAL BODIES

Consider a small element of material in a solid body as shown in Figure 13.1. The element is in the shape of a rectangular parallelepiped with sides $\mathrm{d}x$, $\mathrm{d}y$, and $\mathrm{d}z$. The energy balance equation can be stated as follows [13.1]:

$$\begin{matrix}\text{Heat inflow} & & \text{Heat generated} & & \text{Heat outflow} & & \text{Change in} \\ \text{during time } \mathrm{d}t & + & \text{by internal} & = & \text{during } \mathrm{d}t & + & \text{internal} \\ & & \text{sources during } \mathrm{d}t & & & & \text{energy during } \mathrm{d}t\end{matrix} \tag{13.7}$$

With the help of rate equations, Eq. (13.7) can be expressed as

$$(q_x + q_y + q_z)\,\mathrm{d}t + \dot{q}\,\mathrm{d}x\,\mathrm{d}y\,\mathrm{d}z\,\mathrm{d}t = (q_{x+\mathrm{d}x} + q_{y+\mathrm{d}y} + q_{z+\mathrm{d}z})\,\mathrm{d}t + \rho c\,\mathrm{d}T\,\mathrm{d}x\,\mathrm{d}y\,\mathrm{d}z \tag{13.8}$$

where

$$\begin{aligned} q_x &= \text{heat inflow rate into the face located at } x \\ &= -k_x A_x \frac{\partial T}{\partial x} = -k_x \frac{\partial T}{\partial x}\,\mathrm{d}y\,\mathrm{d}z \end{aligned} \tag{13.9}$$

$$\begin{aligned} q_{x+\mathrm{d}x} &= \text{heat outflow rate from the face located at } x + \mathrm{d}x \\ &= q|_{x+\mathrm{d}x} \approx q_x + \frac{\partial q_x}{\partial x}\,\mathrm{d}x \\ &= -k_x A_x \frac{\partial T}{\partial x} - \frac{\partial}{\partial x}\left(k_x A_x \frac{\partial T}{\partial x}\right)\mathrm{d}x \\ &= -k_x \frac{\partial T}{\partial x}\,\mathrm{d}y\,\mathrm{d}z - \frac{\partial}{\partial x}\left(k_x \frac{\partial T}{\partial x}\right)\mathrm{d}x\,\mathrm{d}y\,\mathrm{d}z \end{aligned} \tag{13.10}$$

k_x is the thermal conductivity of the material in x direction, A_x is the area normal to the x direction through which heat flows $= \mathrm{d}y\,\mathrm{d}z$, T is the temperature, $\dot{q}$ is the rate of heat generated per unit volume (per unit time), ρ is the density of the material, and c is the specific heat of the material. By substituting Eqs. (13.9) and (13.10) and similar expressions for q_y, $q_{y+\mathrm{d}y}$, q_z, and $q_{z+\mathrm{d}z}$ into Eq. (13.8) and dividing each term by $\mathrm{d}x\,\mathrm{d}y\,\mathrm{d}z\,\mathrm{d}t$, we obtain

$$\frac{\partial}{\partial x}\left(k_x \frac{\partial T}{\partial x}\right) + \frac{\partial}{\partial y}\left(k_y \frac{\partial T}{\partial y}\right) + \frac{\partial}{\partial z}\left(k_z \frac{\partial T}{\partial z}\right) + \dot{q} = \rho c \frac{\partial T}{\partial t} \tag{13.11}$$

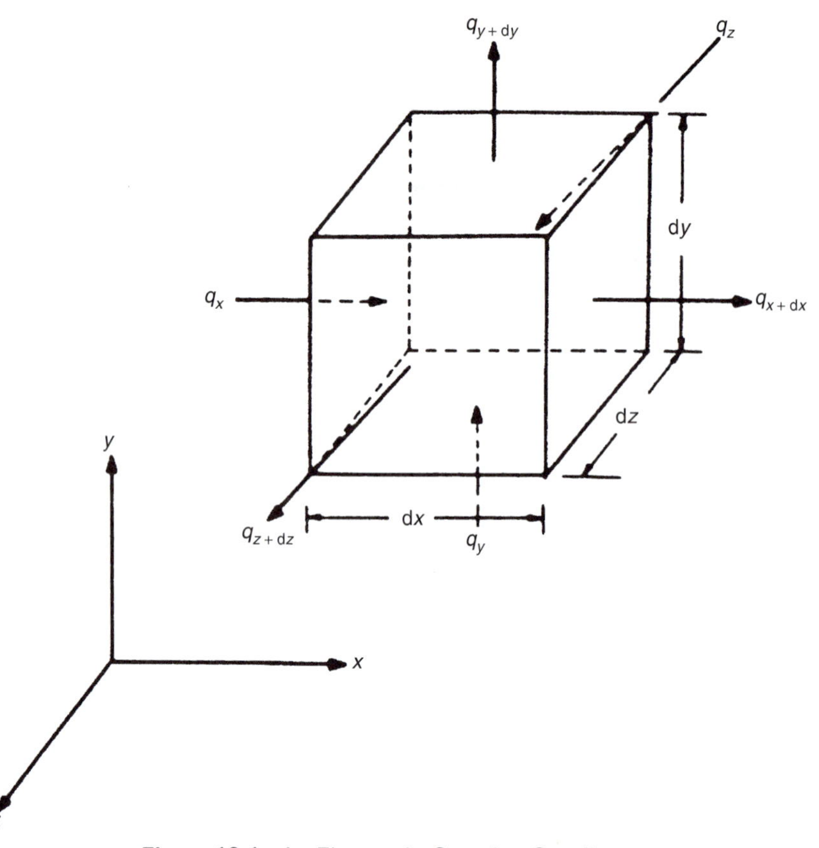

Figure 13.1. An Element in Cartesian Coordinates.

Equation (13.11) is the differential equation governing the heat conduction in an orthotropic solid body. If the thermal conductivities in x, y, and z directions are assumed to be the same, $k_x = k_y = k_z = k =$ constant, Eq. (13.11) can be written as

$$\frac{\partial^2 T}{\partial x^2} + \frac{\partial^2 T}{\partial y^2} + \frac{\partial^2 T}{\partial z^2} + \frac{\dot{q}}{k} = \frac{1}{\alpha}\frac{\partial T}{\partial t} \tag{13.12}$$

where the constant $\alpha = (k/\rho c)$ is called the thermal diffusivity. Equation (13.12) is the heat conduction equation that governs the temperature distribution and the conduction heat flow in a solid having uniform material properties (isotropic body). If heat sources are absent in the body, Eq. (13.12) reduces to the Fourier equation

$$\frac{\partial^2 T}{\partial x^2} + \frac{\partial^2 T}{\partial y^2} + \frac{\partial^2 T}{\partial z^2} = \frac{1}{\alpha}\frac{\partial T}{\partial t} \tag{13.13}$$

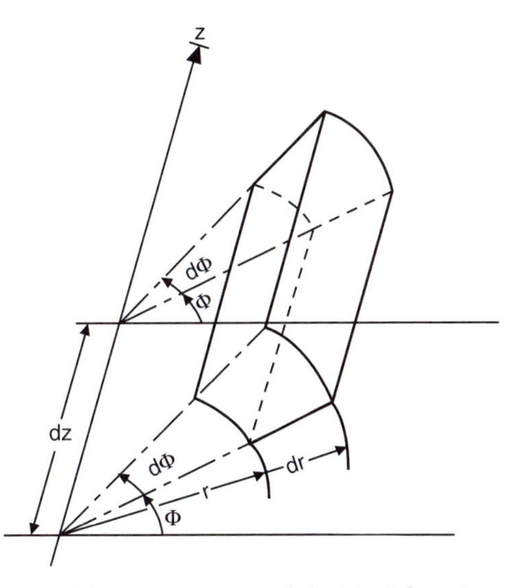

Figure 13.2. An Element in Cylindrical Coordinates.

If the body is in a steady state (with heat sources present), Eq. (13.12) becomes the Poisson's equation

$$\frac{\partial^2 T}{\partial x^2} + \frac{\partial^2 T}{\partial y^2} + \frac{\partial^2 T}{\partial z^2} + \frac{\dot{q}}{k} = 0 \tag{13.14}$$

If the body is in a steady state without any heat sources, Eq. (13.12) reduces to the Laplace equation

$$\frac{\partial^2 T}{\partial x^2} + \frac{\partial^2 T}{\partial y^2} + \frac{\partial^2 T}{\partial z^2} = 0 \tag{13.15}$$

13.3.1 Governing Equation in Cylindrical Coordinate System

If a cylindrical coordinate system (with r, ϕ, z coordinates) is used instead of the Cartesian x, y, z system, Eq. (13.12) takes the form

$$\frac{\partial^2 T}{\partial r^2} + \frac{1}{r}\frac{\partial T}{\partial r} + \frac{1}{r^2}\frac{\partial^2 T}{\partial \phi^2} + \frac{\partial^2 T}{\partial z^2} + \frac{\dot{q}}{k} = \frac{1}{\alpha}\frac{\partial T}{\partial t} \tag{13.16}$$

This equation can be derived by taking the element of the body as shown in Figure 13.2.

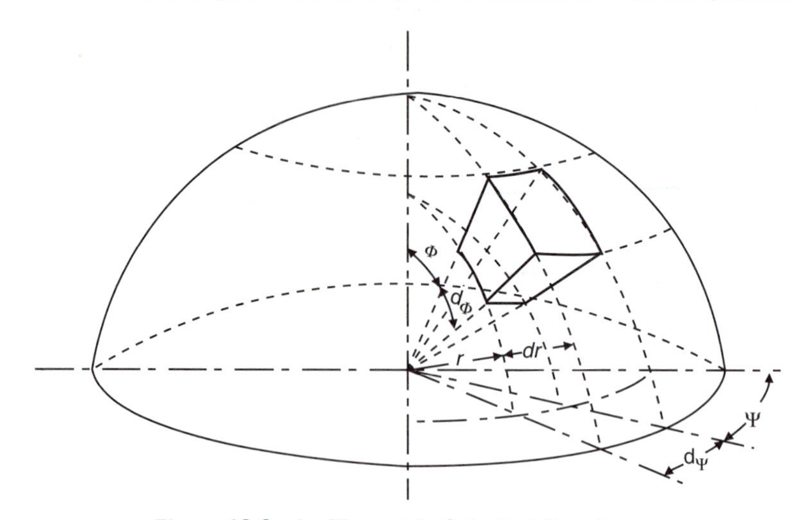

Figure 13.3. An Element in Spherical Coordinates.

13.3.2 Governing Equation in Spherical Coordinate System

By considering an element of the body in a spherical r, ϕ, ψ coordinate system as indicated in Figure 13.3, the general heat conduction equation (13.12) becomes

$$\frac{1}{r^2}\frac{\partial}{\partial r}\left(r^2\frac{\partial T}{\partial r}\right)+\frac{1}{r^2\cdot\sin\phi}\cdot\frac{\partial}{\partial\phi}\left(\sin\phi\cdot\frac{\partial T}{\partial\phi}\right)+\frac{1}{r^2\cdot\sin^2\phi}\cdot\frac{\partial^2 T}{\partial\psi^2}+\frac{\dot{q}}{k}=\frac{1}{\alpha}\frac{\partial T}{\partial t} \quad (13.17)$$

13.3.3 Boundary and Initial Conditions

Since the differential equation, Eq. (13.11) or (13.12), is second order, two boundary conditions need to be specified. The possible boundary conditions are

$$T(x,y,z,t)=T_0 \quad \text{for} \quad t>0 \text{ on } S_1 \quad (13.18)$$

$$k_x\cdot\frac{\partial T}{\partial x}\cdot l_x+k_y\cdot\frac{\partial T}{\partial y}\cdot l_y+k_z\cdot\frac{\partial T}{\partial z}\cdot l_z+q=0 \quad \text{for} \quad t>0 \text{ on } S_2 \quad (13.19)$$

$$k_x\cdot\frac{\partial T}{\partial x}\cdot l_x+k_y\cdot\frac{\partial T}{\partial y}\cdot l_y+k_z\cdot\frac{\partial T}{\partial z}\cdot l_z+h(T-T_\infty)=0 \quad \text{for} \quad t>0 \text{ on } S_3 \quad (13.20)$$

where q is the boundary heat flux, h is the convection heat transfer coefficient, T_∞ is the surrounding temperature, l_x, l_y, l_z are the direction cosines of the outward drawn normal to the boundary, S_1 is the boundary on which the value of temperature is specified as $T_0(t)$, S_2 is the boundary on which the heat flux q is specified, and S_3 is the boundary on which the convective heat loss $h(T-T_\infty)$ is specified. The boundary condition stated in Eq. (13.18) is known as the Dirichlet condition and those stated in Eqs. (13.19) and (13.20) are called Neumann conditions.

Furthermore, the differential equation, Eq. (13.11) or (13.12), is first order in time t and hence it requires one initial condition. The commonly used initial condition is

$$T(x, y, z, t = 0) = \bar{T}_0(x, y, z) \text{ in } V \tag{13.21}$$

where V is the domain (or volume) of the solid body, and $\bar{T}_0$ is the specified temperature distribution at time zero.

13.4 STATEMENT OF THE PROBLEM

13.4.1 In Differential Equation Form

The problem of finding the temperature distribution inside a solid body involves the solution of Eq. (13.11) or Eq. (13.12) subject to the satisfaction of the boundary conditions of Eqs. (13.18)–(13.20) and the initial condition given by Eq. (13.21).

13.4.2 In Variational Form

The three-dimensional heat conduction problem can be stated in an equivalent variational form as follows [13.2]:

Find the temperature distribution $T(x, y, z, t)$ inside the solid body that minimizes the integral

$$I = \frac{1}{2} \iiint_V \left[k_x \left(\frac{\partial T}{\partial x} \right)^2 + k_y \left(\frac{\partial T}{\partial y} \right)^2 + k_z \left(\frac{\partial T}{\partial z} \right)^2 - 2 \left(\dot{q} - \rho c \frac{\partial T}{\underset{\sim}{\partial t}} \right) T \right] \mathrm{d}V \tag{13.22}$$

and satisfies the boundary conditions of Eqs. (13.18)–(13.20) and the initial condition of Eq. (13.21). Here, the term $(\partial T / \underset{\sim}{\partial t})$ must be considered fixed while taking the variations. It can be verified that Eq. (13.11) is the Euler–Lagrange equation corresponding to the functional of Eq. (13.22). Generally it is not difficult to satisfy the boundary condition of Eq. (13.18), but Eqs. (13.19) and (13.20) present some difficulty. To overcome this difficulty, an integral pertaining to the boundary conditions of Eqs. (13.19) and (13.20) is added to the functional of Eq. (13.22) so that when the combined functional is minimized, the boundary conditions of Eqs. (13.19) and (13.20) would be automatically satisfied. The integral pertaining to Eqs. (13.19) and (13.20) is given by

$$\iint_{S_2} qT \, \mathrm{d}S_2 + \iint_{S_3} \frac{1}{2} h (T - T_\infty)^2 \, \mathrm{d}S_3$$

Thus, the combined functional to be minimized will be

$$\begin{aligned} I = {} & \frac{1}{2} \iiint_V \left[k_x \left(\frac{\partial T}{\partial x} \right)^2 + k_y \left(\frac{\partial T}{\partial y} \right)^2 + k_z \left(\frac{\partial T}{\partial z} \right)^2 - 2 \left(\dot{q} - \rho c \frac{\partial T}{\underset{\sim}{\partial t}} \right) T \right] \mathrm{d}V \\ & + \iint_{S_2} qT \, \mathrm{d}S_2 + \frac{1}{2} \iint_{S_3} h (T - T_\infty)^2 \, \mathrm{d}S_3 \end{aligned} \tag{13.23}$$

13.5 DERIVATION OF FINITE ELEMENT EQUATIONS

The finite element equations for the heat conduction problem can be derived either by using a variational approach or by using a Galerkin approach. We shall derive the equations using both the approaches in this section.

13.5.1 Variational Approach

In this approach, we consider the minimization of the functional I given by Eq. (13.23) subject to the satisfaction of the boundary conditions of Eq. (13.18) and the initial conditions of Eq. (13.21). The step-by-step procedure involved in the derivation of finite element equations is given below.

Step 1: Divide the domain V into E finite elements of p nodes each.

Step 2: Assume a suitable form of variation of T in each finite element and express $T^{(e)}(x, y, z, t)$ in element e as

$$T^{(e)}(x,y,z,t) = [N(x,y,z)]\vec{T}^{(e)} \tag{13.24}$$

where

$$[N(x,y,z)] = [N_1(x,y,z) \quad N_2(x,y,z) \quad \dots \quad N_p(x,y,z)]$$

$$\vec{T}^{(e)} = \begin{Bmatrix} T_1(t) \\ T_2(t) \\ \vdots \\ T_p(t) \end{Bmatrix}^{(e)}$$

$T_i(t)$ is the temperature of node i, and $N_i(x, y, z)$ is the interpolation function corresponding to node i of element e.

Step 3: Express the functional I as a sum of E elemental quantities $I^{(e)}$ as

$$I = \sum_{e=1}^{E} I^{(e)} \tag{13.25}$$

where

$$I^{(e)} = \frac{1}{2} \iiint\limits_{V^{(e)}} \left[k_x \left(\frac{\partial T^{(e)}}{\partial x} \right)^2 + k_y \left(\frac{\partial T^{(e)}}{\partial y} \right)^2 + k_z \left(\frac{\partial T^{(e)}}{\partial z} \right)^2 - 2 \left(\dot{q} - \rho c \frac{\partial T^{(e)}}{\underset{\sim}{\partial t}} \right) T^{(e)} \right] \mathrm{d}V$$
$$+ \iint\limits_{S_2^{(e)}} q T^{(e)} \, \mathrm{d}S_2 + \frac{1}{2} \iint\limits_{S_3^{(e)}} h (T^{(e)} - T_\infty)^2 \, \mathrm{d}S_3 \tag{13.26}$$

For the minimization of the functional I, use the necessary conditions

$$\frac{\partial I}{\partial T_i} = \sum_{e=1}^{E} \frac{\partial I^{(e)}}{\partial T_i} = 0, \qquad i = 1, 2, \dots, M$$

where M is the total number of nodal temperature unknowns. From Eq. (13.26), we have

$$\frac{\partial I^{(e)}}{\partial T_i} = \iiint\limits_{V^{(e)}} \left[k_x \frac{\partial T^{(e)}}{\partial x} \frac{\partial}{\partial T_i}\left(\frac{\partial T^{(e)}}{\partial x}\right) + k_y \frac{\partial T^{(e)}}{\partial y} \frac{\partial}{\partial T_i}\left(\frac{\partial T^{(e)}}{\partial y}\right) + k_z \frac{\partial T^{(e)}}{\partial z} \frac{\partial}{\partial T_i}\left(\frac{\partial T^{(e)}}{\partial z}\right) - \left(\dot{q} - \rho c \frac{\partial T^{(e)}}{\partial \underset{\sim}{t}}\right) \frac{\partial T^{(e)}}{\partial T_i} \right] \mathrm{d}V + \iint\limits_{S_2^{(e)}} q \frac{\partial T^{(e)}}{\partial T_i} \,\mathrm{d}S_2 + \iint\limits_{S_3^{(e)}} h(T^{(e)} - T_\infty) \frac{\partial T^{(e)}}{\partial T_i} \cdot \mathrm{d}S_3 \tag{13.27}$$

Note that the surface integrals do not appear in Eq. (13.27) if node i does not lie on S_2 and S_3. Equation (13.24) gives

$$\left.\begin{aligned} \frac{\partial T^{(e)}}{\partial x} &= \left[\frac{\partial N_1}{\partial x}\frac{\partial N_2}{\partial x}\cdots\frac{\partial N_p}{\partial x}\right]\vec{T}^{(e)} \\ \frac{\partial}{\partial T_i}\left(\frac{\partial T^{(e)}}{\partial x}\right) &= \frac{\partial N_i}{\partial x} \\ \frac{\partial T^{(e)}}{\partial T_i} &= N_i \\ \frac{\partial T^{(e)}}{\partial \underset{\sim}{t}} &= [N]\dot{\vec{T}}^{(e)} \end{aligned}\right\} \tag{13.28}$$

where

$$\dot{\vec{T}}^{(e)} = \begin{Bmatrix} \partial T_1/\partial t \\ \vdots \\ \partial T_p/\partial t \end{Bmatrix}$$

Thus, Eq. (13.27) can be expressed as

$$\frac{\partial I^{(e)}}{\partial \vec{T}^{(e)}} = [K_1^{(e)}]\vec{T}^{(e)} - \vec{P}^{(e)} + [K_2^{(e)}]\vec{T}^{(e)} + [K_3^{(e)}]\dot{\vec{T}}^{(e)} \tag{13.29}$$

where the elements of $[K_1^{(e)}]$, $[K_2^{(e)}]$, $[K_3^{(e)}]$, and $\vec{P}^{(e)}$ are given by

$$K1_{ij}^{(e)} = \iiint\limits_{V^{(e)}} \left(k_x \frac{\partial N_i}{\partial x}\frac{\partial N_j}{\partial x} + k_y \frac{\partial N_i}{\partial y}\frac{\partial N_j}{\partial y} + k_z \frac{\partial N_i}{\partial z}\frac{\partial N_j}{\partial z} \right) \cdot \mathrm{d}V \tag{13.30}$$

$$K2_{ij}^{(e)} = \iint\limits_{S_3^{(e)}} h N_i N_j \cdot \mathrm{d}S_3 \tag{13.31}$$

$$K3_{ij}^{(e)} = \iiint\limits_{V^{(e)}} \rho c N_i N_j \cdot \mathrm{d}V \tag{13.32}$$

and

$$P_i^{(e)} = \iiint\limits_{V^{(e)}} \dot{q} N_i \, \mathrm{d}V - \iint\limits_{S_2^{(e)}} q N_i \, \mathrm{d}S_2 + \iint\limits_{S_3^{(e)}} h T_\infty N_i \, \mathrm{d}S_3 \tag{13.33}$$

Step 4: Rewrite Eqs. (13.29) in matrix form as

$$\frac{\partial I}{\partial \underset{\sim}{\vec{T}}} = \sum_{e=1}^{E} \frac{\partial I^{(e)}}{\partial \vec{T}^{(e)}} = \sum_{e=1}^{E} \left(\left[[K_1^{(e)}] + [K_2^{(e)}] \right] T^{(e)} - \vec{P}^{(e)} + [K_3^{(e)}] \dot{\vec{T}}^{(e)} \right) = \vec{0} \tag{13.34}$$

where $\underset{\sim}{\vec{T}}$ is the vector of nodal temperature unknowns of the system:

$$\underset{\sim}{\vec{T}} = \begin{Bmatrix} T_1 \\ T_2 \\ \vdots \end{Bmatrix}$$

By using the familiar assembly process, Eq. (13.34) can be expressed as

$$[\underset{\sim}{K_3}] \dot{\underset{\sim}{\vec{T}}} + [\underset{\sim}{K}] \underset{\sim}{\vec{T}} = \underset{\sim}{\vec{P}} \tag{13.35}$$

where

$$[\underset{\sim}{K_3}] = \sum_{e=1}^{E} [K_3^{(e)}] \tag{13.36}$$

$$[\underset{\sim}{K}] = \sum_{e=1}^{E} \left[[K_1^{(e)}] + [K_2^{(e)}] \right] \tag{13.37}$$

and

$$\underset{\sim}{\vec{P}} = \sum_{e=1}^{E} \vec{P}^{(e)} \tag{13.38}$$

Step 5: Equations (13.35) are the desired equations that have to be solved after incorporating the boundary conditions specified over S_1 [Eq. (13.18) and the initial conditions stated in Eq. (13.21)].

13.5.2 Galerkin Approach

The finite element procedure using the Galerkin method can be described by the following steps.

Step 1: Divide the domain V into E finite elements of p nodes each.

Step 2: Assume a suitable form of variation of T in each finite element and express $T^{(e)}(x, y, z, t)$ in element e as

$$T^{(e)}(x, y, z, t) = [N(x, y, z)]\vec{T}^{(e)} \tag{13.39}$$

Step 3: In the Galerkin method, the integral of the weighted residue over the domain of the element is set equal to zero by taking the weights same as the interpolation functions N_i. Since the solution in Eq. (13.39) is not exact, substitution of Eq. (13.39) into the differential Eq. (13.11) gives a nonzero value instead of zero. This nonzero value will be the residue. Hence, the criterion to be satisfied at any instant of time is

$$\iiint_{V^{(e)}} N_i \left[\frac{\partial}{\partial x}\left(k_x \frac{\partial T^{(e)}}{\partial x}\right) + \frac{\partial}{\partial y}\left(k_y \frac{\partial T^{(e)}}{\partial y}\right) + \frac{\partial}{\partial z}\left(k_z \frac{\partial T^{(e)}}{\partial z}\right) + \dot{q} - \rho c \frac{\partial T^{(e)}}{\partial t} \right] \mathrm{d}V = 0, \qquad i = 1, 2, \ldots, p \tag{13.40}$$

By noting that the first integral term of Eq. (13.40) can be written as

$$\iiint_{V^{(e)}} N_i \frac{\partial}{\partial x}\left(k_x \frac{\partial T^{(e)}}{\partial x}\right) \mathrm{d}V = -\iiint_{V^{(e)}} \frac{\partial N_i}{\partial x} k_x \frac{\partial T^{(e)}}{\partial x} \mathrm{d}V + \iint_{S^{(e)}} N_i k_x \frac{\partial T^{(e)}}{\partial x} l_x \,\mathrm{d}S \tag{13.41}$$

where l_x is the x-direction cosine of the outward drawn normal, and with similar expressions for the second and third integral terms, Eq. (13.40) can be stated as

$$\begin{aligned} &-\iiint_{V^{(e)}} \left[k_x \frac{\partial N_i}{\partial x}\frac{\partial T^{(e)}}{\partial x} + k_y \frac{\partial N_i}{\partial y}\frac{\partial T^{(e)}}{\partial y} + k_z \frac{\partial N_i}{\partial z}\frac{\partial T^{(e)}}{\partial z} \right] \mathrm{d}V \\ &+ \iint_{S^{(e)}} N_i \left[k_x \frac{\partial T^{(e)}}{\partial x} l_x + k_y \frac{\partial T^{(e)}}{\partial y} l_y + k_z \frac{\partial T^{(e)}}{\partial z} l_z \right] \mathrm{d}S \\ &+ \iiint_{V^{(e)}} N_i \left(\dot{q} - \rho c \frac{\partial T^{(e)}}{\partial t} \right) \mathrm{d}V = 0, \qquad i = 1, 2, \ldots, p \end{aligned} \tag{13.42}$$

Since the boundary of the element $S^{(e)}$ is composed of $S_1^{(e)}$, $S_2^{(e)}$, and $S_3^{(e)}$, the surface integral of Eq. (13.42) over $S_1^{(e)}$ would be zero (since $T^{(e)}$ is prescribed to be a constant T_0 on $S_1^{(e)}$, the derivatives of $T^{(e)}$ with respect to x, y, and z would be zero). On the surfaces $S_2^{(e)}$ and $S_3^{(e)}$, the boundary conditions given by Eqs. (13.19) and (13.20) are to be satisfied. For this, the surface integral in Eq. (13.42) over $S_2^{(e)}$ and $S_3^{(e)}$ is written in equivalent form as

$$\begin{aligned} \iint_{S_2^{(e)} + S_3^{(e)}} N_i \left[k_x \frac{\partial T^{(e)}}{\partial x} l_x + k_y \frac{\partial T^{(e)}}{\partial y} l_y + k_z \frac{\partial T^{(e)}}{\partial z} l_z \right] \mathrm{d}S = &-\iint_{S_2^{(e)}} N_i q \,\mathrm{d}S_2 \\ &-\iint_{S_3^{(e)}} h(T^{(e)} - T_\infty) \,\mathrm{d}S_3 \end{aligned} \tag{13.43}$$

By using Eqs. (13.39) and (13.43), Eq. (13.42) can be expressed in matrix form as

$$[K_1^{(e)}]\vec{T}^{(e)} + [K_2^{(e)}]\vec{T}^{(e)} + [K_3^{(e)}]\dot{\vec{T}}^{(e)} - \vec{P}^{(e)} = \vec{0} \tag{13.44}$$

where the elements of the matrices $[K_1^{(e)}]$, $[K_2^{(e)}]$, $[K_3^{(e)}]$, and $\vec{P}^{(e)}$ are the same as those given in Eqs. (13.30)–(13.33).

Step 4: The element equation (13.44) can be assembled in the usual manner to obtain the overall equations as

$$[\underset{\sim}{K_3}]\dot{\underset{\sim}{\vec{T}}} + [\underset{\sim}{K}]\underset{\sim}{\vec{T}} = \underset{\sim}{\vec{P}} \tag{13.45}$$

where $[\underset{\sim}{K_3}]$, $[\underset{\sim}{K}]$, and $\underset{\sim}{\vec{P}}$ are the same as those defined by Eqs. (13.36)–(13.38). It can be seen that the same final equations, Eq. (13.35), are obtained in both the approaches.

Step 5: Equations (13.35) have to be solved after incorporating the boundary conditions specified over S_1 and the initial conditions.

Notes:

1. The expressions for $[K_1^{(e)}]$, $[K_2^{(e)}]$, $[K_3^{(e)}]$, and $\vec{P}^{(e)}$ can be stated using matrix notation as

$$[K_1^{(e)}] = \iiint_{V^{(e)}} [B]^T[D][B]\,\mathrm{d}V \tag{13.46}$$

$$[K_2^{(e)}] = \iint_{S_3^{(e)}} h[N]^T[N]\,\mathrm{d}S_3 \tag{13.47}$$

$$[K_3^{(e)}] = \iiint_{V^{(e)}} \rho c[N]^T[N]\,\mathrm{d}V \tag{13.48}$$

$$\vec{P}^{(e)} = \vec{P}_1^{(e)} - \vec{P}_2^{(e)} + \vec{P}_3^{(e)} \tag{13.49}$$

where

$$\vec{P}_1^{(e)} = \iiint_{V^{(e)}} \dot{q}[N]^T\,\mathrm{d}V \tag{13.50}$$

$$\vec{P}_2^{(e)} = \iint_{S_2^{(e)}} q[N]^T\,\mathrm{d}S_2 \tag{13.51}$$

$$\vec{P}_3^{(e)} = \iint_{S_3^{(e)}} hT_\infty[N]^T\,\mathrm{d}S_3 \tag{13.52}$$

$$[D] = \begin{bmatrix} k_x & 0 & 0 \\ 0 & k_y & 0 \\ 0 & 0 & k_z \end{bmatrix} \tag{13.53}$$

$$[B] = \begin{bmatrix} \frac{\partial N_1}{\partial x} & \frac{\partial N_2}{\partial x} & \cdots & \frac{\partial N_p}{\partial x} \\ \frac{\partial N_1}{\partial y} & \frac{\partial N_2}{\partial y} & \cdots & \frac{\partial N_p}{\partial y} \\ \frac{\partial N_1}{\partial z} & \frac{\partial N_2}{\partial z} & \cdots & \frac{\partial N_p}{\partial z} \end{bmatrix} \tag{13.54}$$

2. When all the three modes of heat transfer are considered, the governing differential equation becomes nonlinear (due to the inclusion of radiation term). An iterative procedure is presented in Section 14.7 for the solution of heat transfer problems involving radiation.

REFERENCES

13.1 F.P. Incropera and D.P. DeWitt: *Fundamentals of Heat and Mass Transfer*, 4th Ed., Wiley, New York, 1996.

13.2 G.E. Myers: *Analytical Methods in Conduction Heat Transfer*, McGraw-Hill, New York, 1971.

PROBLEMS

13.1 Derive the heat conduction equation in cylindrical coordinates, Eq. (13.16), from Eq. (13.12).

Hint: Use the relations $x = r\cos\theta$, $y = r\sin\theta$, and $z = z$ in Eq. (13.12).

13.2 Derive the heat conduction equation in spherical coordinates, Eq. (13.17), from Eq. (13.12).

Hint: Use the relations $x = r\sin\phi\cos\psi$, $y = r\sin\phi\sin\psi$, and $z = r\cos\phi$ in Eq. (13.12).

13.3 The steady-state one-dimensional heat conduction equation is given by:

In Cartesian coordinate system: $\frac{\mathrm{d}}{\mathrm{d}x}\left[k\frac{\mathrm{d}T}{\mathrm{d}x}\right] = 0$

In cylindrical coordinate system: $\frac{\mathrm{d}}{\mathrm{d}r}\left[rk\frac{\mathrm{d}T}{\mathrm{d}r}\right] = 0$

In spherical coordinate system: $\frac{\mathrm{d}}{\mathrm{d}r}\left[kr^2\frac{\mathrm{d}T}{\mathrm{d}r}\right] = 0$

Suggest a suitable temperature distribution model, for each case, for use in the finite element analysis.

13.4 Express the boundary conditions of Figure 13.4 in the form of equations.

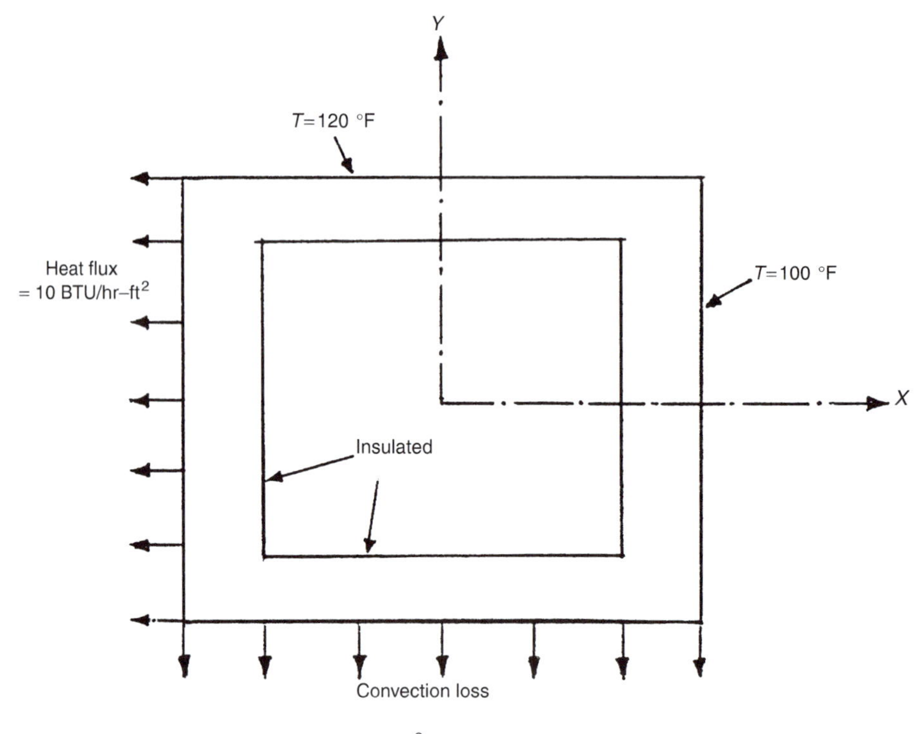

Figure 13.4.

13.5 The thermal equilibrium equation for a one-dimensional problem, including conduction, convection, and radiation, can be expressed as

$$\frac{\mathrm{d}}{\mathrm{d}x}\left[kA\frac{\mathrm{d}T}{\mathrm{d}x}\right] - hP(T - T_\infty) - \varepsilon\sigma P(T^4 - T_\infty^4) + \dot{q}A = 0; \qquad 0 \le x \le L \qquad (\mathrm{E}_1)$$

where k is the conductivity coefficient, h is the convection coefficient, ε is the emissivity of the surface, σ is the Stefan–Boltzman constant, $\dot{q}$ is the heat generated per unit volume, A is the cross-sectional area, P is the perimeter, $T(x)$ is the temperature at location x, T_∞ is the ambient temperature, and L is the length of the body. Show that the variational functional I corresponding to Eq. (E_1) is given by

$$I = \int_{x=0}^{L} \left\{ \dot{q}AT - \frac{1}{2}hPT^2 + hPT_\infty T - \frac{1}{5}\varepsilon\sigma PT^5 + \varepsilon\sigma PT_\infty^4 T - \frac{1}{2}kA\left[\frac{\mathrm{d}T}{\mathrm{d}x}\right]^2 \right\} \mathrm{d}x \qquad (\mathrm{E}_2)$$

13.6 Derive the finite element equations corresponding to Eq. (E_1) of Problem 13.5 using the Galerkin approach.

13.7 Derive the finite element equations corresponding to Eqs. (E_1) and (E_2) of Problem 13.5 using a variational approach.

13.8 Heat transfer takes place by convection and radiation from the inner and outer surfaces of a hollow sphere. If the radii are R_i and R_o, the fluid (ambient) temperatures are T_i and T_o, convection heat transfer coefficients are h_i and h_o, and emissivities are ε_i and ε_o at the inner (i) and outer (o) surfaces of the hollow sphere, state the governing differential equation and the boundary conditions to be satisfied in finding the temperature distribution in the sphere, $T(r)$.

14

ONE-DIMENSIONAL PROBLEMS

14.1 INTRODUCTION

For a one-dimensional heat transfer problem, the governing differential equation is given by

$$k\frac{\mathrm{d}^2T}{\mathrm{d}x^2} + \dot{q} = 0 \tag{14.1}$$

The boundary conditions are

$$T(x = 0) = T_0 \text{ (temperature specified)} \tag{14.2}$$

$$k\frac{\mathrm{d}T}{\mathrm{d}x}l_x + h(T - T_\infty) + q = 0 \text{ on the surface} \tag{14.3}$$

$$\text{(combined heat flux and convection specified)}$$

A fin is a common example of a one-dimensional heat transfer problem. One end of the fin is connected to a heat source (whose temperature is known) and heat will be lost to the surroundings through the perimeter surface and the end. We now consider the analysis of uniform and tapered fins.

14.2 STRAIGHT UNIFORM FIN ANALYSIS

Step 1: Idealize the rod into several finite elements as shown in Figure 14.1(b).

Step 2: Assume a linear temperature variation inside any element "e" as

$$T^{(e)}(x) = a_1 + a_2 x = [N(x)]\vec{q}^{\,(e)} \tag{14.4}$$

where

$$[N(x)] = [N_i(x) \quad N_j(x)] \tag{14.5}$$

$$N_i(x) \equiv N_1(x) = 1 - \frac{x}{l^{(e)}} \tag{14.6}$$

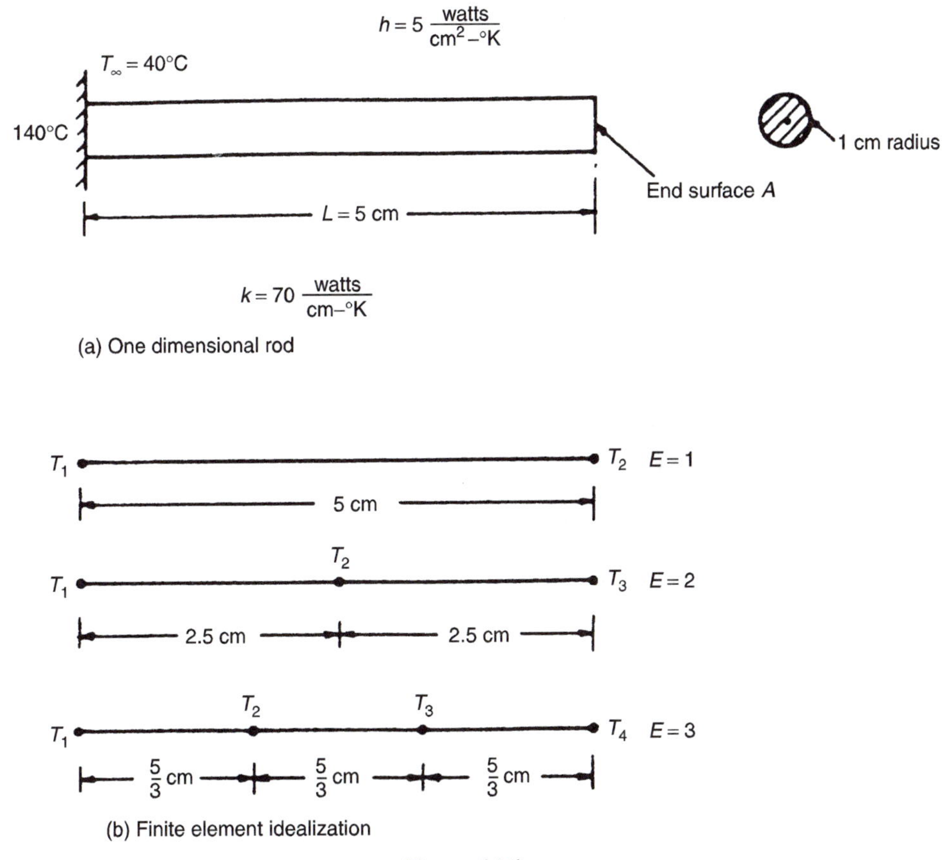

Figure 14.1.

$$N_j(x) \equiv N_2(x) = \frac{x}{l^{(e)}} \tag{14.7}$$

$$\vec{q}^{\,(e)} = \begin{Bmatrix} q_1 \\ q_2 \end{Bmatrix} \equiv \begin{Bmatrix} T_i \\ T_j \end{Bmatrix} \tag{14.8}$$

i and j indicate the global node numbers corresponding to the left- and right-hand-side nodes, and $l^{(e)}$ is the length of element e.

Step 3: Derivation of element matrices:

Since this is a one-dimensional problem, Eqs. (13.53) and (13.54) reduce to

$$[D] = [k] \text{ and } [B] = \begin{bmatrix} \dfrac{\partial N_i}{\partial x} & \dfrac{\partial N_j}{\partial x} \end{bmatrix} = \begin{bmatrix} -\dfrac{1}{l^{(e)}} & \dfrac{1}{l^{(e)}} \end{bmatrix} \tag{14.9}$$

Equations (13.46)–(13.49) become

$$[K_1^{(e)}] = \iiint_{V^{(e)}} [B]^T[D][B]\,\mathrm{d}V = \int_{x=0}^{l^{(e)}} \begin{Bmatrix} -\frac{1}{l^{(e)}} \\ \frac{1}{l^{(e)}} \end{Bmatrix} [k] \begin{Bmatrix} -\frac{1}{l^{(e)}} & \frac{1}{l^{(e)}} \end{Bmatrix} A\,\mathrm{d}x$$

$$= \frac{Ak}{l^{(e)}} \begin{bmatrix} 1 & -1 \\ -1 & 1 \end{bmatrix} \tag{14.10}$$

$$[K_2^{(e)}] = \iint_{S_3^{(e)}} h[N]^T[N]\,\mathrm{d}S_3^{(e)} = \int_{x=0}^{l^{(e)}} h \begin{Bmatrix} 1 - x/l^{(e)} \\ x/l^{(e)} \end{Bmatrix} \left\{ \left(1 - \frac{x}{l^{(e)}}\right) \frac{x}{l^{(e)}} \right\} P\,\mathrm{d}x$$

$$= \frac{hPl^{(e)}}{6} \begin{bmatrix} 2 & 1 \\ 1 & 2 \end{bmatrix} \tag{14.11}$$

since $\mathrm{d}S_3 = P\,\mathrm{d}x$, where P is the perimeter.

$$[K_3^{(e)}] = [0] \text{ since this is a steady-state problem} \tag{14.12}$$

$$\vec{P}^{(e)} = \int_{x=0}^{l^{(e)}} \dot{q} \begin{Bmatrix} 1 - \frac{x}{l^{(e)}} \\ \frac{x}{l^{(e)}} \end{Bmatrix} A\,\mathrm{d}x - \int_{x=0}^{l^{(e)}} q \begin{Bmatrix} 1 - \frac{x}{l^{(e)}} \\ \frac{x}{l^{(e)}} \end{Bmatrix} P\,\mathrm{d}x + \int_{x=0}^{l^{(e)}} hT_\infty \begin{Bmatrix} 1 - \frac{x}{l^{(e)}} \\ \frac{x}{l^{(e)}} \end{Bmatrix} P\,\mathrm{d}x$$

$$= \frac{\dot{q}Al^{(e)}}{2} \begin{Bmatrix} 1 \\ 1 \end{Bmatrix} - \frac{qPl^{(e)}}{2} \begin{Bmatrix} 1 \\ 1 \end{Bmatrix} + \frac{hT_\infty Pl^{(e)}}{2} \begin{Bmatrix} 1 \\ 1 \end{Bmatrix} \tag{14.13}$$

Step 4: Assembled equations: The element matrices can be assembled to obtain the overall equations as [Eq. (13.45)]

$$[\underset{\sim}{K}]\underset{\sim}{\vec{T}} = \underset{\sim}{\vec{P}} \tag{14.14}$$

where

$$[\underset{\sim}{K}] = \sum_{e=1}^{E} \left(\frac{Ak}{l^{(e)}} \begin{bmatrix} 1 & -1 \\ -1 & 1 \end{bmatrix} + \frac{hPl^{(e)}}{6} \begin{bmatrix} 2 & 1 \\ 1 & 2 \end{bmatrix} \right) \tag{14.15}$$

and

$$\underset{\sim}{\vec{P}} = \sum_{e=1}^{E} \vec{P}^{(e)} = \sum_{e=1}^{E} \frac{1}{2} \left(\dot{q}Al^{(e)} - qPl^{(e)} + hT_\infty Pl^{(e)} \right) \begin{Bmatrix} 1 \\ 1 \end{Bmatrix} \tag{14.16}$$

Step 5: The assembled equations (14.14) are to be solved, after incorporating the boundary conditions stated in Eq. (14.2), to find the nodal temperatures.

Example 14.1 Find the temperature distribution in the one-dimensional fin shown in Figure 14.1.

Solution

(i) With one element

Here, $\dot{q} = q = 0$, and hence, for $E = 1$, Eq. (14.14) gives

$$\begin{bmatrix} \left(\frac{Ak}{L} + \frac{2hPL}{6}\right) & \left(-\frac{Ak}{L} + \frac{hPL}{6}\right) \\ \left(-\frac{Ak}{L} + \frac{hPL}{6}\right) & \left(\frac{Ak}{L} + \frac{2hPL}{6}\right) \end{bmatrix} \begin{Bmatrix} T_1 \\ T_2 \end{Bmatrix} = \frac{hPT_\infty L}{2} \begin{Bmatrix} 1 \\ 1 \end{Bmatrix} \tag{E$_1$}$$

By dividing throughout by (Ak/L), Eq. (E_1) can be written as

$$\begin{bmatrix} \left(1 + \frac{2hPL^2}{6kA}\right) & \left(-1 + \frac{hPL^2}{6kA}\right) \\ \left(-1 + \frac{hPL^2}{6kA}\right) & \left(1 + \frac{2hPL^2}{6kA}\right) \end{bmatrix} \begin{Bmatrix} T_1 \\ T_2 \end{Bmatrix} = \frac{hPT_\infty L^2}{2kA} \begin{Bmatrix} 1 \\ 1 \end{Bmatrix} \tag{E$_2$}$$

For the data given, $P = 2\pi$ cm and $A = \pi$ cm^2, and hence

$$\frac{hPL^2}{kA} = \frac{(5)(2\pi)(5^2)}{(70)(\pi)} = \frac{25}{7} \quad \text{and} \quad \frac{hPT_\infty L^2}{2kA} = \frac{(5)(2\pi)(40)(5^2)}{2(70)(\pi)} = \frac{500}{7}$$

Thus, Eq. (E_1) becomes

$$\begin{bmatrix} 92 & -17 \\ -17 & 92 \end{bmatrix} \begin{Bmatrix} T_1 \\ T_2 \end{Bmatrix} = \begin{Bmatrix} 3000 \\ 3000 \end{Bmatrix} \tag{E$_3$}$$

In order to incorporate the boundary condition $T_1 = 140\,^\circ$C, we replace the first equation of (E_3) by $T_1 = 140$ and rewrite the second equation of (E_3) as

$$92T_2 = 3000 + 17T_1 = 3000 + 17(140) = 5380$$

from which the unknown temperature T_2 can be found as $T_2 = 58.48\,^\circ$C.

While solving Eq. (E_3) on a computer, the boundary condition $T_1 = 140$ can be incoporated by modifying Eq. (E_3) as

$$\begin{bmatrix} 1 & 0 \\ 0 & 92 \end{bmatrix} \begin{Bmatrix} T_1 \\ T_2 \end{Bmatrix} = \begin{Bmatrix} 140 \\ 3000 + 17 \times 140 \end{Bmatrix} = \begin{Bmatrix} 140 \\ 5380 \end{Bmatrix} \tag{E$_4$}$$

(ii) With two elements

In this case, Eq. (14.14) represents assembly of two element equations and leads to

$$\begin{bmatrix} a_1 & a_2 & 0 \\ a_2 & 2a_1 & a_2 \\ 0 & a_2 & a_1 \end{bmatrix} \begin{Bmatrix} T_1 \\ T_2 \\ T_3 \end{Bmatrix} = \begin{Bmatrix} b \\ 2b \\ b \end{Bmatrix} \tag{E$_5$}$$

where

$$a_1 = 1 + \frac{2hPL^2}{24kA} = 1 + \frac{2}{24}\left(\frac{25}{7}\right) = \frac{109}{84}$$

$$a_2 = -1 + \frac{hPL^2}{24kA} = -1 + \frac{1}{24}\left(\frac{25}{7}\right) = -\frac{143}{168}$$

$$b = \frac{hPT_\infty L^2}{8kA} = \frac{500}{28}$$

As before, we modify Eq. (E_5) to incorporate the boundary condition $T_1 = 140$ as follows:

$$\begin{bmatrix} 1 & 0 & 0 \\ 0 & 2a_1 & a_2 \\ 0 & a_2 & a_1 \end{bmatrix} \begin{Bmatrix} T_1 \\ T_2 \\ T_3 \end{Bmatrix} = \begin{Bmatrix} 140 & \\ 2b & -a_2 \times T_1 \\ b & -0 \times T_1 \end{Bmatrix} = \begin{Bmatrix} 140 & \\ 2b & -140a_2 \\ b & \end{Bmatrix}$$

or

$$\begin{bmatrix} 1 & 0 & 0 \\ 0 & \frac{218}{84} & -\frac{143}{168} \\ 0 & -\frac{143}{168} & \frac{109}{84} \end{bmatrix} \begin{Bmatrix} T_1 \\ T_2 \\ T_3 \end{Bmatrix} = \begin{Bmatrix} 140 \\ \frac{13010}{84} \\ \frac{1500}{84} \end{Bmatrix} \tag{E_6}$$

The solution of Eq. (E_6) gives

$$T_1 = 140\,^\circ\text{C}, \qquad T_2 = 81.77\,^\circ\text{C}, \quad \text{and} \quad T_3 = 67.39\,^\circ\text{C}$$

Note

In the previous solution, it is assumed that the convection heat loss occurs only from the perimeter surface and not from the end surface A (Figure 14.1). If the convection loss from the end surface is also to be considered, the following method can be adopted. Let the convection heat loss occur from the surface at the right-hand-side node of the element e. Then the surface integral of Eq. (13.47) or Eq. (14.11) should extend over this surface also. Thus, in Eq. (14.11), the following term should also be included:

$$\iint_{S_3^{(e)}} h[N]^T[N]\,\mathrm{d}S_3 = \iint_A h \begin{Bmatrix} N_1 \\ N_2 \end{Bmatrix} \{N_1 \quad N_2\}\,\mathrm{d}S_3 \tag{14.17}$$

corresponding to the surface at the right-side node 2.

Since we are interested in the surface at node 2 (right-side node), we substitute $N_1(x = l^{(e)}) = 0$ and $N_2(x = l^{(e)}) = 1$ in Eq. (14.17) to obtain

$$\iint_A h \begin{Bmatrix} 0 \\ 1 \end{Bmatrix} \{0 \quad 1\} \, \mathrm{d}S_3 = \iint_A h \begin{bmatrix} 0 & 0 \\ 0 & 1 \end{bmatrix} \mathrm{d}S_3 = hA \begin{bmatrix} 0 & 0 \\ 0 & 1 \end{bmatrix} \tag{14.18}$$

Similarly, the surface integral over S_3 in Eq. (13.52) or Eq. (14.13) should extend over the free end surface also. Thus, the additional term to be included in the vector $\vec{P}^{(e)}$ is given by

$$\iint_{S_3^{(e)}} hT_\infty [N]^T \, \mathrm{d}S_3 = hT_\infty \iint_A \begin{Bmatrix} 0 \\ 1 \end{Bmatrix} \mathrm{d}S_3 = hT_\infty A \begin{Bmatrix} 0 \\ 1 \end{Bmatrix} \tag{14.19}$$

Example 14.2 Find the temperature distribution in the fin shown in Figure 14.1 by including the effect of convection from the end surface A.

Solution

(i) With one element

Equation (E_1) of Example 14.1 will be modified as

$$\begin{bmatrix} \left(\frac{Ak}{L} + \frac{2hPL}{6} + 0\right) & \left(-\frac{Ak}{L} + \frac{hPL}{6} + 0\right) \\ \left(-\frac{Ak}{L} + \frac{hPL}{6} + 0\right) & \left(\frac{Ak}{L} + \frac{hPL}{6} + hA\right) \end{bmatrix} \begin{Bmatrix} T_1 \\ T_2 \end{Bmatrix} = \begin{Bmatrix} \frac{hPT_\infty L}{2} + 0 \\ \frac{hPT_\infty L}{2} + hAT_\infty \end{Bmatrix} \tag{E$_1$}$$

For the given data, Eq. (E_1) reduces to [after multiplying throughout by (L/Ak)]

$$\begin{bmatrix} \left(1 + \frac{25}{21}\right) & \left(-1 + \frac{25}{42}\right) \\ \left(-1 + \frac{25}{42}\right) & \left(1 + \frac{25}{21} + \frac{5}{14}\right) \end{bmatrix} \begin{Bmatrix} T_1 \\ T_2 \end{Bmatrix} = \begin{Bmatrix} \frac{500}{7} \\ \frac{500}{7} + \frac{100}{7} \end{Bmatrix}$$

or

$$\begin{bmatrix} 92 & -17 \\ -17 & 107 \end{bmatrix} \begin{Bmatrix} T_1 \\ T_2 \end{Bmatrix} = \begin{Bmatrix} 3000 \\ 3600 \end{Bmatrix} \tag{E$_2$}$$

After incorporating the boundary condition, $T_1 = 140$, Eq. (E_2) becomes

$$\begin{bmatrix} 1 & 0 \\ 0 & 107 \end{bmatrix} \begin{Bmatrix} T_1 \\ T_2 \end{Bmatrix} = \begin{Bmatrix} T_1 \\ 3600 + 17T_1 \end{Bmatrix} = \begin{Bmatrix} 140 \\ 5980 \end{Bmatrix} \tag{E$_3$}$$

from which the solution can be obtained as

$$T_1 = 140\,^\circ\text{C} \quad \text{and} \quad T_2 = 55.89\,^\circ\text{C}$$

(ii) With two elements

The element matrices and vectors are given by

$$[K_1^{(1)}] = \frac{(\pi)(70)}{(2.5)} \begin{bmatrix} 1 & -1 \\ -1 & 1 \end{bmatrix} = 28\pi \begin{bmatrix} 1 & -1 \\ -1 & 1 \end{bmatrix}$$

$$[K_2^{(1)}] = \frac{(5)(2\pi)(2.5)}{6} \begin{bmatrix} 2 & 1 \\ 1 & 2 \end{bmatrix} = 4.1667\pi \begin{bmatrix} 2 & 1 \\ 1 & 2 \end{bmatrix}$$

$$\vec{P}^{(1)} = \frac{1}{2}(5)(40)(2\pi)(2.5) \begin{Bmatrix} 1 \\ 1 \end{Bmatrix} = 500\pi \begin{Bmatrix} 1 \\ 1 \end{Bmatrix}$$

$$[K_1^{(2)}] = \frac{(\pi)(70)}{2.5} \begin{bmatrix} 1 & -1 \\ -1 & 1 \end{bmatrix} = 28\pi \begin{bmatrix} 1 & -1 \\ -1 & 1 \end{bmatrix}$$

$$[K_2^{(2)}] = \frac{(5)(2\pi)(2.5)}{6} \begin{bmatrix} 2 & 1 \\ 1 & 2 \end{bmatrix} + (5)(\pi) \begin{bmatrix} 0 & 0 \\ 0 & 1 \end{bmatrix} = \begin{bmatrix} 8.3334\pi & 4.1667\pi \\ 4.1667\pi & 13.3334\pi \end{bmatrix}$$

$$\vec{P}^{(2)} = \frac{1}{2}(5)(40)(2\pi)(2.5) \begin{Bmatrix} 1 \\ 1 \end{Bmatrix} + (5)(40)\pi \begin{Bmatrix} 0 \\ 1 \end{Bmatrix} = \begin{Bmatrix} 500\pi \\ 700\pi \end{Bmatrix}$$

The assembled equations can be written as

$$\begin{bmatrix} 36.3334\pi & -23.8333\pi & 0 \\ -23.8333\pi & 72.6668\pi & -23.8333\pi \\ 0 & -23.8333\pi & 41.3334\pi \end{bmatrix} \begin{Bmatrix} T_1 \\ T_2 \\ T_3 \end{Bmatrix} = \begin{Bmatrix} 500\pi \\ 1000\pi \\ 700\pi \end{Bmatrix} \tag{E$_4$}$$

After incorporating the boundary condition $T_1 = 140$, Eq. (E$_4$) becomes

$$\begin{bmatrix} 1 & 0 & 0 \\ 0 & 72.6668 & -23.8333 \\ 0 & -23.8333 & 41.3334 \end{bmatrix} \begin{Bmatrix} T_1 \\ T_2 \\ T_3 \end{Bmatrix} = \begin{Bmatrix} T_1 \\ 1000 + 23.8333\ T_1 \\ 700 \end{Bmatrix} = \begin{Bmatrix} 140 \\ 4336 \\ 700 \end{Bmatrix} \tag{E$_5$}$$

The solution of Eq. (E$_5$) gives

$$T_1 = 140\,^\circ\text{C}, \qquad T_2 = 80.44\,^\circ\text{C}, \quad \text{and} \quad T_3 = 63.36\,^\circ\text{C}$$

14.3 COMPUTER PROGRAM FOR ONE-DIMENSIONAL PROBLEMS

A subroutine called HEAT1 is given for the solution of one-dimensional (fin-type) heat transfer problems. The arguments of this subroutine are as follows:

NN = number of nodes (input).
NE = number of elements (input).

NB = semibandwidth of the overall matrix GK (input).
IEND = 0: means no heat convection from free end.
= any nonzero integer: means that heat convection occurs from the free end (input).
CC = thermal conductivity of the material, k (input).
H = convection heat transfer coefficient, h (input).
TINF = atmospheric temperature, T_∞ (input).
QD = strength of heat source, $\dot{q}$ (input).
Q = boundary heat flux, q (input).
NØDE = array of size NE $\times$ 2; NØDE (I, J) = global node number corresponding to Jth (right-hand side) end of element I (input).
P = array of size NN used to store the vector $\vec{\underset{\sim}{P}}$.
PLØAD = array of size NN $\times$ 1.
XC = array of size NN; XC(I) = x coordinate of node I (input).
A = array of size NE; A(I) = area of cross section of element I (input).
GK = array of size NN $\times$ NB to store the matrix $[\underset{\sim}{K}]$.
EL = array of size NE; EL(I) = length of element I.
PERI = array of size NE; PERI (I) = perimeter of element I (input).
TS = array of size NN; TS (I) = prescribed value of temperature of node I (input). If the temperature of node I is not specified, then the value of TS (I) is to be given as 0.0.

This subroutine requires the subroutines ADJUST, DECOMP, and SOLVE. To illustrate the use of the program HEAT1, consider the problem of Example 14.2 with two finite elements. The main program for solving this problem and the numerical results given by the program are given below.

```
C==========================================================================
C
C     ONE-DIMENSIONAL HEAT CONDUCTION
C
C==========================================================================
      DIMENSION NODE(2,2),P(3),PLOAD(3,1),XC(3),A(2),GK(3,2),EL(2),TS(3)
     2 ,PERI(2)
      DATA NN,NE,NB,IEND,CC,H,TINF,QD,Q/3,2,2,1,70.0,5.0,40.0,0.0,0.0/
      DATA NODE/1,2,2,3/
      DATA XC/0.0,2.5,5.0/
      DATA A/3.1416,3.1416/
      DATA PERI/6.2832,6.2832/
      DATA TS/140.0,0.0,0.0/
      CALL HEAT1(NN,NE,NB,IEND,CC,H,TINF,QD,Q,NODE,P,PLOAD,XC,A,GK,EL,
     2 PERI,TS)
      PRINT 10
   10 FORMAT(19H NODAL TEMPERATURES,/)
      DO 20 I=1,NN
   20 PRINT 30,I,PLOAD(I,1)
   30 FORMAT(I4,E15.8)
      STOP
      END
```

```
NODAL TEMPERATURES

  1 0.14000000E+03
  2 0.80447556E+02
  3 0.63322582E+02
```

14.4 TAPERED FIN ANALYSIS

In a tapered fin, the area of cross section A varies with x. By assuming a linear variation of area from node i (local node 1) to node j (local node 2) of element e, the area of cross section at a distance x from node i can be expressed as

$$A(x) = A_i + \frac{(A_j - A_i)x}{l^{(e)}} = A_i N_i(x) + A_j N_j(x) \tag{14.20}$$

where N_i and N_j are the linear shape functions defined in Eq. (14.5), and A_i and A_j are the cross-sectional areas of element e at nodes i and j, respectively.

The matrices $[K_1^{(e)}]$, $[K_2^{(e)}]$, $[K_3^{(e)}]$, and $\vec{P}^{(e)}$ can be obtained as [from Eqs. (13.46)–(13.49)]

$$[K_1^{(e)}] = \iiint\limits_{V^{(e)}} [B]^T[D][B]\,\mathrm{d}V = \int\limits_{x=0}^{l^{(e)}} \begin{Bmatrix} \left(-\frac{1}{l^{(e)}}\right) \\ \left(\frac{1}{l^{(e)}}\right) \end{Bmatrix} [k] \left\{ \left(-\frac{1}{l^{(e)}}\right) \left(\frac{1}{l^{(e)}}\right) \right\} A(x)\,\mathrm{d}x$$

$$= \frac{k}{l^{(e)}} \left(\frac{A_i + A_j}{2}\right) \begin{bmatrix} 1 & -1 \\ -1 & 1 \end{bmatrix} = \frac{k\bar{A}^{(e)}}{l^{(e)}} \begin{bmatrix} 1 & -1 \\ -1 & 1 \end{bmatrix} \tag{14.21}$$

where $\bar{A}^{(e)}$ is the average area of the element e. Since the evaluation of the integral in $[K_2^{(e)}]$ involves the perimeter P, we can use a similar procedure. By writing P as

$$P(x) = P_i N_i(x) + P_j N_j(x) \tag{14.22}$$

where P_i and P_j are the perimeters of element e at nodes i and j, respectively, we obtain

$$[K_2^{(e)}] = h \iint\limits_{S_3^{(e)}} [N]^T[N]\,\mathrm{d}S_3 = h \int\limits_{x=0}^{l^{(e)}} \begin{bmatrix} N_1^2 & N_1 N_2 \\ N_1 N_2 & N_2^2 \end{bmatrix} P\,\mathrm{d}x \tag{14.23}$$

The integrals of Eq. (14.23) can be evaluated as

$$\int\limits_{x=0}^{l^{(e)}} N_i^2(x) P(x)\,\mathrm{d}x = \frac{l^{(e)}}{12}(3P_i + P_j) \tag{14.24}$$

$$\int\limits_{x=0}^{l^{(e)}} N_i(x) N_j(x) P(x)\,\mathrm{d}x = \frac{l^{(e)}}{12}(P_i + P_j) \tag{14.25}$$

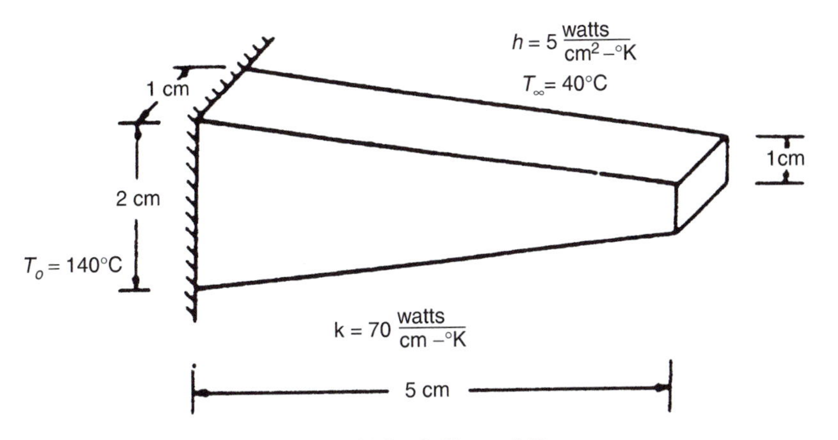

Figure 14.2. A Tapered Fin.

$$\int_{x=0}^{l^{(e)}} N_j^2(x)P(x)\,\mathrm{d}x = \frac{l^{(e)}}{12}(P_i + 3P_j) \tag{14.26}$$

$$\therefore\ [K_2^{(e)}] = \frac{hl^{(e)}}{12}\begin{bmatrix}(3P_i + P_j) & (P_i + P_j)\\ (P_i + P_j) & (P_i + 3P_j)\end{bmatrix}^{(e)} \tag{14.27}$$

Since this is a steady-state problem with $\dot{q} = q = 0$, we have

$$[K_3^{(e)}] = [0] \tag{14.28}$$

$$\vec{P}^{(e)} = \iint_{S_3^{(e)}} hT_\infty [N]^T\,\mathrm{d}S_3 = \frac{hT_\infty l^{(e)}}{6}\begin{Bmatrix}2P_i + P_j\\ P_i + 2P_j\end{Bmatrix}^{(e)} \tag{14.29}$$

Once the element matrices are available, the overall equations can be obtained using Eq. (13.35).

Example 14.3 Find the temperature distribution in the tapered fin shown in Figure 14.2.

Solution

(i) For one element

If $E = 1$, $l^{(1)} = 5$ cm, $A_i = 2$ cm^2, $A_j = 1$ cm^2, $\bar{A} = 1.5$ cm^2, $P_i = 6$ cm, and $P_j = 4$ cm. Thus, Eqs. (14.21), (14.27), and (14.29) give

$$[K_1^{(e)}] = \frac{(70)(1.5)}{5}\begin{bmatrix}1 & -1\\ -1 & 1\end{bmatrix} = \begin{bmatrix}21 & -21\\ -21 & 21\end{bmatrix}$$

$$[K_2^{(e)}] = \frac{(5)(5)}{12}\begin{bmatrix}(3\times 6 + 4) & (6+4)\\ (6+4) & (6 + 3\times 4)\end{bmatrix} = \frac{1}{6}\begin{bmatrix}275 & 125\\ 125 & 225\end{bmatrix}$$

$$\vec{P}^{(e)} = \frac{(5)(40)(5)}{6}\begin{Bmatrix} 2\times 6+4 \\ 6+2\times 4\end{Bmatrix} = \frac{1}{3}\begin{Bmatrix} 8000 \\ 7000\end{Bmatrix}$$

Hence, Eq. (13.35) gives

$$\begin{bmatrix} 401 & -1 \\ -1 & 351\end{bmatrix}\begin{Bmatrix} T_1 \\ T_2\end{Bmatrix} = \begin{Bmatrix} 16000 \\ 14000\end{Bmatrix} \tag{E$_1$}$$

When the boundary condition, $T_1 = 140\,^\circ$C, is incorporated, Eq. (E_1) gets modified to

$$\begin{bmatrix} 1 & 0 \\ 0 & 351\end{bmatrix}\begin{Bmatrix} T_1 \\ T_2\end{Bmatrix} = \begin{Bmatrix} T_1 \\ 14000+T_1\end{Bmatrix} = \begin{Bmatrix} 140 \\ 14140\end{Bmatrix} \tag{E$_2$}$$

The solution of Eq. (E_2) gives

$$T_1 = 140\,^\circ\text{C} \quad \text{and} \quad T_2 = 40.28\,^\circ\text{C}$$

(ii) For two elements

For $e = 1$: $l^{(1)} = 2.5$ cm, $A_i = 2$ cm^2, $A_j = 1.5$ cm^2, $\bar{A} = 1.75$ cm^2, $P_i = 6$ cm, and $P_j = 5$ cm.

$$[K_1^{(1)}] = \frac{(70)(1.75)}{(2.5)}\begin{bmatrix} 1 & -1 \\ -1 & 1\end{bmatrix} = \begin{bmatrix} 49 & -49 \\ -49 & 49\end{bmatrix}$$

$$[K_2^{(1)}] = \frac{(5)(2.5)}{12}\begin{bmatrix} (3\times 6+5) & (6+5) \\ (6+5) & (6+3\times 5)\end{bmatrix} = \begin{bmatrix} 23.95 & 11.45 \\ 11.45 & 21.90\end{bmatrix}$$

$$\vec{P}^{(1)} = \frac{(5)(40)(2.5)}{6}\begin{Bmatrix} 2\times 6+5 \\ 6+2\times 5\end{Bmatrix} = \frac{1}{6}\begin{Bmatrix} 8500 \\ 8000\end{Bmatrix}$$

For $e = 2$: $l^{(2)} = 2.5$ cm, $A_i = 1.5$ cm^2, $A_j = 1$ cm^2, $\bar{A} = 1.25$ cm^2, $P_i = 5$ cm, and $P_j = 4$ cm.

$$[K_1^{(2)}] = \frac{(70)(1.25)}{(2.5)}\begin{bmatrix} 1 & -1 \\ -1 & 1\end{bmatrix} = \begin{bmatrix} 35 & -35 \\ -35 & 35\end{bmatrix}$$

$$[K_2^{(2)}] = \frac{(5)(2.5)}{12}\begin{bmatrix} (3\times 5+4) & (5+4) \\ (5+4) & (5+3\times 4)\end{bmatrix} = \begin{bmatrix} 19.8 & 9.4 \\ 9.4 & 17.7\end{bmatrix}$$

$$\vec{P}^{(2)} = \frac{(5)(40)(2.5)}{6}\begin{Bmatrix} 2\times 5+4 \\ 5+2\times 4\end{Bmatrix} = \frac{1}{6}\begin{Bmatrix} 7000 \\ 6500\end{Bmatrix}$$

The overall equations (13.35) will be

$$\begin{bmatrix} 72.95 & -37.55 & 0 \\ -37.55 & 125.70 & -25.60 \\ 0 & -25.60 & 52.70\end{bmatrix}\begin{Bmatrix} T_1 \\ T_2 \\ T_3\end{Bmatrix} = \begin{Bmatrix} \dfrac{8500}{6} \\ \dfrac{15000}{6} \\ \dfrac{6500}{6}\end{Bmatrix} \tag{E$_3$}$$

Equation (E_3), when modified to incorporate the boundary condition $T_1 = 140\,^\circ\text{C}$, appears as

$$\begin{bmatrix} 1 & 0 & 0 \\ 0 & 125.70 & -25.60 \\ 0 & -25.60 & 52.70 \end{bmatrix} \begin{Bmatrix} T_1 \\ T_2 \\ T_3 \end{Bmatrix} = \begin{Bmatrix} T_1 \\ \dfrac{15000}{6} + 37.55T_1 \\ \dfrac{6500}{6} \end{Bmatrix} = \begin{Bmatrix} 140.0 \\ 7757.0 \\ 1083.3 \end{Bmatrix} \tag{E$_4$}$$

The solution of Eq. (E_4) gives

$$T_1 = 140\,^\circ\text{C}, \quad T_2 = 73.13\,^\circ\text{C}, \quad \text{and} \quad T_3 = 56.07\,^\circ\text{C}$$

14.5 ANALYSIS OF UNIFORM FINS USING QUADRATIC ELEMENTS

The solution of uniform one-dimensional heat transfer problems is considered using a quadratic model for the variation of temperature in the element. The step-by-step procedure is given below.

Step 1: Idealize the fin into E finite elements as shown in Figure 14.3.

Step 2: Assume a quadratic variation of temperature inside any element e as

$$T^{(e)}(x) = a_1 + a_2 x + a_3 x^2 = [N(x)]\vec{q}^{\,(e)} \tag{14.30}$$

where

$$[N(x)] = [N_i(x) \quad N_j(x) \quad N_k(x)] \tag{14.31}$$

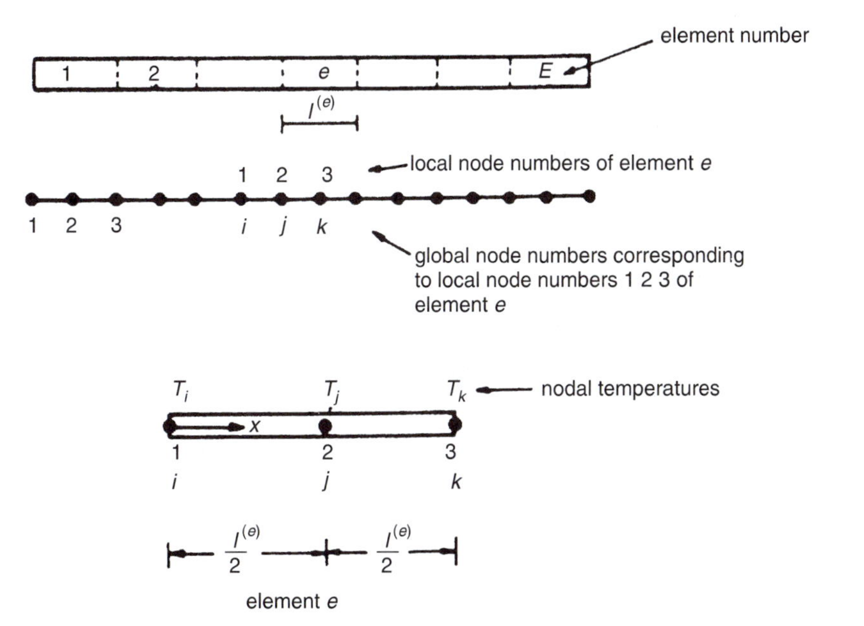

Figure 14.3. A Quadratic Element.

$$N_i(x) = \left(1 - \frac{2x}{l^{(e)}}\right)\left(1 - \frac{x}{l^{(e)}}\right) \tag{14.32}$$

$$N_j(x) = \frac{4x}{l^{(e)}}\left(1 - \frac{x}{l^{(e)}}\right) \tag{14.33}$$

$$N_k(x) = -\frac{x}{l^{(e)}}\left(1 - \frac{2x}{l^{(e)}}\right) \tag{14.34}$$

and

$$\vec{q}^{\,(e)} = \begin{Bmatrix} q_1 \\ q_2 \\ q_3 \end{Bmatrix} = \begin{Bmatrix} T_i \\ T_j \\ T_k \end{Bmatrix}$$

where i, j, and k denote the global node numbers corresponding to local nodes 1 (left end), 2 (middle), and 3 (right end), respectively.

Step 3: Derivation of element matrices:

For the quadratic element, we have

$$[D] = [k] \tag{14.35}$$

$$[B] = \begin{bmatrix} \dfrac{\partial N_i}{\partial x} & \dfrac{\partial N_j}{\partial x} & \dfrac{\partial N_k}{\partial x} \end{bmatrix} = \begin{bmatrix} \dfrac{4x}{l^{(e)2}} - \dfrac{3}{l^{(e)}} & \dfrac{4}{l^{(e)}} - \dfrac{8x}{l^{(e)2}} & \dfrac{4x}{l^{(e)2}} - \dfrac{1}{l^{(e)}} \end{bmatrix} \tag{14.36}$$

The definitions of $[K_1^{(e)}]$, $[K_2^{(e)}]$, $[K_3^{(e)}]$, and $\vec{P}^{(e)}$ remain the same as those given in Eqs. (13.46)–(13.49), but the integrals have to be reevaluated using the quadratic displacement model of Eq. (14.30). This gives

$$[K_1^{(e)}] = kA \int_{x=0}^{l^{(e)}} \mathrm{d}x \times \begin{bmatrix} \left(\dfrac{4x}{l^{(e)2}} - \dfrac{3}{l^{(e)}}\right)^2 & \left(\dfrac{4x}{l^{(e)2}} - \dfrac{3}{l^{(e)}}\right)\left(\dfrac{4}{l^{e)}} - \dfrac{8x}{l^{(e)2}}\right) & \left(\dfrac{4x}{l^{(e)2}} - \dfrac{3}{l^{(e)}}\right)\left(\dfrac{4x}{l^{(e)2}} - \dfrac{1}{l^{(e)}}\right) \\ & \left(\dfrac{4x}{l^{(e)2}} - \dfrac{3}{l^{(e)}}\right)^2 & \left(\dfrac{4}{l^{(e)2}} - \dfrac{8x}{l^{(e)}}\right)\left(\dfrac{4x}{l^{(e)2}} - \dfrac{1}{l^{(e)}}\right) \\ \text{Symmetric} & & \left(\dfrac{4x}{l^{(e)2}} - \dfrac{1}{l^{(e)}}\right)^2 \end{bmatrix}$$

$$= \frac{kA}{3l^{(e)}} \begin{bmatrix} 7 & -8 & 1 \\ -8 & 16 & -8 \\ 1 & -8 & 7 \end{bmatrix} \tag{14.37}$$

where A is the cross-sectional area of the element.

$$[K_2^{(e)}] = h \cdot P \int_{x=0}^{l^{(e)}} \begin{bmatrix} N_i^2(x) & N_i(x) \cdot N_j(x) & N_i(x) \cdot N_k(x) \\ N_i(x) \cdot N_j(x) & N_j^2(x) & N_j(x) \cdot N_k(x) \\ N_i(x) \cdot N_k(x) & N_j(x) \cdot N_k(x) & N_k^2(x) \end{bmatrix} \mathrm{d}x$$

$$= \frac{hPl^{(e)}}{30} \begin{bmatrix} 4 & 2 & -1 \\ 2 & 16 & 2 \\ -1 & 2 & 4 \end{bmatrix} \tag{14.38}$$

where P is the perimeter of the element.

$$[K_3^{(e)}] = [0] \text{ for steady-state problems} \tag{14.39}$$

$$\vec{P}^{(e)} = \dot{q}A \int_{x=0}^{l^{(e)}} \begin{Bmatrix} N_i(x) \\ N_j(x) \\ N_k(x) \end{Bmatrix} \mathrm{d}x - qP \int_{x=0}^{l^{(e)}} \begin{Bmatrix} N_i(x) \\ N_j(x) \\ N_k(x) \end{Bmatrix} \mathrm{d}x + hT_\infty P \int_{x=0}^{l^{(e)}} \begin{Bmatrix} N_i(x) \\ N_j(x) \\ N_k(x) \end{Bmatrix} \mathrm{d}x \tag{14.40}$$

where $\mathrm{d}V^{(e)}$, $\mathrm{d}S_2^{(e)}$, and $\mathrm{d}S_3^{(e)}$ were replaced by $A\,\mathrm{d}x$, $P\,\mathrm{d}x$, and $P\,\mathrm{d}x$, respectively. With the help of Eqs. (14.32)–(14.34), Eq. (14.40) can be expressed as

$$\vec{P}^{(e)} = \frac{\dot{q}Al^{(e)}}{6} \begin{Bmatrix} 1 \\ 4 \\ 1 \end{Bmatrix} - \frac{qPl^{(e)}}{6} \begin{Bmatrix} 1 \\ 4 \\ 1 \end{Bmatrix} + \frac{kT_\infty Pl^{(e)}}{6} \begin{Bmatrix} 1 \\ 4 \\ 1 \end{Bmatrix} \tag{14.41}$$

If convection occurs from the free end of the element, for example, node i, then $N_i(x = o) = 1$, $N_j(x = o) = N_k(x = o) = 0$, and hence the additional surface integral term to be added to the matrix $[K_2^{(e)}]$ will be

$$\iint_{S_3^{(e)}} h[N]^T[N]\,\mathrm{d}S_3$$

$$= h \iint_{S_3^{(e)}} \begin{bmatrix} N_i^2(x=o) & N_i(x=o) \cdot N_j(x=o) & N_i(x=o) \cdot N_k(x=o) \\ N_i(x=o) \cdot N_j(x=o) & N_j^2(x=o) & N_j(x=o) \cdot N_k(x=o) \\ N_i(x=o) \cdot N_k(x=o) & N_j(x=o) \cdot N_k(x=o) & N_k^2(x=o) \end{bmatrix} \mathrm{d}S_3$$

$$= hA_i \begin{bmatrix} 1 & 0 & 0 \\ 0 & 0 & 0 \\ 0 & 0 & 0 \end{bmatrix} \tag{14.42}$$

where A_i is the cross-sectional area of the rod at node i. Similarly, the additional surface integral term to be added to $\vec{P}^{(e)}$ due to convection from the free end (e.g., node i) will be

$$\iint_{S_3^{(e)}} hT_\infty [N]^T\,\mathrm{d}S_3 = hT_\infty \iint_{S_3^{(e)}} \begin{Bmatrix} N_i(x=o) \\ N_j(x=o) \\ N_k(x=o) \end{Bmatrix} \mathrm{d}S_3 = hT_\infty A_i \begin{Bmatrix} 1 \\ 0 \\ 0 \end{Bmatrix} \tag{14.43}$$

Example 14.4 Find the temperature distribution in the fin shown in Figure 14.1 using quadratic elements.

Solution We use one element only and neglect convection from the free end. Thus, Eqs. (14.37), (14.38), and (14.41) become

$$[K_1^{(1)}] = \frac{(70)(\pi)}{3(5)} \begin{bmatrix} 7 & -8 & 1 \\ -8 & 16 & -8 \\ 1 & -8 & 7 \end{bmatrix} = \frac{\pi}{3} \begin{bmatrix} 98 & -112 & 14 \\ -112 & 224 & -112 \\ 14 & -112 & 98 \end{bmatrix}$$

$$[K_2^{(1)}] = \frac{(5)(2\pi)(5)}{30} \begin{bmatrix} 4 & 2 & -1 \\ 2 & 16 & 2 \\ -1 & 2 & 4 \end{bmatrix} = \frac{\pi}{3} \begin{bmatrix} 20 & 10 & -5 \\ 10 & 80 & 10 \\ -5 & 10 & 20 \end{bmatrix}$$

$$\vec{P}^{(1)} = \frac{(5)(40)(2\pi)(5)}{6} \begin{Bmatrix} 1 \\ 4 \\ 1 \end{Bmatrix} = \frac{\pi}{3} \begin{Bmatrix} 1000 \\ 4000 \\ 1000 \end{Bmatrix}$$

The governing equations can be expressed as

$$\begin{bmatrix} 118 & -102 & 9 \\ -102 & 304 & -102 \\ 9 & -102 & 118 \end{bmatrix} \begin{Bmatrix} T_1 \\ T_2 \\ T_3 \end{Bmatrix} = \begin{Bmatrix} 1000 \\ 4000 \\ 1000 \end{Bmatrix} \tag{E$_1$}$$

where $T_1 = T(x = 0)$, $T_2 = T(x = 2.5)$, and $T_3 = T(x = 5.0)$. By incorporating the boundary condition $T_1 = 140$, Eq. (E$_1$) can be modified as

$$\begin{bmatrix} 1 & 0 & 0 \\ 0 & 304 & -102 \\ 0 & -102 & 118 \end{bmatrix} \begin{Bmatrix} T_1 \\ T_2 \\ T_3 \end{Bmatrix} = \begin{Bmatrix} 140 \\ 4000 + 102T_1 \\ 1000 - 9T_1 \end{Bmatrix} = \begin{Bmatrix} 140 \\ 18280 \\ -260 \end{Bmatrix} \tag{E$_2$}$$

The solution of Eq. (E$_2$) gives $T_1 = 140.0\,^\circ\text{C}$, $T_2 = 83.64\,^\circ\text{C}$, and $T_3 = 70.09\,^\circ\text{C}$.

14.6 UNSTEADY STATE PROBLEMS

Time-dependent or unsteady state problems are very common in heat transfer. For some of these time-dependent problems, the transient period occurs between the starting of the physical process and the reaching of the steady-state condition. For some problems, the steady-state condition is never obtained and in these cases the transient period includes the entire life of the process. Several finite element procedures have been suggested for the solution of transient heat transfer problems [14.1, 14.2]. We consider the finite element solution of time-dependent heat transfer problems briefly in this section. The governing differential equation for an unsteady state heat transfer problem is given by Eq. (13.11) and the associated boundary and initial conditions are given by Eqs. (13.18)–(13.21). In general all the parameters k_x, k_y, k_z, $\dot{q}$, and ρc will be time dependent. The finite element solution of this problem leads to a set of first-order linear differential equations,

Eq. (13.35). It can be seen that the term $[K_3]\dot{\underset{\sim}{T}}$ is the additional term that appears because of the unsteady state. The associated element matrix is defined as

$$[K_3^{(e)}] = \iiint\limits_{V^{(e)}} \rho c[N]^T[N]\,\mathrm{d}V \tag{14.44}$$

which is also known as the element capacitance matrix.

14.6.1 Derivation of Element Capacitance Matrices

For the straight uniform one-dimensional element considered in Section 14.2, the shape function matrix is given by Eq. (14.5). By writing the element volume as $\mathrm{d}V = A^{(e)}\mathrm{d}x$, where $A^{(e)}$ is the cross-sectional area of the element e, Eq. (14.44) can be expressed as

$$\begin{aligned}[K_3^{(e)}] &= (\rho c)^{(e)} \int\limits_{x=0}^{l^{(e)}} \begin{Bmatrix} \left(1 - \dfrac{x}{l^{(e)}}\right) \\ \dfrac{x}{l^{(e)}} \end{Bmatrix} \left\{ \left(1 - \frac{x}{l^{(e)}}\right) \;\; \frac{x}{l^{(e)}} \right\} \mathrm{d}x \\ &= \frac{(\rho c)^{(e)} A^{(e)} l^{(e)}}{6} \begin{bmatrix} 2 & 1 \\ 1 & 2 \end{bmatrix} \end{aligned} \tag{14.45}$$

where $(\rho c)^{(e)}$ is assumed to be a constant for the element e. For the linearly tapered fin element considered in Section 14.4, $\mathrm{d}V = A(x)\,\mathrm{d}x = [A_i + ((A_j - A_i)/l^{(e)})x]\,\mathrm{d}x$ and hence Eq. (14.44) becomes

$$\begin{aligned}[K_3^{(e)}] &= (\rho c)^{(e)} \int\limits_{x=0}^{l^{(e)}} \begin{Bmatrix} -\dfrac{1}{l^{(e)}} \\ \dfrac{1}{l^{(e)}} \end{Bmatrix} \left\{ -\frac{1}{l^{(e)}} \;\; \frac{1}{l^{(e)}} \right\} \cdot \left[A_i + \left(\frac{A_j - A_i}{l^{(e)}} \right) x \right] \mathrm{d}x \\ &= \frac{(\rho c)^{(e)} \bar{A}^{(e)}}{l^{(e)}} \begin{bmatrix} 1 & -1 \\ -1 & 1 \end{bmatrix} \end{aligned} \tag{14.46}$$

where $\bar{A}^{(e)}$ is the average cross-sectional area of the element e. For the straight uniform fin considered in Section 14.5 using a quadratic model [defined by Eq. (14.30)], Eq. (14.44) becomes

$$\begin{aligned}[K_3^{(e)}] &= (\rho c)^{(e)} A^{(e)} \int\limits_{x=0}^{l^{(e)}} \begin{bmatrix} N_i^2 & N_i N_j & N_i N_k \\ N_i N_j & N_j^2 & N_j N_k \\ N_i N_k & N_j N_k & N_k^2 \end{bmatrix} \mathrm{d}x \\ &= \frac{(\rho c)^{(e)} \bar{A}^{(e)} l^{(e)}}{30} \begin{bmatrix} 4 & 2 & -1 \\ 2 & 16 & 2 \\ -1 & 2 & 4 \end{bmatrix} \end{aligned} \tag{14.47}$$

Example 14.5 Find the time-dependent temperature distribution in a plane wall that is insulated on one face and is subjected to a step change in surface temperature on the other face as shown in Figure 14.4.

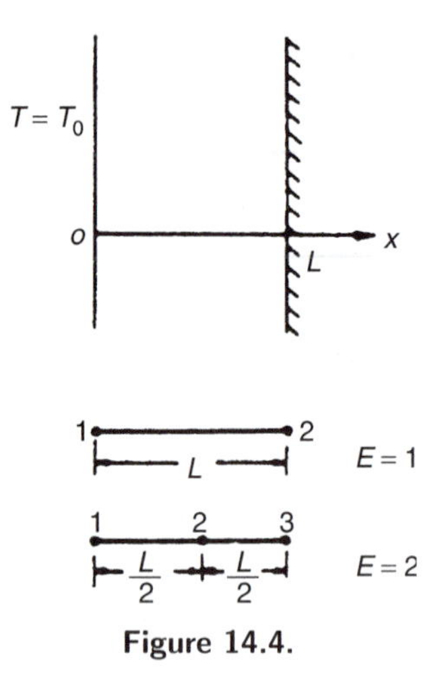

Figure 14.4.

Solution The finite element equations for this one-dimensional transient problem are given by (Eq. 13.35):

$$[\underset{\sim}{K_3}]\,\dot{\underset{\sim}{\vec{T}}}+[\underset{\sim}{K}]\underset{\sim}{\vec{T}}=\underset{\sim}{\vec{P}} \tag{E$_1$}$$

where the element matrices, with the assumption of linear temperature variation, are given by

$$[K_1^{(e)}]=\frac{A^{(e)}k^{(e)}}{l^{(e)}}\begin{bmatrix}1 & -1\\ -1 & 1\end{bmatrix} \tag{E$_2$}$$

$$[K_2^{(e)}]=\begin{bmatrix}0 & 0\\ 0 & 0\end{bmatrix} \text{ since no convection condition is specified} \tag{E$_3$}$$

$$[K_3^{(e)}]=\frac{(\rho c)^{(e)}A^{(e)}l^{(e)}}{6}\begin{bmatrix}2 & 1\\ 1 & 2\end{bmatrix} \tag{E$_4$}$$

$$\vec{P}^{(e)}=\begin{Bmatrix}0\\ 0\end{Bmatrix} \text{ since no } \dot{q}, q, \text{ and } h \text{ are specified in the problem} \tag{E$_5$}$$

(i) Solution with one element

If $E = 1$, T_1 and T_2 denote the temperatures of nodes 1 and 2, and Eq. (E_1) becomes

$$\frac{\rho cAL}{6}\begin{bmatrix}2 & 1\\ 1 & 2\end{bmatrix}\begin{Bmatrix}\dfrac{dT_1}{dt}\\[2mm] \dfrac{dT_2}{dt}\end{Bmatrix}+\frac{Ak}{L}\begin{bmatrix}1 & -1\\ -1 & 1\end{bmatrix}\begin{Bmatrix}T_1\\ T_2\end{Bmatrix}=\begin{Bmatrix}0\\ 0\end{Bmatrix} \tag{E$_6$}$$

Equation (E_6) is to be modified to satisfy the boundary condition at $x = 0$. Since T_2 is the only unknown in the problem, we can delete the first equation of (E_6) and set $T_1 = T_0$

and $(\mathrm{d}T_1/\mathrm{d}t) = 0$ in Eq. (E_6) to obtain

$$\frac{\mathrm{d}T_2}{\mathrm{d}t} = -\frac{3k}{\rho c L^2}(T_2 - T_0) \tag{E_7}$$

By defining $\theta = T_2 - T_0$, Eq. (E_7) can be written as

$$\frac{\mathrm{d}\theta}{\mathrm{d}t} + \alpha\theta = 0 \tag{E_8}$$

where $\alpha = (3k/\rho c L^2)$. The solution of Eq. (E_8) is given by

$$\theta(t) = a_1\,\mathrm{e}^{-\alpha t} \tag{E_9}$$

where a_1 is a constant whose value can be determined from the known initial condition, $T_2(t = 0) = \bar{T}_0$:

$$\theta(t = 0) = \bar{T}_0 - T_0 = a_1 \cdot \mathrm{e}^{-\alpha(0)} = a_1 \tag{E_{10}}$$

Thus, the solution of Eq. (E_7) is

$$T_2(t) = T_0 + (\bar{T}_0 - T_0)\cdot \mathrm{e}^{-(3k/\rho c L^2)t} \tag{E_{11}}$$

(ii) Solution with two elements

If $E = 2$, T_1, T_2, and T_3 indicate the temperatures of nodes 1, 2, and 3 and Eq. (E_1) becomes

$$\begin{bmatrix} 2 & 1 & 0 \\ 1 & 4 & 1 \\ 0 & 1 & 2 \end{bmatrix} \begin{Bmatrix} \mathrm{d}T_1/\mathrm{d}t \\ \mathrm{d}T_2/\mathrm{d}t \\ \mathrm{d}T_3/\mathrm{d}t \end{Bmatrix} + \frac{24k}{\rho c L^2} \begin{bmatrix} 1 & -1 & 0 \\ -1 & 2 & -1 \\ 0 & -1 & 1 \end{bmatrix} \begin{Bmatrix} T_1 \\ T_2 \\ T_3 \end{Bmatrix} = \begin{Bmatrix} 0 \\ 0 \\ 0 \end{Bmatrix} \tag{E_{12}}$$

As before, we delete the first equation from Eq. (E_{12}) and substitute the boundary condition $T_1 = T_0$ [and hence $(\mathrm{d}T_1/\mathrm{d}t) = 0$] in the remaining two equations to obtain

$$\left.\begin{aligned} &4\frac{\mathrm{d}T_2}{\mathrm{d}t} + \frac{\mathrm{d}T_3}{\mathrm{d}t} + \frac{24k}{\rho c L^2}(-T_0 + 2T_2 - T_3) = 0 \\ &\frac{\mathrm{d}T_2}{\mathrm{d}t} + 2\frac{\mathrm{d}T_3}{\mathrm{d}t} + \frac{24k}{\rho c L^2}(-T_2 + T_3) = 0 \end{aligned}\right\} \tag{E_{13}}$$

By defining $\theta_2 = T_2 - T_0$ and $\theta_3 = T_3 - T_0$, Eqs. (E_{13}) can be expressed as

$$\left.\begin{aligned} &4\frac{\mathrm{d}\theta_2}{\mathrm{d}t} + \frac{\mathrm{d}\theta_3}{\mathrm{d}t} + \frac{24k}{\rho c L^2}(2\theta_2 - \theta_3) = 0 \\ &\frac{\mathrm{d}\theta_2}{\mathrm{d}t} + 2\frac{\mathrm{d}\theta_3}{\mathrm{d}t} + \frac{24k}{\rho c L^2}(-\theta_2 + \theta_3) = 0 \end{aligned}\right\} \tag{E_{14}}$$

These equations can be solved using the known initial conditions.

14.6.2 Finite Difference Solution in Time Domain

The solution of the unsteady state equations, namely Eqs. (13.35), based on the fourth-order Runge–Kutta integration procedure was given in Chapter 7. We now present an alternative approach using the finite difference scheme for solving these equations. This scheme is based on approximating the first-time derivative of T as

$$\left.\frac{\mathrm{d}T}{\mathrm{d}t}\right|_t = \frac{T_1 - T_0}{\Delta t} \tag{14.48}$$

where $T_1 = T(t + (\Delta t/2))$, $T_0 = T(t - (\Delta t/2))$, and Δt is a small time step. Thus, $\dot{\vec{T}} = (\mathrm{d}\vec{T}/\mathrm{d}t)$ can be replaced by

$$\left.\frac{\mathrm{d}\vec{T}}{\mathrm{d}t}\right|_t = \frac{1}{\Delta t}(\vec{T}_1 - \vec{T}_0) \tag{14.49}$$

Since $\dot{\vec{T}}$ is evaluated at the middle point of the time interval Δt, the quantities $\vec{T}$ and $\vec{P}$ involved in Eq. (13.35) are also to be evaluated at this point. These quantities can be approximated as

$$\left.\vec{T}\right|_t = \tfrac{1}{2}(\vec{T}_1 + \vec{T}_0) \tag{14.50}$$

and

$$\left.\vec{P}\right|_t = \tfrac{1}{2}(\vec{P}_1 + \vec{P}_0) \tag{14.51}$$

where

$$\vec{P}_1 = \vec{P}\left(t + \frac{\Delta t}{2}\right) \quad \text{and} \quad \vec{P}_0 = \vec{P}\left(t - \frac{\Delta t}{2}\right) \tag{14.52}$$

By substituting Eqs. (14.49)–(14.51) into Eq. (13.35), we obtain

$$\frac{1}{\Delta t}[K_3](\vec{T}_1 - \vec{T}_0) + \frac{1}{2}[K](\vec{T}_1 + \vec{T}_0) = \vec{P}|_t$$

or

$$\left([K] + \frac{2}{\Delta t}[K_3]\right)\vec{T}_1 = \left(-[K] + \frac{2}{\Delta t}[K_3]\right)\vec{T}_0 + (\vec{P}_1 + \vec{P}_0) \tag{14.53}$$

This equation shows that the nodal temperatures $\vec{T}$ at time $t + \Delta t$ can be computed once the nodal temperatures at time t are known since $\vec{P}_1$ can be computed before solving Eq. (14.53). Thus, the known initial conditions (on nodal temperatures) can be used to find the solution at subsequent time steps.

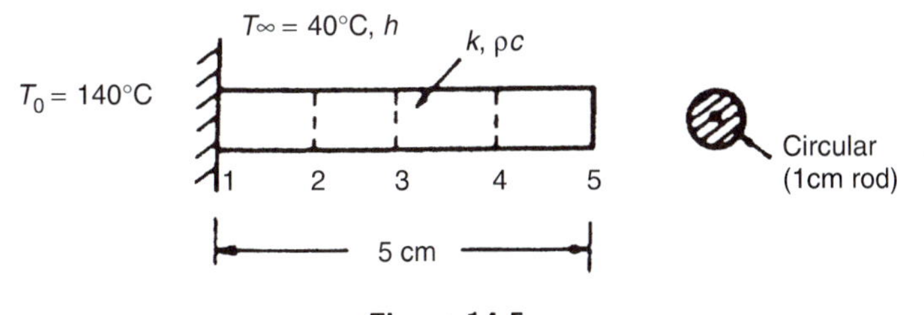

Figure 14.5.

Note

Equation (14.53) has been derived by evaluating the derivative at the middle point of the time interval. The nodal values (i.e., at time t) of $\vec{\underset{\sim}{T}}$ can be computed after solving Eq. (14.53) using Eq. (14.50). In fact, by using Eq. (14.50), Eq. (14.53) can be rewritten as

$$\left([\underset{\sim}{K}] + \frac{2}{\Delta t}[\underset{\sim}{K_3}]\right)\vec{\underset{\sim}{T}}\bigg|_t = \frac{2}{\Delta t}[\underset{\sim}{K_3}]\vec{\underset{\sim}{T}}_0 - \vec{\underset{\sim}{P}}\bigg|_t \tag{14.54}$$

where the nodal temperatures $\vec{\underset{\sim}{T}}|_t$ can be directly obtained.

Example 14.6 Derive the recursive relations, Eq. (14.53), for the one-dimensional fin shown in Figure 14.5 with the following data:

$$k = 70\frac{\text{watts}}{\text{cm–°K}},\qquad h = 10\frac{\text{watts}}{\text{cm}^2\text{–°K}},\qquad T_\infty = 40\,°\text{C},\qquad T_0 = 140\,°\text{C},$$

$$\rho c = 20\frac{\text{Joules}}{\text{cm–°K}},\qquad \Delta t = 2 \text{ minutes}$$

Solution We divide the fin into four finite elements ($E = 4$) so that the element matrices and vectors become

$$[K_1^{(e)}] = \frac{(\pi)(70)}{(1.25)}\begin{bmatrix} 1 & -1 \\ -1 & 1 \end{bmatrix} = \begin{bmatrix} 175.93 & -175.93 \\ -175.93 & 175.93 \end{bmatrix}$$

$$[K_2^{(e)}] = \frac{(10)(2\pi)(1.25)}{6}\begin{bmatrix} 2 & 1 \\ 1 & 2 \end{bmatrix} = \begin{bmatrix} 39.26 & 19.63 \\ 19.63 & 39.26 \end{bmatrix}$$

$$[K_3^{(e)}] = \frac{(20)(\pi)(1.25)}{6}\begin{bmatrix} 2 & 1 \\ 1 & 2 \end{bmatrix} = \begin{bmatrix} 39.26 & 19.63 \\ 19.63 & 39.26 \end{bmatrix}$$

$$\vec{P}^{(e)} = \vec{P}_1^{(e)} - \vec{P}_2^{(e)} + \vec{P}_3^{(e)} = \vec{P}_3^{(e)} = \frac{(10)(40)(2\pi)(1.25)}{2}\begin{Bmatrix} 1 \\ 1 \end{Bmatrix} = \begin{Bmatrix} 1570.80 \\ 1570.80 \end{Bmatrix}$$

The assembled matrices and vectors are

$$[\underset{\sim}{K_1}] = \begin{bmatrix} 175.93 & -175.93 & 0 & 0 & 0 \\ -175.93 & 351.86 & -175.93 & 0 & 0 \\ 0 & -175.93 & 351.86 & -175.93 & 0 \\ 0 & 0 & -175.93 & 351.86 & -175.93 \\ 0 & 0 & 0 & -175.93 & 175.93 \end{bmatrix} \tag{E$_1$}$$

$$[\underset{\sim}{K_3}] = [\underset{\sim}{K_2}] = \begin{bmatrix} 39.26 & 19.63 & 0 & 0 & 0 \\ 19.63 & 78.52 & 19.63 & 0 & 0 \\ 0 & 19.63 & 78.52 & 19.63 & 0 \\ 0 & 0 & 19.63 & 78.52 & 19.63 \\ 0 & 0 & 0 & 19.63 & 39.26 \end{bmatrix} \tag{E$_2$}$$

$$\underset{\sim}{\vec{P}} = \begin{Bmatrix} 1570.80 \\ 3141.60 \\ 3141.60 \\ 3141.60 \\ 1570.80 \end{Bmatrix} \tag{E$_3$}$$

Since $\Delta t = 1/30$ hour, we have

$$[A] = [\underset{\sim}{K}] + \frac{2}{\Delta t}[\underset{\sim}{K_3}] = [\underset{\sim}{K_1}] + [\underset{\sim}{K_2}] + 60[\underset{\sim}{K_3}]$$

$$= \begin{bmatrix} 2570.79 & 1021.50 & 0 & 0 & 0 \\ 1021.50 & 5141.58 & 1021.50 & 0 & 0 \\ 0 & 1021.50 & 5141.58 & 1021.50 & 0 \\ 0 & 0 & 1021.50 & 5141.58 & 1021.50 \\ 0 & 0 & 0 & 1021.50 & 2570.79 \end{bmatrix} \tag{E$_4$}$$

and

$$[B] = -[\underset{\sim}{K}] + \frac{2}{\Delta t}[\underset{\sim}{K_3}] = -[\underset{\sim}{K_1}] - [\underset{\sim}{K_2}] + 60[\underset{\sim}{K_3}]$$

$$= \begin{bmatrix} 2140.41 & 1334.10 & 0 & 0 & 0 \\ 1334.10 & 4280.82 & 1334.10 & 0 & 0 \\ 0 & 1334.10 & 4280.82 & 1334.10 & 0 \\ 0 & 0 & 1334.10 & 4280.82 & 1334.10 \\ 0 & 0 & 0 & 1334.10 & 2140.41 \end{bmatrix} \tag{E$_5$}$$

Hence, the desired recursive relation is

$$[A]\underset{\sim}{\vec{T}}_1 = [B]\underset{\sim}{\vec{T}}_0 + \underset{\sim}{\vec{P}} \tag{E$_6$}$$

where $[A]$, $[B]$, and $\underset{\sim}{\vec{P}}$ are given by Eqs. (E$_4$), (E$_5$), and (E$_3$), respectively.

14.7 HEAT TRANSFER PROBLEMS WITH RADIATION

The rate of heat flow by radiation (q) is governed by the relation

$$q = \sigma \varepsilon A(T^4 - T_\infty^4) \tag{14.55}$$

where σ is the Stefan–Boltzmann constant, ε is the emissivity of the surface, A is the surface area of the body through which heat flows, T is the absolute surface temperature of the body, and T_∞ is the absolute surrounding temperature. Thus, the inclusion of the radiation boundary condition makes a heat transfer problem nonlinear due to the nonlinear relation of Eq. (14.55). Hence, an iterative procedure is to be adopted to find the finite element solution of the problem. For example, for a one-dimensional problem, the governing differential equation is

$$k\frac{\partial^2 T}{\partial x^2} + \dot{q} = \rho c\frac{\partial T}{\partial t} \tag{14.56}$$

If heat flux is specified on the surface of the rod and if both convection and radiation losses take place from the surface, the boundary conditions of the problem can be expressed as

$$T(x = 0, t) = T_0 \tag{14.57}$$

and

$$\left.\begin{aligned} k\frac{\partial T}{\partial x} l_x + h(T - T_\infty) + q + \sigma\varepsilon(T^4 - T_\infty^4) = 0 \\ \text{on the surface} \end{aligned}\right\} \tag{14.58}$$

The initial conditions can be specified as

$$T(x, t = 0) = \bar{T}_0 \tag{14.59}$$

For convenience, we define a radiation heat transfer coefficient (h_r) as

$$h_r = \sigma\varepsilon(T^2 + T_\infty^2)(T + T_\infty) \tag{14.60}$$

so that Eq. (14.58) can be expressed as

$$\left.\begin{aligned} k\frac{\partial T}{\partial x} l_x + h(T - T_\infty) + q + h_r(T - T_\infty) = 0 \\ \text{on the surface} \end{aligned}\right\} \tag{14.61}$$

The inclusion of the convection term $h(T - T_\infty)$ in the finite element analysis resulted in the matrix (Eqs. 13.31)

$$[K_2^{(e)}] = \iint_{S_3^{(e)}} h[N]^T[N]\,\mathrm{d}S_3 \tag{14.62}$$

and the vector (Eq. 13.33)

$$\vec{P}_3^{(e)} = \iint_{S_3^{(e)}} hT_\infty[N]^T\,\mathrm{d}S_3 \tag{14.63}$$

Assuming, for the time being, that h_r is independent of the temperature T, and proceeding as in the case of the term $h(T - T_\infty)$, we obtain the additional matrix

$$[K_4^{(e)}] = \iint_{S_4^{(e)}} h_r[N]^T[N]\,\mathrm{d}S_4 \tag{14.64}$$

and the additional vector

$$\vec{P}_4^{(e)} = \iint_{S_4^{(e)}} h_r T_\infty [N]^T \, \mathrm{d}S_4 \tag{14.65}$$

to be assembled in generating the matrix $[\underset{\sim}{K}]$ and the vector $\underset{\sim}{\vec{P}}$, respectively. In Eqs. (14.64) and (14.65), $S_4^{(e)}$ denotes the surface of the element e from which radiation loss takes place. Since h_r was assumed to be a constant in deriving Eqs. (14.64) and (14.65), its value needs to be changed subsequently. Since the correct solution $(\underset{\sim}{\vec{T}})$ cannot be found unless the correct value of h_r is used in Eqs. (14.64) and (14.65), the following iterative procedure can be adopted:

1. Set the iteration number as $n = 1$ and assume $h_r^{(e)} = 0$.
2. Generate $[K_4^{(e)}]$ and $\vec{P}_4^{(e)}$ using Eqs. (14.64) and (14.65) using the latest values of $h_r^{(e)}$.
3. Assemble the element matrices and vectors to obtain the overall equation (13.35) with $[\underset{\sim}{K}] = \Sigma_{e=1}^{E}[[K_1^{(e)}] + [K_2^{(e)}] + [K_4^{(e)}]]$ and $\underset{\sim}{\vec{P}} = \Sigma_{e=1}^{E}[\vec{P}_1^{(e)} - \vec{P}_2^{(e)} + \vec{P}_3^{(e)} + \vec{P}_4^{(e)}]$.
4. Solve Eqs. (13.35) and find $\underset{\sim}{\vec{T}}$.
5. From the known nodal temperatures $\underset{\sim}{\vec{T}}$, find the new value of $h_r^{(e)}$ using Eq. (14.60) [the average of the two nodal temperatures of the element, $(T_i + T_j)/2$, can be used as $T_{av}^{(e)}$ in place of T]:

$$h_r^{(e)} = \sigma\varepsilon({T_{av}^{(e)}}^2 + T_\infty^2)(T_{av}^{(e)} + T_\infty) \tag{14.66}$$

If $n > 1$, test for the convergence of the method. If

$$\left| \frac{\left[h_r^{(e)}\right]_n - \left[h_r^{(e)}\right]_{n-1}}{\left[h_r^{(e)}\right]_{n-1}} \right| \le \delta_1 \tag{14.67}$$

and

$$\left\| (\underset{\sim}{\vec{T}}^{(\text{total})})_n - (\underset{\sim}{\vec{T}}^{(\text{total})})_{n-1} \right\| \le \delta_2 \tag{14.68}$$

where δ_1 and δ_2 are specified small numbers, the method is assumed to have converged. Hence, stop the method by taking $\underset{\sim}{\vec{T}}|_{\text{correct}} = (\underset{\sim}{\vec{T}})_n$. On the other hand, if either of the inequalities of Eqs. (14.67) and (14.68) is not satisfied, set the new iteration number as $n = n + 1$, and go to step 2.

Example 14.7 Find the steady-state temperature distribution in the one-dimensional fin shown in Figure 14.1 by considering both convection and radiation losses from its perimeter surface. Assume $\varepsilon = 0.1$ and $\sigma = 5.7 \times 10^{-8}$ W/cm^2–K^4.

Solution For linear temperature variation inside the element, the matrix $[K_4^{(e)}]$ and the vector $\vec{P}_4^{(e)}$ can be obtained as

$$[K_4^{(e)}] = \frac{h_r P l^{(e)}}{6} \begin{bmatrix} 2 & 1 \\ 1 & 2 \end{bmatrix}$$

$$\vec{P}_4^{(e)} = \frac{h_r T_\infty P l^{(e)}}{2} \begin{Bmatrix} 1 \\ 1 \end{Bmatrix}$$

By using one-element idealization ($E = 1$), the matrices $[K_1^{(e)}]$, $[K_2^{(e)}]$, and $\vec{P}_3^{(e)}$ can be derived as

$$[K_1^{(1)}] = \frac{(\pi)(70)}{(5)} \begin{bmatrix} 1 & -1 \\ -1 & 1 \end{bmatrix} = \pi \begin{bmatrix} 14 & -14 \\ -14 & 14 \end{bmatrix}$$

$$[K_2^{(1)}] = \frac{(5)(2\pi)(5)}{6} \begin{bmatrix} 2 & 1 \\ 1 & 2 \end{bmatrix} = \pi \begin{bmatrix} 16.67 & 8.33 \\ 8.33 & 16.67 \end{bmatrix}$$

$$\vec{P}_3^{(1)} = \frac{(5)(40)(2\pi)(5)}{2} \begin{Bmatrix} 1 \\ 1 \end{Bmatrix} = \pi \begin{Bmatrix} 1000 \\ 1000 \end{Bmatrix}$$

Iteration 1:

By using $h_r^{(1)} = 0$, the matrices $[K_4^{(1)}]$ and $\vec{P}_4^{(1)}$ can be obtained as $[K_4^{(1)}] = \begin{bmatrix} 0 & 0 \\ 0 & 0 \end{bmatrix}$ and $\vec{P}_4^{(1)} = \begin{Bmatrix} 0 \\ 0 \end{Bmatrix}$. The overall equation (13.35) becomes

$$\begin{bmatrix} 30.67 & -5.67 \\ -5.67 & 30.67 \end{bmatrix} \begin{Bmatrix} T_1 \\ T_2 \end{Bmatrix} = \begin{Bmatrix} 1000 \\ 1000 \end{Bmatrix} \tag{E$_1$}$$

After incorporating the boundary condition $T_1 = 140$, Eq. (E$_1$) becomes

$$\begin{bmatrix} 1 & 0 \\ 0 & 30.67 \end{bmatrix} \begin{Bmatrix} T_1 \\ T_2 \end{Bmatrix} = \begin{Bmatrix} T_1 \\ 1000 + 5.67 T_1 \end{Bmatrix} = \begin{Bmatrix} 140.0 \\ 1793.8 \end{Bmatrix} \tag{E$_2$}$$

from which the solution can be obtained as

$$\underset{\sim}{\vec{T}} = \begin{Bmatrix} T_1 \\ T_2 \end{Bmatrix} = \begin{Bmatrix} 140.00 \\ 58.48 \end{Bmatrix} \tag{E$_3$}$$

The average temperature of the nodes of the element can be computed as

$$T_{av}^{(1)} = \frac{T_1 + T_2}{2} = 99.24\,^\circ\text{C}$$

Thus, the values of $T_{av}^{(1)}$ and T_∞ to be used in the computation of $h_r^{(1)}$ are 372.24 and 313 °K, respectively. The solution of Eq. (14.66) gives the value of

$$h_r^{(1)} = (5.7 \times 10^{-8})(0.1)(372.24^2 + 313^2)(372.24 + 313) = 0.9234$$

Iteration 2:
By using the current value of $h_r^{(1)}$, we can derive

$$[K_4^{(1)}] = \frac{(0.9234)(2\pi)(5)}{6}\begin{bmatrix}2 & 1\\ 1 & 2\end{bmatrix} = \pi\begin{bmatrix}3.078 & 1.539\\ 1.539 & 3.078\end{bmatrix}$$

$$\vec{P}_4^{(1)} = \frac{(0.9234)(40)(2\pi)(5)}{2}\begin{Bmatrix}1\\ 1\end{Bmatrix} = \pi\begin{Bmatrix}184.68\\ 184.68\end{Bmatrix}$$

Thus, the overall equation (13.35) can be written as

$$\begin{bmatrix}33.745 & -4.128\\ -4.128 & 33.745\end{bmatrix}\begin{Bmatrix}T_1\\ T_2\end{Bmatrix} = \begin{Bmatrix}1184.68\\ 1184.68\end{Bmatrix} \tag{E$_4$}$$

The application of the boundary condition ($T_1 = 140$) leads to

$$\begin{bmatrix}1 & 0\\ 0 & 33.745\end{bmatrix}\begin{Bmatrix}T_1\\ T_2\end{Bmatrix} = \begin{Bmatrix}140.00\\ 1762.60\end{Bmatrix} \tag{E$_5$}$$

Equation (E$_5$) gives the solution

$$\underset{\sim}{\vec{T}} = \begin{Bmatrix}T_1\\ T_2\end{Bmatrix} = \begin{Bmatrix}140.00\\ 52.25\end{Bmatrix} \tag{E$_6$}$$

Thus, $T_{av}^{(1)} = 96.125\,^\circ\text{C} = 369.125\,^\circ\text{K}$ and the new value of $h_r^{(1)}$ can be obtained as

$$\begin{aligned} h_r^{(1)} &= (5.7 \times 10^{-8})(0.1)(369.125^2 + 313^2)(369.125 + 313)\\ &= 0.9103 \end{aligned}$$

Iteration 3:
With the present value of $h_r^{(1)}$, we can obtain

$$[K_4^{(1)}] = \frac{(0.9103)(2\pi)(5)}{6}\begin{bmatrix}2 & 1\\ 1 & 2\end{bmatrix} = \pi\begin{bmatrix}3.034 & 1.517\\ 1.517 & 3.034\end{bmatrix}$$

and

$$\vec{P}_4^{(e)} = \frac{(0.9103)(40)(2\pi)(5)}{2}\begin{Bmatrix}1\\ 1\end{Bmatrix} = \pi\begin{Bmatrix}182.06\\ 182.06\end{Bmatrix}$$

Thus, the overall equations (13.35) can be expressed as

$$\begin{bmatrix}33.701 & -4.150\\ -4.150 & 33.701\end{bmatrix}\begin{Bmatrix}T_1\\ T_2\end{Bmatrix} = \begin{Bmatrix}1182.06\\ 1182.06\end{Bmatrix} \tag{E$_7$}$$

After incorporating the known condition, $T_1 = 140$, Eq. (E_7) gives

$$\begin{bmatrix} 1 & 0 \\ 0 & 33.701 \end{bmatrix} \begin{Bmatrix} T_1 \\ T_2 \end{Bmatrix} = \begin{Bmatrix} 140.00 \\ 1763.06 \end{Bmatrix} \qquad (E_8)$$

from which the solution can be obtained as

$$\vec{\underset{\sim}{T}} = \begin{Bmatrix} T_1 \\ T_2 \end{Bmatrix} = \begin{Bmatrix} 140.00 \\ 52.31 \end{Bmatrix} \qquad (E_9)$$

This solution gives $T_{av}^{(1)} = (T_1 + T_2)/2 = 96.155\,°\text{C} = 369.155\,°\text{K}$ and $h_r^{(1)} = (5.7 \times 10^{-8})(0.1)(369.155^2 + 313^2)(369.155 + 313) = 0.9104$.

Since the difference between this value of $h_r^{(1)}$ and the previous value is very small, we assume convergence and hence the solution of Eq. (E_9) can be taken as the correct solution of the problem.

14.8 COMPUTER PROGRAM FOR PROBLEMS WITH RADIATION

A subroutine called RADIAT is given for the solution of the one-dimensional (fin-type) heat transfer problem, including the effect of convection and radiation losses. The arguments NN, NE, NB, IEND, CC, H, TINF, QD, NØDE, P, PLØAD, XC, A, GK, EL, PERI, and TS have the same meaning as in the case of the subroutine HEAT1. The remaining arguments have the following meaning:

EPSIL = emissivity of the surface (input).
EPS = a small number of the order of 10^{-6} for testing the convergence of the method (input).
SIG = Stefan–Boltzmann constant = 5.7×10^{-8} W/m^2–K^4 (input).
HR = array of dimension NE.
HRN = array of dimension NE.
ITER = number of iterations used for obtaining the convergence of the solution (output).

To illustrate the use of the subroutine RADIAT, the problem of Example 14.7 is considered. The main program and the output of the program are given below:

```
C======================================================================
C
C       ONE-DIMENSIONAL HEAT RADIATION
C
C======================================================================
        DIMENSION NODE(1,2),P(2),PLOAD(2,1),XC(2),A(1),GK(2,2),EL(1),
       2 PERI(1),TS(2),HR(1),HRN(1)
        DATA NN,NE,NB,IEND,CC,H,TINF,QD,Q/2,1,2,0,70.0,5.0,40.0,0.0,0.0/
        DATA EPSIL,EPS,SIG/0.1,0.0001,5.7E-08/
        DATA NODE/1,2/
        DATA XC/0.0,5.0/
        DATA A/3.1416/
        DATA PERI/6.2832/
        DATA TS/140.0,0.0/
```

```
      CALL RADIAT(NN,NE,NB,IEND,CC,H,TINF,QD,Q,NODE,P,PLOAD,XC,A,GK,EL,
     2 PERI,TS,EPSIL,EPS,SIG,HR,HRN,ITER)
      PRINT 10
10    FORMAT (//,30H CONVERGED NODAL TEMPERATURES,/)
      DO 20 I=1,NN
20    PRINT 30, I,PLOAD(I,1)
30    FORMAT (2X,I6,2X,E15.8)
      STOP
      END
```

```
ITERATION      PLOAD(1,1)          PLOAD(2,1)
     1       0.14000000E+03     0.58478268E+02
     2       0.14000000E+03     0.52229244E+02
     3       0.14000000E+03     0.52310612E+02
     4       0.14000000E+03     0.52309559E+02

CONVERGED NODAL TEMPERATURES

     1    0.14000000E+03
     2    0.52309559E+02
```

REFERENCES

14.1 L.G. Tham and Y.K. Cheung: Numerical solution of heat conduction problems by parabolic time–space element, *International Journal for Numerical Methods in Engineering, 18*, 467–474, 1982.

14.2 J.R. Yu and T.R. Hsu: Analysis of heat conduction in solids by space–time finite element method, *International Journal for Numerical Methods in Engineering, 21*, 2001–2012, 1985.

PROBLEMS

14.1 A composite wall, made up of two materials, is shown in Figure 14.6. The temperature on the left side of the wall is specified as 80 °F while convection takes place on the right side of the wall. Find the temperature distribution in the wall using two linear elements.

14.2 A fin, of size 1 × 10 × 50 in., extends from a wall as shown in Figure 14.7. If the wall temperature is maintained at 500 °F and the ambient temperature is 70 °F, determine the temperature distribution in the fin using three one-dimensional elements in the x direction. Assume $k = 40$ BTU/hr–ft–°F and $h = 120$ BTU/hr–ft^2–°F.

14.3 Determine the amount of heat transferred from the fin considered in Problem 14.2.

14.4 One side of a brick wall, of width 5 m, height 4 m, and thickness 0.5 m is exposed to a temperature of −30 °C, while the other side is maintained at 30 °C. If the thermal conductivity (k) is 0.75 W/m–°C and the heat transfer coefficient on the colder side of the wall (h) is 5 W/m^2–°C, determine the following:

(a) Temperature distribution in the wall using two one-dimensional elements in the thickness.

(b) Heat loss from the wall.

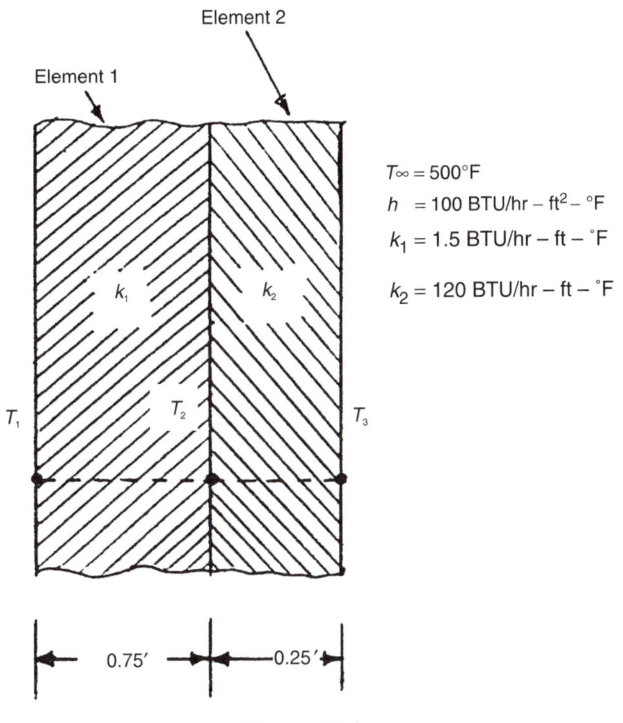

Figure 14.6.

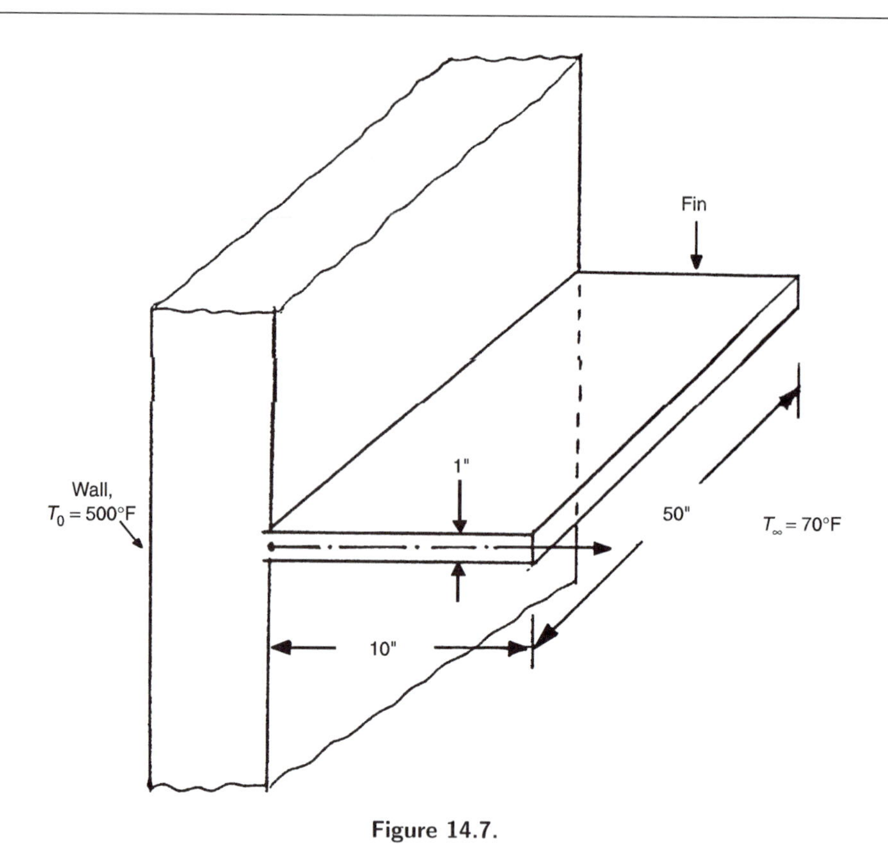

Figure 14.7.

14.5 Figure 14.8 shows a uniform aluminum fin of diameter 2 cm. The root (left end) of the fin is maintained at a temperature of $T_0 = 100\,°\text{C}$ while convection takes place from the lateral (circular) surface and the right (flat) edge of the fin. Assuming $k = 200$ W/m–°C, $h = 1000$ W/m^2–°C, and $T_\infty = 20\,°\text{C}$, determine the temperature distribution in the fin using a two-element idealization.

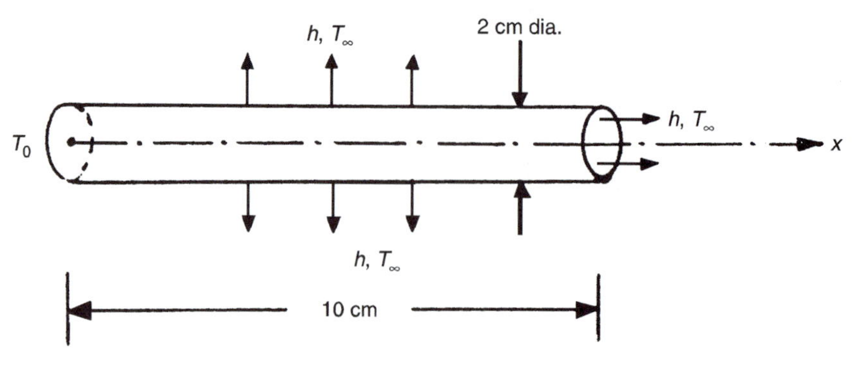

Figure 14.8.

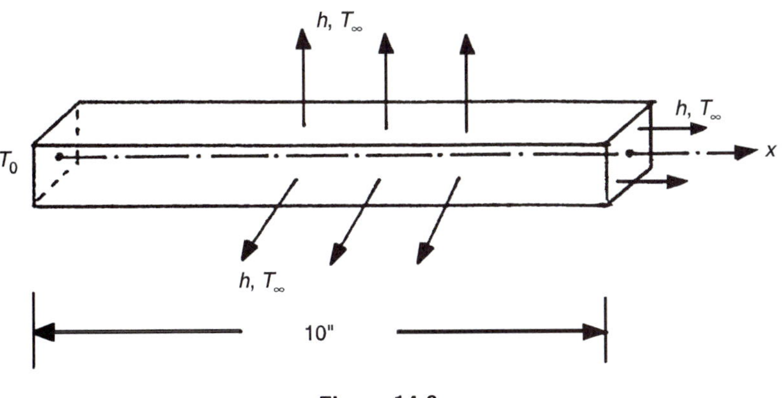

Figure 14.9.

14.6 Solve Problem 14.5 by neglecting heat convection from the right-hand edge of the fin.

14.7 Solve Problem 14.5 by assuming the fin diameter to be varying linearly from 4 cm at the root to 1 cm at the right end.

14.8 A uniform steel fin of length 10 in., with a rectangular cross section 2 × 1 in., is shown in Figure 14.9. If heat transfer takes place by convection from all the surfaces while the left side (root) of the fin is maintained at $T_0 = 500\,^\circ$F, determine the temperature distribution in the fin. Assume that $k = 9$ BTU/hr–ft–°F, $h = 2500$ BTU/hr–ft^2–°F, and $T_\infty = 50$°F. Use two finite elements.

14.9 Solve Problem 14.8 using three finite elements.

14.10 Derive the finite element equations corresponding to Eqs. (14.56)–(14.59) without assuming the radiation heat transfer coefficient (h_r) to be a constant.

14.11 A wall consists of 4-cm-thick wood, 10-cm-thick glass fiber insulation, and 1-cm-thick plaster. If the temperatures on the wood and plaster faces are 20 ° and −20 °C, respectively, determine the temperature distribution in the wall. Assume thermal conductivities of wood, glass fiber, and plaster as 0.17, 0.035, and 0.5 W/m–°C, respectively, and the heat transfer coefficient on the colder side of the wall as 25 W/m^2–°C.

14.12 The radial temperature distribution in an annular fin (Figure 14.10) is governed by the equation

$$\frac{\mathrm{d}}{\mathrm{d}r}\left[ktr\frac{\mathrm{d}T}{\mathrm{d}r}\right] - 2hr(T - T_\infty) = 0$$

with boundary conditions

$$T(r_i) = T_0 \text{ (temperature specified)}$$

$$\frac{\mathrm{d}T}{\mathrm{d}r}(r_0) = 0 \text{ (insulated)}$$

Derive the finite element equations corresponding to this problem.

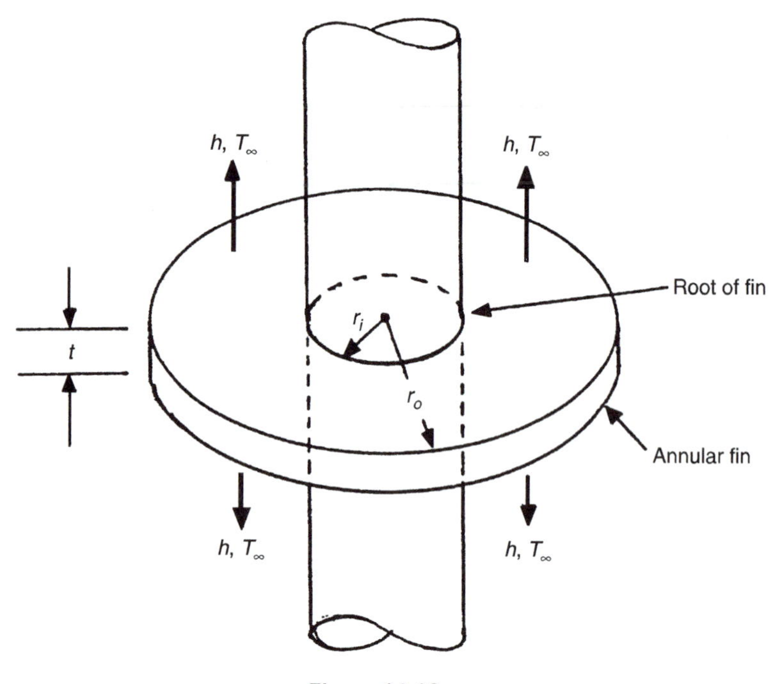

Figure 14.10.

14.13 Derive the element matrix $[K^{(e)}]$ and the vector $\vec{P}^{(e)}$ for a one-dimensional element for which the thermal conductivity k varies linearly between the two nodes.

14.14 Using the finite element method, find the tip temperature and the heat loss from the tapered fin shown in Figure 14.11. Assume that (i) the temperature is uniform

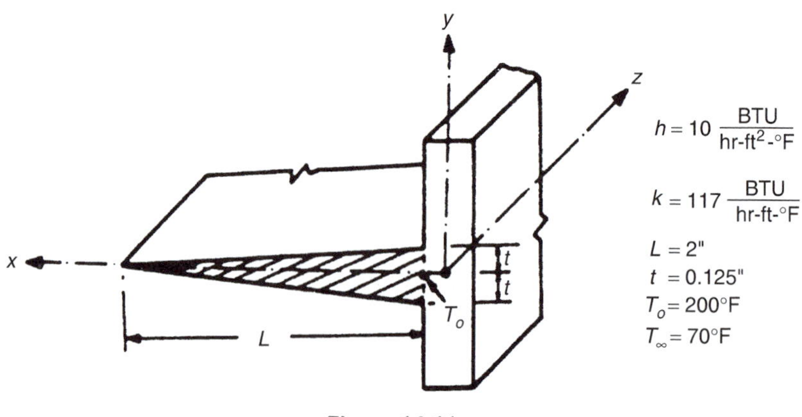

Figure 14.11.

in the y direction, (ii) the heat transfer from the fin edges (one is shown hatched) is negligible, and (iii) there is no temperature variation in the z direction.

14.15 A plane wall of thickness 15 cm has an initial temperature distribution given by $T(x, t = 0) = 500 \sin(\pi x/L)$, where $x = 0$ and $x = L$ denote the two faces of the wall. The temperature of each face is kept at zero and the wall is allowed to approach thermal equilibrium as time increases. Find the time variation of temperature distribution in the wall for $\alpha = (k/\rho c) = 10$ cm^2/hr using the finite element method.

14.16 Derive the matrix $[K_4^{(e)}]$ corresponding to radiation heat transfer for a tapered one-dimensional element.

14.17 Derive the matrix $[K_4^{(e)}]$ corresponding to radiation heat transfer for a one-dimensional element using quadratic temperature variation within the element.

14.18 Find the steady-state temperature distribution in the tapered fin shown in Figure 14.2 by considering both convection and radiation from its perimeter surface. Take $\varepsilon = 0.1$ and $\sigma = 5.7 \times 10^{-8}$ W/m^2–$^\circ K^4$.

14.19 Modify the subroutine RADIAT so that it can be used to find the steady-state temperature distribution in a one-dimensional tapered fin with convection and radiation losses.

14.20 Write a subroutine UNSTDY to find the unsteady temperature distribution in a one-dimensional fin using a linear temperature model.

15

TWO-DIMENSIONAL PROBLEMS

15.1 INTRODUCTION

For a two-dimensional steady-state problem, the governing differential equation is (Figure 15.1(a))

$$\frac{\partial}{\partial x}\left(k_x \frac{\partial T}{\partial x}\right) + \frac{\partial}{\partial y}\left(k_y \frac{\partial T}{\partial y}\right) + \dot{q} = 0 \tag{15.1}$$

and the boundary conditions are

$$T = T_o(x, y) \text{ on } S_1 \tag{15.2}$$

$$k_x \frac{\partial T}{\partial x} l_x + k_y \frac{\partial T}{\partial y} l_y + q = 0 \text{ on } S_2 \tag{15.3}$$

$$k_x \frac{\partial T}{\partial x} l_x + k_y \frac{\partial T}{\partial y} l_y + h(T - T_\infty) = 0 \text{ on } S_3 \tag{15.4}$$

where k_x and k_y are thermal conductivities in the principal (x and y) directions, $\dot{q}$ is the strength of heat source, q is the magnitude of boundary heat flux, $h(T - T_\infty)$ is the surface heat flow due to convection, and l_x and l_y are the direction cosines of the outward normal to the surface.

15.2 SOLUTION

The problem stated in Eqs. (15.1)–(15.4) is equivalent to finding $T(x, y)$, which minimizes the functional

$$I = \frac{1}{2} \iint\limits_A \left[k_x \left(\frac{\partial T}{\partial x}\right)^2 + k_y \left(\frac{\partial T}{\partial y}\right)^2 - 2\dot{q}T \right] \mathrm{d}A + \frac{1}{2}\int_{S_2} qT \, \mathrm{d}S_2$$
$$+ \int\limits_{S_3} h(T^2 - 2TT_\infty)\, \mathrm{d}S_3 \tag{15.5}$$

and satisfies Eq. (15.2). The finite element solution of this problem can be obtained as follows.

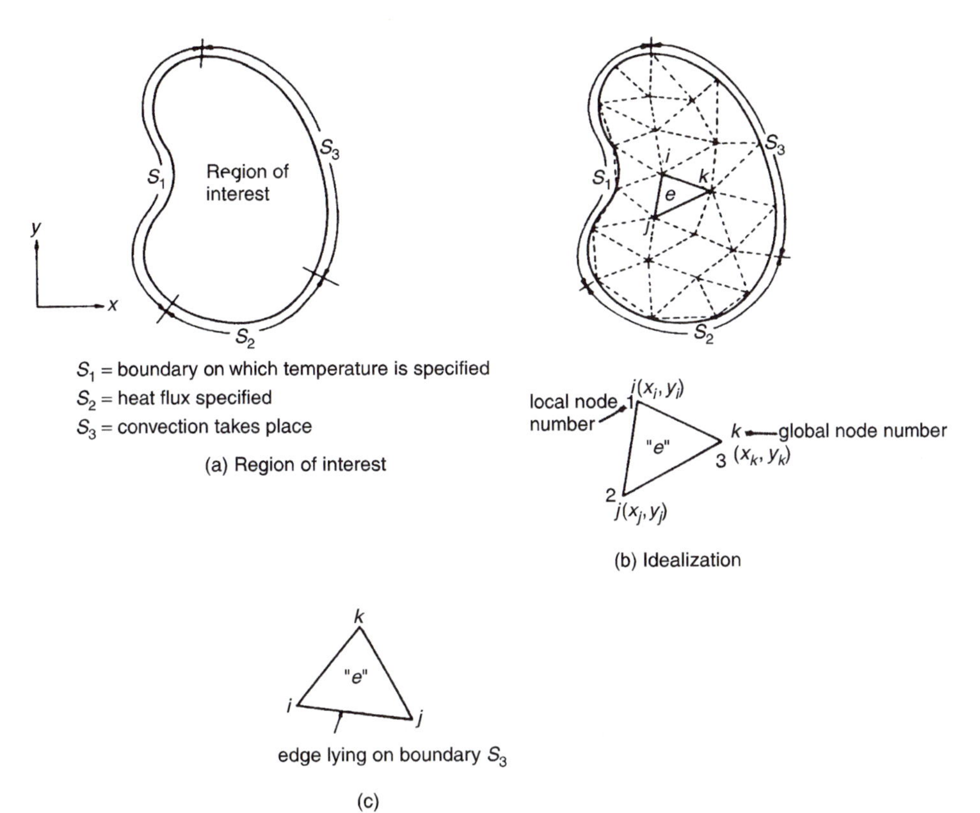

Figure 15.1. Two-Dimensional Problem.

Step 1: Idealize the solution region with triangular elements as shown in Figure 15.1(b).

Step 2: Assuming a linear variation of temperature $T^{(e)}$ inside the finite element "e,"

$$T^{(e)}(x,y) = \alpha_1 + \alpha_2 x + \alpha_3 y = [N(x,y)]\vec{q}^{(e)} \tag{15.6}$$

where

$$[N(x,y)] = \begin{Bmatrix} N_i(x,y) \\ N_j(x,y) \\ N_k(x,y) \end{Bmatrix}^T = \begin{Bmatrix} (a_i + xb_i + yc_i)/2A^{(e)} \\ (a_j + xb_j + yc_j)/2A^{(e)} \\ (a_k + xb_k + yc_k)/2A^{(e)} \end{Bmatrix}^T \tag{15.7}$$

$$\vec{q}^{(e)} = \begin{Bmatrix} q_1 \\ q_2 \\ q_3 \end{Bmatrix} \equiv \begin{Bmatrix} T_i \\ T_j \\ T_k \end{Bmatrix} \tag{15.8}$$

and $A^{(e)}$ is the area and T_i, T_j, and T_k are the nodal temperatures of element e. The expressions for a_i, b_i, c_i, and $A^{(e)}$ are given by Eqs. (3.32) and (3.31), respectively.

Step 3: Derivation of element matrices:

Once the matrix $[N(x,y)]$ is defined, Eq. (15.7), the matrix $[B]$ of Eq. (13.54) can be computed as

$$[B] = \begin{bmatrix} \frac{\partial N_i}{\partial x} & \frac{\partial N_j}{\partial x} & \frac{\partial N_k}{\partial x} \\ \frac{\partial N_i}{\partial y} & \frac{\partial N_j}{\partial y} & \frac{\partial N_k}{\partial y} \end{bmatrix} = \frac{1}{2A^{(e)}} \begin{bmatrix} b_i & b_j & b_k \\ c_i & c_j & c_k \end{bmatrix} \tag{15.9}$$

Using

$$[D] = \begin{bmatrix} k_x & 0 \\ 0 & k_y \end{bmatrix} \tag{15.10}$$

Equation (13.46) gives

$$[K_1^{(e)}] = \frac{1}{4A^{(e)2}} \iiint_{V^{(e)}} \begin{bmatrix} b_i & c_i \\ b_j & c_j \\ b_k & c_k \end{bmatrix} \begin{bmatrix} k_x & 0 \\ 0 & k_y \end{bmatrix} \begin{bmatrix} b_i & b_j & b_k \\ c_i & c_j & c_k \end{bmatrix} \cdot \mathrm{d}V \tag{15.11}$$

Assuming a unit thickness, the elemental volume can be expressed as $\mathrm{d}V = \mathrm{d}A$. Thus, Eq. (15.11) becomes

$$[K_1^{(e)}] = \frac{k_x}{4A^{(e)}} \begin{bmatrix} b_i^2 & b_i b_j & b_i b_k \\ b_i b_j & b_j^2 & b_j b_k \\ b_i b_k & b_j b_k & b_k^2 \end{bmatrix} + \frac{k_y}{4A^{(e)}} \begin{bmatrix} c_i^2 & c_i c_j & c_i c_k \\ c_i c_j & c_j^2 & c_j c_k \\ c_i c_k & c_j c_k & c_k^2 \end{bmatrix} \tag{15.12}$$

For an isotropic material with $k_x = k_y = k$, Eq. (15.12) reduces to

$$[K_1^{(e)}] = \frac{k}{4A^{(e)}} \begin{bmatrix} (b_i^2 + c_i^2) & (b_i b_j + c_i c_j) & (b_i b_k + c_i c_k) \\ & (b_j^2 + c_j^2) & (b_j b_k + c_j c_k) \\ \text{Symmetric} & & (b_k^2 + c_k^2) \end{bmatrix} \tag{15.13}$$

To determine the matrix $[K_2^{(e)}]$, integration over the surface $S_3^{(e)}$ is to be performed:

$$[K_2^{(e)}] = h \iint_{S_3^{(e)}} \begin{bmatrix} N_i^2 & N_i N_j & N_i N_k \\ N_i N_j & N_j^2 & N_j N_k \\ N_i N_k & N_j N_k & N_k^2 \end{bmatrix} \mathrm{d}S_3 \tag{15.14}$$

Thus, the surface $S_3^{(e)}$ experiencing convection phenomenon must be known for evaluating the integrals of Eq. (15.14). Let the edge ij of element e lie on the boundary S_3 as shown in Figure 15.1(c) so that $N_k = 0$ along this edge. Then Eq. (15.14) becomes

$$[K_2^{(e)}] = h \iint_{S_3^{(e)}} \begin{bmatrix} N_i^2 & N_i N_j & 0 \\ N_i N_j & N_j^2 & 0 \\ 0 & 0 & 0 \end{bmatrix} \cdot \mathrm{d}S_3 \tag{15.15}$$

Note that if the edge ik (or jk) is subjected to convection instead of the edge ij, $N_j = 0$ (or $N_i = 0$) in Eq. (15.14). To evaluate the integrals of Eq. (15.15) conveniently, we can use the triangular or area coordinates introduced in Section 3.9.2. Because the temperature is assumed to vary linearly inside the element, we have $N_i = L_1$, $N_j = L_2$, $N_k = L_3$. Along the edge ij, $N_k = L_3 = 0$ and hence Eq. (15.15) becomes

$$[K_2^{(e)}] = h \int_{s=s_i}^{s_j} \begin{bmatrix} L_1^2 & L_1 L_2 & 0 \\ L_1 L_2 & L_2^2 & 0 \\ 0 & 0 & 0 \end{bmatrix} \mathrm{d}s \tag{15.16}$$

where s denotes the direction along the edge ij, and $\mathrm{d}S_3$ was replaced by $t \cdot \mathrm{d}s = \mathrm{d}s$ since a unit thickness has been assumed for the element e. The integrals of Eq. (15.16) can be evaluated using Eq. (3.77) to find

$$[K_2^{(e)}] = \frac{h s_{ji}}{6} \begin{bmatrix} 2 & 1 & 0 \\ 1 & 2 & 0 \\ 0 & 0 & 0 \end{bmatrix} \tag{15.17}$$

The integrals involved in Eq. (13.49) can be evaluated using triangular coordinates as follows:

$$\vec{P}_1^{(e)} = \iiint_{V^{(e)}} \dot{q} [N]^T \, \mathrm{d}V = \dot{q}_o \iint_{A^{(e)}} \begin{Bmatrix} L_1 \\ L_2 \\ L_3 \end{Bmatrix} \mathrm{d}A = \frac{\dot{q}_o A^{(e)}}{3} \begin{Bmatrix} 1 \\ 1 \\ 1 \end{Bmatrix} \tag{15.18}$$

The integral in

$$\vec{P}_2^{(e)} = \iint_{S_2^{(e)}} q [N]^T \, \mathrm{d}S_2 \tag{15.19}$$

depends on the edge that lies on the heat flux boundary S_2. If the edge ij lies on S_2, $N_k = L_3 = 0$ and $\mathrm{d}S_2 = t \, \mathrm{d}s = \mathrm{d}s$ as in Eq. (15.16) and hence

$$\vec{P}_2^{(e)} = q \int_{s=s_i}^{s_j} \begin{Bmatrix} L_1 \\ L_2 \\ 0 \end{Bmatrix} \mathrm{d}s = \frac{q s_{ji}}{2} \begin{Bmatrix} 1 \\ 1 \\ 0 \end{Bmatrix} \tag{15.20}$$

Similarly, the vector $\vec{P}_3^{(e)}$ can be obtained as

$$\vec{P}_3^{(e)} = \iint_{S_3^{(e)}} hT_\infty [N]^T \, dS_3 = \frac{hT_\infty s_{ji}}{2} \begin{Bmatrix} 1 \\ 1 \\ 0 \end{Bmatrix} \text{ if the edge } ij \text{ lies on } S_3 \tag{15.21}$$

Note that if the heat flux (q) or the convection heat transfer (h) occurs from two sides of the element e, then the surface integral becomes a sum of the integral for each side.

Step 4: The assembled equations (13.35) can be expressed as

$$[\underset{\sim}{K}]\,\underset{\sim}{\vec{T}} = \underset{\sim}{\vec{P}} \tag{15.22}$$

where

$$[\underset{\sim}{K}] = \sum_{e=1}^{E} \left([K_1^{(e)}] + [K_2^{(e)}] \right) \tag{15.23}$$

and

$$\underset{\sim}{\vec{P}} = \sum_{e=1}^{E} \left(\vec{P}_1^{(e)} - \vec{P}_2^{(e)} + \vec{P}_3^{(e)} \right) \tag{15.24}$$

Step 5: The overall equations (15.22) are to be solved, after incorporating the boundary conditions, to obtain the values of nodal temperatures.

Example 15.1 Compute the element matrices and vectors for the element shown in Figure 15.2 when the edges jk and ki experience convection heat loss.

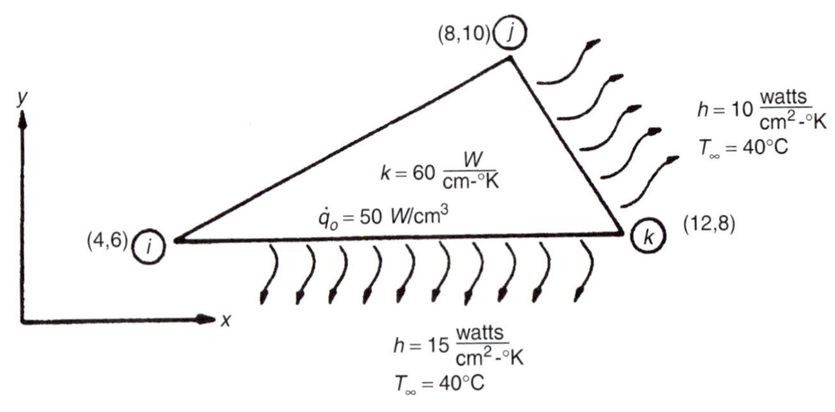

Figure 15.2.

Solution From the data given in Figure 15.2, we can compute the required quantities as

$$
\begin{aligned}
b_i &= (y_j - y_k) = (10 - 8) = 2 \\
b_j &= (y_k - y_i) = (8 - 6) = 2 \\
b_k &= (y_i - y_j) = (6 - 10) = -4 \\
c_i &= x_k - x_j = 12 - 8 = 4 \\
c_j &= x_i - x_k = 4 - 12 = -8 \\
c_k &= x_j - x_i = 8 - 4 = 4 \\
A^{(e)} &= \tfrac{1}{2}|[(-4)(2) - (-4)(-4)]| = \tfrac{1}{2}|(-8 - 16)| = 12 \\
s_{kj} &= s_k - s_j = \text{length of edge } jk = [(x_k - x_j)^2 + (y_k - y_j)^2]^{1/2} = 4.47 \\
s_{ik} &= s_i - s_k = \text{length of edge } ki = [(x_i - x_k)^2 + (y_i - y_k)^2]^{1/2} = 8.25
\end{aligned}
$$

Substitution of these values in Eqs. (15.13) and (15.17)–(15.21) gives

$$
[K_1^{(e)}] = \frac{60}{4 \times 12} \begin{bmatrix} (4+16) & (4-32) & (-8+16) \\ & (4+64) & (-8-32) \\ \text{Symmetric} & & (16+16) \end{bmatrix} = \begin{bmatrix} 25 & -35 & 10 \\ -35 & 85 & -50 \\ 10 & -50 & 40 \end{bmatrix}
$$

$$
\begin{aligned}
[K_2^{(e)}] &= \frac{h_{ik} s_{ik}}{6} \begin{bmatrix} 2 & 0 & 1 \\ 0 & 0 & 0 \\ 1 & 0 & 2 \end{bmatrix} + \frac{h_{kj} s_{kj}}{6} \begin{bmatrix} 0 & 0 & 0 \\ 0 & 2 & 1 \\ 0 & 1 & 2 \end{bmatrix} \\
&= \frac{(15)(8.25)}{6} \begin{bmatrix} 2 & 0 & 1 \\ 0 & 0 & 0 \\ 1 & 0 & 2 \end{bmatrix} + \frac{(10)(4.47)}{6} \begin{bmatrix} 0 & 0 & 0 \\ 0 & 2 & 1 \\ 0 & 1 & 2 \end{bmatrix} \\
&= \begin{bmatrix} 41.250 & 0 & 20.625 \\ 0 & 14.900 & 7.450 \\ 20.625 & 7.450 & 56.150 \end{bmatrix}
\end{aligned}
$$

$$
\vec{P}_1^{(e)} = \frac{\dot{q}_o A^{(e)}}{3} \begin{Bmatrix} 1 \\ 1 \\ 1 \end{Bmatrix} = \frac{(50)(12)}{3} \begin{Bmatrix} 1 \\ 1 \\ 1 \end{Bmatrix} = \begin{Bmatrix} 200 \\ 200 \\ 200 \end{Bmatrix}
$$

$\vec{P}_2^{(e)} = \vec{0}$ since no boundary heat flux is specified

$$
\begin{aligned}
P_3^{(e)} &= \frac{(hT_\infty)_{kj} s_{kj}}{2} \begin{Bmatrix} 0 \\ 1 \\ 1 \end{Bmatrix} + \frac{(hT_\infty)_{ik} s_{ik}}{2} \begin{Bmatrix} 1 \\ 0 \\ 1 \end{Bmatrix} \\
&= \frac{(10)(40)(4.47)}{2} \begin{Bmatrix} 0 \\ 1 \\ 1 \end{Bmatrix} + \frac{(15)(40)(8.25)}{2} \begin{Bmatrix} 1 \\ 0 \\ 1 \end{Bmatrix} = \begin{Bmatrix} 2475 \\ 894 \\ 3369 \end{Bmatrix}
\end{aligned}
$$

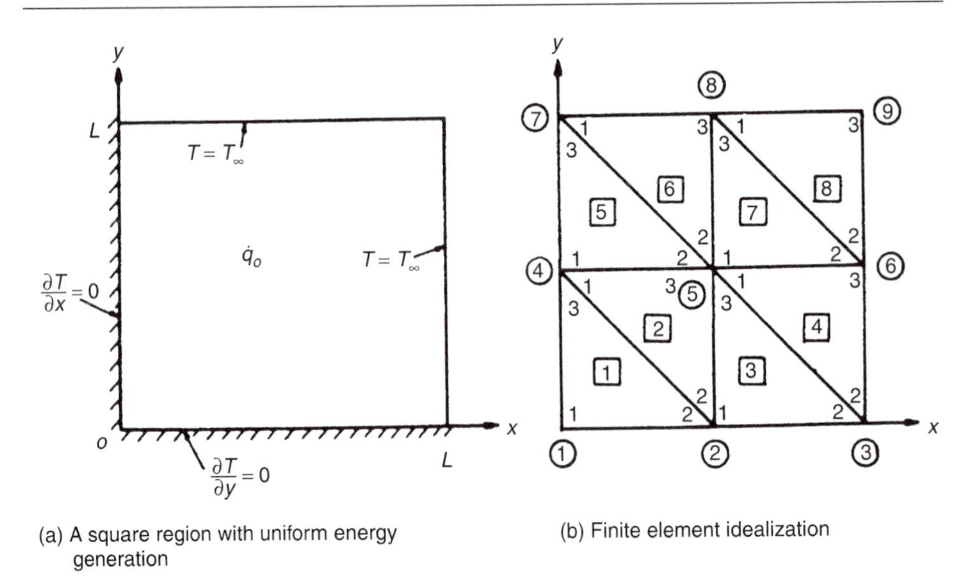

(a) A square region with uniform energy generation

(b) Finite element idealization

Figure 15.3.

Example 15.2 Find the temperature distribution in a square region with uniform energy generation as shown in Figure 15.3(a). Assume that there is no temperature variation in the z direction. Take $k = 30$ W/cm-°K, $L = 10$ cm, $T_\infty = 50\,°$C, and $\dot{q} = \dot{q}_0 = 100$ W/cm^3.

Solution

Step 1: Divide the solution region into eight triangular elements as shown in Figure 15.3(b). Label the local corner numbers of the elements counterclockwise starting from the lower left corner (only for convenience). The information needed for subsequent calculations is given below:

Node number (i)	1	2	3	4	5	6	7	8	9
Coordinates of node i (x_i, y_i)	(0,0)	$\left(\frac{L}{2}, 0\right)$	$(L, 0)$	$\left(0, \frac{L}{2}\right)$	$\left(\frac{L}{2}, \frac{L}{2}\right)$	$\left(L, \frac{L}{2}\right)$	$(0, L)$	$\left(\frac{L}{2}, L\right)$	(L, L)

Element number (e)		1	2	3	4	5	6	7	8
Global node numbers i, j, and k corresponding to local nodes 1, 2, and 3	i	1	4	2	5	4	7	5	8
	j	2	2	3	3	5	5	6	6
	k	4	5	5	6	7	8	8	9

Step 2: Computation of $[N(x, y)]$ of Eq. (15.7) for various elements:

The information needed for the computation of $[N(x, y)]$ is given below (a_i, a_j, and a_k are not computed because they are not needed in the computations):

Step 3: Derivation of element matrices and vectors:

(a) $[K_1^{(e)}]$ matrices (Eq. 15.13):

$$[K_1^{(1)}] = \frac{2k}{L^2}\begin{bmatrix} \left(\frac{L^2}{4}+\frac{L^2}{4}\right) & \left(-\frac{L^2}{4}+0\right) & \left(0-\frac{L^2}{4}\right) \\ & \left(\frac{L^2}{4}+0\right) & (0+0) \\ \text{Symmetric} & & \left(0+\frac{L^2}{4}\right) \end{bmatrix}$$

$$= \frac{k}{2}\begin{bmatrix} 2 & -1 & -1 \\ -1 & 1 & 0 \\ -1 & 0 & 1 \end{bmatrix} \begin{matrix} T_1 \\ T_2 \\ T_4 \end{matrix} \quad \text{(columns } T_1 \; T_2 \; T_4\text{)}$$

$$[K_1^{(2)}] = \frac{2k}{L^2}\begin{bmatrix} \left(\frac{L^2}{4}+0\right) & (0+0) & \left(-\frac{L^2}{4}+0\right) \\ & \left(0+\frac{L^2}{4}\right) & \left(0-\frac{L^2}{4}\right) \\ \text{Symmetric} & & \left(\frac{L^2}{4}+\frac{L^2}{4}\right) \end{bmatrix}$$

$$= \frac{k}{2}\begin{bmatrix} 1 & 0 & -1 \\ 0 & 1 & -1 \\ -1 & -1 & 2 \end{bmatrix} \begin{matrix} T_4 \\ T_2 \\ T_5 \end{matrix} \quad \text{(columns } T_4 \; T_2 \; T_5\text{)}$$

$$[K_1^{(3)}] = \frac{2k}{L^2}\begin{bmatrix} \left(\frac{L^2}{4}+\frac{L^2}{4}\right) & \left(-\frac{L^2}{4}+0\right) & \left(0-\frac{L^2}{4}\right) \\ & \left(\frac{L^2}{4}+0\right) & (0+0) \\ \text{Symmetric} & & \left(0+\frac{L^2}{4}\right) \end{bmatrix}$$

$$= \frac{k}{2}\begin{bmatrix} 2 & -1 & -1 \\ -1 & 1 & 0 \\ -1 & 0 & 1 \end{bmatrix} \begin{matrix} T_2 \\ T_3 \\ T_5 \end{matrix} \quad \text{(columns } T_2 \; T_3 \; T_5\text{)}$$

Element number e	x_i	x_j	x_k	y_i	y_j	y_k	$c_k = x_j - x_i$	$c_i = x_k - x_j$	$c_j = x_i - x_k$	$b_k = y_i - y_j$	$b_i = y_j - y_k$	$b_j = y_k - y_i$	$A^{(e)} = \frac{1}{2}\lvert(x_{ij}y_{jk} - x_{jk}y_{ij})\rvert$
1	0	$\frac{L}{2}$	0	0	0	$\frac{L}{2}$	$\frac{L}{2}$	$-\frac{L}{2}$	0	0	$-\frac{L}{2}$	$\frac{L}{2}$	$\frac{1}{2}\left\lvert\frac{L^2}{4} - 0\right\rvert = \frac{L^2}{8}$
2	0	$\frac{L}{2}$	$\frac{L}{2}$	$\frac{L}{2}$	0	$\frac{L}{2}$	$\frac{L}{2}$	0	$-\frac{L}{2}$	$\frac{L}{2}$	$-\frac{L}{2}$	0	$\frac{1}{2}\left\lvert\frac{L^2}{4} - 0\right\rvert = \frac{L^2}{8}$
3	$\frac{L}{2}$	L	$\frac{L}{2}$	0	0	$\frac{L}{2}$	$\frac{L}{2}$	$-\frac{L}{2}$	0	0	$-\frac{L}{2}$	$\frac{L}{2}$	$\frac{1}{2}\left\lvert\frac{L^2}{4} - 0\right\rvert = \frac{L^2}{8}$
4	$\frac{L}{2}$	L	L	$\frac{L}{2}$	0	$\frac{L}{2}$	$\frac{L}{2}$	0	$-\frac{L}{2}$	$\frac{L}{2}$	$-\frac{L}{2}$	0	$\frac{1}{2}\left\lvert\frac{L^2}{4} - 0\right\rvert = \frac{L^2}{8}$
5	0	$\frac{L}{2}$	0	$\frac{L}{2}$	$\frac{L}{2}$	L	$\frac{L}{2}$	$-\frac{L}{2}$	0	0	$-\frac{L}{2}$	$\frac{L}{2}$	$\frac{1}{2}\left\lvert\frac{L^2}{4} - 0\right\rvert = \frac{L^2}{8}$
6	0	$\frac{L}{2}$	$\frac{L}{2}$	L	$\frac{L}{2}$	L	$\frac{L}{2}$	0	$-\frac{L}{2}$	$\frac{L}{2}$	$-\frac{L}{2}$	0	$\frac{1}{2}\left\lvert\frac{L^2}{4} - 0\right\rvert = \frac{L^2}{8}$
7	$\frac{L}{2}$	L	$\frac{L}{2}$	$\frac{L}{2}$	$\frac{L}{2}$	L	$\frac{L}{2}$	$-\frac{L}{2}$	0	0	$-\frac{L}{2}$	$\frac{L}{2}$	$\frac{1}{2}\left\lvert\frac{L^2}{4} - 0\right\rvert = \frac{L^2}{8}$
8	$\frac{L}{2}$	L	L	L	$\frac{L}{2}$	L	$\frac{L}{2}$	0	$-\frac{L}{2}$	$\frac{L}{2}$	$-\frac{L}{2}$	0	$\frac{1}{2}\left\lvert\frac{L^2}{4} - 0\right\rvert = \frac{L^2}{8}$

$$[K_1^{(4)}] = \frac{2k}{L^2}\begin{bmatrix} \left(\frac{L^2}{4}+0\right) & (0+0) & \left(-\frac{L^2}{4}+0\right) \\ & \left(0+\frac{L^2}{4}\right) & \left(0-\frac{L^2}{4}\right) \\ \text{Symmetric} & & \left(\frac{L^2}{4}+\frac{L^2}{4}\right) \end{bmatrix}$$

$$= \frac{k}{2}\begin{matrix} T_5 & T_3 & T_6 \end{matrix} \begin{bmatrix} 1 & 0 & -1 \\ 0 & 1 & -1 \\ -1 & -1 & 2 \end{bmatrix} \begin{matrix} T_5 \\ T_3 \\ T_6 \end{matrix}$$

$$[K_1^{(5)}] = \frac{2k}{L^2}\begin{bmatrix} \left(\frac{L^2}{4}+\frac{L^2}{4}\right) & \left(-\frac{L^2}{4}+0\right) & \left(0-\frac{L^2}{4}\right) \\ & \left(\frac{L^2}{4}+0\right) & (0+0) \\ \text{Symmetric} & & \left(0+\frac{L^2}{4}\right) \end{bmatrix}$$

$$= \frac{k}{2}\begin{matrix} T_4 & T_5 & T_7 \end{matrix} \begin{bmatrix} 2 & -1 & -1 \\ -1 & 1 & 0 \\ -1 & 0 & 1 \end{bmatrix} \begin{matrix} T_4 \\ T_5 \\ T_7 \end{matrix}$$

$$[K_1^{(6)}] = \frac{2k}{L^2}\begin{bmatrix} \left(\frac{L^2}{4}+0\right) & (0+0) & \left(-\frac{L^2}{4}+0\right) \\ & \left(0+\frac{L^2}{4}\right) & \left(0-\frac{L^2}{4}\right) \\ \text{Symmetric} & & \left(\frac{L^2}{4}+\frac{L^2}{4}\right) \end{bmatrix}$$

$$= \frac{k}{2}\begin{matrix} T_7 & T_5 & T_8 \end{matrix} \begin{bmatrix} 1 & 0 & -1 \\ 0 & 1 & -1 \\ -1 & -1 & 2 \end{bmatrix} \begin{matrix} T_7 \\ T_5 \\ T_8 \end{matrix}$$

$$[K_1^{(7)}] = \frac{2k}{L^2}\begin{bmatrix} \left(\frac{L^2}{4}+\frac{L^2}{4}\right) & \left(-\frac{L^2}{4}+0\right) & \left(0-\frac{L^2}{4}\right) \\ & \left(\frac{L^2}{4}+0\right) & (0+0) \\ \text{Symmetric} & & \left(0+\frac{L^2}{4}\right) \end{bmatrix}$$

$$= \frac{k}{2}\begin{matrix} T_5 & T_6 & T_8 \end{matrix} \begin{bmatrix} 2 & -1 & -1 \\ -1 & 1 & 0 \\ -1 & 0 & 1 \end{bmatrix} \begin{matrix} T_5 \\ T_6 \\ T_8 \end{matrix}$$

$$[K_1^{(8)}] = \frac{2k}{L^2}\begin{bmatrix} \left(\frac{L^2}{4}+0\right) & (0+0) & \left(-\frac{L^2}{4}+0\right) \\ & \left(0+\frac{L^2}{4}\right) & \left(0-\frac{L^2}{4}\right) \\ \text{Symmetric} & & \left(\frac{L^2}{4}+\frac{L^2}{4}\right) \end{bmatrix}$$

$$= \frac{k}{2}\begin{bmatrix} 1 & 0 & -1 \\ 0 & 1 & -1 \\ -1 & -1 & 2 \end{bmatrix}\begin{matrix} T_8 \\ T_6 \\ T_9 \end{matrix} \quad (\text{columns } T_8 \; T_6 \; T_9)$$

(b) $[K_2^{(e)}]$ matrices (Eq. 15.17) and $\vec{P}_3^{(e)}$ vectors (Eq. 15.21):
Because no convective boundary condition is specified in the problem, we have

$$[K_2^{(e)}] = \begin{bmatrix} 0 & 0 & 0 \\ 0 & 0 & 0 \\ 0 & 0 & 0 \end{bmatrix}, \qquad \vec{P}_3^{(e)} = \begin{Bmatrix} 0 \\ 0 \\ 0 \end{Bmatrix} \quad \text{for} \quad e = 1, 2, \ldots, 8$$

(c) $\vec{P}_1^{(e)}$ vectors (Eq. 15.18):
Since $A^{(e)}$ is the same for $e = 1$–8, we obtain

$$\vec{P}_1^{(e)} = \frac{\dot{q}_0 L^2}{24}\begin{Bmatrix} 1 \\ 1 \\ 1 \end{Bmatrix}, \qquad e = 1, 2, \ldots, 8$$

(d) $\vec{P}_2^{(e)}$ vectors (Eq. 15.20):
Since no boundary heat flux is specified in the problem, we have

$$\vec{P}_2^{(e)} = \begin{Bmatrix} 0 \\ 0 \\ 0 \end{Bmatrix}, \qquad e = 1, 2, \ldots, 8$$

Step 4: The element matrices and vectors derived in Step 3 are assembled to obtain the overall system matrices and vectors as follows:

$$[\underset{\sim}{K}] = [\underset{\sim}{K_1}] = \sum_{e=1}^{8} [K_1^{(e)}]$$

	T_1	T_2	T_3	T_4	T_5	T_6	T_7	T_8	T_9	
	2	-1		-1						T_1
	-1	$1+1+2$	-1	$0+0$	$-1-1$					T_2
		-1	$1+1$		$0+0$	-1				T_3
	-1	$0+0$		$1+1+2$	$-1-1$		-1			T_4
$=\frac{k}{2}$		$-1-1$	$0+0$	$-1-1$	$2+1+1$ $+1+1+2$	$-1-1$	$0+0$	$-1-1$		T_5
			-1		$-1-1$	$2+1+1$		$0+0$	-1	T_6
				-1	$0+0$		$1+1$	-1		T_7
					$-1-1$	$0+0$	-1	$+2+1+1$	-1	T_8
						-1		-1	2	T_9

$$
= \frac{k}{2} \begin{bmatrix}
2 & -1 & 0 & -1 & & & & \mathrm{O} & \\
-1 & 4 & -1 & 0 & -2 & & & & \\
0 & -1 & 2 & 0 & 0 & -1 & & & \\
-1 & 0 & 0 & 4 & -2 & 0 & -1 & & \\
 & -2 & 0 & -2 & 8 & -2 & 0 & -2 & \\
 & & -1 & 0 & -2 & 4 & 0 & 0 & -1 \\
 & & & -1 & 0 & 0 & 2 & -1 & 0 \\
 & & & & -2 & 0 & -1 & 4 & -1 \\
 & \mathrm{O} & & & & -1 & 0 & -1 & 2
\end{bmatrix} \tag{E$_1$}
$$

$$
\vec{\underset{\sim}{P}} = \vec{\underset{\sim}{P}}_1 = \sum_{e=1}^{8} \vec{P}_1^{(e)} = \frac{\dot{q}_o L^2}{24} \left\{ \begin{matrix} 1 \\ 1+1+1 \\ 1+1 \\ 1+1+1 \\ 1+1+1+1+1+1 \\ 1+1+1 \\ 1+1 \\ 1+1+1 \\ 1 \end{matrix} \right\} \begin{matrix} T_1 \\ T_2 \\ T_3 \\ T_4 \\ T_5 \\ T_6 \\ T_7 \\ T_8 \\ T_9 \end{matrix} = \frac{\dot{q}_o L^2}{24} \left\{ \begin{matrix} 1 \\ 3 \\ 2 \\ 3 \\ 6 \\ 3 \\ 2 \\ 3 \\ 1 \end{matrix} \right\} \tag{E$_2$}
$$

Thus, the overall system equations are given by Eq. (15.22), where $[\underset{\sim}{K}]$ and $\vec{\underset{\sim}{P}}$ are given by Eqs. (E_1) and (E_2), and

$$
\vec{\underset{\sim}{T}} = \{T_1 \ T_2 \ldots \ T_9\}^T \tag{E$_3$}
$$

Step 5: The boundary conditions to be incorporated are $T_3 = T_6 = T_7 = T_8 = T_9 = T_\infty$. The following procedure can be adopted to incorporate these boundary conditions in

Eq. (15.22) without destroying the symmetry of the matrix. To incorporate the condition $T_3 = T_\infty$, for example, transfer all the off-diagonal elements of the third column (that get multiplied by T_3) to the right-hand side of the equation. These elements are then set equal to zero on the left-hand side. Then, in the third row of $[\underset{\sim}{K}]$, the off-diagonal elements are set equal to zero and the diagonal element is set equal to one. Replace the third component of the new right-hand side by T_∞ (value of T_3). Thus, after the incorporation of the boundary condition $T_3 = T_\infty$, Eq. (15.22) will appear as follows:

$$\begin{bmatrix} 2 & -1 & 0 & -1 & 0 & 0 & 0 & 0 & 0 \\ -1 & 4 & 0 & 0 & -2 & 0 & 0 & 0 & 0 \\ 0 & 0 & 1 & 0 & 0 & 0 & 0 & 0 & 0 \\ -1 & 0 & 0 & 4 & -2 & 0 & -1 & 0 & 0 \\ 0 & -2 & 0 & -2 & 8 & -2 & 0 & -2 & 0 \\ 0 & 0 & 0 & 0 & -2 & 4 & 0 & 0 & -1 \\ 0 & 0 & 0 & -1 & 0 & 0 & 2 & -1 & 0 \\ 0 & 0 & 0 & 0 & -2 & 0 & -1 & 4 & -1 \\ 0 & 0 & 0 & 0 & 0 & -1 & 0 & -1 & 2 \end{bmatrix} \begin{Bmatrix} T_1 \\ T_2 \\ T_3 \\ T_4 \\ T_5 \\ T_6 \\ T_7 \\ T_8 \\ T_9 \end{Bmatrix}$$

$$= \frac{\dot{q}_o L^2}{12k} \begin{Bmatrix} 1 \\ 3 \\ 0 \\ 3 \\ 6 \\ 3 \\ 2 \\ 3 \\ 1 \end{Bmatrix} - T_\infty \begin{Bmatrix} 0 \\ -1 \\ -1 \\ 0 \\ 0 \\ -1 \\ 0 \\ 0 \\ 0 \end{Bmatrix} \tag{E$_4$}$$

It can be observed that the third equation of (E_4) is now decoupled from the remaining equations and has the desired solution $T_3 = T_\infty$ as specified by the boundary condition. After incorporating the remaining boundary conditions, namely $T_6 = T_7 = T_8 = T_9 = T_\infty$, the final equations will appear as follows:

$$\begin{bmatrix} 2 & -1 & 0 & -1 & 0 & 0 & 0 & 0 & 0 \\ -1 & 4 & 0 & 0 & -2 & 0 & 0 & 0 & 0 \\ 0 & 0 & 1 & 0 & 0 & 0 & 0 & 0 & 0 \\ -1 & 0 & 0 & 4 & -2 & 0 & 0 & 0 & 0 \\ 0 & -2 & 0 & -2 & 8 & 0 & 0 & 0 & 0 \\ 0 & 0 & 0 & 0 & 0 & 1 & 0 & 0 & 0 \\ 0 & 0 & 0 & 0 & 0 & 0 & 1 & 0 & 0 \\ 0 & 0 & 0 & 0 & 0 & 0 & 0 & 1 & 0 \\ 0 & 0 & 0 & 0 & 0 & 0 & 0 & 0 & 1 \end{bmatrix} \begin{Bmatrix} T_1 \\ T_2 \\ T_3 \\ T_4 \\ T_5 \\ T_6 \\ T_7 \\ T_8 \\ T_9 \end{Bmatrix} = \frac{\dot{q}_o L^2}{12k} \begin{Bmatrix} 1 \\ 3 \\ 0 \\ 3 \\ 6 \\ 0 \\ 0 \\ 0 \\ 0 \end{Bmatrix} + T_\infty \begin{Bmatrix} 0 \\ 1 \\ 1 \\ 1 \\ 4 \\ 1 \\ 1 \\ 1 \\ 1 \end{Bmatrix} \tag{E$_5$}$$

The solution of Eq. (E_5) gives the following result:

$$T_1 = 133.3\,^\circ\text{C}, \quad T_2 = 119.4\,^\circ\text{C}, \quad T_3 = 50.0\,^\circ\text{C}, \quad T_4 = 119.4\,^\circ\text{C}, \quad T_5 = 105.6\,^\circ\text{C},$$
$$T_6 = 50.0\,^\circ\text{C}, \quad T_7 = 50.0\,^\circ\text{C}, \quad T_8 = 50.0\,^\circ\text{C}, \quad T_9 = 50.0\,^\circ\text{C}$$

15.3 COMPUTER PROGRAM

A subroutine called HEAT2 is given for the solution of two-dimensional heat transfer problems. The arguments of this subroutine are given below:

NN = number of nodes (input).
NE = number of triangular elements (input).
NB = semibandwidth of the overall matrix (input).
NØDE = array of size NE × 3; NØDE (I,J) = global node number corresponding to the Jth corner of element I (input).
XC, YC = array of size NN, XC(I), YC(I) = x and y coordinates of node I (input).
CC = thermal conductivity of the material, k (input).
QD = array of size NE; QD (I) = value of q for element I (input).
GK = array of size NN × NB used to store the matrix $[\underset{\sim}{K}]$.
P = array of size NN used to store the vector $\underset{\sim}{\vec{P}}$.
A = array of size NE; A(I) = area of element I .
ICØN = array of size NE; ICØN = 1 if element I lies on convection boundary and = 0 otherwise (input).
NCØN = array of size NE×2; NCØN (I,J) = Jth node of element I that lies on convection boundary (input). Need not be given if ICØN (I) = 0 for all I.
Q = array of size NE; Q(I) = magnitude of heat flux for element I (input).
TS = array of size NN; TS(I) = specified temperature for node I (input). If the temperature of node I is not specified, then the value of TS (I) is to be set equal to 0.0.
H = array of size NE; H(I) = convective heat transfer coefficient for element I (input).
TINF = array of size NE; TINF (I) = ambient temperature for element I (input).
PLØAD = array of size NN × 1 used to store the final right-hand-side vector. It represents the solution vector (nodal temperatures) upon return from the subroutine HEAT2.

To illustrate the use of the subroutine HEAT2, the problem of Example 15.2 is considered. The main program to solve this problem along with the results are given below.

```
C==================================================================
C
C     TWO-DIMENSIONAL HEAT CONDUCTION
C
C==================================================================
      DIMENSION NODE(8,3),XC(9),YC(9),QD(8),GK(9,4),P(9),A(8),ICON(8),
     2 NCON(8,2),Q(8),TS(9),H(8),TINF(8),PLOAD(9,1)
      DATA NN,NE,NB,CC/9,8,4,30.0/
      DATA NODE/1,4,2,5,4,7,5,8,2,2,3,3,5,5,6,6,4,5,5,6,7,8,8,9/
      DATA XC/0.0,5.0,10.0,0.0,5.0,10.0,0.0,5.0,10.0/
      DATA YC/0.0,0.0,0.0,5.0,5.0,5.0,10.0,10.0,10.0/
      DATA QD/100.0,100.0,100.0,100.0,100.0,100.0,100.0,100.0/
      DATA ICON/0,0,0,0,0,0,0,0/
      DATA Q/0.0,0.0,0.0,0.0,0.0,0.0,0.0,0.0/
      DATA TS/0.0,0.0,50.0,0.0,0.0,50.0,50.0,50.0,50.0/
      DATA H/0.0,0.0,0.0,0.0,0.0,0.0,0.0,0.0/
      DATA TINF/0.0,0.0,0.0,0.0,0.0,0.0,0.0,0.0/
```

```
      CALL HEAT2(NN,NE,NB,NODE,XC,YC,CC,QD,GK,P,A,ICON,NCON,Q,TS,H,
     2 TINF,PLOAD)
      PRINT 10
   10 FORMAT(19H NODAL TEMPERATURES,/)
      DO 20 I=1,NN
   20 PRINT 30,I,PLOAD(I,1)
   30 FORMAT(I4,E15.8)
      STOP
      END
```

```
NODAL TEMPERATURES

 1 0.13333331E+03
 2 0.11944444E+03
 3 0.50000000E+02
 4 0.11944444E+03
 5 0.10555555E+03
 6 0.50000000E+02
 7 0.50000000E+02
 8 0.50000000E+02
 9 0.50000000E+02
```

15.4 UNSTEADY STATE PROBLEMS

The finite element equations governing the unsteady state problem are given by Eqs. (13.35). It can be seen that the term $[\underset{\sim}{K_3}]\dot{\vec{\underset{\sim}{T}}}$ represents the unsteady state part. The element matrix $[K_3^{(e)}]$ can be evaluated using the definition given in Eq. (13.48). Since the shape function matrix used for the triangular element (in terms of natural coordinates) is

$$[N(x,y)] = [L_1 \quad L_2 \quad L_3] \tag{15.25}$$

for unit thickness of the element, we obtain from Eq. (13.48),

$$\begin{aligned} [K_3^{(e)}] &= (\rho c)^{(e)} \iint\limits_{A^{(e)}} \begin{bmatrix} L_1^2 & L_1L_2 & L_1L_3 \\ L_1L_2 & L_2^2 & L_2L_3 \\ L_1L_3 & L_2L_3 & L_3^2 \end{bmatrix} \cdot \mathrm{d}A \\ &= \frac{(\rho c)^{(e)} A^{(e)}}{12} \begin{bmatrix} 2 & 1 & 1 \\ 1 & 2 & 1 \\ 1 & 1 & 2 \end{bmatrix} \end{aligned} \tag{15.26}$$

REFERENCES

15.1 H-C. Huang: *Finite Element Analysis for Heat Transfer: Theory and Software*, Springer-Verlag, London, 1994.

15.2 K.H. Huebner and E.A. Thornton: *The Finite Element Method for Engineers*, 3rd Ed., Wiley, New York, 1995.

15.3 F.L. Stasa: *Applied Finite Element Analysis for Engineers*, Holt, Rinehart & Winston, New York, 1985.

PROBLEMS

15.1 Find the temperature distribution in the square plate shown in Figure 15.4.

15.2 If convection takes place from the triangular faces rather than the edges for the element ijk shown in Figure 15.5, evaluate the surface integrals that contribute to the matrix $[K^{(e)}]$ and the vector $\vec{P}^{(e)}$.

15.3 The temperature distribution in an isotropic plate of thickness t is given by the equation

$$\frac{\partial}{\partial x}\left[kt\frac{\partial T}{\partial x}\right]+\frac{\partial}{\partial y}\left[kt\frac{\partial T}{\partial y}\right]+\dot{q}=0 \tag{E$_1$}$$

with boundary conditions (including radiation heat transfer)

$$T=T_0(x,y) \text{ on } S_1 \tag{E$_2$}$$

$$k\frac{\partial T}{\partial x}l_x+k\frac{\partial T}{\partial y}l_y+q=0 \text{ on } S_2 \tag{E$_3$}$$

$$k\frac{\partial T}{\partial x}l_x+k\frac{\partial T}{\partial y}l_y+h(T-T_\infty)=0 \text{ on } S_3 \tag{E$_4$}$$

$$k\frac{\partial T}{\partial x}l_x+k\frac{\partial T}{\partial y}l_y+\varepsilon\sigma(T^4-T_\infty^4)=0 \text{ on } S_4 \tag{E$_5$}$$

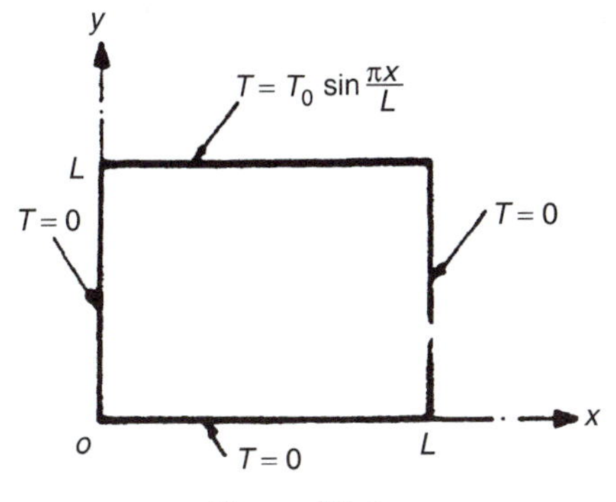

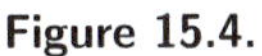

Figure 15.4.

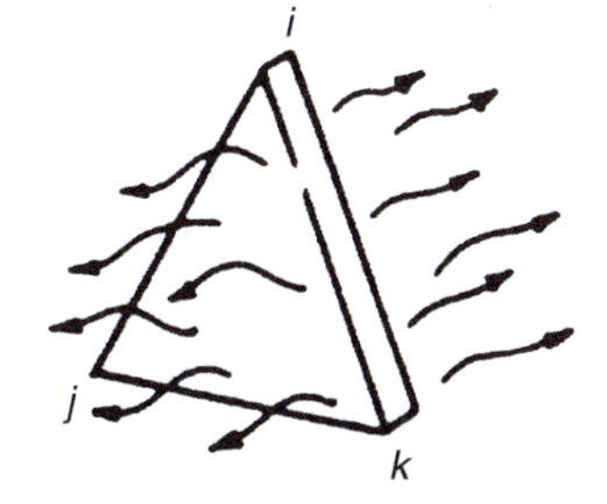

Figure 15.5.

where S_4 denotes the surface from which radiation heat-transfer takes place. Derive the variational functional I corresponding to Eqs. (E_1)–(E_5).

15.4 Derive the finite element equations corresponding to Eqs. (E_1)–(E_5) of Problem 15.3 using the Galerkin method.

15.5 Evaluate the integrals in Eq. (15.14) and derive the matrix $[K_2^{(e)}]$ assuming that convection takes place along the edge jk of element e.

15.6 Evaluate the integrals in Eq. (15.14) and derive the matrix $[K_2^{(e)}]$ assuming that convection takes place along the edge ki of element e.

15.7 If heat flux and convection heat transfer take place from the edge jk of element e, derive the corresponding vectors $\vec{P}_2^{(e)}$ and $\vec{P}_3^{(e)}$.

15.8 If heat flux and convection heat transfer take place from the edge ki of element e, derive the corresponding vectors $\vec{P}_2^{(e)}$ and $\vec{P}_3^{(e)}$.

15.9 Explain why the element matrices resulting from conduction and boundary convection, $[K_1^{(e)}]$ and $[K_2^{(e)}]$, are always symmetric.

15.10 Evaluate the conduction matrix, $[K_1^{(e)}]$, for an isotropic rectangular element with four nodes. Use linear temperature variation in x and y directions.

15.11 A three-noded triangular plate element from a finite element grid is shown in Figure 15.6. The element has a thickness of 0.2 in. and is made up of aluminum with $k = 115$ BTU/hr-ft-°F. Convection heat transfer takes place from all three edges and the two triangular faces of the element to an ambient temperature of 70 °F with a convection coefficient of 100 BTU/hr-ft^2-°F. Determine the characteristic matrices $[K_1^{(e)}]$ and $[K_2^{(e)}]$ of the element.

15.12 If an internal heat source of $\dot{q} = 1000$ BTU/hr-ft^3 is present at the centroid and a heat flux of 50 BTU/hr-ft^2 is imposed on each of the three faces of the triangular element considered in Problem 15.11, determine the characteristic vectors $\vec{P}_1^{(e)}$, $\vec{P}_2^{(e)}$, and $\vec{P}_3^{(e)}$ of the element.

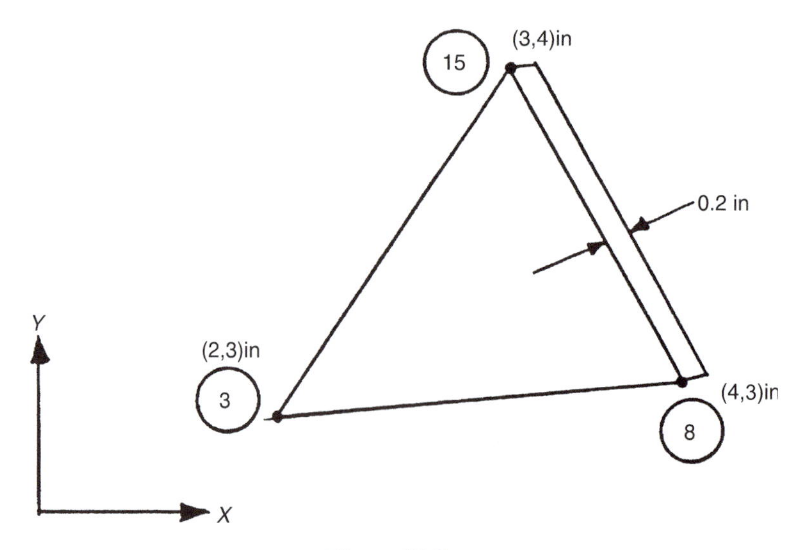

Figure 15.6.

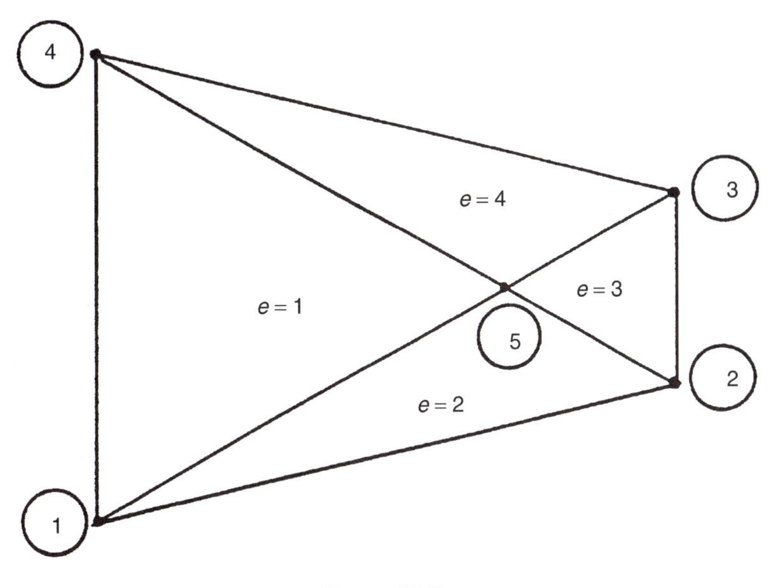

Figure 15.7.

15.13 Consider the trapezoidal plate discretized into four elements and five nodes as shown in Figure 15.7. If $[K_{ij}^{(e)}] \equiv [K_1^{(e)}]$ denotes the characteristic (conduction) matrix of element e ($e = 1, 2, 3, 4$), express the global (assembled) characteristic matrix. Can the bandwidth of the global matrix be reduced by renumbering the nodes? If so, give the details.

15.14 Consider a rectangular element of sides a and b and thickness t idealized as two triangular elements and one rectangular element as shown in Figures 15.8(a) and 15.8(b), respectively.

(a) Derive the assembled characteristic (conduction) matrix, $[K_1]$, for the rectangle.

(b) Compare the result of (a) with the characteristic (conduction) matrix of a rectangular element given by

$$[K_1]_{\text{rect}} = \frac{ktb}{4a}\begin{bmatrix} 1 & 1 & -1 & -1 \\ 1 & 1 & -1 & -1 \\ -1 & -1 & 1 & 1 \\ -1 & -1 & 1 & 1 \end{bmatrix}$$

15.15 The (X, Y) coordinates of the nodes of a triangular element of thickness 0.2 cm are shown in Figure 15.9. Convection takes place from all three edges of the element. If $\dot{q} = 200$ W/cm^3, $k = 100$ W/m-°C, $h = 150$ W/cm^2-°C, and $T_\infty = 30\,°$C, determine the following:

(a) Element matrices $[K_1^{(e)}]$ and $[K_2^{(e)}]$.

(b) Element vectors $\vec{P}_1^{(e)}$ and $\vec{P}_3^{(e)}$.

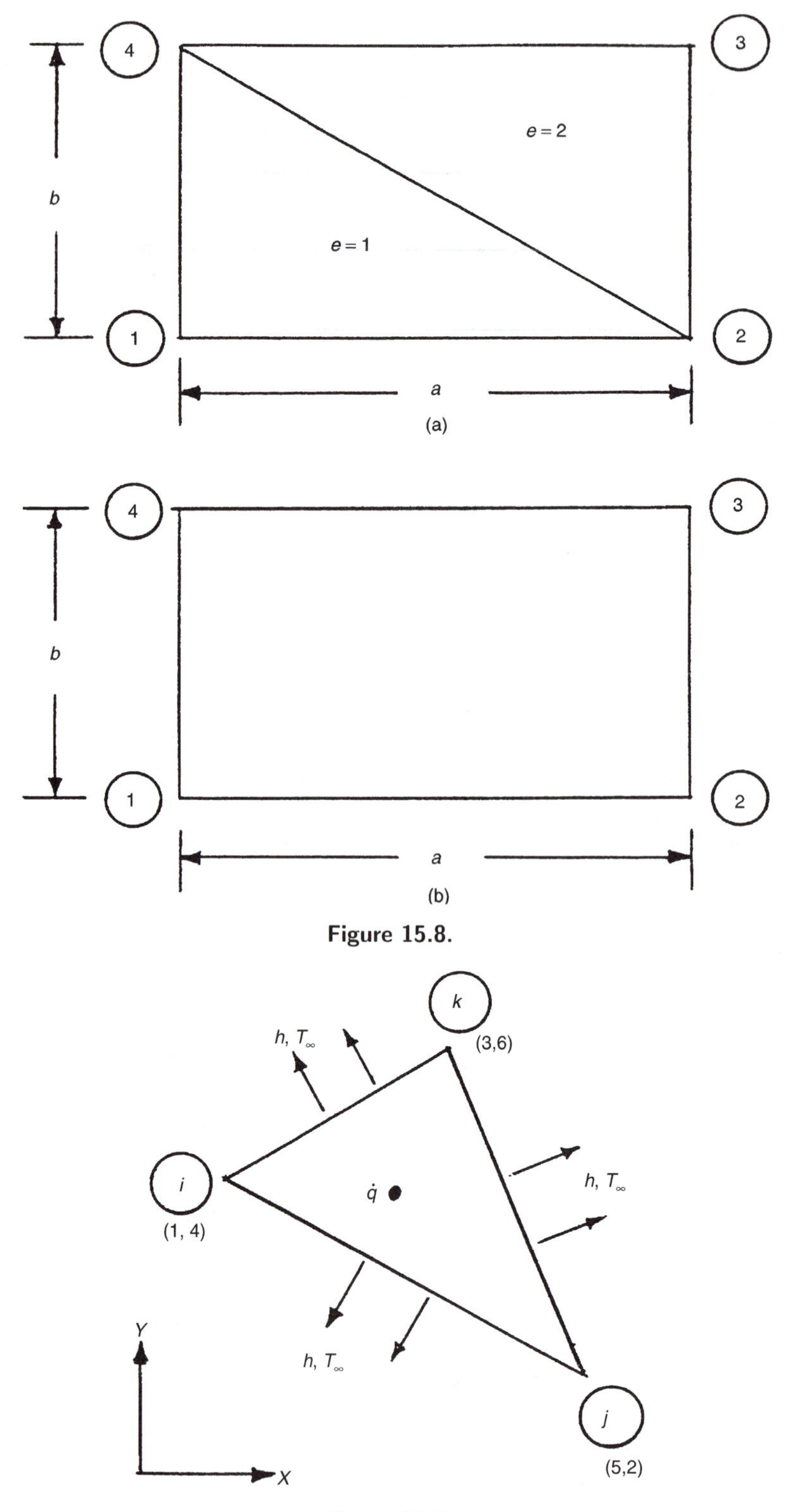

Figure 15.8.

Figure 15.9.

16

THREE-DIMENSIONAL PROBLEMS

16.1 INTRODUCTION

The equations governing heat transfer in three-dimensional bodies were given in Section 13.3. Certain types of three-dimensional problems, namely the axisymmetric problems, can be modeled and solved using ring elements. The solution of axisymmetric problems using triangular ring elements and three-dimensional problems using tetrahedron elements is considered in this chapter. For simplicity, linear interpolation functions, in terms of natural coordinates, are used in the analysis.

16.2 AXISYMMETRIC PROBLEMS

The differential equation of heat conduction for an axisymmetric case, in cylindrical coordinates, is given by [see Eq. (13.16)]

$$\frac{\partial}{\partial r}\left(rk_r\frac{\partial T}{\partial r}\right)+\frac{\partial}{\partial z}\left(rk_z\frac{\partial T}{\partial z}\right)+r\dot{q}=0 \tag{16.1}$$

The boundary conditions associated with the problem are

1. $T = T_0(r,z)$ on S_1 (16.2)

   (temperature specified on surface S_1)

2. $\dfrac{\partial T}{\partial n} =$ on S_2 (16.3)

   (insulated boundary condition on surface S_2)

3. $k_r r\dfrac{\partial T}{\partial r}l_r + k_z r\dfrac{\partial T}{\partial z}l_z + rh(T-T_\infty) = 0$ on S_3 (16.4)

   (convective boundary condition on surface S_3)

4. $k_r r\dfrac{\partial T}{\partial r}l_r + k_z r\dfrac{\partial T}{\partial z}l_z + rq = 0$ on S_4 (16.5)

   (heat flux input specified on surface S_4)

Here, k_r and k_z indicate the thermal conductivities of the solid in r and z directions, n represents the normal direction to the surface, l_r and l_z denote the direction cosines of the outward drawn normal (n), and $hr(T-T_\infty)$ is the surface heat flow due to convection.

The problem defined by Eqs. (16.1)–(16.5) can be stated in variational form as follows: Find the temperature distribution $T(r, z)$ that minimizes the functional

$$I = \frac{1}{2} \iiint\limits_{V} \left[k_r r \left(\frac{\partial T}{\partial r} \right)^2 + k_z r \left(\frac{\partial T}{\partial z} \right)^2 - 2\dot{q} r T \right] \mathrm{d}V + \frac{1}{2} \iint\limits_{S_3} h r (T^2 - 2T_\infty T) \, \mathrm{d}S_3 + \iint\limits_{S_4} r q T \, \mathrm{d}S_4 \tag{16.6}$$

and satisfies the boundary conditions specified by Eqs. (16.2) and (16.3). The finite element solution of the problem is given by the following steps.

Step 1: Replace the solid body of revolution by an assembly of triangular ring elements as shown in Figure 16.1.

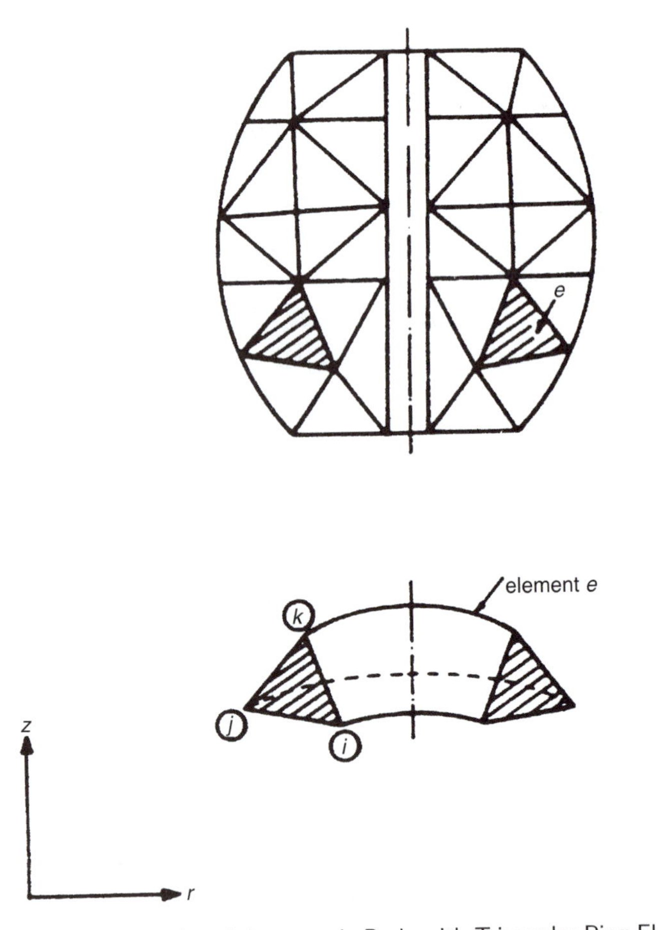

Figure 16.1. Idealization of an Axisymmetric Body with Triangular Ring Elements.

Step 2: We use a natural coordinate system and assume linear variation of temperature inside an element e so that the temperature $T^{(e)}$ can be expressed as

$$T^{(e)} = [N]\vec{q}^{(e)} \tag{16.7}$$

where

$$[N] = \begin{bmatrix} N_i & N_j & N_k \end{bmatrix} \equiv \begin{bmatrix} L_1 & L_2 & L_3 \end{bmatrix} \tag{16.8}$$

and

$$\vec{q}^{(e)} = \begin{Bmatrix} T_i \\ T_j \\ T_k \end{Bmatrix} \tag{16.9}$$

The natural coordinates L_1, L_2, and L_3 are related to the global cylindrical coordinates (r, z) of nodes i, j, and k as

$$\begin{Bmatrix} 1 \\ r \\ z \end{Bmatrix} = \begin{bmatrix} 1 & 1 & 1 \\ r_i & r_j & r_k \\ z_i & z_j & z_k \end{bmatrix} \begin{Bmatrix} L_1 \\ L_2 \\ L_3 \end{Bmatrix} \tag{16.10}$$

or, equivalently,

$$\begin{Bmatrix} L_1 \\ L_2 \\ L_3 \end{Bmatrix} = \frac{1}{2A^{(e)}} \begin{bmatrix} a_1 & b_1 & c_1 \\ a_2 & b_2 & c_2 \\ a_3 & b_3 & c_3 \end{bmatrix} \begin{Bmatrix} 1 \\ r \\ z \end{Bmatrix} \tag{16.11}$$

where

$$\begin{aligned} a_1 &= r_j z_k - r_k z_j \\ a_2 &= r_k z_i - r_i z_k \\ a_3 &= r_i z_j - r_j z_i \\ b_1 &= z_j - z_k \\ b_2 &= z_k - z_i \\ b_3 &= z_i - z_j \\ c_1 &= r_k - r_j \\ c_2 &= r_i - r_k \\ c_3 &= r_j - r_i \end{aligned} \tag{16.12}$$

and $A^{(e)}$ is the area of triangle ijk given by

$$A^{(e)} = \frac{1}{2}[r_i(z_j - z_k) + r_j(z_k - z_i) + r_k(z_i - z_j)] \tag{16.13}$$

Step 3: The element matrices and vectors can be derived using Eqs. (13.46)–(13.49) as follows:
Noting that

$$[D] = \begin{bmatrix} rk_r & 0 \\ 0 & rk_z \end{bmatrix} \tag{16.14}$$

and

$$[B] = \begin{bmatrix} \dfrac{\partial N_i}{\partial r} & \dfrac{\partial N_j}{\partial r} & \dfrac{\partial N_k}{\partial r} \\[2mm] \dfrac{\partial N_i}{\partial z} & \dfrac{\partial N_j}{\partial z} & \dfrac{\partial N_k}{\partial z} \end{bmatrix} = \frac{1}{2A^{(e)}} \begin{bmatrix} b_1 & b_2 & b_3 \\ c_1 & c_2 & c_3 \end{bmatrix} \tag{16.15}$$

and by writing $\mathrm{d}V^{(e)}$ as $2\pi r \cdot \mathrm{d}A$, where $\mathrm{d}A$ is the differential area of the triangle ijk, Eq. (13.46) gives

$$\begin{aligned}[K_1^{(e)}] &= 2\pi \iint_{A^{(e)}} r[B]^T[D][B]\,\mathrm{d}A \\ &= \frac{2\pi k_r}{4A^{(e)2}} \begin{bmatrix} b_1^2 & b_1 b_2 & b_1 b_3 \\ b_1 b_2 & b_2^2 & b_2 b_3 \\ b_1 b_3 & b_2 b_3 & b_3^2 \end{bmatrix} \iint_{A^{(e)}} r^2\,\mathrm{d}A \\ &+ \frac{2\pi k_z}{4A^{(e)2}} \begin{bmatrix} c_1^2 & c_1 c_2 & c_1 c_3 \\ c_1 c_2 & c_2^2 & c_2 c_3 \\ c_1 c_3 & c_2 c_3 & c_3^2 \end{bmatrix} \iint_{A^{(e)}} r^2\,\mathrm{d}A \end{aligned} \tag{16.16}$$

The radial distance r can be expressed in terms of the natural coordinates L_1, L_2, and L_3 as

$$r = r_i L_1 + r_j L_2 + r_k L_3 \tag{16.17}$$

Thus, the integral term in Eq. (16.16) can be expressed as

$$\bar{R}^2 \equiv \iint_{A^{(e)}} r^2 \mathrm{d}A = \iint_{A^{(e)}} (r_i\ r_j\ r_k) \begin{bmatrix} L_1^2 & L_1 L_2 & L_1 L_3 \\ L_1 L_2 & L_2^2 & L_2 L_3 \\ L_1 L_3 & L_2 L_3 & L_3^2 \end{bmatrix} \begin{Bmatrix} r_i \\ r_j \\ r_k \end{Bmatrix} \mathrm{d}A \tag{16.18}$$

By using the integration formula for natural coordinates, Eq. (3.78), Eq. (16.18) can be written as

$$\bar{R}^2 = \iint_{A^{(e)}} r^2\,\mathrm{d}A = \frac{1}{12}(r_i\ r_j\ r_k) \begin{bmatrix} 2 & 1 & 1 \\ 1 & 2 & 1 \\ 1 & 1 & 2 \end{bmatrix} \begin{Bmatrix} r_i \\ r_j \\ r_k \end{Bmatrix} \tag{16.19}$$

and hence

$$[K_1^{(e)}] = \frac{\pi k_r \bar{R}^2}{2A^{(e)}} \begin{bmatrix} b_1^2 & b_1 b_2 & b_1 b_3 \\ b_1 b_2 & b_2^2 & b_2 b_3 \\ b_1 b_3 & b_2 b_3 & b_3^2 \end{bmatrix} + \frac{\pi k_z \bar{R}^2}{2A^{(e)}} \begin{bmatrix} c_1^2 & c_1 c_2 & c_1 c_3 \\ c_1 c_2 & c_2^2 & c_2 c_3 \\ c_1 c_3 & c_2 c_3 & c_3^2 \end{bmatrix} \tag{16.20}$$

For isotropic materials with $k_r = k_z = k$, Eq. (16.20) becomes

$$[K_1^{(e)}] = \frac{\pi k \bar{R}^2}{2A^{(e)}} \begin{bmatrix} (b_1^2 + c_1^2) & (b_1 b_2 + c_1 c_2) & (b_1 b_3 + c_1 c_3) \\ (b_1 b_2 + c_1 c_2) & (b_2^2 + c_2^2) & (b_2 b_3 + c_2 c_3) \\ (b_1 b_3 + c_1 c_3) & (b_2 b_3 + c_2 c_3) & (b_3^2 + c_3^2) \end{bmatrix} \tag{16.21}$$

To evaluate the surface integral of Eq. (13.47), we assume that the edge ij lies on the surface S_3 from which heat convection takes place. Along this edge, $L_3 = 0$ and $\mathrm{d}S_3 = 2\pi r\,\mathrm{d}s$ so that Eq. (13.47) gives

$$[K_2^{(e)}] = 2\pi h \int_{s=s_i}^{s_j} \begin{Bmatrix} L_1 \\ L_2 \\ 0 \end{Bmatrix} \{L_1 \quad L_2 \quad 0\} r\,\mathrm{d}s = 2\pi h \int_{s=s_i}^{s_j} \begin{bmatrix} rL_1^2 & rL_1 L_2 & 0 \\ rL_1 L_2 & rL_2^2 & 0 \\ 0 & 0 & 0 \end{bmatrix} \mathrm{d}s \tag{16.22}$$

By substituting Eq. (16.17) for r and by using the relation

$$\int_{s=s_i}^{s_j} L_1^p L_2^q \,\mathrm{d}s = s_{ji} \frac{p!q!}{(p+q+1)!} \tag{16.23}$$

where $s_{ji} = s_j - s_i$ = length of the edge ij, Eq. (16.22) gives

$$[K_2^{(e)}] = \frac{\pi h s_{ji}}{6} \begin{bmatrix} (3r_i + r_j) & (r_i + r_j) & 0 \\ (r_i + r_j) & (r_i + 3r_j) & 0 \\ 0 & 0 & 0 \end{bmatrix} \tag{16.24}$$

To evaluate the volume integral for $\vec{P}_1^{(e)}$ as

$$\vec{P}_1^{(e)} = \iiint_{V^{(e)}} r\dot{q}[N]^T \,\mathrm{d}V \tag{16.25}$$

we use the approximation $r\dot{q} = r_c\dot{q}$ = constant, where $r_c = (r_i + r_j + r_k)/3$, and the relation $\mathrm{d}V = 2\pi r \cdot \mathrm{d}A$ to obtain

$$\vec{p}_1^{(e)} = 2\pi r_c \dot{q} \iint_{A^{(e)}} \begin{Bmatrix} rL_1 \\ rL_2 \\ rL_3 \end{Bmatrix} \mathrm{d}A \tag{16.26}$$

With the help of Eq. (16.17), Eq. (16.26) can be evaluated to obtain

$$\vec{P}_1^{(e)} = \frac{\pi r_c \dot{q} A^{(e)}}{6} \begin{Bmatrix} (2r_i + r_j + r_k) \\ (r_i + 2r_j + r_k) \\ (r_i + r_j + 2r_k) \end{Bmatrix} \tag{16.27}$$

The surface integral involved in the definition of $\vec{P}_3^{(e)}$ can be evaluated as in the case of Eq. (16.24). Thus, if the edge ij lies on the surface S_3,

$$\vec{P}_3^{(e)} = \iint\limits_{S_3^{(e)}} hT_\infty [N]^T \, dS_3 = 2\pi h T_\infty \int\limits_{s=s_i}^{s_j} \begin{Bmatrix} rL_1 \\ rL_2 \\ 0 \end{Bmatrix} ds = \frac{\pi h T_\infty s_{ji}}{3} \begin{Bmatrix} (2r_i + r_j) \\ (r_i + 2r_j) \\ 0 \end{Bmatrix} \tag{16.28}$$

Similarly, expressions for $\vec{P}_2^{(e)}$ can be obtained as

$$\vec{P}_2^{(e)} = \iint\limits_{S_2^{(e)}} q[N]^T \, dS_2 = \frac{\pi q s_{ji}}{3} \begin{Bmatrix} (2r_i + r_j) \\ (r_i + 2r_j) \\ 0 \end{Bmatrix} \quad \text{if the edge } ij \text{ lies on } S_2 \tag{16.29}$$

Step 4: Once the element matrices and vectors are available, the overall or system equations can be derived as

$$[\underset{\sim}{K}]\underset{\sim}{\vec{T}} = \underset{\sim}{\vec{P}} \tag{16.30}$$

where

$$[\underset{\sim}{K}] = \sum_{e=1}^{E} \left[[K_1^{(e)}] + [K_2^{(e)}] \right] \tag{16.31}$$

and

$$\underset{\sim}{\vec{P}} = \sum_{e=1}^{E} (\vec{P}_1^{(e)} - \vec{P}_2^{(e)} + \vec{P}_3^{(e)}) \tag{16.32}$$

Step 5: The solution of the problem can be obtained by solving Eq. (16.30) after the incorporation of the known boundary conditions.

Example 16.1 Derive the element matrices and vectors for the element shown in Figure 16.2.

Solution From the data shown in Figure 16.2, the required element properties can be computed as

$$b_1 = z_j - z_k = 2 - 6 = -4$$

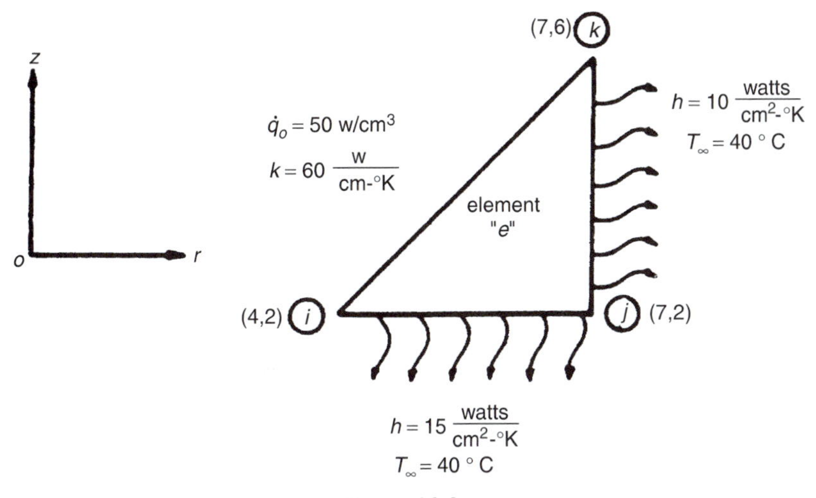

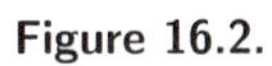

Figure 16.2.

$$b_2 = z_k - z_i = 6 - 2 = 4$$

$$b_3 = z_i - z_j = 2 - 2 = 0$$

$$c_1 = r_k - r_j = 7 - 7 = 0$$

$$c_2 = r_i - r_k = 4 - 7 = -3$$

$$c_3 = r_j - r_i = 7 - 4 = 3$$

$$A^{(e)} = \frac{1}{2}[4(2-6) + 7(6-2) + 7(2-2)] = 6$$

$$\bar{R}^2 = \frac{1}{12}\begin{pmatrix} 4 & 7 & 7 \end{pmatrix} \begin{bmatrix} 2 & 1 & 1 \\ 1 & 2 & 1 \\ 1 & 1 & 2 \end{bmatrix} \begin{Bmatrix} 4 \\ 7 \\ 7 \end{Bmatrix} = \frac{438}{12} = 36.5$$

$$r_c = (r_i + r_j + r_k)/3 = (4 + 7 + 7)/3 = 6$$

$$s_{kj} = [(r_k - r_j)^2 + (z_k - z_j)^2]^{1/2} = [(7-7)^2 + (6-2)^2]^{1/2} = 4$$

$$s_{ji} = [(r_j - r_i)^2 + (z_j - z_i)^2]^{1/2} = [(7-4)^2 + (2-2)^2]^{1/2} = 3$$

$[K_1^{(e)}]$ can be obtained as

$$[K_1^{(e)}] = \begin{bmatrix} 9175 & -9175 & 0 \\ -9175 & 14330 & -5160 \\ 0 & -5160 & 5160 \end{bmatrix} \text{ from Eq. (16.21)}$$

Since convection occurs along the two edges ij and jk, the $[K_2^{(e)}]$ matrix can be written as

$$[K_2^{(e)}] = \frac{\pi(h)_{ij}s_{ji}}{6}\begin{bmatrix}(3r_i+r_j) & (r_i+r_j) & 0\\ (r_i+r_j) & (r_i+3r_j) & 0\\ 0 & 0 & 0\end{bmatrix} + \frac{\pi(h)_{jk}s_{kj}}{6}\begin{bmatrix}0 & 0 & 0\\ 0 & (3r_j+r_k) & (r_j+r_k)\\ 0 & (r_j+r_k) & (r_j+3r_k)\end{bmatrix}$$

$$= \frac{\pi(15)(3)}{6}\begin{bmatrix}(12+7) & (4+7) & 0\\ (4+7) & (4+21) & 0\\ 0 & 0 & 0\end{bmatrix} + \frac{\pi(10)(4)}{6}\begin{bmatrix}0 & 0 & 0\\ 0 & (21+7) & (7+7)\\ 0 & (7+7) & (7+21)\end{bmatrix}$$

$$= \begin{bmatrix}447.7 & 259.2 & 0\\ 259.2 & 1176.0 & 293.2\\ 0 & 293.2 & 586.4\end{bmatrix}$$

Equation (16.27) gives

$$\vec{P}_1^{(e)} = \frac{\pi(60)(50)(6)}{6}\begin{Bmatrix}(8+7+7)\\ (4+14+7)\\ (4+7+14)\end{Bmatrix} = \begin{Bmatrix}20734.5\\ 23561.9\\ 23561.9\end{Bmatrix}$$

Because no boundary heat flux is specified, $\vec{P}_2^{(e)} = \vec{0}$. From Eq. (16.28) and a similar equation for the edge jk, we obtain

$$\vec{P}_3^{(e)} = \frac{\pi(hT_\infty)_{ij}s_{ji}}{3}\begin{Bmatrix}(2r_i+r_j)\\ (r_i+2r_j)\\ 0\end{Bmatrix} + \frac{\pi(hT_\infty)_{jk}s_{kj}}{3}\begin{Bmatrix}0\\ (2r_j+r_k)\\ (r_j+2r_k)\end{Bmatrix}$$

$$= \frac{\pi(15)(40)(3)}{3}\begin{Bmatrix}(8+7)\\ (4+14)\\ 0\end{Bmatrix} + \frac{\pi(10)(40)(4)}{3}\begin{Bmatrix}0\\ (14+7)\\ (7+14)\end{Bmatrix}$$

$$= \begin{Bmatrix}28274.3\\ 69115.0\\ 35185.8\end{Bmatrix}$$

Thus,

$$[K^{(e)}] = [K_1^{(e)}] + [K_2^{(e)}] = \begin{bmatrix}9622.7 & -8915.8 & 0\\ -8915.8 & 15506.0 & -4866.8\\ 0 & -4866.8 & 5746.4\end{bmatrix}$$

and

$$\vec{P}^{(e)} = \vec{P}_1^{(e)} - \vec{P}_2^{(e)} + \vec{P}_3^{(e)} = \begin{Bmatrix} 49008.8 \\ 92676.9 \\ 58747.7 \end{Bmatrix}$$

16.3 COMPUTER PROGRAM FOR AXISYMMETRIC PROBLEMS

A subroutine called HEATAX is given for the solution of axisymmetric heat transfer problems. The arguments NN, NE, NB, TINF, H, Q, and QD have the same meaning as in the case of the subroutine HEAT2. The remaining arguments have the following meaning:

R = array of dimension NN; R(I) = r coordinate of node I (input).
Z = array of dimension NN; Z(I) = z coordinate of node I (input).
LØC = array of dimension NE × 3; LØC(I,J) = global node number corresponding to Jth corner of element I (input).
EDGE = array of dimension NE: EDGE(I) = boundary condition specified for element I (input). EDGE(I) = 0 if no boundary condition is specified for element I; otherwise, its value lies between 1 and 6 as explained in the comments of the main program.
TS = array of dimension NN; TS(I) = specified temperature at node I (input). If the temperature of node I is not specified, then the value of TS(I) is to be set equal to −1.0 E6.
GS = array of dimension NN × NB used to store the matrix $[\underset{\sim}{K}]$.
TEMP = a dummy array of dimension NN × 1.
CK = thermal conductivity (k) of the material (input).

To illustrate the use of the subroutine HEATAX, an infinitely long hollow cylinder of inner radius 1 m and outer radius 2 m is considered. The temperatures of inner and outer surfaces are prescribed as 1000 and 0 °C, respectively. Since the temperature distribution remains constant along the length of the cylinder, an annular disc of axial thickness 0.05 m is considered for the finite element analysis. The idealization is shown in Figure 11.7, where each triangle represents an axisymmetric ring element. The total number of nodes (NN) is 42 and the number of elements (NE) is 40. Only conduction heat transfer is considered in this problem so that H(I), TINF(I), Q(I), and QD(I) are set equal to zero for all elements I. The bandwidth of the overall matrix $[\underset{\sim}{K}]$ can be seen to be 4. The value of CK (thermal conductivity) is taken as 1.0. The main program, in which the data are given/generated, and the results given by the program are shown below.

```
C==============================================================================
C
C     TEMPERATURE DISTRIBUTION IN AXISYMMETRIC SOLIDS
C
C==============================================================================
      INTEGER EDGE(40)
      DIMENSION LOC(40,3),R(42),Z(42),TS(42),TINF(40),H(40),Q(40),QD(40)
     2,GS(42,4),TEMP(42,1)
      DATA NE,NN,NB,CK/40,42,4,1.0/
```

```
      LOC(1,1)=1
      LOC(1,2)=4
      LOC(1,3)=2
      LOC(2,1)=4
      LOC(2,2)=1
      LOC(2,3)=3
      DO 10 J=1,3
      DO 10 I=3,NE
      JJ=I-2
      LOC(I,J)=LOC(JJ,J)+2
 10   CONTINUE
      R(1)=1.0
      R(2)=1.0
      DO 20 I=3,NN,2
      JJ=I-2
      JK=I+1
      R(I)=R(JJ)+0.05
      R(JK)=R(I)
 20   CONTINUE
      DO 30 I=1,NN,2
      Z(I)=0.0
      KK=I+1
      Z(KK)=0.05
 30   CONTINUE
      DO 40 I=1,NN
 40   TS(I)=-1.0E+6
      TS(1)=1000.0
      TS(2)=1000.0
      TS(41)=0.0
      TS(42)=0.0
C     EDGE(I)=1 IF BOUNDARY CONDITION IS SPECIFIED ON EDGE 1-2
C     EDGE(I)=2 IF BOUNDARY CONDITION IS SPECIFIED ON EDGE 2-3
C     EDGE(I)=3 IF BOUNDARY CONDITION IS SPECIFIED ON EDGE 3-1
C     EDGE(I)=4 IF BOUNDARY CONDITION IS SPECIFIED ON EDGES 1-2 AND 2-3
C     EDGE(I)=5 IF BOUNDARY CONDITION IS SPECIFIED ON EDGES 2-3 AND 3-1
C     EDGE(I)=6 IF BOUNDARY CONDITION IS SPECIFIED ON EDGES 3-1 AND 1-2
      DO 50 I=1,NE
 50   EDGE(I)=0
      DO 60 I=1,NE
      Q(I)=0.0
      QD(I)=0.0
      H(I)=0.0
      TINF(I)=0.0
 60   CONTINUE
      CALL HEATAX(LOC,R,Z,NN,NE,NB,CK,Q,QD,H,TINF,TS,EDGE,GS,TEMP)
      PRINT 70
 70   FORMAT(39(1H-)/2X,'NODE',2X,'RADIAL',3X,'AXIAL',3X,'TEMPERATURE'
     2,/,3X,'NO.',2X,'COORD.',3X,'COORD.',/,39(1H-)/)
      DO 80 I=1,NN
      PRINT 90,I,R(I),Z(I),TEMP(I,1)
```

```
80    CONTINUE
90    FORMAT(3X,I2,2X,F6.2,3X,F6.2,3X,F10.4)
      STOP
      END
```

```
--------------------------------------
 NODE  RADIAL   AXIAL   TEMPERATURE
  NO.  COORD.   COORD.
--------------------------------------
    1    1.00    0.00     999.9999
    2    1.00    0.05    1000.0000
    3    1.05    0.00     904.8005
    4    1.05    0.05     904.7729
    5    1.10    0.00     818.2379
    6    1.10    0.05     818.2104
    7    1.15    0.00     739.1964
    8    1.15    0.05     739.1725
    .
    .
    .
   38    1.90    0.05      52.6421
   39    1.95    0.00      25.6482
   40    1.95    0.05      25.6458
   41    2.00    0.00       0.0000
   42    2.00    0.05       0.0000
```

16.4 THREE-DIMENSIONAL HEAT TRANSFER PROBLEMS

The governing differential equation for the steady-state heat conduction in a solid body is given by Eq. (13.11) with the right-hand-side term zero and the boundary conditions by Eqs. (13.18)–(13.20). The finite element solution of these equations can be obtained by using the following procedure.

Step 1: Divide the solid body into E tetrahedron elements.

Step 2: We use a natural coordinate system and assume linear variation of temperature inside an element e so that the temperature $T^{(e)}$ can be expressed as

$$T^{(e)} = [N]\vec{q}^{(e)} \tag{16.33}$$

where

$$[N] = \begin{bmatrix} N_i & N_j & N_k & N_l \end{bmatrix} \equiv \begin{bmatrix} L_1 & L_2 & L_3 & L_4 \end{bmatrix} \tag{16.34}$$

and

$$\vec{q}^{(e)} = \begin{Bmatrix} T_i \\ T_j \\ T_k \\ T_l \end{Bmatrix} \tag{16.35}$$

The natural coordinates L_1, L_2, L_3, and L_4 are related to the global Cartesian coordinates of the nodes i, j, k, and l by Eq. (3.84).

Step 3: The element matrices and vectors can be derived using Eqs. (13.46)–(13.49) as follows:

$$[D] = \begin{bmatrix} k_x & 0 & 0 \\ 0 & k_y & 0 \\ 0 & 0 & k_z \end{bmatrix} \tag{16.36}$$

$$[B] = \begin{bmatrix} \frac{\partial N_i}{\partial x} & \frac{\partial N_j}{\partial x} & \frac{\partial N_k}{\partial x} & \frac{\partial N_l}{\partial x} \\ \frac{\partial N_i}{\partial y} & \frac{\partial N_j}{\partial y} & \frac{\partial N_k}{\partial y} & \frac{\partial N_l}{\partial y} \\ \frac{\partial N_i}{\partial z} & \frac{\partial N_j}{\partial z} & \frac{\partial N_k}{\partial z} & \frac{\partial N_l}{\partial z} \end{bmatrix} = \frac{1}{6V^{(e)}} \begin{bmatrix} b_1 & b_2 & b_3 & b_4 \\ c_1 & c_2 & c_3 & c_4 \\ d_1 & d_2 & d_3 & d_4 \end{bmatrix} \tag{16.37}$$

$$[K_1^{(e)}] = \iiint\limits_{V^{(e)}} [B]^T [D][B]\, \mathrm{d}V = \frac{k_x}{36V^{(e)}} \begin{bmatrix} b_1^2 & b_1 b_2 & b_1 b_3 & b_1 b_4 \\ b_1 b_2 & b_2^2 & b_2 b_3 & b_2 b_4 \\ b_1 b_3 & b_2 b_3 & b_3^2 & b_3 b_4 \\ b_1 b_4 & b_2 b_4 & b_3 b_4 & b_4^2 \end{bmatrix}$$

$$+ \frac{k_y}{36V^{(e)}} \begin{bmatrix} c_1^2 & c_1 c_2 & c_1 c_3 & c_1 c_4 \\ c_1 c_2 & c_2^2 & c_2 c_3 & c_2 c_4 \\ c_1 c_3 & c_2 c_3 & c_3^2 & c_3 c_4 \\ c_1 c_4 & c_2 c_4 & c_3 c_4 & c_4^2 \end{bmatrix}$$

$$+ \frac{k_z}{36V^{(e)}} \begin{bmatrix} d_1^2 & d_1 d_2 & d_1 d_3 & d_1 d_4 \\ d_1 d_2 & d_2^2 & d_2 d_3 & d_2 d_4 \\ d_1 d_3 & d_2 d_3 & d_3^2 & d_3 d_4 \\ d_1 d_4 & d_2 d_4 & d_3 d_4 & d_4^2 \end{bmatrix} \tag{16.38}$$

For an isotropic material with $k_x = k_y = k_z = k$, Eq. (16.38) becomes

$$[K_1^{(e)}] = \frac{k}{36V^{(e)}}$$

$$\times \begin{bmatrix} (b_1^2 + c_1^2 + d_1^2) & (b_1 b_2 + c_1 c_2 + d_1 d_2) & (b_1 b_3 + c_1 c_3 + d_1 d_3) & (b_1 b_4 + c_1 c_4 + d_1 d_4) \\ & (b_2^2 + c_2^2 + d_2^2) & (b_2 b_3 + c_2 c_3 + d_2 d_3) & (b_2 b_4 + c_2 c_4 + d_2 d_4) \\ & & (b_3^2 + c_3^2 + d_3^2) & (b_3 b_4 + c_3 c_4 + d_3 d_4) \\ \text{Symmetric} & & & (b_4^2 + c_4^2 + d_4^2) \end{bmatrix} \tag{16.39}$$

The matrix $[K_2^{(e)}]$ is given by

$$[K_2^{(e)}] = \iint_{S_3^{(e)}} h \begin{bmatrix} N_i^2 & N_iN_j & N_iN_k & N_iN_l \\ & N_j^2 & N_jN_k & N_jN_l \\ & & N_k^2 & N_kN_l \\ \text{Symmetric} & & & N_l^2 \end{bmatrix} dS_3 \tag{16.40}$$

If the face ijk of the element experiences convection, $N_l = 0$ along this face and hence Eq. (16.40) gives

$$[K_2^{(e)}] = \frac{hA_{ijk}}{12} \begin{bmatrix} 2 & 1 & 1 & 0 \\ 1 & 2 & 1 & 0 \\ 1 & 1 & 2 & 0 \\ 0 & 0 & 0 & 0 \end{bmatrix} \tag{16.41}$$

where A_{ijk} is the surface area of the face ijk. There are three other forms of Eq. (16.41), one for each of the other faces jkl, kli, and lij. In each case the value of the diagonal terms will be two and the values of the nonzero off-diagonal terms will be one. The coefficients in the row and the column associated with the node not lying on the surface will be zero.

$$\vec{P}_1^{(e)} = \iiint_{V^{(e)}} \dot{q} \begin{Bmatrix} N_i \\ N_j \\ N_k \\ N_l \end{Bmatrix} dV = \frac{\dot{q}_o V^{(e)}}{4} \begin{Bmatrix} 1 \\ 1 \\ 1 \\ 1 \end{Bmatrix} \tag{16.42}$$

If the face ijk lies on the surface S_2 on which heat flux is specified,

$$\vec{P}_2^{(e)} = \iint_{S_2^{(e)}} q \begin{Bmatrix} N_i \\ N_j \\ N_k \\ N_l \end{Bmatrix} dS_2 = q \iint_{S_2^{(e)}} \begin{Bmatrix} L_1 \\ L_2 \\ L_3 \\ 0 \end{Bmatrix} dS_2 = \frac{qA_{ijk}}{3} \begin{Bmatrix} 1 \\ 1 \\ 1 \\ 0 \end{Bmatrix} \tag{16.43}$$

and similarly, if convection loss occurs from the face ijk,

$$\vec{P}_3^{(e)} = \iint_{S_3^{(e)}} hT_\infty \begin{Bmatrix} N_i \\ N_j \\ N_k \\ N_l \end{Bmatrix} dS_3 = hT_\infty \iint_{S_3^{(e)}} \begin{Bmatrix} L_1 \\ L_2 \\ L_3 \\ 0 \end{Bmatrix} dS_3 = \frac{hT_\infty A_{ijk}}{3} \begin{Bmatrix} 1 \\ 1 \\ 1 \\ 0 \end{Bmatrix} \tag{16.44}$$

There are three other forms of Eqs. (16.43) and (16.44). In these equations, the zero coefficient will be located in the row corresponding to the node not lying on the face.

Example 16.2 Derive the element equations for the element shown in Figure 16.3.

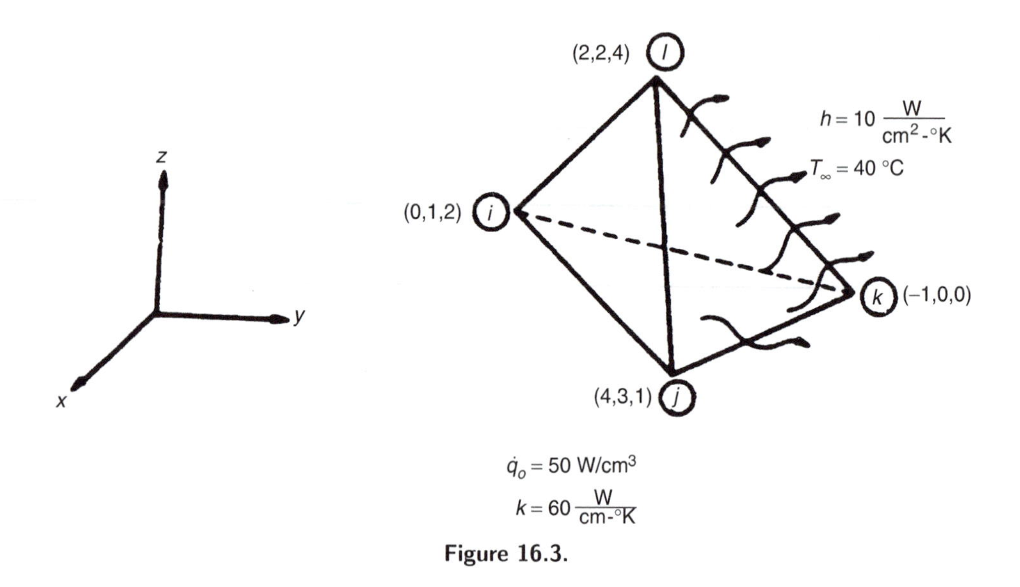

Figure 16.3.

Solution From the given data, the required element properties can be computed as follows:

$$V^{(e)} = \frac{1}{6}\begin{vmatrix} 1 & 0 & 1 & 2 \\ 1 & 4 & 3 & 1 \\ 1 & -1 & 0 & 0 \\ 1 & 2 & 2 & 4 \end{vmatrix} = \frac{5}{6},$$

$$b_1 = -\begin{vmatrix} 1 & 3 & 1 \\ 1 & 0 & 0 \\ 1 & 2 & 4 \end{vmatrix} = 10, \qquad c_1 = -\begin{vmatrix} 4 & 1 & 1 \\ -1 & 1 & 0 \\ 2 & 1 & 4 \end{vmatrix} = -17,$$

$$d_1 = -\begin{vmatrix} 4 & 3 & 1 \\ -1 & 0 & 1 \\ 2 & 2 & 1 \end{vmatrix} = 1,$$

$$b_2 = -\begin{vmatrix} 1 & 0 & 0 \\ 1 & 2 & 4 \\ 1 & 1 & 2 \end{vmatrix} = 0, \qquad c_2 = -\begin{vmatrix} -1 & 1 & 0 \\ 2 & 1 & 4 \\ 0 & 1 & 2 \end{vmatrix} = 2,$$

$$d_2 = -\begin{vmatrix} -1 & 0 & 1 \\ 2 & 2 & 1 \\ 0 & 1 & 1 \end{vmatrix} = -1,$$

$$b_3 = -\begin{vmatrix} 1 & 2 & 4 \\ 1 & 1 & 2 \\ 1 & 3 & 1 \end{vmatrix} = -5, \qquad c_3 = -\begin{vmatrix} 2 & 1 & 4 \\ 0 & 1 & 2 \\ 4 & 1 & 1 \end{vmatrix} = 10,$$

$$d_3 = -\begin{vmatrix} 2 & 2 & 1 \\ 0 & 1 & 1 \\ 4 & 3 & 1 \end{vmatrix} = 0,$$

$$b_4 = -\begin{vmatrix} 1 & 1 & 2 \\ 1 & 3 & 1 \\ 1 & 0 & 0 \end{vmatrix} = 5, \qquad c_4 = -\begin{vmatrix} 0 & 1 & 2 \\ 4 & 1 & 1 \\ -1 & 1 & 0 \end{vmatrix} = -11,$$

$$d_4 = -\begin{vmatrix} 0 & 1 & 1 \\ 4 & 3 & 1 \\ -1 & 0 & 1 \end{vmatrix} = 2$$

To compute the area A_{jkl}, we use the formula

$$A_{jkl} = [s(s-\alpha)(s-\beta)(s-\gamma)]^{1/2}$$

where α, β, γ are the lengths of the sides of the triangle:

$$\begin{aligned}
\alpha = \ & \text{length } jk = [(x_k - x_j)^2 + (y_k - y_j)^2 + (z_k - z_j)^2]^{1/2} \\
& = (25 + 9 + 1)^{(1/2)} = 5.916 \\
\beta = \ & \text{length } kl = [(x_l - x_k)^2 + (y_l - y_k)^2 + (z_l - z_k)^2]^{1/2} \\
& = (9 + 4 + 16)^{(1/2)} = 5.385 \\
\gamma = \ & \text{length } lj = [(x_j - x_l)^2 + (y_j - y_l)^2 + (z_j - z_l)^2]^{1/2} \\
& = (4 + 1 + 9)^{(1/2)} = 3.742 \\
s & = \frac{1}{2}(\alpha + \beta + \gamma) = \frac{1}{2}(5.916 + 5.385 + 3.742) = 7.522 \\
A_{jkl} & = [7.522(7.522 - 5.916)(7.522 - 5.385)(7.522 - 3.742)]^{(1/2)} \\
& = 9.877
\end{aligned}$$

Equation (16.39) gives

$$[K_1^{(e)}] = \frac{60 \times 6}{36 \times 5}\begin{bmatrix} (100 + 289 + 1) & (0 - 34 - 1) & (-50 - 170 + 0) & (50 + 187 + 2) \\ & (0 + 4 + 1) & (0 + 20 + 0) & (0 - 22 - 2) \\ & & (25 + 100 + 0) & (-25 - 110 + 0) \\ \text{Symmetric} & & & (25 + 121 + 4) \end{bmatrix}$$

$$= \begin{bmatrix} 780 & -70 & -440 & 478 \\ & 10 & 40 & -48 \\ & & 250 & -270 \\ \text{Symmetric} & & & 300 \end{bmatrix}$$

The matrix $[K_2^{(e)}]$ will be a modification of Eq. (16.41):

$$[K_2^{(e)}] = \frac{h \cdot A_{jkl}}{12} = \begin{bmatrix} 0 & 0 & 0 & 0 \\ 0 & 2 & 1 & 1 \\ 0 & 1 & 2 & 1 \\ 0 & 1 & 1 & 2 \end{bmatrix} = \frac{(10)(9.877)}{12} \begin{bmatrix} 0 & 0 & 0 & 0 \\ 0 & 2 & 1 & 1 \\ 0 & 1 & 2 & 1 \\ 0 & 1 & 1 & 2 \end{bmatrix}$$

$$= \begin{bmatrix} 0 & 0 & 0 & 0 \\ 0 & 16.462 & 8.231 & 8.231 \\ 0 & 8.231 & 16.462 & 8.231 \\ 0 & 8.231 & 8.231 & 16.462 \end{bmatrix}$$

$$\vec{P}_1^{(e)} = \frac{50 \times 5}{6 \times 4} \begin{Bmatrix} 1 \\ 1 \\ 1 \\ 1 \end{Bmatrix} = \begin{Bmatrix} 10.42 \\ 10.42 \\ 10.42 \\ 10.42 \end{Bmatrix}$$

$$\vec{P}_2^{(e)} = \begin{Bmatrix} 0 \\ 0 \\ 0 \\ 0 \end{Bmatrix} \text{ since no boundary heat flux is specified}$$

$$\vec{P}_3^{(e)} = \frac{10 \times 40 \times 9.877}{3} \begin{Bmatrix} 0 \\ 1 \\ 1 \\ 1 \end{Bmatrix} = \begin{Bmatrix} 0 \\ 1316.92 \\ 1316.92 \\ 1316.92 \end{Bmatrix}$$

$$[K^{(e)}] = [K_1^{(e)}] + [K_2^{(e)}] = \begin{bmatrix} 780.000 & -70.000 & -440.000 & 478.000 \\ & 26.462 & 48.231 & -39.769 \\ & & 266.462 & -261.769 \\ \text{Symmetric} & & & 316.462 \end{bmatrix} \quad (E_1)$$

$$\vec{P}^{(e)} = \vec{P}_1^{(e)} - \vec{P}_2^{(e)} + \vec{P}_3^{(e)} = \begin{Bmatrix} 10.42 \\ 1327.34 \\ 1327.34 \\ 1327.34 \end{Bmatrix} \quad (E_2)$$

The element equations are

$$[K^{(e)}]\vec{q}^{(e)} = \vec{P}^{(e)}$$

where $[K^{(e)}]$ and $\vec{P}^{(e)}$ are given by Eqs. (E_1) and (E_2), respectively, and

$$\vec{q}^{(e)} = \begin{Bmatrix} T_i \\ T_j \\ T_k \\ T_l \end{Bmatrix}$$

16.5 UNSTEADY STATE PROBLEMS

The finite element equations governing the unsteady state problems are given by Eqs. (13.35). It can be seen that the term $[\underset{\sim}{K}_3]\,\dot{\underset{\sim}{T}}$ represents the unsteady state part. The element matrix $[K_3^{(e)}]$ can be evaluated using the definition given in Eq. (13.48).

16.5.1 Axisymmetric Problems

For a triangular ring element, the matrix $[N(r,z)]$, in terms of natural coordinates, is given by

$$[N] = \begin{bmatrix} L_1 & L_2 & L_3 \end{bmatrix} \tag{16.45}$$

By expressing $\mathrm{d}V = 2\pi r\,\mathrm{d}A$, Eq. (13.48) can be written as

$$[K_3^{(e)}] = \iiint\limits_{V^{(e)}} \rho c [N]^T [N]\,\mathrm{d}V$$

$$= (\rho c)^{(e)} 2\pi \iint\limits_{A^{(e)}} \begin{bmatrix} L_1^2 & L_1L_2 & L_1L_3 \\ L_1L_2 & L_2^2 & L_2L_3 \\ L_1L_3 & L_2L_3 & L_3^2 \end{bmatrix} (r_iL_1 + r_jL_2 + r_kL_3)\,\mathrm{d}A \tag{16.46}$$

where Eq. (16.17) has been substituted for r. By carrying out the area integrations indicated in Eq. (16.46), we obtain

$$[K_3^{(e)}] = \frac{\pi(\rho c)^{(e)} A^{(e)}}{30} \begin{bmatrix} (6r_i + 2r_j + 2r_k) & (2r_i + 2r_j + r_k) & (2r_i + r_j + 2r_k) \\ & (2r_i + 6r_j + 2r_k) & (r_i + 2r_j + 2r_k) \\ \text{Symmetric} & & (2r_i + 2r_j + 6r_k) \end{bmatrix} \tag{16.47}$$

16.5.2 Three-Dimensional Problems

The shape function matrix for the tetrahedron element is given by

$$[N(x,y,z)] = \begin{bmatrix} L_1 & L_2 & L_3 & L_4 \end{bmatrix} \tag{16.48}$$

With this, the $[K_3^{(e)}]$ matrix can be derived as

$$[K_3^{(e)}] = (\rho c)^{(e)} \iiint\limits_{V^{(e)}} \begin{bmatrix} L_1^2 & L_1L_2 & L_1L_3 & L_1L_4 \\ L_1L_2 & L_2^2 & L_2L_3 & L_2L_4 \\ L_1L_3 & L_2L_3 & L_3^2 & L_3L_4 \\ L_1L_4 & L_2L_4 & L_3L_4 & L_4^2 \end{bmatrix} \mathrm{d}V$$

$$= \frac{(\rho c)^{(e)} \cdot V^{(e)}}{20} \begin{bmatrix} 2 & 1 & 1 & 1 \\ 1 & 2 & 1 & 1 \\ 1 & 1 & 2 & 1 \\ 1 & 1 & 1 & 2 \end{bmatrix} \tag{16.49}$$

REFERENCES

16.1 H-C. Huang: *Finite Element Analysis for Heat Transfer: Theory and Software*, Springer-Verlag, London, 1994.
16.2 G. Comini: *Finite Element Analysis in Heat Transfer: Basic Formulation and Linear Problems*, Taylor & Francis, Washington, DC, 1994.
16.3 J.N. Reddy: *The Finite Element Method in Heat Transfer and Fluid Dynamics*, CRC Press, Boca Raton, FL, 1994.

PROBLEMS

16.1 If radiation takes place on surface S_5 of an axisymmetric problem, state the boundary condition and indicate a method of deriving the corresponding finite element equations.

16.2 Derive the finite element equations corresponding to Eqs. (16.1)–(16.5) using the Galerkin approach.

16.3 If convection heat transfer takes place from the face corresponding to edge jk of a triangular ring element, derive the matrix $[K_2^{(e)}]$ and the vector $\vec{P}_3^{(e)}$.

16.4 If convection heat transfer takes place from the face corresponding to edge ki of a triangular ring element, derive the matrix $[K_2^{(e)}]$ and the vector $\vec{P}_3^{(e)}$.

16.5 Evaluate the conduction matrix, $[K_1^{(e)}]$, for an isotropic, axisymmetric ring element of rectangular cross section with four nodes. Use linear temperature variation in r and z directions.

16.6 A three-noded axisymmetric aluminum triangular ring element from a finite element grid is shown in Figure 16.4. Convection heat transfer takes place from all the faces (edges) of the triangle with a convection coefficient of 100 BTU/hr–ft^2–°F. If k = 115 BTU/hr–ft–°F, determine the characteristic matrices $[K_1^{(e)}]$ and $[K_2^{(e)}]$ of the element.

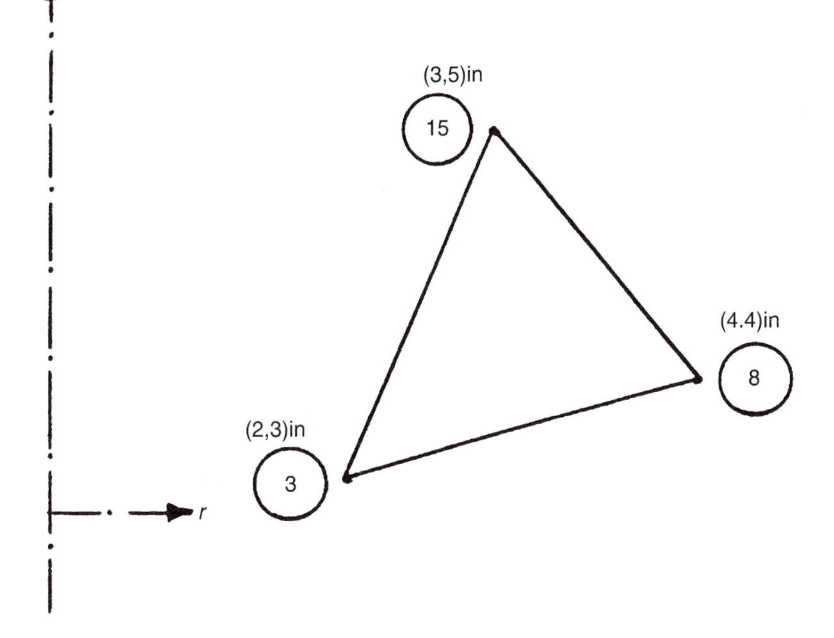

Figure 16.4.

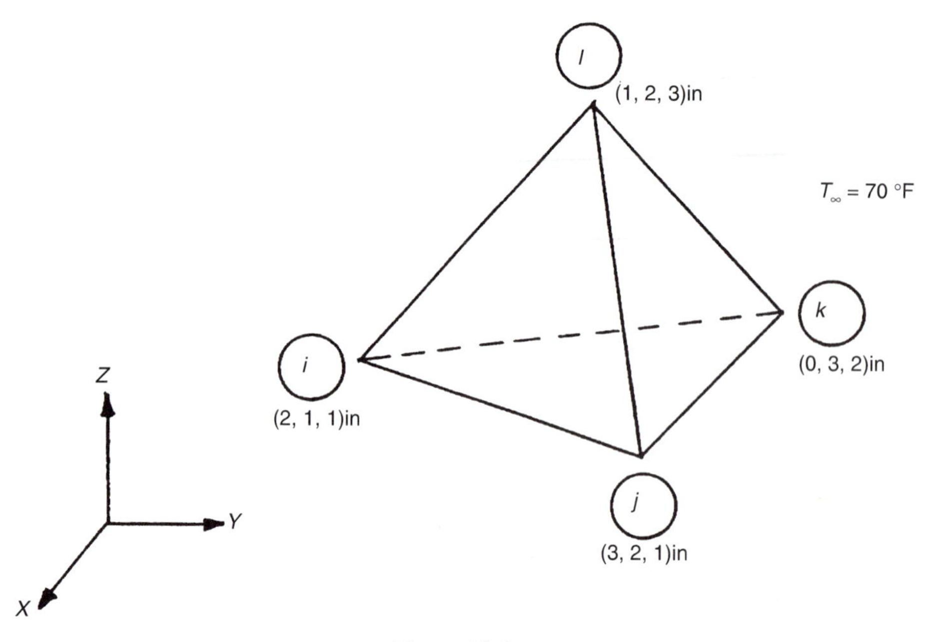

Figure 16.5.

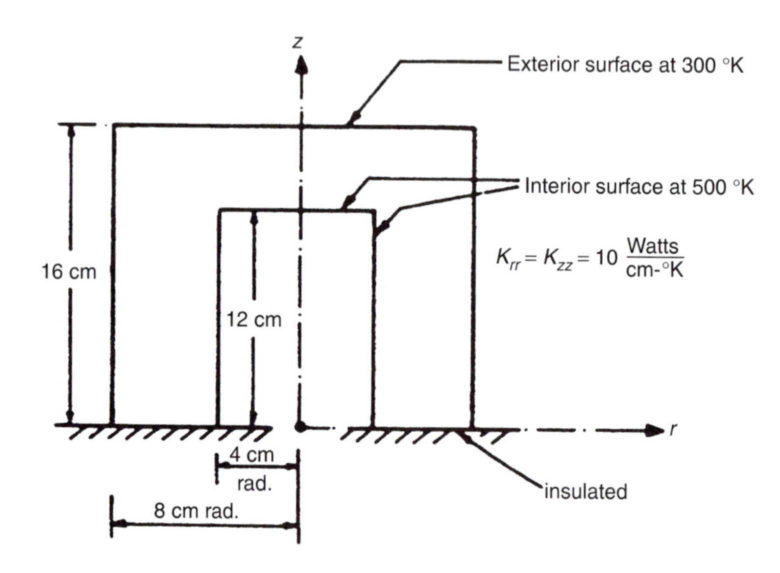

Figure 16.6.

16.7 If convection takes place from the face ijl of a tetrahedron element in a solid body, derive the matrix $[K_2^{(e)}]$ and the vector $\vec{P}_3^{(e)}$.

16.8 If convection takes place from the face jkl of a tetrahedron element in a solid body, derive the matrix $[K_2^{(e)}]$ and the vector $\vec{P}_3^{(e)}$.

16.9 If convection takes place from the face ikl of a tetrahedron element in a solid body, derive the matrix $[K_2^{(e)}]$ and the vector $\vec{P}_3^{(e)}$.

16.10 Derive the element equations for the tetrahedron element of a three-dimensional body shown in Figure 16.5. Assume $k = 100$ BTU/hr–ft–°F, $h = 150$ BTU/hr–ft^2–°F from face ijk, and $\dot{q}_0 = 500$ BTU/hr–ft^3.

16.11 Evaluate the matrix $[K_3^{(e)}]$ for the triangular ring element shown in Figure 16.2. Assume that $\rho c = 20$ Joules/cm-°C.

16.12 Use the subroutine HEATAX to find the temperature distribution in the axisymmetric problem shown in Figure 16.6.

APPLICATION TO FLUID MECHANICS PROBLEMS

17

BASIC EQUATIONS OF FLUID MECHANICS

17.1 INTRODUCTION

Although the finite element method was extensively developed for structural and solid mechanics problems, it was not considered a powerful tool for the solution of fluid mechanics problems until recently. One of the reasons is the success achieved with the more traditional finite difference procedures in solving fluid flow problems. In recent years, significant contributions have been made in the solution of different types of fluid flow problems using the finite element method. This chapter presents a summary of the basic concepts and equations of fluid mechanics.

17.2 BASIC CHARACTERISTICS OF FLUIDS

A fluid is a substance (gas or liquid) that will deform continuously under the action of applied surface (shearing) stresses. The magnitude of the stress depends on the *rate of angular deformation*. On the other hand, a solid can be defined as a substance that will deform by an amount proportional to the stress applied after which static equilibrium will result. Here, the magnitude of the shear stress depends on the *magnitude of angular deformation*.

Different fluids show different relations between stress and the rate of deformation. Depending on the nature of relation followed between stress and rate of deformation, fluids can be classified as Newtonian and non-Newtonian fluids. A Newtonian fluid is one in which the shear stress is directly proportional to the rate of deformation starting with zero stress and zero deformation. The constant of proportionality is defined as μ, the absolute or dynamic viscosity. Common examples of Newtonian fluids are air and water. A non-Newtonian fluid is one that has a variable proportionality between stress and rate of deformation. Common examples of non-Newtonian fluids are some plastics, colloidal suspensions, and emulsions. Fluids can also be classified as compressible and incompressible. Usually, liquids are treated as incompressible, whereas gases and vapors are assumed to be compressible.

A flow field is described in terms of the velocities and accelerations of fluid particles at different times and at different points throughout the fluid-filled space. For the graphical representation of fluid motion, it is convenient to introduce the concepts of streamlines and path lines. A *streamline* is an imaginary line that connects a series of points in space at a given instant in such a manner that all particles falling on the line at that instant have velocities whose vectors are tangent to the line. Thus, the streamlines represent the direction of motion at each point along the line at the given instant. A *path line* is the locus of points through which a fluid particle of fixed identity passes as it moves in space. For a steady flow the streamlines and path lines are identical, whereas they are, in general, different for an unsteady flow.

A flow may be termed as inviscid or viscous depending on the importance of consideration of viscosity of the fluid in the analysis. An *inviscid flow* is a frictionless flow characterized by zero viscosity. A *viscous flow* is one in which the fluid is assumed to have nonzero viscosity. Although no real fluid is inviscid, there are several flow situations in which the effect of viscosity of the fluid can be neglected. For example, in the analysis of a flow over a body surface, the viscosity effects are considered in a thin region close to the flow boundary (known as boundary layer), whereas the viscosity effect is neglected in the rest of the flow.

Depending on the dynamic macroscopic behavior of the fluid flow, we have laminar, transition, and turbulent motion. A *laminar flow* is an orderly state of flow in which macroscopic fluid particles move in layers. A *turbulent flow* is one in which the fluid particles have irregular, fluctuating motions and erratic paths. In this case, macroscopic mixing occurs both lateral to and in the direction of the main flow. A *transition flow* occurs whenever a laminar flow becomes unstable and approaches a turbulent flow.

17.3 METHODS OF DESCRIBING THE MOTION OF A FLUID

The motion of a group of particles in a fluid can be described by either the Lagrangian method or the Eulerian method. In the Lagrangian method, the coordinates of the moving particles are represented as functions of time. This means that at some arbitrary time t_0, the coordinates of a particle (x_0, y_0, z_0) are identified and that thereafter we follow that particle through the fluid flow. Thus, the position of the particle at any other instant is given by a set of equations of the form

$$x = f_1(x_0, y_0, z_0, t), \qquad y = f_2(x_0, y_0, z_0, t), \qquad z = f_3(x_0, y_0, z_0, t)$$

The Lagrangian approach is not generally used in fluid mechanics because it leads to more cumbersome equations. In the Eulerian method, we observe the flow characteristics in the vicinity of a fixed point as the particles pass by. Thus, in this approach the velocities at various points are expressed as functions of time as

$$u = f_1(x, y, z, t), \qquad v = f_2(x, y, z, t), \qquad w = f_3(x, y, z, t)$$

where u, v, and w are the components of velocity in x, y, and z directions, respectively.

The velocity change in the vicinity of a point in the x direction is given by

$$\mathrm{d}u = \frac{\partial u}{\partial t}\mathrm{d}t + \frac{\partial u}{\partial x}\mathrm{d}x + \frac{\partial u}{\partial y}\mathrm{d}y + \frac{\partial u}{\partial z}\mathrm{d}z \tag{17.1}$$

(total derivative expressed in terms of partial derivatives).

The small distances moved by a particle in time $\mathrm{d}t$ can be expressed as

$$\mathrm{d}x = u\ \mathrm{d}t, \qquad \mathrm{d}y = v\ \mathrm{d}t, \qquad \mathrm{d}z = w\ \mathrm{d}t \tag{17.2}$$

Thus, dividing Eq. (17.1) by $\mathrm{d}t$ and using Eq. (17.2) leads to the *total* or *substantial derivative* of the velocity u (x component of acceleration) as

$$a_x = \frac{\mathrm{d}u}{\mathrm{d}t} \equiv \frac{\mathrm{D}u}{\mathrm{D}t} = \frac{\partial u}{\partial t} + u\frac{\partial u}{\partial x} + v\frac{\partial u}{\partial y} + w\frac{\partial u}{\partial z} \tag{17.3a}$$

The other components of acceleration can be expressed in a similar manner as

$$a_y = \frac{\mathrm{d}v}{\mathrm{d}t} \equiv \frac{\mathrm{D}v}{\mathrm{D}t} = \frac{\partial v}{\partial t} + u\frac{\partial v}{\partial x} + v\frac{\partial v}{\partial y} + w\frac{\partial v}{\partial z} \tag{17.3b}$$

$$a_z = \frac{\mathrm{d}w}{\mathrm{d}t} \equiv \frac{\mathrm{D}w}{\mathrm{D}t} = \frac{\partial w}{\partial t} + u\frac{\partial w}{\partial x} + v\frac{\partial w}{\partial y} + w\frac{\partial w}{\partial z} \tag{17.3c}$$

17.4 CONTINUITY EQUATION

To derive the continuity equation, consider a differential control volume of size $\mathrm{d}x\ \mathrm{d}y\ \mathrm{d}z$ as shown in Figure 17.1. Assuming that the density and the velocity are functions of space and time, we obtain the flux of mass per second for the three directions x, y, and z, respectively, as $-\frac{\partial}{\partial x}(\rho u)\cdot \mathrm{d}y\ \mathrm{d}z$, $-\frac{\partial}{\partial y}(\rho v)\cdot \mathrm{d}x\ \mathrm{d}z$, and $-\frac{\partial}{\partial z}(\rho w)\cdot \mathrm{d}x\ \mathrm{d}y$. From the principle of conservation of matter, the sum of these must be equal to the time rate of change

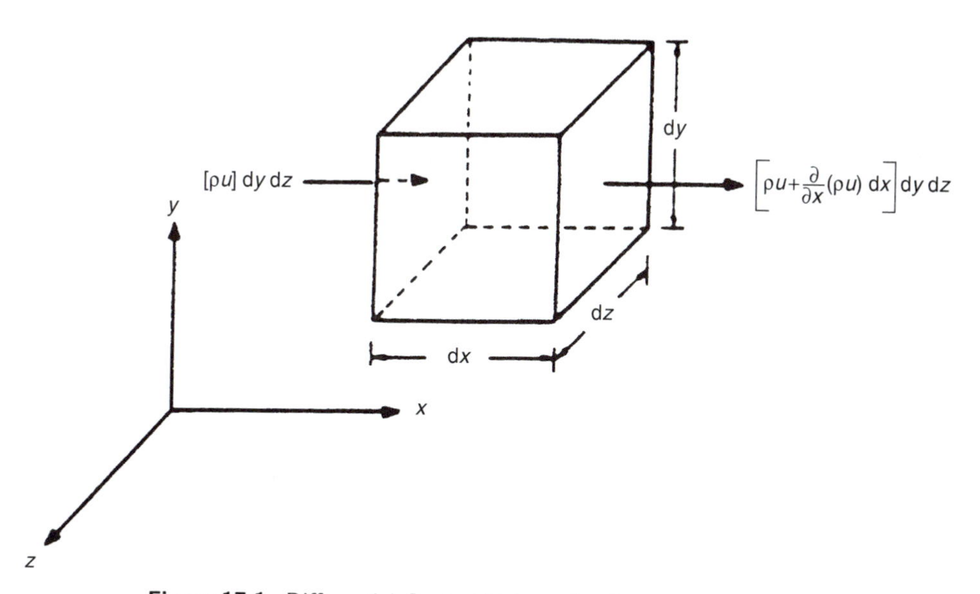

Figure 17.1. Differential Control Volume for Conservation of Mass.

of mass, $\frac{\partial}{\partial t}(\rho \mathrm{d}x \; \mathrm{d}y \; \mathrm{d}z)$. Since the control volume is independent of time, we can cancel $\mathrm{d}x \; \mathrm{d}y \; \mathrm{d}z$ from all the terms and obtain

$$\frac{\partial \rho}{\partial t} + \frac{\partial}{\partial x}(\rho u) + \frac{\partial}{\partial y}(\rho v) + \frac{\partial}{\partial z}(\rho w) = 0 \tag{17.4a}$$

where ρ is the mass density; u, v, and w are the x, y, and z components of velocity, respectively; and t is the time. By using the vector notation

$$\vec{V} = u\vec{i} + v\vec{j} + w\vec{k} = \text{velocity vector}$$

and

$$\vec{\nabla} = \frac{\partial}{\partial x}\vec{i} + \frac{\partial}{\partial y}\vec{j} + \frac{\partial}{\partial z}\vec{k} = \text{gradient vector}$$

where, $\vec{i}$, $\vec{j}$, and $\vec{k}$ represent the unit vectors in x, y, and z directions, respectively. Equation (17.4a) can be expressed as

$$\frac{\partial \rho}{\partial t} + \vec{\nabla} \cdot \rho \vec{V} = 0 \tag{17.4b}$$

This equation can also be written as

$$\frac{\partial u}{\partial x} + \frac{\partial v}{\partial y} + \frac{\partial w}{\partial z} = -\frac{1}{\rho}\frac{\mathrm{D}(\rho)}{\mathrm{D}t} \tag{17.4c}$$

where $(\mathrm{D}/\mathrm{D}t)$ represents the total or substantial derivative with respect to time. Equation (17.4) represents the general three-dimensional continuity equation for a fluid in unsteady flow. If the fluid is incompressible, the time rate of volume expansion of a fluid element will be zero and hence the continuity equation, for both steady and unsteady flows, becomes

$$\vec{\nabla} \cdot \vec{V} = \frac{\partial u}{\partial x} + \frac{\partial v}{\partial y} + \frac{\partial w}{\partial z} = 0 \tag{17.5}$$

17.5 EQUATIONS OF MOTION OR MOMENTUM EQUATIONS

17.5.1 State of Stress in a Fluid

The state of stress in a fluid is characterized, as in the case of a solid, by the six stress components σ_{xx}, σ_{yy}, σ_{zz}, σ_{xy}, σ_{yz}, and σ_{zx}. However, if the fluid is at rest, the shear stress components σ_{xy}, σ_{yz}, and σ_{zx} will be zero and all the normal stress components will be the same and equal to the negative of the hydrostatic pressure, p—that is, $\sigma_{xx} = \sigma_{yy} = \sigma_{zz} = -p$.

17.5.2 Relation between Stress and Rate of Strain for Newtonian Fluids

Stress–Strain Relations for Solids

In solids the stresses σ_{ij} are related to the strains ε_{ij} according to Hooke's law, Eq. (8.7):

$$\left.\begin{aligned} \varepsilon_{xx} &= \frac{1}{E}\left[\sigma_{xx} - v(\sigma_{yy} + \sigma_{zz})\right], \ldots \\ \varepsilon_{xy} &= \frac{\sigma_{xy}}{G}, \ldots \end{aligned}\right\} \tag{17.6}$$

If an element of solid having sides dx, dy, and dz is deformed into an element having sides $(1+\varepsilon_{xx})dx, (1+\varepsilon_{yy})dy$, and $(1+\varepsilon_{zz})dz$, the volume dilation of the element (e) is defined as

$$\begin{aligned} e &= \frac{\text{change in volume of the element}}{\text{original volume of the element}} \\ &= \frac{(1+\varepsilon_{xx})(1+\varepsilon_{yy})(1+\varepsilon_{zz})dx\ dy\ dz - dx\ dy\ dz}{dx\ dy\ dz} \\ &= \varepsilon_{xx} + \varepsilon_{yy} + \varepsilon_{zz} \end{aligned} \tag{17.7}$$

Using Eq. (17.6), Eq. (17.7) can be expressed as

$$e = \frac{1-2v}{E}(\sigma_{xx} + \sigma_{yy} + \sigma_{zz}) = \frac{1-2v}{E}3\bar{\sigma} \tag{17.8}$$

where $\bar{\sigma}$ is the arithmetic mean of the three normal stresses defined as

$$\bar{\sigma} = (\sigma_{xx} + \sigma_{yy} + \sigma_{zz})/3 \tag{17.9}$$

The Young's modulus and Poisson's ratio are related as

$$G = \frac{E}{2(1+v)} \quad \text{or} \quad 2G = \frac{E}{1+v} \tag{17.10}$$

The first equation of Eq. (17.6) can be rewritten as

$$\varepsilon_{xx} = \frac{1}{E}\left[\sigma_{xx} - v(\sigma_{xx} + \sigma_{yy} + \sigma_{zz}) + v\sigma_{xx}\right] = \frac{E}{1+v}\varepsilon_{xx} + \frac{3v}{1+v}\bar{\sigma} \tag{17.11}$$

Substituting for $3\bar{\sigma}$ from Eq. (17.9) into Eq. (17.11) and using the relation (17.10), we obtain

$$\sigma_{xx} = \frac{E}{1+v}\varepsilon_{xx} + \frac{v}{1+v}\frac{E \cdot e}{1-2v} = 2G\,\varepsilon_{xx} + \frac{2Gv}{1-2v}\cdot e \tag{17.12}$$

By subtracting $\bar{\sigma}$ from both sides of Eq. (17.12) we obtain

$$\sigma_{xx} - \bar{\sigma} = 2G\,\varepsilon_{xx} + \frac{2Gv}{1-2v}e - \bar{\sigma} \tag{17.13}$$

Using Eq. (17.8), Eq. (17.13) can be expressed as

$$\sigma_{xx} - \bar{\sigma} = 2G\,\varepsilon_{xx} + \frac{2Gv}{1-2v}e - \frac{E}{3(1-2v)}e$$

that is,

$$\sigma_{xx} - \bar{\sigma} = \left(2G\,\varepsilon_{xx} - \frac{e}{3}\right) \tag{17.14}$$

In a similar manner, the following relations can be derived:

$$\sigma_{yy} - \bar{\sigma} = 2G\left(\varepsilon_{yy} - \frac{e}{3}\right) \tag{17.15}$$

$$\sigma_{zz} - \bar{\sigma} = 2G\left(\varepsilon_{zz} - \frac{e}{3}\right) \tag{17.16}$$

The shear stress–shear strain relations can be written, from Eq. (17.6), as

$$\sigma_{xy} = G\,\varepsilon_{xy} \tag{17.17}$$

$$\sigma_{yz} = G\,\varepsilon_{yz} \tag{17.18}$$

$$\sigma_{zx} = G\,\varepsilon_{zx} \tag{17.19}$$

Stress–Rate of Strain Relations for Newtonian Fluids

Experimental results indicate that the stresses in a fluid are related to the time rate of strain instead of strain itself. Thus, the stress–rate of strain relations can be derived by analogy from Eqs. (17.14)–(17.19). As an example, consider Eq. (17.14). By replacing the shear modulus (G) by a quantity expressing its dimensions, we obtain

$$\sigma_{xx} - \bar{\sigma} = 2\left(\frac{F}{L^2}\right)\left(\varepsilon_{xx} - \frac{e}{3}\right) \tag{17.20}$$

where F is the force, and L is the length. Since the stresses are related to the time rates of strain for a fluid, Eq. (17.20) can be used to obtain the following relation valid for a Newtonian fluid:

$$\sigma_{xx} - \bar{\sigma} = 2\left(\frac{FT}{L^2}\right)\frac{\partial}{\partial t}\left(\varepsilon_{xx} - \frac{e}{3}\right) \tag{17.21}$$

In Eq. (17.21), the dimension of time (T) is added to the proportionality constant in order to preserve the dimensions. The proportionality constant in Eq. (17.21) is taken as the dynamic viscosity μ having dimensions FT/L^2. Thus, Eq. (17.21) can be expressed as

$$\sigma_{xx} - \bar{\sigma} = 2\mu\frac{\partial \varepsilon_{xx}}{\partial t} - \frac{2}{3}\frac{\partial e}{\partial t} \tag{17.22}$$

$$\sigma_{yy} - \bar{\sigma} = 2\mu\frac{\partial \varepsilon_{yy}}{\partial t} - \frac{2}{3}\frac{\partial e}{\partial t} \tag{17.23}$$

$$\sigma_{zz} - \bar{\sigma} = 2\mu\frac{\partial \varepsilon_{zz}}{\partial t} - \frac{2}{3}\frac{\partial e}{\partial t} \tag{17.24}$$

$$\sigma_{xy} = \mu\varepsilon_{xy} \tag{17.25}$$

$$\sigma_{yz} = \mu\varepsilon_{yz} \tag{17.26}$$

$$\sigma_{zx} = \mu\varepsilon_{zx} \tag{17.27}$$

If the coordinates of a point before deformation are given by x, y, z and after deformation by $x+\xi$, $y+\eta$, $z+\zeta$, the strains are given by

$$\left.\begin{aligned} &\varepsilon_{xx} = \frac{\partial \xi}{\partial x}, \quad \varepsilon_{yy} = \frac{\partial \eta}{\partial y}, \quad \varepsilon_{zz} = \frac{\partial \zeta}{\partial z}, \\ &\varepsilon_{xy} = \frac{\partial \xi}{\partial y} + \frac{\partial \eta}{\partial x}, \quad \varepsilon_{yz} = \frac{\partial \eta}{\partial z} + \frac{\partial \zeta}{\partial y}, \quad \varepsilon_{zx} = \frac{\partial \zeta}{\partial x} + \frac{\partial \xi}{\partial z} \end{aligned}\right\} \tag{17.28}$$

The rate of strain involved in Eq. (17.22) can be expressed as

$$\frac{\partial \varepsilon_{xx}}{\partial t} = \frac{\partial}{\partial t}\left(\frac{\partial \xi}{\partial x}\right) = \frac{\partial}{\partial x}\left(\frac{\partial \xi}{\partial t}\right) = \frac{\partial u}{\partial x} \tag{17.29}$$

where u is the component of velocity in x direction, and

$$\frac{\partial e}{\partial t} = \frac{\partial}{\partial t}(\varepsilon_{xx} + \varepsilon_{yy} + \varepsilon_{zz}) = \frac{\partial u}{\partial x} + \frac{\partial v}{\partial y} + \frac{\partial w}{\partial z} = \vec{\nabla}\cdot\vec{V} \tag{17.30}$$

The mean stress $\bar{\sigma}$ is generally taken as $-p$, where p is the mean fluid pressure. Thus, Eqs. (17.22)–(17.27) can also be expressed as

$$\sigma_{xx} = -p + 2\mu\frac{\partial u}{\partial x} - \frac{2}{3}\mu\vec{\nabla}\cdot\vec{V} \tag{17.31}$$

$$\sigma_{yy} = -p + 2\mu\frac{\partial v}{\partial y} - \frac{2}{3}\mu\vec{\nabla}\cdot\vec{V} \tag{17.32}$$

$$\sigma_{zz} = -p + 2\mu\frac{\partial w}{\partial z} - \frac{2}{3}\mu\vec{\nabla}\cdot\vec{V} \tag{17.33}$$

$$\sigma_{xy} = \mu\left(\frac{\partial v}{\partial x} + \frac{\partial u}{\partial y}\right) \tag{17.34}$$

$$\sigma_{yz} = \mu\left(\frac{\partial w}{\partial y} + \frac{\partial v}{\partial z}\right) \tag{17.35}$$

$$\sigma_{zx} = \mu\left(\frac{\partial u}{\partial z} + \frac{\partial w}{\partial x}\right) \tag{17.36}$$

17.5.3 Equations of Motion [17.1]

The equations of motion can be derived by applying Newton's second law to a differential volume $(\mathrm{d}x\,\mathrm{d}y\,\mathrm{d}z)$ of a fixed mass $\mathrm{d}m$. If the body forces acting on the fluid per unit mass are given by the vector

$$\vec{B} = B_x\vec{\imath} + B_y\vec{\jmath} + B_z\vec{k} \tag{17.37}$$

the application of Newton's law in x direction gives

$$\mathrm{d}F_x = \mathrm{d}m\ a_x = (\rho\,\mathrm{d}x\,\mathrm{d}y\,\mathrm{d}z)a_x \tag{17.38}$$

where $\mathrm{d}F_x$ is the differential force acting in x direction, and a_x is the acceleration of the fluid in x direction. Using a figure similar to that of Figure 8.2, Eq. (17.38) can be rewritten as

$$\mathrm{d}F_x = (\rho \mathrm{d}x\ \mathrm{d}y\ \mathrm{d}z)B_x - \sigma_{xx}\mathrm{d}y\ \mathrm{d}z + \left(\sigma_{xx} + \frac{\partial \sigma_{xx}}{\partial x}\mathrm{d}x\right)\mathrm{d}y\,\mathrm{d}z - \sigma_{yx}\mathrm{d}x\ \mathrm{d}z$$
$$+ \left(\sigma_{yx} + \frac{\partial \sigma_{yx}}{\partial y}\mathrm{d}y\right)\mathrm{d}x\,\mathrm{d}z - \sigma_{zx}\mathrm{d}x\ \mathrm{d}y + \left(\sigma_{zx} + \frac{\partial \sigma_{zx}}{\partial z}\mathrm{d}z\right)\mathrm{d}x\,\mathrm{d}y$$

Dividing this equation throughout by the volume of the element gives

$$\rho B_x + \frac{\partial \sigma_{xx}}{\partial x} + \frac{\partial \sigma_{yx}}{\partial y} + \frac{\partial \sigma_{zx}}{\partial z} = \rho a_x \tag{17.39a}$$

Similarly, we can obtain for the y and z directions,

$$\rho B_y + \frac{\partial \sigma_{xy}}{\partial x} + \frac{\partial \sigma_{yy}}{\partial y} + \frac{\partial \sigma_{zy}}{\partial z} = \rho a_y \tag{17.39b}$$

$$\rho B_z + \frac{\partial \sigma_{xz}}{\partial x} + \frac{\partial \sigma_{yz}}{\partial y} + \frac{\partial \sigma_{zz}}{\partial z} = \rho a_z \tag{17.39c}$$

Equations (17.39) are general and are applicable to any fluid with gravitational-type body forces. For Newtonian fluids with a single viscosity coefficient, we substitute Eqs. (17.31)–(17.36) into Eqs. (17.39) and obtain the equations of motion in x, y, and z directions as

$$\rho a_x = \rho B_x - \frac{\partial p}{\partial x} + \frac{\partial}{\partial x}\left(2\mu\frac{\partial u}{\partial x} - \frac{2}{3}\mu\vec{\nabla}\cdot\vec{V}\right) + \frac{\partial}{\partial y}\left[\mu\left(\frac{\partial v}{\partial x} + \frac{\partial u}{\partial y}\right)\right] + \frac{\partial}{\partial z}\left[\mu\left(\frac{\partial u}{\partial z} + \frac{\partial w}{\partial x}\right)\right] \tag{17.40a}$$

$$\rho a_y = \rho B_y - \frac{\partial p}{\partial y} + \frac{\partial}{\partial x}\left[\mu\left(\frac{\partial u}{\partial y} + \frac{\partial v}{\partial x}\right)\right] + \frac{\partial}{\partial y}\left(2\mu\frac{\partial v}{\partial y} - \frac{2}{3}\mu\vec{\nabla}\cdot\vec{V}\right) + \frac{\partial}{\partial z}\left[\mu\left(\frac{\partial v}{\partial z} + \frac{\partial w}{\partial y}\right)\right] \tag{17.40b}$$

$$\underbrace{\rho a_z}_{\text{Inertia force}} = \underbrace{\rho B_z}_{\text{Body force}} \underbrace{- \frac{\partial p}{\partial z}}_{\text{Pressure force}} + \underbrace{\frac{\partial}{\partial x}\left[\mu\left(\frac{\partial w}{\partial x} + \frac{\partial u}{\partial z}\right)\right] + \frac{\partial}{\partial y}\left[\mu\left(\frac{\partial v}{\partial z} + \frac{\partial w}{\partial y}\right)\right] + \frac{\partial}{\partial z}\left(2\mu\frac{\partial w}{\partial z} - \frac{2}{3}\mu\vec{\nabla}\cdot\vec{V}\right)}_{\text{Viscous force}} \tag{17.40c}$$

Equations (17.40) are called the *Navier–Stokes equations* for compressible Newtonian fluids in Cartesian form. For incompressible fluids, the Navier–Stokes equations of motion, Eqs. (17.40), become

$$a_x = \frac{\partial u}{\partial t} + u\frac{\partial u}{\partial x} + v\frac{\partial u}{\partial y} + w\frac{\partial u}{\partial z} = B_x - \frac{l}{\rho}\frac{\partial p}{\partial x} + \frac{\mu}{\rho}\left(\frac{\partial^2 u}{\partial x^2} + \frac{\partial^2 u}{\partial y^2} + \frac{\partial^2 u}{\partial z^2}\right) \tag{17.41a}$$

$$a_y = \frac{\partial v}{\partial t} + u\frac{\partial v}{\partial x} + v\frac{\partial v}{\partial y} + w\frac{\partial v}{\partial z} = B_y - \frac{l}{\rho}\frac{\partial p}{\partial y} + \frac{\mu}{\rho}\left(\frac{\partial^2 v}{\partial x^2} + \frac{\partial^2 v}{\partial y^2} + \frac{\partial^2 v}{\partial z^2}\right) \tag{17.41b}$$

$$a_z = \frac{\partial w}{\partial t} + u\frac{\partial w}{\partial x} + v\frac{\partial w}{\partial y} + w\frac{\partial w}{\partial z} = B_z - \frac{l}{\rho}\frac{\partial p}{\partial z} + \frac{\mu}{\rho}\left(\frac{\partial^2 w}{\partial x^2} + \frac{\partial^2 w}{\partial y^2} + \frac{\partial^2 w}{\partial z^2}\right) \tag{17.41c}$$

Furthermore, when viscosity μ is zero, Eqs. (17.41) reduce to the *Euler equations*:

$$a_x = B_x - \frac{l}{\rho}\frac{\partial p}{\partial x} \tag{17.42a}$$

$$a_y = B_y - \frac{l}{\rho}\frac{\partial p}{\partial y} \tag{17.42b}$$

$$a_z = B_z - \frac{l}{\rho}\frac{\partial p}{\partial z} \tag{17.42c}$$

For steady flow, all derivatives with respect to time will be zero in Eqs. (17.40)–(17.42).

17.6 ENERGY, STATE, AND VISCOSITY EQUATIONS

17.6.1 Energy Equation

When the flow is nonisothermal, the temperature of the fluid will be a function of x, y, z, and t. Just as the continuity equation represents the law of conservation of mass and gives the velocity distribution in space, the energy equation represents the conservation of energy and gives the temperature distribution in space. To derive the energy equation, we consider a differential control volume of fluid of size $\mathrm{d}x\,\mathrm{d}y\,\mathrm{d}z$ and write the energy balance equation as

$$\text{Energy input} = \text{energy output} + \text{energy accumulation} \tag{17.43}$$

The energy input to the element per unit time is given by

$$\begin{aligned}
&\Bigg\langle \underbrace{\left\{\rho u E - \frac{\partial}{\partial x}(\rho u E)\cdot\frac{\mathrm{d}x}{2}\right\}}_{\text{Internal energy}} + \frac{1}{2}\underbrace{\left\{\rho u(u^2+v^2+w^2) - \frac{\partial}{\partial x}[\rho u(u^2+v^2+w^2)]\cdot\frac{\mathrm{d}x}{2}\right\}}_{\text{Kinetic energy}} \\
&+ \underbrace{\left\{pu - \frac{\partial}{\partial x}(pu)\frac{\mathrm{d}x}{2}\right\}}_{\text{Pressure–volume work}} - \underbrace{\left\{k\frac{\partial T}{\partial x} - \frac{\partial}{\partial x}\left(k\frac{\partial T}{\partial x}\right)\cdot\frac{\mathrm{d}x}{2}\right\}}_{\text{Heat conduction}}\Bigg\rangle \cdot \mathrm{d}y\mathrm{d}z \\
&+ \text{similar terms for } y \text{ and } z \text{ directions} \\
&+ \frac{\partial Q}{\partial t}\,\mathrm{d}x\,\mathrm{d}y\,\mathrm{d}z + \Phi\,\mathrm{d}x\,\mathrm{d}y\,\mathrm{d}z
\end{aligned}$$

where T is the temperature, k is the thermal conductivity, Q is the heat generated in the fluid per unit volume, and Φ is the dissipation function—that is, time rate of energy dissipated per unit volume due to the action of viscosity.

Similarly, the energy output per unit time is given by

$$\left\langle \left[\rho u E + \frac{\partial}{\partial x}(\rho u E) \cdot \frac{\mathrm{d}x}{2}\right] + \frac{1}{2}\left\{\rho u(u^2+v^2+w^2) + \frac{\partial}{\partial x}\left[\rho u(u^2+v^2+w^2)\right] \cdot \frac{\mathrm{d}x}{2}\right\}\right.$$
$$\left. + \left[pu + \frac{\partial}{\partial x}(pu)\frac{\mathrm{d}x}{2}\right] - \left\{k\frac{\partial T}{\partial x} + \frac{\partial}{\partial x}\left(k\frac{\partial T}{\partial x}\right) \cdot \frac{\mathrm{d}x}{2}\right\}\right\rangle \cdot \mathrm{d}y\ \mathrm{d}z$$

+ similar terms for y and z directions

The energy accumulated in the element is given by

$$\Big[\underbrace{\frac{\partial}{\partial x}(\rho E)}_{\text{Internal energy}} + \underbrace{\frac{1}{2}\frac{\partial}{\partial t}\left\{\rho(u^2+v^2+w^2)\right\}}_{\text{Kinetic energy}}\Big]\ \mathrm{d}x\ \mathrm{d}y\ \mathrm{d}z$$

By making the energy balance as per Eq. (17.43), we obtain, after some manipulation,

$$\frac{\partial}{\partial x}\left(k\frac{\partial T}{\partial x}\right) + \frac{\partial}{\partial y}\left(k\frac{\partial T}{\partial y}\right) + \frac{\partial}{\partial z}\left(k\frac{\partial T}{\partial z}\right) + \frac{\partial Q}{\partial t} + \Phi$$
$$= \frac{\partial}{\partial x}(pu) + \frac{\partial}{\partial y}(pv) + \frac{\partial}{\partial z}(pw) + \frac{\rho}{2}\frac{\mathrm{D}}{\mathrm{D}t}(u^2+v^2+w^2) + \rho\frac{\mathrm{D}(E)}{\mathrm{D}t} \tag{17.44}$$

By using the relation

$$c_v = \left.\frac{\partial E}{\partial T}\right|_{\text{at constant volume}} = \text{specific heat at constant volume}$$

we can substitute $c_v \cdot \frac{\mathrm{D}T}{\mathrm{D}t}$ in place of $\frac{\mathrm{D}(E)}{\mathrm{D}t}$ in Eq. (17.44).

For inviscid and incompressible fluids, $\vec{\nabla} \cdot \vec{V} = 0$, and the application of Eq. (17.43) leads to

$$\rho c_v \frac{\mathrm{D}T}{\mathrm{D}t} = k\nabla^2 T + \frac{\partial Q}{\partial t} + \Phi \tag{17.45}$$

where c_v is the specific heat at constant volume, T is the temperature, k is the thermal conductivity, Q is the heat generated in the fluid per unit volume, and Φ is the dissipation function (i.e., time rate of energy dissipated per unit volume due to the action of viscosity) given by [17.2]

$$\Phi = -\frac{2}{3}\mu\left(\frac{\partial u}{\partial x} + \frac{\partial v}{\partial y} + \frac{\partial w}{\partial z}\right)^2 + 2\mu\left[\left(\frac{\partial u}{\partial x}\right)^2 + \left(\frac{\partial v}{\partial y}\right)^2 + \left(\frac{\partial w}{\partial z}\right)^2\right]$$
$$+ \mu\left[\left(\frac{\partial w}{\partial y} + \frac{\partial v}{\partial z}\right)^2 + \left(\frac{\partial u}{\partial z} + \frac{\partial w}{\partial x}\right)^2 + \left(\frac{\partial v}{\partial x} + \frac{\partial u}{\partial y}\right)^2\right] \tag{17.46}$$

It can be seen that Φ has a value of zero for inviscid fluids.

17.6.2 State and Viscosity Equations

The variations of density and viscosity with pressure and temperature can be stated in the form of equations of state and viscosity as

$$\rho = \rho(p, T) \tag{17.47}$$

$$\mu = \mu(p, T) \tag{17.48}$$

17.7 SOLUTION PROCEDURE

For a general three-dimensional flow problem, the continuity equation, the equations of motion, the energy equation, the equation of state, and the viscosity equation are to be satisfied. The unknowns are the velocity components (u, v, w), pressure (p), density (ρ), viscosity (μ), and the temperature (T). Thus, there are seven governing equations in seven unknowns, and hence the problem can be solved once the flow boundaries and the boundary and initial conditions for the governing equations are known. The general governing equations are valid at any instant of time and are applicable to laminar, transition, and turbulent flows. Note that the solution of the complete set of equations has not been obtained even for laminar flows. However, in many practical situations, the governing equations get simplified considerably, and hence the mathematical solution would not be very difficult. In a turbulent flow, the unknown variables fluctuate about their mean values randomly and the solution of the problem becomes extremely complex.

For a three-dimensional inviscid fluid flow, five unknowns, namely, u, v, w, p, and ρ, will be there. In this case, Eqs. (17.4) and (17.40) are used along with the equation of state (expressing ρ in terms of pressure p only) to find the unknowns. In the solution of these equations, constants of integration appear that must be evaluated from the boundary conditions of the specific problem.

17.8 INVISCID FLUID FLOW

In a large number of fluid flow problems (especially those with low-viscosity fluids, such as water and the common gases) the effect of viscosity will be small compared to other quantities, such as pressure, inertia force, and field force; hence, the fluid can be treated as an inviscid fluid. Typical problems in which the effect of viscosity of the fluid can be neglected are flow through orifices, flow over weirs, flow in channel and duct entrances, and flow in converging and diverging nozzles. In these problems, the conditions very near to the solid boundary, where the viscosity has a significant effect, are not of much interest and one would normally be interested in the movement of the main mass of the fluid. In any fluid flow problem, we would be interested in determining the fluid velocity and fluid pressure as a function of the spatial coordinates and time. This solution will be greatly simplified if the viscosity of the fluid is assumed to be zero.

The equations of motion (Euler's equations) for this case are

$$\left.\begin{aligned}
\frac{\mathrm{D}u}{\mathrm{D}t} &= \frac{\partial u}{\partial t} + u\frac{\partial u}{\partial x} + v\frac{\partial u}{\partial y} + w\frac{\partial u}{\partial z} = B_x - \frac{1}{\rho}\frac{\partial p}{\partial x}\\
\frac{\mathrm{D}v}{\mathrm{D}t} &= \frac{\partial v}{\partial t} + u\frac{\partial v}{\partial x} + v\frac{\partial v}{\partial y} + w\frac{\partial v}{\partial z} = B_y - \frac{1}{\rho}\frac{\partial p}{\partial y}\\
\frac{\mathrm{D}w}{\mathrm{D}t} &= \frac{\partial w}{\partial t} + u\frac{\partial w}{\partial x} + v\frac{\partial w}{\partial y} + w\frac{\partial w}{\partial z} = B_z - \frac{1}{\rho}\frac{\partial p}{\partial z}
\end{aligned}\right\} \tag{17.49}$$

The continuity equation is given by Eq. (17.4). Thus, the unknowns in Eqs. (17.49) and (17.4) are u, v, w, p, and ρ. Since the density ρ can be expressed in terms of the pressure p by using the equation of state, the four equations represented by Eqs. (17.49) and (17.4) are sufficient to solve for the four unknowns u, v, w, and p. While solving these equations, the constants of integration that appear are to be evaluated from the boundary conditions of the specific problem.

17.9 IRROTATIONAL FLOW

Let a point A and two perpendicular lines AB and AC be considered in a two-dimensional fluid flow. These lines, which are fixed to the fluid, are assumed to move with the fluid and assume the positions $A'B'$ and $A'C'$ after time Δt as shown in Figure 17.2. If the original lines AB and AC are taken parallel to the x and y axes, the angular rotation of the fluid immediately adjacent to point A is given by $\frac{1}{2}(\beta_1 + \beta_2)$, and hence the rate of rotation of the fluid about the z axis (ω_z) is defined as

$$\omega_z = \frac{1}{2}\frac{\beta_1 + \beta_2}{\Delta t} \tag{17.50}$$

If the velocities of the fluid at the point A in x and y directions are u and v, respectively, the velocity components of the point C are $u + (\partial u/\partial y) \cdot \Delta y$ and $v + \partial v/\partial y \cdot \Delta y$ in x and y directions, respectively, where $\Delta y = AC$. Since β_2 is small,

$$\tan \beta_2 = \beta_2 = \frac{C'C_2}{A'C_2} = \frac{A_1A' - CC_1}{A'C_2} = \frac{u\Delta t - \left(u + \dfrac{\partial u}{\partial y}\Delta y\right)\Delta t}{\Delta y} = -\frac{\partial u}{\partial y}\Delta t \tag{17.51}$$

where it was assumed that $A'C_2 \approx AC = \Delta y$. Similarly,

$$\tan \beta_1 = \beta_1 = \frac{B_1B'}{A'B_1} = \frac{\left(v + \dfrac{\partial v}{\partial x}\Delta x\right)\Delta t - v\Delta t}{\Delta x} = \frac{\partial v}{\partial x}\Delta t \tag{17.52}$$

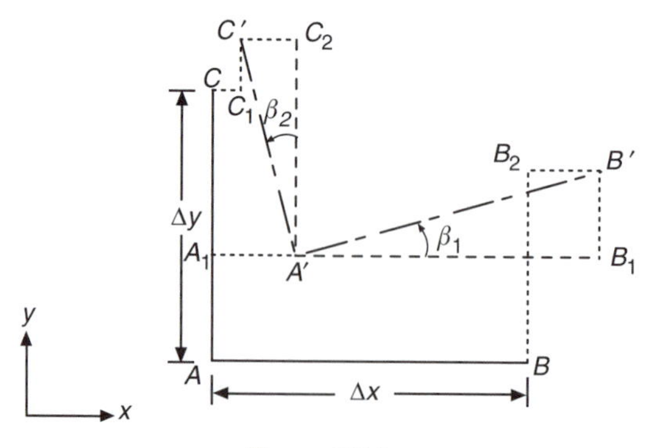

Figure 17.2.

Thus, the rate of rotation (also called rotation) can be expressed as

$$\omega_z = \frac{1}{2}\left(\frac{\dfrac{\partial v}{\partial x}\Delta t - \dfrac{\partial u}{\partial y}\Delta t}{\Delta t}\right) = \frac{1}{2}\left(\frac{\partial v}{\partial x} - \frac{\partial u}{\partial y}\right) \tag{17.53}$$

By proceeding in a similar manner, the rates of rotation about the x, y, and z axes in a three-dimensional fluid flow can be derived as

$$\omega_x = \frac{1}{2}\left(\frac{\partial w}{\partial y} - \frac{\partial v}{\partial z}\right) \tag{17.54a}$$

$$\omega_y = \frac{1}{2}\left(\frac{\partial u}{\partial z} - \frac{\partial w}{\partial x}\right) \tag{17.54b}$$

$$\omega_z = \frac{1}{2}\left(\frac{\partial v}{\partial x} - \frac{\partial u}{\partial y}\right) \tag{17.54c}$$

When the particles of the fluid are not rotating, the rotation is zero and the fluid is called *irrotational.* The physical meaning of irrotational flow can be seen from Figure 17.3. In Figure 17.3(a), the particle maintains the same orientation everywhere along the streamline without rotation. Hence, it is called irrotational flow. On the other hand, in Figure 17.3(b), the particle rotates with respect to fixed axes and maintains the same orientation with respect to the streamline. Hence, this flow is not irrotational. The vorticity or fluid rotation vector ($\vec{\omega}$) is defined as the average angular velocity of any two mutually perpendicular line segments of the fluid whose x, y, and z components are given by Eq. (17.54).

Note

In both Figures 17.3(a) and 17.3(b), the particles can undergo deformation without affecting the analysis. For example, in the flow of a nonviscous fluid between convergent boundaries, the elements of the fluid deform as they pass through the channel, but there is no rotation about the z axis as shown in Figure 17.4.

17.10 VELOCITY POTENTIAL

It is convenient to introduce a function ϕ, called the potential function or velocity potential, in integrating Eqs. (17.49). This function ϕ is defined in such a way that its partial

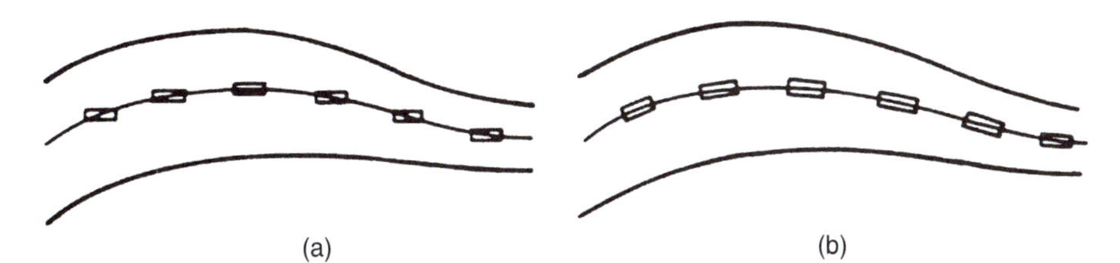

Figure 17.3. Irrotational and Rotational Flows.

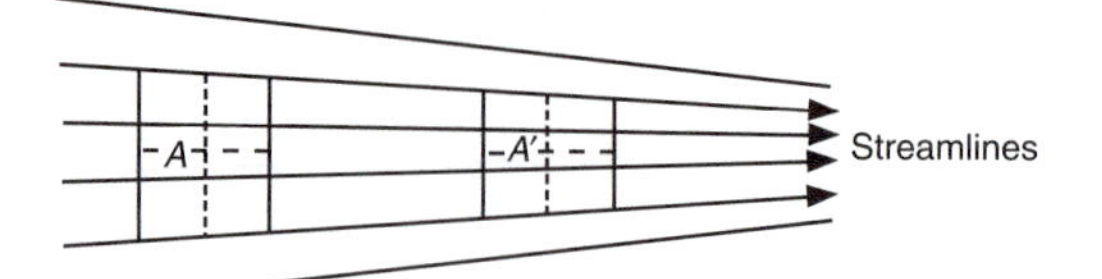

Figure 17.4. Irrotational Flow between Convergent Boundaries.

derivative in any direction gives the velocity in that direction; that is,

$$\frac{\partial \phi}{\partial x} = u, \quad \frac{\partial \phi}{\partial y} = v, \quad \frac{\partial \phi}{\partial z} = w \tag{17.55}$$

Substitution of Eqs. (17.55) into Eq. (17.4) gives

$$-\rho\left(\frac{\partial^2 \phi}{\partial x^2} + \frac{\partial^2 \phi}{\partial y^2} + \frac{\partial^2 \phi}{\partial z^2}\right) = u\frac{\partial \rho}{\partial x} + v\frac{\partial \rho}{\partial y} + w\frac{\partial \rho}{\partial z} + \frac{\partial \rho}{\partial t} = \frac{\mathrm{D}\rho}{\mathrm{D}t} \tag{17.56}$$

For incompressible fluids, Eq. (17.56) becomes

$$\nabla^2 \phi = \frac{\partial^2 \phi}{\partial x^2} + \frac{\partial^2 \phi}{\partial y^2} + \frac{\partial^2 \phi}{\partial z^2} = 0 \tag{17.57}$$

By differentiating u and v with respect to y and x, respectively, we obtain

$$\frac{\partial u}{\partial y} = \frac{\partial^2 \phi}{\partial y \partial x}, \quad \frac{\partial v}{\partial x} = \frac{\partial^2 \phi}{\partial x \partial y} \tag{17.58}$$

from which we can obtain

$$\frac{\partial u}{\partial y} - \frac{\partial v}{\partial x} = 0 \tag{17.59}$$

$$\frac{\partial u}{\partial z} - \frac{\partial w}{\partial x} = 0 \tag{17.60}$$

$$\frac{\partial w}{\partial y} - \frac{\partial v}{\partial z} = 0 \tag{17.61}$$

The terms on the left-hand side of Eqs. (17.59)–(17.61) can be seen to be equal to twice the rates of rotation of the fluid element. Thus, the assumption of a velocity potential defined by Eq. (17.55) requires the flow to be irrotational.

17.11 STREAM FUNCTION

The motion of the fluid, at every point in space, can be represented by means of a velocity vector showing the direction and magnitude of the velocity. Since representation by vectors is unwieldy, we can use streamlines, which are the lines drawn tangent to the velocity vector at every point in space. Since the velocity vectors meet the streamlines tangentially for all points on a streamline, no fluid can cross the streamline.

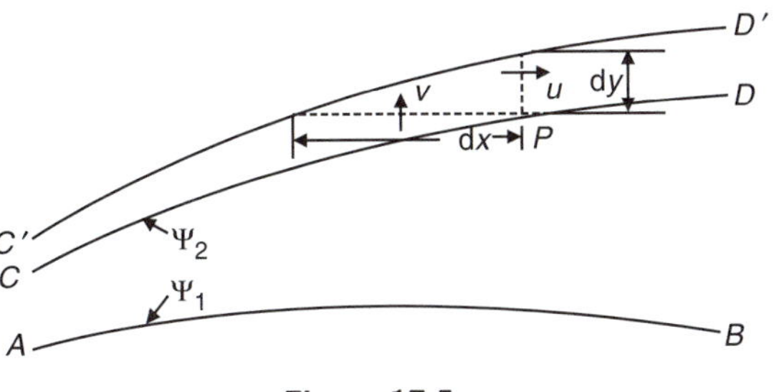

Figure 17.5.

For a two-dimensional flow, the streamlines can be represented in a two-dimensional plane. A stream function ψ may be defined (which is related to the velocity of the fluid) on the basis of the continuity equation and the nature of the streamlines. Let the streamlines AB and CD denote the stream functions ψ_1 and ψ_1, respectively, in Figure 17.5. If a unit thickness of the fluid is considered, $\psi_2 - \psi_1$ is defined as the volume rate of fluid flow between the streamlines AB and CD. Let the streamline $C'D'$ lie at a small distance away from CD and let the flow between the streamlines CD and $C'D'$ be $\mathrm{d}\psi$. At a point P on CD, the distance between CD and $C'D'$ is denoted by the components of distance $-dx$ and dy. Let the velocity of the fluid at point P be u and v in x and y directions, respectively. Since no fluid crosses the streamlines, the volume rate of flow across the element $\mathrm{d}y$ is $u\ \mathrm{d}y$ and the volume rate of flow across the element $-\mathrm{d}x$ is $-v\ \mathrm{d}x$. If the flow is assumed to be incompressible, this volume rate of flow must be equal to $\mathrm{d}\psi$

$$\therefore \mathrm{d}\psi = u\ \mathrm{d}y = -v\ \mathrm{d}x \tag{17.62}$$

Because ψ is a function of both x and y, we use partial derivatives and rewrite Eq. (17.62) as

$$\frac{\partial \psi}{\partial y} = u, \qquad \frac{\partial \psi}{\partial x} = -v \tag{17.63}$$

Equation (17.63) defines the stream function ψ for a two-dimensional incompressible flow. Physically, the stream function denotes the volume rate of flow per unit distance normal to the plane of motion between a streamline in the fluid and an arbitrary reference or base streamline. Hence, the volume rate of flow between any two adjacent streamlines is given by

$$Q = \psi_2 - \psi_1 \tag{17.64}$$

where ψ_1 and ψ_2 are the values of the adjacent streamlines, and Q is the flow rate per unit depth in the z direction. The streamlines also possess the property that there is no flow perpendicular to their direction. For a two-dimensional incompressible flow, the continuity equation is given by

$$\frac{\partial u}{\partial x} + \frac{\partial v}{\partial y} = 0 \tag{17.65}$$

which is automatically satisfied by the stream function ψ—that is, by Eq. (17.63). If the flow is irrotational, the equation to be satisfied is

$$\frac{\partial u}{\partial y} - \frac{\partial v}{\partial x} = 0 \tag{17.66}$$

By substituting Eq. (17.63) into Eq. (17.66), we obtain

$$\frac{\partial^2 \psi}{\partial x^2} + \frac{\partial^2 \psi}{\partial y^2} = 0 \tag{17.67}$$

It can be seen that in a two-dimensional irrotational and incompressible flow, the solution of the Laplace's equation gives either stream functions or velocity potentials depending on the choice.

17.12 BERNOULLI EQUATION

Bernoulli equation can be derived by integrating Euler's equations (17.49) with the help of Eqs. (17.55) and (17.59)–(17.61). By substituting the relations $\partial v/\partial x$ and $\partial w/\partial x$ for $\partial u/\partial y$ and $\partial u/\partial z$, respectively (from Eqs. 17.59 and 17.60), and $\partial \phi/\partial x$ for u (from Eq. 17.55) into the first equation of (17.49), and assuming that a body force potential (Ω), such as gravity, exists, we have $B_x = -(\partial\Omega/\partial x)$, $B_y = -(\partial\Omega/\partial y)$, and $B_z = -(\partial\Omega/\partial z)$, and hence

$$\frac{\partial^2 \phi}{\partial x \partial t} + u\frac{\partial u}{\partial x} + v\frac{\partial v}{\partial x} + w\frac{\partial w}{\partial x} + \frac{\partial \Omega}{\partial x} + \frac{1}{\rho}\frac{\partial p}{\partial x} = 0 \tag{17.68}$$

By integrating Eq. (17.68) with respect to x, we obtain

$$\frac{\partial \phi}{\partial t} + \frac{u^2}{2} + \frac{v^2}{2} + \frac{w^2}{2} + \Omega + \frac{p}{\rho} = f_1(y, z, t) \tag{17.69}$$

where f_1 cannot be a function of x because its partial derivative with respect to x must be zero. Similarly, the second and third equations of (17.49) lead to

$$\frac{\partial \phi}{\partial t} + \frac{u^2}{2} + \frac{v^2}{2} + \frac{w^2}{2} + \Omega + \frac{p}{\rho} = f_2(x, z, t) \tag{17.70}$$

$$\frac{\partial \phi}{\partial t} + \frac{u^2}{2} + \frac{v^2}{2} + \frac{w^2}{2} + \Omega + \frac{p}{\rho} = f_3(x, y, t) \tag{17.71}$$

Since the left-hand sides of Eqs. (17.69)–(17.71) are the same, we have

$$f_1(y, z, t) = f_2(x, z, t) = f_3(x, y, t) = f(t) \tag{17.72}$$

where $f(t)$ is a function of t alone. Since the magnitude of the velocity vector $\vec{V}$ is given by

$$|\vec{V}| = (u^2 + v^2 + w^2)^{1/2} \tag{17.73}$$

Eqs. (17.69)–(17.71) can be expressed as

$$\frac{\partial \phi}{\partial t} + \frac{|\vec{V}|^2}{2} + \Omega + \frac{p}{\rho} = f(t) \tag{17.74}$$

where $f(t)$ is a function of time. For a steady flow, Eq. (17.74) reduces to

$$\frac{|\vec{V}|^2}{2} + \Omega + \frac{p}{\rho} = \text{constant} \tag{17.75}$$

If the body force is due to gravity, $\Omega = gz$, where g is the acceleration due to gravity and z is the elevation. By substituting this expression of Ω into Eq. (17.75) and dividing throughout by g, we obtain a more familiar form of the Bernoulli equation for steady flows as

$$\underbrace{\frac{\vec{V}^2}{2g}}_{\substack{\text{Velocity}\\ \text{head}}} + \underbrace{z}_{\substack{\text{Eleva-}\\ \text{tion}\\ \text{head}}} + \underbrace{\frac{p}{\gamma}}_{\substack{\text{Pres-}\\ \text{sure}\\ \text{head}}} = \text{constant} \tag{17.76}$$

where $\gamma = \rho g$.

REFERENCES

17.1 J.W. Daily and D.R.F. Harleman: *Fluid Dynamics*, Addison-Wesley, Reading, MA, 1966.

17.2 J.G. Knudsen and D.L. Katz: *Fluid Dynamics and Heat Transfer*, McGraw-Hill, New York, 1958.

PROBLEMS

17.1 Derive the continuity equation in polar coordinates for an ideal fluid by equating the flow into and out of the polar element of area $r\,dr\,d\theta$.

17.2 If the x component of velocity in a two-dimensional flow is given by $u = x^2 + 2x - y^2$, find the y component of velocity that satisfies the continuity equation.

17.3 The potential function for a two-dimensional flow is given by $\phi = a_1 + a_2x + a_3y + a_4x^2 + a_5xy + a_6y^2$, where a_i ($i = 1$–6) are constants. Find the expression for the stream function.

17.4 The velocity components in a two-dimensional flow are $u = -2x^2 + 3y$ and $v = 3x + 2y$. Determine whether the flow is incompressible or irrotational or both.

17.5 The potential function for a two-dimensional fluid flow is $\phi = 8xy + 6$. Determine whether the flow is incompressible or irrotational or both. Find the paths of some of the particles and plot them.

17.6 In the steady irrotational flow of a fluid at point P, the pressure is 15 kg/m^2 and the velocity is 10 m/sec. At point Q, which is located 5 m vertically above P, the velocity is 5 m/sec. If ρ of the fluid is 0.001 kg/cm^3, find the pressure at Q.

17.7 A nozzle has an inlet diameter of $d_1 = 4$ in. and an outlet diameter of $d_2 = 2$ in. (Figure 17.6). Determine the gage pressure of water required at the inlet of the nozzle in order to have a steady flow rate of 2 ft^3/sec. Assume the density of water as 1.94 slug/ft^3.

17.8 The x and y components of velocity of a fluid in a steady, incompressible flow are given by $u = 4x$ and $v = -4y$. Find the stream function corresponding to this flow.

17.9 The velocity components of a three-dimensional fluid flow are given by

$$u = a_1x + a_2y + a_3z, \qquad v = a_4x + a_5y + a_6z, \qquad w = a_7x + a_8y + a_9z$$

Determine the relationship between the constants $a_1, a_2, \ldots, a_9$ in order for the flow to be an incompressible flow.

17.10 The stream function corresponding to a fluid flow is given by $\Psi(x, y) = 10(x^2 - y^2)$.

(a) Determine whether the flow is irrotational.
(b) Find the velocity potential of the flow.
(c) Find the velocity components of the flow.

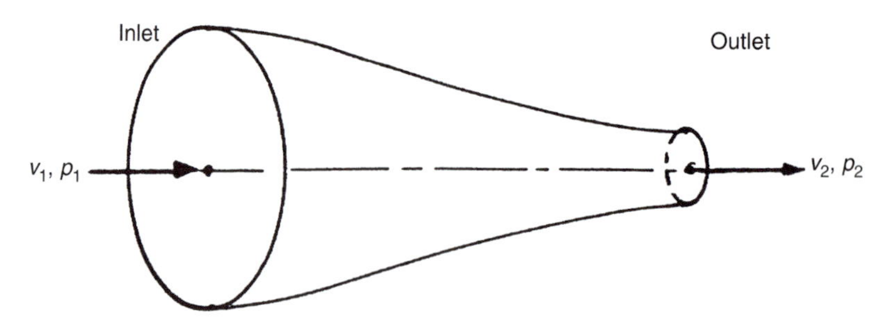

Figure 17.6.

18

INVISCID AND INCOMPRESSIBLE FLOWS

18.1 INTRODUCTION

In this chapter, we consider the finite element solution of ideal flow (inviscid incompressible flow) problems. Typical examples that fall in this category are flow around a cylinder, flow out of an orifice, and flow around an airfoil. The two-dimensional potential flow (irrotational flow) problems can be formulated in terms of a velocity potential function (ϕ) or a stream function (ψ). In terms of the velocity potential, the governing equation for a two-dimensional problem is given by [obtained by substituting Eq. (17.55) into Eq. (17.5)]

$$\frac{\partial^2 \phi}{\partial x^2} + \frac{\partial^2 \phi}{\partial y^2} = 0 \tag{18.1}$$

where the velocity components are given by

$$u = \frac{\partial \phi}{\partial x}, \qquad v = \frac{\partial \phi}{\partial y} \tag{18.2}$$

In terms of stream function, the governing equation is (Eq. 17.67)

$$\frac{\partial^2 \psi}{\partial x^2} + \frac{\partial^2 \psi}{\partial y^2} = 0 \tag{18.3}$$

and the flow velocities can be determined as

$$u = \frac{\partial \psi}{\partial y}, \qquad v = -\frac{\partial \psi}{\partial x} \tag{18.4}$$

In general, the choice between velocity and stream function formulations in the finite element analysis depends on the boundary conditions, whichever is easier to specify. If the geometry is simple, no advantage of one over the other can be claimed.

If the fluid is ideal, its motion does not penetrate into the surrounding body or separate from the surface of the body and leave empty space. This gives the boundary condition that the component of the fluid velocity normal to the surface must be equal to the component of the velocity of the surface in the same direction. Hence,

$$\vec{V} \cdot \vec{n} = \vec{V}_B \cdot \vec{n}$$

or

$$u l_x + v l_y = u_B l_x + v_B l_y \tag{18.5}$$

where $\vec{V}$ is the velocity of the fluid, $\vec{V}_B$ is the velocity of the boundary, and $\vec{n}$ is the outward drawn normal to the boundary whose components (direction cosines) are l_x and l_y. If the boundary is fixed ($\vec{V}_B = \vec{0}$), there will be no flow and hence no velocity perpendicular to the boundary. This implies that all fixed boundaries can be considered as streamlines because there will be no fluid velocity perpendicular to a streamline. If there is a line of symmetry parallel to the direction of flow, it will also be a streamline. If $\vec{V}_B = \vec{0}$, Eqs. (18.5), (18.4), and (18.2) give the conditions

$$\frac{\partial \psi}{\partial s} = \frac{\partial \psi}{\partial y} l_x - \frac{\partial \psi}{\partial x} l_y = 0 \tag{18.6}$$

$$\frac{\partial \phi}{\partial n} = \frac{\partial \phi}{\partial x} l_x + \frac{\partial \phi}{\partial y} l_y = 0 \tag{18.7}$$

Equation (18.6) states that the tangential derivative of the stream function along a fixed boundary is zero, whereas Eq. (18.7) indicates that the normal derivative of the potential function (i.e., velocity normal to the fixed boundary) is zero.

The finite element solution of potential flow problems is illustrated in this chapter with reference to the problem of flow over a circular cylinder between two parallel plates as shown in Figure 18.1(a). Both potential and stream function formulations are considered.

18.2 POTENTIAL FUNCTION FORMULATION

The boundary value problem for potential flows can be stated as follows.

18.2.1 Differential Equation Form

Find the velocity potential $\phi(x, y)$ in a given region S surrounded by the curve C such that

$$\nabla^2 \phi = \frac{\partial^2 \phi}{\partial x^2} + \frac{\partial^2 \phi}{\partial y^2} = 0 \text{ in } S \tag{18.8}$$

with the boundary conditions

Dirichlet condition: $\quad \phi = \phi_0 \text{ on } C_1 \qquad (18.9)$

Neumann condition: $\quad V_n = \frac{\partial \phi}{\partial n} = \frac{\partial \phi}{\partial x} l_x + \frac{\partial \phi}{\partial y} l_y = V_0 \text{ on } C_2 \qquad (18.10)$

where $C = C_1 + C_2$, and V_0 is the prescribed value of the velocity normal to the boundary surface.

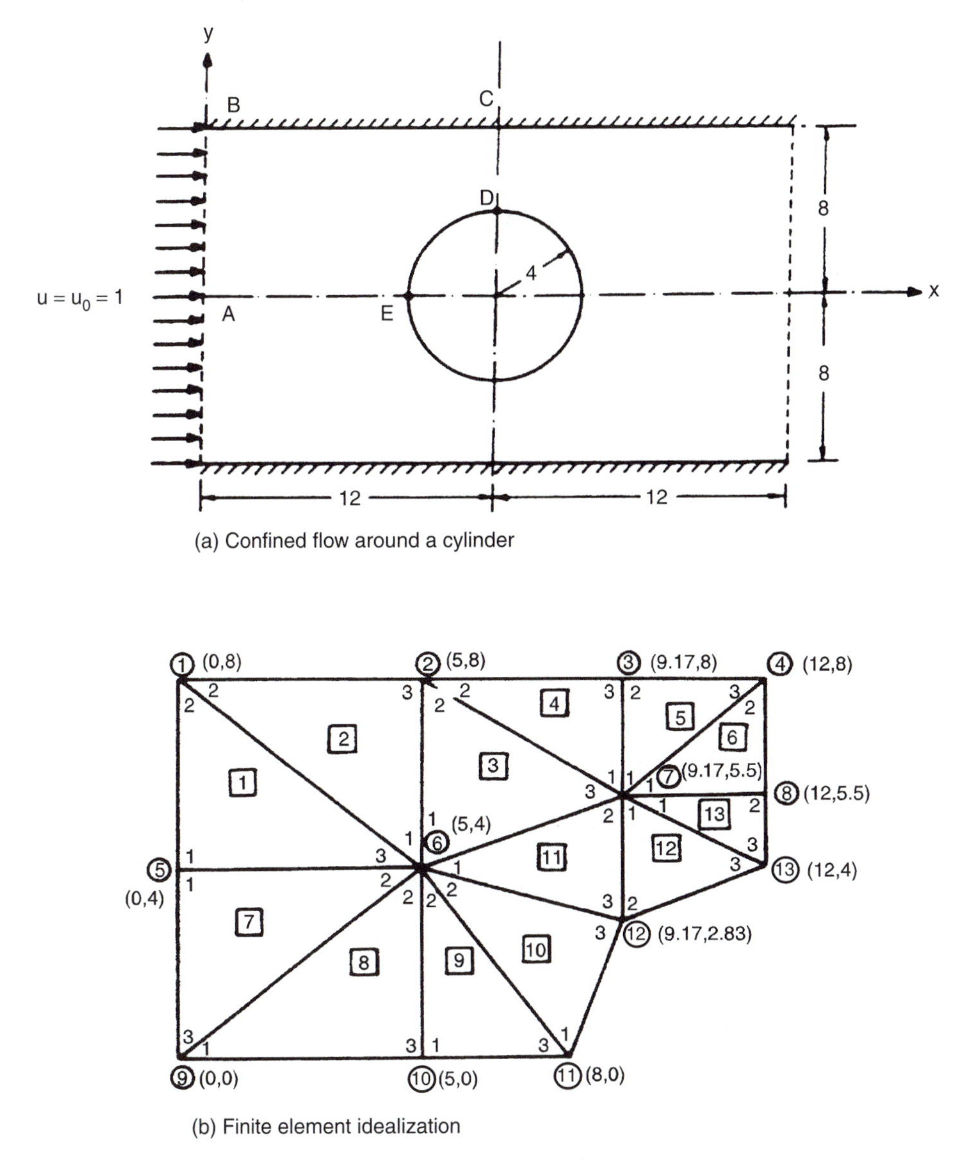

Figure 18.1.

18.2.2 Variational Form

Find the velocity potential $\phi(x, y)$ that minimizes the functional

$$I = \frac{1}{2} \iint_S \left[\left(\frac{\partial \phi}{\partial x} \right)^2 + \left(\frac{\partial \phi}{\partial y} \right)^2 \right] \cdot \mathrm{d}S - \int_{C_2} V_0 \, \phi \, \mathrm{d}C_2 \tag{18.11}$$

with the boundary condition

$$\phi = \phi_0 \text{ on } C_1 \tag{18.12}$$

18.3 FINITE ELEMENT SOLUTION USING THE GALERKIN APPROACH

The finite element procedure using the Galerkin method can be stated by the following steps:

Step 1: Divide the region S into E finite elements of p nodes each.

Step 2: Assume a suitable interpolation model for $\phi^{(e)}$ in element e as

$$\phi^{(e)}(x,y) = [N(x,y)]\vec{\Phi}^{(e)} = \sum_{i=1}^{p} N_i(x,y)\Phi_i^{(e)} \tag{18.13}$$

Step 3: Set the integral of the weighted residue over the region of the element equal to zero by taking the weights same as the interpolation functions N_i. This yields

$$\iint\limits_{S^{(e)}} N_i \left[\frac{\partial^2 \phi^{(e)}}{\partial x^2} + \frac{\partial^2 \phi^{(e)}}{\partial y^2}\right] \mathrm{d}S = 0, \quad i = 1, 2, \ldots, p \tag{18.14}$$

The integrals in Eq. (18.14) can be written as (see Appendix A)

$$\iint\limits_{S^{(e)}} N_i \frac{\partial^2 \phi^{(e)}}{\partial x^2} \mathrm{d}S = -\iint\limits_{S^{(e)}} \frac{\partial N_i}{\partial x}\frac{\partial \phi^{(e)}}{\partial x} \mathrm{d}S + \int\limits_{C^{(e)}} N_i \frac{\partial \phi^{(e)}}{\partial x} l_x \,\mathrm{d}C \tag{18.15}$$

Similarly,

$$\iint\limits_{S^{(e)}} N_i \frac{\partial^2 \phi^{(e)}}{\partial y^2} \mathrm{d}S = -\iint\limits_{S^{(e)}} \frac{\partial N_i}{\partial y}\frac{\partial \phi^{(e)}}{\partial y} \mathrm{d}S + \int\limits_{C^{(e)}} N_i \frac{\partial \phi^{(e)}}{\partial y} l_y \,\mathrm{d}C \tag{18.16}$$

Thus, Eq. (18.14) can be expressed as

$$-\iint\limits_{S^{(e)}} \left(\frac{\partial N_i}{\partial x}\frac{\partial \phi^{(e)}}{\partial x} + \frac{\partial N_i}{\partial y}\frac{\partial \phi^{(e)}}{\partial y}\right) \mathrm{d}S + \int\limits_{C^{(e)}} N_i \left(\frac{\partial \phi^{(e)}}{\partial x} l_x + \frac{\partial \phi^{(e)}}{\partial y} l_y\right) \mathrm{d}C = 0,$$
$$i = 1, 2, \ldots, p \tag{18.17}$$

Since the boundary of the element $C^{(e)}$ is composed of $C_1^{(e)}$ and $C_2^{(e)}$, the line integral of Eq. (18.17) would be zero on $C_1^{(e)}$ (since $\phi^{(e)}$ is prescribed to be a constant ϕ_0 on $C_1^{(e)}$, the derivatives of $\phi^{(e)}$ with respect to x and y would be zero). On the boundary $C_2^{(e)}$, Eq. (18.10) is to be satisfied. For this, the line integral of Eq. (18.17) can be rewritten as

$$\int_{C_1^{(e)}+C_2^{(e)}} N_i \left(\frac{\partial \phi^{(e)}}{\partial x} l_x + \frac{\partial \phi^{(e)}}{\partial y} l_y \right) \mathrm{d}C = \int_{C_2^{(e)}} V_0 N_i \, \mathrm{d}C_2 \tag{18.18}$$

By using Eqs. (18.13) and (18.18), Eq. (18.17) can be expressed in matrix form as

$$[K^{(e)}]\vec{\Phi}^{(e)} = \vec{P}^{(e)} \tag{18.19}$$

where

$$[K^{(e)}] = \iint_{S^{(e)}} [B]^T [D][B] \cdot \mathrm{d}S \tag{18.20}$$

$$\vec{P}^{(e)} = -\int_{C_2^{(e)}} V_0 [N]^T \, \mathrm{d}C_2 \tag{18.21}$$

$$[B] = \begin{bmatrix} \dfrac{\partial N_1}{\partial x} & \dfrac{\partial N_2}{\partial x} & \cdots & \dfrac{\partial N_p}{\partial x} \\ \dfrac{\partial N_1}{\partial y} & \dfrac{\partial N_2}{\partial y} & \cdots & \dfrac{\partial N_p}{\partial y} \end{bmatrix} \tag{18.22}$$

and

$$[D] = \begin{bmatrix} 1 & 0 \\ 0 & 1 \end{bmatrix} \tag{18.23}$$

Step 4: Assemble the element equations (18.19) to obtain the overall equations as

$$[\underset{\sim}{K}]\underset{\sim}{\vec{\Phi}} = \underset{\sim}{\vec{P}} \tag{18.24}$$

Step 5: Incorporate the boundary conditions specified over C_1 and solve Eqs. (18.24).

Example 18.1 (*Confined Flow around a Cylinder*) Find the velocity distribution along the vertical center line CD in Figure 18.1(a).

Solution Due to symmetry, we can consider only the portion $ABCDEA$ in the finite element analysis. The boundary condition is that ϕ is constant along CD. This constant can be taken as zero for convenience.

Step 1: Idealize the solution region using triangular elements. In the current case, 13 elements are used for modeling the region as shown in Figure 18.1(b). This idealization, although crude in representing the cylindrical boundary, is considered for simplicity. The local corner numbers of the elements are labeled in an arbitrary manner. The information needed for subsequent calculations is given in Table 18.1.

Step 2: Determine the nodal interpolation functions. The variation of $\phi^{(e)}$ inside the element e is assumed to be linear as

$$\phi^{(e)}(x,y) = [N(x,y)]\vec{\Phi}^{(e)} \tag{E$_1$}$$

where

$$[N(x,y)] = \begin{Bmatrix} N_i(x,y) \\ N_j(x,y) \\ N_k(x,y) \end{Bmatrix}^T = \begin{Bmatrix} (a_i + xb_i + yc_i)/2A^{(e)} \\ (a_j + xb_j + yc_j)/2A^{(e)} \\ (a_k + xb_k + yc_k)/2A^{(e)} \end{Bmatrix}^{(e)}$$

$$\vec{\Phi}^{(e)} = \begin{Bmatrix} \Phi_i \\ \Phi_j \\ \Phi_k \end{Bmatrix}$$

and the constants $a_i, a_j, \ldots, c_k$ are defined by Eqs. (3.32). The information needed for the computation of $[N(x,y)]$ is given in Table 18.1 (the constants a_i, a_j, and a_k are not given because they are not required in the computations).

Step 3: Derive the element matrices using the known values of $A^{(e)}, b_i, b_j, \ldots, c_k$. The element characteristic matrix is given by

$$[K^{(e)}] = \iint_{A^{(e)}} [B]^T[D][B] \cdot \mathrm{d}x\,\mathrm{d}y$$

$$= \frac{1}{4A^{(e)}} \begin{bmatrix} (b_i^2 + c_i^2) & (b_ib_j + c_ic_j) & (b_ib_k + c_ic_k) \\ & (b_j^2 + c_j^2) & (b_jb_k + c_jc_k) \\ \text{Symmetric} & & (b_k^2 + c_k^2) \end{bmatrix} \tag{E$_2$}$$

Thus we obtain in this case,

$$[K^{(1)}] = \begin{array}{c} \begin{matrix} \quad 5 & \quad\quad 1 & \quad\quad 6 \end{matrix} \\ \begin{bmatrix} 1.025 & -0.625 & -0.400 \\ -0.625 & 0.625 & 0.000 \\ -0.400 & 0.000 & 0.400 \end{bmatrix} \begin{matrix} 5 \\ 1 \\ 6 \end{matrix} \end{array}$$

$$[K^{(2)}] = \begin{array}{c} \begin{matrix} \quad 6 & \quad\quad 1 & \quad\quad 2 \end{matrix} \\ \begin{bmatrix} 0.625 & 0.000 & -0.625 \\ 0.000 & 0.400 & -0.400 \\ -0.625 & -0.400 & 1.025 \end{bmatrix} \begin{matrix} 6 \\ 1 \\ 2 \end{matrix} \end{array}$$

Table 18.1.

Node number (i)	1	2	3	4	5	6	7	8	9	10	11	12	13
Global coordinates of node i (x_i, y_i)	(0,8)	(5,8)	(9.17,8)	(12,8)	(0,4)	(5,4)	(9.17,5.5)	(12,5.5)	(0,0)	(5,0)	(8,0)	(9.17,2.83)	(12,4)

Element number (e)		1	2	3	4	5	6	7	8	9	10	11	12	13
Global node numbers i, j, and k corresponding to local nodes 1, 2, and 3	i	5	6	6	7	7	7	5	9	10	11	6	7	7
	j	1	1	2	2	3	4	6	6	6	6	7	12	8
	k	6	2	7	3	4	8	9	10	11	12	12	13	13

Element number (e)	x_i	x_j	x_k	y_i	y_j	y_k	$c_k = x_j - x_i$	$c_i = x_k - x_j$	$c_j = x_i - x_k$	$b_i = y_j - y_k$	$b_j = y_k - y_i$	$b_k = y_i - y_j$	$A^{(e)} = \frac{1}{2}\lvert x_{ij}y_{jk} - x_{jk}y_{ij}\rvert$
1	0	0	5	4	8	4	0	5	−5	4	0	−4	$\frac{1}{2}\lvert 0 \times (-4) - 5 \times 4\rvert = 10$
2	5	0	5	4	8	8	−5	5	0	0	4	−4	$\frac{1}{2}\lvert -5 \times 0 - 5 \times 4\rvert = 10$
3	5	5	9.17	4	8	5.5	0	4.17	−4.17	2.5	1.5	−4	$\frac{1}{2}\lvert 0 \times (-2.5) - 4.17 \times 4\rvert = 8.34$
4	9.17	5	9.17	5.5	8	8	−4.17	4.17	0	0	2.5	−2.5	$\frac{1}{2}\lvert -4.17 \times 0 - 4.17 \times 2.5\rvert = 5.2125$
5	9.17	9.17	12	5.5	8	8	0	2.83	−2.83	0	2.5	−2.5	$\frac{1}{2}\lvert 0 \times 0 - 2.83 \times 2.5\rvert = 3.5375$
6	9.17	12	12	5.5	8	5.5	2.83	0	−2.83	2.5	0	−2.5	$\frac{1}{2}\lvert 2.83 \times (-2.5) - 0 \times 2.5\rvert = 3.5375$
7	0	5	0	4	4	0	5	−5	0	4	−4	0	$\frac{1}{2}\lvert 5 \times (-4) - (-5) \times 0\rvert = 10$
8	0	5	5	0	4	0	5	0	−5	4	0	−4	$\frac{1}{2}\lvert 5 \times (-4) - 0 \times 4\rvert = 10$
9	5	5	8	0	4	0	0	3	−3	4	0	−4	$\frac{1}{2}\lvert 0 \times (-4) - 3 \times 4\rvert = 6$
10	8	5	9.17	0	4	2.83	−3	4.17	−1.17	1.17	2.83	−4	$\frac{1}{2}\lvert -3 \times (-1.17) - 4.17 \times 4\rvert = 6.585$
11	5	9.17	9.17	4	5.5	2.83	4.17	0	−4.17	2.67	−1.17	−1.5	$\frac{1}{2}\lvert 4.17 \times (-2.67) - 0 \times 1.5\rvert = 5.56695$
12	9.17	9.17	12	5.5	2.83	4	0	2.83	−2.83	−1.17	−1.5	2.67	$\frac{1}{2}\lvert 0 \times 1.17 - 2.83 \times (-2.67)\rvert = 3.77805$
13	9.17	12	12	5.5	5.5	4	2.83	0	−2.83	1.5	−1.5	0	$\frac{1}{2}\lvert 2.83 \times (-1.5) - 0 \times 0\rvert = 2.1225$

$$[K^{(3)}] = \begin{array}{c} \begin{matrix} 6 & \quad 2 & \quad 7 \end{matrix} \\ \begin{bmatrix} 0.7086 & -0.4088 & -0.2998 \\ -0.4088 & 0.5887 & -0.1799 \\ -0.2998 & -0.1799 & 0.4796 \end{bmatrix} \end{array} \begin{matrix} \\ 6 \\ 2 \\ 7 \end{matrix}$$

$$[K^{(4)}] = \begin{array}{c} \begin{matrix} 7 & \quad 2 & \quad 3 \end{matrix} \\ \begin{bmatrix} 0.8340 & 0.0000 & -0.8340 \\ 0.0000 & 0.2998 & -0.2998 \\ -0.8340 & -0.2998 & 1.1338 \end{bmatrix} \end{array} \begin{matrix} \\ 7 \\ 2 \\ 3 \end{matrix}$$

$$[K^{(5)}] = \begin{array}{c} \begin{matrix} 7 & \quad 3 & \quad 4 \end{matrix} \\ \begin{bmatrix} 0.5660 & -0.5660 & 0.0000 \\ -0.5660 & 1.0077 & -0.4417 \\ 0.0000 & -0.4417 & 0.4417 \end{bmatrix} \end{array} \begin{matrix} \\ 7 \\ 3 \\ 4 \end{matrix}$$

$$[K^{(6)}] = \begin{array}{c} \begin{matrix} 7 & \quad 4 & \quad 8 \end{matrix} \\ \begin{bmatrix} 0.4417 & 0.0000 & -0.4417 \\ 0.0000 & 0.5660 & -0.5660 \\ -0.4417 & -0.5660 & 1.0077 \end{bmatrix} \end{array} \begin{matrix} \\ 7 \\ 4 \\ 8 \end{matrix}$$

$$[K^{(7)}] = \begin{array}{c} \begin{matrix} 5 & \quad 6 & \quad 9 \end{matrix} \\ \begin{bmatrix} 1.025 & -0.400 & -0.625 \\ -0.400 & 0.400 & 0.000 \\ -0.625 & 0.000 & 0.625 \end{bmatrix} \end{array} \begin{matrix} \\ 5 \\ 6 \\ 9 \end{matrix}$$

$$[K^{(8)}] = \begin{array}{c} \begin{matrix} 9 & \quad 6 & \quad 10 \end{matrix} \\ \begin{bmatrix} 0.400 & 0.000 & -0.400 \\ 0.000 & 0.625 & -0.625 \\ -0.400 & -0.625 & 1.025 \end{bmatrix} \end{array} \begin{matrix} \\ 9 \\ 6 \\ 10 \end{matrix}$$

$$[K^{(9)}] = \begin{array}{c} \begin{matrix} 10 & \quad 6 & \quad 11 \end{matrix} \\ \begin{bmatrix} 1.0417 & -0.3750 & -0.6667 \\ -0.3750 & 0.3750 & 0.0000 \\ -0.6667 & 0.0000 & 0.6677 \end{bmatrix} \end{array} \begin{matrix} \\ 10 \\ 6 \\ 11 \end{matrix}$$

$$[K^{(10)}] = \begin{array}{c} \begin{matrix} 11 & \quad 6 & \quad 12 \end{matrix} \\ \begin{bmatrix} 0.7121 & -0.0595 & -0.6526 \\ -0.0595 & 0.3560 & -0.2965 \\ -0.6526 & -0.2965 & 0.9491 \end{bmatrix} \end{array} \begin{matrix} \\ 11 \\ 6 \\ 12 \end{matrix}$$

$$[K^{(11)}] = \begin{array}{c} \begin{matrix} 6 & \quad 7 & \quad 12 \end{matrix} \\ \begin{bmatrix} 0.3201 & -0.1403 & -0.1799 \\ -0.1403 & 0.8424 & -0.7021 \\ -0.1799 & -0.7021 & 0.8819 \end{bmatrix} \end{array} \begin{matrix} \\ 6 \\ 7 \\ 12 \end{matrix}$$

$$[K^{(12)}] = \begin{array}{c} \begin{matrix} 7 & 12 & 13 \end{matrix} \\ \begin{bmatrix} 0.6505 & -0.4138 & -0.2067 \\ -0.4138 & 0.6788 & -0.2650 \\ -0.2067 & -0.2650 & 0.4717 \end{bmatrix} \end{array} \begin{matrix} 7 \\ 12 \\ 13 \end{matrix}$$

$$[K^{(13)}] = \begin{array}{c} \begin{matrix} 7 & 8 & 13 \end{matrix} \\ \begin{bmatrix} 0.2650 & -0.2650 & 0.0000 \\ -0.2650 & 1.2084 & -0.9433 \\ 0.0000 & -0.9433 & 0.9433 \end{bmatrix} \end{array} \begin{matrix} 7 \\ 8 \\ 13 \end{matrix}$$

For the computation of the element characteristic vectors we use Eq. (18.21) and obtain

$$\vec{P}^{(e)} = -\int_{C_1} V_0 [N]^T \cdot dC_2 = -V_0 \int_{s_i}^{s_j} \begin{Bmatrix} N_1 \\ N_2 \\ 0 \end{Bmatrix} ds = -\frac{V_0 s_{ji}}{2} \begin{Bmatrix} 1 \\ 1 \\ 0 \end{Bmatrix} \tag{E$_3$}$$

if the velocity of the fluid leaving the edge ij is specified as V_0. Similarly, we obtain

$$\vec{P}^{(e)} = -\frac{V_0 s_{kj}}{2} \begin{Bmatrix} 0 \\ 1 \\ 1 \end{Bmatrix} \quad \text{if the velocity of the fluid leaving the edge } jk \text{ is specified as } V_0, \text{ and} \tag{E$_4$}$$

$$\vec{P}^{(e)} = -\frac{V_0 s_{ik}}{2} \begin{Bmatrix} 1 \\ 0 \\ 1 \end{Bmatrix} \quad \text{if the velocity of the fluid leaving the edge } ki \text{ is specified as } V_0. \tag{E$_5$}$$

In Eqs. (E$_3$)–(E$_5$), s_{ji}, s_{kj}, and s_{ik} denote the lengths of the edges ij, jk, and ki, respectively.

In the current case, the velocity entering the boundary AB is prescribed as $u_0 = 1$ (or $V_0 = -1$), and hence the vectors $\vec{P}^{(e)}$ will be nonzero only for elements 1 and 7. These nonzero vectors can be computed as follows:

$$\vec{P}^{(1)} = \frac{1 \times 4}{2} \begin{Bmatrix} 1 \\ 1 \\ 0 \end{Bmatrix} = \begin{Bmatrix} 2 \\ 2 \\ 0 \end{Bmatrix} \begin{matrix} 5 \\ 1 \\ 6 \end{matrix} \quad \leftarrow \text{global node number}$$

(for element 1, the specified velocity is along the edge ij, i.e., 12.)

$$\vec{P}^{(7)} = \frac{1 \times 4}{2} \begin{Bmatrix} 1 \\ 0 \\ 1 \end{Bmatrix} = \begin{Bmatrix} 2 \\ 0 \\ 2 \end{Bmatrix} \begin{matrix} 5 \\ 6 \\ 9 \end{matrix} \quad \leftarrow \text{global node number}$$

(for element 7, the specified velocity is along the edge ik, i.e., 13.)

Step 4: Assemble the element matrices and vectors to obtain the overall system matrix and vector as follows:

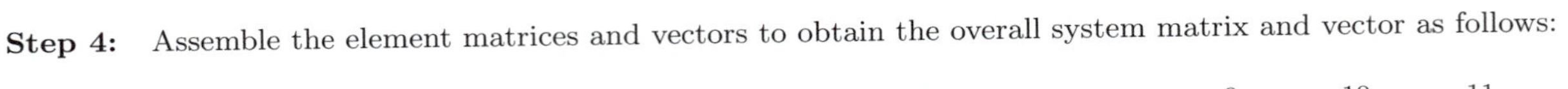

$$
[\underset{\sim}{K}] = \begin{array}{c} \\ \\ \end{array}
\begin{array}{ccccccccccccc}
1 & 2 & 3 & 4 & 5 & 6 & 7 & 8 & 9 & 10 & 11 & 12 & 13
\end{array}
$$

$$
[\underset{\sim}{K}] = \left[\begin{array}{cccccccc:ccccc}
1.0250 & -0.4000 & 0 & 0 & -0.6250 & 0 & 0 & 0 & 0 & 0 & 0 & 0 & 0 \\
-0.4000 & 1.9135 & -0.2998 & 0 & 0 & -1.0338 & -0.1799 & 0 & 0 & 0 & 0 & 0 & 0 \\
0 & -0.2998 & 2.1415 & -0.4417 & 0 & 0 & -1.4000 & 0 & 0 & 0 & 0 & 0 & 0 \\
0 & 0 & -0.4417 & 1.0077 & 0 & 0 & 0 & -0.5660 & 0 & 0 & 0 & 0 & 0 \\
-0.6250 & 0 & 0 & 0 & 2.0500 & -0.800 & 0 & 0 & -0.6250 & 0 & 0 & 0 & 0 \\
0 & -1.0338 & 0 & 0 & -0.800 & 3.8097 & -0.4401 & 0 & 0 & -1.0000 & -0.0595 & -0.4764 & 0 \\
0 & -0.1799 & -1.4000 & 0 & 0 & -0.4401 & 4.0492 & -0.7067 & 0 & 0 & 0 & -1.1159 & -0.2067 \\
0 & 0 & 0 & -0.5660 & 0 & 0 & -0.7067 & 2.2161 & 0 & 0 & 0 & 0 & -0.9433 \\
\hdashline
0 & 0 & 0 & 0 & -0.6250 & 0 & 0 & 0 & 1.0250 & -0.4000 & 0 & 0 & 0 \\
0 & 0 & 0 & 0 & 0 & -1.0000 & 0 & 0 & -0.4000 & 2.0667 & -0.6667 & 0 & 0 \\
0 & 0 & 0 & 0 & 0 & -0.0595 & 0 & 0 & 0 & -0.6667 & 1.3788 & -0.6526 & 0 \\
0 & 0 & 0 & 0 & 0 & -0.4764 & -1.1159 & 0 & 0 & 0 & -0.6526 & 2.5098 & -0.2650 \\
0 & 0 & 0 & 0 & 0 & 0 & -0.2067 & -0.9433 & 0 & 0 & 0 & -0.2650 & 1.4150
\end{array}\right]
\begin{array}{c} 1 \\ 2 \\ 3 \\ 4 \\ 5 \\ 6 \\ 7 \\ 8 \\ 9 \\ 10 \\ 11 \\ 12 \\ 13 \end{array},
$$

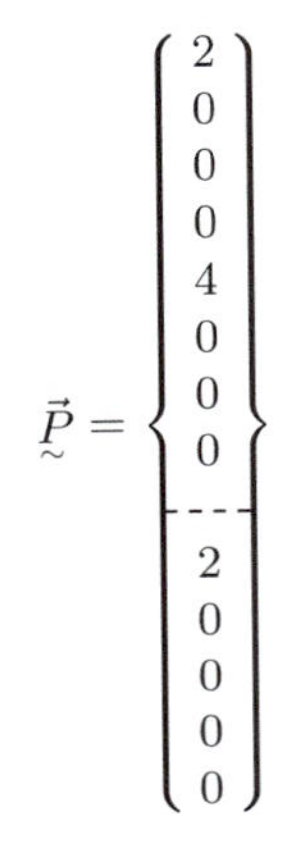

$$
\vec{\underset{\sim}{P}} = \left\{\begin{array}{c} 2 \\ 0 \\ 0 \\ 0 \\ 4 \\ 0 \\ 0 \\ 0 \\ \hdashline 2 \\ 0 \\ 0 \\ 0 \\ 0 \end{array}\right\}
$$

Step 5: Solve the assembled equations

$$[\underset{\sim}{K}]\underset{\sim}{\vec{\Phi}} = \underset{\sim}{\vec{P}} \tag{E$_6$}$$

after incorporating the boundary conditions specified on C_1. In the current case, the value of ϕ is set equal to zero along CD. Thus, the boundary conditions to be satisfied are $\Phi_4 = \Phi_8 = \Phi_{13} = 0$. One way of incorporating these boundary conditions is to delete the rows and columns corresponding to these degrees of freedom from Eqs. E(6). Another method is to modify the matrix $[\underset{\sim}{K}]$ and vector $\underset{\sim}{\vec{P}}$ as indicated in Section 6.5 (with the help of the subroutine ADJUST). The solution of the resulting equations is given by

$$\underset{\sim}{\vec{\Phi}} = \begin{Bmatrix} \Phi_1 \\ \Phi_2 \\ \Phi_3 \\ \Phi_4 \\ \Phi_5 \\ \Phi_6 \\ \Phi_7 \\ \Phi_8 \\ \Phi_9 \\ \Phi_{10} \\ \Phi_{11} \\ \Phi_{12} \\ \Phi_{13} \end{Bmatrix} = \begin{Bmatrix} 14.900 \\ 9.675 \\ 4.482 \\ 0.000 \\ 15.044 \\ 10.011 \\ 4.784 \\ 0.000 \\ 15.231 \\ 10.524 \\ 8.469 \\ 6.229 \\ 0.000 \end{Bmatrix}$$

From these nodal values of ϕ, the average value of the u component of velocity between the nodes 7 and 8 can be computed as

$$(u)_{7-8} = \frac{\partial \phi}{\partial x} \approx \frac{\Phi_8 - \Phi_7}{x_8 - x_7} = \frac{0.000 - 4.784}{9.17 - 12.00} = 1.690$$

18.4 STREAM FUNCTION FORMULATION

In the stream function formulation the problem can be stated as follows.

18.4.1 Differential Equation Form

Find the stream function $\psi(x, y)$ in the given region S surrounded by the curve C such that

$$\nabla^2 \psi = \frac{\partial^2 \psi}{\partial x^2} + \frac{\partial^2 \psi}{\partial y^2} = 0 \text{ in } S \tag{18.25}$$

with the boundary conditions

Dirichlet condition: $\psi = \psi_0$ on C_1 (18.26)

Neumann condition: $V_s = -\dfrac{\partial \psi}{\partial n} = V_0$ on C_2 (18.27)

where V_0 is the velocity of the fluid parallel to the boundary C_2.

18.4.2 Variational Form

Find the stream function $\psi(x, y)$ that minimizes the functional

$$I = \frac{1}{2} \iint_S \left[\left(\frac{\partial \psi}{\partial x} \right)^2 + \left(\frac{\partial \psi}{\partial y} \right)^2 \right] \mathrm{d}S - \int_{C_2} V_0 \psi \, \mathrm{d}C_2 \tag{18.28}$$

with the boundary condition

$$\psi = \psi_0 \text{ on } C_1 \tag{18.29}$$

18.4.3 Finite Element Solution

Since the governing equations (18.25)–(18.27) and (18.28)–(18.29) are similar to Eqs. (18.8)–(18.10) and (18.11)–(18.12), the finite element equations will also be similar. These can be expressed as

$$[K^{(e)}] \vec{\Psi}^{(e)} = \vec{P}^{(e)} \tag{18.30}$$

where $[K^{(e)}]$ and $\vec{P}^{(e)}$ are given by Eqs. (18.20) and (18.21), respectively, and

$$\vec{\Psi}^{(e)} = \begin{Bmatrix} \Psi_1 \\ \Psi_2 \\ \vdots \\ \Psi_p \end{Bmatrix} \tag{18.31}$$

Example 18.2 (*Confined Flow around a Cylinder*) Find the velocity distribution along the vertical center line CD in Figure 18.1(a).

Solution Here also, we consider only the quadrant $ABCDEA$ for analysis. The boundaries AED and BC can be seen to be streamlines. We assume the value of the streamline along AED as zero as a reference value. Because the velocity component entering the face AB is constant, we obtain

$$\int_{\psi_A}^{\psi_B} \mathrm{d}\psi = \psi_B - \psi_A \equiv \int_{y=y_A}^{y=y_B} u_0 \, \mathrm{d}y = u_0(y_B - y_A) \tag{E$_1$}$$

Because $u_0 = 1$ and the streamline passing through A, namely ψ_A, is taken as 0, Eq. (E_1) gives

$$\psi_B - 0 = 1(4 - 0) \text{ or } \psi_B = 4 \tag{E$_2$}$$

Since u_0 is constant along AB, the value of ψ varies linearly along AB, and hence the value of ψ at node 5 (Ψ_5) will be equal to 2. Thus, the Dirichlet boundary conditions are given by

$$\Psi_1 = \Psi_2 = \Psi_3 = \Psi_4 = 4, \qquad \Psi_5 = 2, \qquad \Psi_9 = \Psi_{10} = \Psi_{11} = \Psi_{12} = \Psi_{13} = 0 \qquad (E_3)$$

Because there is no velocity specified parallel to any boundary, the element characteristic vectors $\vec{P}^{(e)}$ will be $\vec{0}$ for all the elements. The element characteristic matrices $[K^{(e)}]$ and the assembled matrix $[\underset{\sim}{K}]$ will be same as those derived in Example 18.1. Thus, the final matrix equations to be solved are given by

$$\underset{13\times 13}{[\underset{\sim}{K}]}\ \underset{13\times 1}{\vec{\underset{\sim}{\Psi}}} = \underset{13\times 1}{\vec{\underset{\sim}{P}}} \qquad (E_4)$$

where $[\underset{\sim}{K}]$ is the same as the one derived in Example 18.1, $\vec{\underset{\sim}{\Psi}}^T = (\Psi_1\ \Psi_2\ \Psi_3 \ldots\ \Psi_{13}) =$ vector of nodal unknowns, and $\vec{\underset{\sim}{P}}^T = (0\,0\,0 \ldots\ 0) =$ vector of nodal actions. The boundary conditions to be satisfied are given by Eq. (E_3). The solution of this problem can be obtained as

$$\vec{\underset{\sim}{\Psi}} = \begin{Bmatrix} \Psi_1 \\ \Psi_2 \\ \Psi_3 \\ \Psi_4 \\ \Psi_5 \\ \Psi_6 \\ \Psi_7 \\ \Psi_8 \\ \Psi_9 \\ \Psi_{10} \\ \Psi_{11} \\ \Psi_{12} \\ \Psi_{13} \end{Bmatrix} = \begin{Bmatrix} 4.0 \\ 4.0 \\ 4.0 \\ 4.0 \\ 2.0 \\ 1.741 \\ 2.042 \\ 1.673 \\ 0.0 \\ 0.0 \\ 0.0 \\ 0.0 \\ 0.0 \end{Bmatrix} \qquad (E_5)$$

Since the stream function has been assumed to vary linearly within each element, the velocity will be a constant in each element. Thus, the u component of velocity, for example, can be computed between any two nodes i and j as

$$u = \frac{\partial \Psi}{\partial y} \approx \frac{\Psi_j - \Psi_i}{y_j - y_i}$$

where y_i and y_j denote the y coordinates of nodes i and j, respectively. By using this formula, we can obtain the value of u along the line 4–8–13 as

$$(u)_{4-8} = \frac{\Psi_8 - \Psi_4}{y_8 - y_4} = \frac{1.673 - 4.0}{5.5 - 8.0} = 0.9308$$

$$(u)_{8-13} = \frac{\Psi_{13} - \Psi_8}{y_{13} - y_8} = \frac{0 - 1.673}{4.0 - 5.5} = 1.1153$$

18.5 COMPUTER PROGRAM FOR POTENTIAL FUNCTION APPROACH

A subroutine called PHIFLØ is written for the solution of the problem of confined flow around a cylinder based on the potential function formulation. The finite element idealization consists of triangular elements. The arguments of this subroutine are as follows:

NN = number of nodes (input).
NE = number of elements (input).
NB = semibandwidth of the overall matrix GK (input).
XC, YC = array of size NN; XC(I), YC(I) = x and y coordinates of node I (input).
NØDE = array of size NE $\times$ 3; NØDE (I, J) = global node number corresponding to Jth corner of element I (input).
GK = array of size NN $\times$ NB used to store the matrix $[\underset{\sim}{K}]$.
P = array of size NN used to store the vector $\underset{\sim}{\vec{P}}$.
Q = array of size NE; Q(I) = velocity of the fluid leaving the element I through one of its edges (input).
A = array of size NE; A(I) = area of element I.
PS = array of size NN; PS(I) = specified value of ϕ at node I. If the value of ϕ is not specified at node I, then the value of PS(I) is to be set equal to -1000.0 (input).
ICØN = array of size NE; ICØN(I) = 1 if element lies along the boundary on which the velocity is specified, and = 0 otherwise (input).
NCØN = array of size NE $\times$ 2; NCØN(I,J) = Jth node of element I that lies on the boundary on which the velocity is specified. Need not be given if ICØN(I) = 0 for all I (input).
PLØAD = array of size NN $\times$ 1 used to store the final right-hand-side vector. It represents the solution vector (nodal values of ϕ) upon return from the subroutine PHIFLØ.

To illustrate the use of the subroutine PHIFLØ, the problem of Example 18.1 is considered. The main program that calls the subroutine PHIFLØ and the results given by the program are listed below.

```
C======================================================================
C
C     MAIN PROGRAM TO CALL PHIFLO
C
C======================================================================
      DIMENSION XC(13),YC(13),NODE(13,3),ICON(13),NCON(13,2),GK(13,7),
     2 A(13),PS(13),PLOAD(13,1),Q(13),P(13)
      DATA NN,NE,NB/13,13,7/
      DATA(NODE(I,1),I=1,13)/5,6,6,7,7,7,5,9,10,11,6,7,7/
      DATA(NODE(I,2),I=1,13)/1,1,2,2,3,4,6,6,6,6,7,12,8/
      DATA(NODE(I,3),I=1,13)/6,2,7,3,4,8,9,10,11,12,12,13,13/
      DATA XC/0.0,5.0,9.17,12.0,0.0,5.0,9.17,12.0,0.0,5.0,8.0,9.17,12.0/
      DATA YC/8.0,8.0,8.0,8.0,4.0,4.0,5.5,5.5,0.0,0.0,0.0,2.83,4.0/
      DATA ICON/1,0,0,0,0,0,1,0,0,0,0,0,0/
      DATA(NCON(I,1),I=1,13)/5,0,0,0,0,0,9,0,0,0,0,0,0/
      DATA(NCON(I,2),I=1,13)/1,0,0,0,0,0,5,0,0,0,0,0,0/
      DO 10 I=1,NN
```

```
10    PS(I) = -1000.0
      PS(4) = 0.0
      PS(8) = 0.0
      PS(13) = 0.0
      DATA Q/-1.0,0.0,0.0,0.0,0.0,0.0,-1.0,0.0,0.0,0.0,0.0,0.0,0.0/
      CALL PHIFLO(NN,NE,NB,NODE,XC,YC,GK,P,A,ICON,NCON,Q,PS,PLOAD)
      PRINT 20
20    FORMAT(2X,'VALUES OF POTENTIAL FUNCTION AT VARIOUS NODES',/)
      DO 30 I=1,NN
30    PRINT 40,I,PLOAD(I,1)
40    FORMAT(I6,E15.8)
      STOP
      END
```

```
VALUES OF POTENTIAL FUNCTION AT VARIOUS NODES

     1 0.14900365E+02
     2 0.96753922E+01
     3 0.44818130E+01
     4 0.00000000E+00
     5 0.15044345E+02
     6 0.10010671E+02
     7 0.47837925E+01
     8 0.00000000E+00
     9 0.15231425E+02
    10 0.10523736E+02
    11 0.84687223E+01
    12 0.62288446E+01
    13 0.00000000E+00
```

REFERENCES

18.1 G.F. Pinder and W.G. Gray: *Finite Element Simulation in Surface and Subsurface Hydrology*, Academic Press, New York, 1977.

18.2 J.C. Connor and C.A. Brebbia: *Finite Element Techniques for Fluid Flow*, Butterworths, London, 1976.

PROBLEMS

18.1 The unsteady fluid flow through a porous medium (seepage flow) is governed by the equation

$$\frac{\partial}{\partial x}\left[k_x \frac{\partial \phi}{\partial x}\right] + \frac{\partial}{\partial y}\left[k_y \frac{\partial \phi}{\partial y}\right] + \frac{\partial}{\partial z}\left[k_z \frac{\partial \phi}{\partial z}\right] + \dot{q} = \alpha \frac{\partial \phi}{\partial t} \tag{E$_1$}$$

with boundary conditions

$$\phi = \phi_0 \text{ on } S_1 \tag{E$_2$}$$

$$k_x \frac{\partial \phi}{\partial x} l_x + k_y \frac{\partial \phi}{\partial y} l_y + k_z \frac{\partial \phi}{\partial z} l_z + q(t) = 0 \text{ on } S_2 \tag{E$_3$}$$

where k_x, k_y, k_z are the coefficients of permeability in $x, y,$ and z directions; $\dot{q}$ is the quantity of fluid added (recharge) per unit time; α is the specific storage (for a confined flow); ϕ is the fluid potential; l_x, l_y, l_z are the direction cosines of the outward normal to surface S_2; ϕ_0 is the specified value of ϕ on the boundary S_1; and $q(t)$ is the specified value of velocity of the fluid normal to the surface S_2. Derive the finite element equations of the flow using the Galerkin approach.

18.2 Consider the steady-state confined seepage through a rectangular soil mass subject to a specified fluid pressure head on the left side as indicated in Figure 18.2. Assuming the permeabilities of the soil in the horizontal and vertical directions

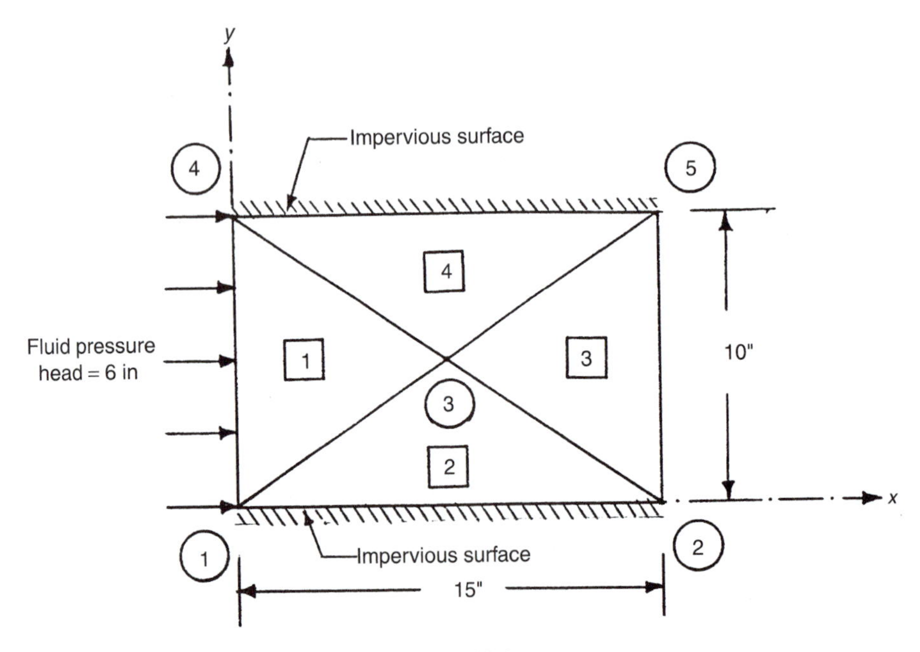

Figure 18.2.

as $k_x = k_y = 2$ in./sec, determine the distribution of the potential in the soil mass.

Hint: The governing equation is given by

$$k_x \frac{\partial^2 \phi}{\partial x^2} + k_y \frac{\partial^2 \phi}{\partial y^2} = 0$$

subject to $\phi_1 = \phi_4 = 6$ in.

18.3 Determine the velocity components of the fluid for the seepage flow considered in Problem 18.2 using the nodal values of ϕ and Darcy's law:

$$u = -k_x \frac{\partial \phi}{\partial x}, \qquad v = -k_y \frac{\partial \phi}{\partial y}$$

18.4 The dam shown in Figure 18.3 retains water at a height of 12 ft on the upstream side and 2 ft on the downstream side. If the permeability of the soil, considered isotropic, is $k = 10$ ft/hr, indicate a procedure for determining the following:

(a) Equipotential lines
(b) Quantity of water seeping into the soil per hour per 1-ft thickness of the dam (in z direction)

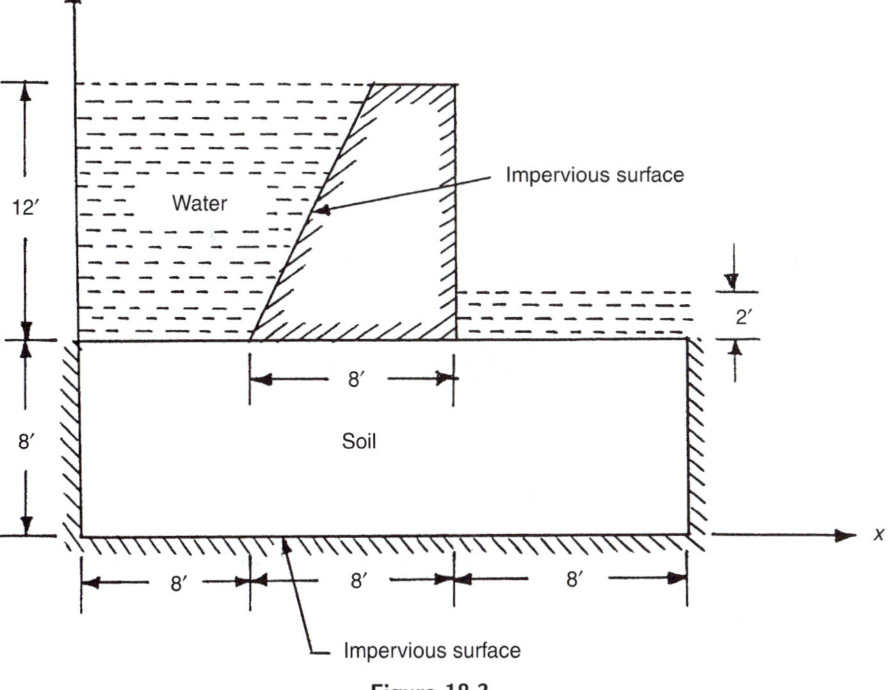

Figure 18.3.

18.5 In the finite element analysis of a two-dimensional flow using triangular elements, the velocity components u and v are assumed to vary linearly within an element (e) as

$$u(x,y) = a_1 U_i^{(e)} + a_2 U_j^{(e)} + a_3 U_k^{(e)}$$
$$v(x,y) = a_1 V_i^{(e)} + a_2 V_j^{(e)} + a_3 V_k^{(e)}$$

where $(U_i^{(e)}, V_i^{(e)})$ denote the values of (u, v) at node i. Find the relationship between $U_i^{(e)}, V_i^{(e)}, \ldots, V_k^{(e)}$ that is to be satisfied for the flow to be incompressible.

18.6 Develop the necessary finite element equations for the analysis of two-dimensional steady flow (seepage) toward a well using linear rectangular isoparametric elements.

18.7 Write a subroutine called PSIFLØ for the analysis of two-dimensional flow using triangular elements based on stream function formulation. Solve the problem of Example 18.2 using this subroutine.

18.8 The fluid flow in a duct is governed by the equation

$$\frac{\partial^2 W}{\partial x^2} + \frac{\partial^2 W}{\partial y^2} + 1 = 0$$

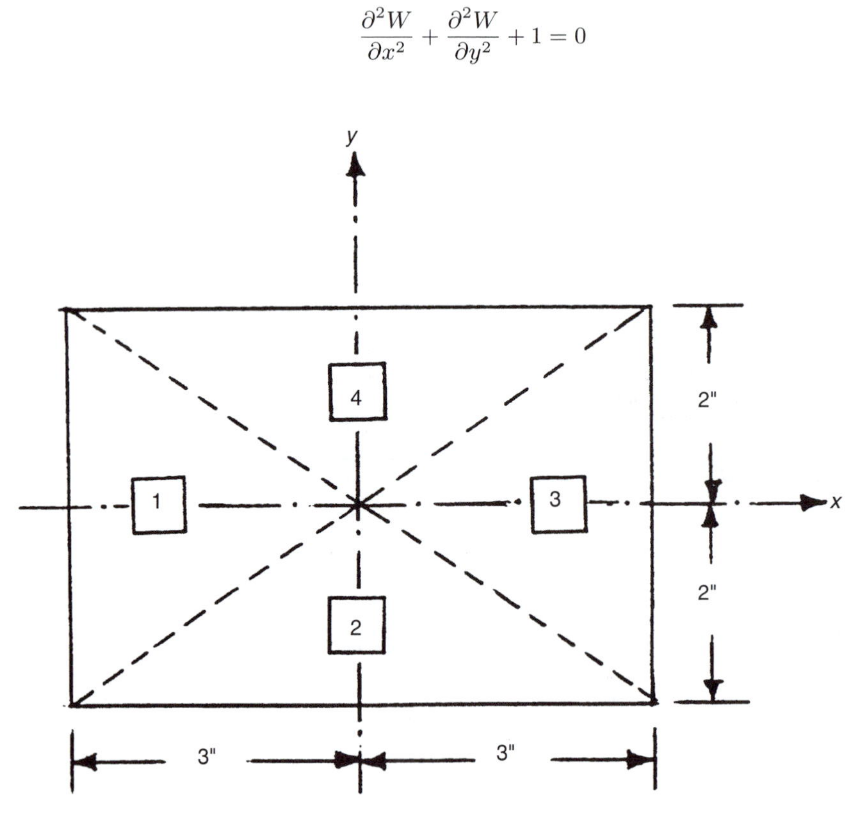

Figure 18.4.

with $W = 0$ on the boundary, and $W(x, y)$ is the nondimensional velocity of the fluid in the axial direction given by

$$W(x, y) = \frac{w(x, y)}{2w_0 f R_e}$$

where $w(x, y)$ is the axial velocity of the fluid, w_0 is the mean value of $w(x, y)$, f is the Fanning friction factor, R_e is the Reynolds number $[R_e = (w_0\, d_h/v)]$, v is the kinematic viscosity of the fluid, and d_h is the hydraulic diameter of the cross section $[d_h = (4.\ \text{area}/\text{perimeter})]$.

(a) Determine the distribution of $W(x, y)$ in a rectangular duct using four linear triangles for idealization as shown in Figure 18.4.
(b) Suggest a method of finding the value of the Fanning friction factor f in each triangular element using the known nodal values of W.
(c) Find the Fanning friction factor f for a flow with $R_e = 200$.

18.9 Find the velocity distribution along the vertical center line CD in Figure 18.1 using a different finite element grid.

18.10 Write a subroutine called *PSIFLO* for the solution of the problem of confined flow around a cylinder based on the stream function formulation. Using this subroutine, find the velocity distribution along the vertical center line CD in Figure 18.1(a).

19

VISCOUS AND NON-NEWTONIAN FLOWS

19.1 INTRODUCTION

The basic equations governing the two-dimensional steady incompressible Newtonian flow can be obtained from Eqs. (17.41) and (17.5) as

Conservation of momentum in x direction:
$$\frac{\partial p}{\partial x} + \rho \left(u \frac{\partial u}{\partial x} + v \frac{\partial u}{\partial y} \right) = X + \mu \left(\frac{\partial^2 u}{\partial x^2} + \frac{\partial^2 u}{\partial y^2} \right) \tag{19.1}$$

Conservation of momentum in y direction:
$$\frac{\partial p}{\partial y} + \rho \left(u \frac{\partial v}{\partial x} + v \frac{\partial v}{\partial y} \right) = Y + \mu \left(\frac{\partial^2 v}{\partial x^2} + \frac{\partial^2 v}{\partial y^2} \right) \tag{19.2}$$

Continuity equation:
$$\frac{\partial u}{\partial x} + \frac{\partial v}{\partial y} = 0 \tag{19.3}$$

where X and Y denote, respectively, the x and y components of the body force per unit volume ($X = \rho B_x$, $Y = \rho B_y$). When the convective terms (terms involving ρ) in Eqs. (19.1) and (19.2) are neglected, we obtain

$$\mu \left(\frac{\partial^2 u}{\partial x^2} + \frac{\partial^2 u}{\partial y^2} \right) - \frac{\partial p}{\partial x} + X = 0 \tag{19.4}$$

$$\mu \left(\frac{\partial^2 v}{\partial x^2} + \frac{\partial^2 v}{\partial y^2} \right) - \frac{\partial p}{\partial y} + Y = 0 \tag{19.5}$$

which are known as Stokes equations. The boundary conditions for the problem may be specified in terms of pressure, velocity, and velocity gradient. Three different formulations can be used for solving Eqs. (19.1)–(19.3). These are (i) stream function formulation, in which the stream function (ψ) is treated as the unknown function [19.1]; (ii) velocity–pressure formulation, in which u, v, and p are treated as unknowns [19.2]; and (iii) stream function–vorticity formulation, in which the stream function (ψ) and vorticity (ω) are taken as unknown field variables [19.3]. We consider all these formulations in this chapter.

19.2 STREAM FUNCTION FORMULATION (USING VARIATIONAL APPROACH)

By introducing the stream function ψ defined by Eq. (17.63), the continuity equation (19.3) can be satisfied exactly and the momentum equations can be combined to obtain a single equation in terms of ψ as (assuming the body forces to be zero)

$$\nu \nabla^4 \psi + \frac{\partial \psi}{\partial x} \nabla^2 \left(\frac{\partial \psi}{\partial y} \right) - \frac{\partial \psi}{\partial y} \nabla^2 \left(\frac{\partial \psi}{\partial x} \right) = 0 \tag{19.6}$$

where $\nu = \mu/\rho$, $\nabla^2 = (\partial^2/\partial x^2) + (\partial^2/\partial y^2)$ = harmonic operator, and $\nabla^4 = (\partial^4/\partial x^4) + 2(\partial^4/\partial x^2 \partial y^2) + (\partial^4/\partial y^4)$ = biharmonic operator. The nonlinear terms in Eq. (19.6), which come from the convective terms of Eqs. (19.1) and (19.2), are the ones that make this equation very difficult to solve. However, approximate numerical solutions may be obtained using the finite element method if Eq. (19.6) is recast as an equivalent variational problem.

Although no universally accepted variational principle is available for the Navier–Stokes equations or for Eq. (19.6), Olson [19.1] developed a pseudo-variational principle for solving Eq. (19.6) using Cowper's 18 degrees-of-freedom triangular element (conforming bending element stated in Section 10.8). For a typical triangular element shown in Figure 19.1, if the edge 1–2 is a boundary, then the boundary conditions along the edge 1–2 will be

$$\text{either} \qquad \psi = \text{constant} \quad \text{or} \quad \frac{\partial p}{\partial \xi} = 0 \tag{19.7}$$

and

$$\text{either} \qquad \frac{\partial \psi}{\partial \eta} = \text{constant} \quad \text{or} \quad \sigma_{\xi\eta} = \mu \left(\frac{\partial u}{\partial \eta} + \frac{\partial v}{\partial \xi} \right) = 0 \tag{19.8}$$

where (ξ, η) represents the local coordinate system, and $\sigma_{\xi\eta}$ denotes the shear stress (for a Newtonian fluid). The functional I, which on minimization gives the governing differential equation (19.6) and the boundary conditions of Eqs. (19.7) and (19.8), is given by

$$\begin{aligned} I(\psi) = & \iint\limits_{A^{(e)}} \left[\frac{\nu}{2} (\nabla^2 \psi)^2 + \left(\underline{\frac{\partial \psi}{\partial \eta} \nabla^2 \psi} \right) \frac{\partial \psi}{\partial \xi} - \left(\underline{\frac{\partial \psi}{\partial \xi} \nabla^2 \psi} \right) \frac{\partial \psi}{\partial \eta} \right] d\xi \, d\eta \\ & + \int_{-b}^{a} \left[2\nu \underline{\frac{\partial^2 \psi}{\partial \xi^2}} \frac{\partial \psi}{\partial \eta} - \left(\underline{\frac{\partial \psi}{\partial \xi} \frac{\partial^2 \psi}{\partial \xi^2} + \frac{\partial \psi}{\partial \eta} \frac{\partial^2 \psi}{\partial \xi \partial \eta}} \right) \psi \right] \Bigg|_{\eta=0} d\xi \end{aligned} \tag{19.9}$$

where $A^{(e)}$ is the area of the triangular element, and the underlined terms are to be taken as constants while taking the variation of I. Note that the boundary integral appears in Eq. (19.9) only when the triangular element (edge 1–2) lies on a boundary where some of the conditions of Eqs. (19.7) and (19.8) are to be satisfied.

Since the functional $I(\psi)$ contains derivatives up to order two, the continuity of ψ and its first derivatives between elements is to be maintained to guarantee convergence to the correct solution as the number of elements is increased. The triangular bending element due to Cowper *et al.* [19.4] does provide this continuity. This element employs the

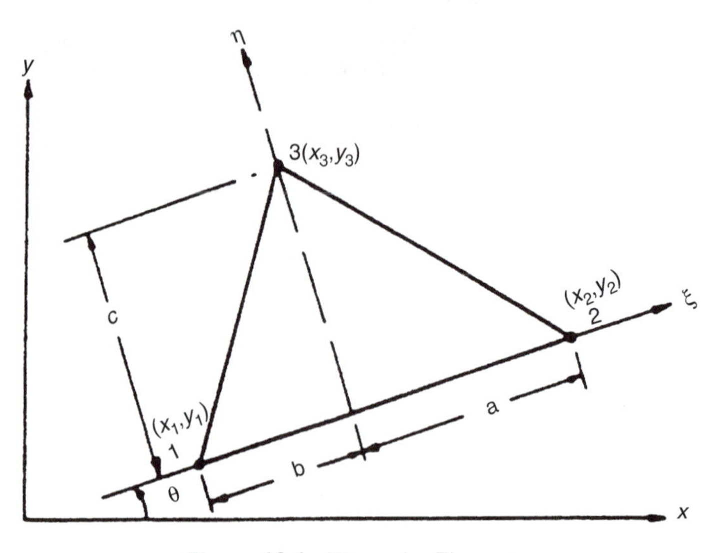

Figure 19.1. Triangular Element.

values of ψ, $(\partial\psi/\partial x)$, $(\partial\psi/\partial y)$, $(\partial^2\psi/\partial x^2)$, $(\partial^2\psi/\partial x\partial y)$, and $(\partial^2\psi/\partial y^2)$ at each of the three vortices as nodal variables. The interpolation function is taken as a full fifth-degree polynomial with 21 parameters as

$$\begin{aligned}\psi(\xi,\eta) &= \alpha_1 + \alpha_2\xi + \alpha_3\eta + \alpha_4\xi^2 + \alpha_5\xi\eta + \alpha_6\eta^2 + \alpha_7\xi^3 + \alpha_8\xi^2\eta + \alpha_9\xi\eta^2 \\ &\quad + \alpha_{10}\eta^3 + \alpha_{11}\xi^4 + \alpha_{12}\xi^3\eta + \alpha_{13}\xi^2\eta^2 + \alpha_{14}\xi\eta^3 + \alpha_{15}\eta^4 + \alpha_{16}\xi^5 \\ &\quad + \alpha_{17}\xi^3\eta^2 + \alpha_{18}\xi^2\eta^3 + \alpha_{19}\xi\eta^4 + \alpha_{20}\eta^5 + \alpha_{21}\xi^4\eta \\ &= \sum_{i=1}^{21} \alpha_i \xi^{m_i}\eta^{n_i} \end{aligned} \tag{19.10}$$

Since the element has only 18 degrees of freedom (six at each node), the 21 constants (α_i) of Eq. (19.10) are evaluated by using 18 nodal conditions and 3 additional conditions. The additional conditions are taken as that the variation of the derivative of ψ normal to an edge (called normal slope) be a cubic function of the edgewise coordinate. It can be seen that the condition that the normal slope $(\partial\psi/\partial\eta)$ be a cubic equation in ξ along the edge $\eta = 0$ (edge 12) can be satisfied if we set $\alpha_{21} = 0$ in Eq. (19.10). With this, Eq. (19.10) can be expressed as

$$\underset{1\times 1}{\psi}(\xi,\eta) = \underset{1\times 20}{[\beta]}\ \underset{20\times 1}{\vec{\alpha}} \tag{19.11}$$

where

$$[\beta] = [1 \quad \xi \quad \eta \quad \xi^2 \quad \dots \quad \eta^5] \tag{19.12}$$

and

$$\vec{\alpha}^T = \{\alpha_1 \quad \alpha_2 \quad \dots \quad \alpha_{20}\} \tag{19.13}$$

The conditions for cubic variation of normal slope along the remaining two edges are somewhat more complicated and are given by [19.4]

For cubic variation of normal slope along the edge 13:

$$5b^4 c\alpha_{16} + (3b^2c^3 - 2b^4c)\alpha_{17} + (2bc^4 - 3b^3c^2)\alpha_{18} + (c^5 - 4b^2c^3)\alpha_{19} - 5bc^4\alpha_{20} = 0 \tag{19.14}$$

For cubic variation of normal slope along the edge 23:

$$5a^4 c\alpha_{16} + (3a^2c^3 - 2a^4c)\alpha_{17} + (-2ac^4 + 3a^3c^2)\alpha_{18} + (c^5 - 4a^2c^3)\alpha_{19} + 5ac^4\alpha_{20} = 0 \tag{19.15}$$

where the dimensions a, b, and c (indicated in Figure 19.1) are given by

$$\left.\begin{aligned} a &= [(x_2 - x_3)(x_2 - x_1) + (y_2 - y_3)(y_2 - y_1)]/r \\ b &= [(x_3 - x_1)(x_2 - x_1) + (y_3 - y_1)(y_2 - y_1)]/r \\ c &= [(x_2 - x_1)(y_3 - y_1) - (x_3 - x_1)(y_2 - y_1)]/r \end{aligned}\right\} \tag{19.16}$$

where

$$r = [(x_2 - x_1)^2 + (y_2 - y_1)^2]^{1/2} \tag{19.17}$$

and (x_i, y_i) are the global (x, y) coordinates of node $i (i = 1, 2, 3)$. The 18 nodal unknowns of the element are given by

$$\begin{aligned} \vec{\psi}^{(e)^T} = \{&\psi_1 \quad \psi_{\xi_1} \quad \psi_{\eta_1} \quad \psi_{\xi\xi_1} \quad \psi_{\xi\eta_1} \quad \psi_{\eta\eta_1} \quad \psi_2 \quad \psi_{\xi_2} \quad \psi_{\eta_2} \\ &\psi_{\xi\xi_2} \quad \psi_{\xi\eta_2} \quad \psi_{\eta\eta_2} \quad \psi_3 \quad \psi_{\xi_3} \quad \psi_{\eta_3} \quad \psi_{\xi\xi_3} \quad \psi_{\xi\eta_3} \quad \psi_{\eta\eta_3}\} \end{aligned} \tag{19.18}$$

where $\psi_i = \psi$ at node i, $\psi_{\xi\eta_i} = (\partial^2\psi/\partial\xi\partial\eta)$ at node i, etc. Using Eqs. (19.11) and (19.18), we can obtain

$$\underset{18 \times 1}{\vec{\psi}^{(e)}} = \underset{18 \times 20}{[\beta]} \underset{20 \times 1}{\vec{\alpha}} \tag{19.19}$$

where $[\beta]$ can be derived without much difficulty. By using Eqs. (19.14), (19.15), and (19.19), we get 20 equations in 20 unknown coefficients $\alpha_i, i = 1, 2, \dots, 20$. These equations can be expressed in matrix form as

$$\underset{20 \times 1}{\begin{Bmatrix} \vec{\psi}^{(e)} \\ 0 \\ 0 \end{Bmatrix}} = \underset{20 \times 20}{[\underset{\sim}{\beta}]} \underset{20 \times 1}{\vec{\alpha}} \tag{19.20}$$

where $[\underset{\sim}{\beta}]$ can be identified without much difficulty.

Equation (19.20) can be inverted to obtain

$$\underset{20\times 1}{\vec{\alpha}} = \underset{20\times 20}{[\underset{\sim}{\beta}]^{-1}} \underset{20\times 1}{\begin{Bmatrix} \vec{\psi}^{(e)} \\ 0 \\ 0 \end{Bmatrix}} \tag{19.21}$$

and

$$\underset{20\times 1}{\vec{\alpha}} = \underset{20\times 18}{[R]} \underset{18\times 1}{\vec{\psi}^{(e)}} \tag{19.22}$$

where the 20×18 matrix $[R]$ consists of the first 18 columns of $[\underset{\sim}{\beta}]^{-1}$. The element properties can be obtained first with respect to the polynomial coefficients α_i by substituting Eq. (19.10) into Eq. (19.9). For an interior element, this gives

$$I = \sum_{i=1}^{21}\sum_{j=1}^{21} k_{ij}\alpha_i\alpha_j + \sum_{i=1}^{21}\sum_{j=1}^{21}\sum_{l=1}^{21} f_{ijl}\bar{\alpha}_i\bar{\alpha}_j\alpha_l \tag{19.23}$$

where

$$\begin{aligned} k_{ij} &= \frac{\nu}{2}\{m_i m_j(m_i-1)(m_j-1)g(m_i+m_j-4, n_i+n_j-1) + [m_i n_j(m_i-1)(n_i-2) \\ &\quad + m_j n_i(m_j-1)(n_j-2)]\cdot g(m_i+m_j-2, n_i+n_j-3) \\ &\quad + n_i n_j(n_i-2)(n_j-2)\cdot g(m_i+m_j, n_i+n_j-5)\} \end{aligned} \tag{19.24}$$

$$\begin{aligned} f_{ijl} &= (n_i m_l - m_i n_l)[m_j(m_j-1)\cdot g(m_i+m_j+m_l-3, n_i+n_j+n_l-3) \\ &\quad + n_j(n_j-2)g(m_i+m_j+m_l-1, n_i+n_j+n_l-5)] \end{aligned} \tag{19.25}$$

$$g(m,n) = \iint\limits_{A^{(e)}} x^m y^n \,\mathrm{d}x\,\mathrm{d}y \tag{19.26}$$

and the bars over α_i in Eq. (19.23) indicate that those terms are not varied when the variation of I is taken. The element properties are transformed to the global coordinate system by using the basic transformation relation

$$\vec{\psi}_i^{(e)} = [\lambda_i]\vec{\Psi}_i^{(e)} \tag{19.27}$$

where

$$\vec{\psi}_i^{(e)} = \begin{Bmatrix} \psi_i \\ \psi_{\xi_i} \\ \psi_{\eta_i} \\ \psi_{\xi\xi_i} \\ \psi_{\xi\eta_i} \\ \psi_{\eta\eta_i} \end{Bmatrix}, \qquad \vec{\Psi}_i^{(e)} = \begin{Bmatrix} \psi_i \\ \psi_{x_i} \\ \psi_{y_i} \\ \psi_{xx_i} \\ \psi_{xy_i} \\ \psi_{yy_i} \end{Bmatrix}, \text{ and}$$

$$[\lambda_i] = \begin{bmatrix} 1 & 0 & 0 & 0 & 0 & 0 \\ 0 & \cos\theta & \sin\theta & 0 & 0 & 0 \\ 0 & -\sin\theta & \cos\theta & 0 & 0 & 0 \\ 0 & 0 & 0 & \cos^2\theta & 2\sin\theta\cos\theta & \sin^2\theta \\ 0 & 0 & 0 & -\sin\theta\cos\theta & (\cos^2\theta - \sin^2\theta) & \sin\theta\cos\theta \\ 0 & 0 & 0 & \sin^2\theta & -2\sin\theta\cos\theta & \cos^2\theta \end{bmatrix} \tag{19.28}$$

The final element equations can be expressed in the familiar form as

$$[K^{(e)}]\vec{\Psi}^{(e)} + [F^{(e)}(\vec{\Psi}^{(e)})]\vec{\Psi}^{(e)} = \vec{0} \tag{19.29}$$

where the elements of the matrix $[F^{(e)}]$ are nonlinear functions of the nodal unknowns $\vec{\Psi}^{(e)}$ with

$$\vec{\Psi}^{(e)} = \begin{Bmatrix} \vec{\Psi}_1^{(e)} \\ \vec{\Psi}_2^{(e)} \\ \vec{\Psi}_3^{(e)} \end{Bmatrix} \tag{19.30}$$

The overall or assembled equations can be obtained as

$$[\underset{\sim}{K}]\,\underset{\sim}{\vec{\Psi}} + [\underset{\sim}{F}(\underset{\sim}{\vec{\Psi}})]\,\underset{\sim}{\vec{\Psi}} = \vec{0} \tag{19.31}$$

These equations represent a set of N nonlinear algebraic equations in the N global variables $\Psi_i, i = 1, 2, \ldots, N$. Olson [19.1] solved these equations using the Newton–Raphson method.

Example 19.1 The finite element formulation presented previously is applied for the solution of the flow over a circular cylinder as shown in Figure 19.2(a) [19.1]. The following boundary conditions are imposed on the problem:

Along the upstream edge: $\psi = y, \quad \dfrac{\partial\psi}{\partial x} = 0$

Along the x axis: $\psi = 0$

Along the upper edge: $\dfrac{\partial\psi}{\partial y} = 1$

Along the downstream edge: $\dfrac{\partial\psi}{\partial x} = 0$

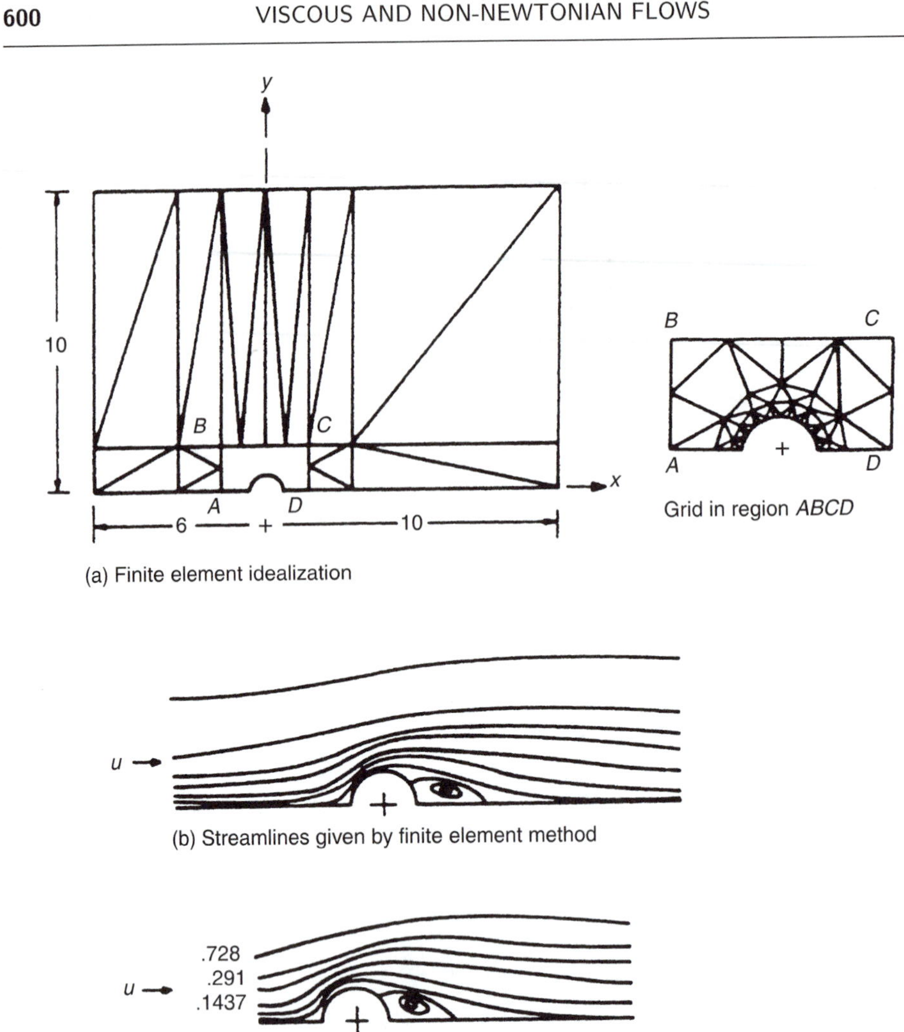

Figure 19.2. Flow over a Circular Cylinder ($R_e = 20$) [19.1].

On the surface of the cylinder: $\psi = 0,\quad \dfrac{\partial \psi}{\partial n} = 0$ (n denotes the normal to the cylinder)

The natural boundary conditions of zero shear stress along the symmetry line (bottom edge) and zero pressure gradient along the top and downstream edges are not imposed but are left for the program to approximate. The pattern of streamlines obtained from the solution for a Reynolds number of 20 are shown in Figure 19.2(b). These results compare well with those given by an accurate finite difference solution [shown in Figure 19.2(c)]. It can be seen that even the separation phenomenon behind the cylinder is predicted accurately.

19.3 VELOCITY–PRESSURE FORMULATION (USING GALERKIN APPROACH)

First, we consider the solution of Stokes equations, Eqs. (19.4) and (19.5), and the continuity equation (19.3). We consider the pressure (p), the velocity component parallel to the x axis (u), and the velocity component parallel to the y axis (v) as unknowns in the formulation. For a typical finite element inside the region S, the unknowns p, u, and v are assumed to vary as

$$\begin{aligned} p^{(e)}(x,y) &= \sum_i N_i^p(x,y)P_i^{(e)} = [N^p(x,y)]\vec{P}^{(e)} \\ u^{(e)}(x,y) &= \sum_i N_i^u(x,y)U_i^{(e)} = [N^u(x,y)]\vec{U}^{(e)} \\ v^{(e)}(x,y) &= \sum_i N_i^v(x,y)V_i^{(e)} = [N^v(x,y)]\vec{V}^{(e)} \end{aligned} \tag{19.32}$$

where N_i^p, N_i^u, and N_i^v are the interpolation functions, which need not necessarily be of the same order, and $\vec{P}^{(e)}$, $\vec{U}^{(e)}$, and $\vec{V}^{(e)}$ are the vectors of nodal unknowns of element e. For simplicity, we assume $N_i^u = N_i^v \equiv N_i$ in the following development.

By using Galerkin's weighted residual method, we can write at any node i of element e three sets of equations. The first one is given by

$$\iint\limits_{A^{(e)}} N_i(x,y)\left[X - \frac{\partial p^{(e)}}{\partial x} + \mu\left(\frac{\partial^2 u^{(e)}}{\partial x^2} + \frac{\partial^2 u^{(e)}}{\partial y^2}\right)\right] \mathrm{d}A^{(e)} = 0 \tag{19.33}$$

where the shape function corresponding to node i, $N_i(x,y)$, is used as the weighting function. Integrating Eq. (19.33) by parts, we obtain

$$\begin{aligned} &\iint\limits_{A^{(e)}} \left[-N_i X - \frac{\partial N_i}{\partial x}p^{(e)} + \mu\frac{\partial N_i}{\partial x}\frac{\partial u^{(e)}}{\partial x} + \mu\frac{\partial N_i}{\partial y}\frac{\partial u^{(e)}}{\partial y}\right]\mathrm{d}A + \int\limits_{C^{(e)}} N_i p^{(e)} l_x \cdot \mathrm{d}C \\ &- \int\limits_{C^{(e)}} \mu N_i\left(\frac{\partial u^{(e)}}{\partial x}l_x + \frac{\partial u^{(e)}}{\partial y}l_y\right)\mathrm{d}C = 0 \end{aligned} \tag{19.34}$$

where l_x and l_y indicate the direction cosines of the outward drawn normal at the boundary of the element e, $C^{(e)}$. By substituting Eqs. (19.32) into Eq. (19.34), we obtain

$$\begin{aligned} &\iint\limits_{A^{(e)}} \left[-N_i X - \frac{\partial N_i}{\partial x}[N^p]\vec{P}^{(e)} + \left(\mu\frac{\partial N_i}{\partial x}\frac{\partial [N]}{\partial x} + \mu\frac{\partial N_i}{\partial y}\frac{\partial [N]}{\partial y}\right)\vec{U}^{(e)}\right]\mathrm{d}A \\ &- \int\limits_{C^{(e)}} \mu N_i\left(\frac{\partial [N]}{\partial x}l_x + \frac{\partial [N]}{\partial y}l_y\right)\vec{U}^{(e)}\,\mathrm{d}C + \int\limits_{C^{(c)}} N_i[N^p]\vec{P}^{(e)} l_x\,\mathrm{d}C = 0 \end{aligned} \tag{19.35}$$

The second equation is similar to Eq. (19.35) and can be obtained by interchanging x and y, U and V, and X and Y in Eq. (19.35). The third equation arising from Eq. (19.3), by using N_i^p as the weighting function, is

$$\iint\limits_{A^{(e)}} N_i^p \frac{\partial [N]}{\partial x} \vec{U}^{(e)} \, dA + \iint\limits_{A^{(e)}} N_i^p \frac{\partial [N]}{\partial y} \vec{V}^{(e)} \, dA = 0 \tag{19.36}$$

By assuming a quadratic variation for the velocity components u and v and a linear variation for the pressure inside a triangular element, the element equations can be written in matrix form as

$$[K^{(e)}] \vec{\Phi}^{(e)} = \vec{\underline{P}}^{(e)} \tag{19.37}$$

where

$$\underset{15 \times 15}{[K^{(e)}]} = \left[\begin{array}{c:c:c} \underset{6\times 6}{[K_1{}^{(e)}]} & \underset{6\times 6}{[0]} & \underset{6\times 3}{[K_2{}^{(e)}]} \\ \hdashline \underset{6\times 6}{[0]} & \underset{6\times 6}{[K_1{}^{(e)}]} & \underset{6\times 3}{[K_3{}^{(e)}]} \\ \hdashline \underset{3\times 6}{-[K_2{}^{(e)}]^T} & \underset{3\times 6}{-[K_3{}^{(e)}]^T} & \underset{3\times 3}{[0]} \end{array} \right] \tag{19.38}$$

$$\underset{15 \times 1}{\vec{\Phi}^{(e)}} = \{ U_1^{(e)} \quad U_2^{(e)} \ \dots \ U_6^{(e)} \ \vdots \ V_1^{(e)} \quad V_1^{(e)} \ \dots \ V_6^{(e)} \ \vdots \ -P_1^{(e)} \quad -P_2^{(e)} \quad -P_3^{(e)} \}^T \tag{19.39}$$

$$\underset{15 \times 1}{\vec{\underline{P}}^{(e)}} = \{ \underline{P}_1^{(e)} \quad \underline{P}_2^{(e)} \ \dots \ \underline{P}_6^{(e)} \ \vdots \ \underline{P}_7^{(e)} \quad \underline{P}_8^{(e)} \ \dots \ \underline{P}_{12}^{(e)} \ \vdots \ 0 \quad 0 \quad 0 \}^T \tag{19.40}$$

$$K_{1\,ij}^{(e)} = \iint\limits_{A^{(e)}} \mu \left(\frac{\partial N_i}{\partial x} \frac{\partial N_j}{\partial x} + \frac{\partial N_i}{\partial y} \frac{\partial N_j}{\partial y} \right) dA \tag{19.41}$$

$$K_{2\,ij}^{(e)} = - \iint\limits_{A^{(e)}} \frac{\partial N_i}{\partial x} N_j^p \, dA \tag{19.42}$$

$$K_{3\,ij}^{(e)} = - \iint\limits_{A^{(e)}} \frac{\partial N_i}{\partial y} N_j^p \, dA \tag{19.43}$$

$$\underline{P}_i^{(e)} = \int\limits_{C^{(e)}} N_i \left(\mu \frac{\partial u^{(e)}}{\partial n} - p^{(e)} \right) l_x \, dC - \iint\limits_{A^{(e)}} N_i X \, dA; \qquad i = 1, 2, \dots, 6 \tag{19.44}$$

$$\underline{P}_i^{(e)} = \int\limits_{C^{(e)}} N_i \left(\mu \frac{\partial v^{(e)}}{\partial n} - p^{(e)} \right) l_y \, dC - \iint\limits_{A^{(e)}} N_i Y \, dA; \qquad i = 7, 8, \dots, 12 \tag{19.45}$$

The element equations (19.37) can be assembled to get the overall equations as

$$[\underset{\sim}{K}] \, \underset{\sim}{\vec{\Phi}} = \underset{\sim}{\vec{P}} \tag{19.46}$$

where

$$[\underset{\sim}{K}] = \sum_{e=1}^{E} [K^{(e)}], \qquad \underset{\sim}{\vec{\Phi}} = \sum_{e=1}^{E} \vec{\Phi}^{(e)}, \qquad \underset{\sim}{\vec{P}} = \sum_{e=1}^{E} \underline{\vec{P}}^{(e)} \tag{19.47}$$

and E is the number of elements.

19.4 SOLUTION OF NAVIER–STOKES EQUATIONS

To extend the solution of Stokes equations to the full Navier–Stokes equations, the following iterative procedure can be adopted. Let u_n, v_n, and p_n be an approximate solution (in nth iteration) to the flow problem. Then we introduce this solution into the coefficients of convective terms of Eqs. (19.1) and (19.2) and write the set of equations as

$$\left.\begin{aligned}
&\mu\left(\frac{\partial^2 u}{\partial x^2} + \frac{\partial^2 u}{\partial y^2}\right) - \rho\left(u_n\frac{\partial u}{\partial x} + v_n\frac{\partial u}{\partial y}\right) - \frac{\partial p}{\partial x} + X = 0\\
&\mu\left(\frac{\partial^2 v}{\partial x^2} + \frac{\partial^2 v}{\partial y^2}\right) - \rho\left(u_n\frac{\partial v}{\partial x} + v_n\frac{\partial v}{\partial y}\right) - \frac{\partial p}{\partial y} + Y = 0\\
&\frac{\partial u}{\partial x} + \frac{\partial v}{\partial y} = 0
\end{aligned}\right\} \tag{19.48}$$

To start the iterative process, u_n, v_n, and p_n may be the solution of the Stokes equations. When the Galerkin procedure is applied to Eqs. (19.48), we get

$$\left.\begin{aligned}
&\iint\limits_{A^{(e)}} \left[\mu\left(N_i\frac{\partial^2 u^{(e)}}{\partial x^2} + N_i\frac{\partial^2 u^{(e)}}{\partial y^2}\right) - \rho\left(u_n^{(e)} N_i\frac{\partial u^{(e)}}{\partial x} + v_n^{(e)} N_i\frac{\partial u^{(e)}}{\partial y}\right) - N_i\frac{\partial p^{(e)}}{\partial x} + N_i X\right] \mathrm{d}A = 0\\
&\iint\limits_{A^{(e)}} \left[\mu\left(N_i\frac{\partial^2 v^{(e)}}{\partial x^2} + N_i\frac{\partial^2 v^{(e)}}{\partial y^2}\right) - \rho\left(u_n^{(e)} N_i\frac{\partial v^{(e)}}{\partial x} + v_n^{(e)} N_i\frac{\partial v^{(e)}}{\partial y}\right) - N_i\frac{\partial p^{(e)}}{\partial y} + N_i Y\right] \mathrm{d}A = 0\\
&\iint\limits_{A^{(e)}} \left(N_i\frac{\partial u^{(e)}}{\partial x} + N_i\frac{\partial v^{(e)}}{\partial y}\right) \mathrm{d}A = 0
\end{aligned}\right\} \tag{19.49}$$

We next integrate by parts all terms in Eqs. (19.49) except those involving $u_n^{(e)}$ and $v_n^{(e)}$. In this manner, the natural boundary conditions are kept identical to those for the Stokes equations. This leads to element equations of the same form as Eq. (19.37) except that the submatrix $[K_1^{(e)}]$ will be different in the current case. The elements of the matrix $[K_1^{(e)}]$ in the current case are given by

$$K_{1\,ij}^{(e)} = \iint\limits_{A^{(e)}} \left[\mu\left(\frac{\partial N_i}{\partial x}\frac{\partial N_j}{\partial x} + \frac{\partial N_i}{\partial y}\frac{\partial N_j}{\partial y}\right) - \rho u_n^{(e)} N_i \frac{\partial N_j}{\partial x} - \rho v_n^{(e)} N_i \frac{\partial N_j}{\partial y}\right] \mathrm{d}A \tag{19.50}$$

Note that the total elemental matrix $[K^{(e)}]$ will be unsymmetric because of the presence of the convective terms. The overall system equations can be expressed as

$$[\underset{\sim}{K}(u_n, v_n)]\underset{\sim}{\vec{\Phi}}_{n+1} = \underset{\sim}{\vec{P}}_{n+1} \tag{19.51}$$

where the subscripts n and $n+1$ indicate the successive stages of iteration. Thus, Eq. (19.51) is to be solved successively for $\underset{\sim}{\vec{\Phi}}_{n+1}$ using the nodal values of u and v obtained in the previous iteration. This process is continued until the two vectors $\underset{\sim}{\vec{\Phi}}_n$ and $\underset{\sim}{\vec{\Phi}}_{n+1}$ are sufficiently close.

Example 19.2 Yamada *et al.* [19.5] considered the problem of flow past a circular cylinder to illustrate the previous procedure. As indicated in Figure 19.3, the infinite field of flow is confined by a circle of radius $r/a = 8.0$ (or 20.0) and the region is divided into finite elements. The boundary conditions at these radii are specified as $u = 5.0$, $v = 0$, and $p = 0$ for a Reynolds number of 30.0. The boundary conditions on the surface of the cylinder ($r = a$) are taken as $u = 0$ and $v = 0$. The velocity distribution obtained by solving Stokes equations is given by Figure 19.4. For the Navier–Stokes equations, the velocity distribution becomes unsymmetrical due to the inclusion of convective terms as shown in Figure 19.5. The convergence of Navier–Stokes equations has been obtained in eight iterations starting from the Stokes solution.

19.5 STREAM FUNCTION–VORTICITY FORMULATION

19.5.1 Governing Equations

In the stream function formulation, the final equation governing the incompressible viscous flows is of order four (see Eq. 19.6). If we choose the stream function (ψ) and vorticity

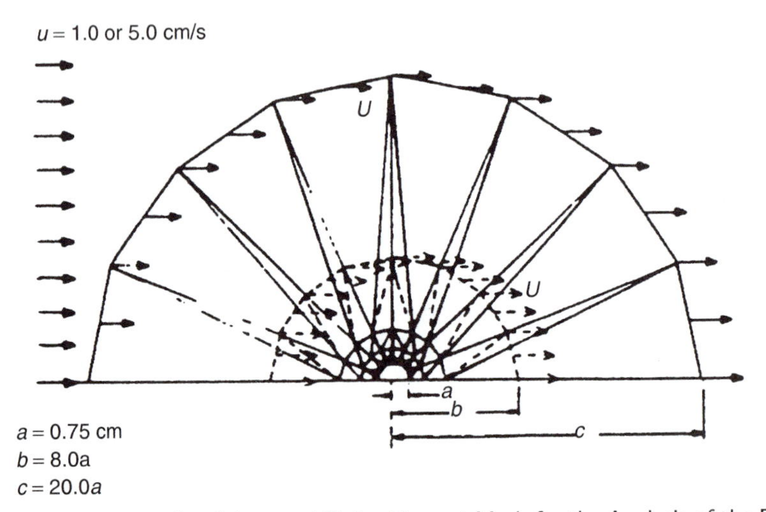

Figure 19.3. Boundary Conditions and Finite Element Mesh for the Analysis of the Flow Past a Circular Cylinder [19.5].

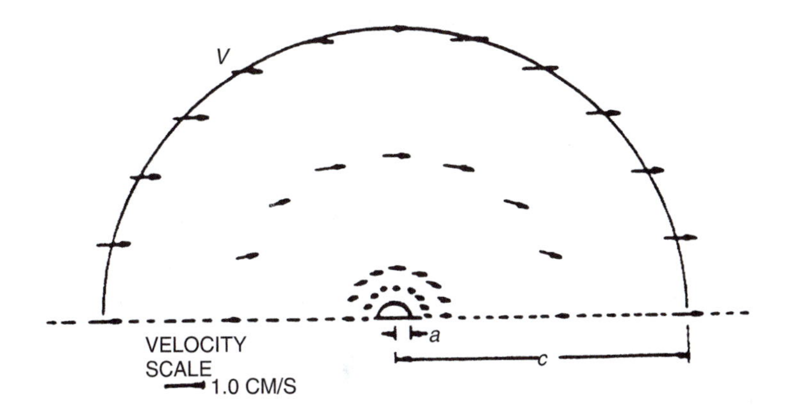

Figure 19.4. Velocity Solution of Stokes Equations [19.5].

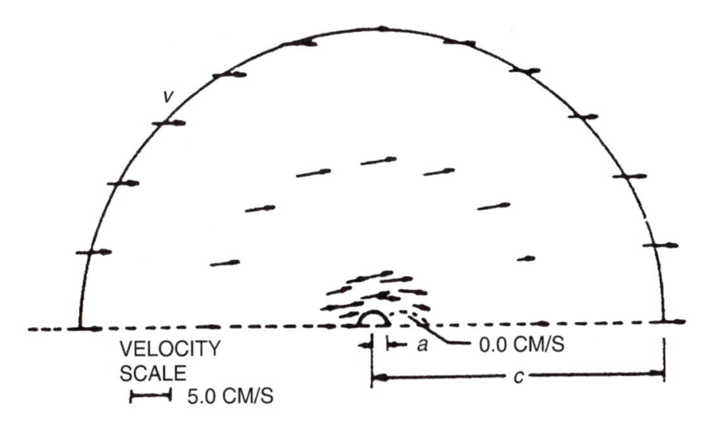

Figure 19.5. Velocity Solution of Navier–Stokes Equations ($R_e = 30$) [19.5].

(ω) as the unknowns, the governing equations will be two second-order equations coupled in ψ and ω as shown below [19.6]. From the definition of vorticity in two dimensions (see Eq. 17.54c),

$$\omega = \omega_z = \frac{1}{2}\left(\frac{\partial v}{\partial x} - \frac{\partial u}{\partial y}\right) \tag{19.52}$$

By substituting the expressions of u and v from Eq. (17.63), Eq. (19.52) gives

$$\omega = -\frac{1}{2}\nabla^2\psi \tag{19.53}$$

Equation (19.6) can be rewritten as

$$\nu\nabla^2(\nabla^2\psi) + \frac{\partial\psi}{\partial x}\frac{\partial}{\partial y}(\nabla^2\psi) - \frac{\partial\psi}{\partial y}\frac{\partial}{\partial x}(\nabla^2\psi) = 0 \tag{19.54}$$

By substituting Eq. (19.53) into Eq. (19.54), we obtain

$$\nu \nabla^2 \omega + \frac{\partial \psi}{\partial x}\frac{\partial \omega}{\partial y} - \frac{\partial \psi}{\partial y}\frac{\partial \omega}{\partial x} = 0 \tag{19.55}$$

For unsteady state problems, Eq. (19.55) will be modified as

$$\nu \nabla^2 \omega + \frac{\partial \psi}{\partial x}\frac{\partial \omega}{\partial y} - \frac{\partial \psi}{\partial y}\frac{\partial \omega}{\partial x} - \frac{\partial \omega}{\partial t} = 0 \tag{19.56}$$

Thus, Eqs. (19.53) and (19.56) represent two coupled second-order equations [Eq. (19.56) is nonlinear] governing the transient incompressible viscous flows.

19.5.2 Finite Element Solution (Using Variational Approach)

Cheng [19.3] presented an iterative procedure based on a quasi-variational approach for the solution of the differential equations (19.53) and (19.56). In this method, the solution at the nth time step (ψ_n, ω_n) is assumed to be known and the solution at the $n+1$th time step is determined by solving the following set of linear differential equations:

$$-\nabla^2 \psi_{n+1} = \omega_n \tag{19.57}$$

$$\nu \nabla^2 \omega_{n+1} + \frac{\partial \psi_{n+1}}{\partial x}\frac{\partial \omega_n}{\partial y} - \frac{\partial \psi_{n+1}}{\partial y}\frac{\partial \omega_n}{\partial x} - \frac{\partial \omega_n}{\partial t} = 0 \tag{19.58}$$

The functionals I_1 and I_2, whose Euler–Lagrange equations yield Eqs. (19.57) and (19.58), respectively, are given by

$$I_1 = \frac{1}{2}\iint_S \left[\left(\frac{\partial \psi_{n+1}}{\partial x}\right)^2 + \left(\frac{\partial \psi_{n+1}}{\partial y}\right)^2 - 2\omega_n \psi_{n+1}\right] \mathrm{d}S \tag{19.59}$$

$$I_2 = \frac{1}{2}\iint_S \left[\nu\left\{\left(\frac{\partial \omega_{n+1}}{\partial x}\right)^2 + \left(\frac{\partial \omega_{n+1}}{\partial y}\right)^2\right\} + 2\left(-\frac{\partial \psi_{n+1}}{\partial x}\frac{\partial \omega_n}{\partial y} + \frac{\partial \psi_{n+1}}{\partial y}\frac{\partial \omega_n}{\partial x} + \underline{\frac{\partial \omega_n}{\partial t}}\right)\omega_{n+1}\right] \mathrm{d}S \tag{19.60}$$

where the underlined term in Eq. (19.60) is to be taken as a constant while taking the variation of I_2. Since the functions I_1 and I_2 involve only the first derivatives of ψ and ω, the interpolation functions need to satisfy only the C^0 continuity. By assuming the variations of ψ and ω inside an element e at the time step $n+1$ as

$$\psi_{n+1}^{(e)}(x, y, t) = [N(x, y)]\vec{\Psi}_{n+1}^{(e)}(t) \tag{19.61}$$

$$\omega_{n+1}^{(e)}(x, y, t) = [N(x, y)]\vec{\Omega}_{n+1}^{(e)}(t) \tag{19.62}$$

the element equations can be derived from the conditions $\delta I_1 = 0$ and $\delta I_2 = 0$ as

$$[K_1^{(e)}]\vec{\Psi}_{n+1}^{(e)} + \vec{P}_{1n}^{(e)} = \vec{0} \tag{19.63}$$

$$[K_2^{(e)}]\dot{\vec{\Omega}}_{n+1}^{(e)} + [K_3^{(e)}]\vec{\Omega}_{n+1}^{(e)} + \vec{P}_{2n}^{(e)} = \vec{0} \tag{19.64}$$

where

$$K_{1\,ij}^{(e)} = \iint\limits_{A^{(e)}} \left(\frac{\partial N_i}{\partial x} \frac{\partial N_j}{\partial x} + \frac{\partial N_i}{\partial y} \frac{\partial N_j}{\partial y} \right) \mathrm{d}A \tag{19.65}$$

$$K_{2\,ij}^{(e)} = \iint\limits_{A^{(e)}} N_i N_j \cdot \mathrm{d}A \tag{19.66}$$

$$K_{3\,ij}^{(e)} = \nu K_{1\,ij}^{(e)} \tag{19.67}$$

$$P_{1n_i} = -\iint\limits_{A^{(e)}} \omega_n^{(e)} N_i \, \mathrm{d}A \tag{19.68}$$

$$P_{2n_i}^{(e)} = \iint\limits_{A^{(e)}} N_i \left(\frac{\partial \psi_{n+1}^{(e)}}{\partial y} \frac{\partial \omega_n^{(e)}}{\partial x} - \frac{\partial \psi_{n+1}^{(e)}}{\partial x} \frac{\partial \omega_n}{\partial y} \right) \cdot \mathrm{d}A \tag{19.69}$$

and $A^{(e)}$ denotes the area of element e. Equations (19.63) and (19.64) can be assembled in the usual manner to obtain the overall system equations as

$$[\underset{\sim}{K_1}]\underset{\sim}{\vec{\Psi}}_{n+1} + \underset{\sim}{\vec{P}}_{1n} = \vec{0} \tag{19.70}$$

$$[\underset{\sim}{K_2}]\,\dot{\underset{\sim}{\vec{\Omega}}}_{n+1} + [\underset{\sim}{K_3}]\underset{\sim}{\vec{\Omega}}_{n+1} + \underset{\sim}{\vec{P}}_{2n} = \vec{0} \tag{19.71}$$

By approximating the vector of time derivatives $\dot{\underset{\sim}{\vec{\Omega}}}_{n+1}$ in Eq. (19.71) as

$$\dot{\underset{\sim}{\vec{\Omega}}}_{n+1} = \begin{Bmatrix} \dfrac{\partial \omega}{\partial t} \text{ (at node 1)} \\ \dfrac{\partial \omega}{\partial t} \text{ (at node 2)} \\ \vdots \end{Bmatrix}_{n+1} = \frac{1}{\Delta t}(\underset{\sim}{\vec{\Omega}}_{n+1} - \underset{\sim}{\vec{\Omega}}_n) \tag{19.72}$$

where the time interval Δt is given by $\Delta t = t_{n+1} - t_n$, Eq. (19.71) can be expressed as

$$\left([\underset{\sim}{K_3}] + \frac{1}{\Delta t}[\underset{\sim}{K_2}] \right) \underset{\sim}{\vec{\Omega}}_{n+1} = \frac{1}{\Delta t}[\underset{\sim}{K_2}]\underset{\sim}{\vec{\Omega}}_n - \underset{\sim}{\vec{P}}_{2n} \tag{19.73}$$

Equations (19.70) and (19.73) represent the final matrix equations to be solved after applying the known boundary conditions. The solution procedure starts with known initial values of $\underset{\sim}{\vec{\Omega}}_n$ (for $n = 0$). Once $\underset{\sim}{\vec{\Omega}}_n$ is known, $\underset{\sim}{\vec{P}}_{1n}$ can be evaluated using Eq. (19.68) and then Eq. (19.70) can be solved for $\underset{\sim}{\vec{\Psi}}_{n+1}$. From the known values of $\underset{\sim}{\vec{\Omega}}_n$ and $\underset{\sim}{\vec{\Psi}}_{n+1}$, the vector $\underset{\sim}{\vec{P}}_{2n}$ can be determined from Eq. (19.69), and hence Eq. (19.73) can be solved for $\underset{\sim}{\vec{\Omega}}_{n+1}$. This recursive procedure is continued until the solution at the specified final time is found.

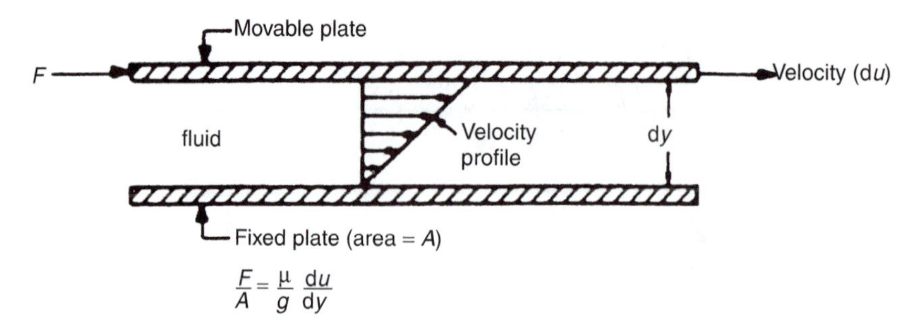

Figure 19.6. Definition of Viscosity for a Newtonian Fluid.

19.6 FLOW OF NON-NEWTONIAN FLUIDS

19.6.1 Governing Equations

(i) Flow curve characteristic

Many practical fluid flows do not follow Newton's law of viscosity and the assumption of a constant viscosity independent of temperature, density, and shear rate does not hold. Such types of fluids are called non-Newtonian fluids. The flow of many crude oils, especially at low temperatures, industrial wastes, slurries, and suspensions of all kinds fall under the category of non-Newtonian fluids. The shear stress τ of a Newtonian fluid in uniaxial flow is given by (see Figure 19.6)

$$\tau = -\frac{\mu}{g}\frac{du}{dy} \tag{19.74}$$

where μ is the coefficient of viscosity [units = mass/(length × time)], g is the acceleration due to gravity, and (du/dy) is the velocity gradient that is equivalent to the shear rate. Here, the quantity (μ/g) is a constant and is independent of the shear rate. Equation (19.74) is represented as a linear curve in Figure 19.7. Certain fluids, known as Bingham plastic fluids, behave as a rigid solid until a certain level of shear stress (τ_0) is attained and behave as a Newtonian fluid afterwards.

Thus, they can be described by

$$\tau = \begin{cases} -\frac{\mu}{g}\frac{du}{dy} + \tau_0 & \text{for} \quad |\tau| > \tau_0 \\ \\ 0 & \text{for} \quad |\tau| < \tau_0 \end{cases} \tag{19.75}$$

True Bingham fluids, which follow Eq. (19.75), are not encountered in practice, but most fluids exhibit pseudo-plastic characteristics. Hence, many empirical equations have been developed to represent their behavior. One widely used such equation, known as the power law, is given by

$$\tau = -\frac{k}{g}\left|\frac{du}{dy}\right|^{n-1}\frac{du}{dy} \tag{19.76}$$

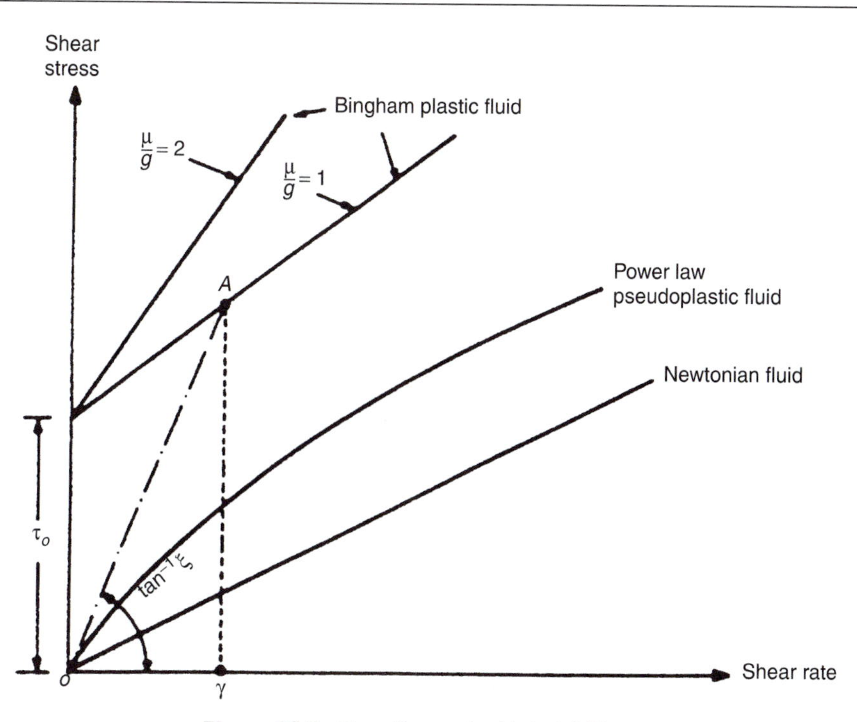

Figure 19.7. Flow Curves for Uniaxial Flow.

where k is the consistency index, and n is the flow behavior index ($n < 1$). Equations (19.75) and (19.76) are also shown plotted in Figure 19.7.

(ii) Equation of motion

By assuming the fluid flow to be incompressible and time independent but possessing a variable viscosity, the steady-state equation of motion for flow through a parallel-sided conduit can be stated in the form of a nonlinear Poisson equation as [19.7, 19.8]

$$\frac{\partial}{\partial x}\left(\frac{\mu}{g}\frac{\partial u}{\partial x}\right) + \frac{\partial}{\partial y}\left(\frac{\mu}{g}\frac{\partial u}{\partial y}\right) + \rho g - \frac{\partial p}{\partial z} = 0 \tag{19.77}$$

where u is the velocity, ρ is the density, and $\partial p/\partial z$ is the pressure gradient. If the pressure gradient is known, the solution of Eq. (19.77) enables us to find the velocity distribution, $u(x, y)$, and the quantity of flow.

19.6.2 Finite Element Equations Using Galerkin Method

In the Galerkin method, the weighted residue must be zero over the region of flow, S. Thus,

$$\iint_S \left[\frac{\partial}{\partial x}\left(\frac{\mu}{g}\frac{\partial u}{\partial x}\right) + \frac{\partial}{\partial y}\left(\frac{\mu}{g}\frac{\partial u}{\partial y}\right) + Q\right] W \cdot \mathrm{d}S = 0 \tag{19.78}$$

where

$$Q = \rho g - \frac{\partial p}{\partial z} \tag{19.79}$$

and W is the weighting function. By applying Green–Gauss theorem of Appendix A, Eq. (19.78) can be expressed as

$$-\iint_S \frac{\mu}{g}\left[\frac{\partial u}{\partial x}\frac{\partial W}{\partial x} + \frac{\partial u}{\partial y}\frac{\partial W}{\partial y}\right]\cdot \mathrm{d}S + \iint_S WQ \cdot \mathrm{d}S + \int_C W\frac{\mu}{g}\left(\frac{\partial u}{\partial x}l_x + \frac{\partial u}{\partial y}l_y\right)\cdot \mathrm{d}C \tag{19.80}$$

where C denotes the boundary of the region S, and l_x and l_y represent the direction cosines of the normal to this boundary at any point. The velocity u is assumed to vary within an element e as

$$u(x,y) = \sum_i N_i U_i^{(e)} = [N]\vec{U}^{(e)} \tag{19.81}$$

where N_i is the shape function corresponding to the ith nodal degree of freedom $U_i^{(e)}$ of element e. It can be seen from Eq. (19.80) that the shape functions N_i need to satisfy only the C^1 continuity. By substituting Eq. (19.81) into Eq. (19.80) and by taking $W = N_i$, we obtain

$$\begin{aligned}&-\iint_{A^{(e)}} \frac{\mu}{g}\left\{\frac{\partial N_i}{\partial x}\frac{\partial}{\partial x}[N]\cdot\vec{U}^{(e)} + \frac{\partial N_i}{\partial y}\frac{\partial}{\partial y}[N]\cdot\vec{U}^{(e)}\right\}\mathrm{d}A + \iint_{A^{(e)}} N_i Q\,\mathrm{d}A\\ &+ \int_{C^{(e)}} \frac{\mu}{g}N_i\frac{\partial[N]}{\partial n}\vec{U}^{(e)}\cdot\mathrm{d}C = 0\end{aligned} \tag{19.82}$$

The boundary conditions applicable are as follows:

(i) $u = u_0 \quad \text{on} \quad C_1$ (19.83)

(known value of velocity on boundary C_1)

(ii) $\alpha u = -\dfrac{\mu}{g}\dfrac{\partial u}{\partial n} \quad \text{on} \quad C_2$ (19.84)

where α is the coefficient of "sliding friction" between the fluid and the boundary (fluid slippage at the boundary is proportional to the velocity gradient normal to the boundary C_2). With the help of Eq. (19.84), the last term on the left-hand side of Eq. (19.82) can be replaced by $-\int_{C_2} \alpha N_i[N]\cdot\mathrm{d}C$. Using Eq. (19.82), the element equations can be stated in matrix form as

$$([K_1^{(e)}] + [K_2^{(e)}])\vec{U}^{(e)} = \vec{P}^{(e)} \tag{19.85}$$

where the elements of the matrices $[K_1^{(e)}]$ and $[K_2^{(e)}]$ and the vector $\vec{P}^{(e)}$ are given by

$$K_{1\,ij}^{(e)} = \iint\limits_{A^{(e)}} \frac{\mu}{g}\left[\frac{\partial N_i}{\partial x}\frac{\partial N_j}{\partial x} + \frac{\partial N_i}{\partial y}\frac{\partial N_j}{\partial y}\right] \mathrm{d}A \tag{19.86}$$

$$K_{2\,ij}^{(e)} = \int\limits_{C_2^{(e)}} \alpha N_i [N] \cdot \mathrm{d}C \tag{19.87}$$

$$P_i^{(e)} = \iint\limits_{A^{(e)}} Q N_i \cdot \mathrm{d}A \tag{19.88}$$

Note that the matrix $[K_2^{(e)}]$ denotes the contribution of slippage at the boundary and is applicable only to elements with such a boundary condition. The assembly of the element equations (19.85) leads to the overall equations

$$[\underset{\sim}{K}]\underset{\sim}{\vec{U}} \equiv [[\underset{\sim}{K_1}] + [\underset{\sim}{K_2}]]\underset{\sim}{\vec{U}} = \underset{\sim}{\vec{P}} \tag{19.89}$$

19.6.3 Solution Procedure

The elements of the matrix $[\underset{\sim}{K_1}]$ can be seen to be functions of viscosity μ and hence are functions of the derivatives of the unknown velocity $u(x, y)$. The viscosity–velocity relationship is nonlinear and is given by Eq. (19.75) or Eq. (19.76). Hence, Eqs. (19.89) represent a set of simultaneous nonlinear equations. These equations can be solved by using an iterative procedure. The procedure starts with the assumption of an apparent viscosity and Eqs. (19.75) and (19.76) are written in the form

$$\tau = -[\xi(\dot{\gamma})]\dot{\gamma} \tag{19.90}$$

This apparent viscosity at any point A shown in Figure 19.7 will be the slope of the secant OA and can be expressed as

$$\frac{\xi_A}{g} = \frac{\tau_A}{\dot{\gamma}_A} \tag{19.91}$$

Thus, for Bingham plastic fluids, the apparent viscosity can be expressed as

$$\xi(\dot{\gamma}) = \frac{\frac{\mu}{g}\dot{\gamma} + \tau_0}{\dot{\gamma}} = \frac{\tau_0}{\dot{\gamma}} + \frac{\mu}{g} \tag{19.92}$$

and for pseudoplastic fluid as

$$\tau = -\frac{k}{g}|\dot{\gamma}|^{n-1}\dot{\gamma} \tag{19.93}$$

or

$$\xi(\dot{\gamma}) = \frac{k}{g}|\dot{\gamma}|^{n-1} \tag{19.94}$$

Then Eq. (19.92) or Eq. (19.94) is substituted into Eq. (19.86) to obtain

$$K_{1\,ij}^{(e)} = \iint\limits_{A^{(e)}} \xi(\dot{\gamma}) \left(\frac{\partial N_i}{\partial x} \frac{\partial N_j}{\partial x} + \frac{\partial N_i}{\partial y} \frac{\partial N_j}{\partial y} \right) \mathrm{d}A \tag{19.95}$$

where

$$\dot{\gamma} = \left[\left(\frac{\partial u}{\partial x} \right)^2 + \left(\frac{\partial u}{\partial y} \right)^2 \right]^{1/2} = \left[\left(\frac{\partial [N]}{\partial x} \vec{U}^{(e)} \right)^2 + \left(\frac{\partial [N]}{\partial y} \vec{U}^{(e)} \right)^2 \right]^{1/2} \tag{19.96}$$

Usually, the initial approximation to the solution of Eq. (19.89) is based on the Newtonian velocity distribution. Equations (19.89) can be written as

$$\underset{\sim}{\vec{U}}_{n+1} = -[\underset{\sim}{K}(\underset{\sim}{\vec{U}}_n)]^{-1} \underset{\sim}{\vec{P}} \tag{19.97}$$

where $\underset{\sim}{\vec{U}}_n$ and $\underset{\sim}{\vec{U}}_{n+1}$ denote the solutions of $\underset{\sim}{\vec{U}}$ in nth and $n+1$th iterations. The solution procedure can be summarized as follows [19.9]:

1. Solve Eqs. (19.89) for Newtonian velocity distribution after applying the boundary conditions of Eq. (19.83).
2. Compute the shear rate and the apparent viscosity.
3. Calculate the matrix $[K_n]$.
4. Solve Eqs. (19.89) for $\underset{\sim}{\vec{U}}_{n+1}$ after applying the boundary conditions of Eq. (19.83).
5. Test for the convergence of the process using the criterion

$$\left| \frac{(U_i)_n - (U_i)_{n-1}}{(U_i)_n} \right| < \varepsilon, \qquad i = 1, 2, \ldots \tag{19.98}$$

where $\varepsilon \approx 0.01$ for each nodal velocity U_i.
If convergence of Eq. (19.98) is not satisfied, repeat steps 2–5.

Example 19.3 The problem of Bingham flow between parallel plates was considered by Lyness *et al.* [19.7]. The problem is shown in Figure 19.8, where the plug dimensions are given by

$$h_p = \tau_0 \Big/ \frac{\mathrm{d}p}{\mathrm{d}z} \tag{E$_1$}$$

where h_p defines the extent of the solid plug. For this problem, the exact solution for the velocity is given by

$$u_p = \frac{1}{\mu} \left[\frac{1}{2} \frac{\mathrm{d}p}{\mathrm{d}z} \left(\frac{H^2}{4} - h_p^2 \right) - \tau_0 \left(\frac{H}{2} - h_p \right) \right] \quad \text{for} \quad h < h_p \tag{E$_2$}$$

$$u = \frac{1}{\mu} \left[\frac{1}{2} \frac{\mathrm{d}p}{\mathrm{d}z} \left(\frac{H^2}{4} - h^2 \right) - \tau_0 \left(\frac{H}{2} - h \right) \right] \quad \text{for} \quad h > h_p \tag{E$_3$}$$

The problem was solved for different values of true viscosity and plug sizes. Five quadratic rectangular elements were used for the idealization as shown in Figure 19.8. The finite

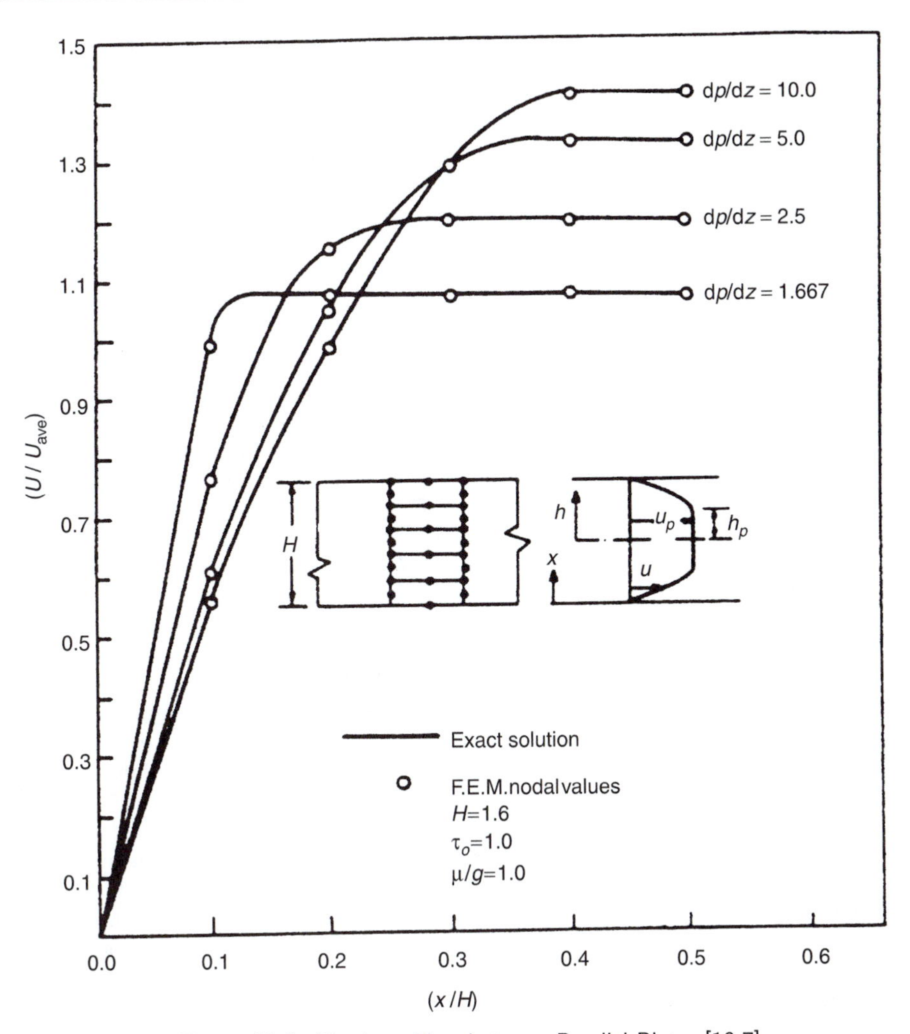

Figure 19.8. Bingham Flow between Parallel Plates [19.7].

element (velocity) solution is compared with the exact solution given by Eqs. (E_2) and (E_3) in Figure 19.8. It can be seen that the finite element solution compares very well with the exact solution. The flow rate through the section can be computed by numerically integrating the velocity over an element and summing over all the elements. The error in the flow rates predicted by the finite element method was found to be less than 1% compared with those obtained by explicit integration of Eqs. (E_2) and (E_3).

19.7 OTHER DEVELOPMENTS

As can be observed from Chapters 18 and 19, a majority of the finite element applications to fluid mechanics problems employed one of the weighted residual criteria for deriving the equations. The problem of laminar boundary layer flow was studied by Lynn [19.10] using the least squares criterion and by Popinski and Baker [19.11] using the Galerkin approach. The transient compressible flow in pipelines was studied by Biggard *et al.* [19.12]. A study of penalty elements for incompressible laminar flows was made by Dhatt and Hubert [19.13]. An optimal control finite element approximation for penalty variational formulation of three-dimensional Navier–Stokes problem was presented by Li *et al.* [19.14]. Numerical experiments with several finite elements were conducted for the study of viscous incompressible flows [19.15]. The problems of finite element mesh generation for arbitrary geometry, efficient solution of equations, derivation of new elements, and development of efficient codes for fluid flow problems have also been investigated [19.16]. An overview of the application of finite elements in computational fluid dynamics was presented by Lohner [19.17].

REFERENCES

19.1 M.D. Olson: Variational finite element methods for two-dimensional and axisymmetric Navier–Stokes equations, in *Finite Elements in Fluids*, Vol. 1, Viscous Flow and Hydrodynamics, R.H. Gallagher *et al.* (Eds.), Wiley, London, 1975.

19.2 C. Taylor and P. Hood: A numerical solution of the Navier–Stokes equations using the finite element technique, *Computers and Fluids*, *1*, 73–100, 1973.

19.3 R.T. Cheng: Numerical solution of the Navier–Stokes equations by the finite element method, *Physics of Fluids*, *15*, 12, 1972.

19.4 G.R. Cowper, E. Kosko, G.M. Lindberg, and M.D. Olson: Static and dynamic applications of a high precision triangular plate bending element, *AIAA Journal*, *7*, 1957–1965, 1969.

19.5 Y. Yamada, K. Ito, Y. Yokouchi, T. Tamano, and T. Ohtsubo: Finite element analysis of steady fluid and metal flow, in *Finite Elements in Fluids*, Vol. 1, Viscous Flow and Hydrodynamics, R.H. Gallagher *et al.* (Eds.), Wiley, London, 1975.

19.6 M.F. Peeters, W.G. Habashi, and E.G. Dueck: Finite element stream function—Vorticity solutions of the incompressible Navier–Stokes equations, *International Journal for Numerical Methods in Fluids*, *7*, 17–28, 1987.

19.7 J.F. Lyness, D.R.J. Owen, and O.C. Zienkiewicz: Finite element analysis of non-Newtonian fluids through parallel sided conduits, in *Finite Element Methods in Flow Problems*, J.T. Oden *et al.* (Eds.), University of Alabama Press, Huntsville, AL, 489–503, 1974.

19.8 R.J. Seeger and G. Temple (Eds.): *Research Frontiers in Fluid Dynamics*, Interscience, New York, 1965.

19.9 P.N. Godbole and O.C. Zienkiewicz: Finite element analysis of steady flow of non-Newtonian fluids, in *Finite Element Methods in Engineering*, V.A. Pulmano and A.P. Kabaila (Eds.), University of New South Wales, Australia, 1974.

19.10 P.O. Lynn: Least squares finite element analysis of laminar boundary layers, *International Journal for Numerical Methods in Engineering*, *8*, 865–876, 1974.

19.11 Z. Popinski and A.J. Baker: An implicit finite element algorithm for the boundary layer equations, *Journal of Computational Physics*, *21*, 55–84, 1976.

19.12 C. Bigaard, H.H. Sorensen, and S. Spangerberg: A finite element method for transient compressible flow in pipelines, *International Journal for Numerical Methods in Fluids*, *7*, 291–304, 1987.

19.13 G. Dhatt and G. Hubert: A study of penalty elements for incompressible laminar flows, *International Journal for Numerical Methods in Fluids*, *6*, 1–20, 1986.

19.14 K.T. Li, A.X. Huag, Y.C. Ma, D. Li, and Z.X. Liu: Optimal control finite element approximation for penalty variational formulation of three-dimensional Navier–Stokes problem, *International Journal for Numerical Methods in Engineering*, *20*, 85–100, 1984.

19.15 M. Fortin and A. Fortin: Experiments with several elements for viscous incompressible flows, *International Journal for Numerical Methods in Fluids*, *5*, 911–928, 1985.

19.16 T.J. Liu: An efficient matrix solver for finite element analysis of non-Newtonian fluid flow problems, *International Journal for Numerical Methods in Fluids*, *5*, 929–938, 1985.

19.17 R. Lohner: Finite elements in CFD—What lies ahead, *International Journal for Numerical Methods in Engineering*, *24*, 1741–1756, 1987.

PROBLEMS

19.1 Consider an incompressible viscous flow using the stream function–vorticity formulation of Section 19.5.1. Derive the corresponding finite element equations using the Galerkin method.

19.2 A fully developed laminar forced flow of a Newtonian, incompressible fluid between two parallel plates is shown in Figure 19.9. Consider the energy balance of an infinitesimal element $dx\,dy$, including the energy transported by the fluid motion, and derive the equation governing the temperature distribution in the fluid, $T(x, y)$, as

$$\rho c u(y) \frac{\partial T}{\partial x} = \frac{\partial}{\partial x}\left[k \frac{\partial T}{\partial x}\right] + \frac{\partial}{\partial y}\left[k \frac{\partial T}{\partial y}\right] \tag{E$_1$}$$

where ρ is the density, c is the specific heat at constant pressure, u is the x component of velocity, and k is the thermal conductivity.

19.3 Using the Galerkin method, derive the finite element equations corresponding to Eq. (E$_1$) of Problem 19.2 by assuming the temperature variation in an element as

$$T(x, y) = [N(x, y)]\vec{T}^{(e)} \tag{E$_1$}$$

Note that the components of velocity of the fluid parallel to the y and x axes are given by 0 and $u(y) = (3/2)u_0\{1 - [y/d]^2\}$, respectively, where u_0 is the average velocity.

19.4 Consider a fully developed laminar forced flow of a Newtonian, incompressible fluid in a circular duct. Axisymmetric temperature and heat fluxes are prescribed on different parts of the boundary (walls) of the duct. Consider the energy balance of an annular element of volume $2\pi r\,dr\,dz$, including the energy transported by

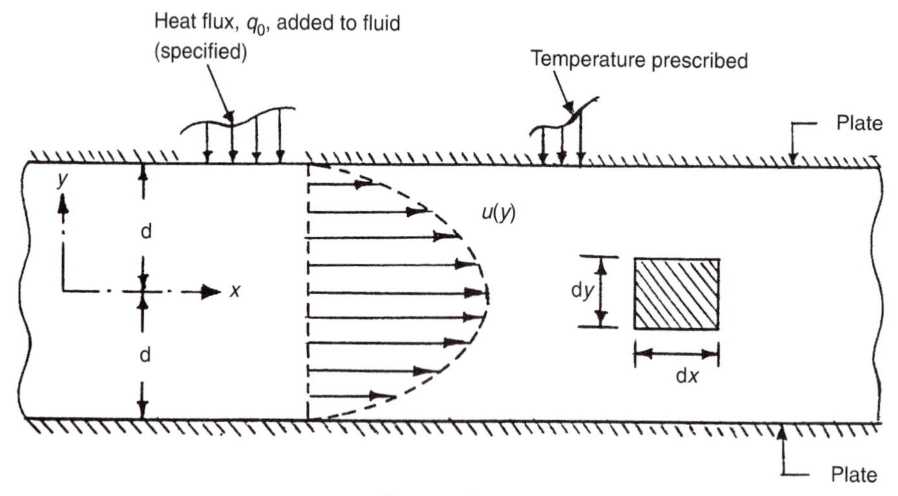

Figure 19.9.

the fluid motion along the axial (z) direction, and derive the equation governing the temperature distribution of the fluid, $T(r, z)$, as

$$\rho c v(r) \frac{\partial T}{\partial z} = \frac{1}{r} \frac{\partial}{\partial r} \left[k r \frac{\partial T}{\partial r} \right] + \frac{\partial}{\partial z} \left[k \frac{\partial T}{\partial z} \right] \tag{E$_1$}$$

where ρ is the density, c is the specific heat at constant pressure, v is the z component of velocity, and k is the thermal conductivity.

19.5 Using the Galerkin method, derive the finite element equations corresponding to Eq. (E_1) of Problem 19.4 by assuming the temperature variation in an element as

$$T(r, z) = [N(r, z)]\vec{T}^{(e)} \tag{E$_1$}$$

Note that the components of velocity of the fluid parallel to the r and z axes are given by 0 and $v(r) = 2v_0[1 - [r/d]^2\}$, respectively, where v_0 is the average velocity and d is the radius of the duct.

ADDITIONAL APPLICATIONS

20

SOLUTION OF QUASI-HARMONIC EQUATIONS

20.1 INTRODUCTION

A significant class of physical problems representing phenomena such as heat conduction, torsion of shafts, and distribution of electric potential are known as field problems. These field problems have the characteristic that they all are governed by similar partial differential equations in terms of the concerned field variable. This permits us to discuss the solution of the governing partial differential equation without identifying the field variable ϕ as a particular physical quantity. The steady-state (time-independent) field problems are governed by the quasi-harmonic equation

$$\frac{\partial}{\partial x}\left(k_x \frac{\partial \phi}{\partial x}\right) + \frac{\partial}{\partial y}\left(k_y \frac{\partial \phi}{\partial y}\right) + \frac{\partial}{\partial z}\left(k_z \frac{\partial \phi}{\partial z}\right) + c = 0 \tag{20.1}$$

where ϕ is an unknown function or field variable (assumed to be single valued in the domain), and k_x, k_y, k_z, and c are known functions of x, y, and z.

The physical interpretation of k_x, k_y, k_z, c, and ϕ depends on the particular physical problem. Table 20.1 lists some typical field problems along with the significance of ϕ and other parameters for each problem. Equation (20.1) assumes that the medium is inhomogeneous and/or anisotropic, and the coordinates x, y, and z coincide with the principal coordinates. If the medium is homogeneous, k_x, k_y, and k_z will be constants, and if it is isotropic, $k_x = k_y = k_z = k =$ constant. The general boundary conditions for Eq. (20.1) are given by

$$\phi = \bar{\phi}; \text{ value of } \phi \text{ prescribed on part of the boundary, } S_1 \text{ (Dirichlet condition)} \tag{20.2}$$

and

$$k_x \frac{\partial \phi}{\partial x} l_x + k_y \frac{\partial \phi}{\partial y} l_y + k_z \frac{\partial \phi}{\partial z} l_z + q + r\phi = 0$$

on the remaining part of the boundary, for example, S_2 (Cauchy condition) (20.3)

Table 20.1. Typical Steady-State Field Problems Governed by Eq. (20.1)

Sr. No.	Physical problem	Field variable, ϕ	Significance of k_x, k_y, k_z	c	Remarks
1	Heat conduction	Temperature	Thermal conductivities	Rate of internal heat generation	$q =$ boundary heat generation $r =$ convective heat transfer coefficient
2	Seepage flow	Pressure	Permeability coefficients	Internal flow source	—
3	Torsion of prismatic shafts	Stress function	$1/G$, where G is shear modulus	$c = 2\theta$, where θ is angle of twist per unit length	—
4	Irrotational flow of ideal fluids	Velocity potential or stream function	—	$c = 0$	$q =$ boundary velocity, $r = 0$
5	Fluid film lubrication	Pressure	k_x and k_y are functions of film thickness and viscosity, $k_z = 0$	Net flow due to various actions	$q =$ boundary flow
6	Distribution of electric potential	Electric potential (voltage)	Specific conductivities	Internal current source	$q =$ externally applied boundary current
7	Electrostatic field	Electric force field intensity	Permittivities	Internal current source	—
8	Magnetostatics	Magnetomotive force	Magnetic permeabilities	Internal magnetic field source	$q =$ externally applied magnetic field intensity

where l_x, l_y, and l_z are direction cosines of the outward drawn normal to the surface. If $q = r = 0$, the Cauchy boundary condition becomes a Neumann condition. The field problem stated in Eqs. (20.1)–(20.3) is called an elliptic* mixed boundary value problem — mixed because some portion of the boundary S has Dirichlet conditions, whereas the remaining one has Cauchy conditions.

If $k_x = k_y = k_z = k =$ constant and $q = r = 0$, Eq. (20.1) reduces to

$$\nabla^2 \phi = -\frac{c(x, y, z)}{k} \tag{20.4}$$

which is called Poisson's equation. In this case, the boundary condition of Eq. (20.3) reduces to

$$\frac{\partial \phi}{\partial n} = 0 \qquad \text{(nonconducting boundary on } S_2\text{)} \tag{20.5}$$

where n indicates the direction of outward drawn normal to the surface. Furthermore, if $c = 0$, Eq. (20.4) becomes

$$\nabla^2 \phi = 0 \tag{20.6}$$

which is known as Laplace equation.

20.2 FINITE ELEMENT EQUATIONS FOR STEADY-STATE PROBLEMS

The finite element solution of the differential equation (20.1) subject to the boundary conditions, Eqs. (20.2) and (20.3), was presented in Sections 13.5.1 and 13.5.2 using variational and Galerkin methods, respectively, in connection with the solution of heat transfer problems. The overall system equations can be stated as

$$[\underset{\sim}{K}]\underset{\sim}{\vec{\Phi}} = \underset{\sim}{\vec{P}} \tag{20.7}$$

---

* Partial differential equations (of second order) can be classified as parabolic, elliptic, hyperbolic, or some combination of these three categories, such as elliptically parabolic, hyperbolically parabolic, etc. To indicate the method of classification, consider the following general partial differential equation in two independent variables:

$$A\frac{\partial^2 \phi}{\partial x^2} + 2B\frac{\partial^2 \phi}{\partial x \partial y} + C\frac{\partial^2 \phi}{\partial y^2} = D\left(\phi, \frac{\partial \phi}{\partial x}, \frac{\partial \phi}{\partial y}, x, y\right)$$

where A, B, and C are functions of x and y, and D is a function of $x, y, \phi, (\partial \phi/\partial x)$, and $(\partial \phi/\partial y)$. The nature of expression for D will decide whether the partial differential equation is linear or nonlinear. Irrespective of the form of D, the partial differential equation is called parabolic if $B^2 - AC = 0$, elliptic if $B^2 - AC < 0$, and hyperbolic if $B^2 - AC > 0$. Usually, the solution domains are defined by closed boundaries for elliptic problems and by open domains for parabolic and hyperbolic problems. Most of the finite element applications so far have been directed toward the solution of elliptic boundary value problems with irregular solution domains.

where the element characteristic matrices $[K^{(e)}]$ and characteristic vectors $\vec{P}^{(e)}$, whose assembly yields Eq. (20.7), can be obtained similar to Eqs. (13.30), (13.31), and (13.33) as

$$K1_{ij}^{(e)} = \iiint\limits_{V^{(e)}} \left(k_x \frac{\partial N_i}{\partial x} \frac{\partial N_j}{\partial x} + k_y \frac{\partial N_i}{\partial y} \frac{\partial N_j}{\partial y} + k_z \frac{\partial N_i}{\partial z} \frac{\partial N_j}{\partial z} \right) \mathrm{d}V \tag{20.8}$$

$$K2_{ij}^{(e)} = \iint\limits_{S_2^{(e)}} r N_i N_j \, \mathrm{d}S_2 \tag{20.9}$$

$$P_i^{(e)} = \iiint\limits_{V^{(e)}} c N_i \, \mathrm{d}V - \iint\limits_{S_2^{(e)}} q N_i \, \mathrm{d}S_2 \tag{20.10}$$

$$K_{ij}^{(e)} = K1_{ij}^{(e)} + K2_{ij}^{(e)} \tag{20.11}$$

The vector of nodal unknowns of element e is denoted by $\vec{\Phi}^{(e)} = \{\Phi_1^{(e)} \Phi_2^{(e)} \ldots\}^T$ and the vector $\underset{\sim}{\vec{\Phi}}$ is given by the assemblage of $\vec{\Phi}^{(e)}$. The solution of Eq. (20.7), after incorporating the boundary conditions given by Eq. (20.2), gives the nodal values of the field variable ϕ. Once the nodal values Φ_i are known, the element resultants, if required, can be evaluated without much difficulty.

20.3 SOLUTION OF POISSON'S EQUATION

It can be seen that Poisson's equation (Eq. 20.4) is a special case of the general field equation (20.1). We consider the solution of the Poisson's equation in the context of torsion of prismatic shafts in this section.

20.3.1 Derivation of the Governing Equation for the Torsion Problem

Consider a solid prismatic shaft having any arbitrary cross section as shown in Figure 20.1. When this shaft is subjected to a twisting moment M_z, it is usual to assume that all stresses except the shear stresses σ_{xz} and σ_{yz} are zero [20.1]. Thus,

$$\sigma_{xx} = \sigma_{yy} = \sigma_{zz} = \sigma_{xy} = 0 \tag{20.12}$$

For this case, the equilibrium equations given by Eq. (8.4) reduce, in the absence of body forces, to

$$\frac{\partial \sigma_{xz}}{\partial x} + \frac{\partial \sigma_{yz}}{\partial y} = 0 \tag{20.13}$$

$$\frac{\partial \sigma_{xz}}{\partial z} = 0 \tag{20.14}$$

$$\frac{\partial \sigma_{yz}}{\partial z} = 0 \tag{20.15}$$

Equations (20.14) and (20.15) indicate that the stresses σ_{xz} and σ_{yz} do not vary with the coordinate z (i.e., along a line taken parallel to the axis of the shaft). Now a stress

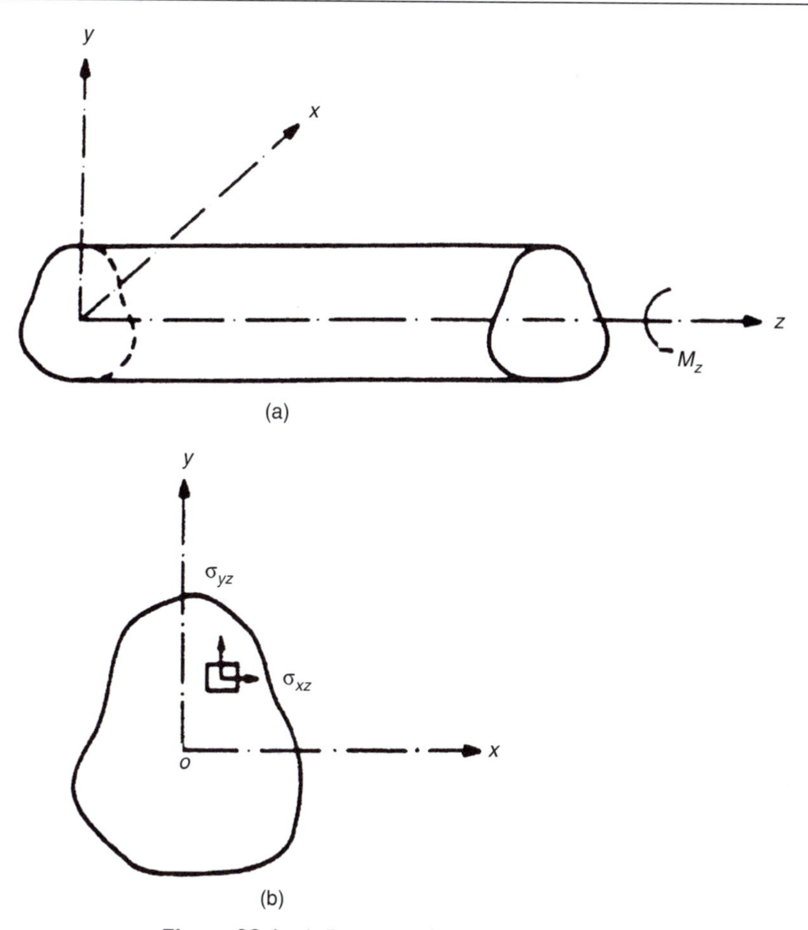

Figure 20.1. A Prismatic Shaft under Torsion.

function ϕ, known as Prandtl's stress function, is defined as

$$\left.\begin{aligned}\sigma_{xz} &= \frac{\partial \phi}{\partial y}\\ \sigma_{yz} &= -\frac{\partial \phi}{\partial x}\end{aligned}\right\} \tag{20.16}$$

so that Eq. (20.13) is satisfied automatically. If we consider strains, Hooke's law (Eq. 8.7) indicates that only ε_{xz} and ε_{yz} will be nonzero. Thus,

$$\left.\begin{aligned}\varepsilon_{xz} &= \frac{\sigma_{xz}}{G}\\ \varepsilon_{yz} &= \frac{\sigma_{yz}}{G}\end{aligned}\right\} \tag{20.17}$$

where G is the modulus of rigidity. Clearly, these shear strains must be independent of the coordinate z. If the displacement components are taken as [20.2]

$$u = -\theta yz \tag{20.18a}$$

$$\nu = \theta xz \tag{20.18b}$$

$$\frac{\partial w}{\partial x} = \frac{1}{G}\frac{\partial \phi}{\partial y} + \theta y \tag{20.18c}$$

$$\frac{\partial w}{\partial y} = -\frac{1}{G}\frac{\partial \phi}{\partial x} - \theta x \tag{20.18d}$$

where θ denotes the angle of twist per unit length, the stress–strain relations, Eq. (20.17), can be satisfied. To ensure continuity of the axial displacement w, we have, by differentiating Eqs. (20.18c) and (20.18d), that

$$\frac{\partial}{\partial x}\left(\frac{1}{G}\frac{\partial \phi}{\partial x}\right) + \frac{\partial}{\partial y}\left(\frac{1}{G}\frac{\partial \phi}{\partial y}\right) = -2\theta \tag{20.19}$$

If G is a constant, Eq. (20.19) reduces to the Poisson equation

$$\nabla^2 \phi = \frac{\partial^2 \phi}{\partial x^2} + \frac{\partial^2 \phi}{\partial y^2} = -2G\theta \tag{20.20}$$

Since the shear stress normal to the external boundary must be zero, it follows from Eq. (20.16) that the stress function on this boundary must have a constant value. This constant can be arbitrarily fixed as zero. Thus, a sufficient and necessary boundary condition on the external boundary becomes

$$\phi = 0 \tag{20.21}$$

To find the torque acting on the section (M_z), we integrate the moment due to shear stresses as

$$M_z = \iint_S (x\sigma_{yz} - y\sigma_{xz})\,dx\,dy \tag{20.22}$$

where S denotes the cross section of the shaft. By substituting Eqs. (20.16) and (20.21), Eq. (20.22) becomes

$$M_z = 2\iint_S \phi\,dx\,dy \tag{20.23}$$

By defining a set of nondimensional quantities as

$$x' = \frac{x}{l}, \qquad y' = \frac{y}{l}, \qquad \phi' = \frac{\phi}{G\theta l^2} \tag{20.24}$$

where l is the length of the shaft, Eqs. (20.20) and (20.21) reduce to

$$\nabla^2 \phi' = -2 \equiv -c \tag{20.25}$$

with

$$\phi' = 0 \quad \text{on external boundary} \tag{20.26}$$

Note

The torsion problem can be formulated using a different approach. In this approach, the warping function $\psi(x, y)$, which represents the movement of the cross section in the z direction per unit twist, is treated as the unknown function. Finally, a Laplace's equation is to be solved for determining $\psi(x, y)$. This approach was used by Herrmann [20.3] for the torsion analysis of irregular shapes.

20.3.2 Finite Element Solution

The functional corresponding to Eq. (20.25) can be written as

$$I(\phi') = \frac{1}{2} \iint_S \left[\left(\frac{\partial \phi'}{\partial x'} \right)^2 + \left(\frac{\partial \phi'}{\partial y'} \right)^2 \right] \mathrm{d}x'\, \mathrm{d}y' + \iint_S c\phi'\, \mathrm{d}x'\, \mathrm{d}y' \tag{20.27}$$

Let the cross section S be idealized by using triangular elements and let the nodal values of ϕ', namely $\Phi'_1, \Phi'_2, \ldots, \Phi'_M$, be taken as unknowns. By choosing a suitable form of variation of ϕ' within each element as

$$\phi'(x', y') = [N(x', y')] \vec{\underset{\sim}{\Phi}}'^{(e)} \tag{20.28}$$

the element matrices and vectors can be obtained as indicated in Eqs. (20.8)–(20.11). The resulting element equations are then assembled to obtain the overall system equations as in Eq. (20.7) and are solved to obtain $\vec{\underset{\sim}{\Phi}}'$. It will be necessary to apply the boundary conditions $\phi' = 0$ on the outer periphery of S before solving Eq. (20.7).

Once $\vec{\underset{\sim}{\Phi}}'$ and hence $\vec{\underset{\sim}{\Phi}}'^{(e)}$ and $\phi'(x, y)$ within all the elements are known, we can find θ by using Eq. (20.23) as

$$M_z = 2 \iint_S \phi\, \mathrm{d}x\, \mathrm{d}y = 2G\theta l^2 \iint_S \phi'(x', y')\, \mathrm{d}x'\, \mathrm{d}y' \tag{20.29}$$

or

$$\theta = \frac{M_z}{2Gl^2 \displaystyle\iint_S \phi'\, \mathrm{d}x'\, \mathrm{d}y'} = \frac{M_z}{2Gl^2 \displaystyle\sum_{e=1}^{E} \begin{pmatrix}\text{area of triangle } e, \\ \text{average of three nodal} \\ \text{values of } \phi' \text{ of element } e\end{pmatrix}} \tag{20.30}$$

Finally, the shear stresses within any element can be computed as

$$\begin{Bmatrix} \sigma_{xz} \\ \sigma_{yz} \end{Bmatrix} = G\theta l^2 \begin{Bmatrix} \dfrac{\partial \phi'}{\partial y'} \\ -\dfrac{\partial \phi'}{\partial x'} \end{Bmatrix} \tag{20.31}$$

where the derivatives of ϕ' can be obtained by differentiating Eq. (20.28).

Example 20.1 Find the stresses developed in a 4×4-cm square shaft when the angle of twist is 2° in a length of 100 cm. The value of G is 0.8×10^6 N/cm^2.

Solution

Step 1: Idealize the region by finite elements. Since the shaft has four axes of symmetry, we consider only one-eighth of the total cross section for analysis. The idealization using four elements is shown in Figure 20.2. The information needed for subsequent computations is given in Table 20.2.

Step 2: Assume a suitable interpolation model for the field variable ϕ [to solve Eq. (20.20)]. By assuming linear variation within any element, $\phi^{(e)}(x, y)$ can be expressed as

$$\phi^{(e)}(x, y) = \alpha_1 + \alpha_2 x + \alpha_3 y \equiv N_i \Phi_i + N_j \Phi_j + N_k \Phi_k \tag{E$_1$}$$

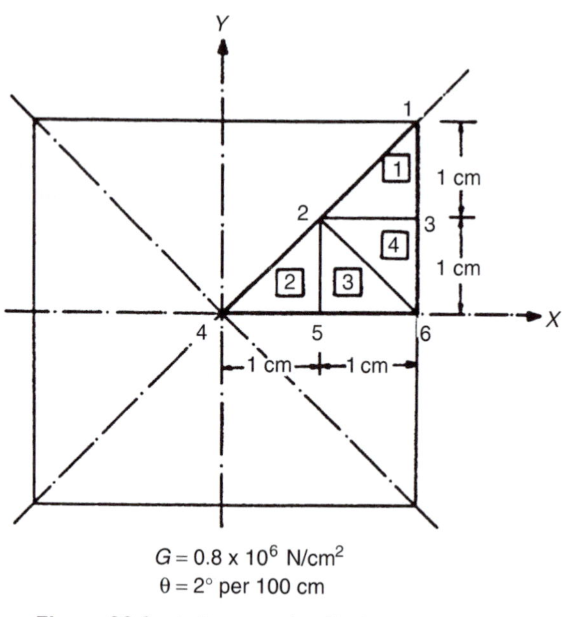

Figure 20.2. A Rectangular Shaft under Torsion.

Table 20.2.

Node	1	2	3	4	5	6
$X(x)$ coordinate (cm)	2.0	1.0	2.0	0.0	1.0	2.0
$Y(y)$ coordinate (cm)	2.0	1.0	1.0	0.0	0.0	0.0

Element number e		1	2	3	4
Global nodes	i	2	4	5	6
corresponding	j	3	5	6	3
to local nodes	k	1	2	2	2

Element number e	x_i	x_j	x_k	y_i	y_j	y_k	$c_k = (x_j - x_i)$	$c_j = (x_k - x_j)$	$c_j = (x_i - x_k)$
1	1.0	2.0	2.0	1.0	1.0	2.0	1.0	0.0	−1.0
2	0.0	1.0	1.0	0.0	0.0	1.0	1.0	0.0	−1.0
3	1.0	2.0	1.0	0.0	0.0	1.0	1.0	−1.0	0.0
4	2.0	2.0	1.0	0.0	1.0	1.0	0.0	−1.0	1.0

Element number e	$b_i = (y_j - y_k)$	$b_j = (y_k - y_i)$	$b_k = (y_i - y_j)$	$A^{(e)} = \frac{1}{2}\lvert x_{ij}y_{jk} - x_{jk}y_{ij}\rvert$
1	−1.0	1.0	0.0	0.5
2	−1.0	1.0	0.0	0.5
3	−1.0	1.0	0.0	0.5
4	0.0	1.0	−1.0	0.5

where α_1, α_2, and α_3 are given by Eq. (3.30) and N_i, N_j, and N_k by Eq. (3.35):

$$\left.\begin{aligned} \alpha_1 &= (a_i\Phi_i + a_j\Phi_j + a_k\Phi_k)/(2A^{(e)}) \\ \alpha_2 &= (b_i\Phi_i + b_j\Phi_j + b_k\Phi_k)/(2A^{(e)}) \\ \alpha_3 &= (c_i\Phi_i + c_j\Phi_j + c_k\Phi_k)/(2A^{(e)}) \end{aligned}\right\} \tag{E$_2$}$$

$$\left.\begin{aligned} N_i &= (a_i + b_i x + c_i y)/(2A^{(e)}) \\ N_j &= (a_j + b_j x + c_j y)/(2A^{(e)}) \\ N_k &= (a_k + b_k x + c_k y)/(2A^{(e)}) \end{aligned}\right\} \tag{E$_3$}$$

and Φ_i, Φ_j, and Φ_k denote the values of ϕ at nodes i, j, and k of element e, respectively.

Step 3: Derive element characteristic matrices and vectors. The finite element equations corresponding to Eq. (20.20) can be derived as a special case of Eq. (20.7) with

$$K_{ij}^{(e)} = \iint\limits_{A^{(e)}} \left[\frac{\partial N_i}{\partial x}\frac{\partial N_j}{\partial x} + \frac{\partial N_i}{\partial y}\frac{\partial N_j}{\partial y} \right] \mathrm{d}A$$

$$= \frac{1}{4A^{(e)}} \begin{bmatrix} (b_i^2 + c_i^2) & (b_ib_j + c_ic_j) & (b_ib_k + c_ic_k) \\ & (b_j^2 + c_j^2) & (b_jb_k + c_jc_k) \\ \text{Symmetric} & & (b_k^2 + c_k^2) \end{bmatrix} \tag{E$_4$}$$

and

$$P_i^{(e)} = \iint\limits_{A^{(e)}} cN_i \, \mathrm{d}A = \frac{cA^{(e)}}{3} \tag{E$_5$}$$

where $c = 2G\theta$. Thus, we can compute the element matrices and vectors as

$$[K^{(1)}] = [K^{(2)}] = [K^{(4)}] = \frac{1}{2}\begin{bmatrix} 1.0 & -1.0 & 0.0 \\ -1.0 & 2.0 & -1.0 \\ 0.0 & -1.0 & 1.0 \end{bmatrix} \tag{E$_6$}$$

$$[K^{(3)}] = \frac{1}{2}\begin{bmatrix} 2.0 & -1.0 & -1.0 \\ -1.0 & 1.0 & 0.0 \\ -1.0 & 0.0 & 1.0 \end{bmatrix} \tag{E$_7$}$$

$$\vec{P}^{(1)} = \vec{P}^{(2)} = \vec{P}^{(3)} = \vec{P}^{(4)} = \frac{G\theta}{3}\begin{Bmatrix} 1 \\ 1 \\ 1 \end{Bmatrix} \tag{E$_8$}$$

Step 4: Derive the overall equations. By using the nodal connectivity information of step 1, the overall matrix $[\underset{\sim}{K}]$ and vector $\underset{\sim}{\vec{P}}$ can be derived as

$$[\underset{\sim}{K}] = \frac{1}{2}\begin{bmatrix} 1.0 & 0.0 & -1.0 & 0.0 & 0.0 & 0.0 \\ 0.0 & 4.0 & -2.0 & 0.0 & -2.0 & 0.0 \\ -1.0 & -2.0 & 4.0 & 0.0 & 0.0 & -1.0 \\ 0.0 & 0.0 & 0.0 & 1.0 & -1.0 & 0.0 \\ 0.0 & -2.0 & 0.0 & -1.0 & 4.0 & -1.0 \\ 0.0 & 0.0 & -1.0 & 0.0 & -1.0 & 2.0 \end{bmatrix} \tag{E$_9$}$$

$$\underset{\sim}{\vec{P}} = \frac{G\theta}{3}\begin{Bmatrix} 1 \\ 4 \\ 2 \\ 1 \\ 2 \\ 2 \end{Bmatrix} \tag{E$_{10}$}$$

Step 5: Solve the system equations after applying the boundary conditions. The boundary conditions are given by $\phi = 0$ on the external boundary; that is, $\Phi_1 = \Phi_3 = \Phi_6 = 0$. By eliminating these variables Φ_1, Φ_3, and Φ_6, the system equations can be written as

$$[K]\vec{\Phi} = \vec{P} \tag{E$_{11}$}$$

where

$$[K] = \frac{1}{2}\begin{bmatrix} 4.0 & 0.0 & -2.0 \\ 0.0 & 1.0 & -1.0 \\ -2.0 & -1.0 & 4.0 \end{bmatrix}, \quad \vec{P} = \frac{G\theta}{3}\begin{Bmatrix} 4 \\ 1 \\ 2 \end{Bmatrix}, \quad \text{and} \quad \vec{\Phi} = \begin{Bmatrix} \Phi_2 \\ \Phi_4 \\ \Phi_5 \end{Bmatrix}$$

The solution of Eqs. (E$_{11}$) gives

$$\Phi_1 = \Phi_3 = \Phi_6 = 0, \qquad \Phi_2 = 3G\theta/2 = 418.8, \qquad \Phi_4 = 7G\theta/3 = 651.5,$$
$$\Phi_5 = 5G\theta/3 = 465.3$$

Step 6: Compute the element resultants.

The shear stresses induced are given by Eq. (20.16):

$$\sigma_{xz} = \frac{\partial\phi}{\partial y} = \alpha_3 = (c_i\Phi_i + c_j\Phi_j + c_k\Phi_k)/(2A^{(e)}) \tag{E$_{12}$}$$

$$\sigma_{yz} = -\frac{\partial\phi}{\partial x} = -\alpha_2 = -(b_i\Phi_i + b_j\Phi_j + b_k\Phi_k)/(2A^{(e)}) \tag{E$_{13}$}$$

For $e = 1$: $i = 2, j = 3$, and $k = 1$:

$$\sigma_{xz} = -\Phi_3 + \Phi_1 = 0, \qquad \sigma_{yz} = \Phi_2 - \Phi_3 = 418.8 \text{ N/cm}^2$$

For $e = 2$: $i = 4, j = 5$, and $k = 2$:

$$\sigma_{xz} = -\Phi_5 + \Phi_2 = -46.5 \text{ N/cm}^2, \qquad \sigma_{yz} = \Phi_4 - \Phi_5 = 186.2 \text{ N/cm}^2$$

For $e = 3$: $i = 5, j = 6$, and $k = 2$:

$$\sigma_{xz} = -\Phi_5 + \Phi_2 = -46.5 \text{ N/cm}^2, \qquad \sigma_{yz} = \Phi_5 - \Phi_6 = 465.3 \text{ N/cm}^2$$

For $e = 4$: $i = 6, j = 3$, and $k = 2$:

$$\sigma_{xz} = -\Phi_6 + \Phi_3 = 0.0, \qquad \sigma_{yz} = -\Phi_3 + \Phi_2 = 418.8 \text{ N/cm}^2$$

Computation of the Twisting Moment (M_Z)

The twisting moment acting on the shaft can be computed, using Eq. (20.23), as

$$\text{Twisting moment} = 2\iint_S \phi \, dx \, dy \simeq \sum_{e=1}^{4} \iint_{A^{(e)}} 2\phi^{(e)} \, dx \, dy \tag{E$_{14}$}$$

Since $\phi^{(e)}$ is given by Eq. (E_1), the integral in Eq. (E_{14}) can be evaluated to obtain

$$\begin{aligned}\text{Twisting moment} &= \sum_{e=1}^{4} \frac{2A^{(e)}}{3}(\Phi_i + \Phi_j + \Phi_k)^{(e)} \\ &= \frac{1}{3}[(418.8 + 0.0 + 0.0) + (651.5 + 465.3 + 418.8) \\ &\quad + (465.3 + 0.0 + 418.8) + (0.0 + 0.0 + 418.8)] \\ &= 1085.67 \text{ N-cm}\end{aligned}$$

Since the region subdivided into finite elements is only one-eighth of the total cross section, the total twisting moment (M_z) is given by

$$M_z = 8(1085.67) = 8685.36 \text{ N-cm}$$

The exact solution for a square shaft ($2a \times 2a$) is given by [20.1]

$$\begin{aligned}M_z &= 0.1406\, G\theta(2a)^4 \\ &= 0.1406(0.8 \times 10^6)\left(\frac{2}{100 \times 57.3}\right)(4)^4 = 10046.0 \text{ N-cm}\end{aligned}$$

Thus, the finite element solution can be seen to be in error by 13.54%. This error can be reduced either by increasing the number of elements or by using higher order elements for idealization.

20.4 COMPUTER PROGRAM FOR TORSION ANALYSIS

A subroutine called TØRSØN is written for the solution of torsion of prismatic shafts using triangular elements. The following input data are required for this subroutine:

NN = number of nodes.
NE = number of elements.
NB = semibandwidth of the overall matrix GK.
XC, YC = array of size NN; XC(I), YC(I) = x and y coordinates of node I.
NFIX = number of nodes lying on the outer boundary (number of nodes at which $\phi = 0$).
NØDE = array of size NE × 3; NØDE(I,J) = global node number corresponding to Jth corner of element I.
G = shear modulus of the material.
THETA = angle of twist in degrees per 100-cm length.
IFIX = array of size NFIX; IFIX(I) = Ith node number at which $\phi = 0$.

The other arguments of the subroutine are as follows:

GK = array of size NN × NB used to store the matrix $[\underset{\sim}{K}]$.

P = array of size NN used to store $\underset{\sim}{\vec{P}}$.

PLØAD = array of size NN × 1 used to store the final right-hand-side vector. It represents the solution vector (nodal values of ϕ) upon return from the subroutine TØRSØN.

To illustrate the use of the subroutine TØRSØN, the problem of Example 20.1 is considered. The main program that calls the subroutine TØRSØN and the results given by the program are given below.

```
C==================================================================
C
C     MAIN PROGRAM FOR CALLING THE SUBROUTINE TORSON
C
C==================================================================
      DIMENSION XC(6),YC(6),NODE(4,3),IFIX(3),GK(6,5),P(6),
     2 PLOAD(6,1)
      PRINT 1
 1    FORMAT(1HT)
      DATA NN,NE,NB,NFIX,G,THETA /6,4,5,3,0.8E6,2.0/
      DATA(NODE(I,1),I=1,4) /2,4,5,6/
      DATA(NODE(I,2),I=1,4) /3,5,6,3/
      DATA(NODE(I,3),I=1,4) /1,2,2,2/
      DATA XC /2.0,1.0,2.0,0.0,1.0,2.0/
      DATA YC /2.0,1.0,1.0,0.0,0.0,0.0/
      DATA IFIX /1,3,6/
      CALL TORSON(NN,NE,NB,NODE,XC,YC,GK,NFIX,IFIX,P,PLOAD,G,THETA)
      PRINT 10
10    FORMAT(2X,' NODE',2X,'VALUE OF PHI',/)
      DO 20 I=1,NN
20    PRINT 30,I,PLOAD(I,1)
30    FORMAT(2X,I3,4X,E12.4)
      STOP
      END
```

```
  NODE    VALUE OF PHI

    1     0.0000E+00
    2     0.4188E+03
    3     0.0000E+00
    4     0.6515E+03
    5     0.4654E+03
    6     0.0000E+00
```

20.5 TRANSIENT FIELD PROBLEMS

20.5.1 Governing Equations

Whenever a field problem involves time as an independent parameter, it is called a propagation, transient, dynamic, or time-dependent problem. The transient field problems are governed by the quasi-harmonic equation with time differentials. For example, in three dimensions, we have

$$\frac{\partial}{\partial x}\left(k_x\frac{\partial\phi}{\partial x}\right)+\frac{\partial}{\partial y}\left(k_y\frac{\partial\phi}{\partial y}\right)+\frac{\partial}{\partial z}\left(k_z\frac{\partial\phi}{\partial z}\right)-c-\alpha\frac{\partial\phi}{\partial t}-\beta\frac{\partial^2\phi}{\partial t^2}=0 \tag{20.32}$$

where k_x, k_y, k_z, c, α, and β will be, in general, functions of x, y, z and time t. If time is not considered as a variable, Eq. (20.32) can be seen to reduce to the steady-state quasi-harmonic equation considered in the previous section. The boundary conditions associated with Eq. (20.32) are

$$\phi = \bar{\phi} \quad \text{for} \quad t > 0 \text{ on } S_1 \tag{20.33}$$

and

$$k_x \frac{\partial \phi}{\partial x} l_x + k_y \frac{\partial \phi}{\partial y} l_y + k_z \frac{\partial \phi}{\partial z} l_z + q + r\phi = 0 \quad \text{for} \quad t > 0 \text{ on } S_2 \tag{20.34}$$

Since this is a time-dependent problem, the initial conditions also have to be specified as

$$\phi(x, y, z, t = 0) = \phi_0(x, y, z) \text{ in } V \tag{20.35}$$

and

$$\frac{\mathrm{d}\phi}{\mathrm{d}t}(x, y, z, t = 0) = \dot{\phi}_0(x, y, z) \text{ in } V \tag{20.36}$$

Equation (20.32) represents a general damped wave equation and has application in phenomena such as electromagnetic waves, acoustic waves, and surface waves. No variational principle (functional I) exists for the problem stated by Eqs. (20.32)–(20.36).

20.5.2 Finite Element Solution

We present the finite element solution of the problem according to the weighted residual (Galerkin) method in this section.

Step 1: Discretize the domain V into E three-dimensional finite elements having p nodes each.

Step 2: Assume the variation of the field variable in a typical element e as

$$\phi(x, y, z, t) = \sum_{i=1}^{p} N_i(x, y, z)\Phi_i^{(e)} = [N]\vec{\Phi}^{(e)} \tag{20.37}$$

where N_i is the interpolation function corresponding to the nodal unknown $\Phi_i^{(e)}$ of element e. The nodal unknowns $\Phi_i^{(e)}$ are assumed to be functions of time.

Step 3: Derive the finite element equations using the Galerkin method.

In this method, the criterion to be satisfied at any instant in time is given by

$$\iiint\limits_{V^{(e)}} N_i \left[\frac{\partial}{\partial x}\left(k_x \frac{\partial \phi}{\partial x}\right) + \frac{\partial}{\partial y}\left(k_y \frac{\partial \phi}{\partial y}\right) + \frac{\partial}{\partial z}\left(k_z \frac{\partial \phi}{\partial z}\right) - c - \alpha \frac{\partial \phi}{\partial t} - \beta \frac{\partial^2 \phi}{\partial t^2} \right] \mathrm{d}V = 0$$

$$i = 1, 2, \ldots, p \tag{20.38}$$

Each of the first three terms in the brackets in Eq. (20.38) can be integrated by parts using the Green–Gauss theorem of Appendix A:

$$\iiint_{V^{(e)}} N_i \frac{\partial}{\partial x}\left(k_x \frac{\partial \phi}{\partial x}\right) \mathrm{d}V = -\iiint_{V^{(e)}} \frac{\partial N_i}{\partial x} k_x \frac{\partial \phi}{\partial x} \mathrm{d}V + \iint_{S^{(e)}} N_i k_x \frac{\partial \phi}{\partial x} \mathrm{d}y\, \mathrm{d}z$$

$$= -\iiint_{V^{(e)}} \frac{\partial N_i}{\partial x} k_x \frac{\partial \phi}{\partial x} \mathrm{d}V + \iint_{S^{(e)}} N_i k_x \frac{\partial \phi}{\partial x} l_x \, \mathrm{d}S \tag{20.39}$$

where l_x is the x-direction cosine of the outward normal. Thus, Eq. (20.38) can be written as

$$-\iiint_{V^{(e)}} \left[k_x \frac{\partial N_i}{\partial x}\frac{\partial \phi}{\partial x} + k_y \frac{\partial N_i}{\partial y}\frac{\partial \phi}{\partial y} + k_z \frac{\partial N_i}{\partial z}\frac{\partial \phi}{\partial z}\right] \mathrm{d}V + \iint_{S^{(e)}} N_i \left[k_x \frac{\partial \phi}{\partial x} l_x + k_y \frac{\partial \phi}{\partial y} l_y + k_z \frac{\partial \phi}{\partial z} l_z\right] \mathrm{d}S$$

$$-\iiint_{V^{(e)}} N_i \left(c + \alpha \frac{\partial \phi}{\partial t} + \beta \frac{\partial^2 \phi}{\partial t^2}\right) \mathrm{d}V = 0, \qquad i = 1, 2, \ldots, p \tag{20.40}$$

Since the boundary of the elements, $S^{(e)}$, is composed of $S_1^{(e)}$ and $S_2^{(e)}$, the surface integral in Eq. (20.40) over $S_1^{(e)}$ would be zero (since ϕ is prescribed to be a constant $\bar{\phi}$ on $S_1^{(e)}$, the derivative of ϕ would be zero). On the surface $S_2^{(e)}$, the boundary condition given by Eq. (20.34) is to be satisfied. For this, we express the surface integral in Eq. (20.40) over $S_2^{(e)}$ in equivalent form as

$$\iint_{S_2^{(e)}} N_i \left[k_x \frac{\partial \phi}{\partial x} l_x + k_y \frac{\partial \phi}{\partial y} l_y + k_z \frac{\partial \phi}{\partial z} l_z\right] \mathrm{d}S_2 = \iint_{S_2^{(e)}} N_i [-q - r\phi] \, \mathrm{d}S_2 \tag{20.41}$$

By using Eqs. (20.37) and (20.41), Eq. (20.40) can be expressed in matrix form as

$$[K^{(e)}]\vec{\Phi}^{(e)} + [K_1^{(e)}]\dot{\vec{\Phi}}^{(e)} + [K_2^{(e)}]\ddot{\vec{\Phi}}^{(e)} + [K_3^{(e)}]\vec{\Phi}^{(e)} + \vec{P}^{(e)} = \vec{0} \tag{20.42}$$

where the elements of the various matrices in Eq. (20.42) are given by

$$K_{ij}^{(e)} = \iiint_{V^{(e)}} \left(k_x \frac{\partial N_i}{\partial x}\frac{\partial N_j}{\partial x} + k_y \frac{\partial N_i}{\partial y}\frac{\partial N_j}{\partial y} + k_z \frac{\partial N_i}{\partial z}\frac{\partial N_j}{\partial z}\right) \mathrm{d}V \tag{20.43}$$

$$K1_{ij}^{(e)} = \iiint_{V^{(e)}} \alpha N_i N_j \, \mathrm{d}V \tag{20.44}$$

$$K2_{ij}^{(e)} = \iiint_{V^{(e)}} \beta N_i N_j \, \mathrm{d}V \tag{20.45}$$

$$K3_{ij}^{(e)} = \iint_{S_2^{(e)}} r N_i N_j \, \mathrm{d}S_2 \tag{20.46}$$

$$P_i^{(e)}(t) = \iiint_{V^{(e)}} c N_i \, \mathrm{d}V + \iint_{S_2^{(e)}} q N_i \, \mathrm{d}S_2 \tag{20.47}$$

Step 4: Assemble the element equations and obtain the overall equations. Equation (20.42) represents the element equations and the assembly of these equations leads to the following type of ordinary differential equations:

$$[\underset{\sim}{K}]\,\underset{\sim}{\vec{\Phi}} + [\underset{\sim}{K_1}]\,\underset{\sim}{\dot{\vec{\Phi}}} + [\underset{\sim}{K_2}]\,\underset{\sim}{\ddot{\vec{\Phi}}} + [\underset{\sim}{K_3}]\,\underset{\sim}{\vec{\Phi}} + \underset{\sim}{\vec{P}} = \vec{0} \tag{20.48}$$

where a dot over $\underset{\sim}{\Phi}$ represents the time derivative.

Step 5: Solve the assembled equations. The system of equations (20.48) can be solved for $\underset{\sim}{\vec{\Phi}}(t)$ with the discretized form of initial conditions stated in Eqs. (20.35) and (20.36) and by incorporating the boundary conditions of Eq. (20.33). The solution procedures outlined in Sections 7.4 and 12.6 can be used to solve Eq. (20.48).

Step 6: Find the element resultants. From the known values of the nodal values of ϕ, the required element resultants can be computed with the help of Eq. (20.37).

20.5.3 Space–Time Finite Elements

In a general time-dependent or propagation problem, three spatial and one time parameters will be involved. Usually, we first use the finite element method to formulate the solution in the physical space. Next, we use a different method, such as finite differences, to find the solution over a period of time. Thus, this procedure involves the idealization of the field variable $\phi(x, y, z, t)$ in any element e (in three-dimensional space) as

$$\phi(x, y, z, t) = [N_1(x, y, z)]\vec{\Phi}^{(e)}(t) \tag{20.49}$$

where $[N_1]$ is the matrix of interpolation or shape functions in space, and $\vec{\Phi}^{(e)}$ is the vector of time-dependent nodal variables. By using Eq. (20.49) and the specified initial conditions, we use a finite difference scheme such as

$$\vec{\Phi}^{(e)}(t) = [N_2(t, \Delta t)]\vec{\Phi}^{(e)}(t - \Delta t) \tag{20.50}$$

where $[N_2]$ indicates the matrix of interpolation functions in the time domain.

Instead of solving the problem using Eqs. (20.49) and (20.50), finite elements can be constructed in four dimensions (x, y, z, and t) and the field variable can be expressed as [20.4–20.6]

$$\phi(x, y, z, t) = [N(x, y, z, t)]\vec{\Phi}^{(e)} \tag{20.51}$$

where $[N]$ represents the matrix of shape functions in x, y, z, and t, and $\vec{\Phi}^{(e)}$ is the vector of nodal values of element e. In this case, the time-dependent problem can be directly solved without using any special techniques.

REFERENCES

20.1 I.H. Shames: *Mechanics of Deformable Solids*, Prentice-Hall, Englewood Cliffs, NJ, 1964.

20.2 Y.C. Fung: *Foundations of Solid Mechanics*, Prentice-Hall, Englewood Cliffs, NJ, 1965.

20.3 L.R. Herrmann: Elastic torsional analysis of irregular shapes, *Journal of Engineering Mechanics Division, 91*, No. EM6, 11–19, 1965.

20.4 O.C. Zienkiewicz and C.J. Parekh: Transient field problems — Two and three dimensional analysis by isoparametric finite elements, *International Journal for Numerical Methods in Engineering, 2*, 61–71, 1970.

20.5 J.H. Argyris and D.W. Scharpf: Finite elements in time and space, *Aeronautical Journal of the Royal Aeronautical Society, 73*, 1041–1044, 1969.

20.6 I. Fried: Finite element analysis of time dependent phenomena, *AIAA Journal, 7*, 1170–1173, 1969.

PROBLEMS

20.1–20.6 Identify whether the given partial differential equation is parabolic, elliptic, or hyperbolic:

(a) Steady-state fluid (seepage) flow under a dam:

$$k_x \frac{\partial^2 h}{\partial x^2} + k_y \frac{\partial^2 h}{\partial y^2} = 0$$

(b) Laminar flow heat exchanger equation:

$$\frac{\partial^2 T}{\partial r^2} + \frac{1}{r}\frac{\partial T}{\partial r} + \frac{\partial^2 T}{\partial z^2} = \frac{\rho u c}{k}\frac{\partial T}{\partial z}$$

(c) Ion exchange equation (for flow of a solution through a packed column containing an ion exchange resin):

$$\frac{\partial u}{\partial t} + \alpha(x,t)\frac{\partial u}{\partial x} = p(x,t)$$

(d) Transient heat conduction in two dimensions:

$$\rho c \frac{\partial T}{\partial t} = k \frac{\partial^2 T}{\partial x^2} + k \frac{\partial^2 T}{\partial y^2} + Q(x,y)$$

(e) Torsion of a prismatic shaft (Poisson's equation):

$$\frac{\partial^2 \phi}{\partial x^2} + \frac{\partial^2 \phi}{\partial y^2} + p(x,y) = 0$$

(f) Vibration of a membrane:

$$\frac{\partial^2 u}{\partial x^2} + \frac{\partial^2 u}{\partial y^2} = \frac{\partial^2 u}{\partial t^2}$$

20.7 A steel shaft, with elliptic cross section and length 1 m, is twisted by an angle of $\theta = 2^\circ$ (Figure 20.3). Assuming the shear modulus as $G = 80$ GPa, determine the maximum shear stress and the torque (M_z) in the shaft by modeling one-quarter of the ellipse using linear triangular finite elements. Compare the finite element solution with the following exact solution:

$$\tau_{\max} = \frac{2M_z}{\pi a b^2} \quad \text{at} \quad x = 0, \qquad y = \pm b$$

$$M_z = \frac{G\theta\pi a^3 b^3}{(a^2 + b^2)}$$

20.8 Consider the torsion of a uniform shaft with an I section (Figure 20.4). Using the finite element method, determine the stresses in the elements and the torque on the shaft for the following data:
Length = 1 m, $G = 80$ GPa, $\theta = 1^\circ$.

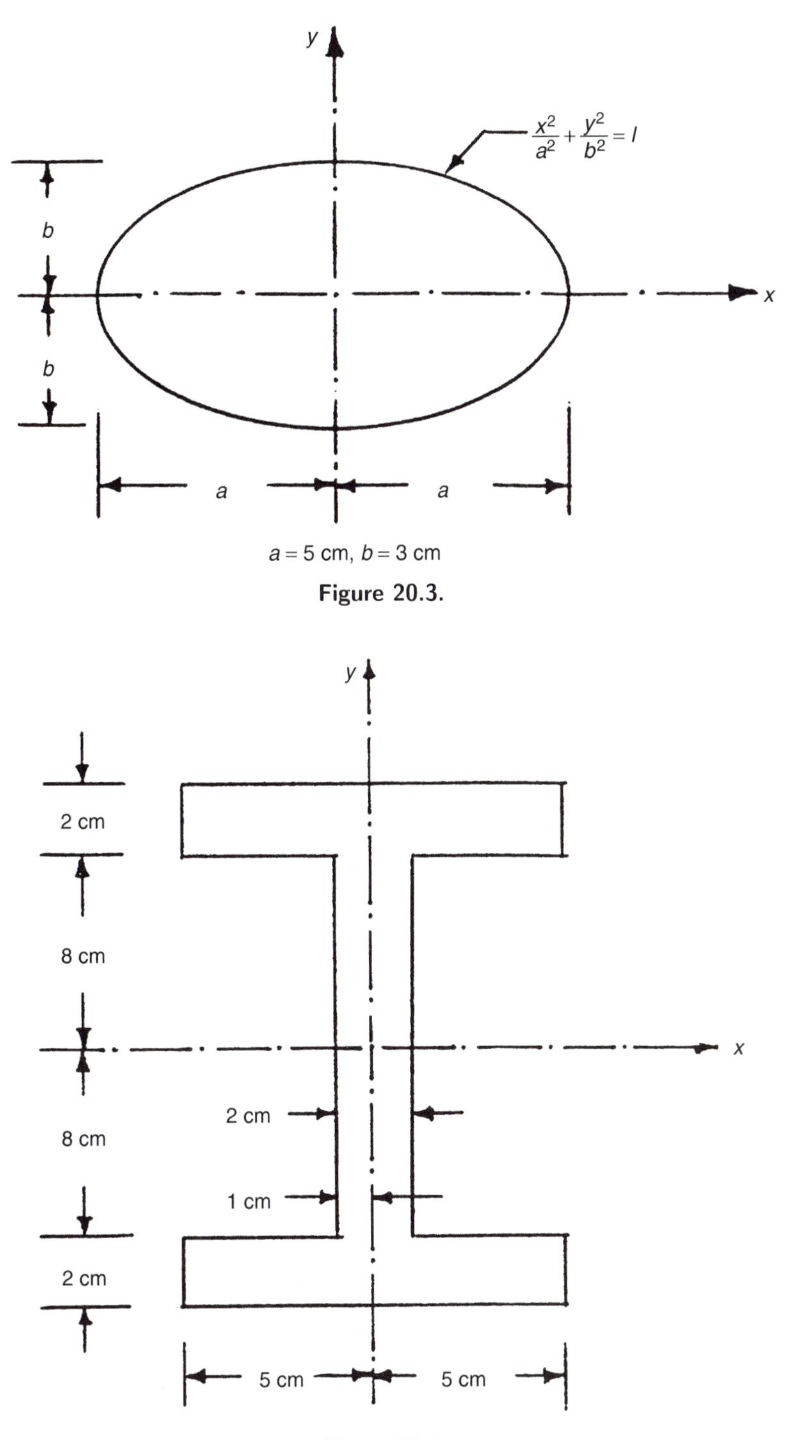

$a = 5$ cm, $b = 3$ cm

Figure 20.3.

Figure 20.4.

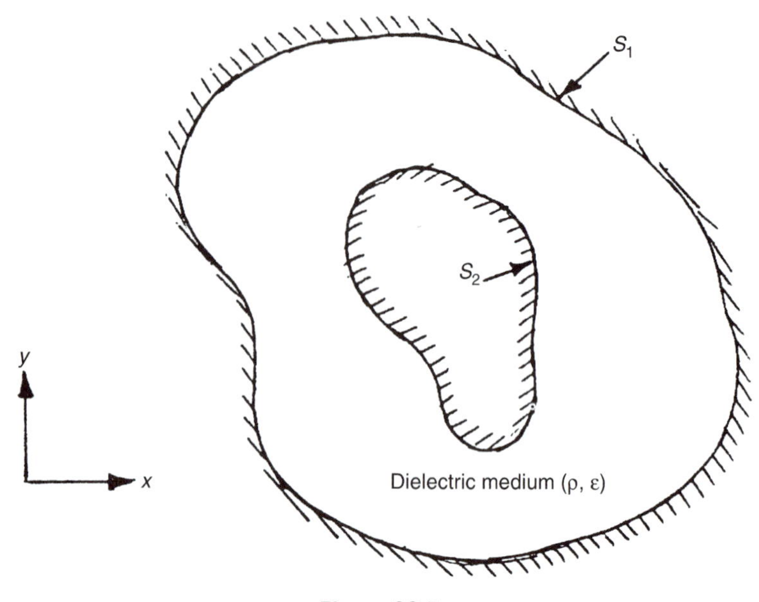

Figure 20.5.

20.9 The torsion of a prismatic shaft is governed by the equation

$$\frac{\partial^2 \psi}{\partial x^2} + \frac{\partial^2 \psi}{\partial y^2} + 2 = 0$$

in the interior cross section of the shaft subject to $\psi = 0$ on the boundary of the cross section. Derive the corresponding finite element equations using the Galerkin method.

20.10 In the electrostatic problem, the electric potential (ϕ) is governed by the equation (Figure 20.5)

$$\frac{\partial^2 \phi}{\partial x^2} + \frac{\partial^2 \phi}{\partial y^2} + \frac{\rho}{\varepsilon} = 0$$

with the boundary conditions

$$\phi = c_1 \text{ on } S_1; \qquad \phi = c_2 \text{ on } S_2$$

where ρ is the charge density (Coulomb/m^3), and ε is the permittivity of the dielectric medium (Farad/m). Derive the finite element equations using a variational approach.

20.11 Determine the distribution of voltage in the circular coaxial cable shown in Figure 20.6 if a voltage of 200 V is applied at the inner surface of the dielectric insulator. Assume the voltage at the outer surface to be zero.

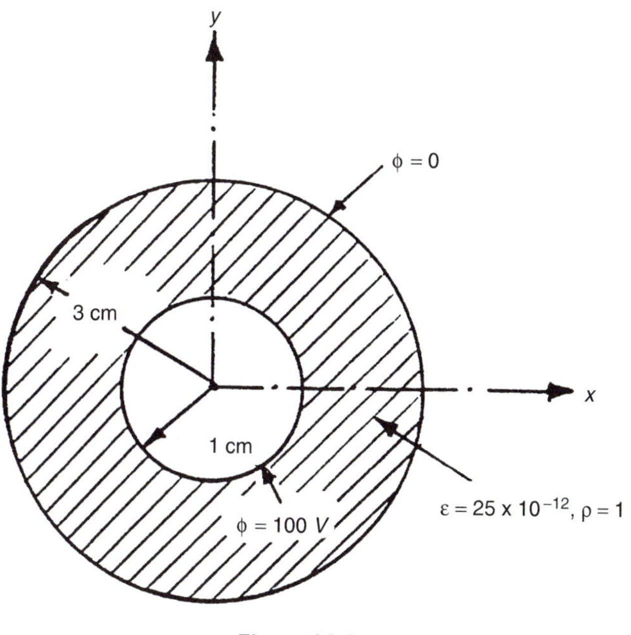

Figure 20.6.

20.12 Consider the following transient field problem:

$$c_1 \frac{\partial^2 \phi}{\partial x^2} + c_2 \frac{\partial \phi}{\partial t} = 0; \qquad 0 \le x \le L; \qquad 0 \le t \le T \tag{E$_1$}$$

with $\phi = \phi_{x0}$ or $(\mathrm{d}\phi/\mathrm{d}x) = 0$ at $x = 0$ and $x = L$, and $\phi = \phi_{t0}(x)$ at $t = 0$. Using $(\partial \phi/\partial t) = (\phi_{i+1} - \phi_i)/\Delta t$ with $\phi_i = \phi(t_i)$ and $\phi_{i+1} = \phi(t_{i+1} = t_i + \Delta t)$, derive the finite element equations using the Galerkin approach.

20.13 Consider the transient field problem described in Problem 20.12. Using linear triangular elements in a space–time coordinate system, derive the finite element equations using the Galerkin method.

21

SOLUTION OF HELMHOLTZ EQUATION

21.1 INTRODUCTION

Physical problems such as the vibration of a membrane, the propagation of electromagnetic waves in a waveguide, the oscillatory or seiche motion of an enclosed mass of water in a lake or harbor, and the acoustic vibrations of a body of fluid enclosed in a room or vehicle are all governed by the Helmholtz equation given by [21.1]

$$\frac{\partial}{\partial x}\left(k_x \frac{\partial \phi}{\partial x}\right) + \frac{\partial}{\partial y}\left(k_y \frac{\partial \phi}{\partial y}\right) + \frac{\partial}{\partial z}\left(k_z \frac{\partial \phi}{\partial z}\right) + \lambda^2 \phi = 0 \tag{21.1}$$

with the boundary conditions

$$\phi = \bar{\phi} \text{ on } S_1 \text{ (Dirichlet condition)} \tag{21.2}$$

and

$$\frac{\partial \phi}{\partial n} = 0 \text{ on } S_2 \text{ (Neumann condition)} \tag{21.3}$$

where n denotes the normal direction. The physical significance of k_x, k_y, k_z, ϕ, and λ for various types of problems is given in Table 21.1. In Eq. (21.1), ϕ and λ are both unknowns.

21.2 FINITE ELEMENT SOLUTION

The functional I, which on minimization gives Eq. (21.1) and the Neumann boundary condition, is given by

$$I(\phi) = \iiint\limits_V \left[k_x \left(\frac{\partial \phi}{\partial x}\right)^2 + k_y \left(\frac{\partial \phi}{\partial y}\right)^2 + k_z \left(\frac{\partial \phi}{\partial z}\right)^2 - \lambda^2 \phi^2\right] \mathrm{d}V \tag{21.4}$$

Thus, the function $\phi(x, y, z)$, which minimizes $I(\phi)$ of Eq. (21.4) and satisfies the given Dirichlet boundary condition, also satisfies the Helmholtz Eq. (21.1). There is no need to

Table 21.1. Typical Problems Governed by Eq. (21.1)

Problem	Physical significance of k_x, k_y, k_z	Physical significance of ϕ	Physical significance of λ^2
1. Propagation of electromagnetic waves in a waveguide filled with a dielectric material [21.2, 21.3]	$k_x = k_y = k_z = \frac{1}{\varepsilon_d}$, where ε_d = permittivity of the dielectric	A component of the magnetic field strength vector (when transverse magnetic modes are considered)	$\lambda^2 = \omega^2 \mu_0 \varepsilon_0$, where ω = wave frequency, μ_0 = permeability of free space, ε_0 = permittivity of free space
2. Seiche motion (oscillations of an enclosed water mass in a lake or harbor) [21.4]	$k_x = k_y = h$, $k_z = 0$, where h = depth from the mean water level to the lake or harbor bed	Elevation of free surface (or standing wave) measured from the mean water level	$\lambda^2 = \frac{4\pi^2}{gT^2}$, where g = acceleration due to gravity, T = period of oscillation
3. Free transverse vibrations of a membrane [21.5]	$k_x = k_y = k_z = 1$	Transverse displacement of the membrane	$\lambda^2 = \frac{\rho\omega^2}{T}$, where ρ = mass per unit area, ω = natural frequency, T = tension (force per unit length) of membrane
4. Vibrations of an enclosed fluid (acoustic vibrations) [21.6, 21.7]	$k_x = k_y = k_z = 1$	Excess pressure above ambient pressure	$\lambda^2 = \omega^2/c^2$, where ω = wave frequency, c = wave velocity in the medium

consider the Neumann boundary condition separately because it is naturally taken into account in the functional I, with the additional term being identically zero. To derive the finite element equations, we divide the domain V in which Eq. (21.1) is to be satisfied into E finite elements. We then express ϕ within each element in terms of its nodal values, $\vec{\Phi}^{(e)}$, as

$$\phi(x, y, z) = \sum_i N_i(x, y, z)\Phi_i^{(e)} = [N(x, y, z)]\vec{\Phi}^{(e)} \tag{21.5}$$

Next, we substitute Eq. (21.5) into the discretized form of Eq. (21.4), namely

$$I = \sum_{e=1}^{E} I^{(e)} \quad \text{and} \quad \text{set} \quad \frac{\partial I}{\partial \Phi_i} = 0 \quad \text{for} \quad i = 1, 2, \ldots, M \tag{21.6}$$

where Φ_i is the ith nodal unknown, and M is the total number of nodal unknowns. This gives the system equations as

$$[[\underset{\sim}{K}] - \lambda^2[\underset{\sim}{M}]]\vec{\underset{\sim}{\Phi}} = \vec{0} \tag{21.7}$$

where

$$[\underset{\sim}{K}] = \sum_{e=1}^{E}[K^{(e)}] \tag{21.8}$$

$$[\underset{\sim}{M}] = \sum_{e=1}^{E}[M^{(e)}] \tag{21.9}$$

$$\vec{\underset{\sim}{\Phi}} = \begin{Bmatrix} \Phi_1 \\ \Phi_2 \\ \vdots \\ \Phi_M \end{Bmatrix} \tag{21.10}$$

$$K_{ij}^{(e)} = \int_{V^{(e)}} \left[k_x \frac{\partial N_i}{\partial x}\frac{\partial N_j}{\partial x} + k_y \frac{\partial N_i}{\partial y}\frac{\partial N_j}{\partial y} + k_z \frac{\partial N_i}{\partial z}\frac{\partial N_j}{\partial z}\right] \mathrm{d}V \tag{21.11}$$

$$M_{ij}^{(e)} = \int_{V^{(e)}} N_i N_j \,\mathrm{d}V \tag{21.12}$$

Equation (21.7) represents an algebraic eigenvalue problem (similar to the free vibration equation of Chapter 7), and, in general, the number of values of λ (eigenvalues) that satisfy Eq. (21.7) is equal to the total number of degrees of freedom M. For each value of λ, there will be a corresponding vector $\vec{\Phi}$ representing the mode or distribution of ϕ. Both the matrices $[\underset{\sim}{K}]$ and $[\underset{\sim}{M}]$ would be symmetric and positive semidefinite, and hence all the nonzero eigenvalues would be distinct and positive with all the corresponding eigenvectors $\vec{\Phi}$ independent [21.8].*

21.3 NUMERICAL EXAMPLES

21.3.1 Propagation of Electromagnetic Waves in a Waveguide

A waveguide is a system of conductors and dielectrics used to guide electromagnetic energy from one point to another. A simple rectangular waveguide shown in Figure 21.1 is considered to illustrate the use of finite element technique for the solution of Helmholtz equation [21.3]. In this case, the Neumann boundary condition that $(\partial\phi/\partial n) = 0$ has to be satisfied for H-waves and the Dirichlet condition that $\phi = 0$ for E-waves. Here, use is made of symmetry and the boundary conditions imposed are shown in Figure 21.1.

* After incorporating the boundary conditions on S_1, the matrices $[\underset{\sim}{K}]$ and $[\underset{\sim}{M}]$ become positive definite, in which case all the eigenvalues will be distinct and positive. This property is true even for structural eigenvalue problems.

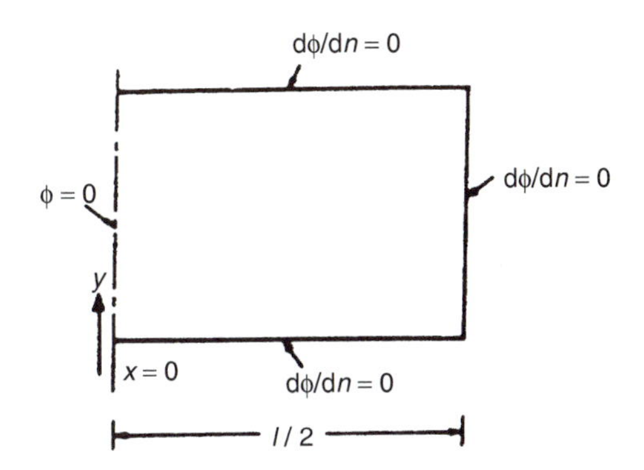

Figure 21.1. Rectangular Waveguide (Half Cross Section).

Table 21.2. Results of Rectangular Waveguide *H*-Mode [21.3]

Method	Finite element	Finite differences	Exact	(x/l)
Value of $(\lambda l)^2$	9.9	9.79	9.87	
Value of field	1.00000	1.00000	1.00000	0.5
variable ϕ	0.95105	0.95105	0.95105	0.4
	0.80902	0.80902	0.80902	0.3
	0.58778	0.58778	0.58778	0.2
	0.30902	0.30902	0.30902	0.1
	0.00000	0.00000	0.00000	0

Table 21.2 gives the results obtained by Arlett *et al.* [21.3] along with those reported from finite differences and exact methods. The exact solution is given by

$$(\lambda l)^2 = \pi, \quad \phi = \sin(\pi x/l)$$

Next, a T septate lunar waveguide shown in Figure 21.2 is considered. The finite element subdivision and the field distribution are also indicated in Figure 21.2. Table 21.3 shows the finite element results given by Arlett *et al.* [21.3] along with those obtained using a regular grid by the finite differences and experimental results.

21.3.2 Frequencies and Modes of Sound Pressure

The problem of noise in the passenger compartment of an automobile is that of an acoustic field in a room with a complex shape. Since the theoretical solution of the governing (Helmholtz) equation is extremely difficult over irregular-shaped domains, the finite element method is used to find the frequencies and modes of sound pressure inside passenger cars [21.7, 21.10]. Shuku and Ishihara [21.10] used a triangular element with linear and cubic interpolations for idealizing a two-dimensional acoustic field enclosed by an irregularly shaped boundary. The values of sound pressure (p) and pressure gradients $[(\partial p/\partial x)$ and $(\partial p/\partial y)]$ are taken as nodal unknowns in the case of the cubic interpolation model. After deriving the element equations in a convenient local xy coordinate system, the

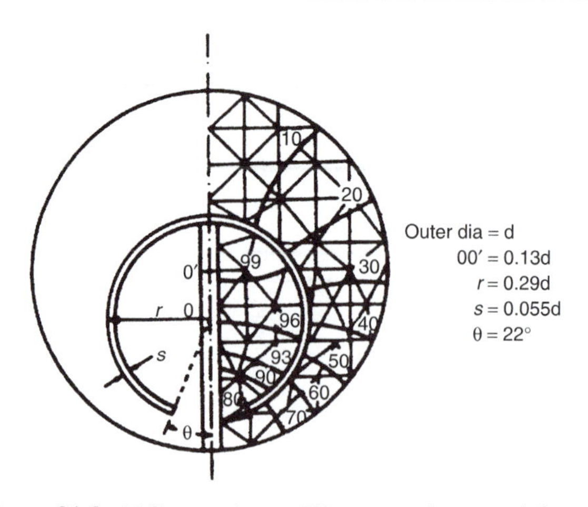

Figure 21.2. *T* Septate Lunar Waveguide (*H*-Mode) [21.3].

Table 21.3. Results of *T* Septate Lunar Waveguide [21.3]

Method	Value of (λd)	Value of (h/d)
Finite element	1.052	$\simeq \frac{1}{10}$ to $\frac{1}{20}$
Finite differences	1.039	$\frac{1}{52}$
Experimental [21.9]	1.042	—

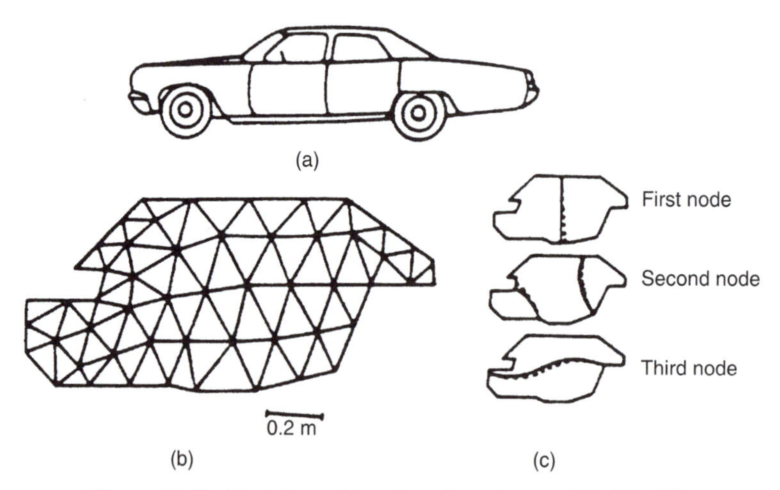

Figure 21.3. Modeling of Interior of an Automobile [21.10].

Table 21.4. Comparison of Normal Frequencies of an Automobile Compartment Model [21.10]

Mode number	Triangular elements: With cubic interpolation (number of nodes = 53)	Triangular elements: With linear interpolation (nodes = 53)	Finite differences method (nodes = 57)	Experimental
1	86.8	86.9	88.0	87.5
2	138.9	142.9	137.7	138.5
3	154.6	159.1	156.4	157.0

transformation matrix, $[\lambda]$, to a global XY coordinate system is established by using the relation

$$\begin{Bmatrix} p \\ \partial p/\partial x \\ \partial p/\partial y \end{Bmatrix} = \begin{bmatrix} 1 & 0 & 0 \\ 0 & \cos\theta & -\sin\theta \\ 0 & \sin\theta & \cos\theta \end{bmatrix} \begin{Bmatrix} p \\ \partial p/\partial X \\ \partial p/\partial Y \end{Bmatrix} \equiv [\lambda] \begin{Bmatrix} p \\ \partial p/\partial X \\ \partial p/\partial Y \end{Bmatrix}$$

where θ is the angle that the x axis makes with the X axis. The interior of an automobile is modeled as shown in Figure 21.3. In Figure 21.3(c), the solid and dotted lines indicate the loci of points where the sound pressure level is minimum (mode shapes). Although in reality, the interior of an automobile has soft boundaries, the boundaries are assumed to be rigid in the current analysis. The normal frequencies predicted by the finite element method are compared with those obtained by other methods in Table 21.4 [21.10].

REFERENCES

21.1 P. Silvester: Tetrahedral polynomial finite elements for the Helmholtz equation, *International Journal for Numerical Methods in Engineering*, *4*, 405–413, 1972.

21.2 W.H. Haytt, Jr.: *Engineering Electromagnetics*, McGraw-Hill, New York, 1958.

21.3 P.L. Arlett, A.K. Bahrani, and O.C. Zienkiewicz: Application of finite elements to the solution of Helmholtz equation, *Proceedings of Institution of Electrical Engineers*, *115*, 1762–1766, 1968.

21.4 C. Taylor, B.S. Patil, and O.C. Zienkiewicz: Harbor oscillation: A numerical treatment for undamped natural modes, *Proceedings of the Institution of Civil Engineers*, *43*, 141–155, 1969.

21.5 L. Meirovitch: *Analytical Methods in Vibrations*, Macmillan, New York, 1967.

21.6 L.E. Kinsler and A.R. Frey: *Fundamentals of Acoustics*, Wiley, New York, 1950.

21.7 A. Craggs: The use of simple three dimensional acoustic finite elements for determining the natural modes and frequencies of complex shaped enclosures, *Journal of Sound and Vibration*, *23*, 331–339, 1972.

21.8 J.H. Wilkinson: *The Algebraic Eigenvalue Problem*, Oxford University Press, London, 1965.

21.9 J.B. Davies and C.A. Muilwyk: Numerical solution of uniform hollow waveguides with boundaries of arbitrary shape, *Proceedings of Institution of Electrical Engineers*, *113*, 277–284, 1966.

21.10 T. Shuku and K. Ishihara: The analysis of the acoustic field in irregularly shaped rooms by the finite element method, *Journal of Sound and Vibration*, *29*, 67–76, 1973.

PROBLEMS

21.1 The equation governing the acoustical vibrations in a closed pipe (Figure 21.4) is given by

$$\frac{d^2\phi}{dx^2} + a\phi = 0; \qquad 0 \le x \le l \tag{E$_1$}$$

subject to

$$\frac{d\phi}{dx} = 0 \quad \text{at} \quad x = 0 \text{ and } l \tag{E$_2$}$$

where $a = (\omega^2/c^2)$, ω is the wave frequency, and c is the wave velocity in the medium.

(a) Derive the finite element equations corresponding to Eq. (E_1) using the Galerkin approach.
(b) Derive the element matrices using a linear two-noded element.

21.2 Using a one-element idealization of the pipe shown in Figure 21.4 of Problem 21.1, derive the eigenvalue problem of the system. Also, find the eigenvalues and eigenvectors of the system.

21.3 Using a two-element idealization of the pipe shown in Figure 21.4 of Problem 21.1, derive the eigenvalue problem of the system. Also, find the eigenvalues and eigenvectors of the system.

21.4 Using a three-element idealization of the pipe shown in Figure 21.4 of Problem 21.1, derive the eigenvalue problem of the system. Also, find the eigenvalues and eigenvectors of the system.

21.5 The equation governing the buckling of a uniform pin–pin column (Figure 21.5) is given by

$$\frac{d^2w}{dx^2} + aw = 0; \qquad 0 \le x \le l \tag{E$_1$}$$

with boundary conditions

$$w = 0 \quad \text{at} \quad x = 0 \quad \text{and} \quad x = l \tag{E$_2$}$$

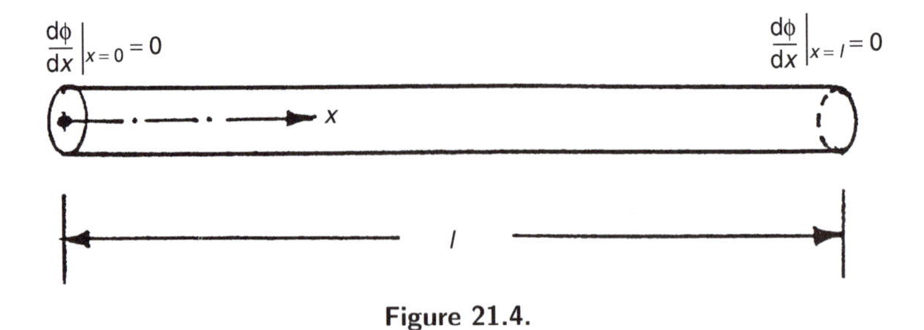

Figure 21.4.

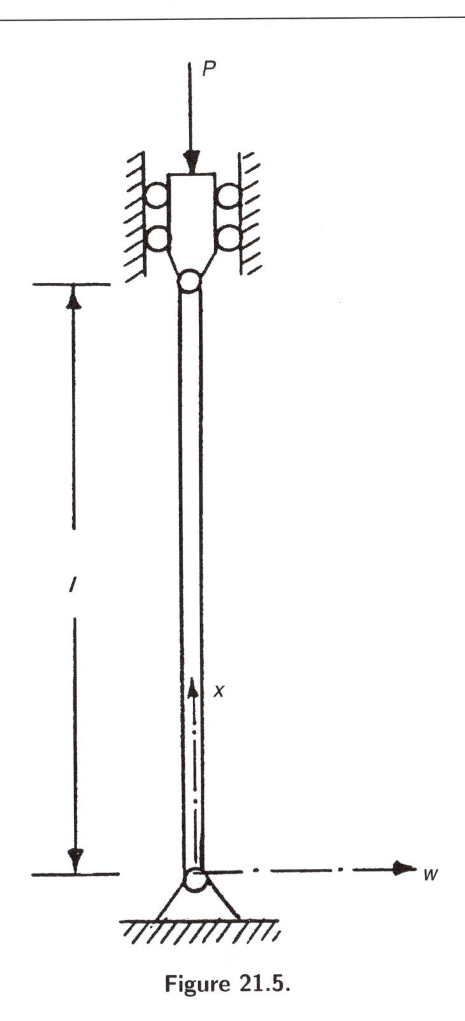

Figure 21.5.

where $a = (P/EI)$, P is the axial (compressive) load, I is the area moment of inertia of the cross section, and E is the Young's modulus of the column. Derive the element matrices corresponding to Eq. (E_1) using a two-noded linear element.

21.6 Using a two-element idealization of the column shown in Figure 21.5 of Problem 21.5, formulate and solve the corresponding eigenvalue problem.

22

SOLUTION OF REYNOLDS EQUATION

22.1 HYDRODYNAMIC LUBRICATION

Effective lubrication requires that the two surfaces should be nearly but not quite parallel as shown in Figure 22.1. The pressure necessary to separate the two surfaces is developed due to the wedge action. The applied (normal) load is resisted by the fluid pressure, thereby reducing friction and wear. This type of lubrication is called "hydrodynamic lubrication" and is governed by the Reynolds equation [22.1]:

$$\nabla\left[\frac{h^3}{12}\nabla p\right] = \nabla\left(h\vec{U} + \frac{h^3}{12\mu}\vec{B}\right) + \frac{\partial h}{\partial t} + v_{\mathrm{d}} \tag{22.1}$$

with boundary conditions

$$p = \bar{p} \text{ on } C_1 \tag{22.2}$$

and

$$\vec{n}\left[\vec{U}h + \vec{B}\frac{h^3}{12\mu} - \frac{h^3}{12\mu}\nabla p\right] = q \text{ on } C_2 \tag{22.3}$$

where h is the distance of separation of the surfaces (film thickness), μ is the viscosity, p is the pressure, $\vec{U}$ is the surface velocity vector, $\vec{B}$ is the body force vector (acting in the film), $(\partial h/\partial t) = \dot{h} =$ squeeze velocity, t is the time, v_d is the net outward velocity of diffusion flow through porous boundary surfaces located at $z = 0$ and $z = h$, $\vec{n}$ is the unit outer normal vector to C_2, q is the lineal normal flow across the boundary C_2, C_1 is the part of the boundary over which pressure is prescribed, and C_2 is the part of the boundary over which flow is prescribed.

The assumptions made in the derivation of Eq. (22.1) are as follows:

1. The fluid is incompressible.
2. The fluid is Newtonian.

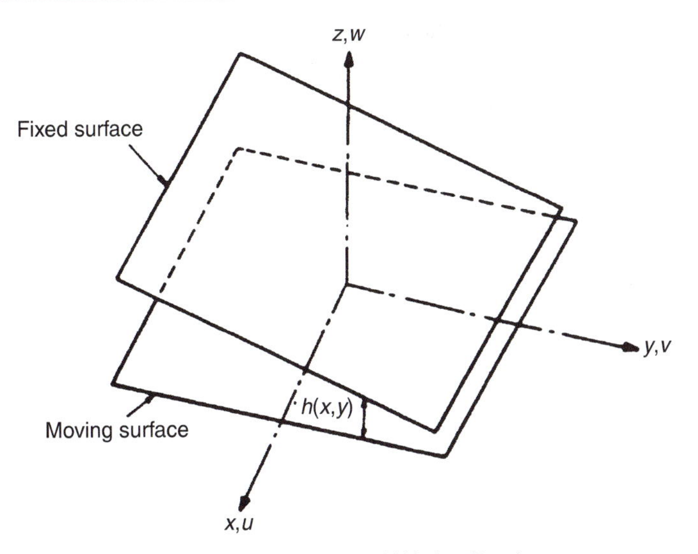

Figure 22.1. A Tapered Wedge Bearing.

3. The curvature of the bearing components introduces only second-order negligible effects in journal bearings.
4. The viscosity remains constant throughout the pressure film.
5. The inertia terms governing the motion of the fluid are negligible.
6. The pressure gradient through the thickness of the film is zero.
7. The motion of the fluid in a direction normal to the surface can be neglected compared to the motion parallel to it.

The problem can be stated as follows: Given the parameters μ, h, $(\partial h/\partial t)$, q, and bearing geometry, find the pressure distribution $p(x, y)$ in the lubricant film.

22.2 FINITE ELEMENT SOLUTION

The functional $I(p)$, which on minimization gives Eqs. (22.1)–(22.3), is given by [22.2–22.4]

$$I(p) = \iint_S \left[\frac{h^3}{24\mu} \nabla p \nabla p - h\vec{U}\nabla p - \frac{h^3}{12\mu}\vec{B}\nabla p + \left(\frac{\partial h}{\partial t} + v_d\right) p \right] \mathrm{d}S + \int_{C_2} qp \, \mathrm{d}C_2 \quad (22.4)$$

Let the region of interest in the xy plane be divided into E finite elements (triangular element is more versatile for two-dimensional regions). By expressing the pressure and the distributions of the various forcing functions in terms of their respective nodal values through interpolation functions $N_i(x, y)$, we obtain

$$\left. \begin{aligned} p(x, y) &= \sum_i N_i(x, y) P_i^{(e)} = [N(x, y)]\vec{P}^{(e)} \\ U_x(x, y) &= u(x, y) = \sum_i N_i U_i^{(e)} = [N(x, y)]\vec{U}^{(e)} \\ &\vdots \\ v_d(x, y) &= \sum_i N_i V_{d_i}^{(e)} = [N(x, y)]\vec{V}_d^{(e)} \end{aligned} \right\} \quad (22.5)$$

where $\vec{P}^{(e)}, \vec{U}^{(e)}, \ldots$ are the vectors of nodal pressures, x component velocities, ..., respectively, of element e. If Eq. (22.5) is substituted into the functional I of Eq. (22.4), we obtain the discretized form of the functional for element e. By writing $I = \sum_{e=1}^{E} I^{(e)}$ and using the conditions for the minimum of I, namely

$$\frac{\partial I}{\partial P_i} = 0, \qquad i = 1, 2, \ldots, M \tag{22.6}$$

where M is the total number of nodal unknowns, we obtain the elemental equations as

$$\begin{aligned}[K_p^{(e)}]\vec{P}^{(e)} &= \vec{Q}^{(e)} - [K_u^{(e)}]\vec{U}^{(e)} - [K_v^{(e)}]\vec{V}^{(e)} - [K_{B_x}^{(e)}]\vec{B}_x^{(e)} - [K_{B_y}^{(e)}]\vec{B}_y^{(e)} \\ &- [K_{\dot{h}}^{(e)}]\vec{\dot{H}}^{(e)} - [K_{v_d}^{(e)}]\vec{V}_d^{(e)} = \vec{F}^{(e)}\end{aligned} \tag{22.7}$$

where the elements of the various matrices are given by

$$K_{p_{ij}}^{(e)} = -\iint_{A^{(e)}} \frac{h^3}{12\mu}\left(\frac{\partial N_i}{\partial x}\frac{\partial N_j}{\partial x} + \frac{\partial N_i}{\partial y}\frac{\partial N_j}{\partial y}\right) \mathrm{d}A \tag{22.8}$$

$$K_{u_{ij}}^{(e)} = \iint_{A^{(e)}} h\frac{\partial N_i}{\partial x} N_j \,\mathrm{d}A \tag{22.9}$$

$$K_{v_{ij}}^{(e)} = \iint_{A^{(e)}} h\frac{\partial N_i}{\partial y} N_j \,\mathrm{d}A \tag{22.10}$$

$$K_{B_{x_{ij}}}^{(e)} = \iint_{A^{(e)}} \frac{h^3}{12\mu}\frac{\partial N_i}{\partial x} N_j \,\mathrm{d}A \tag{22.11}$$

$$K_{B_{y_{ij}}}^{(e)} = \iint_{A^{(e)}} \frac{h^3}{12\mu}\frac{\partial N_i}{\partial y} N_j \,\mathrm{d}A \tag{22.12}$$

$$K_{v_{d_{ij}}}^{(e)} = K_{\dot{h}_{ij}}^{(e)} = -\iint_{A^{(e)}} N_i N_j \,\mathrm{d}A \tag{22.13}$$

$$Q_i^{(e)} = \int_{C_2^{(e)}} q N_i \,\mathrm{d}C_2 \tag{22.14}$$

The element equations (22.7) are assembled in the usual manner to obtain the overall system equations as

$$[\underset{\sim}{K}_p]\,\underset{\sim}{\vec{P}} = \underset{\sim}{\vec{F}} \tag{22.15}$$

where the order of the matrix $[\underset{\sim}{K}_p]$ is equal to M. Equation (22.15) can be solved after applying the specified pressure boundary conditions on S_1.

Once the nodal pressures $\vec{\underset{\sim}{P}}$ and flows $\vec{\underset{\sim}{U}}$ and $\vec{\underset{\sim}{V}}$ are found, the bearing load capacity (W) and the friction force vector ($\vec{f}$) can be computed from the relations

$$W = \iint_S p(x,y)\,\mathrm{d}S \tag{22.16}$$

and

$$\vec{f} = \iint_S \mu \frac{\partial}{\partial z}(u\vec{i} + v\vec{j})\Big|_{z=0,h}\,\mathrm{d}S \tag{22.17}$$

where $\vec{i}$ and $\vec{j}$ indicate unit vectors along x and y directions, respectively.

22.3 NUMERICAL EXAMPLES

22.3.1 Step Bearing

The simple infinite-width step bearing shown in Figure 22.2 is divided into two finite elements (idealized as a one-dimensional problem). The film pressure inside any element is assumed to vary linearly as

$$\underline{p}(\underline{x}) = a_1 + a_2\underline{x} \tag{E$_1$}$$

where $\underline{x} = x/l$, $\underline{p} = (ph_2^2/\mu ul)$, and a_1 and a_2 are constants. The film pressure is assumed to be zero on both sides of the bearing. The element equations for the first element can be derived as

$$-\frac{a^3}{12b_1}\begin{bmatrix} 1 & -1 \\ -1 & 1 \end{bmatrix}\begin{Bmatrix} \underline{p}_1 \\ \underline{p}_2 \end{Bmatrix} + \frac{a}{2}\begin{Bmatrix} -1 \\ 1 \end{Bmatrix} = \begin{Bmatrix} 0 \\ 0 \end{Bmatrix} \tag{E$_2$}$$

and the assembled system of equations as

$$\begin{bmatrix} 1 & 0 & 0 \\ \dfrac{a^3}{12b_1} & -\dfrac{1}{12}\left(\dfrac{a^3}{b_1} + \dfrac{1}{b_2}\right) & \dfrac{1}{12b_2} \\ 0 & 0 & 1 \end{bmatrix}\begin{Bmatrix} \underline{p}_1 \\ \underline{p}_2 \\ \underline{p}_3 \end{Bmatrix} + \begin{Bmatrix} 0 \\ \dfrac{a-1}{2} \\ 0 \end{Bmatrix} = \begin{Bmatrix} 0 \\ 0 \\ 0 \end{Bmatrix} \tag{E$_3$}$$

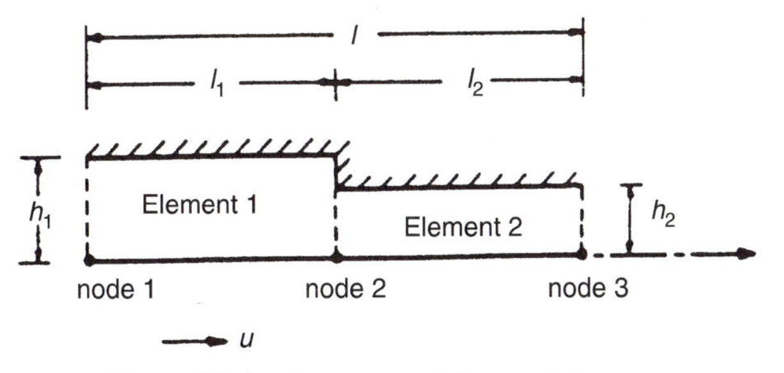

Figure 22.2. Geometry of Stepped Bearing.

where $a = h_1/h_2$, $b_1 = l_1/l$, $b_2 = l_2/l$, and $\underline{p}_1$, $\underline{p}_2$, and $\underline{p}_3$ are the values of $\underline{p}$ at nodes 1, 2, and 3, respectively. Equations (E_3) give the solution

$$\underline{p}_2 = \underline{p}(\underline{x} = b_1) = \frac{6b_1b_2(a-1)}{b_2a^3 + b_1} \tag{E$_4$}$$

The oil flow (Q) at node 2 can be computed as [22.5, 22.6]

$$\underline{Q} = -\frac{\underline{h}^3}{12}\frac{\mathrm{d}\underline{p}}{\mathrm{d}\underline{x}} + (\underline{h}/2) = \frac{\underline{p}_2}{12b_2} + \frac{1}{2} \tag{E$_5$}$$

where $\underline{Q} = Q/(h_2u)$, and $\underline{h} = h/h_2$. The solution given by Eqs. (E_4) and (E_5) coincides with the exact solution for this case.

22.3.2 Journal Bearing

As a second example, the journal bearing shown in Figure 22.3 is considered with

$$\underline{p}(\underline{x}) = a_1 + a_2\underline{x} \tag{22.18}$$

with the boundary conditions $\underline{p}(0) = \underline{p}(\pi) = 0$. The results obtained for eccentricity ratios of $\varepsilon = 0.2$ and 0.8 are shown in Table 22.1. It can be seen that the finite element results are better in the case of $\varepsilon = 0.2$ than in the case of $\varepsilon = 0.8$. The reason for this is that as the eccentricity ratio becomes smaller, the pressure distribution tends to be flatter and hence a linear variation for $\underline{p}$ gives better results. When the eccentricity ratio is large, the film pressure changes abruptly; therefore, the results become slightly inferior. Of course, better results can be obtained by increasing either the number of elements or the degree of polynomial assumed for $\underline{p}$ within any element.

Although numerical results for simple one-dimensional problems are given in this chapter, several complex lubrication problems have been solved by the finite element method in the literature [22.7].

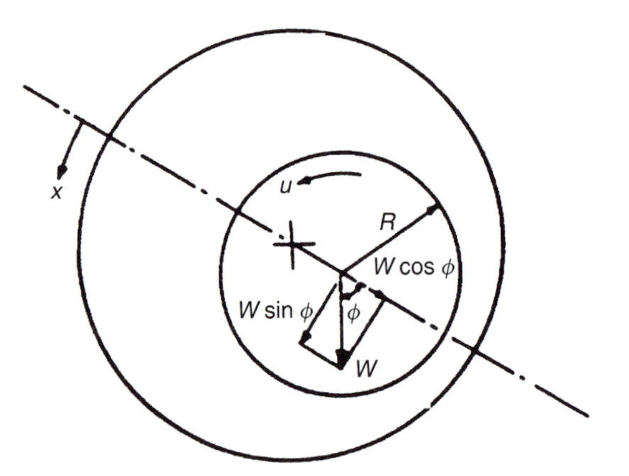

Figure 22.3. Geometry of Journal Bearing.

Table 22.1. Film Pressure of Journal Bearing at $x = 90°$ [22.5]

	For eccentricity ratio $\varepsilon = 0.2$			For eccentricity ratio $\varepsilon = 0.8$		
Number of finite elements	Finite element method	Finite difference method	Exact solution	Finite element method	Finite difference method	Exact solution
2	1.132	1.397		2.449	3.021	
4	1.170	1.233		3.291	3.522	
6	1.174	1.201		3.496	3.606	
12	1.176	1.183	1.176	3.601	3.630	3.636
18	1.176	1.179		3.621	3.634	
24	1.176	1.178		3.627	3.635	
36	1.176	1.177		3.632	3.636	

REFERENCES

22.1 P.R. Trumpler: *Design of Film Bearings*, Macmillan, New York, 1966.

22.2 M.M. Reddi: Finite element solution of the incompressible lubrication problem, *Journal of Lubrication Technology, Transactions of ASME, 91*, 524–533, 1969.

22.3 M.M. Reddi and T.Y. Chu: Finite element solution of the steady-state compressible lubrication problem, *Journal of Lubrication Technology, Transactions of ASME, 92*, 495–503, 1970.

22.4 J.F. Booker and K.H. Huebner: Application of finite element methods to lubrication: An engineering approach, *Journal of Lubrication Technology, Transactions of ASME, 94*, 313, 1972.

22.5 S. Wada, H. Hayashi, and M. Migita: Application of finite element method to hydrodynamic lubrication problems: Part 1, Infinite width bearings, *Bulletin of the Japan Society of Mechanical Engineers, 14*, 1222–1233, 1971.

22.6 S. Wada and H. Hayashi: Application of finite element method to hydrodynamic lubrication problems: Part 2, Finite width bearings, *Bulletin of the Japan Society of Mechanical Engineers, 14*, 1234–1244, 1971.

22.7 K.H. Huebner: Application of finite element methods to thermohydrodynamic lubrication, *International Journal for Numerical Methods in Engineering, 8*, 139–168, 1974.

PROBLEMS

22.1 Solve the infinite-width step-bearing problem indicated in Figure 22.2 using two one-dimensional quadratic elements.

22.2 Derive the finite element equations corresponding to Eqs. (22.1)–(22.3) using the Galerkin approach.

22.3 Find the pressure distribution in the stepped bearing shown in Figure 22.2 using four linear elements. Also, determine the oil flow at node 2 of Figure 22.2.

22.4 Find the film pressure in the journal bearing shown in Figure 22.3 using two linear elements for an eccentricity ratio of $\varepsilon = 0.2$.

22.5 Find the film pressure in the journal bearing shown in Figure 22.3 using two linear elements for an eccentricity ratio of $\varepsilon = 0.8$.

APPENDIX A

GREEN-GAUSS THEOREM

(Integration by parts in two and three dimensions)

In the derivation of finite element equations for two-dimensional problems, we need to evaluate integrals of the type

$$\iint_S \psi \frac{\partial \phi}{\partial x} \, dx \, dy \tag{A.1}$$

where S is the area or region of integration, and C is its bounding curve. We can integrate Eq. (A.1) by parts, first with respect to x, using the basic relation

$$\int_{x_l}^{x_r} u \, dv = -\int_{x_l}^{x_r} v \, du + uv \Big|_{x_l}^{x_r} \tag{A.2}$$

to obtain

$$\iint_S \psi \frac{\partial \phi}{\partial x} \, dx \, dy = -\iint_S \frac{\partial \psi}{\partial x} \phi \, dx \, dy + \int_{y=y_1}^{y_2} (\psi \phi) \Big|_{x_l}^{x_r} \cdot dy \tag{A.3}$$

where (x_l, x_r) and (y_1, y_2) denote the limits of integration for x and y as shown in Figure A.1. However, dy can be expressed as

$$dy = \pm dC \cdot l_x \tag{A.4}$$

where dC is an element of the boundary curve, l_x is the cosine of the angle between the normal n and the x direction, and the plus and minus signs are applicable to the right- and left-hand-side boundary curves (see Figure A.1). Thus, the last term of Eq. (A.3) can be expressed in integral form as

$$\int_{y_1}^{y_2} (\psi \phi) \Big|_{x_l}^{x_r} \cdot dy = \oint_C \psi \phi \, dC l_x \tag{A.5}$$

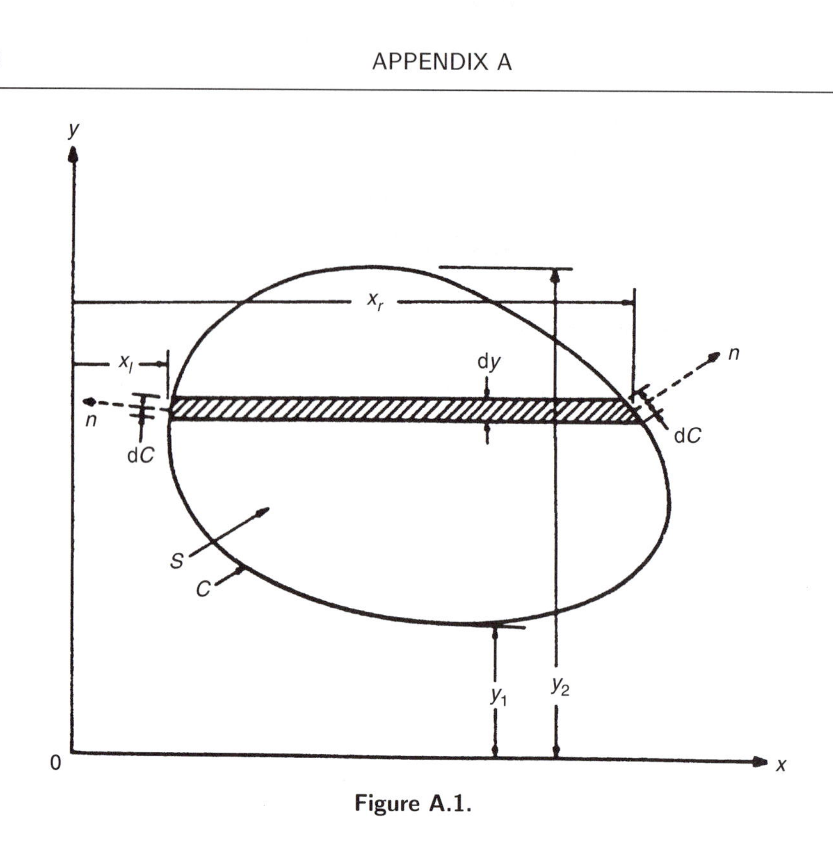

Figure A.1.

Thus, the integral of Eq. (A.1) can be evaluated as

$$\iint_S \psi \frac{\partial \phi}{\partial x} \,\mathrm{d}x\,\mathrm{d}y = -\iint_S \frac{\partial \psi}{\partial x} \phi \,\mathrm{d}x\,\mathrm{d}y + \oint_C \psi \phi l_x \,\mathrm{d}C \tag{A.6}$$

Similarly, if the integral (A.1) contains the term $(\partial\phi/\partial y)$ instead of $(\partial\phi/\partial x)$, it can be evaluated as

$$\iint_S \psi \frac{\partial \phi}{\partial y} \,\mathrm{d}x\,\mathrm{d}y = -\iint_S \frac{\partial \psi}{\partial y} \phi \,\mathrm{d}x\,\mathrm{d}y + \oint_C \psi \phi l_y \,\mathrm{d}C \tag{A.7}$$

where l_y is the cosine of the angle between the normal n and the y direction.

Equations (A.6) and (A.7) can be generalized to the case of three dimensions as

$$\iiint_V \psi \frac{\partial \phi}{\partial x} \,\mathrm{d}x\,\mathrm{d}y\,\mathrm{d}z = -\iiint_V \frac{\partial \psi}{\partial x} \phi \,\mathrm{d}x\,\mathrm{d}y\,\mathrm{d}z + \oint_S \psi \phi l_x \,\mathrm{d}S \tag{A.8}$$

where V is the volume or domain of integration and S is the surface bounding the domain V. Expressions similar to Eq. (A.8) can be written if the quantity $(\partial\phi/\partial y)$ or $(\partial\phi/\partial z)$ appears instead of $(\partial\phi/\partial x)$ in the original integral.

INDEX